New Results in Numerical and Experimental Fluid Mechanics II

Edited by
Wolfgang Nitsche
Hans-Joachim Heinemann and
Reinhard Hilbig

Notes on Numerical Fluid Mechanics (NNFM)   Volume 72

New Results in Numerical and Experimental Fluid Mechanics II

Contributions to the
11th AG STAB/DGLR Symposium
Berlin, Germany 1998

Edited by
Wolfgang Nitsche
Hans-Joachim Heinemann and
Reinhard Hilbig

Springer Fachmedien Wiesbaden GmbH

Die Deutsche Bibliothek – CIP-Einheitsaufnahme

Arbeitsgemeinschaft Strömungen mit Ablösung:
Contributions to the ... AG STAB/DGLR symposium ... –
Braunschweig; Wiesbaden: Vieweg
   (Notes on numerical fluid mechanics; ...)

11. New results in numerical and experimental fluid mechanics
2. Berlin, Germany, 1998 – 1999

New results in numerical and experimental fluid mechanics. –
Braunschweig; Wiesbaden: Vieweg
   Contributions to the ... AG STAB/DGLR symposium ...; ...)

2. Berlin, Germany, 1998 / ed. by Wolfgang Nitsche ... – 1999
(Contributions to the ... AG STAB/DGLR symposium ... ; 11)

   ISBN 978-3-663-10903-7   ISBN 978-3-663-10901-3 (eBook)
   DOI 10.1007/978-3-663-10901-3

http://www.vieweg.de

Produced by W. Langelüddecke, Braunschweig
Printed on acid-free paper

ISSN 0179-9614

FOREWORD

This volume contains the contributions to the 11th DGLR/AG STAB-Symposium held at the Technical University Berlin, November, 10 to 12, 1998. AG STAB is the German Aerospace Aerodynamics Association, founded at the end of the 70', while DGLR is the German Society for Aeronautics and Astronautics (Deutsche Gesellschaft für Luft- und Raumfahrt - Lilienthal Oberth e.V.).

In the AG STAB German scientists and engineers from universities, research-establishments and industry are involved, who are doing research and project work in numerical and experimental fluidmechanics and aerodynamics for aerospace and other applications.

About 20 years ago it became obvious for this community that a joint effort of members of universities, the DLR and industry was necessary to counter-act declining budgets in the field. It was decided to approach high-level persons in industry, ministries and the parliament for help to shift the trend with its negative effects for research and industry. From the begin it was clear that an effort should be built around a central theme. "Flow with Separation" became the topic of the AG STAB (Arbeitsgemeinschaft Strömung mit Ablösung), which developed fast into a lively association, with, however, a larger scope than just flow with separation.

One of the general guidelines of STAB is to concentrate resources and know-how in the institutions involved and to avoid duplication in research work as much as possible. Today, this is more then ever necessary. The experience made in the past makes it easier now, to obtain new knowledge for solving today's and tomorrow's problems.

Strongly involved in STAB from the beginning was Prof. Dr. Hermann L. Jordan, who held leading positions in national and international institutions. From 1973 until 1987 he was Chairman of the Board of Directors of DFVLR, now the DLR. His enthusiasm as well as his managing qualities were very important. He was the first Chairman of the Board of the AG STAB. He died in December 1998. It is the AG STAB who is very grateful to Prof. Dr. Hermann L. Jordan and, therefore, this book is dedicated to him.

Since 1986 the symposia have been organized every two years at different locations in Germany. In between STAB workshops are held. It is now for the second time that the contributions to the symposium are published after having been subjected to peer review. Many of the contributions are giving results from the "Luftfahrtforschungsprogramm der Bundesregierung (German Aeronautical Research Programme) 1995-1998". Some of the papers report on work sponsored by the Deutsche Forschungsgemeinschaft (German Research Council), DFG. Therefore, the volume gives a broad overview of the ongoing work in this field in Germany.

The Review-Board, which is partly identical with the Program-Committee, consisted of J. Ballmann (Aachen), K.A. Bütefisch (Göttingen), G. Cucinelli (München), U. Chr. Dallmann (Göttingen), R. Friedrich (München), F.-R. Grosche (Göttingen), D.K. Hennecke (Darmstadt), P. Hennig (München), R. Hilbig (Bremen), E.H. Hirschel (München), H. Hönlinger (Göttingen), H. Körner (Braunschweig), W. Kordulla (Göttingen), H.-P. Kreplin (Göttingen), D. Kröner (Freiburg), N. Kroll (Braunschweig), A. Leder (Rostock), G.E.A. Meier (Göttingen), R. Niehuis (Aachen), W. Nitsche (Berlin) - Chairman -, H. Olivier (Aachen), J.-J.

Philippe (Châtillon), G. Redeker (Braunschweig), U. Rist (Stuttgart), R.P. Shreeve (Monterey), J. Szodruch (Hamburg), P. Thiede (Bremen), F. Thiele (Berlin), B. Wagner (Oberpfaffenhofen), S. Wagner (Stuttgart) and H.B. Weyer (Köln). Nevertheless, the authors sign responsible for the contents of their contributions.

The editors are also grateful to Prof. Dr. E.H. Hirschel as the general editor of the „Notes on Numerical Fluid Mechanics" and to the Vieweg-Verlag for the opportunity to publish the results of the symposium.

W. Nitsche, Berlin; H.-J. Heinemann, Göttingen; R. Hilbig, Bremen   June 1999

This book is dedicated to the memory of

Prof. Dr. H.L. Jordan

CONTENTS

CONTENTS (continued)

CONTENTS (continued)

CONTENTS (continued)

CONTENTS (continued)

On the application of suction for the stabilization of crossflow instability over perforated walls

C. Abegg, H. Bippes, F. P. Bertolotti

Deutsches Zentrum für Luft- und Raumfahrt
Institut für Strömungsmechanik
Bunsenstraße 10, D-37083 Göttingen
Germany

Summary

Experimental investigations to study the control of crossflow instability in three-dimensional boundary layers with the aid of suction through perforated walls are presented. It is shown that suction can be a powerful tool to stabilize boundary layer flows that are subject to crossflow instability. To elucidate possible improvements, two different suction devices, which differ in the hole diameter of the perforated metal sheet, are used inside a swept flat plate. Stationary and travelling modes have a high receptivity to roughness and suction at their neutral stability point, hence we pay special attention to the chordwise position where the suction is applied. In particular, suction within the stagnation area can lead to an undesired increase of the disturbance content due to a spatial non-uniformity in the suction velocity. The suction device with larger holes in the perforated metal sheet leads to more homogeneous suction velocities and therefore to a significant damping of the travelling modes.

Introduction

In swept wing flow typical for large transport aircraft, transition occurs predominantly in the leading edge area due to leading edge contamination. Therefore, attempts to keep swept wing flows laminar have been concentrated on this wing portion. Recent experiments have demonstrated that suction enables the relaminarisation of the attachment line flow [2], [3]. This success inspired aircraft industry and research to study the feasibility of laminar swept wing flow by damping crossflow instability. In flight and wind–tunnel tests known so far, much less success was achieved than predicted by stability analyses. On the other hand, the stability problem in the leading edge area of swept wings, especially where stagnation flow instability switches over to crossflow instability, is not well understood. Moreover, the strong non-parallelism in this area leads to numerical difficulties. In order to provide a better physical insight into this problem, basic experiments are needed. The present work contributes to this goal. It is specifically aimed at studying the stabilisation of crossflow disturbances by means of suction isolated from stagnation flow instability. The results document both the positive influence and the shortcomings in the suction technique and give some hints on possible improvements and benefits. Previous experiments suggested, that a more uniform suction velocity would improve the stabilization effect. Therefore additional tests on this aspect have been performed.

The Experiment

In favour of a realistic estimate of the stabilizing effect of suction on crossflow instability, the same model as in previous experiments was used for the present study. It is a swept flat plate, where the pressure gradient is induced by a wing-like body outside the boundary layer. Details of the model design and the pressure distribution are reported in [1]. For the suction experiments, the front part of the plate can be equipped with two different suction devices (Fig. 1a). The first device has four suction chambers in the chordwise direction with a spanwise length of 500 mm, and the suction is applied through holes spaced 400 μm apart and with nominal diameters of 50 μm (Fig. 1b). The second device contains only the suction chamber III and IV covered by a perforated metal sheet with holes spaced 960 μm and with diameters of 120 μm (Fig. 1c). On this model the surface from the leading edge up to the beginning of the suction chamber III is solid and polished.

The tests are performed in a wind tunnel with an open section of 1 m $\times$ 0.7 m and a turbulence level of Tu $= 100/Q_\infty \left[1/3 \left(\overline{u'^2} + \overline{v'^2} + \overline{w'^2}\right)\right]^{1/2} = 0.15$ %, measured in the frequency range 2 Hz $< f < 2$ kHz. For unsteady flow measurements of the wall-parallel velocity components, hot-wire anemometry with two-wire probes in a V-arrangement was used.

Results

Preliminary experiments with the four chamber suction unit revealed the largest decrease of the disturbance content when suction was applied to chambers III and IV only. Furthermore, applying suction to chamber I only, which is located near the neutral stability point of cross-flow vortices, increased the disturbance amplitude compared with the no-suction case. Only if the suction surface was covered beyond the stagnation line (see Fig. 1b), the negative effect could be avoided. To illustrate the effectiveness of suction on attenuating the disturbances, Figure 2 (taken from [4]) shows the spanwise velocity distribution in the boundary layer at a dimensionless chordwise position of $x_c/c = 0.8$, far inside the nonlinear regime of the instability process. At the chosen wall distance of $z/\delta = 0.25$, the disturbances reach their maximum amplitudes. On the left-hand side, the spanwise variation of the mean velocity $\overline{U_s}(y_c)/Q_e$ due to the presence of stationary vortices is depicted. On the right-hand side, the size of the travelling modes is directly given by the r.m.s. fluctuations $u_{rms}(y_c)/Q_e$. The spanwise variation of the unsteady disturbances indicates the nonlinear state of the disturbance development. Comparing the measurements without suction (Fig. 2a) and with suction in all four chambers (Fig. 2b) using a suction rate $c_q = \overline{W}_{suct}/Q_\infty = 0.1\%$, it becomes obvious that the suction has considerably damped the growth of the unsteady modes, whereas the amplitudes of the stationary modes are only marginally influenced. In the case where only chambers II, III and IV are activated (Fig. 2c), the amplitudes of the travelling modes and of the stationary modes are decreased below the level observed in the case when all four suction chambers are active. A possible reason for this behaviour is a non-uniformity in the suction distribution caused by manufacturing deficiencies of the perforated metal sheet and by sporadically appearing closed holes. Our calculations employing non-local stability theory revealed a strong increase of receptivity to non-uniformity in suction as the chordwise location of the non-uniformity approached the attachment line. The coupling of increased receptivity and suction non-uniformity can lead to a perturbation similar to surface roughness and therefore stimulate stationary vortices in the same way.

In more extensive investigation using the four chamber unit it became evident that the optimal configuration of suction chambers is a combination of chambers III and IV. Figure 3 shows the spanwise averaged profiles of the mean velocity components U/U_e,

2

V/U_e (averaged over two wavelengths) — the stationary modes — and the r.m.s. values integrated normal to the wall over the whole boundary layer, as well as integrated spanwise over two wavelengths. The depiction documents that — using only one suction chamber — chamber III reduces the travelling modes most effectively. The r.m.s value reaches approximately 75% of that value where chambers II or IV are activated alone. With two active suction chambers, the most effective case for damping the instationary modes is the combination III/IV. Here, the integrated r.m.s. value decreases to 34% compared with the case where only chamber III is activated.

Lastly, an optimized suction unit with only two chambers III and IV and a perforated sheet with larger holes was used (Fig 1b). The diameter of the holes in the perforated sheet is $120\mu m$, which should lead to a more homogeneous suction. The following results show the downstream development of the stationary and travelling modes for the two different suction devices. In the four chamber unit, only chamber III and IV are activated. In both experiments, the suction rate is $c_q = 0.1\%$ and a velocity of the external flow of $Q_e = 19m/s$ is used. Figure 4 depicts the development of the different modes in a streamline oriented system. Again, the integrated r.m.s. value and the amount of the stationary disturbances are shown. When comparing the two different suction devices it becomes evident that the amplification rates of the travelling modes decrease stronger than the stationary modes. At the dimensionless cord position of $x_c/c = 0.9$, the amplitudes of the stationary mode of the two chamber device is 30% lower and the instationary modes reduce by 75%. The high receptivity of the stationary mode to nonuniform suction is the reason for the different behaviour and explains the higher damping of the travelling modes. Although, it is not clear why the stationary mode decreases so strongly when using the two different suction devices with the same suction rate. A possible reason is the manufacturing quality of the two different models, especially of the edge, where the perforated metal sheet of the suction device is connected with the polished surface.

Conclusions

The experiments have shown that suction can be a powerful aid to stabilize three-dimensional boundary layer flow exposed to crossflow instability. In agreement with theory, travelling modes are more efficiently damped than stationary modes. In addition, non-uniformity in the spatial distribution of suction strongly causes streamwise vortices, especially in the proximity of the stagnation area. Therefore, a new model which contains only suction chambers downstream of the attachment line is used to improve the efficiency of the suction technique. Due to a more uniform suction, the travelling modes have been reduced by 70%. In conclusion, an economic and efficient use of suction requires an improvement in the suction technology, especially in the case of swept wing flows, when suction has to be applied already at the stagnation area because of leading edge contamination.

References

[1] Deyhle, H. and H. Bippes: Disturbance growth in an unstable three-dimensional boundary layer and its dependence on environmental conditions. *J. Fluid Mech.* **316**, 73-113.

[2] Juillen, J.C. and D. Arnal (1994): Experimental study of boundary layer suction effects on leading edge contamination along the attachment line on swept wing. *In Laminar-Turbulent Transition, IUTAM Symp., Sendai Japan 1994*, 137-144.

[3] Poll, D. I. A. and M. Danks (1994): Relaminarisation of swept wing attachment-line by surface suction. *In Laminar-Turbulent Transition, IUTAM Symp., Sendai Japan 1994*, 137-144.

[4] Wiedemann, A. (1996): Ermittlung des Einflusses von Absaugung im Nasenbereich eines schiebenden Flügels auf die Stabilität der dreidimensionalen Grenzschichtströmung. Diploma thesis, Georg August University, Göttingen.

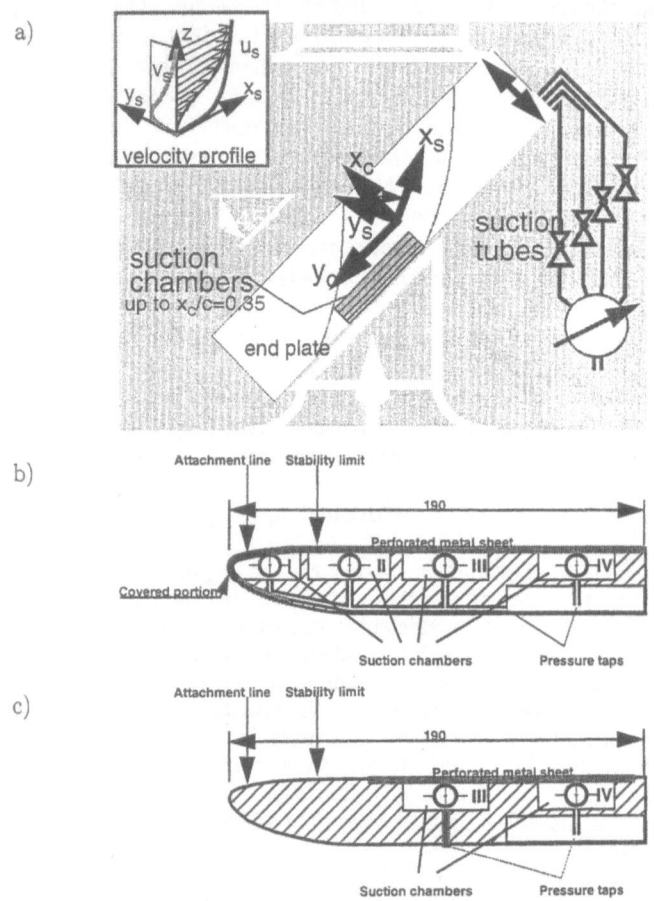

Figure 1: (a) Experimental set-up. (The displacement body which generates the pressure gradient on the flat plate is not shown. For details of the model design we refer to [1].) (b) The arrangement of the suction chambers in the front part of the model for the four chamber model (hole diameter is $50\mu m$). (c) The arrangement of the suction chambers in the front part of the model for the two chamber model (hole diameter is $120\mu m$).

4

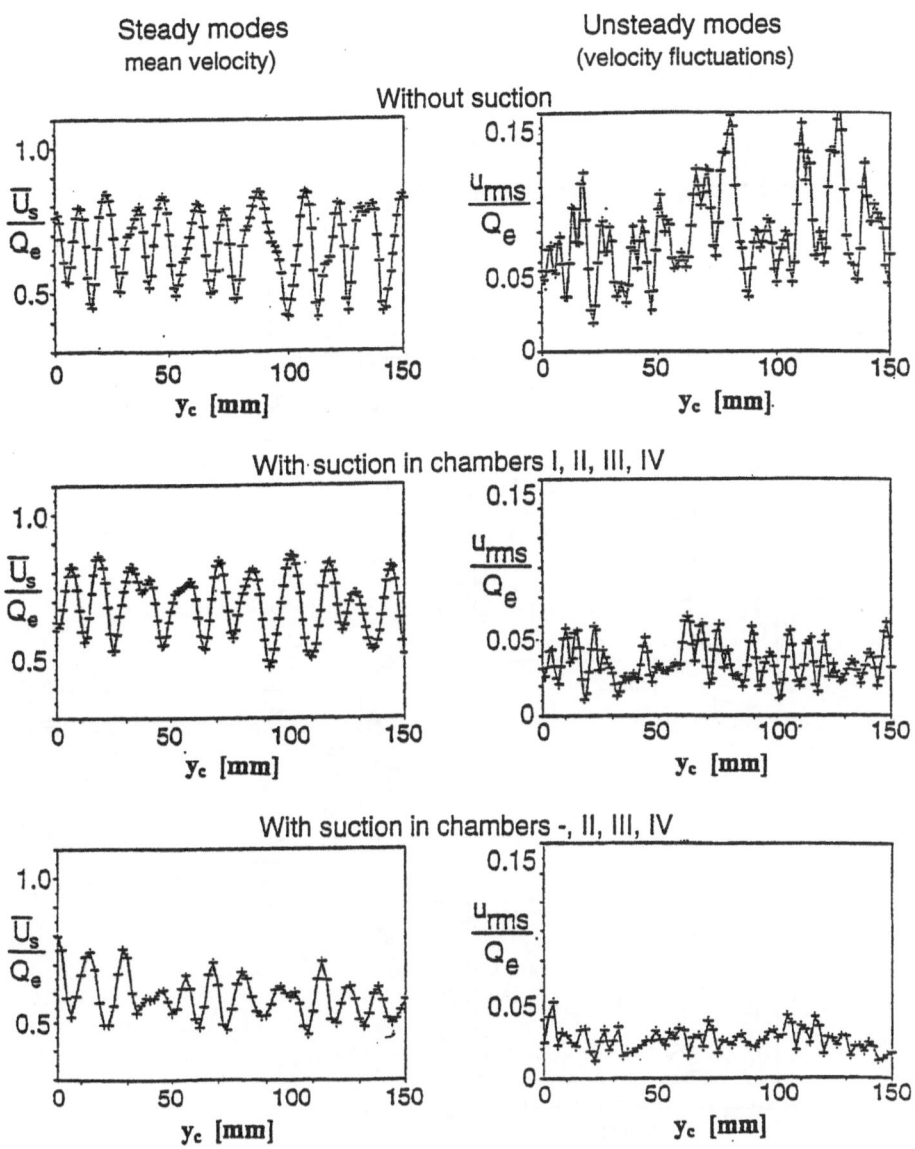

Figure 2: Spanwise velocity distributions for the stationary mode (left-hand side) and the travelling mode (right-hand side) at a chordwise position of $x_c/c = 0.8$ and a wall distance of $z/\delta = 0.25$. To illustrate the optimization of suction, different combinations of active suction chambers using the four chamber device are shown.

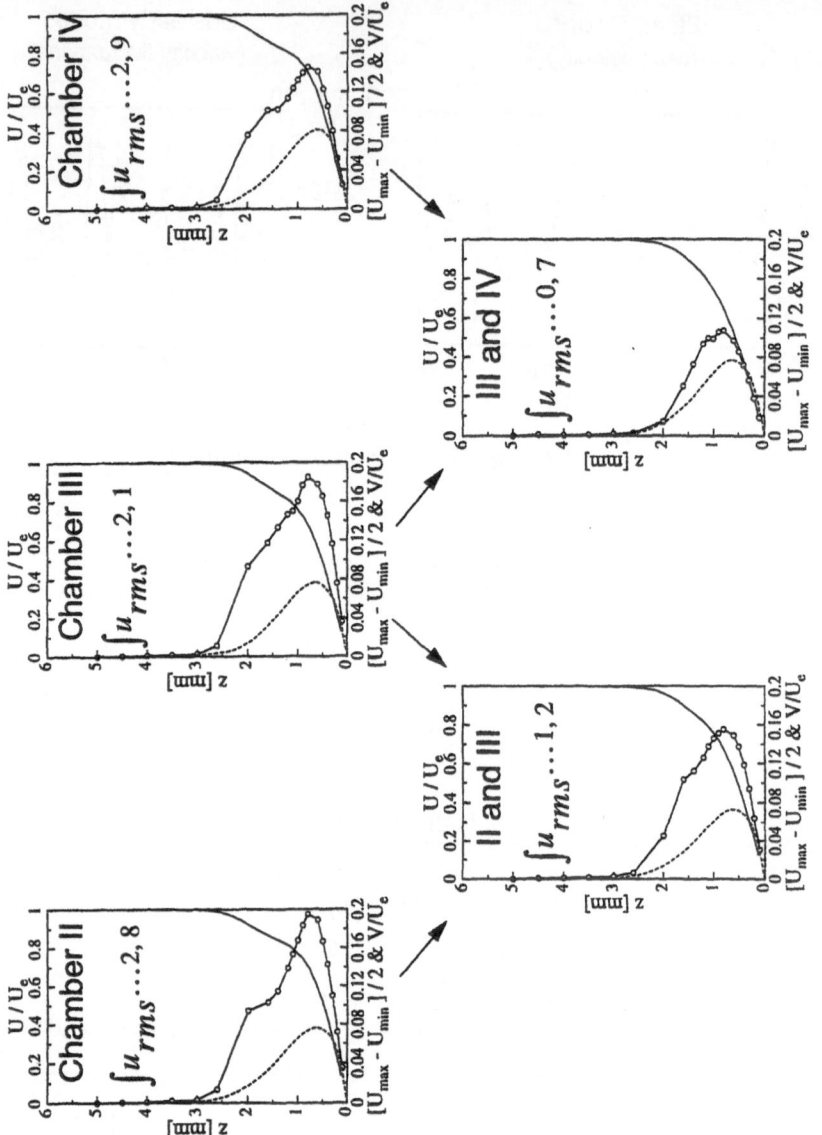

Figure 3: Mean velocity components (dashed: crossflow component, solid: streamwise component) and the stationary disturbance (solid line with circles) at a chordwise position of $x_c/c = 0.85$ for a suction rate of $c_d = 0.1\%$ with different combination of active suction chambers. The r.m.s values are integrated over two wavelength.

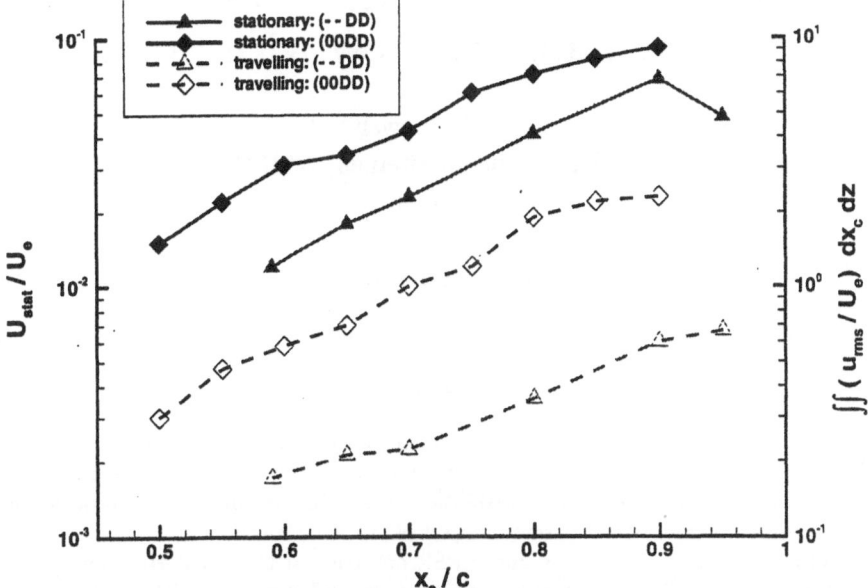

Figure 4: Growths of the stationary and travelling modes for the four chamber suction unit (00DD), where only chamber III and IV are active and for the two chamber unit (--DD) using a suction rate of $c_q = 0.1\%$. The stationary modes are given by the spanwise variation of the mean velocity and the travelling modes by the r.m.s. fluctuations, integrated spanwise over two wavelength and normal to the wall over the whole boundary layer.

Impact of Stringer-Width on the Effect of Hybrid Laminar Flow with Suction

R. Alewelt
University of Bremen, ZARM
Am Fallturm
D-28359 Bremen
Germany

Summary

This paper addresses some realistic requirements for the design of a Hybrid Laminar Flow wing with special attention to the application on production aircraft. After successful wind tunnel and flight tests the capabilities of the HLF technology are proven and the application on a commercial basis is subject of current efforts.

Stability computations with the linear PSE approach of the A320 HLF Fin reveal the aerodynamical drawbacks in test cases with increased stringer width, suction chamber failure and outflow from a suction chamber.

It comes out that within limits concessions to the ease of manufacturing or to other systems for a commercial application might be balanced with higher suction rates and therefore higher power requirements for the HLF system.

Introduction

The reduction of the aerodynamic drag is one of the major challenges for the manufacturers of modern transport aircraft. The most promising approach is the creation or extension of laminar flow regions on wings, nacelles and empennages. Since high cruise Mach numbers, resulting in large sweep angles, and high Reynolds numbers prohibit natural laminar flow, additional efforts have to be made. Keeping the boundary layer laminar with suction panels in the leading edge regions and an appropriate shaping of the wing body further downstream, the Hybrid Laminar Flow concept, turned out to be a suitable method. The application of the HLF technology on the above mentioned components would lead to an estimated 20% saving in fuel consumption and additional benefit regarding pollutant emission for a commercial aircraft.

Recent results from the flight tests with the A320 with HLF Fin show laminar regions up to 50% chord under cruise conditions and prove the feasibility of the technique. The HLF systems for the wind tunnel and free flight test were tailored with respect to a large extension of laminar flow and to provide extensive insight into physical mechanisms leading to laminar turbulent transition [2, 6, 9]. During the design of an HLF system for a production aircraft the bare necessities of other systems such as de-icing facilities and high-lift devices and a reasonable cost to benefit relation have to be considered [8].

Some aspects of realistic HLF design requirements are addressed in this paper. This includes the width of the stringers, that separate the suction chambers, and some off-design cases with blockage of suction chambers and even outflow.

The recent A320 HLF-Fin flight tests provide a wealth of experimental results, including off-design cases with switched off suction chambers and outflow. This makes the fin a favourable candidate for our analysis, providing evidence for later application of the HLF concept on a wing. The test cases are analysed with the linear parabolized stability equations (PSE). The comparison of the amplification rates and N-factors for some relevant disturbance modes give a qualitative rate of the cases. A more quantitative investigation will follow as soon as the flight test data are available.

Layout of the Suction System

Once a suitable wing profile is found, one has to decide on the number of suction chambers and their distribution. The suction chambers with their pipes to the suction pump have to fit into the nose box, limited by the front spar. The position of the front spar in a wing is given by structural needs and required fuel tank capacity. Since other systems like the de-icing, tracks of the high-lift devices and electrical cable share the room in the nose box, the number of suction chambers is limited and hence the possibilities of suction velocity variation are limited too.

Fig. 1 gives an overview of the A320 HLF Fin. The suction pump is located inside the pressure cabin and connected to the suction panel in the fin nose box with a ducting system. The suction velocity for each chamber can be adjusted by valves. The laminar flow regions are detected by infrared cameras, mounted on the horizontal stabilizers. The suction chambers near the leading edge are narrow spaced to allow for a fine suction velocity variation according to the outer pressure gradient. The downstream chambers (no. 1,2,8,9) are larger and supported by a honeycomb substructure to cope with the loads, resulting from aerodynamic forces and pressure differences between outer flow and suction chamber. The substructures provide additional non-sucked areas in the corresponding chambers, however – they are not covered by the current investigation.

The adjustable parameter for the suction system is the pressure inside the suction chamber. The chamber pressures have to be chosen according to the respective flight conditions, in order to achieve maximum extension of laminar flow regions with minimal suction power. The local suction velocity depends on the pressure difference between the outer flow static pressure, varying along the streamwise extension of the chamber, and the pressure inside the suction chamber. From experimental evidence the following quadratic approach for the pressure drop of the flow through the perforated suction panel could be derived [9]:

$$\Delta P = Ac_q^2 + Bc_q \tag{1}$$

The coefficients A and B in equation (1) are adjusted to experimental results and depend mainly on the porosity and hole diameters of the suction panel. The suction parameter c_q is the dimensionless suction velocity, normalized with the free stream velocity.

From the outer pressure distribution and a given chamber pressure the local suction velocities can be derived. The suction velocities are restricted to a lower limit, given by the no outflow condition for the suction chamber, i. e. the pressure inside the suction chamber stays at any position below the outer static pressure. Since the outer pressure varies along the chamber, a small pressure difference has to be applied for no outflow, to prevent a ventilation through the suction chamber. The upper limit for the suction velocities is given by the capabilities of the suction pump, see Fig. 2.

Numerical Procedure

In order to investigate the influence of suction distribution and stringer width of suction chamber walls on the extend of laminar flow numerically, the suction velocity distribution on the surface has to be modelled carefully.

At the edges of the suction chambers suction velocity rises suddenly from a zero value to a certain suction velocity or vice versa. The gradients at these edges are very steep and care has to be taken not to violate the assumptions made in the boundary layer and stability codes. For that reason we applied an exponential rise/descent for the suction velocities over a number of input points at the chamber edges. To resolve the chamber edges and the stringers between subsequent chambers, the input data has to be interpolated with a sufficient resolution in the area of interest. For the cases treated in this paper up to 700 input stations were necessary.

In a first step the geometry information is interpolated to the required degree. Next the edges of the suction chambers are projected onto the geometry, considering the width of the separating stringers.

For the presented test cases a basis suction distribution is chosen, so that a reasonable damping of the dominating stationary crossflow disturbance is achieved, but still some amplification can be found. Since only a symmetric flow is investigated in this paper, results are restricted to one side of the profile with suction chambers 5 to 9 according the numbering scheme in Fig. 1.

The input geometry, pressure distribution and local suction velocities are fed into the LISW boundary layer code, solving for the velocity and temperature profiles and their derivatives [3]. The code has been validated before and found to be reasonable accurate for the subsequent stability computations.

For the stability analysis we used the linear NOLOT/PSE code, derived at DLR Göttingen and FFA Stockholm [5, 4]. This code solves the parabolized stability equations, proposed by Herbert and Bertolotti [1, 7]. In contrary to the classical theory the parabolized stability equations account for boundary layer growth, the streamwise development of the flow and the curvature terms.

From the stability analysis we get the local amplification rates and N-factors, based on the total kinetic disturbance energy, for the investigated disturbances. The amplification rates and N-factors give a rating of the disturbance damping effectiveness of the different test cases. The results are plotted against the surface arclength normal to the leading edge, normalized with the chord length (s/c in the figures). The suction chamber edges are marked with vertical lines in Figures 4 - 7.

The results presented in this paper for a single disturbance hold qualitatively for a variety of spanwise wavenumbers and frequencies.

Test Cases

The following test cases assume cruise flight conditions (M_∞ =0.78, H=31 000ft) for the A320 HLF Fin. The reference case is a zero stringerwidth suction distribution. For all test cases the absolute mass flow per suction chamber is kept constant, so that we get higher suction velocities in the corresponding chambers by increasing stringer width.

The suction chambers 4,5 and 6 in Fig. 1 are small and the impact of increased stringer width is most significant for these chambers. With a total width of 30mm for chambers 4 to 6 a stringer of 10mm width reduces effective suction area by 30%.

The following test cases are discussed in this paper, numbering of suction chambers follows Fig. 1, suction velocities are shown in Fig. 3:

- case1: reference case width zero stringer width
- case2: 10mm stringer width for all chambers
- case3: 10mm stringer width between chambers 5 - 6 and 6 - 7, 20mm all others
- case4: like case 2, failure of chamber 5
- case5: like case 2, failure chamber 6
- case6: like case 2, failure of chamber 5+6
- case7: like case 4, with increased suction in chamber 6,7,8,9
- case8: like case 2, outflow chamber 5.

Variation of Stringer Width

In Fig. 4 the effect of stringers is demonstrated. The investigated crossflow vortices show largest amplification rates in the region of suction chambers 5 and 6. Since these chambers extend to only 30mm on surface arclength, effect of stringer variation is most significant for these chambers. For the reference case 1 amplification of the disturbance starts in chamber 6. Some wiggles can be found where the suction chambers meet and gradients in suction velocity occur. The maximum amplification does not exceed $50m^{-1}$ for case 1 and resulting N-Factor remains below 3 (Fig. 4(a) and 4(b)).

With a 10mm stringer width, case 2, amplification already starts in chamber 5, reaching a maximum of $125m^{-1}$ and a N-Factor of 5.

Increasing stringer width in case 3 to 20mm for chambers 7,8 and 9 has no further significant effect on the investigated disturbance. In the region of chamber number 7 to 9 we get slightly higher N-Factors due to some slightly higher amplification rates.

The impact of the downstream chambers is a bit more for travelling crossflow and TS waves (not shown in this paper), but still far less significant than the effect of the front suction chambers on steady crossflow vortices.

Suction Chamber Failure

Fig. 5 reflects the higher amplification rates and resulting N-Factors for a failure of suction chamber 5, 6 or both in comparison to case 2. The maximum amplification rates for case 4 and 6 are more than $175m^{-1}$ and at $125m^{-1}$ for case 5. The large extension of the regions with high amplification rates lead to a maximum N-Factor of 10 for case 6. This results shows the importance of the chambers close to the attachment line for a swept profile like the HLF Fin. The steady crossflow vortices can best be tackled on the first percent of chord. Within this short distance along surface arc length N-Factors reach high values that may lead to transition.

In case 7 the suction rates for chambers 6 to 9 are increased to compensate the failure of chamber 5, Fig. 6. This does work, when suction rates are increased by 30% in this case for the downstream chambers. The rate of increase may be different in other cases, but it can be stated that lacking suction chambers in the attachment line region on swept wings lead to far stronger demands to the suction system capabilities concerning mass flow.

Outflow

Fig. 7 shows the effect of outflow from suction chamber 5 (case 8). The mass flow is moderate and the effect on the amplification rates is only slightly higher than in case 4, with failure chamber 5. Although this will certainly change with increasing outflow velocities, the effect of blowing compared to the no outflow case 4 is less spectacular than expected. The depicted stronger amplification of the disturbances rests on the additional thickening of the boundary layer only. The effect of additional roughness or disturbances introduced by blowing through the sheet can not be covered by the applied PSE methods. They are rather a topic of receptivity and subject of current research.

Conclusions

The investigations presented in this paper reveal a significant impact of suction chamber blocking near the leading edge due to stringers or chamber failure on the disturbance damping effectiveness. This is particularly true for the damping of crossflow disturbances, that are strongly amplified near the attachment line. These effects can be compensated by higher suction rates for the downstream chambers.

The stringer width of chambers beyond 2% of chord (for the presented cases) is less critical compared to the stringer width of chambers near the leading edge because of the moderate amplification rates of oblique travelling waves, dominant in that region.

The presented qualitative results have to be supported by further more quantitative investigations, based upon the data of the A320 HLF Fin flight test. Corresponding analysis of wing configurations is necessary. Especially receptivity aspects, including disturbances introduced by inhomogeneous suction, have to be treated.

References

[1] F. P. Bertolotti. *Linear and Nonlinear Stability of Boundary Layers with Streamwise Varying Properties*. PhD thesis, The Ohio State University, 1991.

[2] H. Bieler, J. Pfennig, and R. Herrmann. A320 HLF Fin: Interdisciplinary Approach to a Boundary Layer Suction System. *Paper on Second European Forum on Laminar Flow Technology, Bordeaux*, 1996.

[3] E. Elsholz. Ein inverses LISW-Grenzschicht Verfahren. *MBB Technical Report*, TE-1681, 1988.

[4] A. Hanifi, D. Henningson, S. Hein, F.P. Bertolotti, and M. Simen. Linear Nonlocal Instability Analysis, the linear NOLOT code. *FFA*, TN54-1994, Stockholm 1994.

[5] S. Hein, F.P. Bertolotti, M. Simen, A. Hanifi, and D. Henningson. Linear Nonlocal Instability Analysis, the linear NOLOT code. *DLR-IB*, 223-94 A56, 1994, Göttingen.

[6] R. Henke, P. Capbern, A.J. Davies, R. Hinsinger, and J.L. Santana. The A320 HLF Fin-Programme: Objectives and Challenges. *Paper on Second European Forum on Laminar Flow Technology, Bordeaux*, 1996.

[7] Th. Herbert. Boundary-layer transition – Analysis and prediction revisited. *AIAA paper*, 91-0737, 1991.

[8] D.H. Jagger and A.J. Davies. Design and Engineering Issues of a Hybrid Laminar Flow Wing. *Paper on Second European Forum on Laminar Flow Technology, Bordeaux*, 1996.

[9] J. Preist and B. Paluch. Design Specification and Inspection of Perforated Panels for HLFC Suction Systems. *Paper on Second European Forum on Laminar Flow Technology, Bordeaux*, 1996.

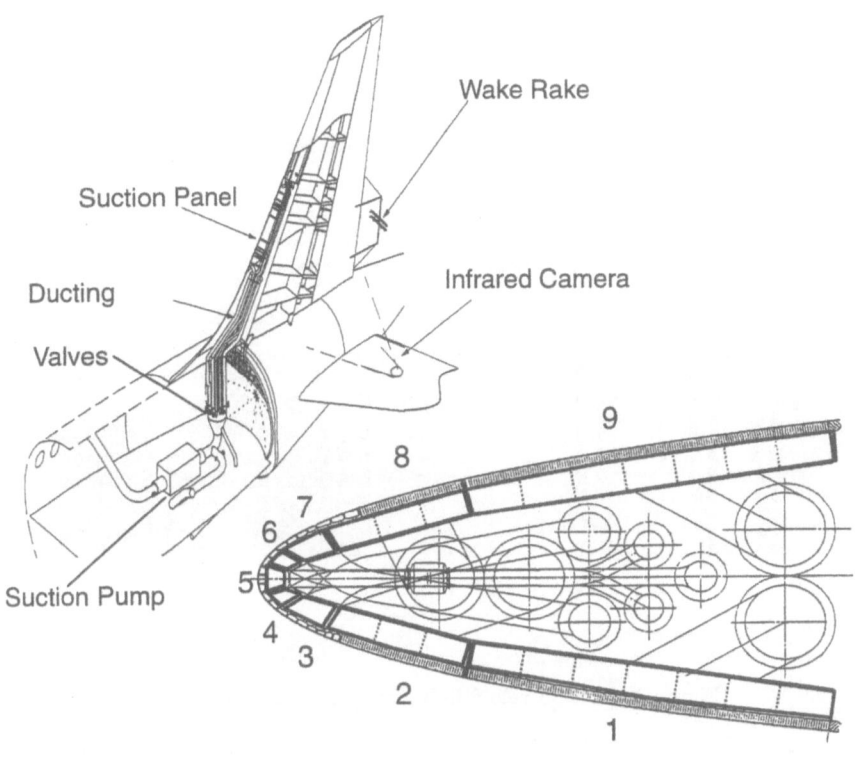

Figure 1: A320 HLF Fin

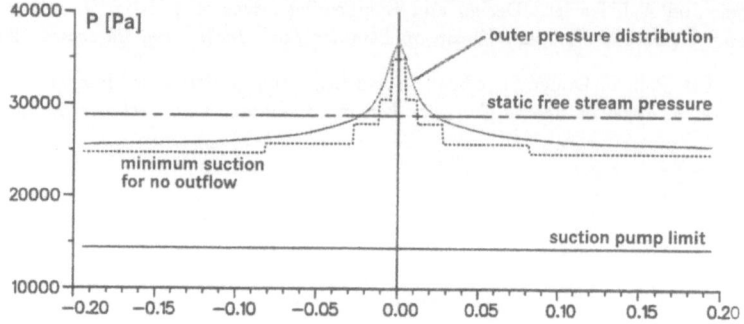

Figure 2: HLF design for A320 Fin

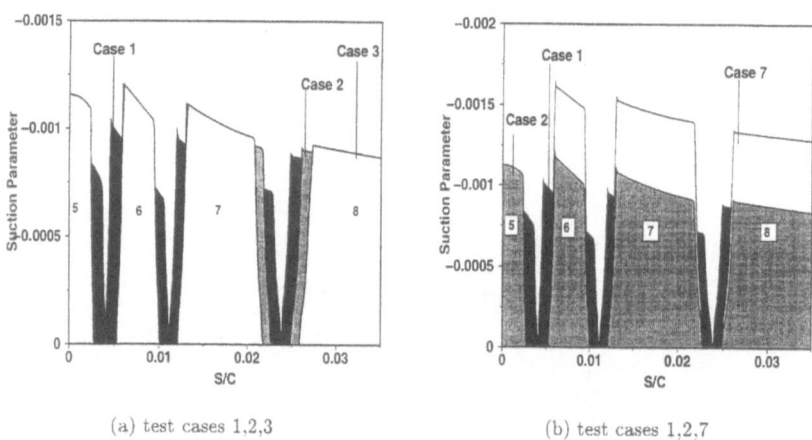

(a) test cases 1,2,3

(b) test cases 1,2,7

Figure 3: Suction velocities for various test cases

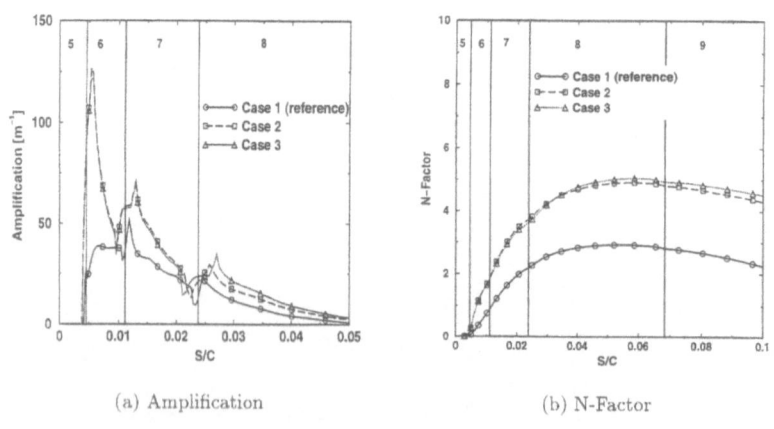

(a) Amplification

(b) N-Factor

Figure 4: Comparison Case 1,2,3

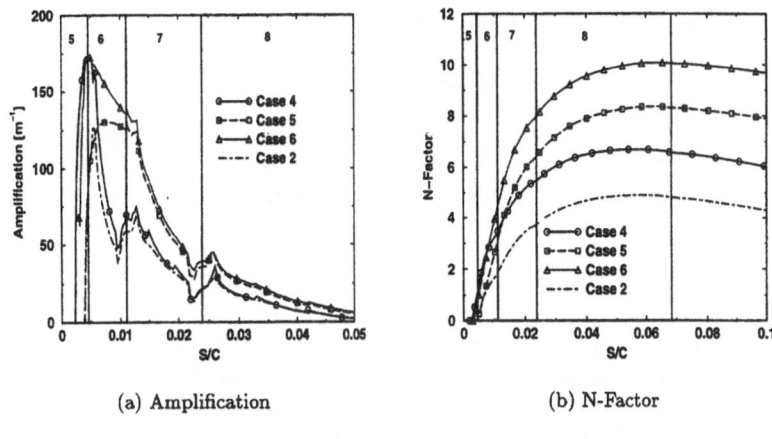

(a) Amplification

(b) N-Factor

Figure 5: Comparison Case 4,5,6,2

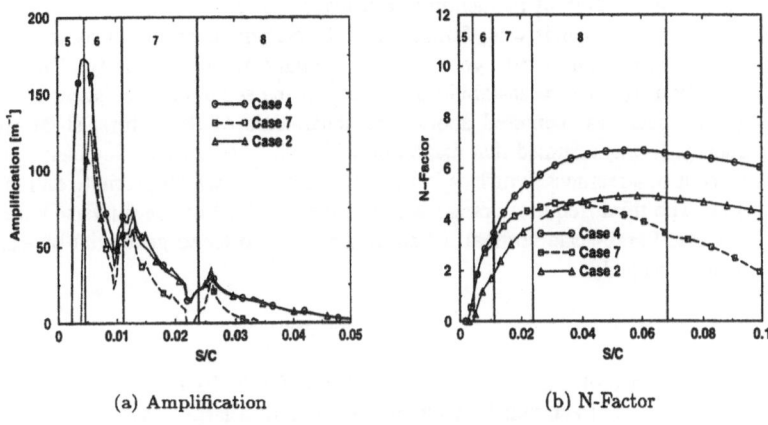

(a) Amplification

(b) N-Factor

Figure 6: Comparison Case 4,7,2

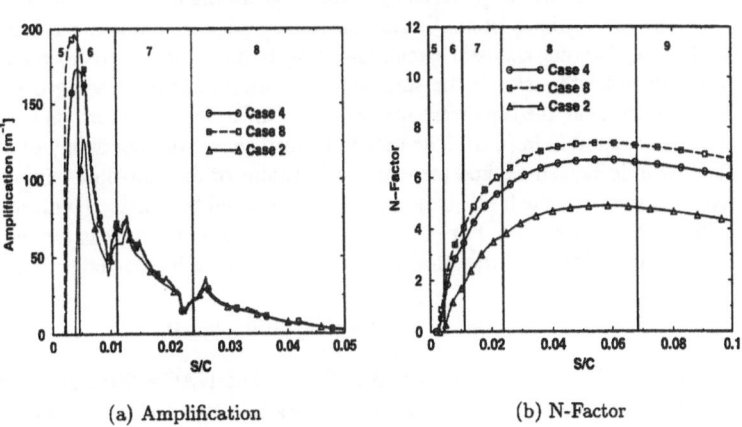

(a) Amplification

(b) N-Factor

Figure 7: Comparison Case 4,8,2

Control of flow separation by dynamic excitation of the free shear layer

Viktor Bader, Friedrich-Reinhard Grosche

DLR Institut für Strömungsmechanik
Bunsenstraße 10
37073 Göttingen, Germany

Summary

Reattachment of a fully separated flow at a NACA 0012 airfoil at angles of incidence above the critical angle could be achieved by periodic suction and blowing through a narrow slot close to the leading edge of the airfoil. A small cavity below the slot was connected to a horn-driver providing periodic pressure fluctuations to drive the flow through the slot. The present research confirms the results of previous studies [1] that the mechanism for such a control consists of stimulation of momentum exchange across the separated free shear layer. Three optimum frequency bands were found where the periodic excitation was very effective (i.e. where low sound pressure levels were required to obtain an attached flow). The first frequency band was about the natural instability frequency of the separated free shear layer. The second frequency band was centered about the subharmonic of the natural shear layer instability frequency. It was assumed that excitation within these two frequency bands promotes the development of streamwise vortices in the free shear layer which promote entrainment of the outer flow. The third frequency range was below about 1/10 of the fundamental free shear layer instability. Large-scale vortical structures are believed to be generated by excitation within this frequency band.

Introduction

A significant problem of the operation of an airfoil is the leading-edge separation occurring at high angles of attack which results in formation of a large-scale separation region, so that the lift cannot increase further with the angle of attack. Our objective is to find control mechanisms which are as effective as possible to attain a higher critical angle of incidence. This paper demonstrates two methods of controlling the leading-edge separation: First, according to previous studies [1] internal acoustic excitation using horndrivers to create pressure pulsations in a cavity underneath a slot in the surface of the airfoil near the leading edge. The optimum frequencies to excite the flow were investigated. Similar investigations on leading edge separation have been reported in [6-8]. The second method shows the possibility of attaining an attached flow by utilizing self-excited resonance-oscillations of the slot-flow without any kind of external energy source. Such self-induced resonances could be obtained with suitable configurations of the slot-cavity system. For both methods the separated flow was excited by periodic suction and blowing through narrow slots in the surface near the leading edge of the airfoil.

Experimental Set-Up

A sketch of the test set-up is shown in Fig.1a. The NACA 0012-airfoil is mounted between two end-plates to obtain optimal two-dimensional flow conditions. The chord length of the airfoil is c = 30 cm and the span is also 30 cm. Pressure holes on the upper side of the model in the symmetry plane are used to measure the pressure distribution. A cylindrical cavity is placed inside the model near the leading edge (Fig. 1b). It has a diameter of 1 cm and a length of 30 cm and was oriented in spanwise direction. Pressure oscillations inside the cavity cause periodic suction and blowing through a slot. This can excite flow instabilities. The pressure fluctuations within the cavity can be generated by an acoustic horndriver connected to both sides of the cavity. The slot has a length of 12 cm and is 0,5 mm wide. It is placed at 5 % of the chord length and symmetrically in spanwise direction. We have found that pressure oscillations can also be generated by self-excited resonances, if the cavity was closed except for the slot and if the flow velocity was high enough. For the experiments with self-excited resonances a second slot at 1,3 % of the chord length is added. It is connected to the same cavity. The sound-pressure level A was measured always without flow by a microphone placed 2 mm above the slot in the plane of symmetry. A single hot-wire probe mounted on a traversing mechanism was used to measure the flow perturbations. The coordinate x' is the distance from the leading edge along the chord of the model, c is the chord length.

Results and Discussion

Fig. 2 shows the variation of the pressure distribution on the airfoil with increasing sound pressure level A. The excitation frequency is 892 Hz and the angle of attack = 15° at $Re = 1,4 \times 10^5$. At this angle of attack and Re-number, the flow is separated completely from the airfoil under natural conditions as shown by the curve without excitation with A = 0 dB. The curve with the sound pressure level A = 95,4 dB exhibits the pressure distribution with the flow reattached again to the model. This curve shows a suction peak near the leading edge instead of a flat distribution. The curve also indicates, with a high plateau of -Cp near the leading edge, the appearance of a separation bubble. With further increase of the sound pressure level A, only a small increase of the negative pressure coefficient takes place. An interesting result is a strong hysteresis with increasing and decreasing sound pressure level. Fig. 3 shows the dependence of the pressure at the chord position x'/c = 0,013 on increasing and decreasing sound pressure level. The highest value of -Cp in the case of attached flow was always found at this pressure hole (compare Fig. 2). In the range 83 dB ≤ A ≤ 93 dB, it depends on the history of the flow whether it is separated or attached. If one comes from a high value of A, the flow remains attached until about 83 dB, although it becomes attached first at about 93 dB if one comes from a low value of excitation. A similar behavior exhibits the Fig. 5 for the excitation frequency of 430 Hz. But, the hysteresis loop is much smaller in this case of lower excitation frequency. Fig. 4 shows, for three different angles of attack, the minimum sound pressure level A required to obtain attached flow as a function of the excitation frequency. In all cases the flow was separated under natural conditions. The slot was again at 5 % of the chord length. It is obvious that there are three frequency bands where only low sound pressure levels are necessary to obtain an attached flow. The first one is about 800 Hz, the second one about half of that and the third one below 1/10 of the first. The highest frequency band is determined by a coupling of the forcing with the linear instability of the separated free shear layer (Kelvin-Helmholtz-Instability). A peak at about 750 Hz is found in the spectra of disturbances in the unexcited separated free shear layer by means of hot-wire measurements. So the second effective frequency band in Fig. 4 is centered at about the subharmonic of the natural free shear layer instability. A reason for the two first effective frequency bands can be the development of streamwise vortices in the free shear layer. In experiments in free shear layers behind splitter plates, structures in the form of spatially stationary counter-rotating streamwise vortices were found in addition to the spanwise coherent structures [3]. It was also found that the vortex merging process of the primary vortices suppresses the activity of the streamwise vortices [4]. However, the further vortex merging process can be delayed by exciting the free shear layer with the natural instability frequency or the first subharmonic. In our case, the streamwise vortices could be responsible for the entrainment of the separated flow by the outer flow and they could deliver a higher momentum flux to the surface than under unexcited conditions. They could thus explain the first two effective frequency ranges. Below the excitation frequency of about 150 Hz, no hysteresis loop like the one in Fig. 3 or 5 was found anymore (compare Fig. 6). The excitation level A given in Fig. 4 for this frequency range is determined by the highest gradient dCp/dA. The third effective, lowest, frequency range could be explained by the development of large-scale vortical structures. In experiments on excited free shear layers behind splitter plates it was found that periodic excitations of the flow with frequencies below 1/10 of the natural instability and with high enough excitation amplitude can cause the formation of large-scale vortical structures. Such an excitation phenomenon was called „collective interaction" [5]. These large-scale vortical structures could deliver a higher momentum flux to the surface and could thus explain the lowest effective frequency range. Fig. 7 compares three effective frequencies for the same sound pressure level of A = 99,6 dB. The plot shows the ratio Cp/C_{nat} of the pressure coefficient with excitation normalized by the pressure coefficient of the flow without excitation as function of the angle of attack. With the excitation frequency of 780 Hz (which is in the range of the highest effective frequency regime) the flow remains attached up to the angle of attack of 15,5° whereas it is separated at the excitation frequency of 430 Hz. The excitation at 90 Hz was the most effective one in the range of 16° to 20°, where large scale vortical structures could probably deliver a higher momentum flux to the surface.

Until now results were presented which were obtained at a Reynolds-Number of $Re = 1,4 \times 10^5$ (which means a flow velocity of U = 7,3 m/s). Above the flow velocity of about U = 8 m/s, self-excited resonances appeared in the configuration of open slot and cavity being closed at both sides (compare Fig. 1b). It was found that the amplitudes of these resonances were high enough to excite the flow sufficiently to control separation. The critical angle of attack and therefore the maximum lift of the airfoil could be increased. Another slot was now positioned at 1,3 % of the chord length, parallel to the first slot, and it was also connected to the same cavity. This slot was also 12 cm long and 0,5 mm wide. Fig. 8 shows the variation of the pressure coefficient at x'/c = 0,013 with increasing flow velocity for different slot configurations, and angle of attack = 15,5°. Under natural conditions the flow became attached at about 14 m/s (or $Re = 2,8 \times 10^5$), as indicated in the curve with the closed slot by a sudden increase of the negative pressure coefficient. The other curves with different slot configurations show a completely different behavior. The flow is already attached in each case at lower flow velocities. The most effective configuration seems to be the combination of both open slots. An explanation for this is that the configuration of one open slot at 5 % of the chord length develops much higher sound pressure levels (> 120 dB in the cavity) of the self-excited resonances (for attached flow) than in the case of one open slot at 1,3 % of the chord length (< 110 dB in the cavity). On the other hand the most effective position to excite the flow is near the separation line [2]. Therefore the configuration „both slots open" should be the most favorable one here because the excitation happens very close to the separation position of the flow with a high sound pressure level. In all cases, several resonance frequency peaks were found in the range of 500 to 6000 Hz [2].

References

[1] Dovgal, A. (1993); „Control of leading-edge separation on an airfoil by localized excitation", DLR-FB 93-16.
[2] Bader, V. (1997); „Steuerung der Vorderkantenablösung an einem Tragflügelprofil mittels interner Schallanregung", Diplomarbeit, Universität Göttingen.

[3] Leboeuf, R.L. (1996); Mehta, R.D.; „Vortical structure morphology in the initial region of a forced mixing layer: roll-up and pairing", Journ. Fluid Mech. 315, pp. 175-221.

[4] Huang, L.-S.; Ho, C.-M. (1990); „Small-scale transition in a plane mixing layer", Journ. Fluid Mech. 210, pp. 475-500.

[5] Ho, C.-M.; Huerre, P. (1984); „Perturbed free shear layers"; Annual Review of Fluid Mechanics 16, pp. 365-424.

[6] Huang, L.S.; Maestrello, L.; Bryant, T.D. (1987); „Separation control over an airfoil at high angles of attack by sound emanating from the surface"; AIAA-Paper 87-1261.

[7] Hsiao, F.-B.; Liu, C.-F.; Shyu, J.-Y. (1990); „Control of wall-separated flow by internal acoustic excitation"; AIAA Journ. 28, pp. 1440-1446.

[8] Erk, P.P. (1997); „Separation control on a post-stall airfoil using acoustically generated perturbations", Number 328 in Fortschr.-Ber. VDI Reihe 7. VDI Verlag, Düsseldorf.

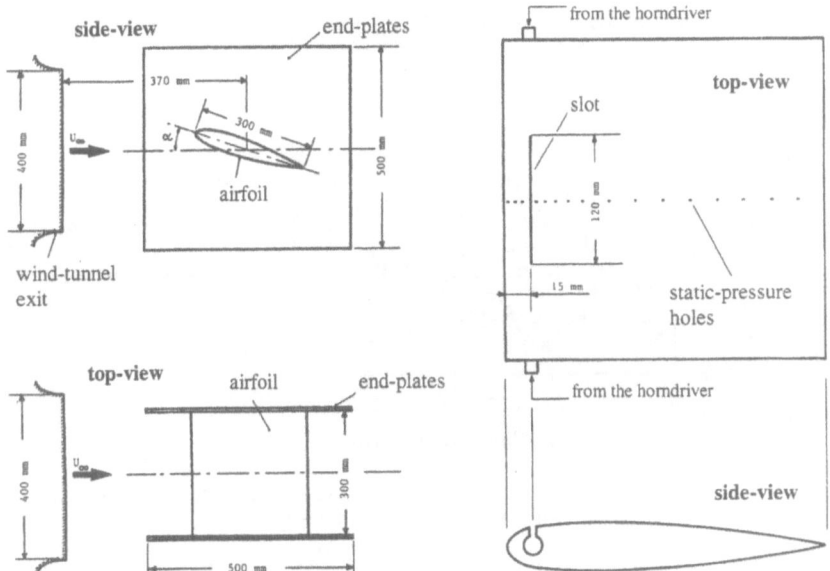

Fig. 1: a) Schematic sketch of the test set-up at the „Small Low Turbulence Wind Tunnel" with open test section of the DLR in Göttingen;
b) Schematic sketch of the NACA 0012-airfoil with a slot at 5 % of the chord length

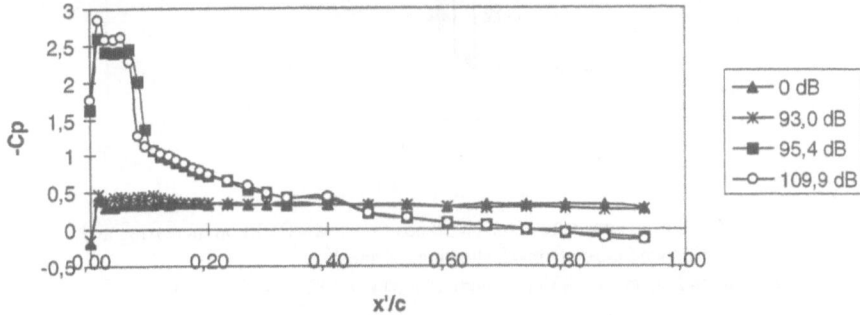

Fig. 2: Dependence of the pressure distribution on increasing sound pressure level A, with a slot at 5 % of the chord length. Excitation frequency = 892 Hz, angle of attack = 15°, Re = $1,4 \times 10^5$

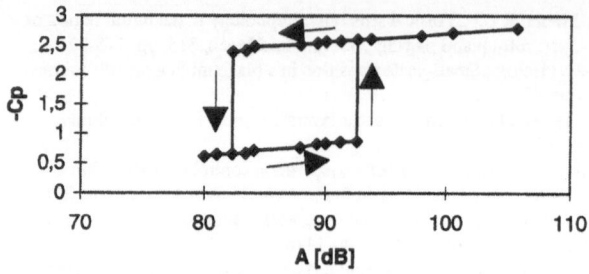

Fig. 3: Dependence of the pressure at the chord position x'/c = 0,013 on increasing and decreasing sound pressure level, showing strong hysteresis.
Excitation frequency = 892 Hz, angle of attack = 15°, Re = 1,4 × 10⁵

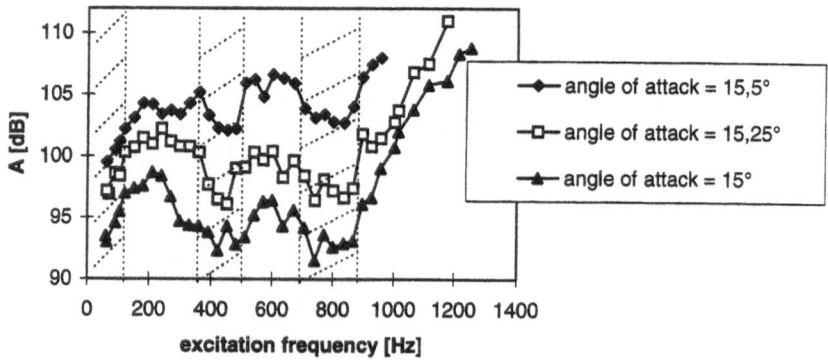

Fig. 4: The minimum sound pressure level A to obtain attached flow, as function of excitation frequency. Slot at 5 % of the chord length, Re = 1,4 × 10⁵

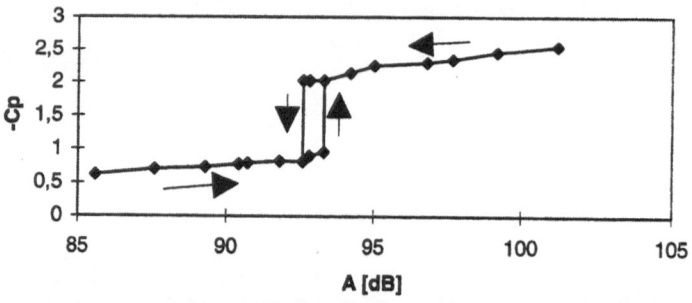

Fig. 5: Dependence of the pressure at the chord position x'/c = 0,013 on increasing and decreasing sound pressure level, showing hysteresis.
Excitation frequency = 430 Hz, angle of attack = 15°, Re = 1,4 × 10⁵

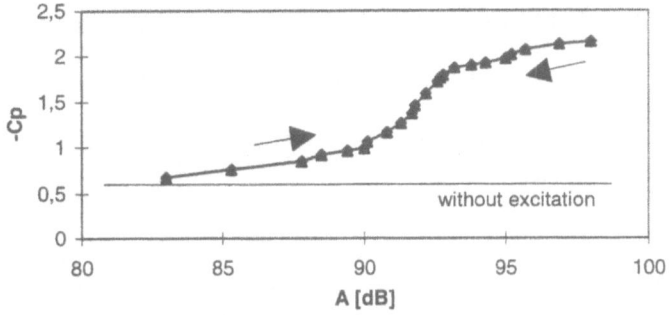

Fig. 6: Dependence of the pressure at the chord position x'/c = 0,013 on increasing and decreasing sound pressure level, showing no hysteresis. Excitation frequency = 90 Hz, angle of attack = 15°, Re = 1,4 × 10⁵

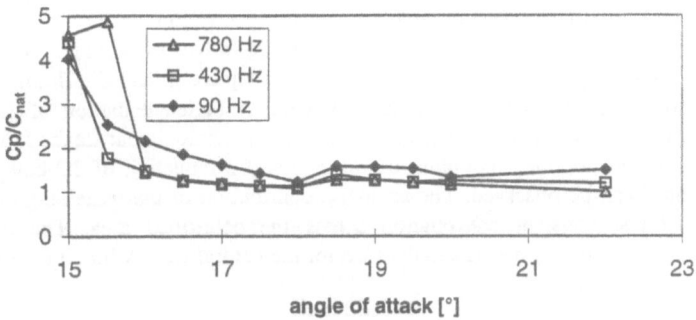

Fig 7: Dependence of the pressure at chord position x'/c = 0,013 on angle of attack for different excitation frequencies at constant sound pressure level of 99,6 dB, Re = 1,4 × 10⁵

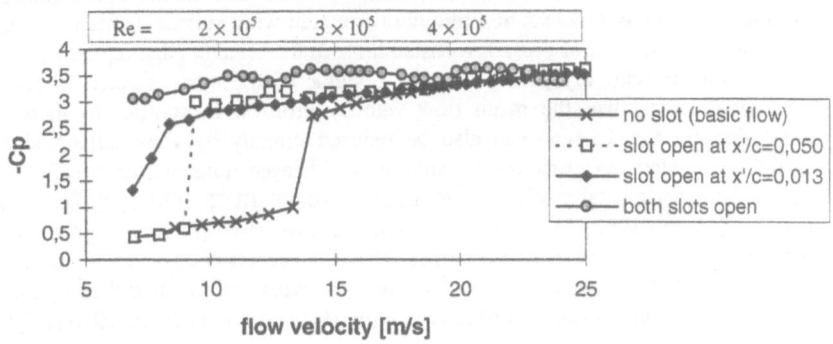

Fig. 8: Variation of the pressure at x'/c = 0,013 with increasing Reynolds-Number for different slot configurations, angle of attack = 15,5°

On Active Control of Boundary Layer Instabilities on a Wing

M. Baumann, D. Sturzebecher, W. Nitsche

Technische Universität Berlin, Institut für Luft- und Raumfahrt, Sekr. F2,
Marchstr. 14, 10587 Berlin, Germany

Summary

Different methods are known to stabilize a laminar boundary layer on a wing in order to achieve a delay of the laminar-turbulent transition resulting in a reduction of the viscous friction drag. One less known approach is the reduction of natural occurring instability waves, so called Tollmien-Schlichting-waves (TS-waves) of a laminar boundary layer by means of superposition with artificially excited canceling waves. For technical applications of this active control method, sensors, actuators and control devices must be integrated in the wing surface. The control devices are processing the sensor signals and drive the actuators to excite the required canceling waves. This method is investigated in basic research on an unswept wing in a low speed wind tunnel at the TU-Berlin since several years. In the present paper an arrangement of 16 surface sensors and an actuator to excite a 2D-canceling-wave is successfully used for attenuation of natural occurring TS-waves on a wing. The actuator is driven by a digital signal processor, executing an adaptive filter algorithm to perform feedforward control of the wave cancellation process. A high attenuation of TS-waves is obtained not only in the early linear stage, but also in the weak nonlinear stage of the TS-wave development. Due to nonlinear coupling of wave modes a reduction of 3D-components by purely 2D-control can be observed. The spanwise distribution of the remaining disturbances indicates an almost constant cancellation across the controlled area. The downstream amplification of attenuated waves is smaller than for the natural wave without control.

Introduction

A laminar boundary layer at high Reynolds numbers of transport aircraft wings can be maintained only by artificial methods, reducing the natural occurring instabilities. The recent research work is primary focused on an indirect reduction of instabilities by modification of the mean velocity profile of the laminar boundary layer. Here the most promising method is a distributed boundary layer suction through perforated wing surface areas, which is especially successful in reduction of crossflow instabilities in the leading edge region of a swept wing. For an aircraft with cryogenic fuel, the cooling of the wing surface is additionally in discussion, to stabilize the mean flow velocity profiles. In contrast to these techniques, instabilities of the TS-type can also be reduced directly by wave superposition with an artificially excited canceling wave resulting in a delayed transition as already demonstrated e.g. in the experimental works of LIEPMANN & NOSENCHUCK 1981 [10], PUPATOR & SARIC 1989 [11], LADD 1990 [9], GROSCHE & YONG-GUANG 1990 [5]. However, the excitation of artificial waves to cancel natural occurring TS-wave packets is difficult, but can effectively be managed employing a sensor actuator system in the wing surface controlled by a digital signal processor executing an adaptive filter algorithm (BAUMANN & NITSCHE 1996 [1,2]).

Wing with sensors, actuator and adaptive controller

The present investigation was carried out on an unswept test wing (modified NACA 0008 airfoil, c=1300 mm) at a free stream velocity of 17m/s. The principle sketch in figure 1 shows the wing with the sensor-actuator-system in the mid chord region. The actuator to excite the canceling wave is located at x/c=0.46 resulting in a Reynolds number of $Re_x=0.7\cdot10^6$ at the actuator. The actuator acts on the boundary layer by periodic suction and blowing through a narrow spanwise slot in the wing surface. Vibrating membranes in the wing surface were successfully tested as an alternative actuator type, but the most experience was collected at the time with the suction blowing slot.

At least one sensor is required upstream of the slot to measure the traveling TS-waves and a second one downstream to control the remaining disturbances and thus the success of the wave cancellation. The controller generates the actuator signal by adaptive filtering (FIR-filter, FIR - Finite Impulse Response) of the upstream sensor signal (disturbance input). The downstream sensor delivers an error signal to be minimized (least mean square - LMS) by adaptation of the FIR-filter. The linear FIR-Filter is a good choice to build an internal model of the physical transfer function between sensor and actuator, because the downstream coherence of the TS-wave development is relatively high compared to the disturbance development in turbulent flows. TS-wave reduction up to 90% could be obtained using the described feedforward LMS-adapted FIR-Filter. The applied algorithm is known from the literature (ELLIOTT 1993 [4]) as filtered-x-LMS or filtered-LMS, because it requires an internal model (second FIR-Filter) of the secondary signal path (error path) between actuator and downstream sensor to perform the LMS-adaptation with a filtered and correlated error signal. This internal reference signal is obtained by multiplying the error signal with the filtered disturbance signal, employing the FIR-filter of the error path.

The newly applied surface sensors of the present investigation are basing on the hot-wire technique. Their basic principle is similar to cavity hot-films, where a reduced heat flux into the wall results in a better signal-to-noise ratio and sensitivity. These flush mounted hot-wires with a slot beneath (0.1 mm wide) were our best choice of known sensors for low speed experiments. The mechanical and thermal roughness effects are negligible, especially compared to the actuator slot (0.3mm wide), which causes a small roughness effect.

To investigate the 2D- and 3D-stage of TS-waves under 2D-control, small spanwise arrays of the cavity hot-wire sensors, as indicated in figure 2, were used instead of single sensors in our previous experiments [1,2]. Each array consists on 8 sensors with a spanwise spacing of 14 mm, which is approximately half the 2D-TS-wavelength under the chosen Reynolds number. To obtain a single input signal for the controller, the signals of each array are averaged, representing only the 2D-part of the TS-waves. With these spanwise averaged sensor signals it was possible to achieve very clean 2D-control with almost constant residual disturbance amplitudes at least across the spanwise dimension of the sensor arrays (98 mm).

Experimental Results of TS-Wave Control

Typical signals of the sensors and the actuator without control and under active wave control (AWC) are shown in figure 3. The left signals indicate the downstream amplification of the natural TS-wave, if the different scaling is taken into account. The right signals are obtained under control, as already indicated by the actuator signal. Here the downstream control sensor indicates a very high attenuation compared to the case without control. The power spectra of the control sensor (figure 4) shows an attenuation of the fundamental TS-disturbances (frequency range: 250-400 Hz) by up to 20 dB (reduction by 90%). Comparative good values are also obtained in the higher harmonic frequency range (550-800 Hz) and can be explained by coupling to the fundamental disturbances.

The downstream amplification of the TS-waves (figure 5) indicates smaller amplitudes due to control and furthermore a reduced amplification of the residual disturbances, as indicated by the different slopes of the approximation lines. Since there is no change in the mean velocity profiles (figure 6) due to active wave control, this effect can be explained by the fact, that the dominant 2D-part of the TS-waves is almost completely removed by control. The remaining fluctuations are 3D-dominated with a lower amplification. A similar behavior was already found in the numerical simulations of LAURIEN & KLEISER 1985 [6,7] and KRAL & FASEL 1989 [8]. The downstream shifted amplitude growth suggests a transition delay of around 120 mm. The u'-fluctuations of the diagram were measured with a hot-wire in constant wall distance of 1 mm and the dimensionless values are u'_{RMS} referenced to the free stream velocity, which differs only slightly from the velocity of the boundary layer edge. The velocity profiles of figure 6 and the u'-profiles (velocity fluctuations) of figure 7 are also obtained from these hot-wire measurements and the u'-component is scaled in the same way. Already by definition, active TS-wave control does not affect the mean velocity profile of the boundary layer. The observable differences in figure 6 between the controlled and uncontrolled case are probably measurement uncertainties. But the u'-profiles are indicating a significant change due to control across the whole boundary layer thickness. At the last x-position, after a certain downstream development, a reduction of the u'-maximum by factor 5 is obvious. The measured u'-components are scattering, but the typical u'-eigenfunction of TS-disturbances can be recognized. The increasing values next to the wall, below y=0.5 mm, are caused by surface roughness of the sensors and especially of the actuator slot, because this phenomena was not observed during measurements on a clean wing surface.

To get an impression of local effects close to the slot during TS-wave cancellation, phase coupled hot-wire measurements with high spatial and temporal resolution were conducted. For these measurements a test case with artificially excited sinusoidal TS-waves was chosen. A reduced angle of attack with very low natural disturbances was adjusted and the TS-wave excitation was performed using an additional excitation slot in the wing surface at x/c=0.2. The hot-wire measurements were triggered with the excitation sine (300 Hz). The obtained data set can be animated to suggest a simultaneous measurement across the whole flowfield. One timestep of the u'-field is shown as iso-surface plot in figure 8. The downstream growing TS-wave is visible until x=600 mm, where the actuator is located. The typical 180° phase jump of the boundary layer motion can be observed around y=2mm, indicated by the sign reversal of the u'-component along the wall distance y. The sign reversal, respectively the values of u'=0 are marked with a thin black line. The antiphase wave, introduced at x=600 mm, causes an immediate cancellation across the whole boundary layer thickness. Already 10 mm downstream of the slot, minimized disturbances can be observed, which grow downstream again. After a very short relaxation distance of 30 mm (at x=630 mm) the typical u'-distribution is recovered with reduced amplitudes. The dimensions have to be compared to the TS-wavelength of λ_x=22 mm. It can additionally be observed that the actuator also produces an upstream enhancement of the disturbances in the vicinity of the slot. This upstream effect of the actuator is of physical nature and is typical for wave cancellation. The effect can be compared with any rapid change in a subsonic flowfield, where the flow prevents discontinuities by an upstream pressure field, which influences the upstream flow field.

To find the limits and the best operating conditions for our active wave control system, the attenuation was measured for different angles of attack under constant freestream velocity (figure 9). The 8 downstream sensor signals were simultaneously sampled to allow a calculation of a spanwise RMS-Value (3D-RMS) and a downstream RMS-Value (2D-RMS) to quantify the 3D- and 2D-part of the TS-waves. The spanwise- or 3D-RMS is the time averaged spanwise RMS-value of the sensor signals and quantifies the three-dimensionality of the passing wave. The value is zero for a plane 2D-wave, where all spanwise sensors indicate the same amplitude and phase. In contrast, the downstream- or 2D-RMS is the RMS-voltage of the previously spanwise averaged signal samples and represents the two-dimensional part of the wave. The diagram (figure 9) shows the attenuation ratios (ratio of the corresponding

RMS-voltages between the uncontrolled and controlled case) divided in 2D- and 3D-attenuation. The additional total attenuation is the ratio of the RMS-values of all signals. The data are plotted over the local pressure gradient in the control region, instead of the angle of attack. An increasing adverse pressure gradient causes an increasing instability of the boundary layer with higher disturbances. For the 2D disturbances the best attenuation of 5 is found around $\Delta cp/\Delta x=0.26$. For further increased $\Delta cp/\Delta x$ a drastic performance loss was observed. This limitation occurs due to instabilities of the controller caused by feedback of the actuator to the upstream sensors. Feedback problems of feedforward adaptive filter controllers are also a general problem of recent active noise control applications, which can be also solved with modifications of the control algorithm.

The natural disturbances at low pressure gradients are small and the signal-to-noise ratio of the system is the other limiting factor of the attenuation. The disturbance amplitudes of the shown operating range are comparatively small ($u'_{AVG}=0.07\%$ up to $u'_{AVG}=0.18\%$), measured at the actuator position without control and averaged across the boundary layer. The u'-maximum is approximately two times the average value. No non-linear effects, like mode coupling were expected at these low amplitudes and almost no 3D-attenuation was observed here. The total attenuation has a maximum of only 2 due to these not affected 3D background disturbances.

To indicate that the stability limit of the controller is primarily related to boundary layer stability causing feedback, and only secondary to the disturbance amplitude itself, an amplitude variation of artificially excited 2D-TS-waves was carried out under constant angle of attack. The operating point with the smallest pressure gradient of figure 9 was chosen and the disturbance level was varied by means of random noise excitation employing a second slot type actuator upstream of the control region at $x/c=0.2$. The plot of figure 10 shows the obtained attenuations over the disturbance amplitude, measured at the control actuator position without control. The indicated u'_{AVG} values are here again averaged across the boundary layer profile. During these measurements no feedback instabilities were observed at all. The best 2D-attenuation of 8 was obtained for disturbance amplitudes around $u'_{AVG}=0.15\%$. For higher amplitudes the 2D-attenuation decreases due to increasing non-linear development of the TS-waves. But in the weakly non-linear stage a surprisingly high 3D attenuation of 2 was observed, indicated by the maximum of the 3D-line (figure 10). The non-linear interaction of 2D- and 3D-modes enables 3D-attenuation by 2D-control. For higher amplitudes the nonlinearities become too strong to be effectively controlled. The attenuation of the total disturbances has a maximum between the 2D- and 3D-case, indicating that the non-linear 3D effects are positive up to a certain stage and the best total attenuation is already in the weak non-linear stage. Usually best wave cancellation by superposition is expected in the early linear stage under theoretical conditions. Non-linear effects of mode coupling under TS-wave control were already found in numerical investigations of BIRINGEN 1984 [3], KLEISER & LAURIEN 1985 [7], and KRAL & FASEL 1989 [8]. The numerical results also indicated a reduction of 3D-waves under pure 2D-control.

Figure 11 shows an example of signals of the downstream sensor array in the weakly non-linear stage, where the best total attenuation was obtained. Without control they are still dominated by 2D-wave packets, whereas in the controlled case only small 3D-dominated disturbances remain. This effect can more detailed be observed, if the wavenumber frequency spectrum is calculated, as shown in figure 12. The RMS-voltages of the sensor signals are shown in logarithmic gray scale (dB V_{RMS} referenced to 1 V_{RMS}). The spanwise wavenumber k_z is the wavelength λ_z referenced with twice the sensor distance $\Delta z=28$ mm, which is approximately a 2D-TS-wavelength. The 2D-wave is here only represented by a gray scaled line at $k_z=0.0$. The most attenuated modes in the control case are below $k_z=0.2$ (small propagation angles), but generally all 3D-modes are affected by 2D-control, also wave modes with large propagation angles ($k_z=0.5 - 1.0$). An attenuation of all modes can be identified also for the higher harmonic frequency range (500 Hz - 800 Hz).

Conclusions

The described method of TS-wave control was very successful in low speed experiments, resulting in an attenuation of the TS-waves by up to 90%. Two additional gaining effects were investigated to show that TS-wave control can be more effective than linear wave superposition would suggest. An attenuation of 3D-TS-wave modes by purely 2D-control was examined, which can be explained with non-linear effects, like mode coupling of TS-waves, as to be expected in the weakly non-linear stage. The second, more important gaining effect is the reduced downstream amplification of the actively attenuated TS-waves. This can be explained by the drastic change of the mode composition in the control region. The 2D-part of the TS-mode spectra is highly reduced by active control and the 3D-modes are the major part of the remaining TS-waves, which have a lower downstream amplification than the 2D-mode. The investigation of the flowfield close to the actuator indicated a very short relaxation distance of 1.5 wavelengths downstream of the actuator until a natural like TS-wave remains and also indicated a typical upstream effect of the actuator related to the wave cancellation. The operating range of active wave control was limited in the natural case by feedback of the actuator to the upstream sensor, causing unstable operation of the control algorithm. For this reason our present work is focused on feedback reduction to avoid this limitation.

Acknowledgments

The work on active wave control was performed under financial support of the DFG (Deutsche Forschungsgemeinschaft) in the priority research program 'Transition'.

References

[1] Baumann, M., Nitsche, W., 1996, Investigations on Active Control of Tollmien Schlichting Waves on a Wing, Transitional Boundary Layers in Aeronautics, Ed's: Henkes, R.A.W.M., van Ingen, J.L., KNAW, Verhandelingen 46, North Holland, Amsterdam 1996, 89-98.

[2] Baumann, M., Nitsche, W., 1996, Experiments on Active Control of Tollmien-Schlichting Waves on a Wing, New Results in Numerical and Experimental Fluid Mechanics, Ed's: Körner, H., Hilbig, R., Braunschweig 1996, 56-63.

[3] Biringen, S., 1984, Active control of transition by periodic suction-blowing, Physics of Fluids 27(6), June 1984, 1345-1347.

[4] Elliott, S.J., Nelson, P.A., 1993, Active Noise Control, IEEE Signal Processing Magazine, October 1993, 12-35.

[5] Grosche, F.R., Yong-guang, T., 1990, Experimente zur Dämpfung von Tollmien-Schlichting-Wellen durch aktive Anregung von Wandschwingungen, IB 222-90 A 46 (internal report of the DLR-Göttingen, Germany).

[6] Kleiser, L., Laurien, E., 1984, Three-Dimensional Numerical Simulation of Laminar-Turbulent Transition and its Control by Periodic Disturbances, IUTAM Symposium on Laminar-Turbulent Transition (2nd. 1984, Novosibirsk, RSFSR) Springer-Verlag Berlin, Heidelberg 1985, 29-37.

[7] Kleiser, L., Laurien, E., 1985, Numerical Investigation of Interactive Transition Control, AIAA Paper 85-0566.

[8] Kral, L. D., Fasel, H., 1989, Numerical Investigation of the Control of the Secondary Instability Process in Boundary Layers, AIAA Paper 89-0984, AIAA 2nd Shear Flow Conference, March 89, Tempe, AZ, USA.

[9] Ladd, D.M., 1990, Control of Natural Laminar Instability Waves on an Axisymmetric Body, AIAA Journal, 28, 367-369.

[10] Liepmann, H.W., Nosenchuck, D.M., 1981, Active control of laminar-turbulent transition, J. Fluid Mech., 118, 201-204.

[11] Pupator, P.T., Saric, W.S., 1989, Control of random disturbances in a laminar boundary layer, AIAA Paper 89-1007.

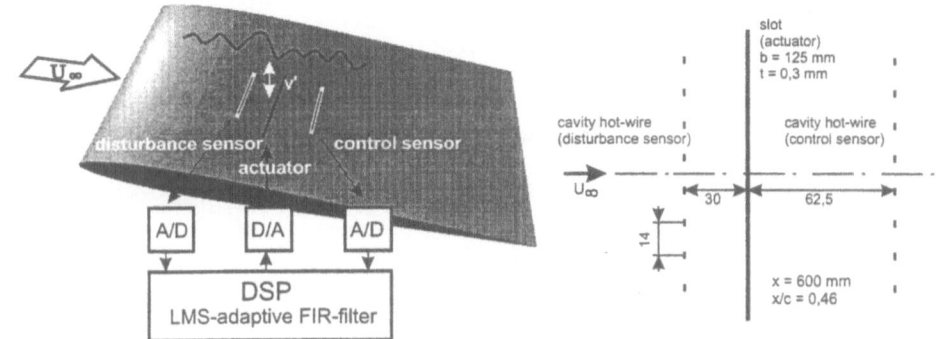

Figure 1: Principle of active wave control (AWC)

Figure 2: Sensor-actuator-layout

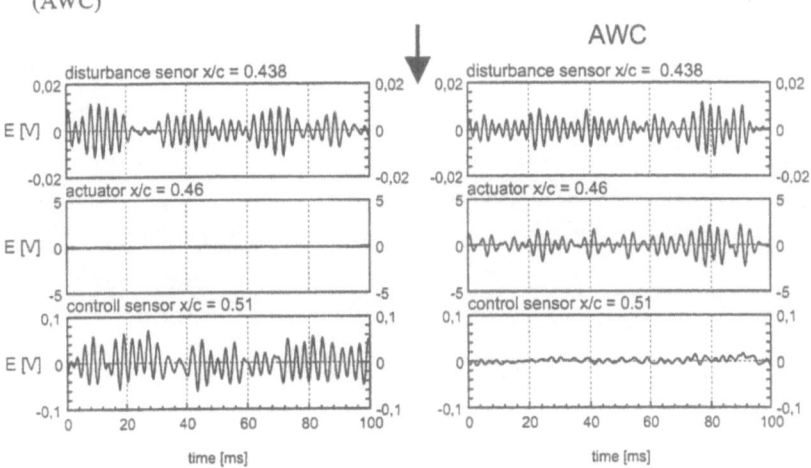

Figure 3: Signals of the sensor-actuator-system without and with AWC

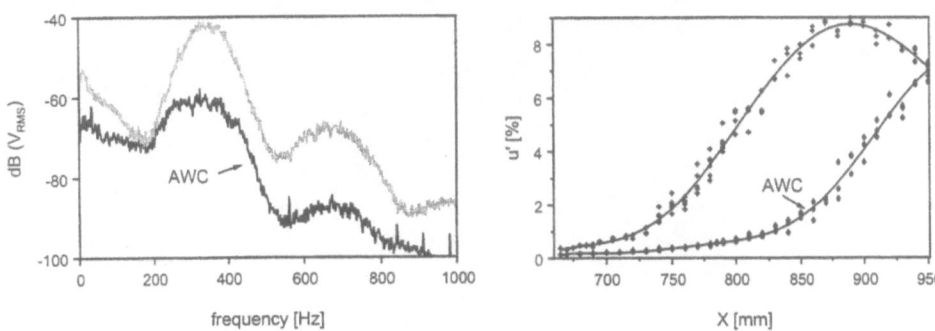

Figure 4: Power spectra of the downstream control sensor

Figure 5: Downstream amplification of u'- turbulence at constant wall distance of 1 mm

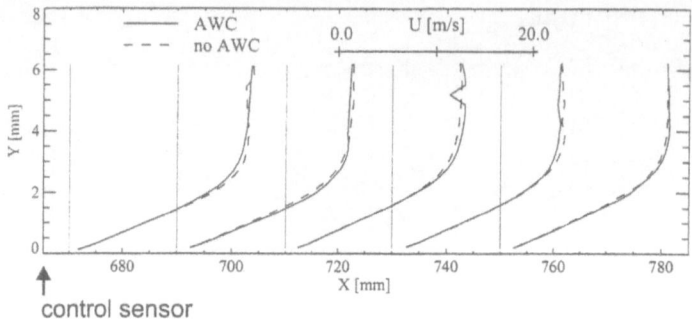

Figure 6: Velocity profiles downstream of the control sensor

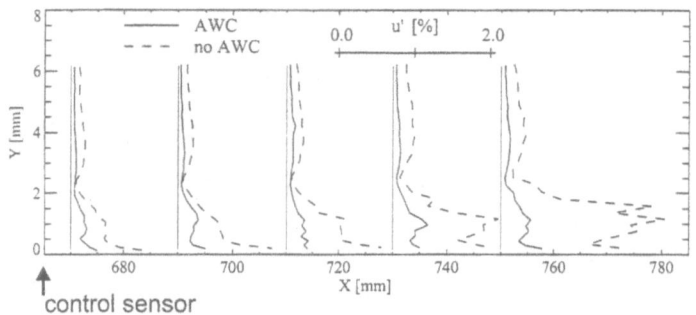

Figure 7: Downstream development of the u'-profiles

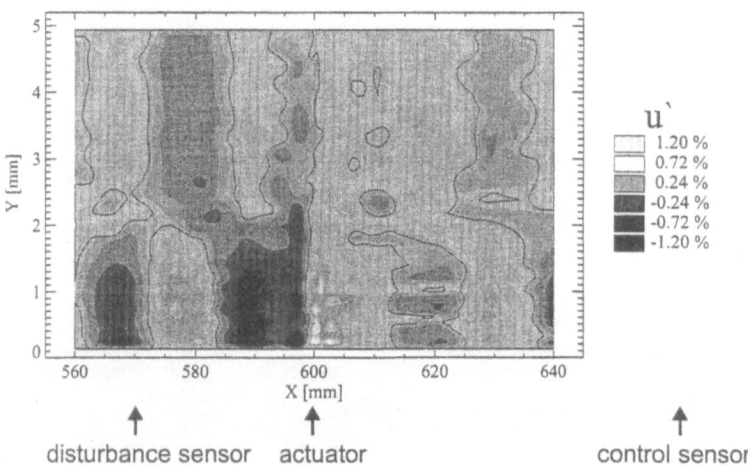

Figure 8: Instantaneous perturbation velocity field of the sensor-actuator-system

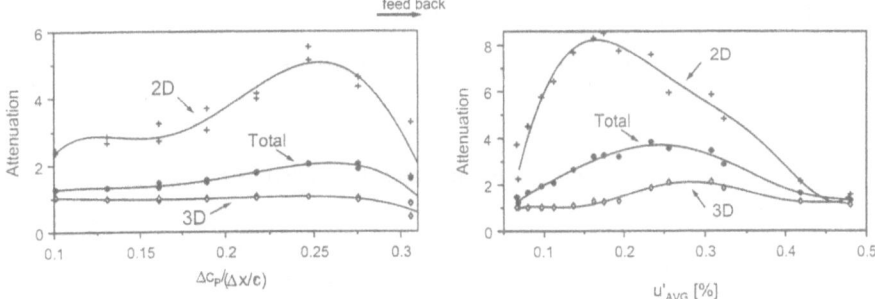

Figure 9: Attenuation of 2D- and 3D-parts versus local pressure gradient (variation of the angle of attack)

Figure 10: Attenuation of 2D- and 3D-parts versus averaged u'(amplitude variation of the random noise excitation)

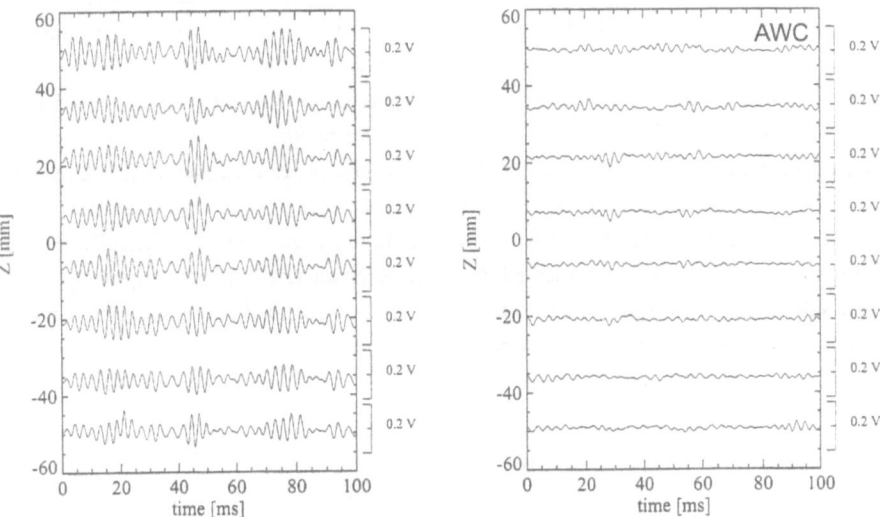

Figure 11: Signals of the spanwise control sensor array

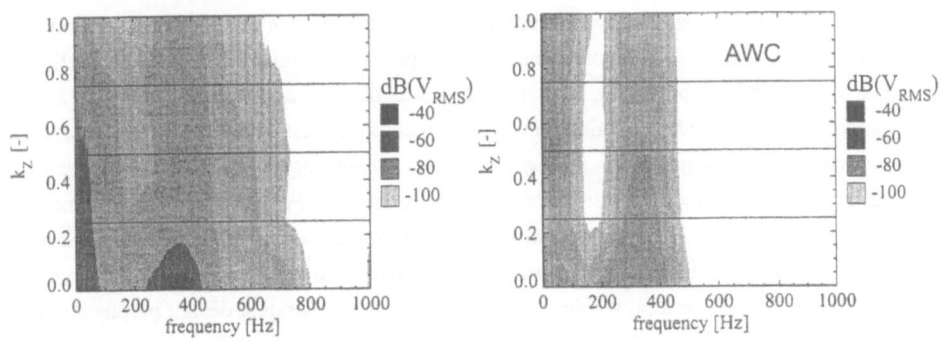

Figure 12: Frequency-wavenumber-spectra of the spanwise control sensor array

In-flight Boundary Layer Investigations on a Laminar Wing using Laser Doppler Anemometry

S. Becker, F. Durst and H. Lienhart

LSTM Erlangen, Institute of Fluid Mechanics, University of Erlangen-Nuremberg,
Cauerstrasse 4, D-91058 Erlangen, Germany

Summary

The development of laminar wing technology requires strategies that combine wind tunnel studies with numerical flow simulations and in-flight measurements. The latter topic is covered by the present paper and it is outlined that instruments are needed to measure local flow velocities in free flight experiments. It is shown that Laser Doppler Anemometry (LDA) is well suited to provide the required local velocity information in the boundary layer but specially adapted LDA systems need to be developed. The paper describes an LDA system constructed for the specific purpose. Free flight measurements studying boundary layer transition on a aeroplane wing downstream of an excitation source were successfully carried out and a summary of results is presented. In order to characterise the development of the Tollmien-Schlichting wave spectral and harmonic analysis were adopted. Finally, suggestions for further advancements of the LDA systems are proposed.

Introduction

Research and development work in aircraft aerodynamics has been heavily based on wind tunnel investigations so far, although it has been known that the test facilities employed did not provide the required flow conditions to yield design information which is directly applicable. The size of the wind tunnel test sections employed were usually too small and the velocities too low to yield Reynolds numbers comparable to those in free flight. Reliable methods to predict the boundary layer transition are not yet available for realistic flight environments and the identification and quantification of the relevant parameters, like pressure gradient, wing geometry, surface curvature, sweep angle, Mach number, heat and mass transfer at the surface, etc. are far from being complete. For the studies of laminar to turbulent transition, individual wind tunnels inherently introduce their own specific spectrum of flow disturbances and hence, their laminar to turbulent transition data can be effected by 'wind tunnel noise'. Therefore, results of wind tunnel studies yielding design parameters for laminar wings have to be verified by flight experiments. For this task there is a pronounced need for advanced measuring techniques that should be preferably nonintrusive and must be reliable in the harsh environment of in-flight tests. Several of those techniques have been developed in past years, e.g., hot film arrays, piezo foils, and infra red camera for transition detection. All of these techniques are only capable of acquiring information directly on the wall surface, whereas LDA may give an insight into the complete boundary layer and the surrounding flow velocity field.

LDA- system

The test aeroplane for the present studies consisted of a GROB 109B, a two seated powered sail plane equipped with a 'wing glove' on its starboard wing. The size and the design of the plane imposed very stringent restrictions on weight and space available to the optical and electronical components as well as on the power consumption of the measuring system. This situation and the demand for a forward scattering arrangement which was adopted for giving maximum signal power from the very small scattering particles available as natural aerosols in the atmosphere resulted in an unconventional probe design. It employed the narrow gap between wing and wing glove for the optical components and the beam path was diverted twice by mirrors for both the transmitting and the receiving path. As sketched in Figure 1 all optics apart from the upper mirrors were placed underneath the wing glove surface and only these mirrors are protruding. Therefore, the distortion of the flow induced by the measuring system was minimal. The optics were mounted on a traversing mechanism that allowed automated measurements of boundary layer profiles. The compact design of the probe, the integral machining of its mechanical structure from titanium and the way it is clamped to the wing glove ensured mechanical and optical stability and, thus, prevented misalignment during flight experiments due to vibration, bending of the wing and stresses due to changes in temperature. Figure 2 shows the LDA system in actual flight tests.

The laser, traversing controller, photomultiplier, power supplies, and signal acquisition and processing unit completed the measuring system. The LDA probe was connected to the laser and the photodetector by monomode and multimode glass fibre cables, respectively. Figure 3 gives an overview of the instrumentation distributed in different locations of the aeroplane. It indicates that the wing glove mounted on the starboard wing is equipped with the LDA probe, the traversing system and the excitation source with the power supplies are located in the underwing station. The laser, photodetector and signal processor (Burst Spectrum Analyzer – BSA) were mounted on the instrumentation platform behind the pilots's seats. The port wing carried the flight data acquisition of the test aircraft which was described in detail by Erb, Ewald and Roth [1] . All systems were controlled by the onboard computer in the cockpit. It ran the different programmes in multitasking operation, in this way simultaneous data acquisition could be performed. The battery that powered the instrumentation lasted for about 30 min of run time per flight.

Results

During the extensive test campaigns the rate of validated velocity data turned out to be approximate by 200 Hz for clear atmospheric weather conditions whereas for hazy weather data rates up to several kHz were obtained. The major atmospheric parameter influencing the validated data rate was the humidity of air. Experiments carried out at flight levels above the inversion layer yielded significantly lower data rates than flights below the inversion. Nevertheless, in all conditions sufficient data could be obtained during flight experiments to provide useful boundary layer information.

In order to demonstrate the performance of the LDA system described to detect Tollmien-Schlichting waves an excitation source for small flow disturbances was installed in the test wing section. This source consisted of a small loudspeaker driver mounted underneath the wing glove upper surface and was located at 27 % of the chord length. The excitation frequency was set to 900 Hz that corresponded to the frequency of maximum amplification according to a stability computation.

A measured boundary layer velocity profile without any excitation is presented in Figure 4. It displays mean velocity distribution and turbulence intensity. This measurement was taken in the boundary layer on the wing glove at a chordwise position of 42.5 %. The position was chosen to be in the region of laminar flow.

Some results of flight tests with a wave train introduced by the excitation source are additionally shown in Figure 4. Whereas the influence of the wave train in the mean velocities could hardly be identified, the turbulence intensities were significantly increased with introduced excitation and showed the characteristic peaks of a transitional stage. There are profiles of mean velocity and turbulence intensity at different spanwise positions. The highest increase of the turbulence intensity could be observed at a distance of 20 mm from the symmetry plane. This corresponded to an angle of lateral growth of the disturbance of about 6° which is good agreement with other observations reported in literature and the Direct Numerical Simulation (DNS) performed by Stemmer [2].

For good weather conditions a data rate achieved that allowed a direct calculation of the power spectral density function from the velocity-time series via Fourier transformation. Figure 5 shows the density spectrum at a distance of about 1 mm above the wing surface. The peak corresponding to the frequency of excitation at 900 Hz could easily be identified. For lower data rates amplitude and phase angle of the Tollmien-Schlichting wave at any location in the boundary layer were analysed using harmonic analysis of the time series. Figure 6 shows an example displaying the characteristic phase change of the Tollmien-Schlichting wave at about 1.5 mm distance from the wall.

Concluding remarks and outlook

The present paper summarises the outcome of the LDA system development yielding a miniaturised optical system optimised to be small, light, robust and able to detect signals from very small scattering particles. Typical results of performed in-flight measurements are presented. The mean streamwise velocities, rms values of turbulent velocity fluctuations and energy spectra of the fluctuating components were measured.

There is still room for further development and related work. Whereas the optical system design dominated in the present research efforts, conventional LDA signal processing equipment was employed. Future work should concentrate on developing LDA electronic systems that are small in size, light in weight and robust, so that they can be reliably operated under flight conditions. It should also aim for a reduction of power consumption of the electronic systems to permit long time measurements in small aeroplanes of the kind employed in this research work.

Acknowledgements

The authors gratefully acknowledge the financial support they received for their research work in the through the DFG (German Science Foundation). The authors are also thankful to the Institute of Aerodynamics of the University of Darmstadt for making the test aeroplane available to accomplish the in-flight measurements.

References

[1] Erb, P., Ewald, B., and Roth, M.: Flight Experiment Guidance Technique for Research on Transition with G109b Aircraft of the Technische Hochschule Darmstadt, New Results in Numerical and Experimental Fluid Mechanics,NNFM60, Vieweg Verlag, 1996, pp. 143 – 150.

[2] Stemmer, C., Kloker, M., and Wagner, S., DNS of Harmonic Point Source Disturbances in an Airfoil Boundary Layer, AIAA Paper 98-2436, 1998, 29th AIAA Fluid Dynamics Conference, Albuquerque, NM.

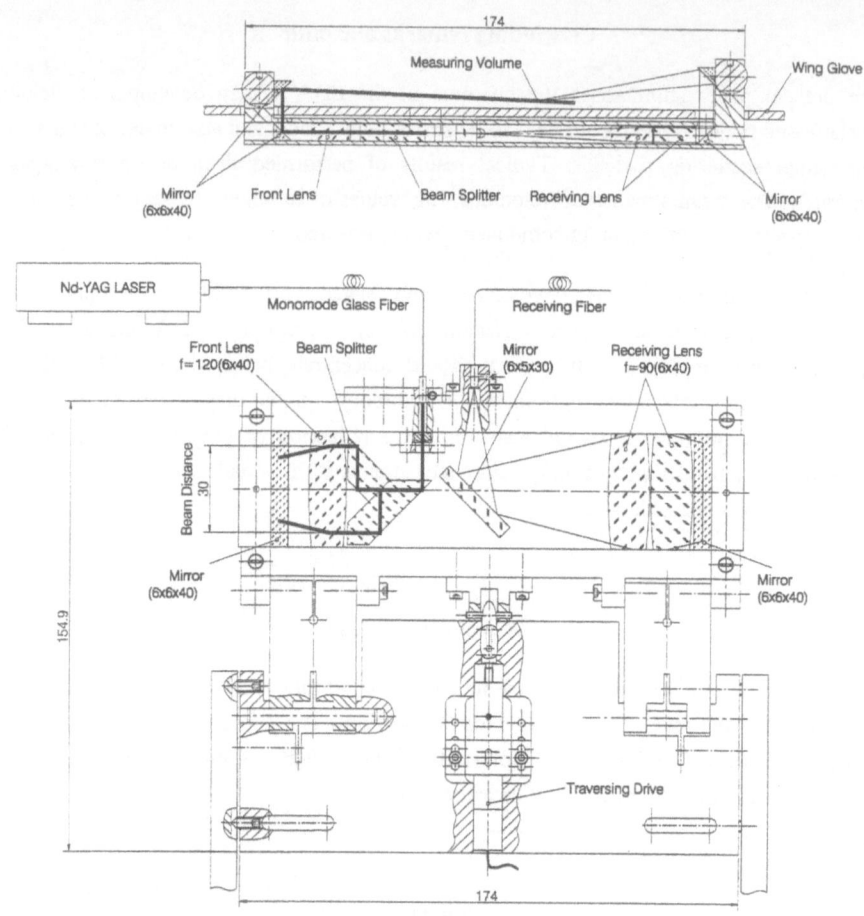

Fig. 1: Cross sectional view of LDA probe for in-flight measurements

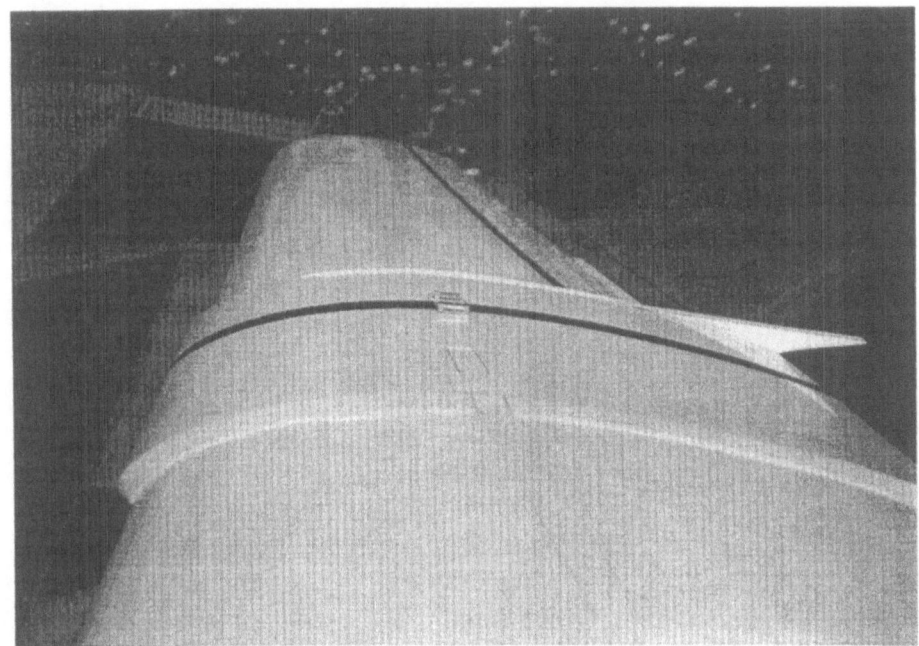

Fig. 2: LDA system during in-flight tests

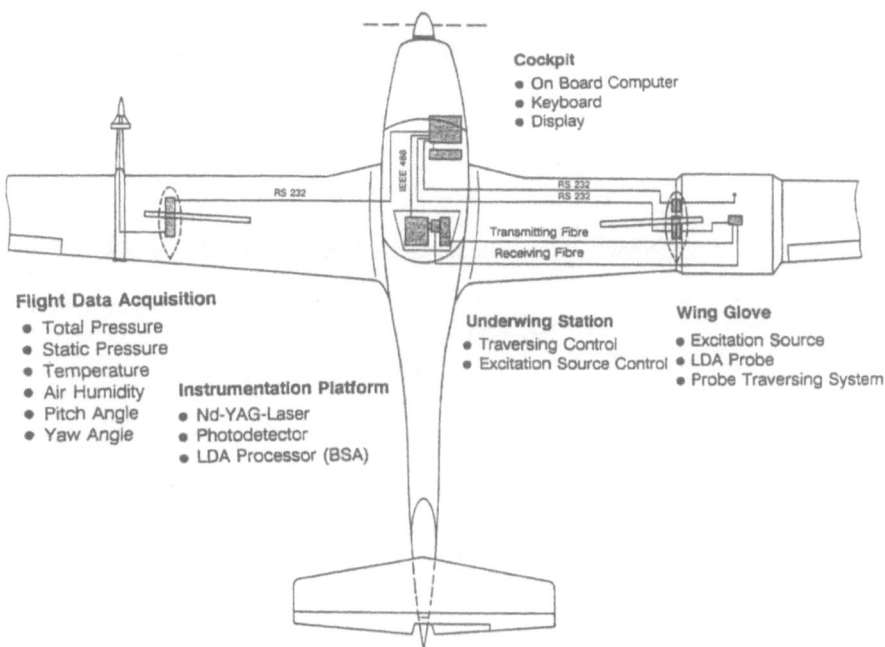

Cockpit
- On Board Computer
- Keyboard
- Display

RS 232

RS 232
RS 232

Transmitting Fibre
Receiving Fibre

Flight Data Acquisition
- Total Pressure
- Static Pressure
- Temperature
- Air Humidity
- Pitch Angle
- Yaw Angle

Instrumentation Platform
- Nd-YAG-Laser
- Photodetector
- LDA Processor (BSA)

Underwing Station
- Traversing Control
- Excitation Source Control

Wing Glove
- Excitation Source
- LDA Probe
- Probe Traversing System

Fig. 3: Arrangement of instrumentation for in-flight tests

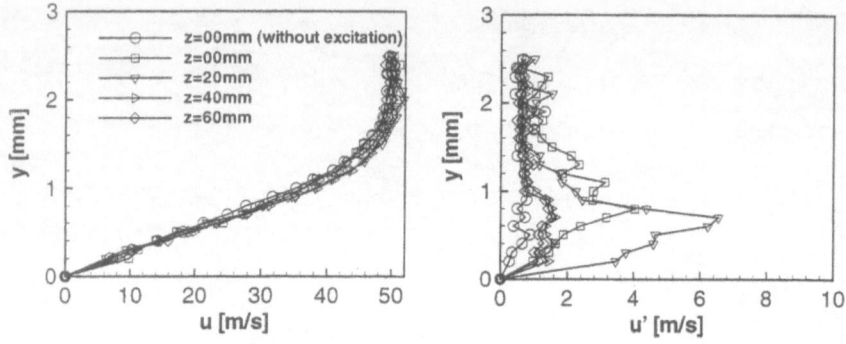

Fig. 4: Mean velocity and turbulence intensities by different spanwide positions with and without excitation

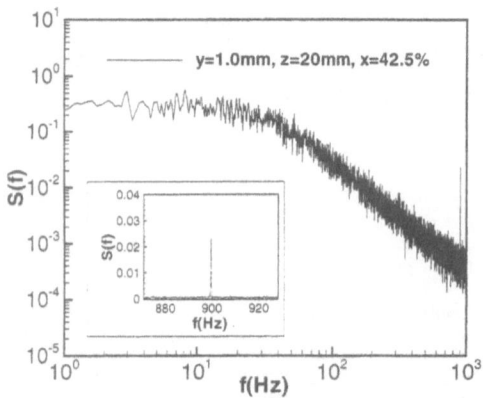

Fig. 5: Power density spectrum of the Tollmien-Schlichting wave by x/c = 42.5 %

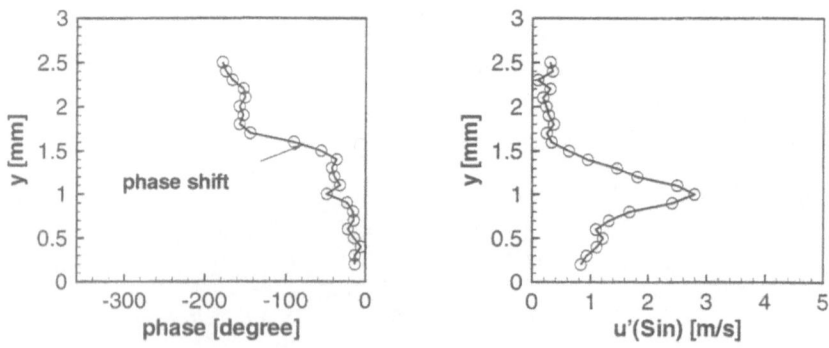

Fig. 6: Phase angle and the turbulence intensity of the Tollmien-Schlichting wave

Effect of mean Flow Accuracy on PSE Results

H. Bieler

Daimler-Benz Aerospace Airbus GmbH, D-28183 Bremen; Germany

R. Alewelt

University of Bremen, ZARM, Am Fallturm

D-28359 Bremen; Germany

Summary

The description of transitional boundary layer phenomena is a key issue for aeronautical technology development. For that purpose a certain code robustness and basic flow computational quality is required. The mean flow prediction tools have been thoroughly validated in industry in the past for their use within the local, linear stability theory. The application of nonlocal methods (PSE: Parabolized Stability Equations) is - however - in sight [9]. With such tools streamwise developments of the basic flow, nonlinear effects and receptivity issues can be taken into account; however more mean flow data with sufficient precision are needed.

We consider here the effect of systematic streamwise changes of the outer boundary condition on the PSE results for a well defined model flow (2D wedge flow). It is shown that boundary layer and stability computations react very sensitive to small changes in the outer boundary condition.

Introduction

For a variety of so-called model flows (e.g. similarity solutions of the boundary layer equations or exact solutions of the Navier-Stokes equations) the outer (inviscid) and inner (mass flow, heat transfer) boundary conditions are often analytically prescribed. The precision of the numerical solution is therefore linked to the precision of the applied code only. The industrial practice - however - is usually different (Fig. 1). The input to boundary layer tools consists of the free stream conditions, the geometry or the pressure distribution information of the considered part of the aircraft (outer boundary condition), and the inner boundary condition (heat and mass transfer). If - for instance - the measured results of flight test campaigns have to be compared with numerical results, the following approach is adopted: The aircraft has to be kept in a more or less constant situation (constant Mach number, angle of incidence, altitude etc.) for the time interval of the measurement. The measured free stream values within the time interval are usually averaged. In addition to these changes of the free stream conditions (which might be very small depending on the length of the measurement), the accuracy of the applied methods to detect the local properties of the flow (e.g. pressure distribution, free stream velocity) has to be considered. The uncertainty for the measured pressure distribution is linked to the available precision to measure the local data and to the available density of pressure sensors.

We investigate here the effect of systematic local changes of the inviscid data (at the boundary layer edge) on PSE results (local and integral amplification rates). For the flow on the wedge (unswept with respect to oncoming flow) an analytical change of the inviscid velocity distribution has been applied to compare also with results from a DNS run. In addition to this case with slowly varying velocities a second case with stochastic changes has been constructed to simulate experimental errors and short term changes (short in relation with the time interval of the measurement; see above).

Numerical Methods

At the University of Bremen, a flexible numerical method has been programmed which solves the laminar, steady, incompressible boundary layer equations on a swept wedge and allows the user to prescribe analytical changes of the inviscid velocity distribution [2]. The code has been validated and tested with emphasis on producing the relevant data necessary for a subsequent PSE analysis. For this parameter configuration (unswept wedge with analytically prescribed wave), a DNS run was also performed at the University of Stuttgart (IAG) [8].

For the second part of our investigation the inviscid data of the 2D wedge flow have been stochastically disturbed. These served as input for the DA in-house boundary layer code [3].

For the stability analysis, the linear NOLOT/PSE code developed at DLR Göttingen and FFA Stockholm [5, 4] has been used. Some NOLOT results were compared with output from the nonlinear PSE code COPS [1, 6, 7], which is applied in industry to derive transition prediction criteria on the basis of a variety of interaction scenarios.

Model Flow (2D Wedge Flow) with Distorted Outer Boundary Condition

Sinusoidal Distortion (outer BC) of 2D Wedge Flow

In order to simulate boundary layer effects which might occur close to the attachment line of a wing, the case $\beta_H = 0.2$ (accelerated flow) has been chosen without sweep. This is for our purpose general enough because we restrict our investigation to running modes of Tollmien-Schlichting type. It has also been verified that the effects produced by the considered wave are typical in the sense that other waves and other TS modes behave similarly. Fig. 2 shows a variety of U_e settings, where at an arbitrary position on the wedge a distortion in form of a wave has been superposed to the U_e variation for $\beta_H = 0.2$. Local surface effects, experimental uncertainties or slow changes of the aircraft speed during the measuring interval might produce such a velocity distribution. The calculated shape factor H_{12} resulting from the boundary layer code agrees fairly well with the corresponding value from a DNS run, Fig. 3. The boundary layer reacts very fast to local changes (both in direction of increasing and decreasing U_e) of the boundary layer conditions as shown in Fig. 4, which contains the N-factors based on PSE and DNS codes.

Random Distortion (outer BC) of 2D Wedge Flow

For the above mentioned case $\beta_H = 0.2$, a stochastic signal (Gaussian distribution) with a 1% deviation around the mean c_p value has been selected. This simulates scatter effects due to experimental errors and slight time-dependent changes of the outer flow conditions. For the same absolute deviation of 1% (around the mean value) an assessment of the gradients between two subsequent points has been done in the following way: the mean flow was calculated with 100 points and with 400 points, respectively, Fig. 5 and Fig. 6. The c_p distortion is so small that it is visible in the shape factor plot only.

The amplification rates using different stepsizes are shown in Fig. 7 and Fig. 8 (100 respectively 400 points in the mean flow direction). The gradient between two points (larger for the case with 400 mean flow points) of the stochastically disturbed mean flow has a larger influence on the amplification rates than the absolute deviation. This is proven by a PSE run with 400 mean flow points and a PSE stepsize which gives the same number of stations (streamwise) as a PSE run with 100 points and reduced PSE stepsize. The local changes for the case with 100 points visible in the local amplification rate are somehow smeared out by integrating the local amplification to the N-factor.

These runs were performed with the NOLOT/PSE code from DLR and FFA. In order to get more experience with these phenomena some runs were done with COPS ([7]), which served to derive some transition scenarios (nonlinear PSE runs) for configurations of industrial interest (ATTAS, F-100 etc.). The principal conclusions - however - (not the quantitative data due to numerical differences and stochastic properties of the input) are the same for both methods. The behavior of the local amplification rates indicates that care is needed if filtering or averaging approaches are necessary in order to do PSE calculations. In order to resolve the information of local changes in the c_p distribution, a method with sufficient resolution has to be applied. This is somewhat less important if integral amplification rates are of interest only.

Conclusions

The case with an analytically prescribed wave superimposed onto the Ue variation shows that the boundary layer and the stability computations reflect local changes immediately. In both cases (U_e variation and stochastic signal overlayed) the basic flow reacts; this is clearly visible in the shape factor development for the latter case. The cases with stochastically overlayed signals show that the number of points in the mean flow direction (with the same 1% deviation of mean data) has an effect on the amplification rates via different gradients between two stations. This was somehow expected because the PSE approach takes streamwise developments of the flow into account. However, our analysis quantified the effect of an 1% inviscid flow deviation onto the amplification rates within the PSE approach.

These results might be used to estimate the effect of e.g. experimental errors and local surface effects (which influence the c_p) on the PSE results in cases of practical interest (wings, empennage etc. of Airbus configurations). The sensitivity analysis of the streamwise mean flow changes is something to be considered before PSE codes can be applied on a project basis.

References

[1] F. P. Bertolotti. *Linear and Nonlinear Stability of Boundary Layers with Streamwise Varying Properties.* PhD thesis, The Ohio State University, 1991.

[2] V. Bode. Numerische Aspekte der laminaren, stationären inkompressiblen und unendlich schiebenden Grenzschichtströmung. *Diplomarbeit*, Universität Bremen, 1998.

[3] E. Elsholz. Ein inverses LISW-Grenzschicht Verfahren. *MBB Technical Report*, TE-1681, 1988.

[4] A. Hanifi, D. Henningson, S. Hein, F.P. Bertolotti, and M. Simen. Linear Nonlocal Instability Analysis, the linear NOLOT code. *FFA*, TN54-1994, Stockholm 1994.

[5] S. Hein, F.P. Bertolotti, M. Simen, A. Hanifi, and D. Henningson. Linear Nonlocal Instability Analysis, the linear NOLOT code. *DLR-IB*, 223-94 A56, 1994, Göttingen.

[6] Th. Herbert. Boundary-layer transition – Analysis and prediction revisited. *AIAA paper*, 91-0737, 1991.

[7] Th. Herbert, D.C. Hill, S. Huang, N. Lin, G.K. Stuckert, and M. Wang. User's Guide to COPS 2.05, software for the analysis of the stability of boundary layer flows. *DynaFlow, Inc Columbus, Ohio, U.S.A*, 1996.

[8] R. Messing. Private Communication. *IAG*, Universität Stuttgart, 1998.

[9] H. Bieler R. Alewelt. Industrial application of PSE codes in the field of HLFC technologies. *Paper at ECCOMAS 1998*, Conference Proceedings, 1998.

Figures

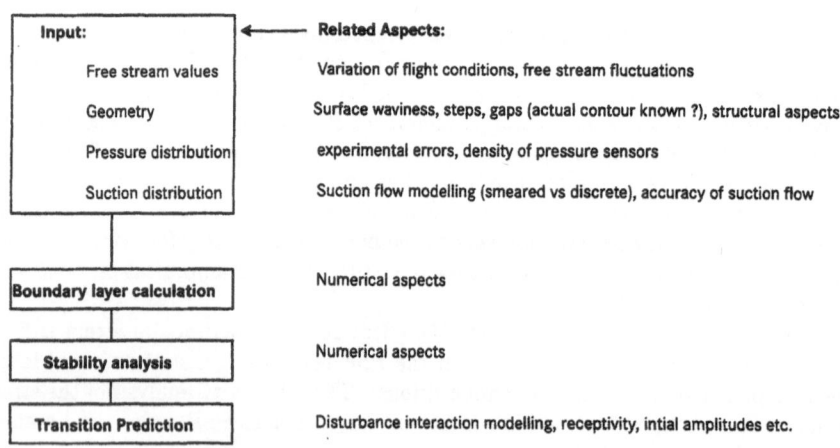

Figure 1: Dependence of mean flow accuracy from input data

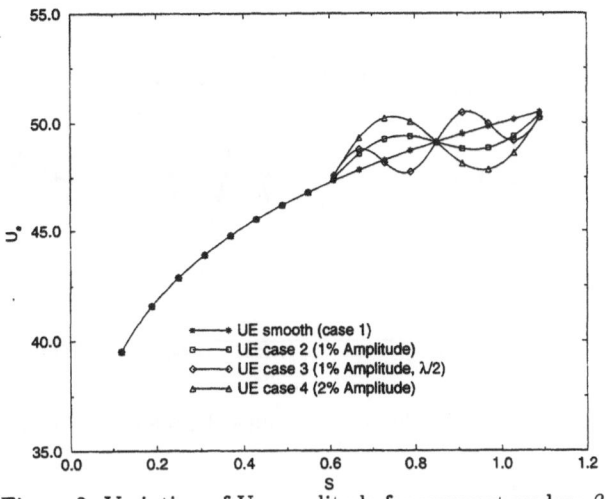

Figure 2: Variation of Ue amplitude for unswept wedge; $\beta_H = 0.2$

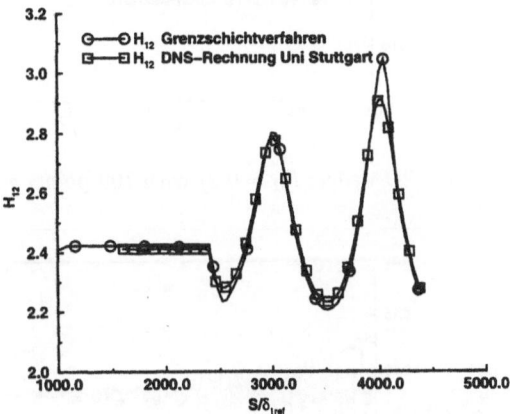

Figure 3: Shape Factor: DNS and Boundary Layer Computation (Ue case 3)

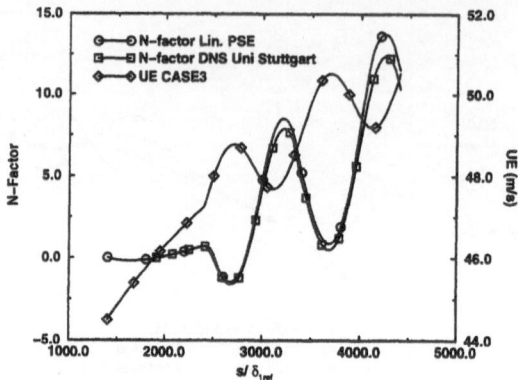

Figure 4: N-Factor Results with linear PSE and DNS (Ue case 3)

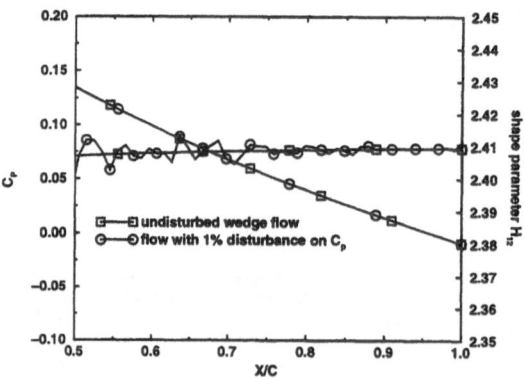

Figure 5: Mean flow (2D wedge; $\beta_H = 0.2$) with 100 points and 1% random deviation

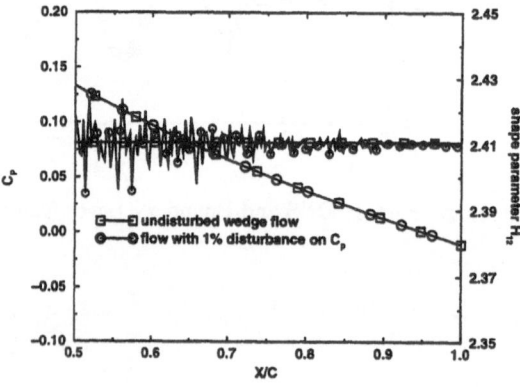

Figure 6: mean flow (2D wedge; $\beta_H = 0.2$) with 400 points and 1% random deviation

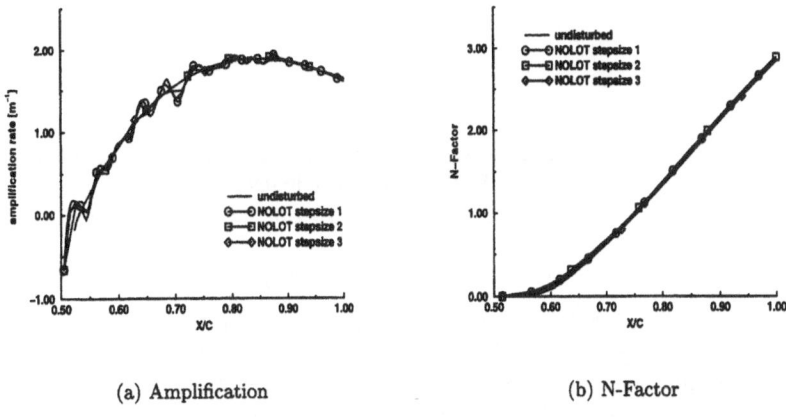

(a) Amplification   (b) N-Factor

Figure 7: NOLOT/PSE-results with different PSE-stepsizes (100 points)

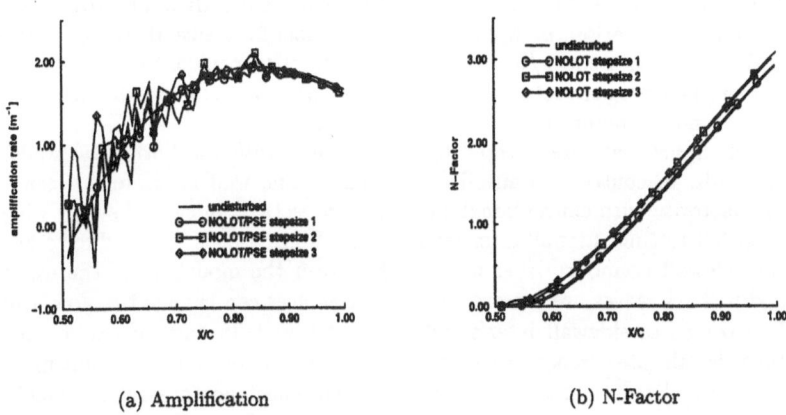

(a) Amplification   (b) N-Factor

Figure 8: NOLOT/PSE-results with different PSE-stepsizes (400 points)

Adapted Wind Tunnel Walls for a Swept Wing Experiment

J.Birkemeyer

Deutsches Zentrum für Luft- und Raumfahrt (DLR) e.V.
Institut für Strömungsmechanik
Bunsenstraße 10, 37073 Göttingen, Germany

Summary

For two reasons infinite-span swept wing flow is of interest: 1) the three-dimensional nature of the boundary layer can be investigated in its most simple form and 2) the quasi-two-dimensional flow field is representative of the mid-section of modern transport aircraft wings. Swept wing wind tunnel tests are affected by strong sidewall interference effects which can be reduced by a contoured lining. The design of such a contour requires the application of numerical methods. A design process is described and the effectiveness of one contour is investigated at different flow conditions. Interferences of the swept wing model with the horizontal walls are present due to a technical restriction of most adaptive test sections, which allow only two-dimensional adaptation. The influence of different horizontal wall adaptation strategies on the wing flow field is shown.

Introduction

The project ADIF (Adaptive Wing) of Daimler-Benz Research and Technology, Daimler-Benz Aerospace Airbus GmbH and DLR includes aerodynamic investigations of a local contour modification (bump) and variable camber for transonic wings. Airfoil experiments with bumps as a shock/boundary layer interaction control measure have shown a large potential for drag reduction. Further experiments with a swept wing (sweep angle $\varphi = 26°$) are planned to determine three-dimensional effects.

Wind tunnel tests are affected by interferences with the horizontal walls and with the sidewalls. In contrast to airfoil wind tunnel tests, wall interferences cannot be reduced satisfactorily with conventional adaptive test sections in case of swept wing flow [4],[5].

In airfoil testing sidewall interferences stem from viscous effects due to an interaction of the sidewall boundary layer with the flow over the model [2]. If the aspect ratio is low and if shock waves are present, these interferences can change the flow field significantly. In most cases sidewall interferences are neglected. In case of swept wing experiments the sidewall interferences are dominated by non-viscous effects, resulting from the three-dimensional nature of the streamlines. Viscous effects are present too and both effects are coupled and cannot be neglected if the flow field of an infinite-span swept wing is to be produced. Therefore, for the present experiment, the wind tunnel sidewalls are covered by a lining, consisting of three parts on each sidewall (Fig. 1). The upstream wedge is followed by the contoured liner segment, which is mounted rotatable together with the wing and the downstream wedge. As the liner contour depends on the wing geometry and the flow conditions, the contour is correct only for one design point. On the other hand, the time- and cost-intensive manufacture of the contoured sidewall segments forbids to produce several ones. For that reason it is desirable that the designed contour is effective also at slightly off-design conditions (wing with bump, different angles of attack).

Interferences with the horizontal walls can be treated as a non-viscous problem in airfoil tests as well as in swept-wing tests and can substantially be reduced by streamlined walls. An infinite-span swept wing flow field requires a swept adaptation of the horizontal walls. This is a technical problem. The adaptive test section of the Transonic Wind Tunnel Göttingen (TWG), where the sheared-wing model is going to be investigated, allows two-dimensional wall adaptation.

Numerical methods

To calculate the 3D- and the 2.5D[1]-flow field the DLR Navier-Stokes code FLOWer has been applied. A chimera boundary condition has been implemented to compute the swept wing/contoured sidewalls configuration. For the calculation of the infinite-span swept wing, a periodic boundary condition has been implemented. The code is equipped with a Baldwin-Lomax turbulence model.

The computational grids for the 2.5D-calculations have a C-structure (176x48 cells). The three-dimensional grids for the design of the sidewall contour have a C-H-type topology with a C-structure in streamwise direction and an H-structure in spanwise direction (176x48x56 cells). The grid is divided into two blocks in spanwise direction. A no-slip wall boundary condition is applied for the sidewalls. The farfield boundary of the 2.5D- and the 3D-grids is set to ten chord lengths. To calculate the wind tunnel flow, including the horizontal walls, a multiblock topology has been used; the grid, shown in Fig. 3, has been stacked in spanwise direction creating 56 cells (Fig. 2). The upper and lower blocks, representing the horizontal wall shape, can easily be exchanged. The horizontal walls have a slip wall boundary condition.

Design process for the sidewall contour

During the design process the wind tunnel flow is to be simulated by Navier-Stokes calculations. The following simplifications have been made: 1) wedges 1 and 2 are not simulated, 2) the horizontal walls are not simulated and 3) the contour of the lining is extended to the farfield boundary.

The design process (Fig. 4) begins with the calculation of the infinite-span swept wing flow. From these results, streamlines are calculated which describe the basic shape for the contoured liner segment. The spanwise deflections of the streamlines inside the wing boundary layer are extremely high. Therefore, these deflections are replaced by the deflections of a streamline close to the wing, but outside the wing boundary layer. This is done separately on the upper and lower side of the wing. This procedure generates a staggered contour in the wake. In the following iteration the displacement thickness of the sidewall boundary layer is evaluated from Navier-Stokes simulations of the wind tunnel flow and is then taken into account for the liner contour. The iteration is stopped when the pressure distribution on the wing is equal to the 2.5D-pressure distribution over the prescribed spanwise extension (here: a half chord length on either side of the center line). The iteration can be accelerated if the displacements on the sidewall are well estimated for the first iteration step.

Results

Plane sidewalls

The pressure distribution of the swept wing/plane sidewalls configuration is shown in Fig. 5. The interference effects are schematically indicated in Fig. 6.

The streamlines of the swept wing are three-dimensional, but near the wall the flow is

---

[1]2.5D is used as an abbreviation for infinite-span swept wing or quasi-two-dimensional.

constrained to follow the wall-direction. This leads to compression waves at the upstream sidewall, while expansion waves are generated at the downstream side. Viscous interference effects are essential in the shock region, where a strong thickening of the sidewall boundary layer is induced. The changing of the flow direction near the wall reduces the shock angle which is then less than the geometrical sweep angle; it generally changes the spanwise direction of the isobars. The shock strengths and the pressure gradients upstream of the shock differ from the infinite-yawed swept-wing values.

Contoured sidewalls

During the design process the wing and the sidewall boundary layers were treated as turbulent boundary layers. The flow conditions of the design point are $Ma = 0.84$, $\alpha = 0.0°$. The design process was stopped after three iteration steps. Then the following calculations were made with a transition location on the wing at $x_{tr}/c = 0.37/0.15$ between $-0.79 \le z/c \le 0.79$.

The pressure distribution of the swept wing/contoured sidewalls configuration in Fig. 7 shows a good agreement with that of the 2.5D-flow. Using the wing with a bump instead of the datum wing, the pressure distribution is in good agreement with the 2.5D-pressure distribution too (Fig. 8). The large deviation in the pressure distribution at $z/c = -1.0$ from the 2.5D-pressure distribution results from the design process for the sidewall contour. Behind the shock, the streamlines, located inside the wing boundary layer, are strongly inclined towards the downstream side. This is not taken into account for the liner contour. The pressure distribution at the center line of the swept wing/contoured sidewalls configuration at different angles of attack (Fig. 9) shows differences to the 2.5D-case mainly in the shock region. At high angles of attack ($\alpha = 0.5°$, $\alpha = 1.0°$) numerically-converged solutions were difficult to obtain due to significant regions of separation and unsteady flow conditions. The 3D-flow differs slightly from the 2.5D-flow, but the effect of a bump on buffet onset can in principle be investigated in the experiment.

Height, length, shape and the position relative to the shock are the characteristic parameters of the bump [1],[3]. The effectiveness of a given bump depends on the shock strength, the relativ shock location and the pressure gradient ahead of the shock as was found by the present author. Therefore, the suitability of the flow for shock control investigations with a bump can be characterized by these three parameters. Ignoring the lowest angle of attack considered here ($\alpha = -1.0°$), the following statements can be made with regard to conditions over a spanwise region of $-0.5 \le z/c \le 0.5$:

- The shock strength, in transonic flow determined by the Mach number or the pressure coefficient upstream of the shock, $c_{p,1}$, differs from the 2.5D-value by less than +7%/-3% (Fig. 10). (The influence of the horizontal wall interferences is shown in Fig. 10 - 12 too, an explanation is given later.)

- The shock location x_1, here defined by the location of the suction peak ahead of the shock, differs from the 2.5D-value by less than +3.5%/-2% c. The shock angle decreases with increasing angle of attack. At $\alpha = 0.0°$ the shock angle is equal to the geometrical sweep angle (Fig. 11).

- The pressure gradient upstream of the shock, here characterized by the average c_p-gradient over 7% c upstream of x_1, shows the largest deviation from the 2.5D-case (Fig. 12). Since the primary effect of a bump is a reduction of the wave drag, the sensitivity of the wave drag reduction by a bump to the changes in the c_p-gradient has been investigated in 2D MSES calculations, where the shock strength and the position of the bump relative to the shock have been kept constant. The influence of the c_p-gradient on the wave drag reduction through a bump is indicated in Fig. 13.

Despite these differences the drag polars of the swept wing/contoured sidewalls configuration are qualitatively similar to the drag polars of the 2.5D-calculations, Fig. 14. The calculated drag reductions due to the bump are underpredicted in all cases due to the coarse mesh.

Horizontal Wall Interferences

Different shapes for the horizontal walls have been investigated at the flow conditions $Ma = 0.84$, $\alpha = 0.0°$ and a transition location on the wing as indicated above. As a reference case a swept adaptation (Fig. 15) has been applied, where a streamline of the infinite-span swept wing is moved along the span shaping the wall. The results (Fig. 16) show details of the pressure distribution in the shock region. The deviation in the pressure distribution at $z/c = -1.0$ from that of the 2.5D-case is even stronger than without horizontal walls (Fig. 7), because the horizontal wall is shaped assuming an infinite-span swept wing.

The two-dimensional horizontal wall shapes that have been considered are shown in Fig. 17: a) a 2.5D-streamline, which is strictly speaking correct for the center line only (Fig. 18), b) a crosswise average of the swept adaptation between $-0.5 \leq z/c \leq 0.5$ (not shown here) and c) a crosswise average of the swept adaptation between $-1.0 \leq z/c \leq 1.0$ (Fig. 19). The two latter strategies mainly lead to a reduction of the maximum wall deflection, as indicated in Fig. 17. This induces an acceleration of the flow, coupled with a movement of the shock further downstream. The parameters essential for the effectiveness of a bump - mentioned above - have been evaluated and the results of case a) have shown the best agreement with the 2.5D-flow (Fig. 10 - 12).

Conclusions

A wind tunnel liner contour has been designed to eliminate sidewall interferences for a swept-wing experiment. The contour is effective at the design point with and without shock control on the wing. The drag polars of the infinite-span swept wing and the swept wing/contoured sidewalls configurations show a qualitatively good agreement. The influence of a bump on buffet onset can be investigated. Interferences with the 2D-adapted horizontal walls are small, when the wall contour is set according to the streamline at the center line.

References

[1] P.R. Ashill, J.L. Fulker, and A. Shires. *A novel technique for controlling shock strength of laminar flow airfoil sections.* DGLR-Bericht 92-06, 1992.

[2] U. Ganzer, E. Stanewsky, and J. Ziemann. *Sidewall effects on airfoil tests.* AIAA Journal, 22(2), 1984.

[3] A. Knauer. *Die Leistungsverbesserung transsonischer Profile durch Konturmodifikationen im Stoßbereich.* DLR FB 98-03, 1998.

[4] G.G. Mateer and A. Bertelrud. *Contouring tunnel walls to achieve free-air flow over a transonic, swept wing.* AIAA Paper 83-1725, 1983.

[5] P.A. Newman, E.C. Anderson, and J.B. Peterson Jr. *Aerodynamic design of the contoured wind-tunnel liner for the NASA supercritical, laminar-flow-control, swept-wing experiment.* NASA TP 2335, 1985.

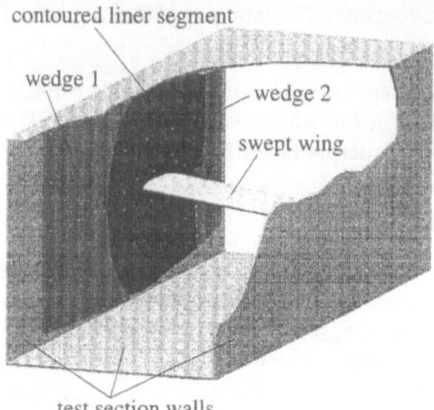

Fig. 1: Experimental setup

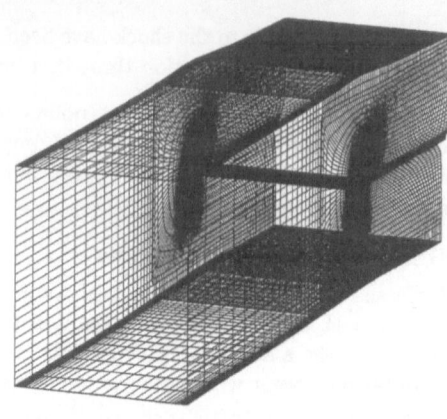

Fig. 2: Computational grid for for the simulation of the wind tunnel flow with adapted walls

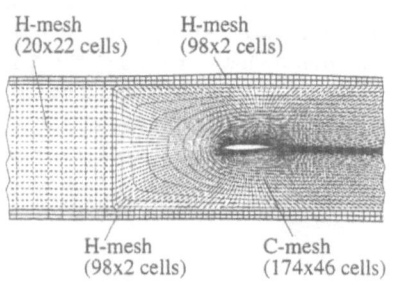

Fig. 3: Grid layer for 3d-grid

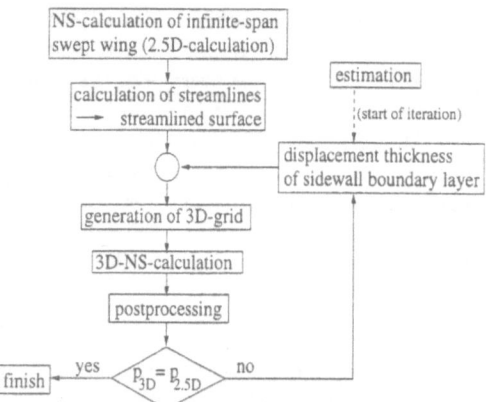

Fig. 4: Design process for the sidewall contour

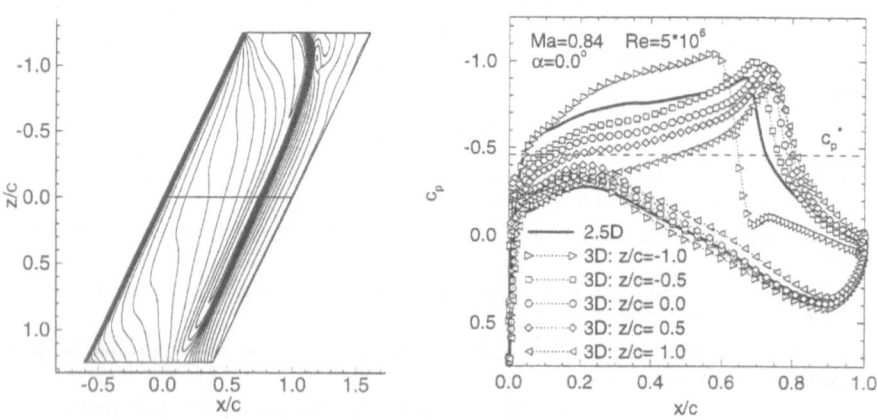

Fig. 5: Swept wing/plane sidewalls; left: isobars on wing upper side, right: c_p-distribution at different spanwise locations

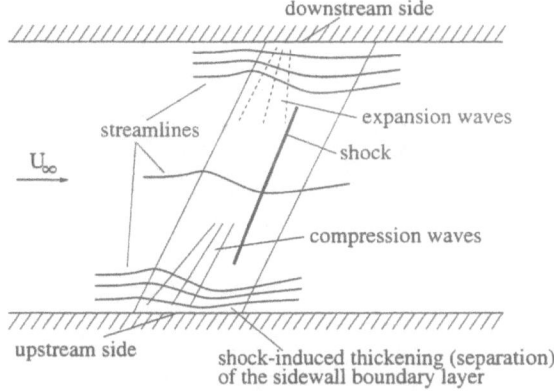

Fig. 6: Sidewall interferences schematically in case of swept wing experiments

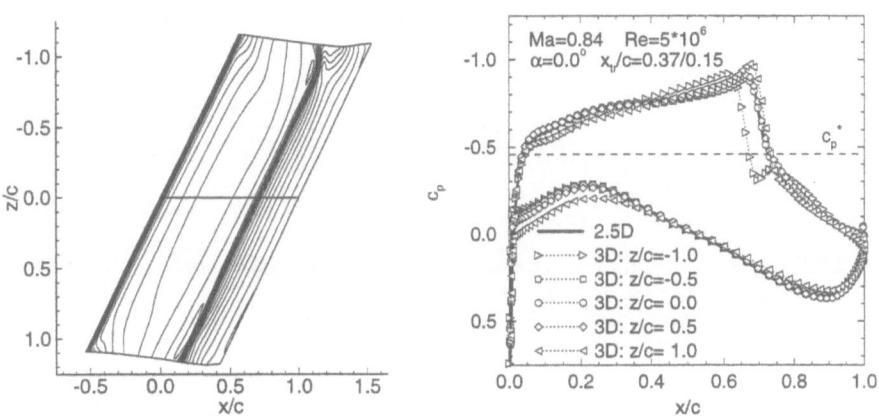

Fig. 7: Swept wing/contoured sidewalls; left: isobars on wing upper side, right: c_p-distribution at different spanwise locations

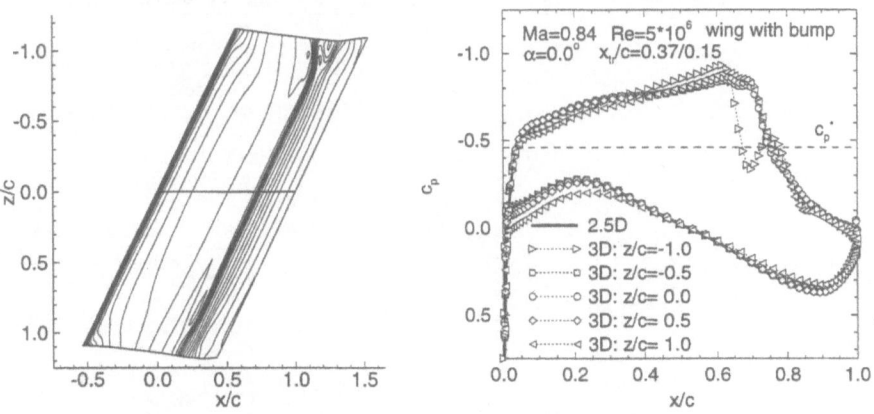

Fig. 8: Swept wing with bump/contoured sidewalls; left: isobars on wing upper side, right: c_p-distribution at different spanwise locations

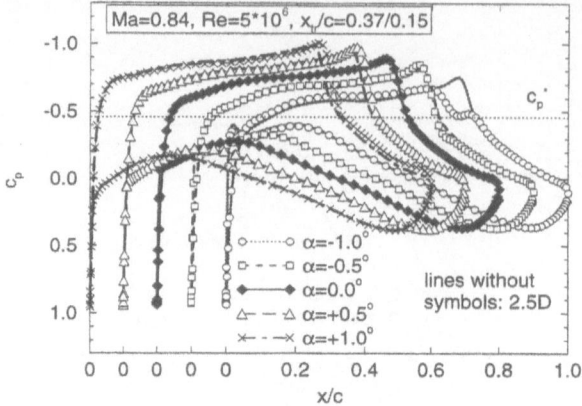

Fig. 9: Swept wing/contoured sidewalls; c_p-distribution at various angles of attack at z/c=0.0

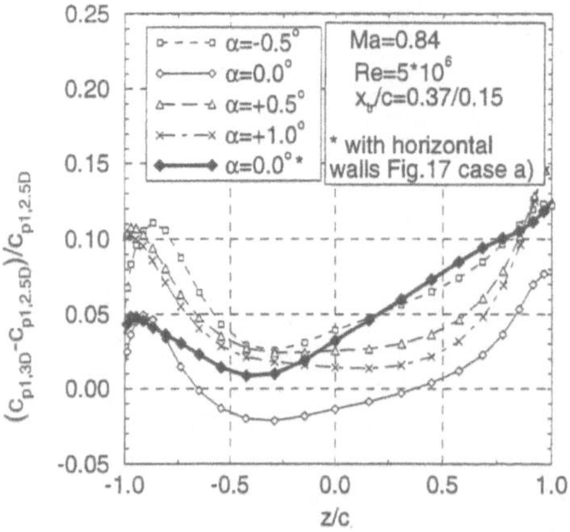

Fig. 10: Shock strength

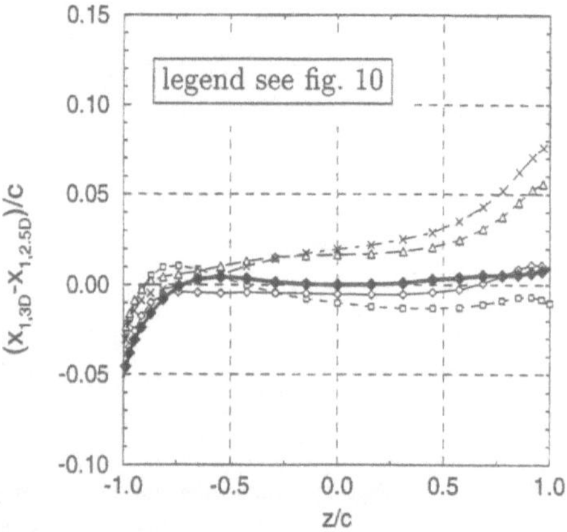

Fig. 11: Shock location

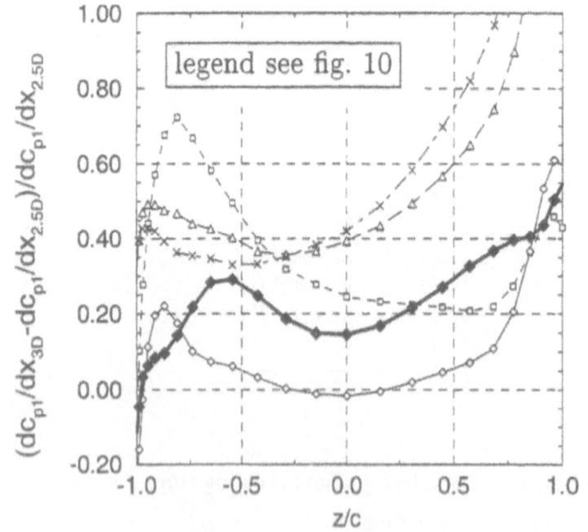

Fig. 12: c_p-gradient

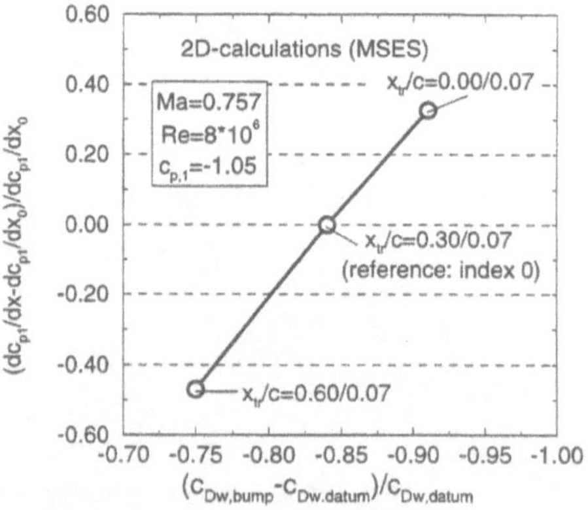

Fig. 13: Influence of c_p-gradient on wave drag reduction

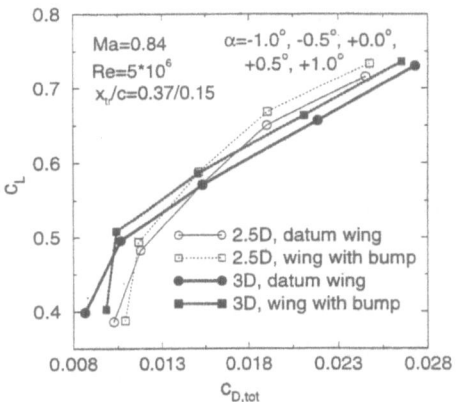

Fig. 14: Drag polar from 2.5D- and 3D-calculations

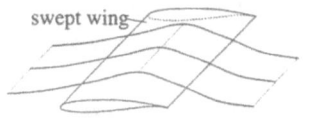

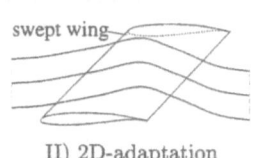

Fig. 15: Types of adaptations for horizontal walls

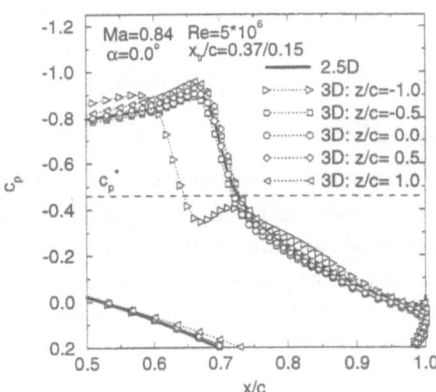

Fig. 16: Swept adaptation

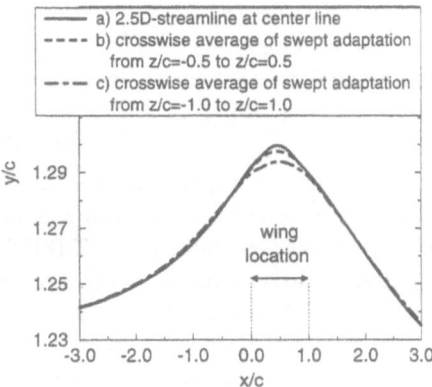

Fig. 17: Shapes of 2D-adaptations

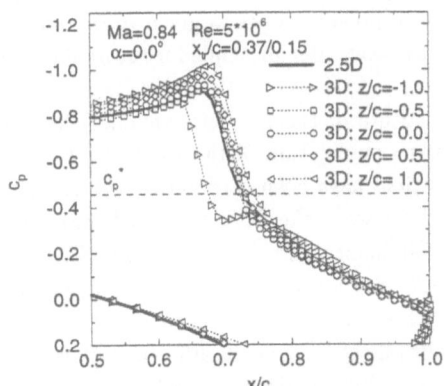

Fig. 18: 2D-adaptation at $z/c=0.0$

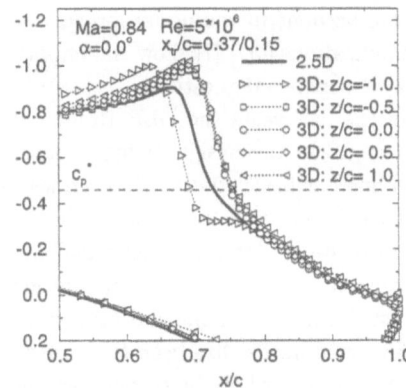

Fig. 19: 2D-adaptation averaged from $z/c=-1.0$ to $z/c=+1.0$

Experimental and numerical investigations on a waverider configuration in incompressible flow [†]

R.C. Blaschke, D. Hummel

Institut für Strömungsmechanik, TU Braunschweig
Bienroder Weg 3, 38106 Braunschweig
Germany

Summary

A waverider configuration named DLR–F8 has been investigated in incompressible flow both experimentally and numerically. A complex vortex system with primary, secondary and fuselage vortex has been found and a strong interaction between fuselage and primary vortex exists. In the primary vortex centre low velocities have been found surrounded by a region with high velocities above and beneath the vortex axis. The flow characteristics on this configuration are well predicted by the solution of the Navier–Stokes equations.

1. Introduction

For hypersonic transport aircrafts and space transportation systems, slender configurations with practicable shapes and reasonable distribution of volume are applied. To generate such slender configurations T.R.F. Nonweiler [1] suggested in 1959 a simple method by using the exact solution for the inviscid hypersonic flow past wedges. The procedure to design flow fields around 3D–bodies with a wedge flow on the lower side and a freestream upper surface is described in more detail in [2]. The shock is attached along the leading edges and due to the fact that such configurations ride on their own shock wave, they are called *waveriders*. In the past, Nonweilers method has been developped extensively. Modern design concepts are founded on the hypersonic flow past a cone. They allow the generation of waveriders with various planform shapes, and viscous effects are also taken into account in the design process. A survey of previous work is given in [3], and new publications [4] – [7] show the importance of the waverider principle for future designs of hypersonic configurations.

In the last years the DLR Braunschweig (Institute of Design Aerodynamics) has been engaged in the design of hypersonic vehicles by application of the waverider principle. Numerous basic studies of waverider configurations have been carried out and a survey is given in [8]. As a result of a close cooperation between the DLR and TU Braunschweig (Institute of Aircraft Design and Structure Mechanics and Institute of Fluid Mechanics) a realistic waverider configuration DLR–F8 with a practicable shape and reasonable distribution of volume and camber was created [8]. The windtunnel model of the DLR–F8 configuration has been the object of detailed investigations in different test facilities, covering the whole Mach number range of a real flight trajectory. In the present paper experimental as well as numerical results for this waverider configuration will be presented for incompressible flow.

[†]The investigations in this paper have been supported by the DFG under grant Hu 254/15

2. Experimental setup

2.1 Windtunnel model

The windtunnel model of the waverider configuration DLR–F8 is shown in Fig. 1. The root chord is $c = 504.4mm$ and the wing span is $b = 330.0mm$. It has an aspect ratio $\Lambda = b^2/S = 1.15$ and a taper ratio $\lambda = 0.05$. For further geometrical data refer to [8]. The model has sharp leading edges and is equipped with pressure taps in two cross sections and one chordwise section. In the experiment the base pressure in the nozzle area is measured by one pressure tap. Therefore, the contribution of the base area to the drag can be estimated and subtracted from the balance–measured drag in order to derive the forebody drag. In the windtunnel the model is fixed by a rear sting with the constant cross section of the nozzle area at the end of the configuration.

2.2 Experimental program

The experiments have been carried out in the $1.3m$ windtunnel of TU Braunschweig (Institute of Fluid Mechanics) for a free stream velocity $U_\infty = 31m/s$, corresponding to a Reynolds number based on the root chord c of $Re_\infty = 1.0 \times 10^6$. They comprise the determination of the aerodynamic forces and moments using an internal strain gauge balance (reference point is the geometric neutral point N_{25}), measurements of surface pressure distributions and flowfield investigations by means of surface oilflow patterns and a conventional five–hole probe. A detailed description of the experimental program is given in [8].

3. Numerical investigations

Navier–Stokes calculations have been carried out on a 2–Block structured C–O grid with 1.2 million grid points [8] by means of the DLR code FLOWER [9]. The solid sting of the experimental setup has been considered with the same geometry as in the experiments. The numerical investigations in the present paper comprise viscous flows with turbulent boundary layer only, using the Baldwin–Lomax turbulence model with the Degani–Shiff modification. Concerning post processing, the aerodynamic forces and moments of the DLR–F8 configuration have been calculated for the forebody only. The base area has not been taken into account, and this is the reason why the base pressure has been measured in the experiment in order to be able to compare the numerical and experimental results.

4. Results

4.1 Aerodynamic forces and pitching moment

The results of the three–component measurements (Fig. 2) show the well known nonlinear dependence of the aerodynamic coefficients on the angle of attack for slender configurations at small and medium angles of attack. For angles of attack $\alpha > 18°$ vortex breakdown takes place over the configuration and lift, drag and nose–down pitching moment are reduced as indicated in Fig. 2 by hatching. In addition, the results of a Navier–Stokes calculation for $\alpha = 12.2°$ is given in Fig. 2 by the filled symbols. The lift coefficient as well as the drag coefficient are very well predicted, but there are some differences between the numerical and experimental result for the pitching moment which will be discussed later.

4.2 Surface pressure distribution

The experimental upper surface pressure distributions in the two cross sections at $x/c = 0.29$ and $x/c = 0.75$ are shown in Fig. 3 for various angles of attack. Lines of constant static pressure coefficient $-C_p = (p - p_\infty)/q_\infty$ are plotted against the dimensionless spanwise coordinate y/y_1 in the upper part and the corresponding cross section area is shown in the lower part of this figure.

Concerning the cross section $x/c = 0.29$ flow separation occurs at the leading edge for $\alpha > 0°$ and a primary vortex is developped over the waverider configuration. For increasing α the minimum static pressure beneath the primary vortex axis moves inward.

In the cross section at $x/c = 0.75$ a primary vortex exists on the wing upper surface already for low negative angles of attack α. This is due to the fact that the leading edge of the wing is inclined against the ridge line of the waverider. For increasing α in the pressure distribution the suction peak beneath the primary vortex is increased up to $\alpha \approx 16°$ followed by a reduction and for $\alpha > 30°$ the suction peak increases again. Concerning the primary vortex axis an inward shift takes place up to $\alpha \approx 22°$ followed by an outward shift for $\alpha > 22°$. Due to the fuselage a corresponding vortex occurs for $\alpha > 10°$. With increasing α this fuselage vortex becomes stronger and its size increases. This leads to a lift up of the primary vortex for $\alpha > 16°$ and to an outward shift for $\alpha > 22°$. This interference mechanism leads to a reduction of the suction peak beneath the primary vortex. At even larger angles of attack vortex breakdown leads to the same behaviour, but both effects cannot be distinguished from one another by the pressure distributions.

The result of the Navier–Stokes calculation for $\alpha = 12.2°$ and $M_\infty = 0.3$ is compared with the experiment in Fig. 4. It is shown that the mean flow characteristics are well predicted by the Navier–Stokes solution, but there are some differences in the pressure quantity between experiment and theory. Thus, the suction peaks beneath the primary vortex axis as well as their locations are different from the experiment in both cross sections. Concerning the rear cross section the Navier–Stokes calculation predicts additional suction at the sidewall of the fuselage similar to the experimental data. The differences shown in Fig. 4 are due to the applied turbulence model and the calculation may also be affected to some extent by the coarse grid in the rear part of the waverider. The described differences of the pressure distributions between theory and experiment are the reason for the deviation of the calculated pitching moment from the experimental result according to Fig. 2.

4.3 Flowfield investigations

For an angle of attack $\alpha = 12.2°$ the results of the probe measurements in three cross sections $x/c = const.$ are shown in Fig. 5 by lines of constant total pressure coefficient. In addition the corresponding solution of the Navier–Stokes equations are given for $M_\infty = 0.3$. The experiment shows all flow characteristics which are well known from slender delta wings. More downstream at the sidewall of the fuselage a very well developped fuselage vortex can be determined, due to the flow separation at the upper fuselage corner. Furthermore, from the experiment a deformation of the primary vortex sheet can be determined which is due to the tip vortex emanating from the kink between the leading edge and the side edge at $x/c = 0.95$.

The solution of the Navier–Stokes equations shows the aforementioned flow characteristics also but there are some differences to the experiment, especially in the rear cross sections at $x/c = 0.75$ and $x/c = 1.01$. Thus the secondary vortex at $x/c = 0.75$ is too small and at $x/c = 1.01$ a secondary flow separation is no longer prescribed. Concerning the fuselage vortex there are differences in shape and strength of this vortex. Furthermore, a tip vortex at $x/c = 1.01$ cannot be observed in the numerical solution.

The described differences between numerical solution and experiment, especially in the rear part of the waverider, may be due to the coarse grid in this region in combination with the numerical viscosity. The results may also be affected by the used turbulence model. The magnitude of the interference effects of the secondary vortex and the fuselage vortex on the primary vortex is slightly different in theory and experiment and this leads to the differences in the pressure distribution discussed earlier.

The experimental dynamic pressure distribution in the cross section at $x/c = 0.29$ (Fig. 6) shows a region with low dynamic pressure in the centre of the primary vortex which is surrounded by a region with high dynamic pressure, especially below and above the vortex axis. This distribution can be explained as a planform effect in the front part of the waverider. The flow separation at the leading edge, starting in the symmetry plane, is similar to the 2D separation on an inclined flat plate. Concerning the axial and circumferential velocity components based on a vortex–fixed coordinate system, there exist no axial velocity components in the symmetry plane but only circumferential ones. Therefore, a region with low velocity is established in the vortex centre. With increasing spanwise distance from the symmetry plane the sweep of the leading edge increases rapidly. A 3D flow separation takes place and the initial region with low dynamic pressure is surrounded by a vortical flow with increasing velocity towards the vortex centre. The corresponding distribution of dynamic pressure is very well predicted by the Navier–Stokes calculation as shown in Fig. 6.

In the plan view of the surface oilflow patterns for $\alpha = 12.2°$ (Fig. 7) the experiment shows in the front part of the waverider a curved attachment line due to the primary vortex. This attachment line approaches the waverider symmetry line more downstream. Close to the leading edge a secondary separation line can be determined which moves outboard at approximately 79 % rootchord position. A kink in the geometry may be the reason for this outboard shift. Towards the fuselage an attachment line can be distinguished which is related to the fuselage vortex. In the side view the experiment shows the corresponding separation line in the vicinity of the upper fuselage corner, and a secondary separation line on the fuselage sidewall can also be observed.

The comparison with the result of the Navier–Stokes calculation indicates that the mean characteristics are well predicted by the theory. But there are some differences with respect to the primary attachment line and the separation line of the secondary vortex. In the rear part the secondary separation line is not predicted correctly. Concerning the fuselage vortex the separation and attachment lines are well predicted as shown in the plan and side view of Fig. 7 but a secondary separation on the sidewall is lacking.

5. Conclusions

The waverider configuration DLR–F8 has been investigated in incompressible flow experimentally and numerically. For this configuration the experiment shows all flow characteristics which are well known from slender delta wings. In addition a fuselage vortex has been found close to the fuselage sidewall and strong interactions have been observed between this vortex and the primary vortex. Furthermore, in the centre of the primary vortex a region with low dynamic pressure has been detected which is surrounded by a region of high dynamic pressure especially above and beneath the vortex centre. This distribution of dynamic pressure is a planform effect. The applied Navier–Stokes code FLOWER predicts all flow characteristics very well. Some differences in comparison with the experiment are due to the grid resolution and to the turbulence model used in the calculations.

6. References

[1] Nonweiler, T.R.F.: Aerodynamic problems of manned space vehicles. J. Roy. Aeron. Soc. Vol. 63 (1959), 521-528.

[2] Nonweiler, T.R.F.: Delta wings of shapes amenable to exact shock–wave theory. J. Roy. Aeron. Soc. Vol 67 (1963), 39–40.

[3] Proceedings of the 1st Internat. Hypersonic Waverider Symposium 1990, University of Maryland.

[4] Pegg, R.J.; Hahne, D.E.; Cockrell Jr., C.E.: Low speed wind tunnel tests of two waverider configuration models. AIAA–Paper 95–6093 (1995).

[5] Cockrell Jr., C.E.; Huebner, L.D.; Finley, D.B.: Aerodynamic characteristics of two waverider–derived hypersonic cruise configurations. NASA TP 3559 (1996).

[6] Hahne, D.E.: Evaluation of the low–speed stability and control characteristics of a Mach 5.5 waverider concept. NASA TM 4756 (1997).

[7] Gillum, M.J.; Lewic, M.J.: Experimental results on a Mach 14 waverider with blunt leading edges. J. Aircraft Vol. 34 (1997), 296–303.

[8] Hummel, D.; Blaschke, R.C.; Eggers, Th.; Strohmeyer, D.: Experimental and numerical investigations on waveriders in different flight regimes. 21st ICAS Congress 1998, Melbourne, Paper ICAS–98–2,1,4 (1998).

[9] Kroll, N., Rossow, C.-C., Becker, K., Thiele, F.: MEGAFLOW - A numerical flow simulation system. 21st ICAS Congress 1998, Melbourne, Paper ICAS–98–2,7,4 (1998).

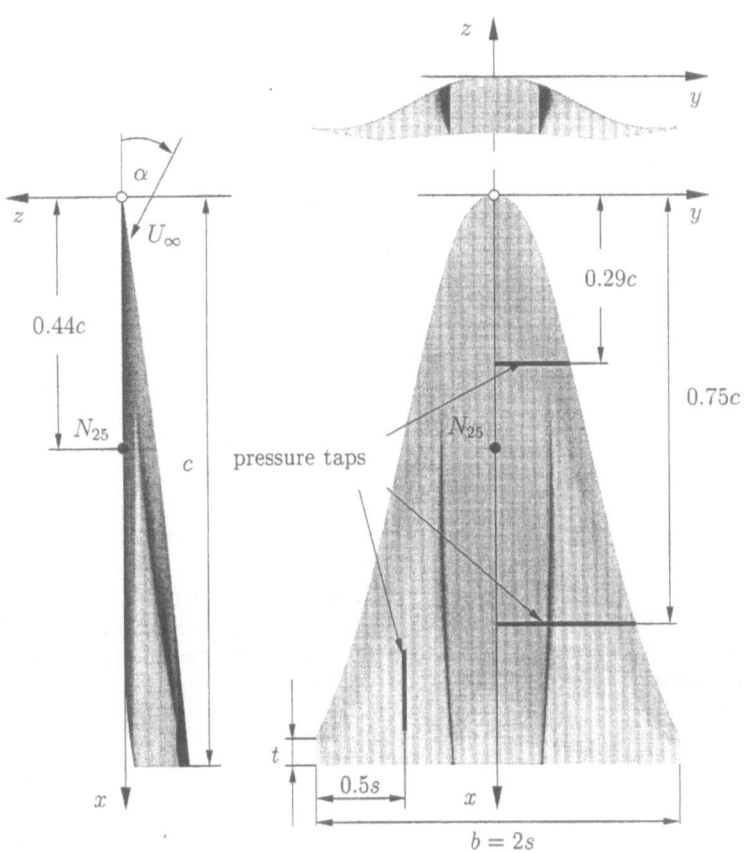

Fig. 1   Windtunnel model DLR–F8

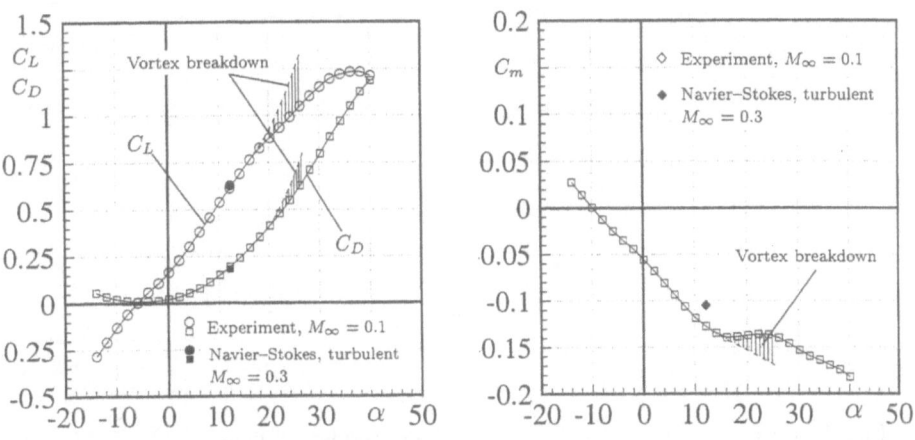

Fig. 2   Lift , drag and pitching moment characteristics for the waverider configuration
DLR–F8 at $Re_\infty = 1.0 \times 10^6$

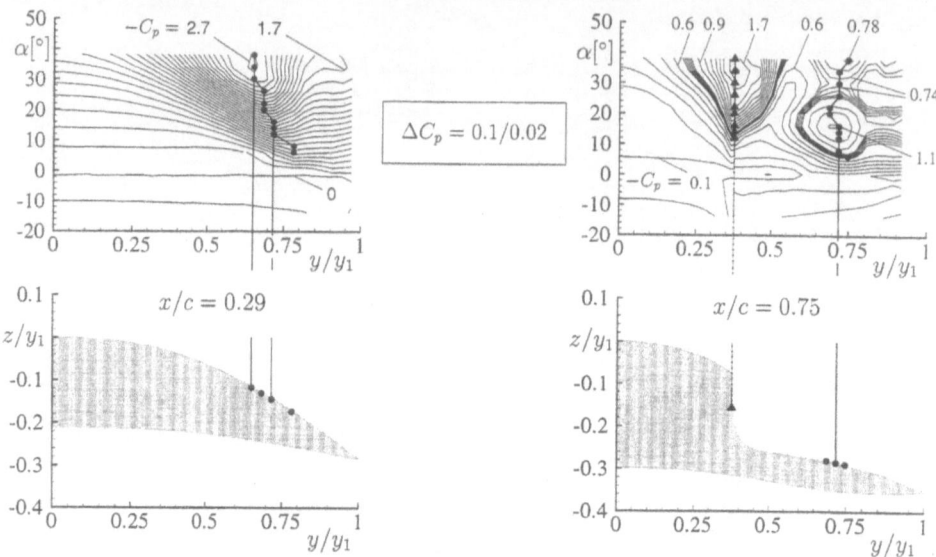

Fig. 3 Experimental upper surface pressure distributions with lines $-C_p = $ const.
in two cross sections $x/c = $ $const.$ on the waverider configuration DLR–F8 for
different angles of attack α and $Re_\infty = 1.0 \times 10^6$. (Pressure minimum due to
primary vortex marked by filled circles, due to fuselage vortex marked by filled
triangles)

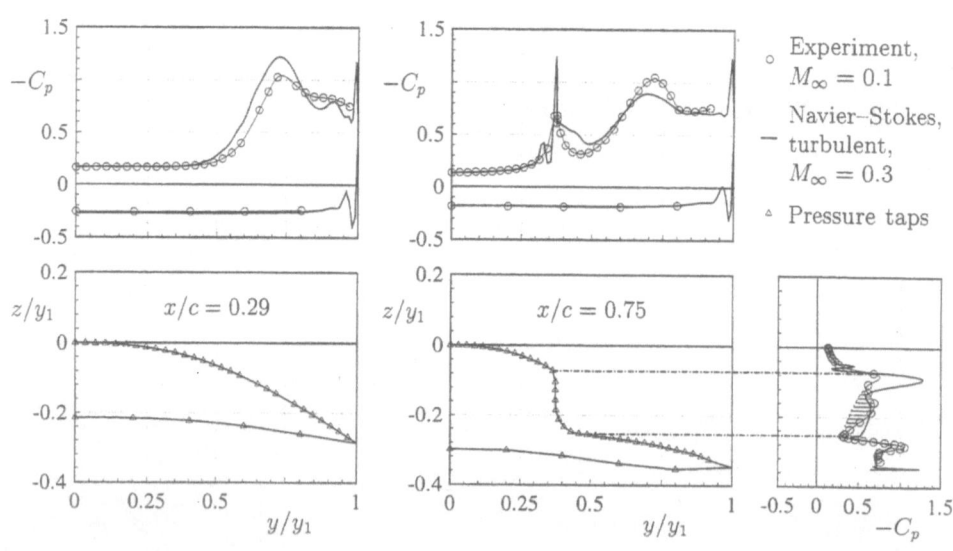

Fig. 4 Surface pressure distributions on the waverider configuration DLR–F8 in two
cross sections $x/c = $ $const.$ for $\alpha = 12.2°$ and $Re_\infty = 1.0 \times 10^6$.
(Effect of fuselage vortex marked by hatching)

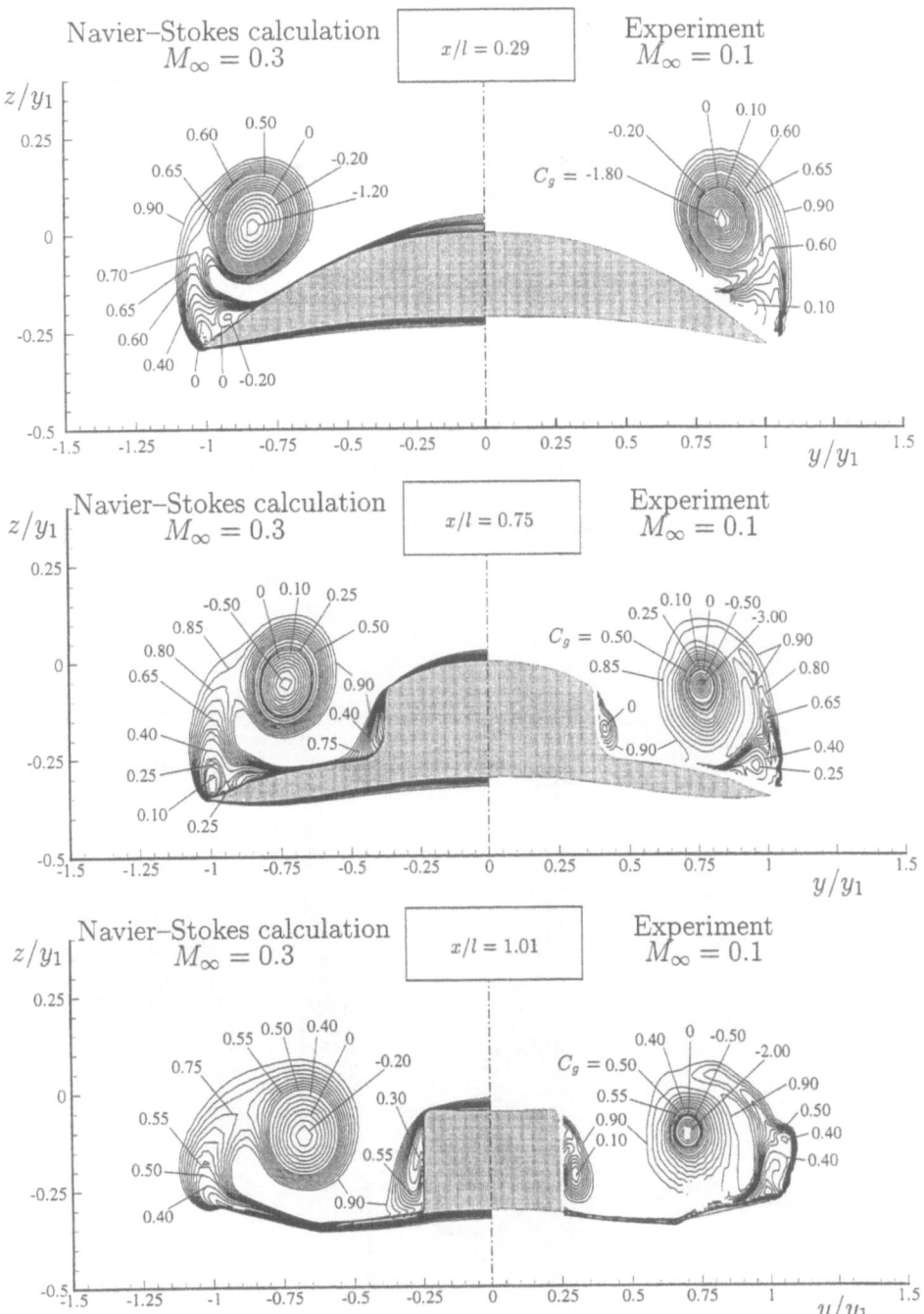

Fig. 5   Flowfield on the waverider configuration DLR–F8 at $\alpha = 12.2°$ and
$Re_\infty = 1.0 \times 10^6$ in different cross sections.(Lines of constant total pressure
coefficient $C_g = (g - p_\infty)/q_\infty$)

Navier-Stokes calculation
$M_\infty = 0.3$
$x/l = 0.29$
Experiment
$M_\infty = 0.1$
z/y_1
$\frac{q}{q_\infty}$
y/y_1
1.90
1.30
1.30
0.90
2.30
1.20
0.80
0.70
=1.20
1.60
1.90
1.30
1.80
0.40
1.20
0.80
0.70
1.10
0.25
0
-0.25
-0.5
-1.5
-1.25
-1
-0.75
-0.5
-0.25
0
0.25
0.5
0.75
1
1.5

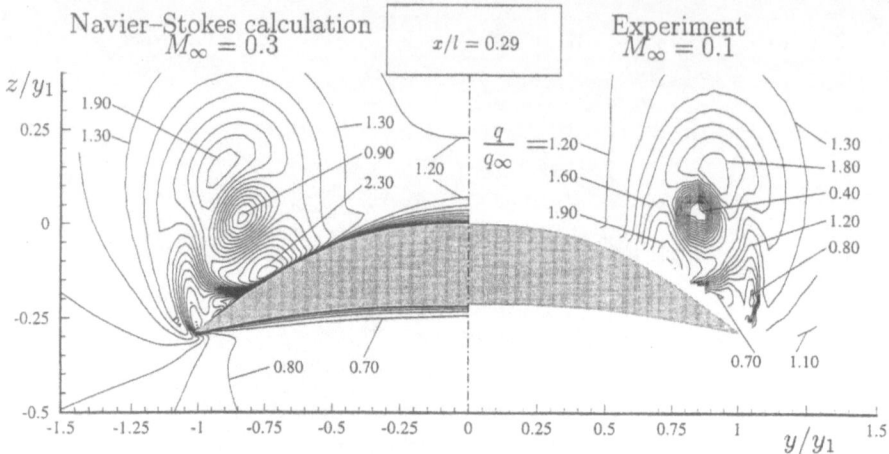

Fig. 6 Flowfield on the waverider configuration DLR–F8 at $\alpha = 12.2°$ and $Re_\infty = 1.0 \times 10^6$ in the cross section at $x/c = 0.29$. (Lines of constant dynamic pressure q/q_∞)

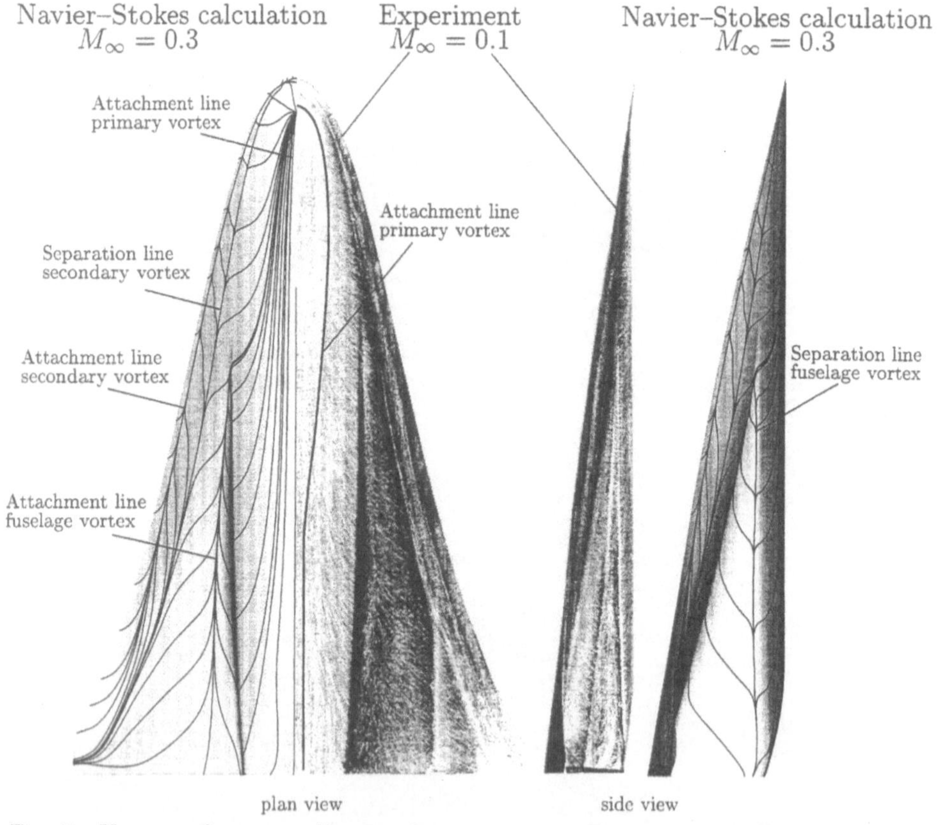

Navier–Stokes calculation
$M_\infty = 0.3$
Experiment
$M_\infty = 0.1$
Navier–Stokes calculation
$M_\infty = 0.3$
Attachment line
primary vortex
Attachment line
primary vortex
Separation line
secondary vortex
Attachment line
secondary vortex
Attachment line
fuselage vortex
Separation line
fuselage vortex
plan view
side view

Fig. 7 Upper surface streamlines on the waverider configuration DLR–F8 at at $\alpha = 12.2°$ and $Re_\infty = 1.0 \times 10^6$. (In the case of the Navier-Stokes calculation, the fuselage is enlarged about 80 % in the side view)

Spatial Navier-Stokes Simulation of Crossflow-Induced Transition in a Three-Dimensional Boundary Layer

G. Bonfigli, M. Kloker

Universität Stuttgart, Institut für Aerodynamik und Gasdynamik (IAG)

Pfaffenwaldring 21, D-70550 Stuttgart, Germany

Summary

Crossflow-dominated transition in the accelerated three-dimensional boundary layer as defined by the canonical DLR-Göttingen Prinzipexperiment is investigated by spatial direct numerical simulation. The numerical scheme is based on a combined 6th-order compact finite-difference/ spanwise Fourier-spectral discretisation with an improved outflow-zone formulation allowing the proper simulation of 3-D non-symmetric flow fields. The implementation of the new outflow-zone is discussed. Results are reported for a parametric study showing the influence of the ratio of the initial amplitudes of stationary and instationary perturbations and for a high-resolution simulation aimed at the investigation of the late stages of transition.

Introduction

The knowledge about transition in three-dimensional accelerating boundary layers (as typically found close to the leading edge of a swept-back wing) is presently limited to the identification of crossflow (CF) waves and vortices as dominating primary perturbations. Still unclear are the interaction mechanisms governing the non-linear stages. However, significant progress would be achieved if the experimental data obtained by Bippes and co-workers at the DLR-Göttingen in the scope of the "Querströmungsprinzipexperiment" (see [5] and Fig. 1 for the most significant parameters of the boundary layer) could be explained and confirmed theoretically.

Direct numerical simulations of the Prinzipexperiment were performed by Kleiser and co-workers (see [8]) using the temporal model, and by W. Müller [6] at IAG using the spatial model. As far as the simulation of W. Müller is regarded no information could be gained with respect to the late stages of transition. Numerical problems and lack of computational power made it impossible to achieve the necessary temporal span and discretisation level.

The work at IAG is continued in the present paper adopting an improved numerical method, exploiting the increase of available computational power, and considering different relative amplitudes for the initial perturbations. Particular attention is paid to the reproduction of events and of flow structures which are supposed to be connected with the high frequency instablility observed within the Prinzipexperiment after saturation of the dominating CF vortices.

Numerical Method

The numerical method relies on the explicit time integration of the complete 3-D incompressible Navier-Stokes equations in the rectangular domain shown in Fig. 1. The steady laminar solution (base flow) is computed as preprocessing before starting the simulation of the unsteady disturbance flow. Hereafter the subscript "b" indicates quantities of the base flow, while a prime identifies perturbation quantities. For the general variable f holds $f = f_b + f'$. A deformation of the meanflow $< f >$ is represented by $< f' > \neq 0$.

Non-dimensional variables are introduced using the reference quantities $\bar{L} = 0.1\ m$, $\bar{U}_\infty = 14\ m/s$ and $\bar{p} = 1.225 kg/m^3$. Indicating dimensional quantities with overlined symbols, the non-dimensionalisation relations may be written as follows:

$$x = \frac{\bar{x}}{\bar{L}}, \quad y = \frac{\bar{y}}{\bar{L}} \cdot \sqrt{Re}, \quad z = \frac{\bar{z}}{\bar{L}}, \quad t = \bar{t} \cdot \frac{\bar{U}_\infty}{\bar{L}},$$

$$u = \frac{\bar{u}}{\bar{U}_\infty}, \quad v = \frac{\bar{v}}{\bar{U}_\infty} \cdot \sqrt{Re}, \quad w = \frac{\bar{u}}{\bar{U}_\infty}, \quad Re = \frac{\bar{U}_\infty \cdot \bar{L}}{\bar{\nu}}. \tag{1}$$

The non-dimensional vorticity vector is correspondingly defined as:

$$\omega_x = \frac{1}{Re}\frac{\partial v}{\partial z} - \frac{\partial w}{\partial y}, \quad \omega_y = \frac{\partial w}{\partial x} - \frac{\partial u}{\partial z}, \quad \omega_z = \frac{\partial u}{\partial y} - \frac{1}{Re}\frac{\partial v}{\partial x}. \tag{2}$$

As far as the spatial discretisation of the perturbation flow is regarded, a Fourier expansion is introduced for the spanwise direction. For the general quantity f' holds

$$f'(x, y, z, t) = \sum_{k=-K}^{+K} \hat{f'}_k(x, y, t)\, e^{ik\gamma z}, \qquad \gamma = \frac{2\pi}{\lambda}, \tag{3}$$

where the fundamental wave length λ follows from the periodicity of the physical flow. Since all physical quantities are real, the relation $\hat{f'}_k = \hat{f'}^*_{-k}$ holds for all variables, and the Fourier coefficents $\hat{f'}_k$ have to be computed only for $0 \leq k \leq K$.

The resolving equations are obtained from the vorticity formulation of the Navier-Stokes equations (see, e.g., [7]) through projection onto the Fourier components of the spanwise discretisation:

$$\frac{\partial \hat{\omega}'_{x\,k}}{\partial t} = \tilde{\Delta}\hat{\omega}'_{x\,k} + \hat{X}_k, \qquad 0 \leq k \leq K, \tag{4a}$$

$$\frac{\partial \hat{\omega}'_{y\,k}}{\partial t} = \tilde{\Delta}\hat{\omega}'_{y\,k} + \hat{Y}_k, \qquad 0 \leq k \leq K, \tag{4b}$$

$$\frac{\partial \hat{\omega}'_{z\,k}}{\partial t} = \tilde{\Delta}\hat{\omega}'_{z\,k} + \hat{Z}_k, \qquad 0 \leq k \leq K, \tag{4c}$$

$$\frac{\partial^2 \hat{u}'_k}{\partial x^2} - k^2\gamma^2 \cdot \hat{u}'_k = -ik\gamma \cdot \hat{\omega}'_{y\,k} - \frac{\partial^2 \hat{v}'_k}{\partial x \partial y}, \qquad 1 \leq k \leq K, \tag{4d}$$

$$\tilde{\Delta}\hat{v}'_k = ik\gamma \cdot \hat{\omega}'_{x\,k} - \frac{\partial \hat{\omega}'_{z\,k}}{\partial x}, \qquad 0 \leq k \leq K, \tag{4e}$$

$$\frac{\partial^2 \hat{w}'_k}{\partial x^2} - k^2\gamma^2 \cdot \hat{w}'_k = \frac{\partial \hat{\omega}'_{y\,k}}{\partial x} - ik\gamma \cdot \frac{\partial \hat{v}'_k}{\partial y}, \qquad 1 \leq k \leq K. \tag{4f}$$

Thereby $\tilde{\Delta}$ is the Laplace operator adapted for the used spectral formulation,

$$\tilde{\Delta} = \frac{1}{Re}\frac{\partial^2}{\partial x^2} + \frac{\partial^2}{\partial y^2} - \frac{k^2\gamma^2}{Re},$$

and $\hat{X}_k$, $\hat{Y}_k$ and $\hat{Z}_k$ represent the spectral components of the non-linear terms of the Navier-Stokes equations (see [7]). The initial value problems (IVPs)

$$\frac{\partial \hat{u}'_0}{\partial x} = -\frac{\partial \hat{v}'_0}{\partial y}, \quad \frac{\partial \hat{w}'_0}{\partial x} = \hat{\omega}_{y\,0} \tag{5}$$

are used for the computation of the harmonic components $k = 0$ of u' and w'.

Dirichlet boundary conditions are set at the inflow boundary requiring the perturbation to be equal to zero for both velocity and vorticity.

At the external boundary the vorticity vector is set to zero, and exponential decay is imposed for v'. No further conditions are given on such boundary for u' and w', which are computed in the same way as in the internal part of the integration domain (Eq. (4d), (4f) and (5)).

No-slip conditions are given for u' and w' at the wall. Steady and unsteady perturbations may be introduced through suction and blowing ($v' \neq 0$) within the perturbation stripe (see Fig. 1). The actual distribution of the wall-normal velocity v_{wall} is given by the expression

$$v'_{wall} = 2\sqrt{Re} \left\{ \sqrt{2}A_{(0,1)} \cdot f_{(0,1)}(x)cos(\gamma z) + A_{(1,1)} \cdot f_{(1,1)}(x)cos(\gamma z - \beta t) \right\} , \qquad (6)$$

where $A_{(0,1)}$ and $A_{(1,1)}$ are amplitude coefficients, β is the frequency of the unsteady perturbation and $f_{(0,1)}(x)$ and $f_{(1,1)}(x)$ are the shape functions sketched in Fig. 1. The wall vorticity components ω'_x and ω'_z are computed solving the following initial or boundary value problems

$$\frac{\partial \hat{\omega}'_{xk}}{\partial x} = -\frac{\partial \hat{\omega}'_{yk}}{\partial y}, \qquad \hat{\omega}'_{xk}|_{in} = 0, \qquad\qquad k = 0, \quad (7a)$$

$$\frac{\partial^2 \hat{\omega}'_{xk}}{\partial x^2} - k^2\gamma^2\hat{\omega}'_{xk} = -\frac{\partial^2 \hat{\omega}'_{yk}}{\partial x\partial y} + ik\gamma\tilde{\Delta}\hat{v}'_k, \qquad \hat{\omega}'_{xk}|_{in} = 0, \quad \frac{\partial^2 \hat{\omega}'_{xk}}{\partial x^2}|_{out} = 0, \quad k \neq 0, \quad (7b)$$

$$\frac{\partial \hat{\omega}'_{zk}}{\partial x} = ik\gamma\hat{\omega}'_{xk} - \tilde{\Delta}\hat{v}'_k, \qquad \hat{\omega}'_{zk}|_{in} = 0, \qquad\qquad \forall k, \quad (7c)$$

where the subscripts in and out indicate boundary conditions to be set at the inflow and at the outflow respectively.

A damping zone, where the disturbance vorticity is artificially suppressed by multiplication with the damping function $G_D(x)$ (see [4]), precedes the outflow boundary as shown in Fig. 1. The region where $G_D(x)$ falls to zero is followed by an equally long region in which the disturbance velocity decays. This way the amplitudes of the perturbations convecting through the outflow boundary are negligibly small and no reflection is observed.

The actual boundary conditions are of minor relevance as far as they do not contrast with the aimed decay of the perturbation amplitudes. In the present implementation, all second order derivatives in x are set to zero at the outflow boundary. This, together with the suppression of the vorticity and equation (4e), forces v' to zero at the outflow. The following compatibility equations are used for u' and w':

$$\frac{\partial^2 \hat{u}'_k}{\partial y^2} = \frac{\partial \hat{\omega}'_{zk}}{\partial y} + \frac{1}{Re} \cdot \frac{\partial^2 \hat{v}'_k}{\partial x\partial y}, \qquad \hat{u}'_k|_{wall} = 0, \quad \frac{\partial \hat{u}'_k}{\partial y}|_{outer} = \frac{1}{Re}\frac{\partial \hat{v}'_k}{\partial x}, \qquad k \neq 0, \quad (8a)$$

$$\frac{\partial^2 \hat{w}'_k}{\partial y^2} = -\frac{\partial \hat{\omega}'_{xk}}{\partial y} + \frac{ik\gamma}{Re} \cdot \frac{\partial \hat{v}'_k}{\partial y}, \qquad \hat{w}'_k|_{wall} = 0, \quad \frac{\partial \hat{w}'_k}{\partial y}|_{outer} = \frac{ik\gamma}{Re}\hat{v}'_k, \qquad k \neq 0. \quad (8b)$$

Once the right hand sides are known, they provide discoupled boundary value problems. The expressions marked with $wall$ and $outer$ assign boundary conditions for the wall and for the upper boundary, respectively.

Sixth-order compact differences are used for the chordwise (x) and wall-normal (y) directions [3]. While the grid step Δx keeps constant over the whole integration domain, Δy is, in the region close to the wall, twice as fine as in the external part of the boundary layer [3]. A 4-step 4th-order Runge-Kutta scheme is applied for the time integration.

The computation of each new Runge-Kutta step starts with the explicit computation of $\underline{\omega}'$ in all grid points except on the plate surface (equations (4a-c)). Thereafter the new vorticity distribution, with the exception of the harmonic component $k = 0$ of ω'_y, is forced to zero within the outflow zone. The wall-normal velocity v' is then computed solving (4e) and the vorticity components ω'_x and ω'_z are evaluated at the wall by integrating equations (7a-c) and superimposing the damping function $G_D(x)$. As a final step, u' and w' are computed solving equations (4d), (4f) and (5) with prior calculation of the boundary values for $k > 0$ using (8a-b).

While most of the mentioned boundary conditions have been directly taken over from the well tested numerical method for symmetrical 3-D flows (here symmetry means mirror symmetry with respect to the plane $z = 0$), crucial modifications were needed with respect to the conditions at the plate surface and the implementation of the damping zone (the formulation used in W. Müllers's work [6] did not work in all cases). Extensive testing showed that the strong asymmetry of the perturbation flow, direct consequence of the asymmetry of the base flow itself, increases the sensitivity of the numerical problem to violations of the solenoidality of the vorticity vector. Symmetry implies the solenoidality of $\underline{\omega}'$ for the spectral component $k = 0$ ($\hat{\omega}'_{x\,0} \equiv 0$, $\hat{\omega}'_{y\,0} \equiv 0$, $\partial\{\hat{\omega}'_{z\,0}cos(0 \cdot \gamma z)\}/\partial z \equiv 0$). In the non-symmetric case, errors in the solenoidality condition may occur also for $k = 0$. The consequences are fatal since the harmonic $k = 0$ is associated with a constant distribution along the z-axis and eventual inconsistencies may not be recovered through relaxation in that direction. As a matter of fact a long-time stable numerical code could be implemented only by enhancing the fulfillment of the solenoidality condition for $k = 0$. This has been achieved with the introduction of the solenoidality condition (7a) as a boundary condition for $\hat{\omega}'_{x\,0}$ and with the removal of the vorticity suppression in the damping zone for $\hat{\omega}'_{y\,0}$.

The effect of inappropriate boundary conditions is shown in Fig. 2A for a simulation case (R1-SHORT) with simultaneous perturbation of steady and unsteady modes (see Table 1 for the simulation parameters). The modal disturbance amplitudes $\tilde{u}'_{(h,k)} = max_y\hat{u}'_{(h,k)}$ resulting from the Fourier expansion of u' in time (index h, fundamental frequency β) and in spanwise direction (index k, fundamental wave number γ) is plotted as a function of the chordwise coordinate x. The plateau P in the curves of modes $(0,0)$ and $(1,0)$ (see Fig. 2A) is clearly unphysical. Its amplitude level grows in time and eventually the computation explodes. On the contrary, no numerical instability appears in Fig. 2B, where the amplitude curves still refer to R1-SHORT but have been computed with the numerical method discussed above. The comparison of results from simulations with different positions of the damping zone (R1-SHORT and R1, also Fig. 2B) shows that noticeable upstream influence is limited to $1 \sim 2$ chordwise wavelengths of the fundamental mode $(1,1)$ ($\lambda \approx 0.07$) or about 25 displacement thicknesses ($\delta_1/\sqrt{Re} \approx 0.007$)). The total chordwise extension of the region, where the solution is unphysically influenced by the damping, amounts typically to about $75\delta_1/\sqrt{Re}$.

Numerical results

The results discussed in this section refer to studies concerning the numerical simulation of transition in the boundary layer of the "Prinzipexperiment". The first set of simulations (R1, R2, R5 and R10) is devoted to the investigation of the influence of the upstream disturbance conditions onto the non-linear downstream disturbance development. The second block of results refers to the high-resolution simulation JERRY, in which the development of the perturbation system also considered in R5 (large amplitude of the CF-vortex) is reproduced up to the late stages of transition. Relevant parameters for the discussed simulations are collected in Table 1.

While linear stability theory and experimental data provide exact indications about the frequency ($\bar{\beta} \approx 135Hz$) and the spanwise wave length ($\bar{\lambda}_z \approx 12.0mm$) for both CF vortex and wave, little evidence is available with respect to the role played by the ratio of their initial amplitudes in determining the transition mechanism. In order to clear this question the simulations R1, R2, R5 and R10 have been driven with simultaneous excitation of the CF-vortex and CF-wave, varying the amplitude of the former and keeping constant that of the latter.

A bifurcation could be detected in the dependence between disturbance and solution. Figures 3A-D show the amplitude development for different modes (h, k) of the velocity component u'_s obtained projecting the vector $\underline{u}'$ onto the direction tangential to the local potential streamline. Two main types of transition are recognized depending on whether the dominant role is played by unsteady (R1, R2) or steady (R5, R10) perturbations. In all considered cases, the CF vortex (mode $(0, 1)$) is the mode with the largest initial amplitude. Nevertheless, the traveling wave (mode $(1, 1)$) undergoes larger amplification and, if the gap between the initial amplitudes is small enough, the curve for $\tilde{u}'_{s\,(1,1)}$ gets over that for $\tilde{u}'_{s\,(0,1)}$. As soon as the dominant mode reaches an amplitude close to $2-3$ per cent of the external velocity, non-linear interactions become relevant and the growth of the other perturbed mode is suppressed.

In R1 and R2 the growth of $(1, 1)$ seems unaffected by the presence of the vortex and the amplification curves overlap with good approximation over the whole integration domain. In R5 and R10, where $(0, 1)$ dominates, saturation for $(1, 1)$ appears earlier and at considerably lower amplitude levels. The amplitudes of modes $(1, 1)$, $(1, -1)$ and $(1, 0)$ level up and the existence of a coupling mechanism reminescent of the secondary instability model of Fischer and Dallmann [1] seems plausible. Indeed the modes considered in [1] for a vortex-oriented coordinate system (VOS) could match with the mentioned modes of the plate-oriented system (POS) for a suitable value of the detuning parameter σ. The POS modes $(1, 0)$, $(1, 1)$ and $(1, -1)$ would correspond to the VOS-modes $(1, -\sigma)$, $(1, 1 - \sigma)$ and $(1, -1 - \sigma)$, respectively. In any case, the presence of a pair of travelling waves, $(1, 1)$ and $(1, -1)$, with similar amplitudes and almost opposite wave vectors, leads to a strong reduction of the unsteady spanwise wave motion, as observed also in the experiment [5].

As expected, the spanwise deformation of the time-averaged velocity profiles increases with the amplitude of the stationary vortex and is maximal in the case R10 (see inserts in Fig. 3A and 3C). This and the experimental evidence, according to which the high-frequency instabilities (HFI) preceeding the final breakdown are spatially localized in regions with marked deformation of the time-averaged flow (marked concavity of the downstream velocity profiles), suggested to adopt the perturbation pair considered in R5 for the simulation JERRY as well. Goal of such simulation is indeed the investigation of the late stages of transition with particular attention to the phenomena observed in Göttingen. Vortex-dominated simulations seem closer to the experimental conditions also with respect to the detected retardation of the spanwise wave motion, which favours the HFI mechanism by rendering the base conditions more steady and localized. Up to the present day the simulation JERRY could not be driven for longer than 9 perturbation periods. Proceeding further in time an explosive growth of the boundary-layer dynamics toward the external region of the integration domain takes place and gives rise to undue interactions between the perturbation and the external boundary, which eventually make the computation explode. A relaxation may be expected by enlarging the height of the itegration domain.

Figures 4A-B show amplification curves for different fundamental modes, computed for the last available simulation period. They provide indications also for the regions where the periodical state is not reached yet or will never be reached. Between $x = 3.2$ and $x = 3.4$, all modes saturate and their amplitudes do not vary significantly. From $x = 3.4$ onwards modes with high spanwise wave number start growing again and at $x = 4.0$ the final breakdown sets on abruptly. Modes with high spanwise wave numbers undergo the greatest relative amplitude jumps growing by a factor of 100 within $x = 4.0$ and $x = 4.1$. At $x = 4.1$, the final stage of chaotic motion is already reached. The amplification plots level off and oscillate irregularly up to the outflow region. Even if the discussed scenario appears compatible with the onset of a HFI as observed in the experiment, absolute certainty is not given yet, since the available DNS data do not allow a proper high frequency analysis.

The visualization of vortical structure by means of the λ_2-criterion [2] is shown in Fig 5A for the time $t = 9T$ (T period of the unsteady perturbation). The comparison with the same visualization for the simulation VORTEX-5, where the stationary mode $(0, 1)$ has been perturbed alone, (see Fig. 5B) highlights the effect of the travelling wave onto the CF-vortex and onto the secondary vortex developing close to the principal one, on the leeward side with respect to the crossflow. A vortical structure climbing toward the external part of the boundary layer is highlighted in Fig. 5D zooming up the region preceding the final breakdown. The same visualization for $t = (8 + \frac{4}{5})T$ (Fig 5C) shows the same structure in a more upstream position and provides a clear proof of its downstream convection. Similar phenomena have been detected also in the scope of temporal DNS. Structures with a very similar shape and undergoing the same convecting movement have been observed by Lerche (see [5], Fig. 76) while visualizing the spatial regions just upstream of the breakdown.

References

[1] Fischer, T. M.; Dallmann, U. (1991): Primary and secondary stability analysis of a three-dimensional boundary-layer flow. *Phys. Fluids A*, 3(10), October 1991.

[2] Jeong, J.; Hussain, F. (1995): On identification of a vortex. *J. Fluid Mech.*, Vol. 285, pp. 69-94.

[3] Kloker, M. (1998): A robust high-resolution split-type compact FD-scheme for spatial direct numerical simulation of boundary-layer transition. *Applied Scientific Research* 59 (4), 1998, pp. 353-377, Kluwer Acad. Publishers.

[4] Kloker, M.; Konzelmann, U.; Fasel, H. (1993): Outflow boundary conditions for spatial Navier-Stokes simulations of transition boundary layers. *AIAA Journal*, Vol. 31, No. 4, April 1993.

[5] Lerche, T. (1997): *Experimentelle Untersuchung nichtlinearer Strukturbildung im Transitionsprozeß einer instabilen dreidimensionalen Grenzschicht.* Fortschrittberichte VDI, Reihe 7: Strömungstechnik, Nr.310.

[6] Müller, W. (1995): *Numerische Untersuchung räumlicher Umschlagvorgänge in dreidimensionalen Grenzschichtströmungen.* Dissertation, Universität Stuttgart.

[7] Wassermann, P.; Kloker, M. (1999): Direct numerical simulation of the development and control of boundary-layer crossflow vortices. In Nitsche, W.; Hilbig, R. (Eds.): *New results in numerical and experimental fluid dynamics.* NNFM, 11.STAB/DGLR Symposium, Berlin, November 1998, Vieweg Verlag.

[8] Wintergerste, T.;Kleiser, L. (1996): DNS of transition in a 3-D boundary layer. In Henkes, R.; Von Ingen, J. (Eds.): *Transitional boundary layers in aeronautics.* North-Holland Publishers.

Table 1: Parameters of the simulations: amplitude of stationary and instationary perturbations, upstream boundary of the damping region, discretisation parameters.

Simulation	$A_{(0,1)}$	$A_{(1,1)}$	$x_{damping}$	Δx	Δy	K
R1	$3.4 \cdot 10^{-4}$	$1.5 \cdot 10^{-4}$	3.88	$1.3089 \cdot 10^{-3}$	$1.392 \cdot 10^{-1}$	5
R2	$6.8 \cdot 10^{-4}$	$1.5 \cdot 10^{-4}$	3.88	$1.3089 \cdot 10^{-3}$	$1.392 \cdot 10^{-1}$	5
R5	$1.7 \cdot 10^{-3}$	$1.5 \cdot 10^{-4}$	3.88	$1.3089 \cdot 10^{-3}$	$1.392 \cdot 10^{-1}$	5
R10	$3.4 \cdot 10^{-3}$	$1.5 \cdot 10^{-4}$	3.88	$1.3089 \cdot 10^{-3}$	$1.392 \cdot 10^{-1}$	5
R1-SHORT	$3.4 \cdot 10^{-4}$	$1.5 \cdot 10^{-4}$	3.48	$1.3089 \cdot 10^{-3}$	$1.392 \cdot 10^{-1}$	5
VORTEX-5	$1.7 \cdot 10^{-3}$	0.0	5.50	$1.3089 \cdot 10^{-3}$	$1.392 \cdot 10^{-1}$	5
JERRY	$1.7 \cdot 10^{-3}$	$1.5 \cdot 10^{-4}$	5.50	$1.3089 \cdot 10^{-3}$	$1.392 \cdot 10^{-1}$	21

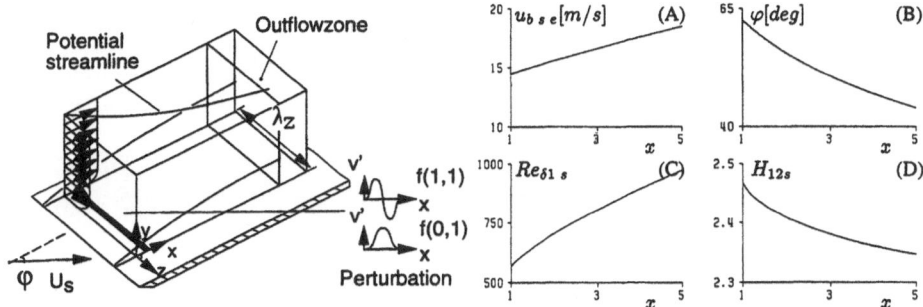

Figure 1: Integration domain and parameters of the base flow: potential velocity (A), local sweep back angle (B), local Reynolds number (C), shape factor (D). Here and in the following figures, the subscript s indicates quantities referring to the direction tangent to the local streamline. The subscipt e is used for quantities of the external potential flow.

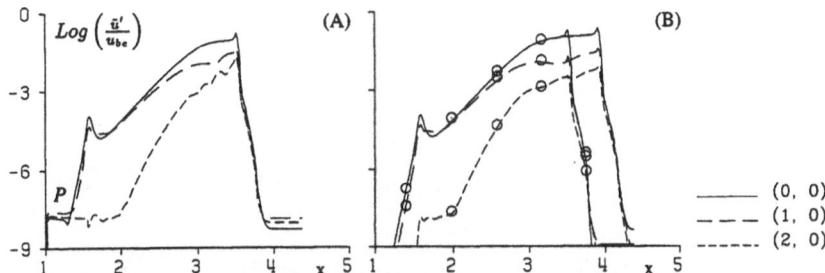

Figure 2: Chordwise amplitude development of the most significant modes for R1-SHORT computed with the original version of the numerical code (A), and for R1-SHORT (B with symbols) and R1 (B without symbols) computed with the final version of the numerical code.

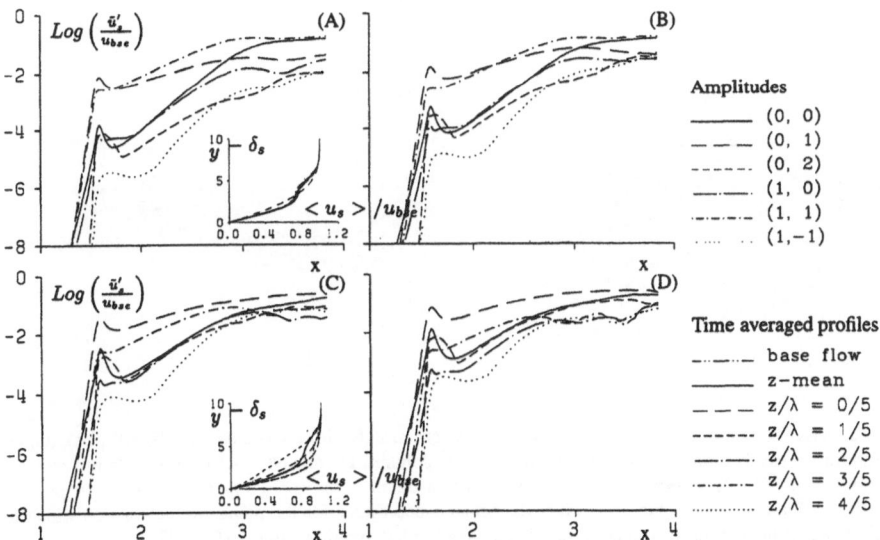

Figure 3: Chordwise amplitude development of the most significant modes, and time-averaged velocity profiles (inserts) at $x = 3.4$ at various spanwise positions for R1 (A), R2 (B), R5 (C) and R10 (D).

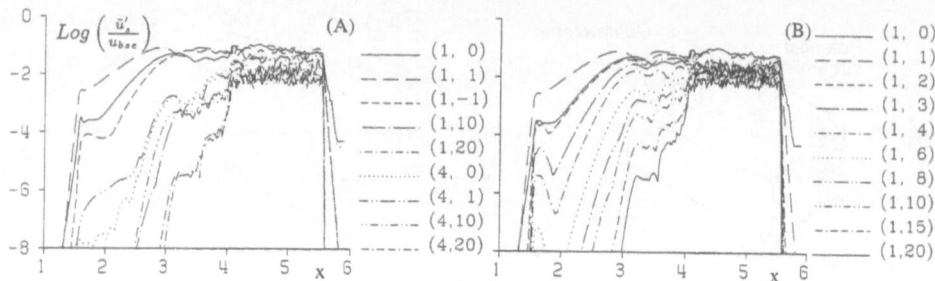

Figure 4: Chordwise amplitude development of the most significant modes for JERRY.

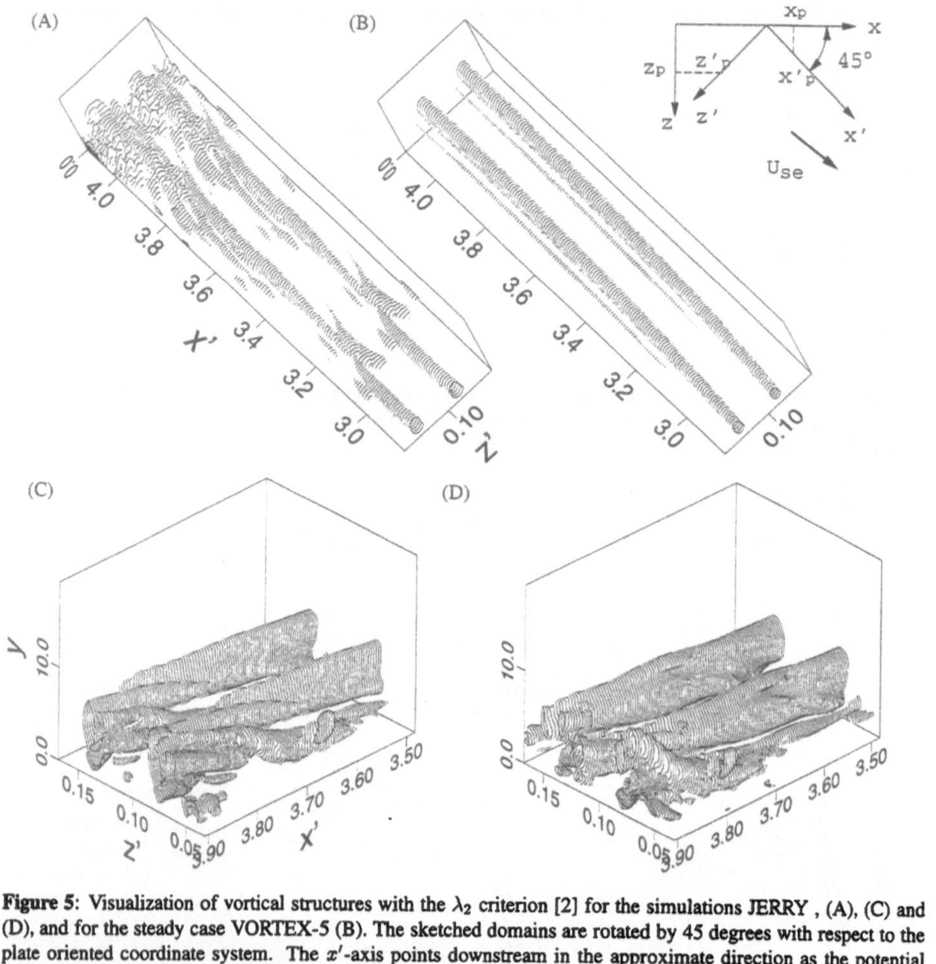

Figure 5: Visualization of vortical structures with the λ_2 criterion [2] for the simulations JERRY , (A), (C) and (D), and for the steady case VORTEX-5 (B). The sketched domains are rotated by 45 degrees with respect to the plate oriented coordinate system. The x'-axis points downstream in the approximate direction as the potential streamlines. The numerical value of x' gives the actual x-coordinate for the points on the edge on which the coordinate labels lie. Figures (A) and (D) refer to the time $t = 9T$, figure (C) to $t = (8 + \frac{4}{5})T$.

Strake Effects On the Turbulent Fin Flowfield Of a High–Performance Fighter Aircraft

C. Breitsamter

Lehrstuhl für Fluidmechanik
Technische Universität München
Boltzmannstr. 15, 85748 Garching, Germany

SUMMARY

Experimental results for the low–speed fin flow environment of a canard–delta fighter configuration are presented. In particular, the investigation deals with the influence of small forebody strakes on the fin buffet situation. Using advanced hot–wire techniques instantaneous velocities are measured resulting in detailed flowfields of mean, rms and spectral quantities. Various angles of attack at nonsymmetric freestream are tested at a Reynolds number of about 1 million. Especially at sideslip and high incidence, a strong interaction between strake and canard vortex systems hampers the inboard movement of the burst windward wing leading–edge vortex. Consequently, the turbulence intensities at a center–line fin station are substantially lower for the strake–on compared to the strake–off configuration. Moreover, the observed narrow–band concentration of turbulent kinetic energy is also less severe that the excitation level on fin structural modes is reduced.

NOMENCLATURE

c_r, c_c	wing root chord, canard root chord, $[m]$
f	frequency, $[Hz]$
k	reduced frequency, $f\, l_\mu/U_\infty$
l_μ	wing mean aerodynamic chord, $[m]$
Re_{l_μ}	Reynolds number based on l_μ, $U_\infty l_\mu/\nu$
rms	root mean square value
S	spectral density, $[1/Hz]$
s, s_F	wing semi–span, fin span, $[m]$
U_∞	freestream velocity, $[m/s]$
u, v, w	axial, lateral and vertical velocity (wind tunnel–axis system), $[m/s]$
$u_{xz_{rms}}$	sum of axial and vertical rms velocity, $\sqrt{1/2\,(\overline{u'^2}+\overline{w'^2})}$
Y, Z	nondimensionalized coordinates of the measurement plane, referred to s
x, y, z	wind tunnel axis system and model coordinate system, resp. $[m]$
α, β	aircraft angle of attack, aircraft angle of sideslip, $[°]$
φ_W, φ_C	wing leading–edge sweep, canard leading–edge sweep, $[°]$
φ_F	fin leading–edge sweep, $[°]$
Λ_W, Λ_C	wing aspect ratio, canard aspect ratio
ν	kinematic viscosity, $[m^2/s]$
Δf	frequency resolution of power spectral density
$\bar{\cdot}\,,\ \cdot'$	time–averaged, fluctuation part

INTRODUCTION

For modern fighter aircraft poststall maneuvering is a key design factor [1]. Consequently, operational flight hours at high angle of attack are significantly increased. Under these conditions the fin buffet problem has become a crucial issue endangering long fatigue life and vibration free control [2]. The buffet loads are attributed to the highly turbulent flow in the wake of burst leading–edge vortices affecting the fin(s). Unsteady flowfield velocity and induced surface pressure show narrow–band peaked distributions linked to a helical mode instability of the breakdown flow [3]. Numerous investigations have been conducted aimed at understanding and reducing the buffet loads [4–7]. Corrective actions include mainly design modifications such as LEX fences at the F/A–18 and stiffening of the fin structure [5]. Recently, concepts of active control are intensively investigated [6].

In this context an extensive research program has been conducted over the last years at the Lehrstuhl für Fluidmechanik of the Technische Universität München. It focuses on the characteristics of the turbulent fin flowfield associated with leading–edge vortices in pre– and post–breakdown stage. Detailed analysis of the very complex flow phenomenology results in general fin buffet prediction parameters [3]. In addition, an evaluation method has been delevoped to determine fin buffet loads based only on the measured unsteady flowfields [8].

MEASUREMENT TECHNIQUE AND TEST PROGRAM

Description of Model and Facility

The wind tunnel model used represents a high–agility aircaft of canard–delta wing type, Fig. 1. Major parts of the model are: nose section, front fuselage including rotatable canards and a single place canopy, center fuselage with delta–wing section and a through–flow air intake underneath, and rear fuselage with nozzle section. A center–line (single) fin can be added which is part of an insert that is bolted to the rear fuselage. The configuration is completed by small strakes mounted at the front fuselage under the canopy, Fig. 1b. All model parts are made of stainless steel. For these tests leading– and trailing–edge flap deflections as well as the canard setting angle were set to zero degree.

The model was sting mounted on its lower surface from a moving support strut. This arrangement enables flowfield measurements to a great extent free from interference. The computer controlled model support provides an incidence range from 0 to 31.5 deg and models may be yawed and rolled 360 deg. The experiments were carried out in a Göttingen type low–speed wind tunnel facility of the Lehrstuhl für Fluidmechanik of the Technische Universität München. The open test section is 1.5 m in diam. and 3 m long. Maximum usable velocity is 55 m/s. Turbulence intensity is less than 0.4%.

Measurement of Unsteady Velocity

Dual–sensor hot–wire probes were used to measure the fluctuating velocity components. The sensors consist of 5–μm–diam. platinum–plated tungsten wires forming a measurement volume of about 1 mm^3. In order to achieve best angular resolution the sensors are arranged perpendicular to each other. The probes were operated by a multi–channel constant–temperature anemometer system (DISA C). Bridge output voltages were low–pass filtered at 1000 Hz before digitization and simultaneously sampled with 12–bit precision at 3000 Hz over 26.24 sec. Thus each sample block contains 78720 points in the time domain, producing 39360 values in the complex frequency domain. The sampling parameters were achieved by preliminary tests to ensure that all significant flowfield phenomena are detected. Statistical accuracy of the calculated quantities was considered as well that the error for the mean and standard deviation and spectral density estimation is less than 0.4, 1 and 3%, respectively. The method to evaluate the instantaneous velocity vector is based on look–up tables derived from the full velocity and flow angle calibration of all sensors [3].

Description of Tests

Flowfields were measured in a plane perpendicular to the model x–axis at a possible single or twin–fin station, Fig. 2. The standard/large plane contains 45/77 points spanwise and 11/16 points vertically. The survey points were evenly spaced giving a grid resolution of 0.014 in the spanwise and 0.020 in the vertical direction referred to the wing span. Tests were made at seven angles of attack, namely at $\alpha = 0°$, $15°$, $20°$, $25°$, $28°$, $30°$ and $31.5°$ at symmetric freestream and at sideslip $\beta = 5°$. The freestream reference velocity U_∞ was kept constant at 40 m/s giving a Reynolds number of $Re_{l_\mu} = 0.97 \times 10^6$ based on the wing mean aerodynamic chord. At all tests turbulent boundary layers were present at wing and control surfaces. The experiments were performed with and without the single fin to identify interference effects. Results given here are for sideslip and fin–off only. The complete studies are reported in Ref. 3.

RESULTS AND DISCUSSION

Vortex Flow Features

The flowfield in the wake of the considered fighter configuration is dominated by several vortex systems, mainly produced by the delta wing and the canard. From laser light sheet tests vortex trajectories and breakdown locations are obtained. In the measurement plane at $x_W/c_r = 1.13$ the burst wing leading–edge vortices (WLV's), the canard vortex systems consisting of leading– and trailing–edge vortices (CLV's and CTV's) and for strake–on the strake vortices (STV's) are present, Fig. 3. Furthermore, wing tip vortices (WTV's) are attached to the outboard part of the wing vortex sheets.

Patterns of Turbulence Intensity

For the strake–off and strake–on configuration, Fig. 4 shows contours of turbulence intensities obtained from axial and vertical rms velocities $u_{xx_{rms}}/U_\infty$ at $\alpha = 20°$, $25°$, $31.5°$ and $\beta = 5°$. At moderate angle of attack, $\alpha = 20°$, the specified vortex systems can be identified by local rms maxima: rms levels of 20–25% at the lower cross section edges are related to the burst WLV's, rms levels of 7–9% around the midsection indicate the CLV's and CTV's, Fig. 4a. Downstream of the canard trailing–edge the dominant CLV induces velocities on the CTV that the CTV is moved upward following a helical path around the center of the CLV. Therefore, in the regarded measurement plane the CTV's are located upward of the CLV's. Because of sideslip, the starboard (windward) vortex systems are shifted inboard and upward while the port (leeward) vortex systems are displaced outboard and downward. At strake–on, CLV's and STV's are strongly connected by their shear layers, Fig. 4b. The STV cores are positioned outboard and slightly downward of the CLV centers. In addition, a vortex pair shed at the canopy can be found near the midsection at $Z = 0$.

With increasing incidence, $\alpha = 25°$, the region of high turbulence intensity associated with the burst WLV's has grown considerably in both size and strength, Fig. 4c. The interaction between wing and canard vortices is such that the CTV's start to move upward away from

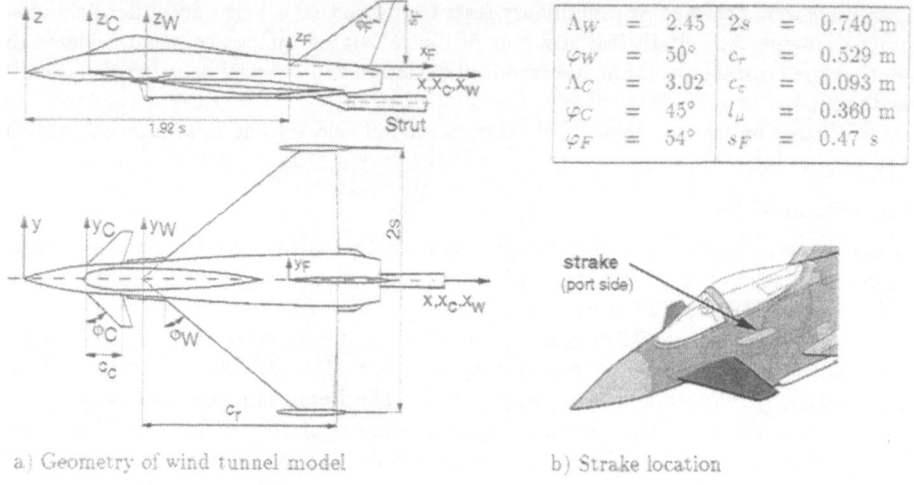

Λ_W	=	2.45	$2s$	=	0.740 m
φ_W	=	50°	c_r	=	0.529 m
Λ_C	=	3.02	c_c	=	0.093 m
φ_C	=	45°	l_μ	=	0.360 m
φ_F	=	54°	s_F	=	0.47 s

a) Geometry of wind tunnel model   b) Strake location

Fig. 1: Delta–Canard Fighter Configuration.

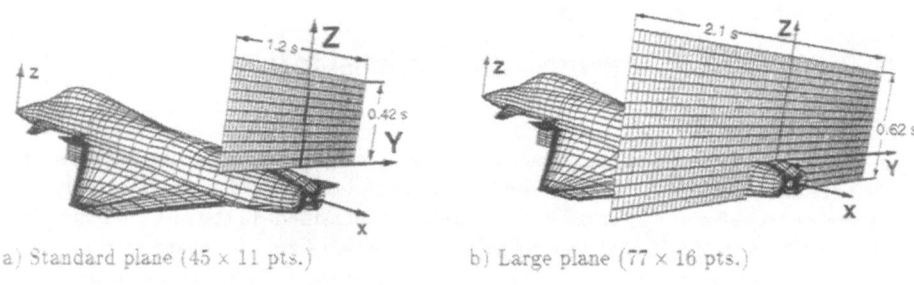

a) Standard plane (45 × 11 pts.)   b) Large plane (77 × 16 pts.)

Fig. 2: Location and discretization of the measurement plane.

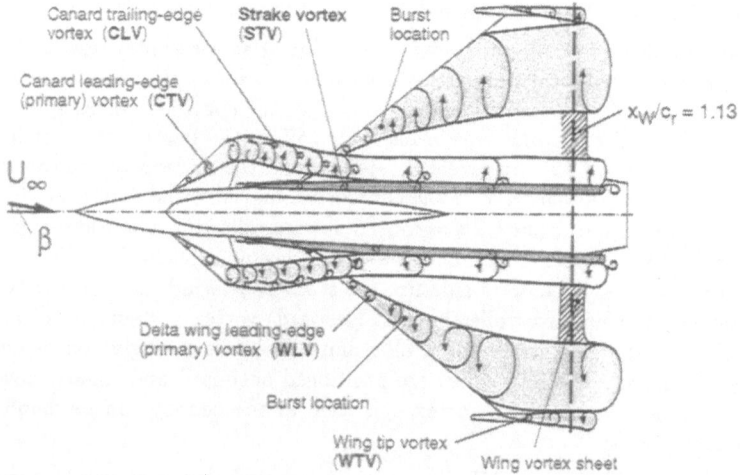

Fig. 3: Schematic representation of vortex flow features for moderate and high angles of attack.

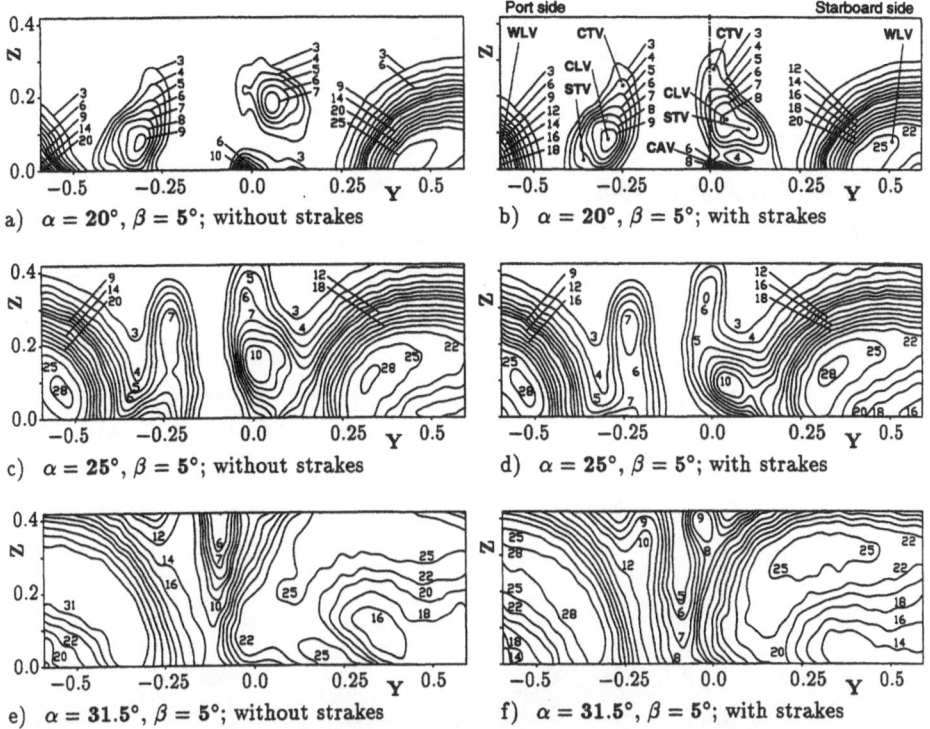

Port side   Starboard side

a) $\alpha = 20°$, $\beta = 5°$; without strakes

b) $\alpha = 20°$, $\beta = 5°$; with strakes

c) $\alpha = 25°$, $\beta = 5°$; without strakes

d) $\alpha = 25°$, $\beta = 5°$; with strakes

e) $\alpha = 31.5°$, $\beta = 5°$; without strakes

f) $\alpha = 31.5°$, $\beta = 5°$; with strakes

Fig. 4: Contours of axial and vertical rms velocity $u_{xz rms}/U_\infty$ in the measurement plane $x_W/c_r = 1.13$ at $\alpha = 20°$, $25°$, $31.5°$ and $\beta = 5°$ for the strake–off and strake–on configuration; $U_\infty = 40 \ m/s$, $Re_{l_\mu} = 0.97 \times 10^6$.

the wing while the CLV's are moved inward and downward to merge with the WLV sheets. Due to the STV downwash the downward shift of the CLV's is stronger for strake–on compared to strake–off, Fig. 4d. Furthermore, the combined CLV's and STV's stunt the expansion of the burst WLV's. In particular, this is evident for the starboard side. Although at high angle of attack, $\alpha = 31.5°$, the maximum rms values have not grown considerably, their region of influence has, Fig. 4e. The CLV's at strake–off as well as the combined CLV's and STV's at strake–on are completely embedded in the inboard part of the wing vortex shear layers, Figs. 4e, f. It can be seen again that at strake–on the inboard shift of the starboard WLV is significantly hampered by the STV interference. Thus, the midsection encounters lower rms levels.

The described sequence of vortex interactions may be also derived from the lateral rms velocity distributions v_{ms}/U_∞, Fig. 5. It is a characteristic feature of the burst WLV that maximum turbulence intensities are concentrated on a limited radial range, Fig 5a, b. Their loci correspond approximately to the points of inflection in the radial profiles of retarded axial flow [3]. At high angle of attack the annular structure of high turbulence intensities is not markedly changed by the embedded STV. However, the inboard rms maxima are decreased, Fig 5c, d.

For a single fin mainly the lateral velocity fluctuations cause buffet whereas for a twin–fin with dihedral both the lateral and vertical components contribute. Fig. 6 summarizes the increase of lateral rms velocity with angle of attack for two vertical stations in the plane of symmetry. It is shown that at sideslip and high-α ($\alpha \geq 25°$), the STV interference leads to a significant reduction in the buffet excitation level of the midsection. This is especially favourable for a center–line fin, although the rms levels are substantially higher than at $\beta = 0°$.

Patterns of Turbulent Shear Stress

Contours of turbulent shear stress $\overline{u'v'}/U_\infty^2$ at $\alpha = 20°$, $25°$, $31.5°$ and $\beta = 5°$ can be taken from Fig. 7. CLV's and CTV's are clearly indicated by local shear stress maxima where

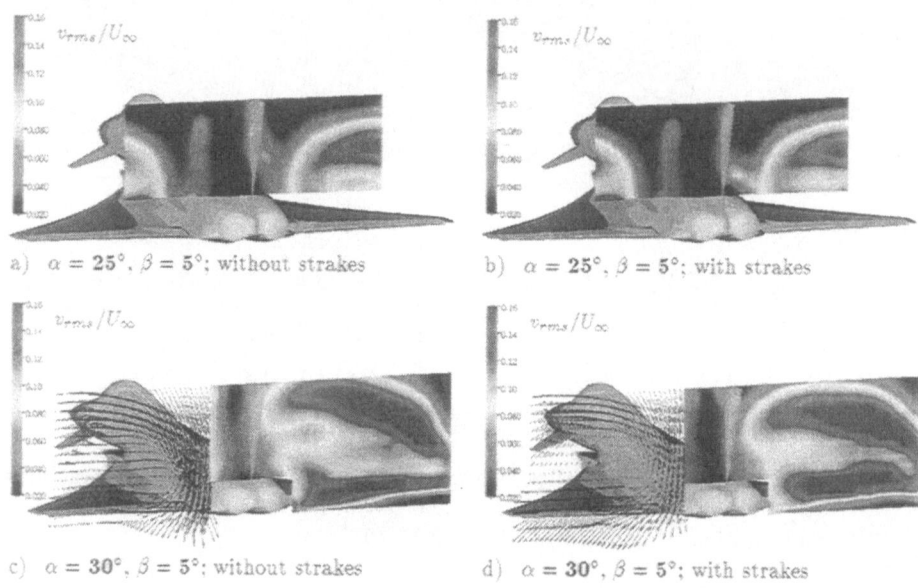

a) $\alpha = 25°$, $\beta = 5°$; without strakes   b) $\alpha = 25°$, $\beta = 5°$; with strakes

c) $\alpha = 30°$, $\beta = 5°$; without strakes   d) $\alpha = 30°$, $\beta = 5°$; with strakes

Fig. 5: Contours of lateral rms velocity v_{rms}/U_∞ in the measurement plane $x_W/c_r = 1.13$ at $\alpha = 25°$, $30°$ and $\beta = 5°$ for the strake–off and strake–on configuration; $U_\infty = 40\ m/s$, $Re_{l_\mu} = 0.97 \times 10^6$.

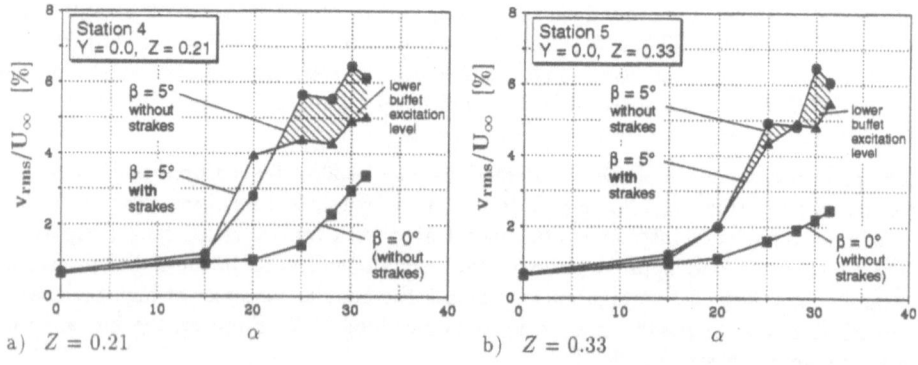

a) $Z = 0.21$ b) $Z = 0.33$

Fig. 6: Lateral turbulence intensity v_{rms}/U_∞ in the plane of symmetry as function of angle of attack for the strake–off and strake–on configuration.

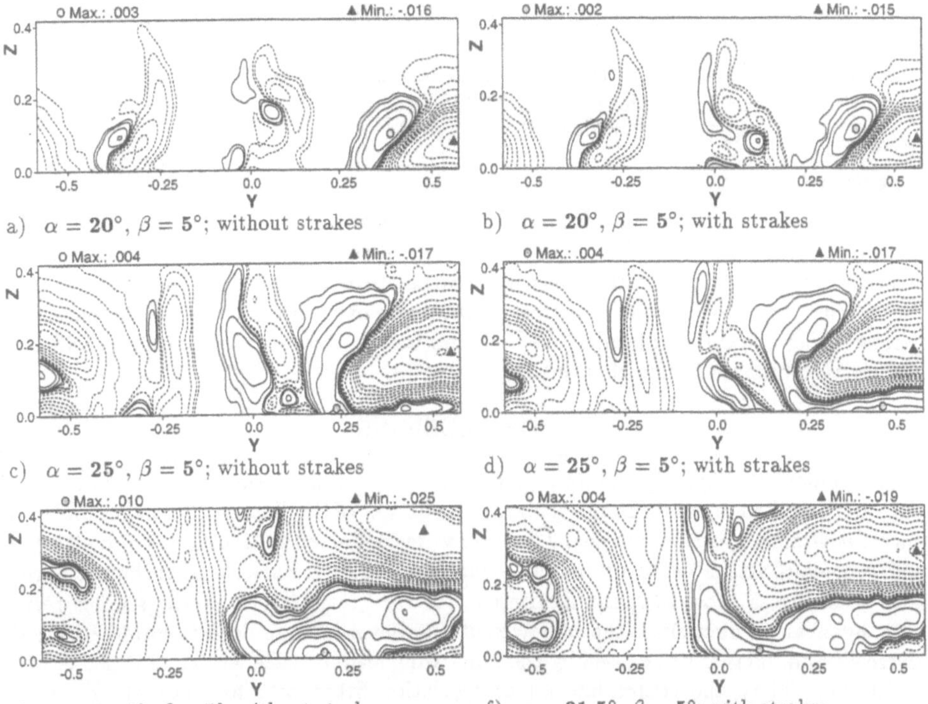

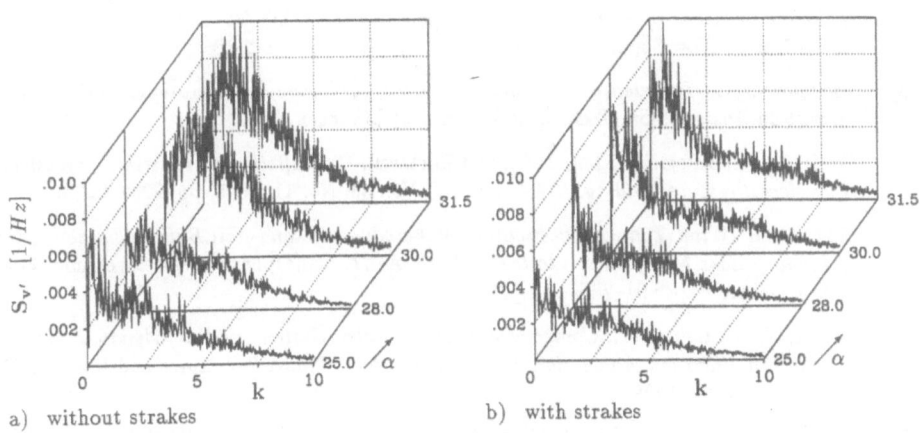

a) $\alpha = 20°$, $\beta = 5°$; without strakes

b) $\alpha = 20°$, $\beta = 5°$; with strakes

c) $\alpha = 25°$, $\beta = 5°$; without strakes

d) $\alpha = 25°$, $\beta = 5°$; with strakes

e) $\alpha = 31.5°$, $\beta = 5°$; without strakes

f) $\alpha = 31.5°$, $\beta = 5°$; with strakes

Fig. 7: Contours of turbulent shear stress $\overline{u'v'}/U_\infty^2$ in the measurement plane $x_W/c_r = 1.13$ at $\alpha = 20°$, $25°$, $31.5°$ and $\beta = 5°$ for the strake–off and strake–on configuration; solid/dashed lines represent positive/negative values; $U_\infty = 40\ m/s$, $Re_{l_\mu} = 0.97 \times 10^6$.

a) without strakes

b) with strakes

Fig. 8: Power spectral densities of lateral velocity fluctuations $S_{v'}$ in the plane of symmetry at station $Z = 0.33$ for the strake–off and strake–on configuration; $\Delta f = 2.93\ Hz$, $U_\infty = 40\ m/s$, $Re_{l_\mu} = 0.97 \times 10^6$.

75

the change from positive to negative values mark the vortex centers, Fig. 7a. Further pairs of positive and negative local shear stress maxima denote the STV's, Fig. 7b. Absolute shear stress maxima exist at the WLV shear layers. Moreover, the shear stress patterns highlight the merging of the starboard CLV and the STV, Figs. 7c, d, as well as the integration process in the WLV sheet, Figs. 7e, f.

Spectral Content

Distinct quasi–periodic oscillations are found within the radial range of maximum turbulence intensities. Because of induction effects the flow of the plane of symmetry exhibit also narrow–band concentrations of turbulent kinetic energy, Fig. 8. The spectral densities of the dominant freqencies increase with angle of attack, Fig 8a. In comparison to strake–off lower peak amplitudes can be found for the strake–on configuration while the frequency content is hardly changed.

CONCLUSIONS

The turbulent flowfields in the fin region of a fighter configuration of canard–delta wing type were studied in detail. In particular, the influence of small forebody strakes on the fin buffet environment was addressed. The tests were conducted at sideslip from moderate to high angles of attack using advance hot–wire anemometry. Patterns of turbulent normal and shear stress show that the interference of the vortices shed at the strakes results in a signifcant reduction of the buffet excitation level in the midsection. Moreover, the narrow–band fluctuations caused by the burst wing leading–edge vortices are also of lower magnitude. Thus, the center–line fin configuration fitted with forebody strakes would encounter less severe fin buffet loads at sideslip.

REFERENCES

[1] Herbst, W. B.: *Future Fighter Technologies.* Journal of Aircraft, Vol. 17, No. 8, Aug. 1980, pp. 561–566.

[2] Ferman, A., Patel, S. R., Zimmermann, N. H., and Gerstenkorn, G.: *A Unified Approach to Buffet Response of Fighter Aircraft Empennage.* AGARD–CP–483, Aircraft Dynamic Loads Due to Flow Separation, Sorrento, Italy, 1990, pp. 2-1-2.18.

[3] Breitsamter, C.: *Turbulente Strömungsstrukturen an Flugzeugkonfigurationen mit Vorderkantenwirbeln.* Dissertation, Technische Universität München, 1997.

[4] Wolfe, S., Canbazoglu, S., Lin, J.-C., and Rockwell, D.: *Buffeting on Fins: An Assessment of Surface Pressure Loading.* AIAA Journal, Vol. 33, No. 11, 1995, pp. 2232–2235.

[5] Shah, G. H: *Wind Tunnel Investigation of Aerodynamic and Tail Buffet Characteristics of Leading-Edge Modifications to the F/A-18.* AIAA Atmospheric Flight Mechanics Conference, AIAA Paper 91–2889, Aug. 1991.

[6] Becker, J., and Luber, W.: *Comparison of Piezoelectric Systems and Aerodynamic Systems for Aircraft Vibration Alleviation.* SPIE Conference " Smart Structures and Materials ", San Diego, March 1998.

[7] Breitsamter, C., and Laschka, B.: *Turbulent Flow Structure Associated with Vortex-Induced Fin Buffeting.* Journal of Aircraft, Vol. 31, No. 4, July-Aug. 1994, pp. 773–781.

[8] Breitsamter, C., and Laschka, B.: *Fin Buffet Pressure Evaluation Based On Measured Flowfield Velocities.* Journal of Aircraft, Vol. 35, No. 5, Sept.-Oct. 1998, pp. 806–815.

Numerical Experiments About the Uniqueness of Aeroelastic Solutions of Elastic Wings in Transonic Flow

G. Britten, J. Ballmann

Lehr- und Forschungsgebiet für Mechanik der RWTH Aachen,
Templergraben 64, 52062 Aachen, Germany

Summary

This work describes the advancements in developing a computational method for the treatment of solid fluid interaction (SOFIA) to analyze and to improve the aerodynamic performance of elastic wings and rotors by direct numerical aeroelastic simulation. The unsteady fluid flow and the structural dynamics are computed simultaneously in a fully coupled manner. The fluid flow is modeled by the Euler equations, or in a new version by the Reynolds-averaged Navier-Stokes equations. The elastic wing is described by a generalized Timoshenko-like beam structure with six degrees of freedom for a material cross-section. Numerical results are shown for the two- and three-dimensional reference configuration of the collaborative research centre SFB401 "Modulation of Flow and Fluid-Structure Interaction at Airplane Wings" of the University of Technology Aachen.

Introduction

Aeroelasticity plays an important role in the design process of large transport aircrafts. Since the transport capacities raise with concurrently reduced structural weights, the deformation of the wings can become as big that their influence on the aerodynamics has to be taken into account already in the aerodynamic design process. In the collaborative research centre SFB401 computational methods are under development as well as experiments focussing on large wings are carried out. The numerical method SOFIA for direct numerical aeroelastic simulation is being developed in the research project B1. In SOFIA the calculation of the flow field and the structural deformation are performed fully coupled and consistent in time. The aerodynamic part consists of either the implicit Euler code IN-FLEX, or in a new version of the Navier-Stokes code FLOWer. ODISA (one-dimensional structural analysis) is a finite-element-based solver for a multi-axial Timoshenko beam model of the wing structure. One might ask the question why we have chosen a simplified model for the structure (and not a full 3D FE code), while on the other hand the fluid flow is modeled by a very complex set of equations. The answer is that it is necessary to solve the Euler or Navier-Stokes equations for the fluid to capture the full non-linearity of aeroelastic problems. Concerning the structure a simplified approach is admissible during the code development and perhaps also justified by the small number of structural model parameters enabling a deeper analysis of fundamental aeroelastic characteristics.

SOFIA is used within the SFB401 to support the experimental projects e.g. in the determination of appropriate pressure gauge positions. When defining the elastic properties of the SFB's 3D cruising flight model the question of unambiguity of the state of aeroelastic equilibrium arised which is discussed in this paper.

Physical Model

Non-Stationary Flow in a Moving Grid

Two different codes for the calculation of the flow field have been implemented in SOFIA, one solving the Euler equations, the other taking into account viscosity and heat transfer in the fluid by integration of the Navier-Stokes equations.

The governing flow equations are solved by finite volume techniques with control volumina dependent on time in order to describe consistently the fluid flow and the motion of the wing's surface. Thereby the nodes on the wing's surface are taken as material points of the wing throughout the motion of the structure, whereas the nodes at the outer boundary, which is the boundary of the computational domain, remain fixed in a rigid body fixed coordinate system. Thus, the grid is deforming with time and the nodes of the finite volumes within the flow field are moved such that a body surface fitting grid is ensured. The three-dimensional Reynolds-averaged Navier-Stokes equations which are implemented in the FLOWer code read in integral form

$$
\frac{\partial}{\partial t} \int_{V(t)} \mathbf{U} dV + \int_{\partial V(t)} \mathbf{F} \mathbf{n} dS + \int_{V(t)} \mathbf{G} dV = 0 \,,
$$

where $\mathbf{U} = (\rho, \rho u, \rho v, \rho w, \rho e)^T$ is the algebraic vector of conserved quantities: density, Cartesian components of momentum and total specific energy density. Contributions due to rigid body motion of the grid, which may be rotating with the angular velocity vector $\omega = (\omega_x, \omega_y, \omega_z)^T$, are contained in the source term $\mathbf{G}$ (see [4]). $\mathbf{F} = \mathbf{F}^c + \mathbf{F}^d$ represents the flux function with its parts $\mathbf{F}^c$ designating the convective and in isentropic flows conservative terms and $\mathbf{F}^d$ denoting the viscous part,

$$
\mathbf{F}^c = \begin{bmatrix} \rho(\mathbf{v} - \mathbf{v}_B) \\ \rho u(\mathbf{v} - \mathbf{v}_B) + p\mathbf{e}_x \\ \rho v(\mathbf{v} - \mathbf{v}_B) + p\mathbf{e}_y \\ \rho w(\mathbf{v} - \mathbf{v}_B) + p\mathbf{e}_z \\ \rho e(\mathbf{v} - \mathbf{v}_B) + p\mathbf{v} \end{bmatrix} \quad \text{and} \quad \mathbf{F}^d = \begin{bmatrix} 0 \\ \tau_{xx}\mathbf{e}_x + \tau_{xy}\mathbf{e}_y + \tau_{xz}\mathbf{e}_z \\ \tau_{yx}\mathbf{e}_x + \tau_{yy}\mathbf{e}_y + \tau_{yz}\mathbf{e}_z \\ \tau_{zx}\mathbf{e}_x + \tau_{zy}\mathbf{e}_y + \tau_{zz}\mathbf{e}_z \\ \psi_x\mathbf{e}_x + \psi_y\mathbf{e}_y + \psi_z\mathbf{e}_z \end{bmatrix} \,.
$$

A detailed description of the diffusive part including the formulation of the symmetrical stress tensor components with Stokes' hypothesis of vanishing pressure viscosity ($\mu_v = \frac{2}{3}\mu$) can be found in [4]. The viscosity is approximated in case of laminar flow by Sutherland's formula as a function of the statical temperature only,

$$
\mu(T) = \frac{\sqrt{\kappa} M_\infty}{Re_\infty} \left(\frac{T}{T_\infty} \right)^{\frac{1}{2}} \frac{T_\infty + 100K}{T + 100K} \,.
$$

The vector of heat conduction is modeled by Fourier's law ($\mathbf{q} = -k \mathrm{grad} T$) with the heat conductivity k given by a constant Prandtl number Pr

$$
k = \frac{\kappa}{\kappa - 1} \frac{\mu}{Pr} \,, \quad Pr = 0.72 \,.
$$

For turbulent flows, the laminar viscosity μ is replaced by $\mu + \mu_t$ with the turbulent part provided by a turbulence model. In version 113 of FLOWer, the algebraic model of Baldwin/Lomax is used. Finally, since the air is assumed to behave as a calorically perfect gas, the pressure is calculated by the equation of state

$$p = (\kappa - 1)\rho \left(e - \tfrac{1}{2}\mathbf{v}^2\right)$$

with κ denoting the ratio of specific heats.

The Euler equations solved by INFLEX are derived by neglecting the dissipative flux vector in the above equations.

Underlying Beam Theory

The elastic wing is modeled in ODISA by a Timoshenko-like beam structure with six degrees of freedom for a material cross-section. The centrelines of mass, bending and torsion are generally non-coinciding. By that assumption, besides the coupling via aerodynamics, all degrees of freedom may be coupled with each other mechanically, too. In contrast to the often used Euler-Bernoulli beam theory which couples bending with translation by kinematic constraint involving anomalous dispersion of deformation energy propagation, the Timoshenko approximation with its two more degrees of freedom concerning the shear deformation exhibits no effects of anomalous dispersion and thus describes unsteady deformation in a physically reasonable way.

Extending Hamilton's functional for the wing by terms for the initial conditions δZ_{IC}, boundary conditions δZ_{BC}, airloads δW^{aero} and additional masses (e.g. winglet, engine) δZ_C leads to the variational equation for the complete initial boundary value problem

$$\delta \mathcal{I}^e(\mathbf{u}_S, \varphi) = \int_{t_a}^{t_e} \delta \int_0^l l_B d\xi + \delta W^{aero} Dt + \delta Z_C + \delta\left(Z_{IC} + Z_{BC}\right)$$

with the secondary conditions $\delta \mathbf{u}(\xi, t_e) = \mathbf{0}$ and $\delta \varphi(\xi, t_e) = \mathbf{0}$. For the Lagrangian density one obtains

$$l_B = \frac{1}{2}\left(\rho A \dot{\mathbf{u}}_S \dot{\mathbf{u}}_S + \dot{\varphi}\Theta_S\dot{\varphi} - GA\gamma\mathbf{K}\gamma - EAu_{B1|1}^2 - \varphi_{|1}\mathbf{C}_{BTW}\varphi_{|1} + \rho A\mathbf{u}_S\mathbf{g}\right)$$

(S and B are the centers of gravity and bending, respectively). Applying Timoshenko's theory the shear angles are represented as

$$\gamma_2 = u_{S2|1} - \varphi_3 - (\zeta_{SD}\varphi_1)_{|1}, \quad \gamma_3 = u_{S3|1} + \varphi_2 + (\eta_{SD}\varphi_1)_{|1}$$

where ζ_{SD} and η_{SD} are the cartesian co-ordinate differences between the center of gravity and the shear center of the cross-section. A detailed description of the remaining terms can be found in [5].

Solution Strategy

Within the SOFIA code, the solution schemes for the flow field and the structure are combined in an iterative process through a strong coupling such that the differential equations of both media can be integrated simultaneously i.e. consistent in time. The central time loop works as follows:

ODISA calculates the velocities and the displacements of the structure under the actual aerodynamic loads.

The grid points on the wing's surface (inner boundary of the aerodynamic grid) are moved corresponding to the structure's change of shape. In contrast to the points at the inner boundary the grid points at the outer boundary remain fixed in a rigid-body fixed co-ordinate system. GRIDGEN (GRID GENerator) computes the point distribution such that the topology of the structured grid is preserved.

Finally the flow field and the new airloads are computed for the current time on the numerical grid, which is updated concerning position and state of velocity. This is done by INFLEX solving the Euler equations or in the new version of SOFIA by FLOWer integrating the Navier-Stokes equations. A subiteration level has been implemented which computes the time step. During repeated calculation an average value of the aerodynamic forces of the current and the preceding time step is used. The iteration is stopped when a certain error bound is reached. Of course, the computer time increases linearly with the (number of) subiteration levels. Numerical experiments have shown that in case of carefully chosen time steps – i.e. all the relevant time scales are resolved properly – just one subiteration is sufficient.

Numerical Methods

The numerical methods used in SOFIA are desribed here only very shortly. Details are presented in the references.

Flow Solver INFLEX

For the numerical integration of the strong conservation form of the Euler equations an implicit relaxation scheme is used [1]. The unfactored Euler equations are solved by applying a Newton iteration method. Relaxation is performed with a point Gauß-Seidel algorithm. The combination of a Newton method with a point Gauß-Seidel algorithm leads to a robust numerical scheme. Concerning the resolution of pressure and shock waves, a characteristic variable splitting technique is employed. Grid generation is done using elliptical smoothing in every time step.

Flow solver FLOWer

The FLOWer code is being developed in the project MEGAFLOW by different german research organisations under the leadership of the DLR/Braunschweig. The results presented in this work have been calculated with version 113 which is two years old and thus does not include some features of newer versions (e.g. upwinding, transport equation turbulence models).

The numerical discretization of the FLOWer code is based on structured grids. Central differences are used for the spatial discretization. Time integration is performed by dual-time stepping. Within each pseudo-time step an explicit multi stage Runge-Kutta method is used which is accelerated by techniques of local time stepping, enthalpy damping and implicit residual smoothing. The solution procedure is embedded into a sophisticated multigrid algorithm. This flow solver is written in a flexible block structured form enabling treatment of complex aerodynamic configurations with any mesh topology.

Our work is focusing on the implementation of our structural solver and a grid generator which makes the FLOWer code applicable to aeroelastic analyses. Starting with an elliptically smoothed grid for the non-deformed reference configuration the grid is adapted algebraically to the current configuration. Thereby, interpolation polynomials are used such that the far field grid points remain fixed in the undisturbed fluid flow, whereas points on the wing's surface are moved fixed to the local cross-section of the wing. This grid generator works very fast and produces good meshes even for large deflections of the wing. However, elliptical smoothing may become necessary to avoid overlapping.

Structural Solver

To determine the generalized deflections, a second order in time system of ordinary differential equations (ODEs) is derived by applying Hamilton's principle and the method of Ritz/Kantorowitsch [5]. Linear damping is included (Rayleigh-damping). Discretization is done by isoparametric, two–noded elements. A reduced integration scheme avoids shear locking. The set of ODEs is integrated by Newmark's method, where the resulting linear system of equations is solved directly with a LU-decomposition. The external forces are assumed to vary linearly during a time-step. Alternatively, the system of ODEs is diagonalized by solving the generalized eigenvalue problem (EVP), and the time integration is done by the evaluation of Duhamel's integral.

Results

All elements of the SOFIA code have been validated seperately with respect to various test cases. A detailed description of the structural part can be found in [5], including examples which demonstrate the necessity of using Timoshenko's model to capture wave propagation correctly. The INFLEX code and the basic FLOWer program are described in [1] and [2, 3, 4], respectively.

A validation test of the extension of the FLOWer code by the algebraic grid generator is presented in Fig. 1. The figure shows the total lift coefficient of the well known AGARD LANN wing configuration ([7], see Fig. 3) which is oscillating about the z-axis. The origin of the co-ordinate system is located in the nose of the first cross-section and the z-axis is perpendicular to it. The solution calculated on the dynamic grid corresponds very well to the one calculated on a grid performing rigid body motion (standard option of FLOWer including validation for AGARD testcase CT5, see [4]).

A first example for coupled fluid/structure analysis using the unsteady SOFIA code is given in Fig. 3. Again, the aerodynamic configuration of the wing is defined by the LANN wing. The figure shows the Navier-Stokes solution for the steady flow field about the rigid wing and an elastic one, respectively. Beginning with the reference configuration, the elastic wing is deformed by the airloads during the iterative solution process. Due to aerodynamic and artificial structural (Rayleigh-) damping, the wing reaches the position depicted in Fig. 3 nearly asymptotically. As can be seen, remarkable differences concerning shock position and strength occur in the tip region.

Fig. 2 shows a comparison of numerical (Navier-Stokes) and experimental results [9] for the two-dimensional cruising flight reference configuration of the SFB401 consisting of the supercritical airfoil BAC 3-11/RES/30/21 [6]. A very good agreement exists for this low speed case. Of course, this is not the true cruise condition, but presently only this erxperiment is available for that configuration.

In a first aeroelastic design study of the SFB's three-dimensional configuration [10] the structural data set for this large wing with 36m half span has been defined by simply scaling up the data of a midrange transport plane [5]. As in the case discussed before, aeroelastic analysis has been carried out starting from the stationary transonic flow ($M_\infty = 0.75, \alpha_\infty = 3°$) about the rigid wing. The history of the wing's tip deflection in vertical direction during the response to the "permission to deform" is shown in Fig. 4. During approximately 3000 time steps only moderate artificial structural damping is applied, in order to avoid too big deflections after the impact in the first steps. As can be seen in the figure, in spite of the structural damping, the wing does not come to rest within 7 cycles but performs a coupled bending/torsion oscillation with an amplitude of about 2m. This test case is an unstable one, since the removal of artificial damping would lead to rapidly rising amplitudes. On the other hand, a doubling of the structural damping diminishes the amplitudes (see Fig. 4): Finally, an apparently steady state aeroelastic solution is reached as depicted in Fig. 5. But reducing the damping would lead to vibration again. Since the torsional and bending deformation behave not monotonously in spanwise direction, the question arises if this apparent equilibrium configuration is the only possible solution. Actually, due to the non-linearities in the flow field and the aeroelastic coupling, non-unique solutions may occur depending on the initial conditions or parameters like the artificial damping. However, for the test case discussed here, all variations lead to the same final situation. For another structural design but the same initial aerodynamic shape a stable equilibrium with monotonous shape of torsional and bending deformation has been found which seems to be also insensitive to variations of the parameters.

Conclusions

In this paper a new version of the SOFIA code for direct numerical simulation of fluid/structure interaction has been presented. Replacing the Euler-based flow solver INFLEX by the DLR-FLOWer code, viscous effects can be taken into account now. A structural part and a grid generator have been added to the Navier-Stokes code. Grid generation is performed algebraically starting with an elliptically smoothed grid for the reference configuration. A comparison of this "dynamic" grid with an oscillating rigid grid for the AGARD-LANN configuration showed that algebraical tracking works well up to considerably large displacements. In SOFIA, the wing is modeled by a generalized Timoshenko beam structure. The relatively small number of parameters in this reduced structural wing model permits fundamental analyses of the aero-structural properties, which become more difficult when more complex 3D finite element discretization is used.

Numerical results are shown for the rigid cruising flight configuration of the SFB401' reference configuration for a low Mach number case. The structural data for an elastic wing were generated by simply scaling up the data of a midrange transport plane for a large 3D configuration with 80m span. The first case is dynamically stable only with high artificial structural damping and comes apparently to rest in an unexpected but uniquely deformed shape. The second one with ten times higher torsional stiffness reaches a stable equilibrium point.

Acknowledgements

This work has been supported by the Deutsche Forschungsgemeinschaft (DFG) in the Collaborative Research Centre SFB 401 "Modulation of Flow and Fluid-Structure Interaction at Airplane Wings" of the RWTH Aachen University of Technology. The experiments concerning the reference configuration of the SFB were accomplished in the Institut für Luft- und Raumfahrt of the university. Computations were performed using the facilities of the Rechenzentrum of the RWTH Aachen. We would like to thank Dr. A. Brenneis and Dr. A. Eberle from DASA for providing the computer code INFLEX as well as Dr. R. Heinrich of the Deutsches Zentrum für Luft- und Raumfahrt (DLR/Braunschweig) and all partners within the project MEGAFLOW developing the FLOWer code.

References

[1] A. Brenneis, A. Eberle; *Evaluation of an Implicit Euler Code Against Two and Three-Dimensional Standard Configurations*, AGARD CP-507: Transonic Unsteady Aerodynamics and Aeroelasticity, Paper No. 10, March 1992.

[2] N. Kroll, R. Radespiel, C.-C. Rossow; *Accurate and Efficient Flow Solvers for 3D Applications on Structured Meshes*, Lecture Series 1994-05 of the von Karman Institute of Fluid Dynamics, 1994.

[3] A. J. Jameson; *Time Dependent Calculation using multigrid, with Applications to Unsteady Flows past Airfoils and Wings*, AIAA-Paper 91-1596, 1991.

[4] R. Heinrich, K. Pahlke, H. Bleecke; *A Three Dimensional Dual-Time Stepping Method for the Solution of the Unsteady Navier-Stokes Equations*, Proc. 'Unsteady Aerodynamics' Conference in London, pp. 5.1-5.12, 1996.

[5] D. Nellessen; *Schallnahe Strömungen um elastische Tragflügel*, Fortschrittsberichte VDI Reihe 7: Strömungstechnik, Nr. 302, 1995.

[6] I.R.M. Moir; *Measurements on a Two-Dimensional Aerofoil with High-Lift Devices*, AGARD AR-303, Test Case A2, 1994.

[7] *Compendium of Unsteady Aerodynamic Measurements*, AGARD AR-702, 1982.

[8] G. Britten, J. Ballmann; *Strömungs-Struktur-Wechselwirkung an einem elastischen Flügel in transsonischer, reibungsbehafteter Strömung*, Proc. Aeroelastik-Tagung der DGLR, Göttingen, pp. 90-97, 1998.

[9] E. Özger, I. Schell, D. Jacob; *Experimental Analysis of the Subsonic Flow about a Two-Dimensional Airfoil*, Private Communications, SFB401, RWTH Aachen, 1998.

[10] J. Ballmann et al.; *SFB401 Strömungsbeeinflussung und Strömungs-Struktur-Wechselwirkung an Tragflügeln*, Finanzierungsantrag 1997-1999, 1996, pp. 60ff.

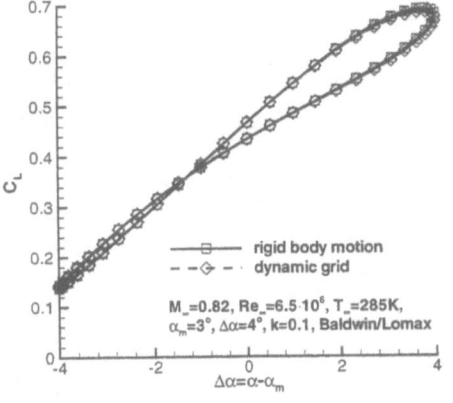

Figure 1. Total lift coefficient for a rotating rigid LANN wing

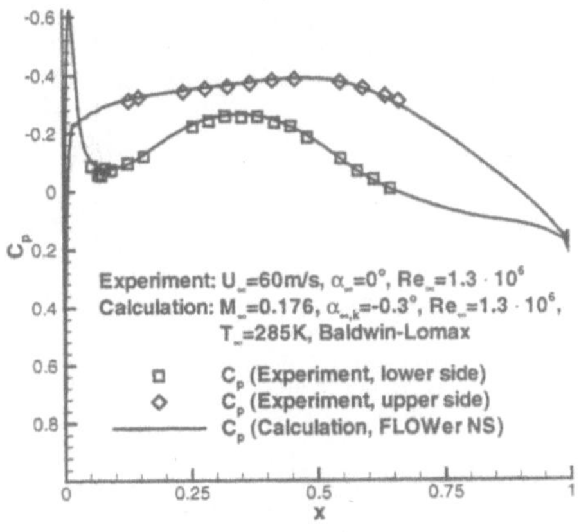

Figure 2. Comparison of numerical and experimental results for the SFB reference airfoil

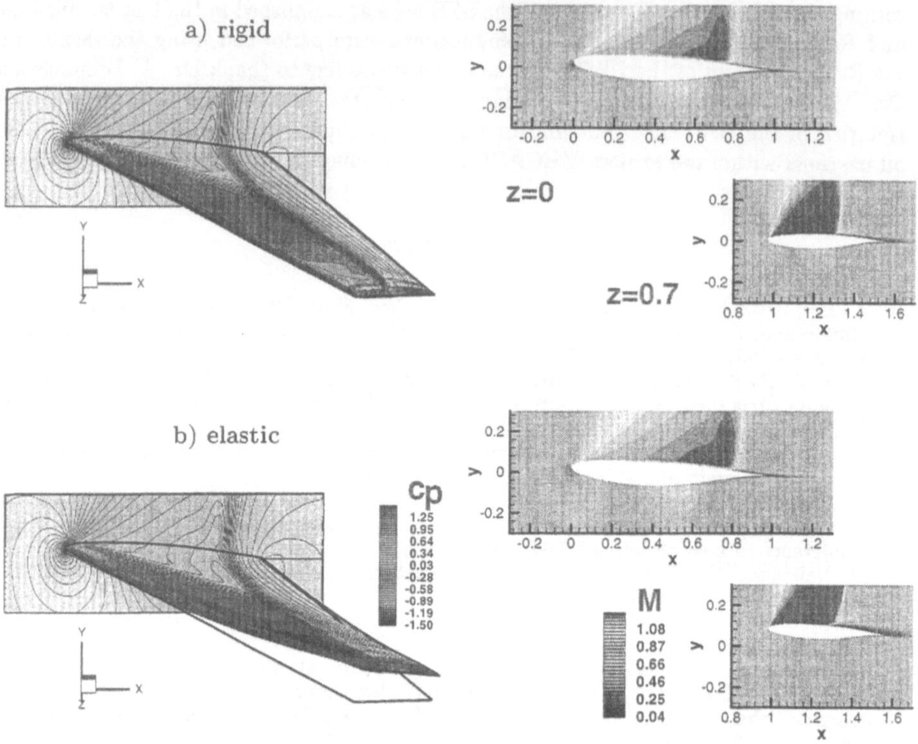

a) rigid

z=0

z=0.7

b) elastic

Cp
1.25
0.95
0.64
0.34
0.03
-0.28
-0.58
-0.89
-1.19
-1.50

M
1.08
0.87
0.66
0.46
0.25
0.04

Figure 3. Pressure distribution on the wing's surface (left) and Mach number distribution in planes at 0% and 70% span (right) of a) a rigid and b) an elastic wing with the aerodynamic configuration of the LANN wing ($M_\infty = 0.82, \alpha_\infty = 3^o, Re = 6.5 \cdot 10^6$, Baldwin-Lomax)

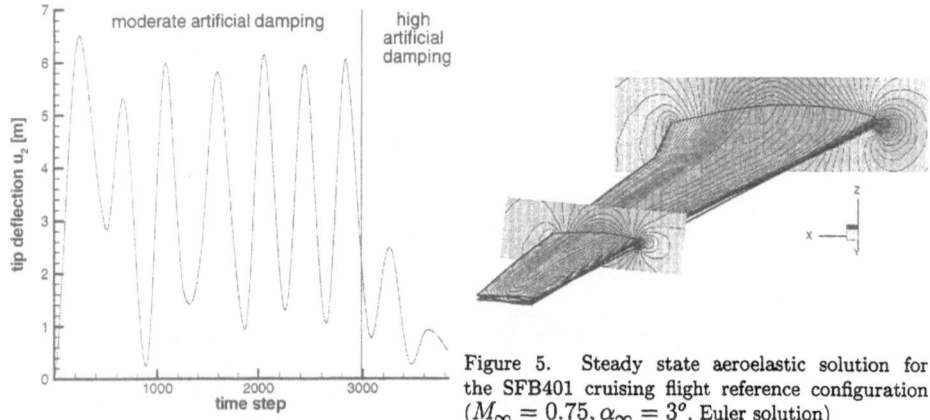

moderate artificial damping

high artificial damping

tip deflection u_2 [m]

time step

Figure 5.   Steady state aeroelastic solution for the SFB401 cruising flight reference configuration ($M_\infty = 0.75, \alpha_\infty = 3^o$, Euler solution)

Figure 4. History of the wing tip deflection during aeroelastic response

84

Computation of Aerodynamic Coefficients for the DLR–F6 Configuration using MEGAFLOW

O. Brodersen, E. Monsen, A. Ronzheimer, R. Rudnik, C.-C. Rossow

DLR, Institute of Design Aerodynamics
Lilienthalplatz 7, D–38108 Braunschweig, Germany

Summary

Navier–Stokes calculations, obtained with software from the MEGAFLOW project, are presented for the Airbus–like DLR–F6 configuration at cruise flight conditions. Results for the wing–fuselage geometry are available for structured grids of different sizes up to 16 million grid cells. The Baldwin–Lomax turbulence model is used and the influence of the numerical dissipation is analyzed. It is demonstrated that even 16 million grid cells are not sufficient to reach a fully grid converged solution for this configuration at transonic flow conditions, if standard non-adapted grids are used. The influence of the numerical dissipation on the pressure distribution is very small for the fine grid. However, the drag coefficient shows a variation of 2.5% for different levels of numerical dissipation. For the configuration with pylon and nacelle, grids with 3.3 and 3.8 million cells are used. In addition to the Baldwin–Lomax model, first computations using the k–ω turbulence model from Wilcox are presented. The pressure distributions as well as the lift and drag coefficients are compared to wind tunnel measurements for both configurations. The comparison shows a good quality of the numerical results.

Introduction

Due to the already high aerodynamic quality of current aircraft, the accurate, reliable and efficient computation of aerodynamic quantities and the analysis of flow phenomena are of increasing importance for the design of new configurations. The capabilities and limits of numerical methods in this area have to be demonstrated before they can be applied in industry with sufficient confidence. Therefore, validation and verification have to be carried out for different realistic configurations. On the one hand, a method has to be validated for different types of flows. In other words, it has to be guaranteed that the 'right equations will be solved'. Otherwise it can not be expected that the flow physics are represented correctly. On the other hand, a verification of the calculation for the given flow conditions, which means 'solving the equations right', has to be performed to minimize numerical discretization and approximation errors. Grid convergence studies and the variation of numerical parameters are an appropriate way to assess calculations [1, 2, 3].

During the last years, the ability to compute lift and drag for wing–fuselage geometries at cruise flight conditions has been demonstrated also by authors from DLR and DASA [4, 5]. The following investigations focus on two aspects. At first, new computational results for a wing–fuselage geometry using the Baldwin–Lomax model and grids with 2.4 and 16 million cells are presented. Secondly, results for the configuration with pylon and nacelle are shown using the Baldwin–Lomax and the k–ω model from Wilcox to demonstrate the capabilities of the MEGAFLOW software to handle complex configurations.

The DLR–F6 configuration has been chosen, because several wind tunnel experiments concerning airframe–engine integration tasks were performed, and a detailed database of different flow conditions is available. Besides that, computations have been carried out for this configuration using Euler or Navier–Stokes methods [6, 7]. Good qualitative results have been obtained with these numerical approaches, but drag coefficients could not be predicted with a sufficient accuracy and reliability. The developments during the last three years in the MEGAFLOW project [8] have significantly improved the Navier–Stokes solver FLOWer, including turbulence modeling [9, 10]. In addition, complex structured multiblock grids can now be generated more easily using MegaCads [11]. The accuracy, which is currently available, and the necessity of a systematic assessment or verification of calculations will be presented for DLR–F6 using FLOWer and MegaCads.

DLR–F6 Configuration

The DLR–F6 configuration is an Airbus–like two–engine civil transport aircraft model for the investigation of airframe–engine interference effects. The design Mach number is $M_\infty = 0.785$ at a lift coefficient of $C_L = 0.5$. The engines are represented by axis-symmetrical throughflow nacelles. The pylon has a symmetrical shape. Since 1990, various experimental investigations have been carried out within the framework of a DLR–ONERA cooperation, using the S2 wind tunnel of the ONERA [6].

Grid Generation

The generation of structured multiblock grids for the F6 configuration has been performed using MegaCads. The grid topology consists of blocks of C–O–type close to the geometry, so that the boundary layers can be resolved adequately. These blocks are surrounded by H–type grid blocks. Two grids are available. Grid 1 contains slightly more than 16 million cells with almost 7.5 million cells located together in the wing and wake block. The wing surface is discretized using 512 cells in streamwise and 176 cells in spanwise direction. 96 cells can be found in the wake region and 64 inside the C–block normal to the wing. In total 90112 cells describe the wing and 53504 cells the fuselage surface. The grid spacings near to the walls are set such that y^+ is nearly 1. Grid 2 consists of 2.4 million cells, and is therefore more suitable for routine applications. Compared to the second level of Grid 1, Grid 2 has a slightly increased number of cells in the near–wall C–blocks (40 instead of 32 cells normal to the walls) and different cell spacings. The objective of using Grid 2 is to reduce the deviation of the solution compared to that one obtainable with the fine grid (Grid 1).

For the configuration with pylon and nacelle (F6–WBNP) two slightly different grids have been created. The first grid, used for the Baldwin–Lomax calculations, has 3.3 million cells. It has been improved for the k–ω investigations by distributing the cells with a smaller stretching ratio and by increasing the number of cells to 3.8 million. The grid topology is similar to that of F6–WB. Due to the higher geometric complexity and the limited computer resources, currently only a coarser discretization is possible compared to the wing–fuselage case. The wing surface is represented by 18016 cells (128 cells in streamwise direction). Figures 1a and 1b give an overview of the topology for selected grid planes.

Numerical Method

All of the presented results are obtained with FLOWer. A detailed description of the numerical method can be found in [12]. Standard parameters are used for most of the computations. Different turbulence models are implemented in FLOWer. Currently, the Baldwin–Lomax and the k–ω model from Wilcox are the most frequently tested models [9, 10], and are therefore used here. Using the so–called 'offcore' capability of FLOWer, only the block being processed has to be stored in main memory. The other blocks are stored temporarily in extended memory or on disk. Otherwise, it would not be possible to perform calculations using Grid 1, because of current memory limitations of the NEC–SX4 computer of DLR.

Results

The analysis of the two configurations has been performed for transonic flow conditions $M_\infty = 0.75$, $Re = 3 \cdot 10^6$. The transition is fixed to the position used in the experiments. The twist of the wing, due to aerodynamic loads, is not considered.

F6–Wing–Fuselage

The results for $\alpha = 0.98^o$ show a good numerical convergence history for Grid 1, as can be seen in Figure 2. Because of the high number of grid cells, Grid 1 is appropriate for a grid convergence study. Figures 3a and 3b present the lift and drag coefficients for three grid levels (16, 2, 0.25 million cells) versus $1/N^{2/3}$ (N = number of cells). Due to the fact that the coefficients nearly form a straight line, it can be stated that FLOWer shows approximately second order accuracy for this configuration. The coefficients for a grid density in the infinite limit can be estimated if C_L and C_D of the fine and medium level of Grid 1 are extrapolated linearly. The lift would slightly increase about 1% and the drag would decrease about 5–6%.

Figures 3a and 3b also include lift and drag of the wing and the fuselage separately, whereby the drag is divided into pressure and friction components. The contribution of the wing to the total lift (85%) is nearly constant for all three grid levels, as shown in Table 1 and Figure 3a. Similar results are observed for the drag coefficients (Figure 3b), where the wing contributes 75%. The grid size has no significant influence on these values. The friction drag $C_{D,F}$ of the fuselage is slightly higher than that of the wing, and the main contribution to the pressure drag $C_{D,P}$, including the wave drag, comes from the wing (62% of total drag). The fuselage contributes only 10% to $C_{D,P}$.

The pressure distributions for the two grids and two levels of numerical dissipation are presented for selected spanwise positions at $\eta = 0.238$, 0.331, 0.377 and 0.635 in Figures 4a to 4d. On the wing's lower side only minor differences occur. The influence of the grid resolution is clearly visible for the wing upper side. With increasing grid density the pressure plateau drops and the shock becomes sharper and moves downstream. Compared with experiments, the shock is located downstream of the measured position using the Baldwin–Lomax model, as is known from previous investigations. The reduction of the numerical dissipation to $k_2 = 1/4$ and $k_4 = 1/64$ shows only a small impact on the pressure distributions.

The Figures 5a to 5d present the solutions for Grid 2 in comparison with the reference calculation on the fine grid (Grid 1). Due to an improvement of grid spacings and a slightly increased number of cells, the calculations show smaller differences to the reference solution than the second level of Grid 1. Only moderate differences to the experimental data are visible. However, it has to be noted, that this is not a grid converged calculation. Compared with the reference calculation the shock is shifted 3–5% upstream. On the contrary, the reduction of the numerical dissipation shows only minor influences on the pressure distributions.

Lift and drag for these computations are plotted in Figure 6a and 6b. Because the pressure distributions are nearly identical, the lift coefficients show only small variations ($\approx$ 1%) for the different grid sizes and numerical dissipation parameters. Regarding drag coefficients, the grid sizes as well as the numerical dissipation still have a significant impact. The solutions using Grid 2 show a variation of 5% for C_D. For Grid 1, approximately 2.5% are computed for the different numerical dissipation values.

F6–Wing–Fuselage–Nacelle–Pylon

Numerical results for the configuration with pylon and nacelle, obtained with the Baldwin–Lomax model on the initial grid ($3.3 \cdot 10^6$ cells) show a satisfactory agreement with experimental data for a range of angles–of–attack from $-3°$ to $2°$, as can be seen in Figures 7a and 7b. The lift has a constant α–shift of approximately $0.5°$. During the investigations, deficiencies in the initial grid in the region of the pylon and nacelle were noted and a refined grid was produced for the k–ω model calculations. Using k–ω, C_L and C_D decrease compared with the solution using the Baldwin–Lomax model.

Figures 8a to 8d show different pressure distributions for $\alpha = 0.98°$ in sections close to the fuselage ($\eta = 0.238$), close to the pylon ($\eta = 0.331, 0.377$), and farther outboard ($\eta = 0.635$). The shock position predicted with the k–ω model is slightly upstream of that predicted with the Baldwin–Lomax model, which is consistent with results of other published computations. The differences in the rear loading portion of the wing, especially oscillations in the Baldwin–Lomax solution, could be attributed to the quality of the initial grid. Lift and drag agree quite well using the k–ω model as can be seen in Fig. 7a and 7b. It must be noted that, in order to achieve sharp shock resolutions on this relatively coarse grid, a numerical dissipation parameter, which takes into account the cell aspect ratios, had to be adapted to reduce numerical dissipation in the wall–near cells.

Conclusion

These Navier–Stokes flow calculations, using structured multiblock grids, have demonstrated that a thourough assessment of numerical results is of significant importance, when aerodynamic coefficients for an aircraft have to be computed. Knowledge about the numerical influences is required to be able to estimate the errors that occur when standard, non–adapted grids with 2–4 million cells are used. In the near future, investigations are necessary to specify the influence of grid resolution and numerical dissipation on the different drag components, especially on the viscous drag, the pressure, and the induced drag.

Acknowledgments

The funding of the project through the German Ministry for Education and Research (BMBF) is gratefully acknowledged.

References

[1] Roache P.J.: *Verification of Codes and Calculations.* AIAA Journal, Vol. 36, No. 5, 1998, pp. 696–702.

[2] Rubbert P.E.: *On Replacing the Concept of CFD Validation with Uncertainty Management.* Presentation at Daimler–Benz, Stuttgart, 10th June, 1998.

[3] Oskam B. Sloof J.W.: *Recent Advances in Computational Aerodynamics at NLR.* AIAA Paper 98–0138, Reno, 1998.

[4] Longo J.M.A.: *Viscous Transonic Flow Simulation Around a Transport Aircraft Configuration.* DGLR–Jahrestagung, Bremen, 1992.

[5] Elsholz E. Longo J.M.A.: *Navier–Stokes Simulation of a Transonic Wing–Body Configuration.* Proc. of 1993 European Forum 'Recent Developments and Applications in Aeronautical CFD', Bristol, Sept. 1993, pp. 4.1–4.12.

[6] Rossow C.–C. Godard J.–L. Hoheisel H. Schmitt V.: *Investigations of Propulsion Integration Interference Effects on a Transport Aircraft Configuration.* AIAA Paper 92–3097, 1992.

[7] Brodersen O. Rossow C.–C.: *Calculation of Interference Phenomena For a Transport Aircraft Configuration Considering Viscous Effects.* Proc. of Europ. Forum: Recent Developments and Applications in Aeronautical CFD, Bristol, 1993, pp. 6.1–6.13.

[8] Kroll N. Rossow C.-C. Becker K. Thiele F.: *MEGAFLOW – A Numerical Flow Simulation System*. 21st ICAS Congress, Melbourne, Australia, Sept. 1998.

[9] Rudnik R.: *Untersuchungen der Leistungsfähigkeit von Zweigleichungs–Turbulenzmodellen bei Profilumströmungen*. DLR–FB 97–49, Braunschweig, 1997.

[10] Monsen E. Rudnik R. Bleeke H.: *Flexibility and Efficiency of a Transport–Equation Turbulence Model for Three–Dimensional Flow*. Notes on Numerical Fluid Mechanics, Ed. Körner H., Hilbig R., Vol. 60, Vieweg Braunschweig, 1996, pp. 237–244.

[11] Brodersen O. Hepperle M. Ronzheimer A. Rossow C.-C. Schöning B.: *The Parametric Grid Generation System MegaCads*. Proc. of 5th Intern. Conf. on Numerical Grid Generation in Comp. Field Simulation. Ed. Soni B.K. et. al., Mississippi, 1996, pp. 353–362.

[12] Raddatz et. al.: *FLOWer 115 – Installation and User Handbook*. DLR Braunschweig, 1998.

Table, Figures

Table 1: F6-WB, % of total lift and drag

| | Baldwin-Lomax (BL) | | | |
| | Grid 1 | | | Grid 2 |
	Lev. 1	Lev. 2	Lev. 3	Lev. 1
$C_L^{Fuselage}$	15.2	15.3	15.3	15.1
C_L^{Wing}	84.8	84.7	84.7	84.9
$C_D^{Fuselage}$	25.1	24.5	25.0	28.7
C_D^{Wing}	74.9	75.5	75.0	71.3
$C_{DP}^{Fuselage}$	9.8	11.8	17.0	10.5
C_{DP}^{Wing}	61.6	64.2	66.8	57.8
$C_{DF}^{Fuselage}$	15.3	12.7	8.0	18.1
C_{DF}^{Wing}	13.4	11.2	8.2	13.5

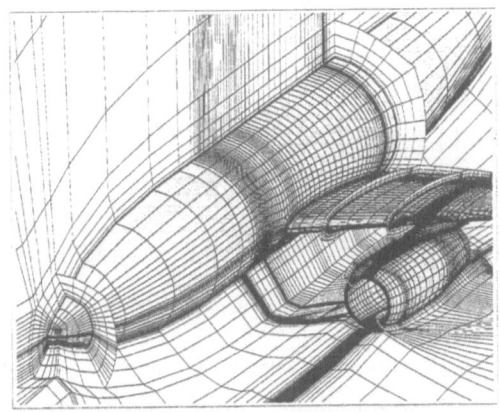

Fig. 1a: Navier-Stokes grid for DLR-F6-WBNP, H-C-O topology, 3.8×10^6 cells, coarse

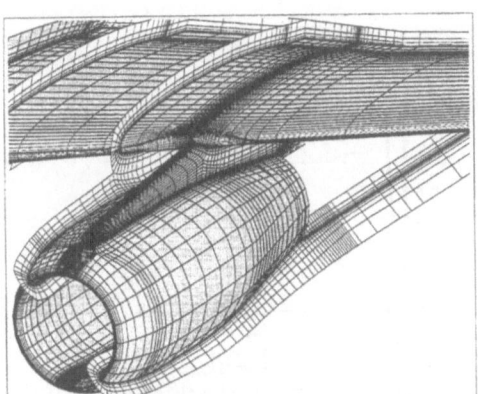

Fig. 1b: Navier-Stokes grid for DLR-F6-WBNP, H-C-O topology, 3.8×10^6 cells, coarse

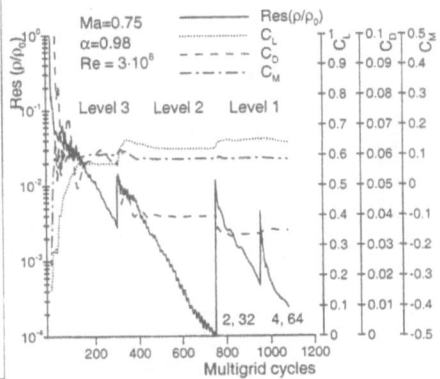

Fig. 2: F6-WB convergence, grid 1, BL, 16×10^6 cells

89

C_L
0.7
0.6
0.5
0.4
0.3
0.2
0.1
0
5E-05
0.0001
0.00015
0.0002
0.00025
0.0003
$1/N^{2/3}$
C_L^{Total}
$C_L^{Fuselage}$
C_L^{Wing}

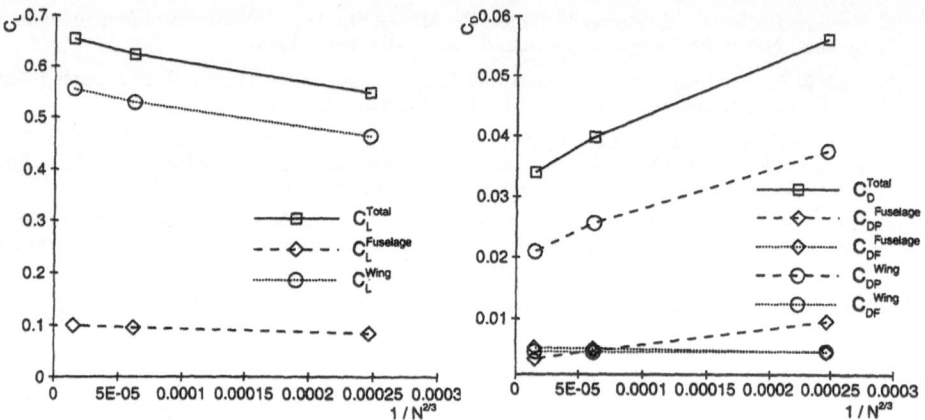

Fig. 3a: F6-WB grid convergence, lift, BL

C_D
0.06
0.05
0.04
0.03
0.02
0.01
0
5E-05
0.0001
0.00015
0.0002
0.00025
0.0003
$1/N^{2/3}$
C_D^{Total}
$C_{DP}^{Fuselage}$
$C_{DF}^{Fuselage}$
C_{DP}^{Wing}
C_{DF}^{Wing}

Fig 3b: F6-WB, grid convg., Cd, pressure, friction, BL

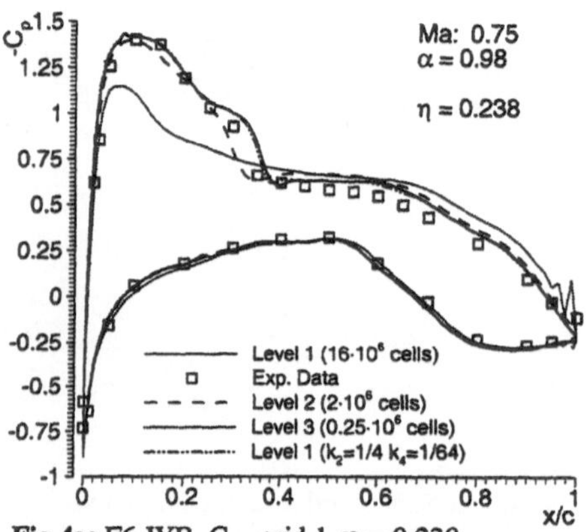

$-C_p$
Ma: 0.75
α = 0.98
η = 0.238
Level 1 (16·10⁶ cells)
Exp. Data
Level 2 (2·10⁶ cells)
Level 3 (0.25·10⁶ cells)
Level 1 (k₂=1/4 k₄=1/64)
x/c

Fig 4a: F6-WB, C_P, grid 1, $\eta = 0.238$

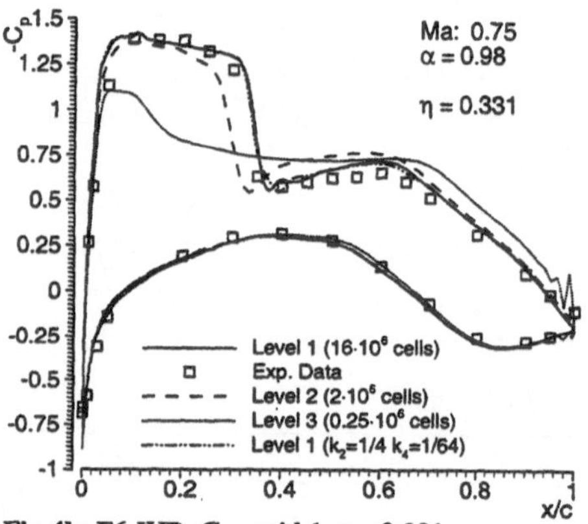

$-C_p$
Ma: 0.75
α = 0.98
η = 0.331
Level 1 (16·10⁶ cells)
Exp. Data
Level 2 (2·10⁶ cells)
Level 3 (0.25·10⁶ cells)
Level 1 (k₂=1/4 k₄=1/64)
x/c

Fig 4b: F6-WB, C_P, grid 1, $\eta= 0.331$

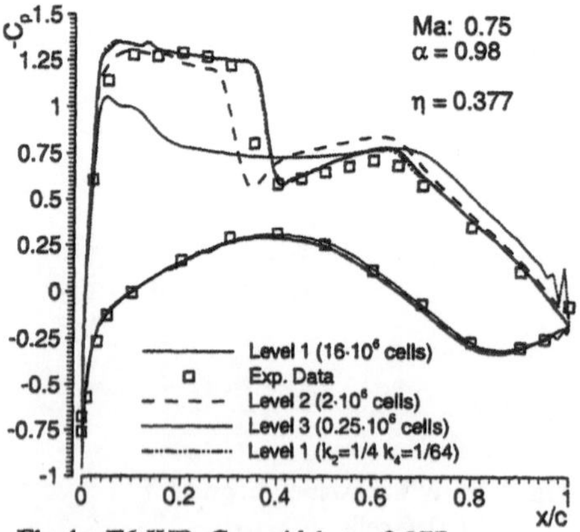

$-C_p$
Ma: 0.75
α = 0.98
η = 0.377
Level 1 (16·10⁶ cells)
Exp. Data
Level 2 (2·10⁶ cells)
Level 3 (0.25·10⁶ cells)
Level 1 (k₂=1/4 k₄=1/64)
x/c

Fig 4c: F6-WB, C_P, grid 1, $\eta= 0.377$

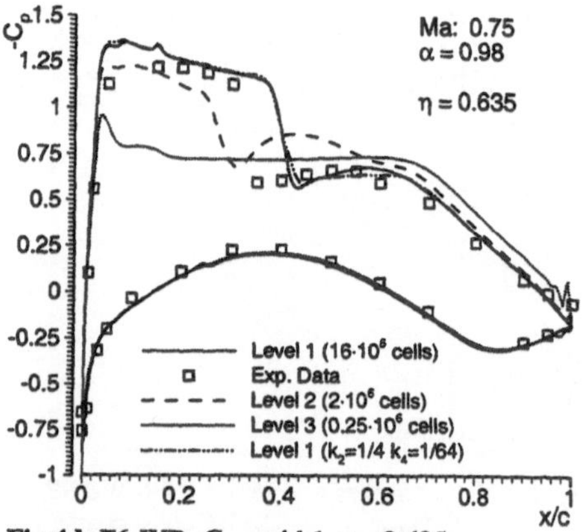

$-C_p$
Ma: 0.75
α = 0.98
η = 0.635
Level 1 (16·10⁶ cells)
Exp. Data
Level 2 (2·10⁶ cells)
Level 3 (0.25·10⁶ cells)
Level 1 (k₂=1/4 k₄=1/64)
x/c

Fig 4d: F6-WB, C_P, grid 1, $\eta= 0.635$

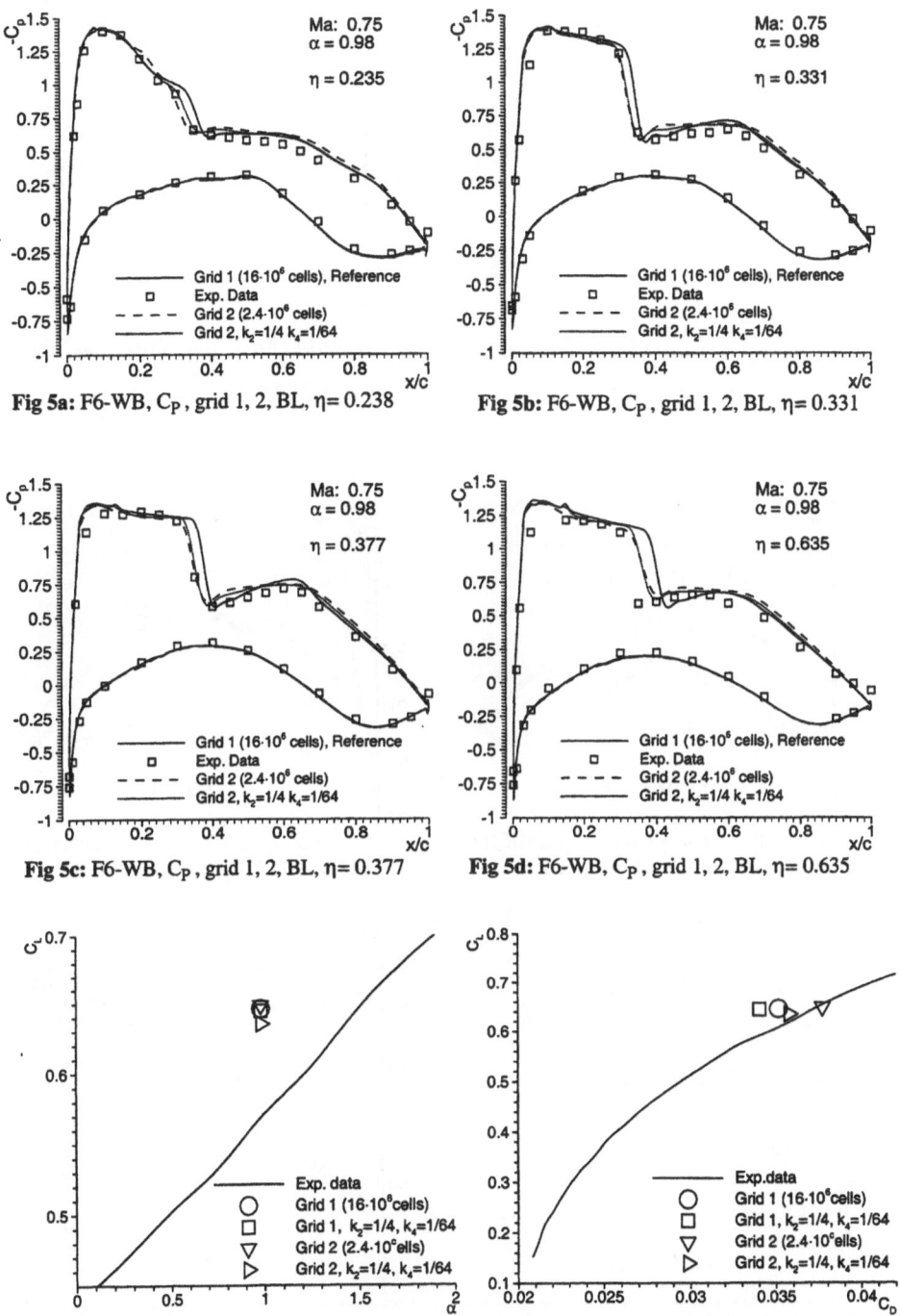

Fig 5a: F6-WB, C_P, grid 1, 2, BL, $\eta = 0.238$

Fig 5b: F6-WB, C_P, grid 1, 2, BL, $\eta = 0.331$

Fig 5c: F6-WB, C_P, grid 1, 2, BL, $\eta = 0.377$

Fig 5d: F6-WB, C_P, grid 1, 2, BL, $\eta = 0.635$

Fig 6a: F6-WB, lift vs. α, grid 1, 2, BL

Fig 6b: F6-WB, drag vs. α, grid 1, 2, BL

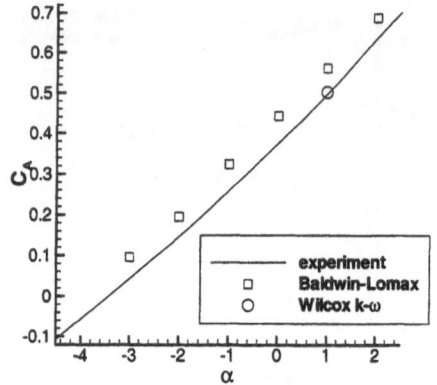

Fig 7a: F6-WBNP, lift vs. α, BL, k-ω

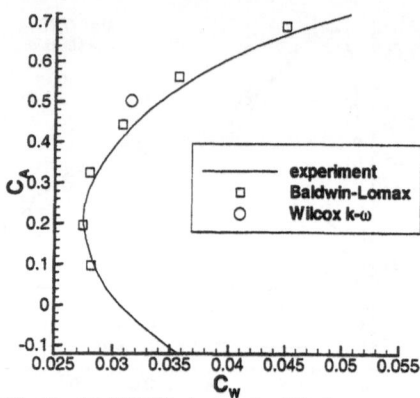

Fig 7b: F6-WBNP, drag polar, BL, k-ω

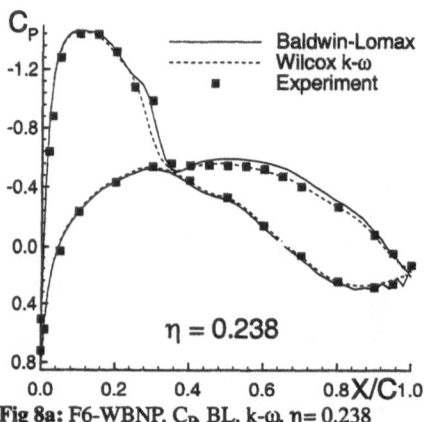

Fig 8a: F6-WBNP, C_P BL, k-ω, η= 0.238

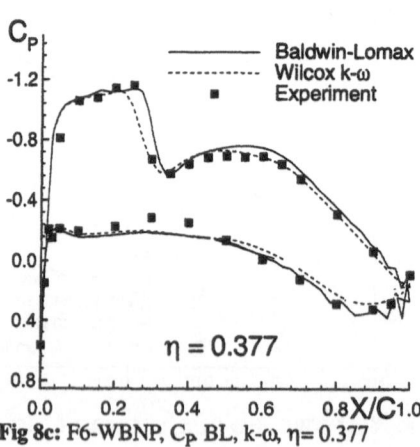

Fig 8c: F6-WBNP, C_P BL, k-ω, η= 0.377

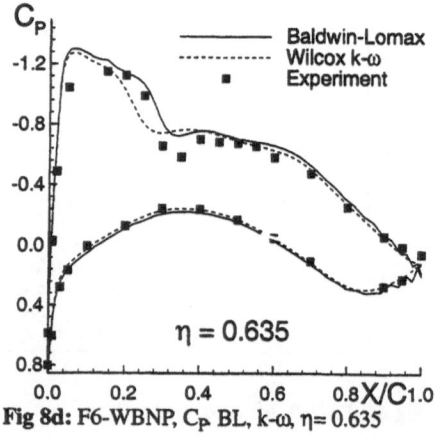

Fig 8b: F6-WBNP, C_P BL, k-ω, η= 0.331

Fig 8d: F6-WBNP, C_P BL, k-ω, η= 0.635

Implementation of the Chimera Method in the Unstructured DLR Finite Volume Code Tau

U. BURGGRAF, M.KUNTZ, B. SCHÖNING

Deutsches Zentrum für Luft- und Raumfahrttechnik e.V.
Institut für Entwurfsaerodynamik
Lilienthalplatz 7 38104 Braunschweig, Germany
email: Udo.Burggraf@dlr.de, Britta.Schoening@dlr.de

Summary

The paper presents a chimera technique for unstructured finite volume schemes which is implemented in the DLR Tau - code, a three-dimensional edge based finite volume scheme solving the Reynolds-averaged Navier-Stokes equations. The chimera technique is assumed to be an efficient way to compute transient multiple body problems, like control surface deflections or store separation. In this paper, we describe the basic algorithms of the implemented chimera technique, which is at present applicable for steady problems only. Steady state solutions for inviscid sub- and supersonic test cases are presented in order to demonstrate the accuracy and efficiency of the implemented algorithms.

Introduction

The intention of the current work is the development of a numerical method, that allows the computation of the flow about multiple moving bodies by solving the unsteady Reynolds averaged Navier-Stokes equations. Applications for this method will be dynamic control surface deflections or store separation problems for example. In these simulations the computational grid has to be adjusted in time according to the changing relative position of the bodies. For inviscid flow calculations approaches like deforming grids and local or global regridding have been successfully applied [4] on unstructured tetrahedral grids. For viscous flow problems, where high aspect ratio cells in the boundary layer are needed, the generation of tetrahedral grids is difficult and it appears that hybrid prismatic - tetrahedral grids provide superior results [5]. However, the automatic adjustment of hybrid grids applying the approaches mentioned above is supposed to be a complex and time consuming task.

An alternative approach, which can be easily applied to hybrid grids, is the chimera technique, usually associated with structured grids[1, 2, 3]. To apply the chimera technique, the vicinity of each independently moving body is discretized by its own local grid. These local grids are designated as blocks in this paper. The blocks are overset to cover the complete domain around the configuration. At the boundaries, that are overlapping with another block, the boundary conditions are computed by linear interpolation of the flow variables provided by this block. To simulate rigid body motions the local grid of the body is moved within the flow field, without changing the elements of the grid and their connectivity. Only the transfer operators at the boundaries must be recomputed.

The present paper reports the current state of the development. It includes a description of the implemented algorithms and and some recent results for inviscid steady flows.

Flow Solver

The DLR Tau - code is a three-dimensional finite volume scheme solving the Reynolds-averaged Navier-Stokes equations in moving frames of reference. Starting from a primary grid of volume elements, a secondary grid of control volumes around each vertex of the primary grid is computed. Due to this so called dual grid approach all kind of initial element types become feasible. Up to now the primary grids can consist of hexahedral, prismatic, pyramidal and tetrahedral elements. The control volumes are constructed by connecting the midpoints of the edges, the face centers and, in 3-D, the centers of the primary elements, see Fig. 2. After the construction of the secondary grid the primary grid is discarded. The flow variables are stored in the vertices of the primary grid, i.e. the midpoints of the control volume.

The inviscid fluxes are computed using an AUSM or a Roe type 2nd order upwind scheme. The gradients of the variables are determined using a Green-Gauß formula. Alternatively, a central discretisation with either scalar or matrix dissipation is employed. The viscous fluxes are discretized with central differences. Time discretisation is done using a multi-step Runge-Kutta scheme. For time accurate calculations, an up to third order accurate dual time stepping approach is implemented.

Convergence to steady state is accelerated by local time stepping, residual smoothing and multigrid. The coarse grids are provided by agglomeration. For parallel computations, subsets of the dual grids are created by domain decomposition. The communication between the subdomains is based on MPI.

The Tau-code is splitted into three independent modules: a preprocessing module, the solver and the grid adaptation module. The construction of the dual grid, the agglomeration of control volumes for the multigrid levels and the grid partitioning for parallel runs are embodied in the preprocessing module. The solver takes the dual grid as input and solves the flow equations. Thus each module can be run separately on the best suited platform. A more detailed description of the code is given in [6].

Implementation

The current method has three principal elements: the creation of the overlapping region, the calculation of the interpolation coefficients and the exchange of the flow variables among the blocks. The first two elements mentioned above are embodied in the preprocessing module, where the primary grid is still known. The third element is part of the solver.

The overlapping region in the presented test cases is constructed as follows. Two blocks sharing a common boundary surface are generated independently using an arbitrary unstructured grid generator, see Fig. 1. The surface triangulation on this boundary is not unique. An additional layer of tetrahedra is created automatically using a three-dimensional advancing front grid generator as proposed by [8]. The front is initialized with the existing chimera boundary triangulation of all blocks. Each front triangle can be imagined as the base a tetrahedron is attached to. If the tetrahedron is generated, the front is updated and the next triangle is chosen from the front. The additional points created in the grid generation process are located in the interior of the neighboring block. The flow variables at these points are interpolated in the solver. The offset of these interpolation points is computed from the average length of all edges connected to the points of the boundary triangle. Hence, the width of the overlapped region is proportional to the local grid spacing. If all vertices of a front triangle are interpolation points, no volume element is created and the triangle is removed from the front. The grid generation process continues until the front is empty. After the grid generation, the connectivity and the metrics of the new primary elements are computed as far as they contribute to the boundary control volumes. This way control volumes similar to those located in the interior domain are achieved at the boundary, see Fig. 2.

The interpolation is based on linear shape-functions well known from finite element methods. At first the element, into which the interpolation point p with coordinates $\vec{x}_p$ is located, has to be determined. A straightforward way is the computation of the linear shape-functions N_i of p with respect to the vertices of the element:

$$\vec{x}_p = \sum_i N_i \vec{x}_i .$$ (1)

This yields together with the sum-property

$$\sum_i N_i = 1$$ (2)

a system of equations

$$
\begin{bmatrix} x_p \\ y_p \\ z_p \\ 1 \end{bmatrix} =
\begin{bmatrix} x_1 & x_2 & x_3 & x_4 \\ y_1 & y_2 & y_3 & y_4 \\ z_1 & z_2 & z_3 & z_4 \\ 1 & 1 & 1 & 1 \end{bmatrix} \cdot
\begin{bmatrix} N_1 \\ N_2 \\ N_3 \\ N_4 \end{bmatrix}
$$ (3)

that allows the evaluation of the four shape-functions needed in case of tetrahedral elements. Other elements are split into tetrahedrons and the shape functions are computed for each of the subelements. A point p is located in the considered element if the criterion

$$\min(N_i, 1 - N_i) \geq 0, \quad \forall i$$ (4)

is satisfied. The fastest method to find this element is an advancing front vicinity algorithm as proposed in [7]. This method has been modified for the edge based data structure of the Tau-code. Consider the interpolation point p and an arbitrary point s of the block the element searched for belongs to. At first all elements surrounding point s are checked. If point p does not fall into one of these elements, the edge connected to point s that encloses the smallest angle with vector $\overrightarrow{sp}$ is determined. We take the other point of this edge as the new starting point and run the algorithm again, until the desired element is found. The performance of this algorithm depends heavily on the chosen starting point. It might even fail if a boundary exists between both points. This is avoided by exploiting the connectivity of the grid. Whenever an element e containing point p is found, we take one of the neighboring points of p as the next interpolation point and one of the points belonging to element e as the next starting point. In the presented test cases, no more than three search steps were needed. For each interpolation point, the vertices of the element satisfying criterion (4) and the corresponding shape functions are transferred to the solver. There a new boundary condition is introduced, in which the flux through the chimera boundary is added to the flux balances of the boundary control volumes. The state of the flow in the interpolation point is computed applying equation (1) to the flow variables and additionally to their gradients in case of the second order discretisation. The flow variables in the boundary points are already known. Hence, central or upwind differences can be applied to compute the fluxes through the chimera boundary faces. The efficiency of the implemented algorithms is demonstrated in Table 1. There, the cpu-time spent for the generation of the additional layer of tetrahedra and the search process in comparison to one four-level W-type multigrid cycle is shown. In this test case, the multigrid cycle takes more than three hundred times longer than the search process and more then thirty times longer than the generation of the additional layer. Keep in mind that the additional layer has to be generated only once in a dynamic simulation, while the search process is run every physical time step .

Table 1: Cpu - time needed on scalar processors

algorithms	factor
grid generation	9.2
search	1
4-W MG-cycle	340

Fig. 5 shows the cpu-time spent for the search algorithm for interpolation point numbers of 386, 1398, and 4608The numbers of elements of these grids are 8269, 79569 and 592186, respectively. It can be easily seen that the cpu-time rises linearly with the number of interpolation points. The data structure of the solver allows to store all blocks in the same array and treat it as one. Thus, splitting the flow domain into several blocks does not degrade the efficiency of vectorisation or parallelisation of the unstructured code.

Applications

As an initial test of the implemented scheme, the incompressible flow about a two-element airfoil is computed, see Fig. 3. The test case is defined by Williams [9]. The airfoils were generated by a double application of the Karman-Trefftz transformation on two lifting circles. The angle of attack of the main airfoil is 0 degree. The trailing edge flap is 30 degree deflected. We defined a chimera boundary around the flap and generated a flap grid composed of 48.000 tetrahedra and a main airfoil grid with 120.000 tetrahedra, respectively. The grid is three-dimensional with a constant width of 0.5 percent of the main airfoil chord. The rectangular farfield boundary is placed in a distance of 10 chord lengths from the main airfoil leading edge. Vortex correction is applied. We chose an inflow Mach number of 0.15 for our calculation. Within 600 multigrid cycles (4-Level W-type), the residual fell about five orders. The computed pressure distribution in the symmetry plane is shown in Fig. 3. It can be seen that the contour lines cross the chimera boundary smoothly. Fig. 4 shows the pressure distribution on the airfoils derived by an exact solution of the plane potential flow [9] in comparison with the solution obtained with the presented method. The results are in excellent agreement.

A supersonic test case is defined to see how the algorithm handles shock waves crossing the chimera boundary. It has been reported in [10] that shocks might dissipate at non-conservative chimera interfaces. The setup of the test case is shown in Fig. 7. The channel is bounded at the bottom and the top by two slip walls. The inlet Mach number is 2.0. The ramp at the bottom produces an oblique shock that crosses the chimera interface at x = 2.7. Reflections occurring at the chimera boundary hit the bottom wall aft of x = 4.1. They can be easily detected in the pressure distribution on the bottom wall. In Fig. 7a, the upper and the lower block have the same grid spacing. The pressure contours in the symmetry plane do not show a significant distortion or disappearance of the shock. Only on the bottom wall a small oscillation in the magnitude of 0.5% of the pressure rise due to the oblique shock becomes evident. Results of the same test case but with a grid spacing ratio of > 2 between the upper and the lower block are shown in fig. 7b. The shock thickness in the upper block is increasing due to the larger spacing. Again no significant reflection is identified in the pressure contours, but the oscillation in the pressure distribution on the lower wall increases by a factor of two. This gives evidence to the assumption that the cell aspect ratio in the overlapping region plays a significant role, when shock waves are crossing chimera boundaries. We propose to employ local grid refinement to keep this difference bounded. The residual was reduced to machine accuracy within 500 multigrid cycles, see Fig. 6.

Conclusion and future research directions

We presented a chimera technique for unstructured, edge based finite volume codes. The method is currently restricted to non-moving grids. The results achieved in the presented sub- and supersonic test case are good. The implemented search and interpolation algorithms are fast enough for time accurate simulations. The cpu time needed increases linearly with the number of interpolation points. We have demonstrated, that the contour lines are crossing the chimera boundary smoothly and that excellent results can be achieved for flows free of discontinuities. We have shown, that the grid spacing ratio in the overlapping region plays a significant role when shocks are crossing the boundary. The best results are achieved, if the local grid spacing ratio of the neighboring blocks is close to 1.

The future work will concentrate on applying the code to unsteady test cases and adding a "hole" definition algorithm in order to simulate arbitrary multiple body motions.

Acknowledgments

The work presented in this paper is part of the AeroSUM project, which is managed by DLR and funded by BMVg.

The authors wish to thank K.D. Pahlke, R. Heinrich and T. Schwarz, DLR, for many helpful discussions.

References

[1] Benek J.A., Buning P.G., Steger J.L.: "A 3-D Chimera Grid Embedding Technique" AIAA paper 85-1523, 1985.

[2] Wang Z.J., Hariharan N., Chen R.: Recent Developments on the conservation property of chimera", AIAA paper 98-0216.

[3] Meakin R.L.: Moving Body Overset Grid Methods for Complete Aircraft Tiltrotor Simulations", AIAA-93-3350 -CP.

[4] Baum J. D., Luo H., Löhner R., Goldberg E., Feldhun A. : Application of Unstructured Adaptive Moving Body Methodology to the Simulation of Fuel Tank Seperation from an F-16 C/D Fighter", AIAA-paper 97-0166, 1997.

[5] Minyard T. , Kallinderis Y. : „A Parallel Navier-Stokes Method and Grid Adapter with Hybrid Prismatic / Tetrahedral Grids", AIAA paper 95-0222.

[6] Gerhold T., Galle M., Friedrich O., Evans J. : "Calculation of Complex Three-Dimensional Configurations Employing the DLR-TAU-Code", AIAA paper 97-0167, 1997.

[7] Rainald Löhner : "Robust, Vectorized Search Algorithms for Interpolation on Unstructured Grids", Journal of Computational Physics 118, 380-387, 1995.

[8] Fleischmann P. ,Selberherr S. : "Three-Dimensional Delaunay Mesh Generation Using a Modified Advancing Front Approach", Proceedings 6th International Meshing Roundtable, (13.-15. 1997, Park City), pp.267-278.

[9] Williams B.R.: "An Exact Test Case for the Plane Potential Flow About Two Adjacent Lifting Aerofoils", A.R.C. Reports and Memoranda No. 3717, 1971.

[10] Wang Z. J. :" A Fully Conservative Interface Algorithm for Overlapped Grids" , Journal of Computational Physics 122, pp. 96-106, 1995.

Figures

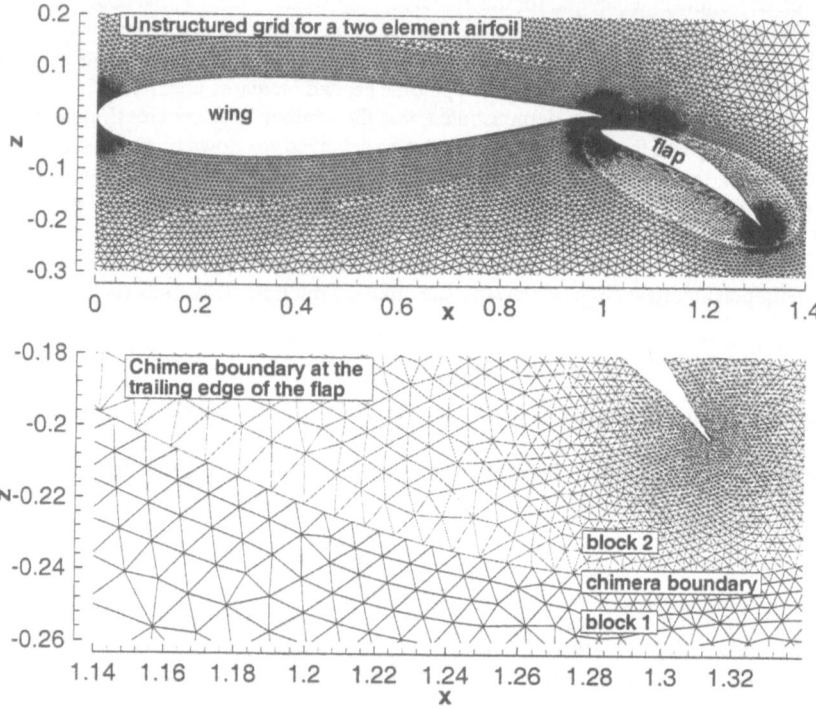

Fig. 1: Two block primary grid for the computation of the flow about a two element airfoil using the chimera technique

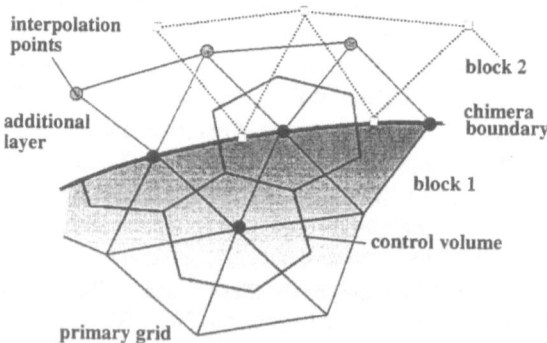

Fig. 2: Primary grid, automatically generated additional layer and secondary grid at the chimera boundary

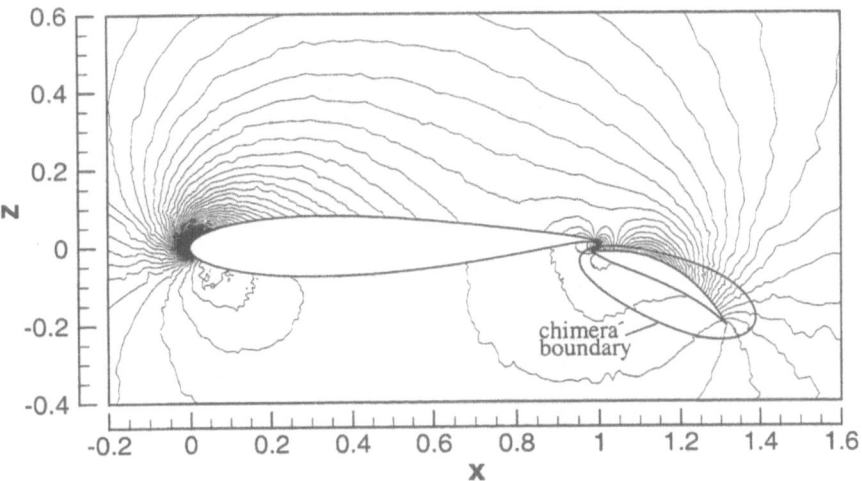

Fig. 3: Pressure contours in the symmetry plane of the two - element airfoil

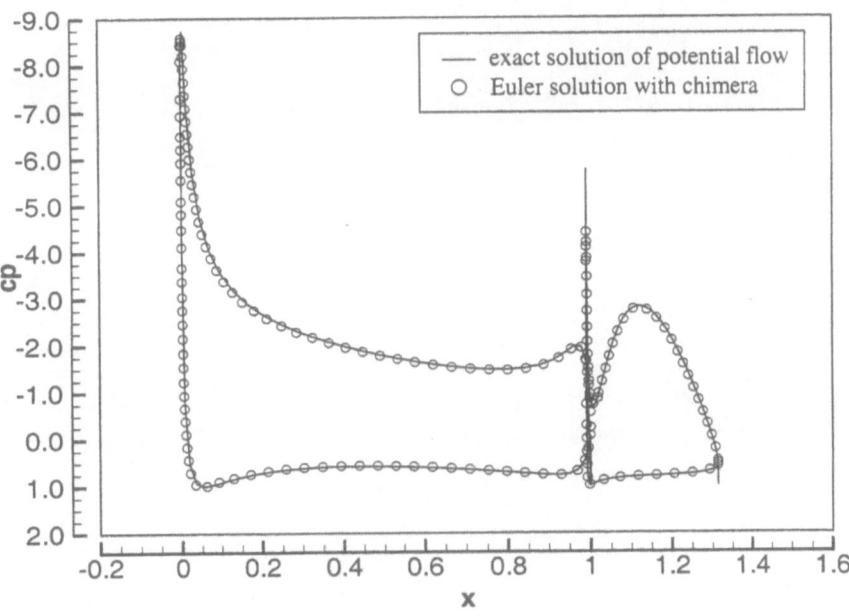

Fig. 4: Distribution of the pressure coefficient on the two - element airfoil. Comparison of the exact solution of potential flow and solution of the Euler equations with chimera

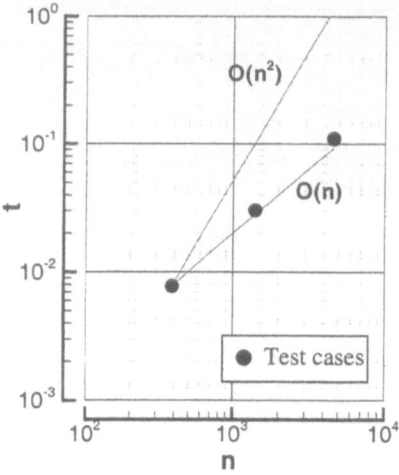

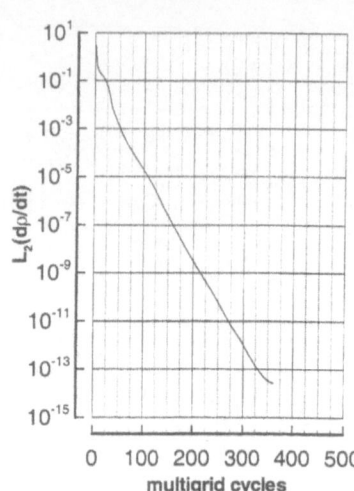

Fig. 5: Cpu-time of the search algorithm versus interpolation point number n

Fig. 6: Convergence history for the supersonic test case

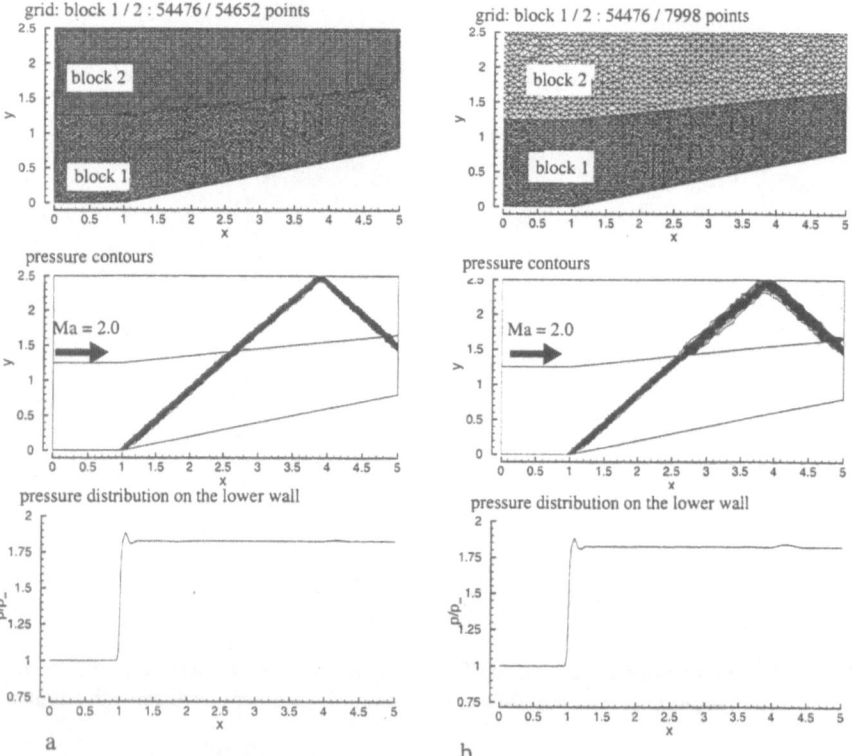

Fig. 7: Grid, pressure contours and pressure distribution on the lower wall for the supersonic flow over a 2D ramp computed on grids with a grid spacing ratio of a) 1 b) 2

Numerical Analyses of an Unsteady Flow Field in a Compressor Cascade with Periodically Changing Inflow Conditions

Damir Delimar, Marius Swoboda, Michael Lötzerich

BMW Rolls-Royce GmbH
Eschenweg 11, 15827 Dahlewitz
Germany

Summary

The presence of unsteady wake-blade interaction and wake transport significantly affects turbomachinary component aerodynamics [1]. This paper presents the results of a CFD analysis which is used to predict the unsteady flow field in a high pressure compressor cascade. In the analyses, the wakes generated by the upstream blade rows are simulated by bars moving in front of the compressor stator blades at various bar speeds and inlet Mach numbers. Particular care has been devoted to the prediction of the wake trajectories and the distribution of the turbulent quantities in the wakes.

Introduction

The process of boundary layer transition on compressor blades depends not only on the adverse pressure gradient but also the influence of unsteady periodic effects due to wakes generated by upstream blade rows. To accurately predict aerodymamic loss, the location and extent of boundary layer transition and the unsteady phenomena due to wakes is a great challange in modern compressor design. In particular, the time dependent wake-airfoil interaction affects the aerodynamic efficiency and structural loading. The simulation of these effects mostly failed in the past due to a lack of transition and appropriate turbulence modelling for periodic unsteady flows. Halstead et al. [3] have given a comprehensive description of the development of the unsteady boundary layers in axial compressors. Part 4 of Halstead's paper discusses some numerical approaches for modelling the transition process. Usually the transition onset must be specified in the models. By giving a detailed view to local flow features, e.g. like shedding vortices behind moving bars, the analyses described in this paper support the ongoing experimental work of Swoboda et al. [6]. The experiment investigates the structure and impact of incoming wakes on the laminar-turbulent transition process in a compressor cascade in compressible flow. The main objective of this study is to predict the wake trajectories and the transport of the wakes through the compressor cascade at various flow conditions. The structure of the turbulence intensity in the wake is shown to play an major role and is an important parameter for accurate flow prediction within the compressor cascade.

The Numerical Method

The commercial CFD package FLUENT 5 applied within this investigation was provided by FLUENT Incorporated. The CFD code solves the Reynolds-averaged Navier-Stokes equations in integral form by applying a finite-volume method to spatially discretize the computational domain. Different numerical methods were employed to solve the discretized equations. At low Mach numbers a segregated scheme was used to discretize the time derivatives with second order accuracy. A point implicit (block Gauss-Seidel) linear equation solver is applied to solve the system of equations. A coupled solver type is considered to be a more appropriate choice at increasing Mach numbers as the compressibilty effects cause a higher degree of coupling between the flow quantities. Therefore, in addition to the segregated solution method, a coupled solver was applied for the analysis at high inlet Mach number and high bar speed. The two-dimensional unsteady calculation was run until a periodic solution was obtained. Both solver types use the implicit approach. Convergence speedup was achieved by the use of the dual-time-stepping method in conjuction with an algebraic multigrid acceleration. The Spalart-Allmaras (S-A) turbulence model [5] was chosen to simulate the turbulent phenomena in the flow field. The Spalart-Allmaras model was designed specifically for turbomachine applications involving wall-bounded flows and has been shown to give good results for boundary layers subjected to adverse pressure gradients.

The Computational Domain and Boundary Conditions

In the experimental work [6], in order to generate a time dependent inlet flow toward the compressor blades, a moving bar device was installed upstream of the cascade. Crossing the core flow in front of the compressor blades the cylindrical bars produce a velocity wake similar to the wake shed by a rotor blade in a real turbomachinary. A detailed description of the moving bar device was given by Acton [1]. Since the maximum bar speed is limited to 40 m/s, the velocity upstream of the moving bars was adjusted to a Mach number of M=0.15 . A bar speed of u_{bar}=20 m/s was chosen. This leads to a Strouhal number typical of high pressure compressors. With respect to the experimental data which were taken at two different inlet Mach numbers (M=0.15 and M=0.7), the calculations were run at the same core flow velocity magnitude. The Reynolds number was 2×10^5 and 4×10^3 based on the blade chord length and bar diameter respectively.

A schematic picture of the two-dimensional computational domain is given in (Fig. 1). The computational domain was subdivided into a moving and a steady frame. The solver uses the sliding mesh interface method in order to transfer the flow quantities from the moving to the steady frame. The bars (d=2 mm) were placed in the moving computational frame with a bar-to-bar-spacing of 40 mm. The downstream blade row has a pitch/chord ratio (chord length=100 mm) value of 0.6 approximating closely the experimental ratio 0.611. This results in a rotor/stator blade count ratio of 3:2. The blade profiles belong to the group of Controlled Diffusion Airfoil (CDA) profiles.

The flow in the mid-span section of the cascade can be considered to be two-dimensional (span/chord ratio=3). Fig. 2 shows the grid structure in the vicinity of the leading edge of the compressor blade. Addressing the issues of near-wall accuracy, the viscous region of the boundary layer was resolved by several quadrilateral grid layers. The initial height of the first layer was adjusted to values resulting in y^+ close to unity. Beyond the near-wall region the computational domain was discretized by triangular elements. The hybrid grid contained in total 74000 nodes.

In this study there were three different cases analysed. Approximating the velocity triangles in a real compressor, a first calculation was performed based on the experimental data for low inlet speed (M=0.17) and low bar speed (u_{bar}=20 m/s). The second case accounts for the compressibilty effects by setting a high Mach number (M=0.72) at the inlet. The boundary conditions were taken from the experiment und were not adjusted to the 2-dimensional calculation. Since the 2-dimensional calculation does not account for the measured 3-dimensional axial velocity density ratio (AVDR). a higher Mach number level was predicted. With respect to the flow conditions in real turbomachines an additional calculation applying high inlet Mach number and high bar speed was carried out. Since there is detailed experimental

102

data available, all CFD calculations were conducted for a Reynolds number of Re=2x10^5. A summary of all inlet/outlet boundary conditions is given in Table 1.

Table 1: Boundary conditions

	Ma=0.17	Ma=0.72	Ma=0.72
Reynolds Number based on Chord Length Re	2x10^5	2x10^5	2x10^5
Reynolds Number based on Bar Diameter Re$_{bar}$	4x10^3	4x10^3	4x10^3
Total Inlet Pressure p_{tin} [Pa]	65056	16977	16977
Total Inlet Temperature T_{tin} [K]	313.14	313.14	313.14
Velocity Inlet Angle over y-direction β_1 [°]	135.1	135.1	135.1
Static Outlet Pressure p_{out} [Pa]	64100	13500	13500
Turbulent Viscosity Ratio μ_t/μ [-]	2	0.5	0.5
Bar Speed u_{bar} [m/s]	20	20	200
Time Step [s]	3.125x10^{-5}	3.125x10^{-6}	3.125x10^{-6}

The Analyses

Low Inlet Mach Number Case. A contour plot of the instantaneous Mach number distribution is depicted in Fig. 3. This gives a qualitative picture of the flow field and shows how the bar wakes convect through the compressor cascade. Since the boundary conditions for the calculation were not adjusted to the 2-dimensional analyses a 20% higher magnitude of the core flow velocity was computed than obtained by the experiment. The time averaged measured velocity distribution and turbulent intensity is shown in Fig. 5. The computational results for the time-averaged velocity distribution at an axial position about half way between the bars and the blades (x=-0.08m) are depicted in Fig. 6. Compared to the experimantal data, the analysis predicts a velocity profile which is less influenced by the bar wakes than detected within the experiment. This is probably caused by a too small value of the turbulent viscosity ratio in the inlet boundary condition. This leads to a lower level of turbulent intensity and thus to a lower mixing rate of the wake profiles. Fig. 7 shows the turbulent viscosity ratio profile.

Regions of higher turbulence indicate the path of the wakes (Fig. 4). Primarily due to variations in the passage velocity field, the wakes deviate toward the suction surface of the blades. The moving bars generate a vortex street shedding at a frequency of about 4 kHz. The calculation shows that in the bar wakes the production of turbulent energy takes place at a lower level than behind the compressor blades.

Fig. 8 shows the instantaneous velocity angle variation along a bar wake cross section. The section is located upstream at a distance of about 32% axial chord length from the blade leading edge (x=-0.063m). Within a wake the velocity angle varies periodically by +2° about the nominal inlet angle β_1=135.1° which leads to an unsteady structural loading of the blades. A more quantitative statement is given by Fig. 11 where the unsteady isentropic Mach number distribution is depicted for a single blade (Blade 1 in Fig. 1) at three various time steps within a single wake-passing cycle. Each individual time step marks a specific wake position relative to the blade surface. As the wake passes the blade leading edge, the periodically changing incidence leads to a fluctuation of the isentropic Mach number which is associated with an unsteady static pressure distribution on the blade surfaces. Curve T1 shows an undisturbed blade leading edge. As the wake covers the leading edge, the blade incidence angle starts rising. This results in a higher acceleration of the particles along the suction surface (Curve T2). Advancing further in time, the complete suction surface shows a higher velocity magnitude (Curve T3). The main result is that, in conjunction with the velocity defect within the wake, the fluctuating velocity angle cause an unsteady loading of the compressor blade.

High Inlet Mach Number Case. Fig. 9 shows the Turbulent Viscosity Ratio contour plot for high inlet Mach number (M=0.72) and low bar speed u_{bar}=20 m/s. Since in this calculation the bar speed is very low compared to the inlet velocity, the wakes pass the cascade nearly following the main stream lines. The bars generate a vortex street with a shedding frequency of about 20 kHz according to the constant Strouhal number.

The region of high turbulent viscosity for the calculation with high bar speed is shown in Fig. 10. Compared to the low inlet speed the bars generate "stronger" wakes and produce turbulent energy at the same level as that behind the blades. The calculation also shows that the chopped wake segments traverse the upper and lower airfoil surface at different speeds. This seperation can be related to the lift, or circulation, associated with the airfoil [4].

Concluding Remarks

Numerical unsteady analyses of wake paths generated by moving bars in a compressor cascade were conducted. The CFD package FLUENT was found to be suitable for predicting the wake structure behind moving bars at different inlet conditions and bar speeds. The analyses provide a general understanding of the bar wake trajectories, their structure and their influence on the unsteady loading of the downstream compressor blades. The results of this study support ongoing experimental investigations for obtaining detailed measurements for the laminar-turbulent transition process. Employing LES, nonlinear Low-Re turbulence models and production term modification models, the unsteady process of transition will be investigated in the future.

Acknowledgment

This reported work was performed within a research project TurboTech II that is part of the national research corporation "AG TURBO". The project has been supported by the German Ministry of Education, Science, Research and Technology (BMBF) and the BMW Rolls Royce GmbH, Dahlewitz. The permission for publication is gratefully acknowledged.

References

[1] P. Acton, *Untersuchungen des Grenzschichtumschlages an einem hochbelastetem Turbinengitter unter inhomogenen und instationären Zuströmbedingungen*, PhD Thesis, Univ. BW Munich, 1998.

[2] N.A. Cumpsty, Y. Dong, Y.S. Li, Compressor Blade Boundary Layers in the Presence of Wakes, McGraw-Hill, New York, 1991.

[3] D.E. Halstead, D.C. Wisler, T.H. Okiishi, G.J. Walker, H.P. Hodson, H. Shin, *Boundary Layer Development in Axial Compressors and Turbines, Part 1: Composite Picture, Part 4: Computation & Analyses*, ASME 95-GT-461.

[4] H.D. Joslyn, J.R. Caspar, R.P. Dring, *Inviscid Modeling of Turbomachinary Wake Transport*, ASME Journal of Propulsion and Power, Vol. 2, No. 2, pp. 175-180, 1986.

[5] P. Spalart and S. Allmaras, *A one-equation turbulence model for aerodynamic flows* , Technical Report AIAA-92-0439, American Institute of Aeronautics and Astronautics, 1992.

[6] M. Swoboda, R. Teusch, L. Fottner, V. Guemmer, U. Wenger, *Experimental Investigation of Boundary Layer Transition in Compressor Cascades at Unsteady Periodic Flow Conditions,* to be presented at 11. DGLR-Fach-Symposium, Strömungen mit Ablösung, Berlin, 1998.

Figures

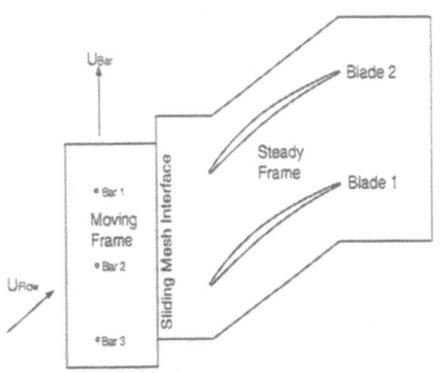

Figure 1: Schematic View of the Computational Domain

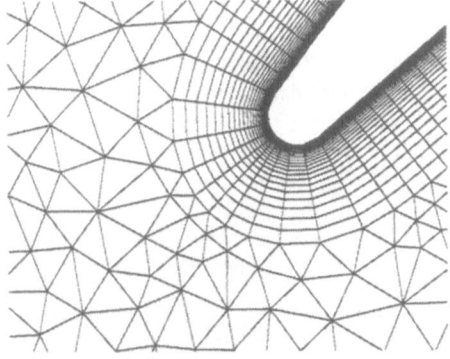

Figure 2: Grid in the Vicinity of the Blade Leading Edge

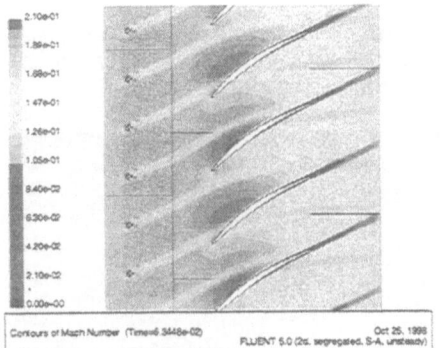

Figure 3: Instantaneous Contours of Mach Number, M=0.17, Re=2x10^5, u_{bar}=20m/s

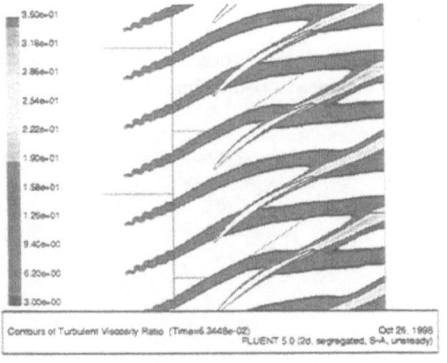

Figure 4: Instantaneous Contours of Turbulent Viscosity Ratio, M=0.17, Re=2x10^5, u_{bar}=20m/s

105

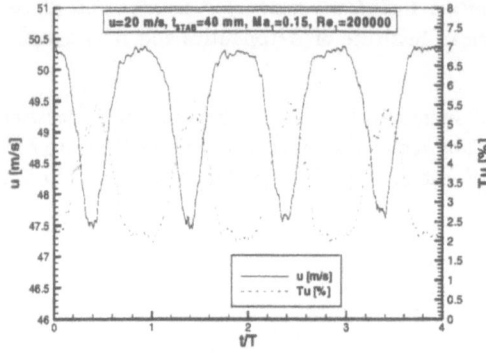

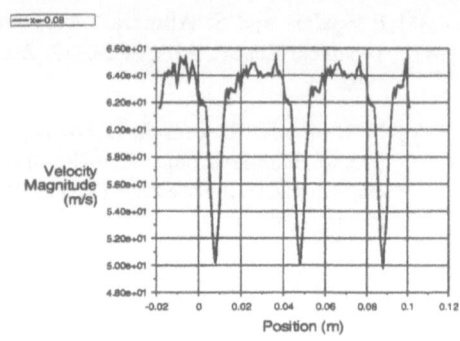

Figure 5: Turbulent Intensity Measurement and Velocity Distribution, M=0.15, Re=2x10⁵, u_{bar}=20m/s

Figure 6: Time Averaged Velocity Magnitude Distribution at x=-0.080m, M=0.17, Re=2x10⁵, u_{bar}=20m/s

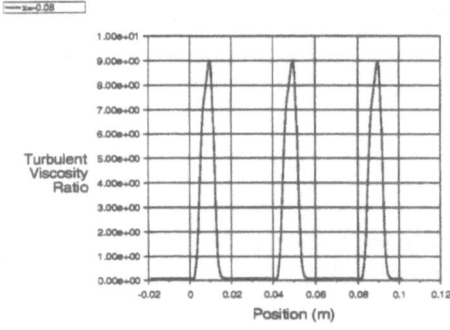

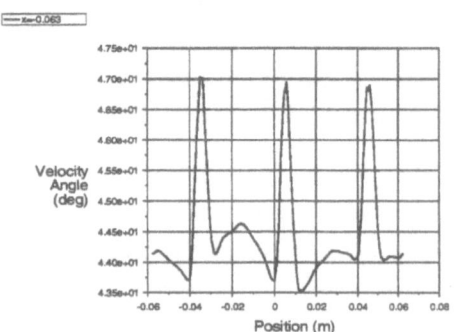

Figure 7: Time Averaged Turbulent Viscosity Ratio Profile at x=-0.080m, M=0.17, Re=2x10⁵, u_{bar}=20m/s

Figure 8: Instantaneous Velocity Angle Distribution at x=-0.063m, M=0.17, Re=2x10⁵, u_{bar}=20m/s

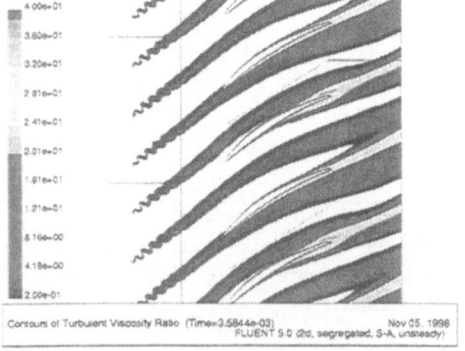

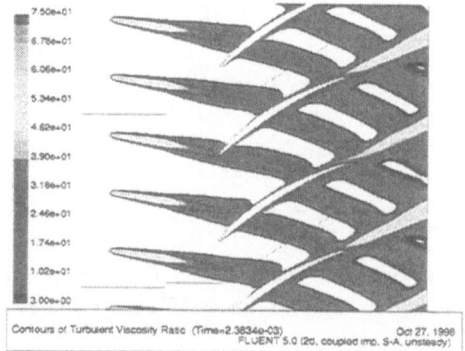

Figure 9: Contours of Turbulent Viscosity Ratio at a certain Time Step, M=0.72, Re=2x10⁵, u_{bar}=20m/s

Figure 10: Contours of Turbulent Viscosity Ratio at a certain Time Step, Coupled Solver, M=0.72, Re=2x10⁵, u_{bar}=200m/s

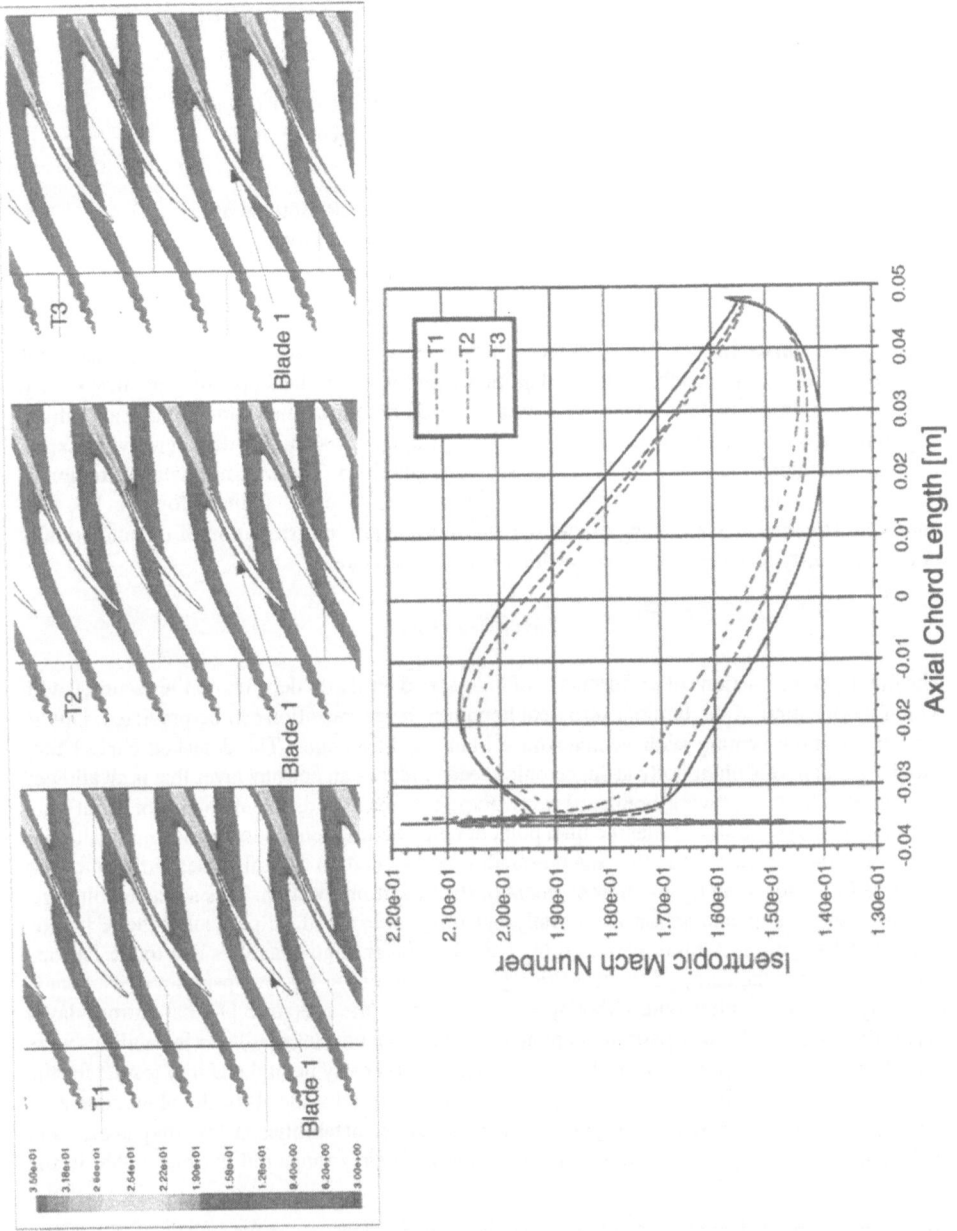

Figure 11: Contours of Turbulent Viscosity Ratio at Various Time Steps and corresponding Unsteady Isentropic Mach Number Curve Plots for Blade 1, M=0.17, Re=2x10^5, u_{bar}=20m/s

Entropy-Layer Instabilities in Plane Supersonic Flow

Guido Dietz, Stefan Mählmann
Aerodynamisches Institut der RWTH Aachen
Wüllnerstraße zw. 5 u. 7, D-52062 Aachen, Germany

Stefan Hein
Institut für Strömungsmechanik des DLR Göttingen
Bunsenstr. 10, D-37073 Göttingen, Germany

Summary

Entropy-layer instabilities in the supersonic flow over a blunt flat plate are investigated. Two types of entropy-layer instabilities are identified with the linear, local, parallel stability theory. The most amplified mode is a two-dimensional one. In order to validate the numerical findings experiments were carried out: The computed steady, laminar mean-flow data agree with experimentally obtained mass-flow and Pitot-pressure distributions. The wall-normal distribution of the mass-flow eigenfunction of the most amplified entropy-layer instability is found in hot-wire data. The streamwise wave number and the wall-normal phase distribution of the numerically obtained density eigenfunctions agree with a Schlieren picture.

Introduction

The quality of estimations of performance of high-speed airplanes depends on the accuracy with which the location of the laminar-turbulent boundary-layer transition can be predicted. Due to the aerodynamic heating such vehicles have blunt leading edges. The detached curved bow shock upstream of a blunt body at supersonic speeds induces an entropy layer that is swallowed downstream by the growing boundary layer. Experiments with different shapes of blunt bodies [1, 2, 3, 4, 5] have shown that the final point of laminar/turbulent boundary-layer transition is displaced downstream with increasing nose radius up to a certain critical value; further increase of the radius leads to an upstream movement of the transition location. This so-called blunting transition reversal phenomenon is currently not fully understood. In particular, the reduction of the local Reynolds number based on the boundary-layer edge quantities due to the entropy layer can explain the delay of the transition [6]. But the reason for the upstream movement is currently not fully understood, although it is assumed in the literature [7] that entropy-layer instabilities [5, 7, 8, 9] may partially explain this phenomenon. Entropy-layer instabilities were first observed in experiments over blunt cones [1]. Besides early neutral stability results for the entropy layer over a blunt plate [7, 8, 9], numerical investigations [10] for the blunt cone case corresponding to experiments [1] identified entropy-layer instabilities at low frequencies and with rather small amplification rates. However, the frequency range did not match that in the experiments.

Herein, we report new results on entropy-layer instabilities observed in the supersonic flow over a blunt flat plate. First numerical results consistent with experimental observations on this type of entropy-layer instability most amplified at zero wave angle were described in [11]. Linear, local, parallel stability analyses of numerically obtained laminar mean-flow results were carried out for free-stream Mach numbers $M_\infty = 2$ to $M_\infty = 4$ at unit Reynolds numbers $Re_\infty/l \approx 10^7/$m. Angles of attack $\alpha = 0^0$ and 7^0 were chosen, while the nose radius amounts $R_n = 2.5$ mm. The numerical results are validated by corresponding experimental data at $M_\infty = 2.5$, $\alpha = 0^0$.

Mathematical models and numerical methods

Mean-flow computation

The Navier-Stokes equations for two-dimensional, compressible flow in generalized orthogonal curvilinear coordinates x^i are the governing equations for the steady, laminar mean flow. Herein x^1 denotes the streamwise and x^3 the wall-normal coordinate in the body-oriented coordinate system x^i ($i=1,(1),3$) with its origin in the stagnation point. A perfect gas with a constant Prandtl number of $Pr = 0.715$ is assumed. The resulting PDE system is closed with Fourier's law of heat conduction and Sutherland's law for the molecular viscosity assuming Stokes's hypothesis for the bulk viscosity. The supersonic inflow boundary condition prescribes the free-stream values. At the outflow boundaries density, mass-fluxes and total enthalpy are extrapolated along gridlines with first order accuracy. Both the hyperbolic outer flow and the parabolic boundary-layer flow are initial-value problems in space, where only the pressure in the subsonic region of the boundary layer transports information upstream. Thus, this simple outflow boundary condition can be applied, provided that only wall-normal profiles are analyzed for stability which are far enough upstream of the outflow boundary. At the solid, adiabatic wall the no-slip condition is applied and according to the boundary-layer approximations the wall-normal pressure gradient vanishes identically. The simulations were initialized specifying a parallel flow with free-stream properties.

The advective terms are discretized using the Advection Upwind Splitting Method in an improved formulation AUSMDV [12]. Second order spatial accuracy is achieved with the MUSCL approach [13]. The diffusive terms are discretized with central differences. Application of an explicit time integration with a five-step Runge-Kutta method yields a numerical method of second-order accuracy in time and space. Convergence to steady state is accelerated with local time stepping and a multi-grid algorithm. Earlier studies [11, 14, 15] have shown that the solutions for the mean flow are sufficiently accurate to perform a reliable instability analysis.

Linear instability analysis

The instability analysis of the laminar mean flow is based on the linear, local stability theory using the DLR/FFA NOLOT-code [16]. The flow quantities in the conservation equations are decomposed into the steady, parallel, laminar mean-flow part and an unsteady disturbance flow component $\bar{q}$, which itself is decomposed into a complex amplitude function $\hat{q}(x^3)$ and a wave function $e^{i(\alpha x^1 + \beta x^2 - \omega t)}$.

In the present work spatial theory is used only, which corresponds to a real circular frequency $\omega = \omega_r$, a complex streamwise wave number $\alpha = \alpha_r + i\alpha_i$ and a real spanwise wave number $\beta = \beta_r$. The indices r and i denote the real and the imaginary part. If the growth rate $\sigma = -\alpha_i$ is positive, the disturbance is amplified which means, that the flow is unstable, and vice versa; neutral instability is obtained for $\sigma = 0$. The wave angle Ψ is defined as $\Psi = \tan^{-1}(\beta_r/\alpha_r)$.

The linear, local, parallel stability equations can be written as a system of eight first order ODEs. These are discretized by fourth order compact finite differences in x^3-direction. According to the no-slip condition the disturbance velocity components vanish at the wall. Even for the adiabatic wall a thermal inertia is assumed such that the wall temperature cannot follow high frequency oscillations. Besides the four boundary conditions at the wall, the asymptotic solution of the stability equations in the approximately constant free stream is applied as far-field boundary condition.

Experimental set-up

Test facility

The suction type trisonic windtunnel is located at the Aerodynamisches Institut of the RWTH Aachen. The tunnel with a measuring chamber cross section size of 40×40 cm^2 is operated intermittently. The adjustable Laval nozzle provides a Mach number range from 0.2 up to 3. The usable testing time for the present experiments depends on the Mach number and amounts about 0.5 s at a free-stream Mach number $M_\infty = 2.5$. The stagnation pressure and the stagnation temperature correspond to their environmental values. So, the free-stream unit Reynolds number depends on the Mach number and here is $Re_\infty/l = 9.9 \cdot 10^6$/m. A description of the tunnel, its instrumentation and its operational characteristics can be found in [17].

Wind tunnel model

The blunt flat plate model is fabricated of polished stainless steal with a semicircular leading edge of 2.5 mm radius and is 36.5 cm long by 40 cm wide. The model was horizontally mounted at zero angle of attack.

Experimental techniques

Flow measurements were performed with application of both constant-temperature hot-wire anemometry and Pitot-pressure technique integrated in a single probe. This probe has been designed and fabricated for supersonic flow investigations at the Aerodynamisches Institut. The body of the probe has an elliptical cross section and is wedge shaped in the front. The Pitot pressure is measured at the tip of this wedge about 0.4 mm downstream of the hot wire. The tungsten hot wire with a diameter of 5 μm and a length of 1 mm was driven by an A.A. Lab Systems AN-1003 CTA bridge with installed high frequency option. The total pressure was measured with a Kulite XCS-062-1barA sensor driven by a ENDEVCO 136 DC amplifier. In order to avoid aliasing the hot-wire anemometer output was low-pass filtered with 100 kHz while the pressure signal was low-pass filtered with 20 kHz. Both filter outputs were decomposed into a mean value (low-pass 50 Hz) and a fluctuating value (high-pass 100 Hz) which all were recorded by a 12 bit ADC with a sampling frequency of 200 kHz. The probe could be moved in wall-normal direction with a traversing unit in steps of 25 μm. At least 16 kilo samples were recorded at each wall-normal position. This measuring equipment was computer controlled and automatically triggered by the static pressure in the measuring chamber, so that about 4 to 5 positions could be sampled within 0.5 s.

The probe was calibrated in the free stream concerning the mass-flow for only one stagnation temperature due to the available test facility. So, the mass-flow values resulting from the hot-wire data obtained in the boundary-layer are affected by an error, because the stagnation temperature there differs from its free-stream value. Thus, the hot-wire was driven at high overheat ratios, where the sensitivity to temperature is small compared to sensitivity to mass-flow. Frequency spectra were obtained with the common Welch method.

Furthermore, visualization of the wall-normal density gradients in flow-field was conducted using a NdYAG laser (512 nm, 8 ns pulse, 170 mJ/pulse) recording conventional Schlieren pictures.

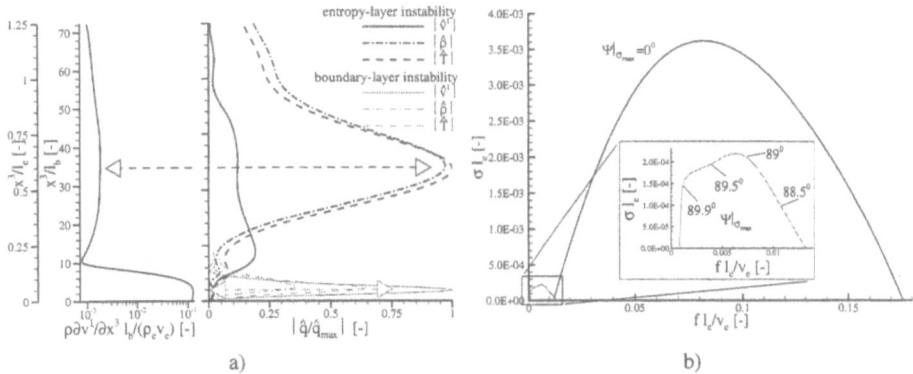

Figure 1 a) Left: Computed $\rho\partial v^1/\partial x^3$ distribution in x^3 direction, Right: Amplitude ratios of the eigenfunctions of an entropy-layer instability (thick, black) compared to eigenfunctions of a first mode boundary-layer instability (thin, gray). b) Growth rates of two different entropy-layer instability modes. $M_\infty = 2.5$, $x^1/R_n = 101.2$, $\alpha=7^0$ windward.

Results

Figure 1 a) on the right shows numerically obtained amplitudes of the eigenfunctions $|\hat{v}^1|$, $|\hat{T}|$ and $|\hat{\rho}|$ of an entropy-layer instability mode compared to those of a first mode boundary-layer instability for $M_\infty = 2.5$, $x^1/R_n = 101.2$, $\alpha=7^0$ windward. The maximum disturbance quantities of both modes are normalized to one, respectively. For the normalization of the x^3 coordinate the two characteristic length scales for the wall-normal non-uniformities in the mean flow are used [8, 9, 11]: One is the common scale for the boundary layer based on a characteristic boundary-layer thickness $l_b = \sqrt{x^1 \nu_e/v_e}$, where v_e denotes the streamwise velocity and ν_e the kinematic viscosity at the boundary-layer edge. The radius of the bow shock measured at the intersection of the shock with the stagnation streamline is used as entropy-layer length scale l_e. It is to be noted, that both the amplitude maxima in $|\hat{T}|$ and $|\hat{\rho}|$ of the entropy-layer instability ($x^3/l_e \approx 0.6$) and the amplitude maximum in $|\hat{v}^1|$ of the first mode boundary-layer instability ($x^3/l_b \approx 2.5$) correspond to generalized inflection points of the mean flow (Fig. 1 a) left) which matches the observations in [1, 7]. In addition to Fig. 1 the data reveal that the corresponding pressure fluctuation amplitudes are rather small for the entropy-layer instability, since there is a phase shift of approximately 180^0 between temperature and density fluctuations. In Fig.1 b) this entropy-layer instability most amplified at zero wave angle is compared to another one, which travels at large wave angles. The former one is about 15 times more amplified at about 10 times larger frequencies than the second one. This second mode corresponds to a mode reported in [10]. Both types are clearly of subsonic character since their streamwise phase velocity $c_r = 2\pi f/\alpha_r/v_e$ is very close to one.

It is shown by Howard in [18] that an inviscid, incompressible, streamwise travelling instability wave associated with an inflection point (Rayleigh wave) has a temporal growth rate $\alpha_r c_i \leq \max |\partial v^1/\partial x^3|/2$ for a bounded shear layer. In Fig. 2 growth rates of the entropy-layer instability are compared to the half of the velocity gradient at the inflection point in the entropy layer, although the flow considered here is a highly compressible one. Gaster's transformation applied to the above stated Howard's semicircle theorem yields the relation $\sigma l_e \leq \max |\partial v^1/\partial x^3| \, l_e/(2\,\partial\omega_r/\partial\alpha_r)$. For all numerically investigated cases the non-dimensional group velocity $\partial\omega_r/\partial\alpha_r/v_e$ of the entropy-layer instability as well as its phase velocity are very close to one as can be seen from Fig. 2 c). Hence, the disturbance energy of this insta-

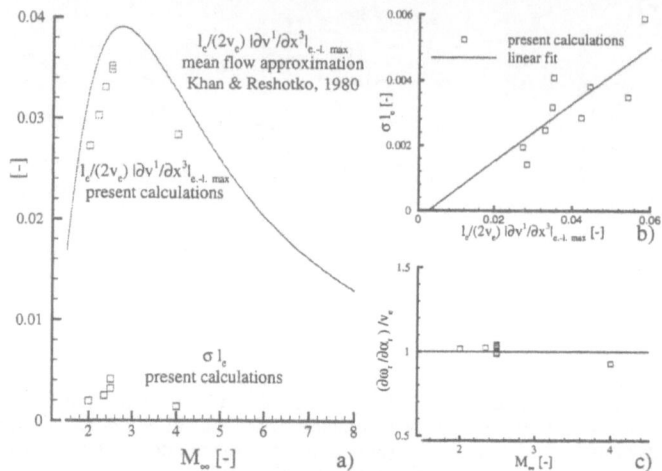

Figure 2 Non-dimensional growth rates σl_e compared to the half of the maximal non-dimensional velocity gradient $|\partial v^1/\partial x^3|\, l_e/2v_e$ in the entropy layer. a) Both values versus free-stream Mach number M_∞, $\alpha = 0^0$. b) Growth rates versus gradient for all evaluated M_∞ and α. c) Non-dimensional group velocity $\partial \omega_r/\partial \alpha_r\, /v_e$ of the most amplified entropy-layer instability versus M_∞ for all evaluated M_∞ and α.

bility mode is convectively transported and the group velocity can be approximately replaced by v_e. In Fig. 2 a) the computed growth rates σl_e are compared to $\partial v^1/\partial x^3\, l_e/(2v_e)$ resulting from the computations and from evaluations of an analytical approximation derived in [9] for the inviscid mean flow in the entropy layer at $\alpha = 0^0$. The approximation predicts the computed values well. For different Mach numbers M_∞ the computed growth rates σl_e are about 10 times smaller than $\partial v^1/\partial x^3\, l_e/(2v_e)$. Nevertheless, a roughly linear dependence of these two quantities can be seen in Fig. 2 b) for all numerically investigated cases. Since the approximation reveals a maximum in $\partial v^1/\partial x^3\, l_e/(2v_e)$ between $M_\infty = 2$ and $M_\infty = 3$ we assume that for plane flow the non-dimensional growth rates of entropy-layer instabilities have a maximum between $2 < M_\infty < 3$. Hence, in the following only the most amplified entropy-layer instability is inspected and the numerical findings are validated by experiments at $M_\infty = 2.5$, $\alpha=0^0$.

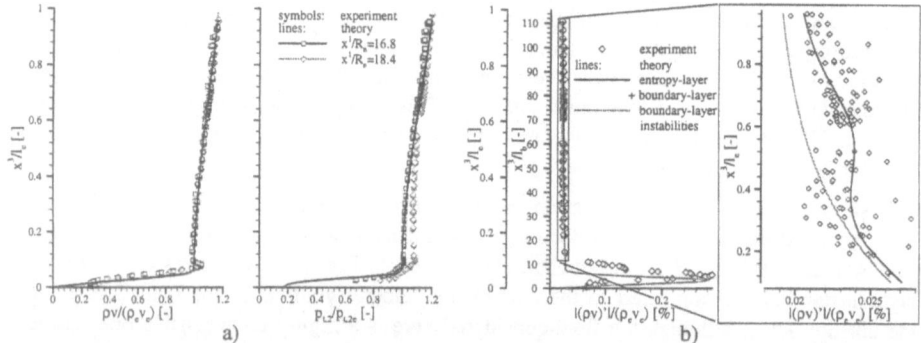

Figure 3 a) Experimentally obtained mean-flow distributions in x^3 direction compared to mean-flow computations at $x^1/R_n = 16.2$ and 18.4. Left: Mass flow, Right: Pitot pressure. b) Experimentally obtained mass-flow fluctuations compared to computed, appropriately scaled mass-flow eigenfunctions at $x^1/R_n = 18.4$. Left: Overall view, Right: Zoom into the entropy-layer region. $M_\infty = 2.5$, $\alpha=0^0$.

Figure 3 a) shows four measured mean-flow profiles compared to the corresponding computational results at two positions $x^1/R_n = 16.2$ and 18.4. In particular, mass-flow (left) and Pitot-pressure (right) distributions are plotted normal to the wall. Outside the boundary layer $(x^3/l_e > 0.15)$ one can recognize the entropy layer, which is characterized by a mass-flow and a Pitot-pressure gradient as well as weakly pronounced inflection points in both distributions at $x^3/l_e \approx 0.6$. The discrepancy between the computed and the measured mass-flow distributions in the boundary-layer region was expected, since the stagnation temperature varies inside of the boundary layer, as explained above. However, the Pitot-pressure distribution on the right-hand side of Fig. 3 a) shows that the computation predicts the boundary-layer thickness very well. In Fig. 3 a) the deviation of the measured pressure data for $x^1/R_n = 18.4$ was caused by a leak in the probe. In Fig. 3 b) the experimentally obtained mass-flow fluctuations are plotted for the frequency of the most amplified entropy-layer instability at $x^1/R_n = 18.4$. The amplitude of the computed mass-flow eigenfunction of the first mode boundary-layer instability at the same frequency is shown in comparison scaled by the measured values. The wall normal position of its maximum in the boundary layer matches the measured one, taking into account that a wall-interference of the probe is likely to occur. On the right-hand side of Fig. 3 b) the plot is zoomed into the entropy-layer region. The amplitude of the computed mass-flow eigenfunction of the entropy-layer instability is superposed to the boundary-layer eigenfunction and scaled analogously. The fluctuation amplitude of the entropy-layer instability is to recognize in the experimental data, although the measured mass-flow fluctuations in this region are about 50 times smaller than those in the boundary layer. The maximum in the amplitude coincides with the generalized inflection point in the entropy layer as stated above.

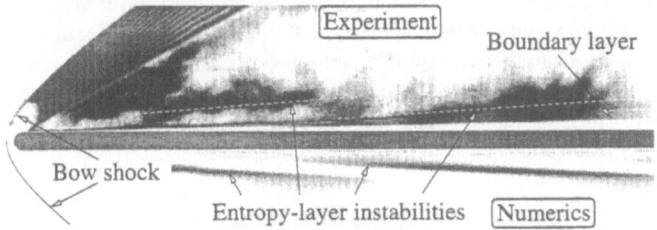

Figure 4 Wall-normal density gradients visualized in a Schlieren picture (upper) compared to those of the numerically obtained perturbation flow of the most amplified entropy-layer instability with a non-dimensional frequency $fl_e/v_e \approx 0.1$ (lower). $M_\infty = 2.5, \alpha = 0^0$.

The density fluctuation amplitudes presented in Fig. 1 a) motivated the visualization of the entropy-layer instability by the Schlieren technique in the upper part of Fig. 4 for $M_\infty = 2.5$, $\alpha = 0^0$. One can see oblique dark areas indicating regions of relatively high wall-normal density gradients. The numerically obtained perturbation flow of the most amplified entropy-layer instability is opposed to the experiment in the lower part, where $\partial\tilde\rho/\partial x^3$ is shown gray-scaled. For a better comparison it has been necessary to adapt the phasing of the calculated eigenfunctions, i.e. the downstream position of the group of lines at a particular time. In addition, the location of the maxima of $\partial\tilde\rho/\partial x^3$ are shown as white dashed lines in the Schlieren picture. Since the streamwise distance of these lines corresponds to the streamwise wave length and their inclination represents the wall-normal phase distribution, the streamwise wave number as well as the wall-normal phase distribution of the density eigenfunction agree with the experimentally determined waves. Furthermore, Fig. 4 supports the theoretical finding that the most amplified entropy-layer instabilities are two-dimensional ones: Their wave angle in the experiment should be close to zero, because the waves are clearly visible in the experimental setup with a plate of 40 cm width.

Conclusion

Two different types of entropy-layer instabilities in the supersonic flow over the windward side of a blunt adiabatic flat plate have been detected numerically. One of them is most amplified at large wave angles and low frequencies. The other exhibits amplification rates at zero wave angle in an order of magnitude higher than the former one. It is shown, that this instability is connected with a generalized inflection point in the entropy layer. From the present computations and the incompressible, inviscid stability theory there is some indication that the non-dimensional growth rates of the entropy-layer instabilities might have a maximum between $2 < M_\infty < 3$. Therefore, the computed steady, laminar mean flow was successfully compared to experimentally determined mass-flow and Pitot-pressure distributions at $M_\infty = 2.5$ and $\alpha = 0^0$. The density perturbation field of the most amplified entropy-layer instability agrees with experimental visualization of the flow-field using the Schlieren technique and its mass-flow eigenfunction could be detected in hot-wire signals. Nevertheless, the mass-flow fluctuations of this entropy-layer instability were much smaller than those of the first mode boundary-layer instability.

Outlook

At first sight one could assume that the entropy-layer instability is not of relevance for the bluntness transition reversal phenomenon, because the amplitudes of the entropy-layer instability mass-flow fluctuations are small compared to that of the boundary-layer instability. However, preliminary computations reveal a damped mode that synchronizes first with the entropy-layer instability and further downstream with the first mode boundary-layer instability while reducing its phase speed from about one to the phase speed of the first mode. So, this mode can transport disturbance energy from the entropy layer into the critical layer of the boundary-layer. Receptivity analyses using adjoint systems e.g. [19] reveal that an exitation in the critical layer is most effective. Thus, an interaction of these three modes might be one reason for the transition reversal phenomenon and is subject of future investigations.

Acknowledgements

We are grateful to Egon Krause (AIA) who initiated this investigation and to Uwe C. Dallmann (DLR) for encouraging the numerical part of this work. We wish to thank Ardeshir Hanifi (FFA) for his contributions to the linear DLR/FFA NOLOT code. This work was supported by the Deutsche Forschungsgemeinschaft, Grant No. II D5–Kr 387/31-1.

References

[1] K. F. Stetson, J. C. Donaldson, L. G. Siler: Laminar boundary layer stability experiments on a cone at Mach 8, Part 2: Blunt cone. AIAA Pap. 84-0006, 1984.

[2] P. F. Brinich: Effects of leading-edge geometry on boundary layer transition at Mach 3.1. NACA TN 3659, 1956.

[3] P. F. Brinich, N. Sands: Effects of bluntness on transition for a cone and a hollow cylinder at Mach 3.1. NACA TN 3679, 1957.

[4] D. W. Jillie, E. J. Hopkins: Effects of Mach-number, leading-edge bluntness, and sweep on boundary-layer transition on a flat plate. NASA TN D-1071, 1961.

[5] V. I. Lysenko: Influence of the entropy layer on the stability of a supersonic shock layer and transition of the laminar boundary layer to turbulence. *J. Appl. Mech. Tech. Phys.*, 31(6):868–873, 1990.

[6] W. E. Moeckel: Some effects of bluntness on boundary-layer transition and heat transfer at supersonic speeds. NACA Rep. 1312, 1957.

[7] A. V. Fedorov: Instability of the entropy layer on a blunt plate in supersonic gas flow. *J. Appl. Mech. Tech. Phys.*, 31(5):722–728, 1990.

[8] E. Reshotko, M. M. S. Khan: Stability of the laminar boundary layer on a blunted plate in supersonic flow. In *Laminar-Turbulent Transition, Symposium Stuttgart, Germany 1979*, Proc. IUTAM, p. 186–200, 1980.

[9] M. M. S. Khan, E. Reshotko: Stability of the laminar boundary layer on a blunted plate in supersonic flow. CWRU FTAS/TR 79-142, 1980.

[10] E. Kufner: Entropy and boundary layer instability of hypersonic cone flows – effects of mean flow variations. In *Laminar-Turbulent Transition, Symposium Sendai, Japan*, Proc. IUTAM, 1994.

[11] G. Dietz, S. Hein: Entropy-layer instabilities over a blunted flat plate in supersonic flow. *Phys. of Fluids*, accepted, publication scheduled in January 1999.

[12] Y. Wada, M.-S. Liou: A flux splitting scheme with high-resolution and robustness for discontinuities. NASA TM 106452 (also AIAA 94-0083), 1994.

[13] B. Van Leer: Towards the ultimate conservative difference scheme v. A second-order sequel to Godunov's method. *J. Comp. Phys.*, 32:101–136, 1979.

[14] A. Meijering, G. Dietz, S. Hein, U. Dallmann: Numerical investigation of the boundary layer instabilities over a blunt flat plate at angle of attack in supersonic flow. DLR IB 223-96 A44, 1996.

[15] G. Dietz, A. Meijering: Numerical investigation of boundary-layer instabilities over a blunt flat plate at angle of attack in supersonic flow. In *Notes on Numerical Fluid Mechanics* 60, p. 103–110, 1996.

[16] S. Hein, F.P. Bertolotti, M. Simen, A. Hanifi, D. Henningson: Linear nonlocal instability analysis -the linear NOLOT code-. DLR IB 223-94 A56, 1994.

[17] M. Jacobs: *Wärmeübergangsmessungen am Modell eines Deltaflügels in supersonischer Strömung.* Diss. Aerodyn. Inst. RWTH-Aachen (DLR FB 97-44), 1997.

[18] P. G. Drazin, W. H. Reid: *Hydrodynamic Stability.* Cambridge University Press, 1997.

[19] D. C. Hill: Adjoint systems and their role in the receptivity problem for boundary layers. *J. Fluid Mech.*, 292:183–204, 1995.

Navier-Stokes Solutions on Wing-Body-Tail Configurations and Viscous Mesh Adaptation Technique

E.Elsholz, H.Steinmeyer, D.John
DaimlerChrysler Aerospace Airbus GmbH, Dept. EFV
D-28183 Bremen
Germany

SUMMARY

A set of 3D Navier-Stokes solutions around a complete wing-body-tail configuration is presented for different mesh resolutions. The local mesh dependent behaviour is investigated and from these experiences, a technique is derived that aims for multi-sweep y+ driven mesh adaptation to wing- and fuselage boundary layer profiles. This technique is described and demonstrated but final results are still under preparation.

INTRODUCTION

It is well known that numerical 3D Navier-Stokes solutions are mesh dependent to a certain degree, i.e. there will be (partially) different solutions on different meshes for the same configuration, turbulence model and flow parameters. In this context, the mesh resolution of the various boundary- and shear layers are of significant importance.

Since the industrial design process becomes more and more dependent on the accuracy and robustness of 3D Navier-Stokes solutions, it is worthwile to optimize the meshes with respect to the necessities of the solution, which results in adaptation to the viscous layers. However, additional mesh adaptation with respect to shocks and geometric curvatures may be also required but will not be included here.

A340-200 PREDICTIONS VS. EXPERIMENTS

Basic solution: medium mesh resolution

A first encouraging Navier-Stokes solution over the complete wing-body-tail configuration (Fig. 1) had been reported in [1] and a preliminary comparison with the ONERA-T2 experiments [2] had been given. The main flow features were already predicted to reasonable satisfaction, i.e. the pressure distribution over the wing and fuselage as well as the skin friction patterns, showing considerable vortex shedding at the lower rear-fuselage area. However, from velocity profile examinations it turned out that the boundary layer resolution over the fuselage was not yet satisfactory in these computations.

These computations have been performed using the FLOWer-code with the Wilcox k-ω turbulence model incorporated.

Enriched and locally condensed mesh solutions

To overcome these deficiencies, the basic mesh which was thought to be of medium resolution with respect to all directions (mesh 'mmm', 560.000 nodes total), has been refined globally in spanwise direction (K- direction), resulting in approximately 1.1 million nodes (mesh 'mmf'), Fig. 2. In a second step, this mesh has been locally redistributed throughout the apparently most critical area at the lower rear-fuselage in order to further improve the boundary layer resolution (mesh 'mmf2'). Comparison of the computed pressure distributions in the 'mmm' and 'mmf2' meshes, Fig. 3, show only marginal changes but the velocity profiles in the critical area respond significantly by steepening of the near-wall gradients, Fig. 4. From this it becomes obvious that the mesh resolution at least in this area has to be properly adapted to the local flow characteristics.

The overall results in the refined mesh are presented in Fig. 5 and Fig. 6. The comparison with the experiments (pressure, skin-friction) show the solution to be of reasonable quality but in a plane downstream of the fuselage tail, the secondary flow and the loss of total pressure still show some deficiencies, Fig. 7.

Forebody refinements

In order to check the influence of the coarse resolution at the fuselage nose with respect to the poor behaviour in the downstream wake plane, another investigation has been performed based on additional local mesh refinement over the canopy and wing leading edge areas, which is shown in Fig. 8. However, there are no significant improvements of this solution observed.

VISCOUS MESH ADAPTATION TECHNIQUE

The mesh modifications described above have been made interactively by hand, using the mesh modification tool [3]. From these experiences it is obvious that an automated adaptation module is needed for improved solutions.

Basic strategy

The basic idea is to adapt the wall-nearest portion of the mesh, say 20-30 layers, to the local boundary layer development and furthermore, to ensure the wall-nearest normal stepsize to be close to $y+ = 1$. Rather than to define the boundary layer edge by searching for the maximum resultant velocity in the solution field, the edge of the logarithmic layer ($y+ = 10^3...10^4$) was chosen to define a unique grid layer to which subsequently all other mesh layers are adapted.

After that, a second adaptation sweep aligns the wall normal distance $y+=10$ (edge of the laminar sublayer) to the 4th grid layer, resulting in $\Delta y+_{wall}$ close to 1 due to the progressive step-size ratio in the normal direction. In both sweeps, the resulting target layers are smoothed out to some degree and the layers of the inner and outer mesh portions are adjusted to these new target layers. Finally, the normal step-size ratios are smoothed out, while the wall-nearest step-size is maintained.

This adaptation process is also performed within the frame of the mesh modification tool [3], which is based on adjustable, non-overshooting interpolation techniques that also may apply on the entire solution field. So after the adaptation process has been completed, a true restart of the Navier-Stokes computations can be performed, starting from the previous solution after interpolation to the adapted mesh nodes.

Wing/wake structures

For the adaptation to a wing boundary layer a linear edge distribution of $y+$ ($10^3...10^4$) is assumed which defines the area of adaptation normal to the wing surface (J-direction). This distribution may roughly account for the typical increase of the boundary layer thickness over a wing. Downstream of the trailing edge, the changes due to adaptation are allowed to decay to the original mesh distribution in order to fit to any downstream block faces.

Fuselage/wake structures

Concerning the fuselage, adaptation is performed along K-direction (normal to the fuselage surface, i.e. the spanwise direction in the present C-O topology) using the techniques described above. Passing the fuselage base, smooth relaxation of the adapted distribution is realized again to fit into the original far-wake mesh structure.

3D corner flow: side-wall boundary layer treatment

Along 3D corners, i.e. in wing-body junction regions or equivalent zones, a special treatment with respect to the side-wall boundary layer(s) becomes necessary, since there is no way to obtain meaningful main-flow boundary layer profiles along mesh lines that are located inside a side-wall boundary layer. As a consequence, any (main-flow) profile evaluation is cancelled when approaching the side-wall up to $y+_{side-wall} = 10^4$.

Results: adapted mesh around A340-200 wing-body-tail configuration

The adapted mesh shown in Fig. 9 and Fig.10, has been realized according to the described adaptation techniques. The figures show the corner-flow adaptation behaviour at a mid-wing cross-section (Fig. 9) and the adapted mesh at the rear-fuselage (Fig. 10), the latter in comparison to the original 'mmf2' mesh version.

It becomes obvious that the adapted mesh ensures considerable improvement of the near-wall resolution while on the other hand, additional mesh refinement might become necessary at the boundary layer edge and in the inviscid near-field.

NEXT STEPS / CONCLUSIONS

From these preliminary experiences, the mesh is planned to be further refined after this first adaptation step and once this is done, a new solution is to be computed including additional, subsequent adaptation processing.

It may be stated in general, that multi-sweep mesh adaptation with respect to the different portions of a velocity profile (in the sense of Clauser plots) should result in very well controlled shear layer resolution. This is seen to be the main improvement in relation to any adaptation technique concentrating solely on the boundary layer thickness development as had been previously proposed in [3].

REFERENCES

[1] E.Elsholz, D.John: Netz-Modifikationstechniken für 3D Navier-Stokes-Netze: Anwendung auf eine Flügel-Rumpf-Leitwerkskonfiguration. STAB Jahresbericht 1997.

[2] E.Coustols, A.Seraudie, A.Mignosi: Rear-Fuselage Transonic Flow Characteristics for a Complete Wing-Body Configuration. J.Aircraft, Vol.34,No.3,pp.337-345.

[3] E.Elsholz, H.Steinmeyer: Mesh Modification Techniques for Navier-Stokes Meshes. Demonstration of Viscous Adaptation, Geometry Change and Mesh Enrichment. In: H.Körner, R.Hilbig (Eds): New Results in Numerical and Experimental Fluid Mechanics. Notes on Numerical Fluid Mechanics, Vol.60, Vieweg 1997.

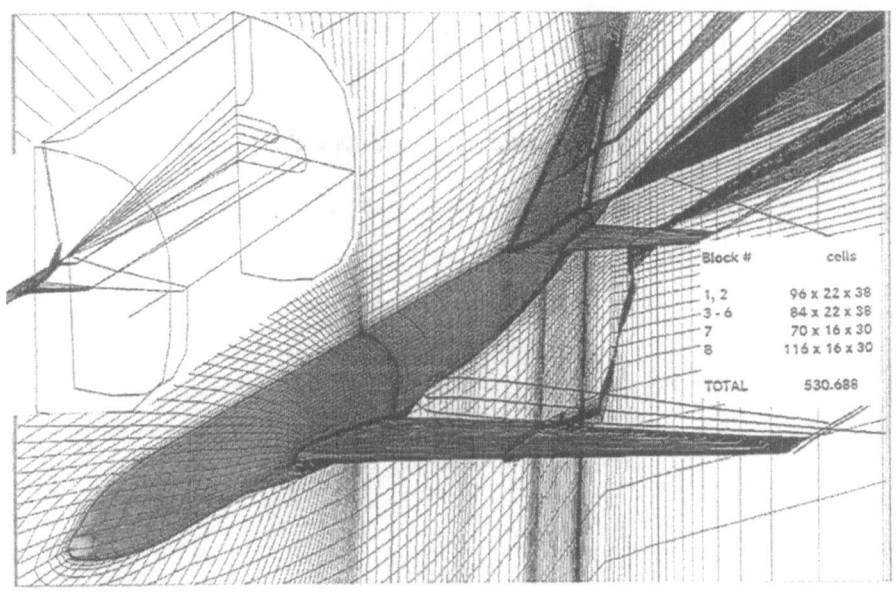

Block #	cells
1, 2	96 x 22 x 38
3 - 6	84 x 22 x 38
7	70 x 16 x 30
8	116 x 16 x 30
TOTAL	530.688

Fig. 1: View on basic mesh

A340-200 Wing-body-tail Configuration
Grid section at downstream wing position

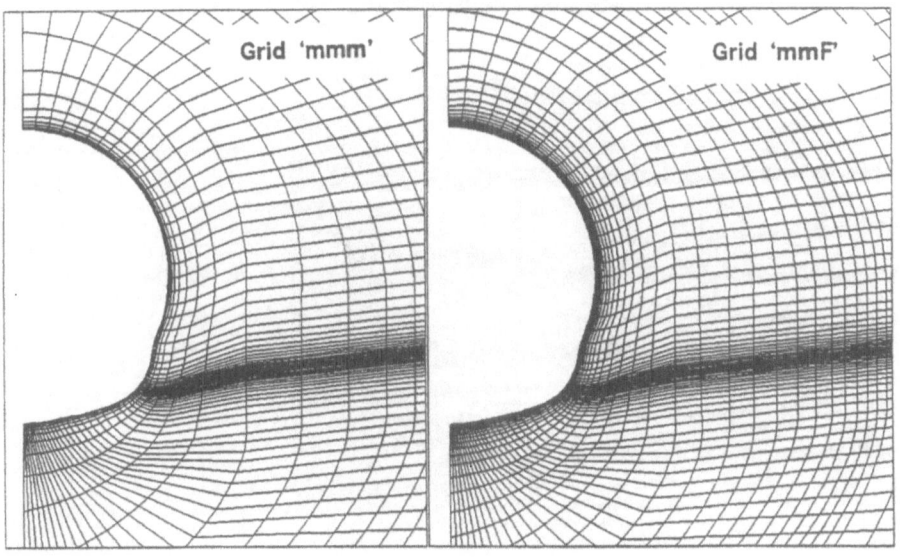

Grid 'mmm'   Grid 'mmF'

Fig. 2: Refinement normal to fuselage surface

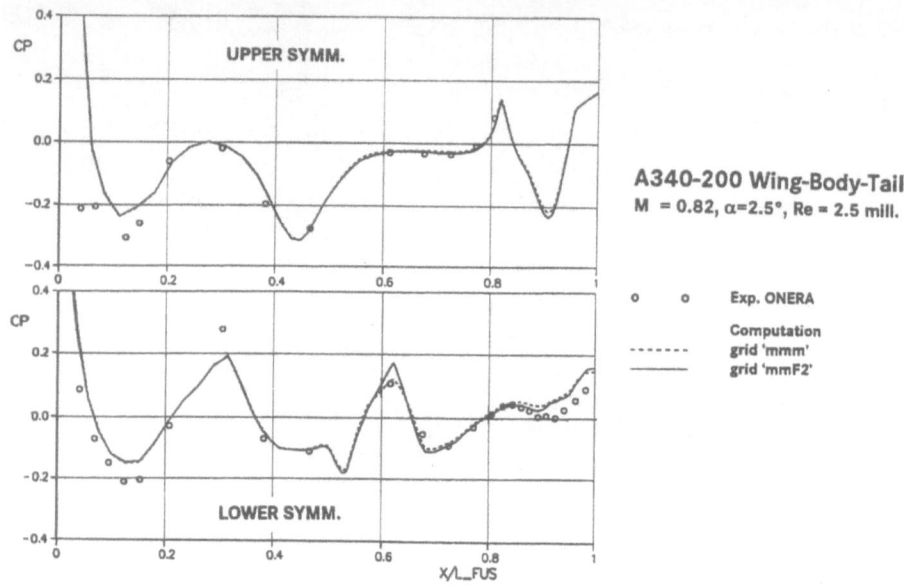

Fig. 3: Pressure distribution in medium and refined mesh

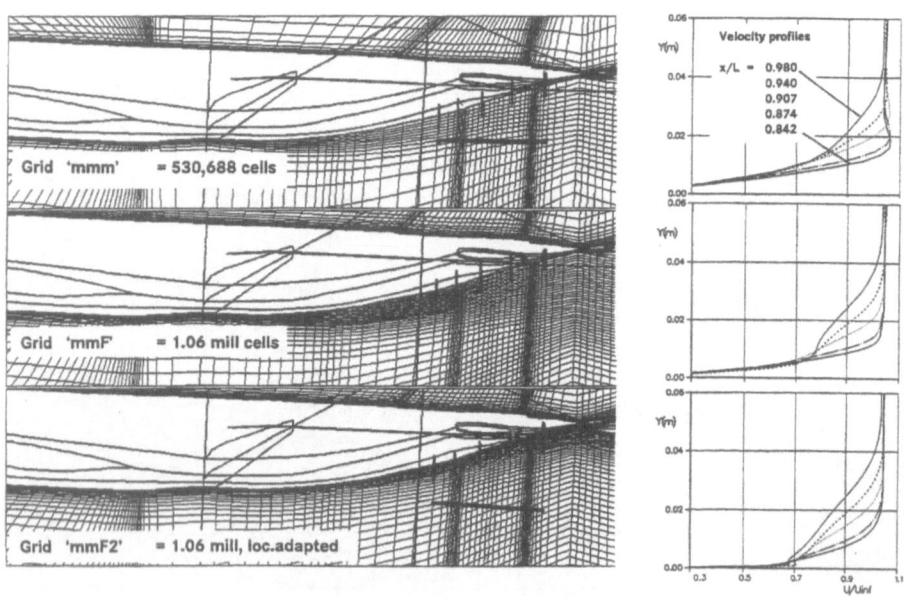

Fig. 4: Mesh dependency on lower rear-fuselage

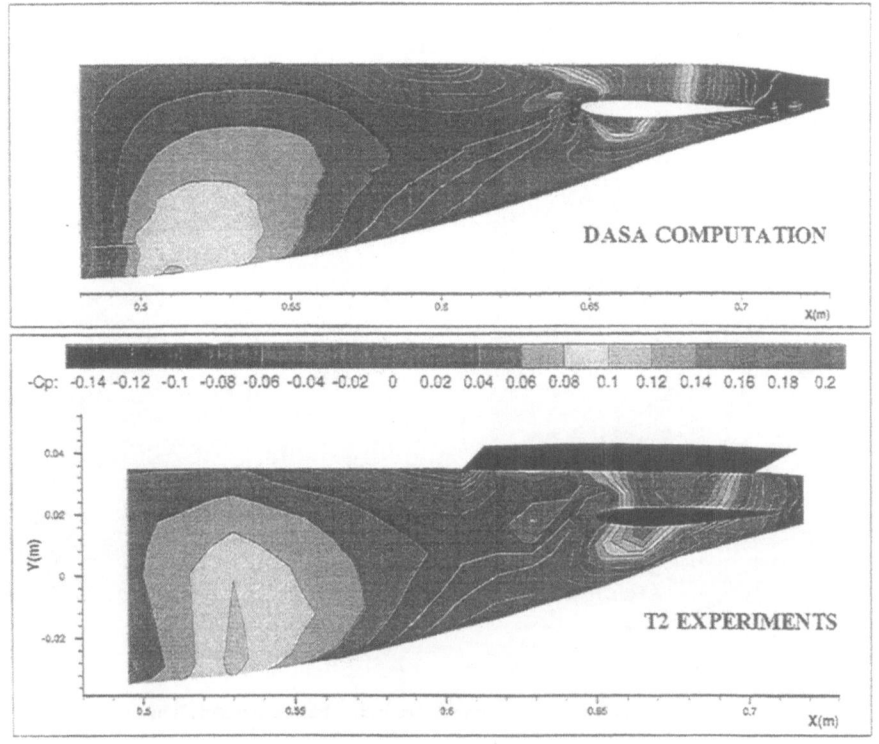

Fig. 5: Pressure distribution   over rear-fuselage

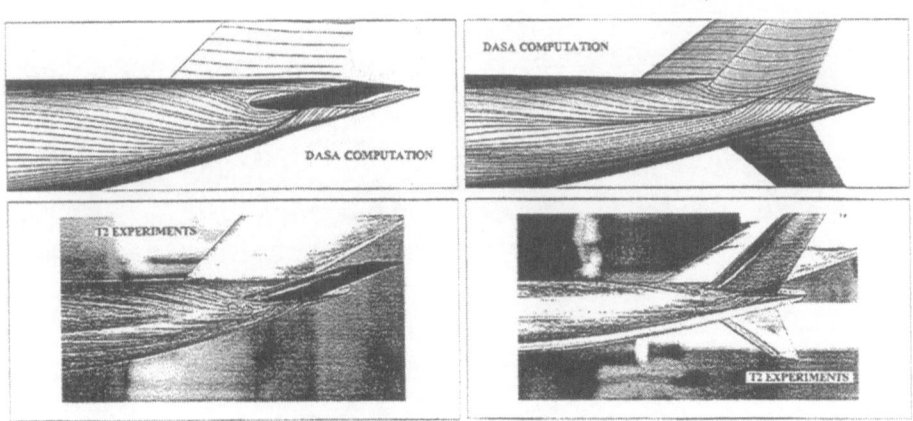

Fig. 6: Skin-friction pattern over rear-fuselage

121

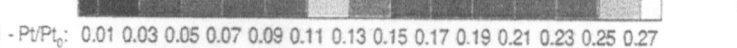

A340-200 WING/BODY/TAIL CONFIGURATION

M=0.82 - α=2.5° - Re_{MAC}=2.5 ·10⁶

1 - Pt/Pt_0: 0.01 0.03 0.05 0.07 0.09 0.11 0.13 0.15 0.17 0.19 0.21 0.23 0.25 0.27

0.1 x V_∞

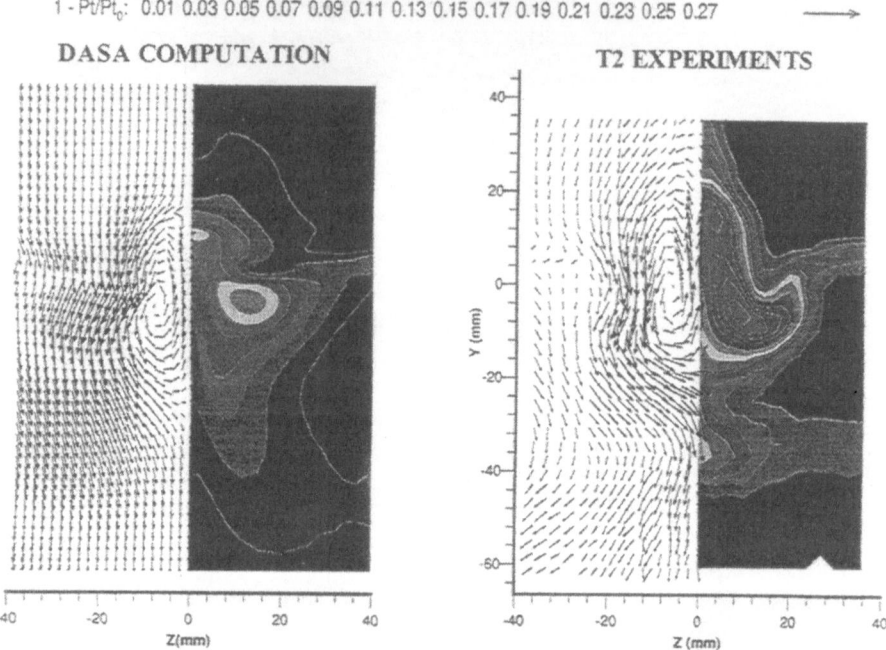

Fig. 7: Secondary flow and total pressure losses in wake plane

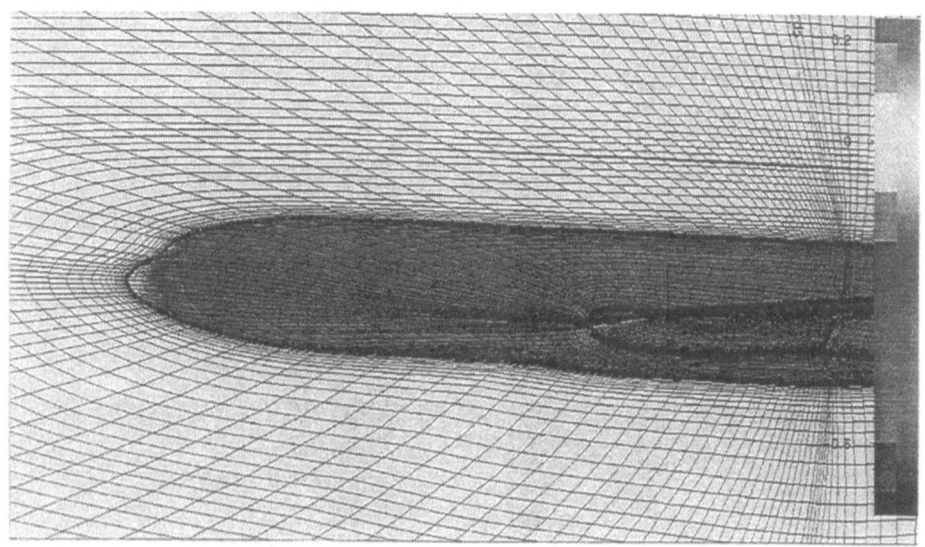

Fig. 8: Mesh refinement at canopy and wing leading edge

Cross section at mid-wing position

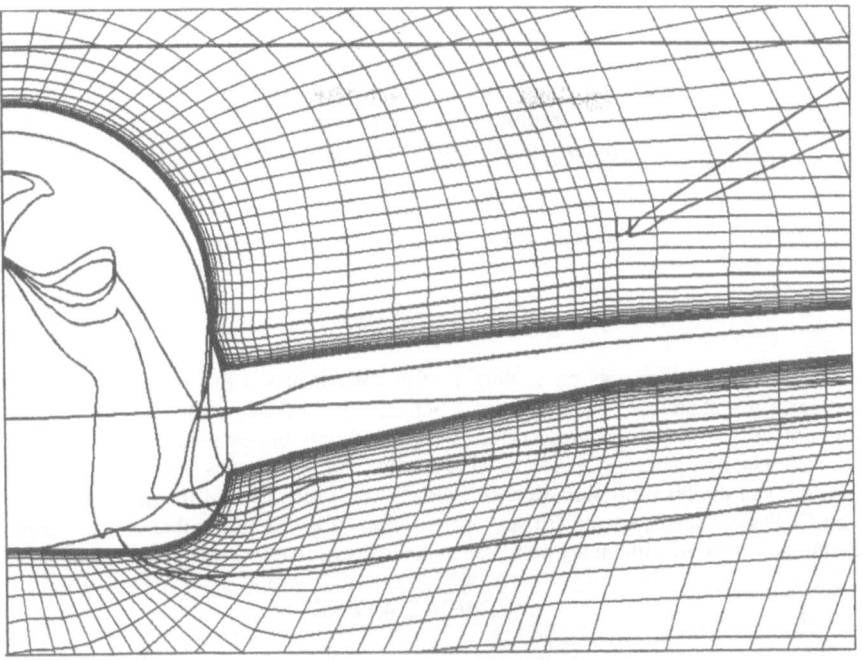

Fig. 9: Cross section of adapted mesh

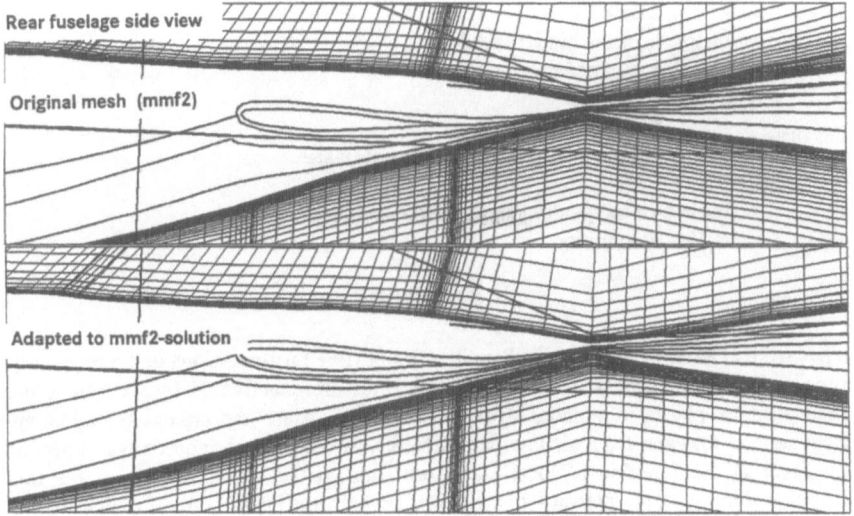

Fig. 10: View on mesh at rear-fuselage

Pressure Sensitive Paint Application in Low-Speed Flows

R. H. Engler, Chr. Klein, P. May*

Deutsches Zentrum für Luft- und Raumfahrt (DLR) e.V.
Institut für Strömungsmechanik
Bunsenstr. 10, 37073 Göttingen, Germany

*Daimler-Benz-Aerospace Airbus GmbH
Hünefeldstr. 1-5, 28199 Bremen, Germany

SUMMARY

Pressure field measurements on a wing profile are conducted to evaluate the accuracy of measurement system using Binary Pressure Sensitive Paint (PSP) for low speed (40-60 m/sec) environments. A part of the wing surface was coated with binary paint B1 from Optrod Ltd. Two different light sources, a Xenon - flash lamp and a nitrogen-laser were used for paint excitation and the resulting fluorescence fields in two spectral band length were acquired by 16 bit CCD slow scan camera with appropriate filters. Results shows that the optical pressure measurement system with binary paint can achieve pressure resolution better than 1 mbar.

NOMENCLATURE

I	= luminescence intensity
K_q	= quenching constant
PO_2	= partial pressure of oxygen
p	= local static pressure
A,B,C	= calibration coefficients of a paint
t	= response time
l	= thickness of the binder layer
D	= diffusion coefficient of oxygen
Λ	= aspect ratio

SUBSCRIPTS

0	= absence of oxygen (vacuum conditions)
ref	= reference condition

INTRODUCTION

The physical basis of all Pressure Sensitive Paint (PSP) formulations is the oxygen-quenching process in which excited luminophore molecules are deactivated by oxygen. Luminophore molecules are placed in a binder layer on the model surface and are excited by an appropriate light source. Oxygen from airflow can diffuse into binder layer and oxygen concentration in the layer is the function of local static pressure on the upper surface of the binder. The resulting luminescent intensity distribution reveals the static pressure distribution on the model [1]. The PSP method gives quantitative as well as qualitative pressure distribution images of the complete observed model surfaces [2] without significant disturbances. This method could also be used for flow visualization and to provide detailed aerodynamic quantitative static pressure information about the models aerodynamics. The PSP techniques have been widely used in wind tunnels for investigations in transonic - and supersonic flow [3]. For these fields of interest, the absolute pressure changes on the model surface are rather large. Therefore changes in the detectable luminescence levels are also larger. Low-speed flows (Ma<0.3) in comparison to the transonic - and supersonic flows produce relatively small pressure changes on the models surface and therefore the variation of luminescence levels due to pressure change is quite small. However, some groups already measured pressure in low-speed flows using the different existing PSP techniques:

Table 1: Previously PSP measurements in low-speed flow

Group	Model	Ma	Δp_{max}[psi]
Morris et al. [4]	Delta wing	0.17	0.3
McLachlan et al. [5]	NACA0012	0.30	1
INTECO [6]	Car-model	0.12	0.12
Sullivan [7]	Jet interaction	-	3

In the described experiment a measurement system and data processing algorithms were optimized to achieve a good signal-to-noise ratio (SNR) and to reduce the influence of possible measurement errors. With this system, reliable experimental results were obtained for flow velocities down to U_∞ = 40m/s. A sample of data is presented to show some aerodynamic effects on the investigated wing profile and to demonstrate the comparison between PSP results and conventional pressure taps data.

PSP BASIS

PSP techniques are based on the deactivation of photochemical-excited organic molecules, so called luminophores, by oxygen molecules. This oxygen-quenching process of luminescence was first discovered and described by Kautsky and Hirsch [8] in 1935 and will be briefly reviewed in this paragraph.

The behavior of photoluminescence of a luminophore quenched by oxygen molecules can be described by the Stern-Volmer relation [1]:

$$I_0/I = 1 + K_q PO_2, \qquad (1)$$

where I_0 is the photoluminescence in the absence of oxygen (vacuum), I the detected photoluminescence, K_q the quenching constant, and PO_2 is the partial pressure of oxygen. K_q is a function of temperature.

In the widely-used pressure sensitive paint formulations, the luminophores (e.g. ruthenium, pyrene) are located in a polymeric binder material (e.g. silicon rubber). The binder compound is permeable to the oxygen molecules. To calculate the static pressure values from a measured intensity distribution for such paint formulations the second order approximation is more useful:

$$p = A(T) + B(T)(I_{ref}/I) + C(T)(I_{ref}/I)^2, \qquad (2)$$

where p is the local static pressure, A,B and C are temperature dependent calibration coefficients of the pressure sensitive paint formulation, which can be determined in laboratory or pressurized windtunnel. I_{ref} is the corresponding intensity value for a constant reference pressure. Thus, using a ratio of images taken at two pressure conditions ("wind on" and "wind off") allows the determination of static pressures over the surface of interest. A problem that occurs when using PSP in this intensity method are the model displacement, uncertainties of the temperature field on the model surface, excitation light non-stability, the spread of exposure time, and the luminescence light scattering on the adjacent model parts or test section walls (self illumination). The position of aerodynamic forces acting on the model changes their position relative to the light source and image acquisition system. Alignment of "wind off" and "wind on" images does not totally eliminate the influence of model displacement and deformation [2] and [9]. Influence of the temperature field uncertainties can be minimized by direct measurements of the temperature distributions on the model surface, by using PSP formulations with small temperature sensitivity or (and) by estimation of the temperature field on the base of measured pressure fields and some assumption about heat flux on the model surface [10]. A more detailed description of different pressure sensitive paint formulations as well as a more complex theoretical background is presented in [10] and [11].

EXPERIMENTAL SETUP

The presented measurements were performed in the Low-Speed-Wind-Tunnel (LSWT) of Daimler-Benz Aerospace in Bremen. This Eiffel type wind tunnel with a test section of 1.8m x 2.0m is a continuously driven facility operating at speeds between U_∞= 20 and 65m/sec. For PSP tests a wall plate element, Fig. 1, was built for the fast implementation of the excitation illuminators and CCD camera observation.

The investigated surface is a part of the suction side of a wing with a constant profile. Fig.2 shows a photo of the PSP coated part of the investigated wing model. Along the pressure taps no PSP coating was applied.

The B1 binary paint formulation from Optrod Ltd., based on pyren as luminophore, was used for the measurements. This paint consists of a screen layer, adhesive layer and an active layer with a total thickness about 50 μm and a response time of 0.5s.

Typical calibration curves for different PSP formulations are presented in Fig.3.

The emission of the PSP depends on the excitation intensity. That means the instabilities of excitation light and model movement or model deformations relative to the light source will lead directly to an error in the pressure measurements. To minimize this error a binary paint composition is necessary. Binary composition provides the possibility to simultaneously acquire an additional reference image, which contains information about local excitation intensity. The excitation intensity fluctuations for Xenon flash lamp or the nitrogen-laser integrated for expose time 1 sec can be estimated as 1% that provides significant input in the measurement error without using binary paint. Fig.4 shows emission spectra of binary paint for different absolute pressures.

The emission band from 425nm to 550nm is used to acquire a pressure sensitive image and the emission band from 610nm to 630nm is used to obtain the "excitation reference" image. Using the above mentioned Xenon flash-lamp light source for excitation and a scientific grade 16bit, 1024x1024 pixel CCD slow-scan camera, exposure times of approximately 10s can be realized for each "pressure/reference" image. These images were acquired sequentially by one camera using a filter shifting system. Such approach does not totally eliminate time non-stability of the light source but is affordable taking long exposure time into consideration. Fig.5 shows the basic PSP system composition for this wind tunnel test. Since pressure sensitive paints are also sensitive to temperature, the temperature of the stream was controlled during the run to minimize the temperature effect on the luminescence painting. The model surface temperature during the run was measured in several points with PT100 thermocouples. The maximum fluctuation of the temperature for all measurements was smaller than 2°C. Required accuracy for pressure measurements in the low speed regime must better than 0.1 mbar. For pyren-based PSP formulations this pressure resolution requires 0.05% of relative intensity resolution. Pyren-based PSP formulations has a temperature coefficient in the range of 0.3 to 0.5%/°C. Uncertainties of 1°C of the temperature distribution on the model surface will create errors of about 10 mbar than is significant for low speed measurements. Therefore one of the most important remaining systematic errors for the PSP measurements comes from temperature influences.

Quantum and read-out noise of used CCD camera referred to maximum dynamic range can be estimated as 0.2%. Taking into account that to calculate relative intensity it is necessary to use six images (three: sensitive, reference intensity and dark images for "wind-off" conditions and the same three for "wind-on" conditions) the error of relative intensity will be not less than 0.4%. Thus there are two possibilities to improve intensity accuracy measurements: to average an appropriate number of each images that is a time consuming operation or to use an additional spatial filtering that finally reduces spatial resolution. Pressure distribution on the wing model is near one-dimensional that gives the possibility to use appropriate spatial filter without significant loss of resolution.

RESULTS

A typical PSP result for the investigated model for U_∞=60m/s, α = 16°is given in Fig. 6. The image in Fig. 6 qualitatively visualizes the pressure distribution on the suction side of the wing.

The comparison for U_∞= 60m/s, α= 16° in Fig.7 of the measured static pressure values using the PSP technique and the conventional pressure tap technique (PSI) shows an acceptable agreement.

The presented comparison in Fig.7 shows rather rough PSP results. This fluctuation gives an impression of the possible SNR of the used PSP setup, for single measurement of "pressure/reference" images. This means that no image averaging method was used to reach a higher SNR.

Because of the more or less one-dimensionality of the pressure distribution on the wing model, it is useful to use a 3 x 3 pixel Gaussian spatial filter. The result of such an approach for the same case than in Fig. 7 is shown in Fig. 8.

The measurements also show a good agreement for the flow speed of U_∞= 60m/s as well as 40m/s. To minimize the noise a spatial filtering-method was used for the comparison of PSP- and PSI data.

CONCLUSION

Obtained results show that the available measurement system can be used for pressure measurements at flow velocities down to 40m/sec. For low-speed measurements the run conditions must be carefully controlled (temperature/dust) and the signal-to-noise ratio must be optimized.

ACKNOWLEDGEMENTS

The test presented here were funded by Daimler-Benz AG. Significant contributions to the success of this effort is attributed to R. Rossmanith from Daimler-Benz AG.

REFERENCES

[1] Vollan, A. & Alati, L. (1991). *A New Optical Pressure Measurement System*, 14th ICIASF Congress, Rockville, MD.

[2] Engler, R.H. & Klein, Chr. (1997). *First Results Using the New DLR PSP System – Intensity and Lifetime Measurements*, Conference "Wind Tunnels and Wind Tunnel Test Techniques", Cambridge UK, ISBN 185768 048 0.

[3] Engler, R.H., Hartmann, K., Troyanovski, I. & Vollan, A. (1992). *Description and assessment of a new optical pressure measurement system (OPMS) demonstrated in the high speed wind tunnel of DLR in Göttingen*, DLR-FB 92-24.

[4] Morris, M.J., Donovan, J.K. Kegelmann, J.T., Schwab, S.D., Levy, R.L. & Crites, R.C. (1993). *Aerodynamic Applications of Pressure Sensitive Paint*, AIAA Journal, Vol. 31, No.3, 419-425.

[5] McLachlan, B.G., Kavandi, J.L., Callis, J.B., Gouterman, M., Green, E. & Khalil, G. (1993). *Surface Pressure Field Mapping Using Luminescent Coatings*, Experiments in Fluids 14, 33-41.

[6] INTECO Report (1994). *OPMS low speed test at Mercedes-Benz, Sindelfingen.*

[7] Torgenson, S.D., Tianshu, L., Sullivan, J.P. (1996). *Use of Pressure Sensitive Paints in Low Speed Flows*, 19th AIAA Advanced Measurement and Ground Testing Technology Conference, 6/1996, New Orleans, LA.

[8] Kautsky, H. & Hirsch, A. (1935). *Nachweis geringster Sauerstoffmengen durch Phosphoreszenztilgung*, Z. f. anorg. Und allg. Chemie, Band 222, 126-134.

[9] Le Sant, Y., Delegise, B., Mebarki, Y. & Merienne, M-C. (1997). *An Automatic Image Alignment Method Applied to Pressure Sensitive Paint Measurements*, 17th ICIASF Congress, Monterey, CA., USA.

[10]   Fonov, S.D., Radchenko, V.N. & Mosharov, V.E. (1997). *Luminescent Pressure Sensors in Aerodynamic Experiments*, TsAGI's books, No. 8005.

[11]   Holmes, J.W. (1997). *Optical Pressure Measurement (PSP)*, Presentation on Modern Optical Flow Measurement Advanced School, Udine, Italy, October 1997.

FIGURES

 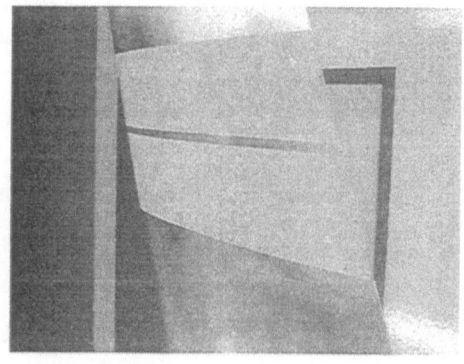

Fig. 1 Observation window, illuminators and camera.

Fig. 2 Investigated model in the test section with PSP coated area

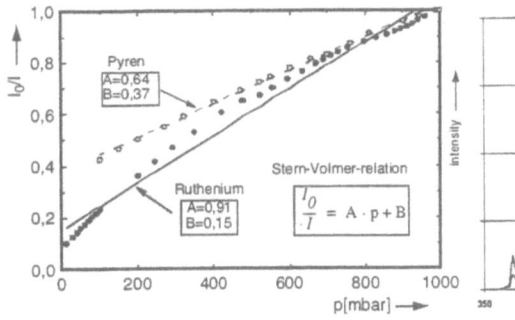

Fig. 3 Calibration curves for different paint

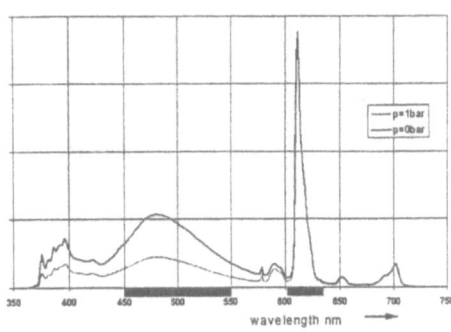

Fig. 4 Emission spectra of binary paint for different absolute pressures and constant temperature.

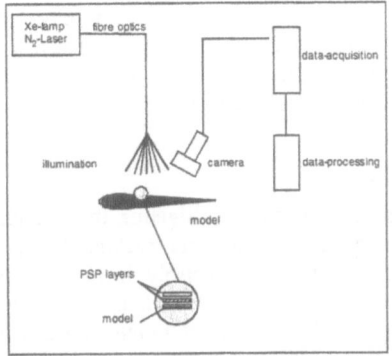

Fig. 5 : Schematic diagram of experimental set-up.

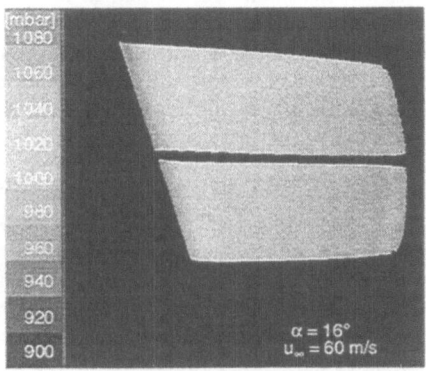

Fig. 6 Visualization of the pressure distribution on the model surface for U_O= 60m/s, α= 16°

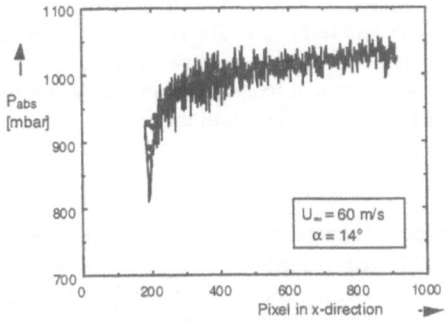

Fig. 7 Comparison of the PSI/PSP pressure distributions on the models suction side surface

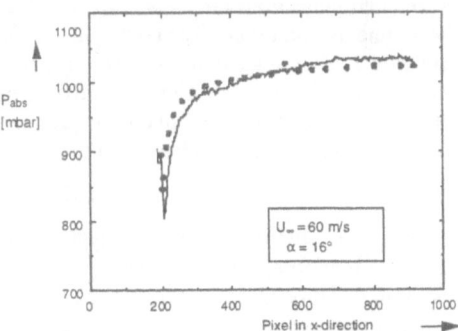

Fig. 8 Comparison of the PSP/PSI pressure distributions on the models suction side surface for U_∞=60m/s, α=16° using spatial filter

129

„In-Situ" calibration procedure for surface-hotfilm-sensors based on a stagnation point flow

P. Erb, B. Ewald, M. Roth, C. Tropea

Technische Universität Darmstadt
Fachgebiet Strömungslehre und Aerodynamik
Flughafenstr. 19, D-64347 Griesheim, Germany

Summary

Novel methods of surface hot-film calibration are required for array sensors and especially for applications in which a removal of the sensors for a channel or pipe calibration is not possible, i.e. in-situ calibration. In the present paper the use of a plane stagnation jet as a 'calibration' flow is explored. Some basic properties of this flow are reviewed and its potential to provide a known shear stress even in an in-situ calibration situation is discussed. First experiments demonstrate the technique for equalizing sensor response in a multi-sensor array. A procedure in principle is outlined for a quantitative shear stress calibration.

Introduction

The focus in this paper is the surface hot-film sensor, used for investigating the boundary layer behavior. The particular application in mind is that of detecting and tracking Tollmien-Schlichting (TS) waves on the surface of a wing in flight. More precisely an instrumented glove is used over the wing, as described more fully in Erb *et al.* (1996), Erb *et al.* (1997). Boundary layer perturbations are introduced using a pulsating air jet mounted flush and normal to the surface [Suttan *et al.* 1996]. The growth of these disturbances are monitored using a 32 sensor hot-film array as illustrated in figure 1. In fact to detect TS instabilities only a qualitative measure of the wall shear stress is required, however even then an equalisation of the sensor response is desirable. A quantitative measure of local and time resolved skin friction is however of even greater interest and then the need for calibration arises.

Several complications are associated with the calibration of such sensor arrays, which can be better understood by briefly viewing the surface hot-film technique. The situation is sketched in figure 2 , showing a thin active film flush with the flow boundary surface and mounted on a capton foil substrate. For a heated active film kept at constant temperature in a CTA bridge, the square of the bridge voltage U^2 is proportional to the heat flux from the film and the heat flux will in some way be related to the wall shear stress

$$\tau_w = \mu \; \partial u_x / \partial y |_{y=0} \qquad (1)$$

regardless of whether the boundary layer is turbulent or laminar [Haselbach, Nitsche 1996]. In principle therefore, the bridge voltage can be calibrated against τ_w. The most complicating factor in this procedure is in fact the heat loss to the substrate, which will depend on substrate material and temperature. This effect becomes even more complicated for sensor arrays, in which mutual interaction may occur [Haselbach 1997]. In the ideal case however the proportionality between convective heat flux and the wall shear stress is given as $Q_C \propto \tau_w^{1/3}$ and this leads to a calibration curve of the form [Bruun 1995]

$$U^2 = A + B\tau_w^{1/3} \qquad (2)$$

The problem of heat losses to the substrate will not be addressed in the present paper but rather attention will be focussed more on generating a flow for which τ_w is known.

Conventionally, surface hot-film sensors are calibrated in a pipe or channel flow in which the wall shear stress can be directly computed from the streamwise pressure drop. These techniques are inappropiate when it is inconvenient to remove the hot-film sensors from their mounting, such as in the present case of in-flight measurements. A "portable" calibration flow is therefore desirable. Early considerations of using a tangential wall jet revealed several practical problems which led to this concept being discarded. These difficulties are summarised as follows:

- exact positioning of jet w.r.t active sensor
- gap between jet and wall
- mechanical damage to upstream sensors in an array
- time consumed to calibrate a large number of sensors in an array.

It appears that these disadvantages can be overcome using a plane stagnation flow as pictured in figure 3. In this concept, the plane jet impinges on the surface and three flow regions can be identified. The first region is the stagnation region in which the Falkner-Skan solution (stagnation flow) is valid [Spurk 1989]. At larger x values the finite width of the jet becomes evident and the flow begins a transition to a wall-jet character [Glauert 1956]. Finally a laminar-turbulent transition will occur and the outer regions (high x) will take on the properties of a turbulent wall jet [Gersten, Herwig 1992]. Each of the above flow regions show a characteristic development of the wall shear stress with x, as noted in figure 3 and described in more detail in the respective references. Clearly however, the mixed behavior of the flow requires a calibration of the wall shear stress as a function of nozzle geometry and position and nozzle exit Reynolds number. This calibration could be either a direct measurement, for instance a measurement of $\partial u/\partial y$ in the near wall layer, or a computation. The latter is however less dependable for the present, since a laminar/turbulent transition can be expected, which cannot be well predicted presently. An experimental determination of the wall shear stress under the impinging plane jet is presently underway using a laser Doppler anemometer with a very small measurement volume.

Plane Impinging Jet

To utilize the properties of the plane impinging jet outlined above, a calibration apparatus as pictured in figure 4 has been conceived. The plane impinging jet was generated using a simple axial fan (50W) mounted on a plenum with flow straighteners and screens and a nozzle with a 30:1 contraction ratio. For an outlet area of 6mm*95mm the mean flow velocity was 24m/s and a laminar flow region in the stagnation point area was detected. The nozzle was designed to yield close to a top-hat exit velocity profile. Hot-wire measurements confirmed that the jet could be considered plane over the middle 75mm section.

The nozzle was mounted on a carriage with rollers, allowing it to be traversed over the hot-film sensor array, even for a curved surface, such as the wing. The curvature introduces deviations of impingement angle and also height variations. The height variation was minimized by mounting the jet exit almost directly over the front rollers. The impinging angle variations remained below ±0.6° from the normal with the particular wing geomety used in this study.

The nozzle is traversed across the hot-film sensor array and thus each sensor is exposed to wall shear stresses varying from zero at the outer edges, achieving a maximum before the jet reaches the sensor position and a minimum when the sensor is at the stagnation point. The situation repeats itself in reverse as the jet moves beyond the sensor. Such a behavior can be recognized directly from the voltage signal of a single sensor during the nozzle traverse, as shown in figure 5. This data was obtained for a nozzle height above the surface of 5 mm.

Before interpreting this signal further it is valuable to also show that the signal response is not strongly dependent on the speed at which the nozzle is traversed, thereby opening up the possibility of a manual traverse in the field. This is documented in figure 6, which compares the minimum and maxima voltages of the signal for various traversing speeds.

Similairly these voltages are quite insensitive to impingement angle, as shown by the data presented in figure 7. Not only that, the difference between the maximum and minimum voltage is even less sensitive if the mean maximum of peak A and B is used, as also shown in this figure.

Finally it is important to quantify the mutual interaction of sensors in an array. To demonstrate this effect the signals from a tightly packed array with a 1mm spacing were recorded, as displayed in figure 8. The substrate material was aluminum giving a nearly constant substrate temperature distribution so that the mutual interaction due to increased substrate temperature is minimized. Both the first and the last sensor exhibit increased outer maxima whereas the inner maxima are significant smaller. This is caused by the smaller heat loss to the strongly heated flow coming from the sensors positioned upstream. To avoid this decrease of maxima due to heated flow, a minimum spacing between sensors greater than the distance between stagnation point to maximum shear stress of the flow is recommended (6mm for the present operating conditions). However this minimum spacing has to be increased if a substrate material with high thermal conductivity is used.

Calibration Procedure

The present concept forsees the jet operated at a constant velocity. The different shear stress calibration points are obtained from the time dependent signal during the traverse. Already the unconditioned signals in figures 5 and 8 indicate that the flow remained laminar between the stagnation point and approximately the point of maximum heat transfer (shear stress). This can be influenced by the exit velocity and the height of the jet above the wall and is the subject of on-going experimental investigations. Thus by knowing the shear stress in this restricted region of the impinging jet, the sensor can be calibrated between a minimum and maximum shear stress.

The minimum value should be zero, however the finite size of the sensor and its induced natural convection precludes this. Thus the sensor voltage never reaches the 'no flow' voltage during the traverse, as seen in figure 5. Despite this, it would be attractive to use this minimum recorded voltage as a zero reference, since this would simplify the calibration procedure to only one measurement during the traverse and immediately allow corrections for temperature drift. Multiple measurements under different operating conditions showed that the observed voltage difference remained at a constant value for each sensor. So this difference is only to be measured once and can be subtracted from the minimum voltage to enable the calibration between 'no flow' and maximum shear stress.

As a preliminiary step to a full quantitative calibration for wall shear stress, the present apparatus can be used directly for response equalization of the sensors in an array. Each sensor has slightly different dimensions and electrical properties, leading to unequal responses given the same imposed flow field. This is demonstrated for the array pictured in figure 1. This array is used to measure the steady flow component and also to detect TS instabilities generated by the pulsating air jet with a given excitation frequency. Hence an equalisation for steady flow conditions and for TS wave frequency is necessary. A single traverse of the nozzle was performed recording each of the sensor signals which are shown in figure 9. The steady flow equalisation can be done using the differences between the signal mean maximum and the minimum voltage. The TS measurements will use the bandpass filtered signals in the frequency domain (e.g. 850Hz±20Hz) of the air jet excitation. For scaling these signals the rms values of the bandpass filtered signals in figure 9 are used. Figure 10 displays these two equalisation factors for every sensor, showing big differences both for maximum-minimum voltage differences and for the rms values. This justifies equalisation of the hot-film sensors in the array.

Conclusions

A new concept for an in-situ calibration of hot-film surface sensors for wall shear stress is introduced based on a traversible plane impinging jet. The technique provides some advantages over alternative methods, the most important being the potential for field use. Qualitative results are presented which show the feasibility of calibrating many sensors of an array in one pass, however at present this is restricted to an equalization of the sensor response. The flow itself should display a unique wall shear stress distribution, whose calibration is the subject of on-going work. Problems arise however with an absolute calibration of hot film surface probes due to heat losses to the substrate, however, this does not lessen the potential of the impinging jet to provide a calibration flow.

The financial support of the Deutsche Forschungsgemeinschaft through grant EW 17/10-4 is gratefully acknowledged.

References

Bruun, H. H.(1995): Hot-wire Anemometry. Principle and Signal Analysis, Oxford University Press, Oxford.

Erb, P.; Ewald, B.; Roth, M. (1996): Flight Experiment Guidance Technique for Research on Transition with Grob G109b Aircraft of the Technische Hochschule Darmstadt, New Results in Numerical and Experimental Fluid Mechanics, Contributions to the 10th AG-STAB/DGLR Symposium, Braunschweig, Germany.

Erb, P.; Ewald, B.; Roth, M.; Stenger, M.(1997): Das Meßflugzeug der TH-Darmstadt - Kalibrierergebnisse und Fähigkeiten, Tagungsband zur DGLR Jahrestagung.

Gersten, K.; Herwig, H. (1992): Strömungsmechanik, Vieweg-Verlag Braunschweig.

Glauert, M. B. (1956): The wall jet, Journal of Fluid Mechanics 1 625-643.

Haselbach F. (1997): Thermalhaushalt und Kalibration von Oberflächenheißfilmen und Heißfilmarrays, Fortschrittsbericht VDI, Reihe 7, Nr. 326 Düsseldorf: VDI Verlag.

Haselbach F.; Nitsche W. (1996): Calibration of single-surface hot-films and in-line hot-film arrays in laminar or turbulent flows, Meas. Sci. Techn. 7 1428-1438.

Spurk, J. H. (1989): Strömungslehre, Springer-Verlag, Berlin Heidelberg.

Suttan, J.; Baumann, M.; Fühling, S.; Becker, S.; Lienhart, H.; Stemmer, C. (1996): In-Flight Research on Boundary Layer Transition- Works of the DFG-University Research Group, New Results in Numerical and Experimental Fluid Mechanics, Contributions to the 10th AG-STAB/DGLR Symposium, Braunschweig, Germany.

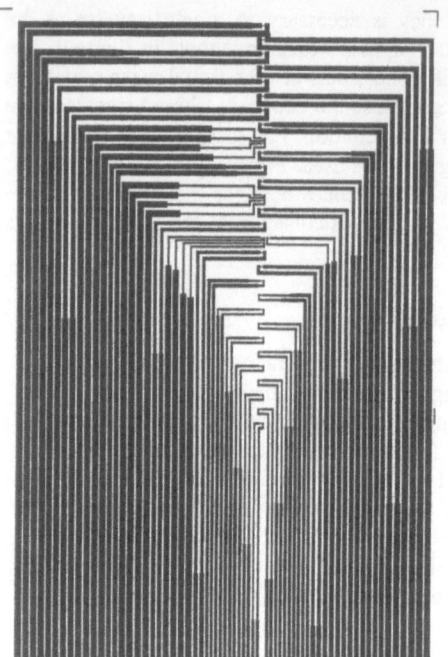

Fig. 1: Layout of 32 sensor hotfilm array.

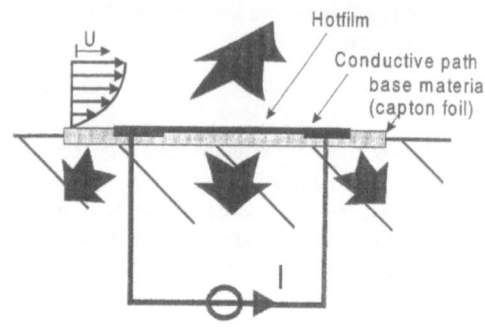

Fig. 2: Surface hotfilm with heat flow into the boundar layer and the substrate.

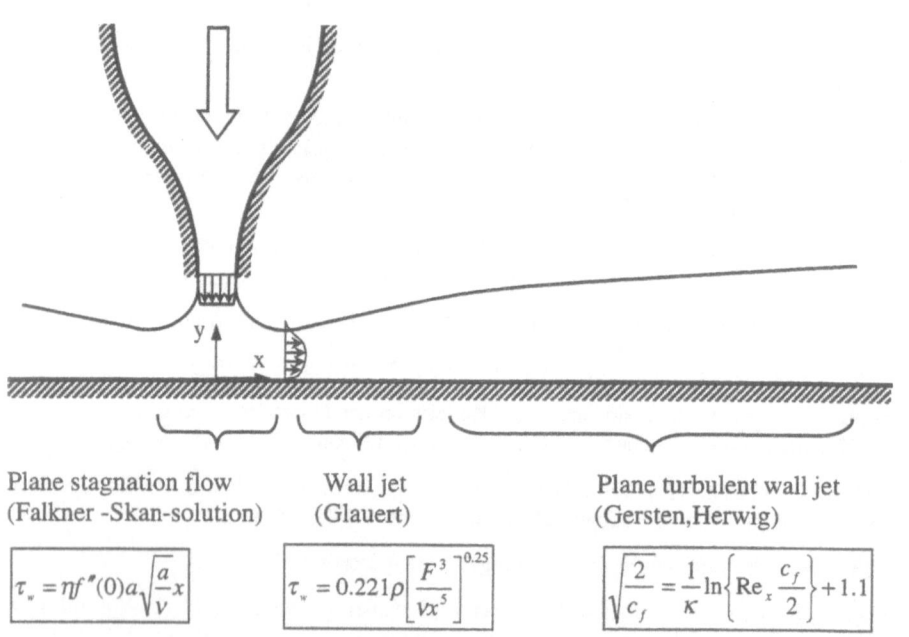

Plane stagnation flow
(Falkner -Skan-solution)

$$\tau_w = \eta f''(0) a \sqrt{\frac{a}{v}} x$$

Wall jet
(Glauert)

$$\tau_w = 0.221 \rho \left[\frac{F^3}{v x^5} \right]^{0.25}$$

Plane turbulent wall jet
(Gersten,Herwig)

$$\sqrt{\frac{2}{c_f}} = \frac{1}{\kappa} \ln \left\{ Re_x \frac{c_f}{2} \right\} + 1.1$$

Fig. 3:Flow regions of a plane impinging jet.

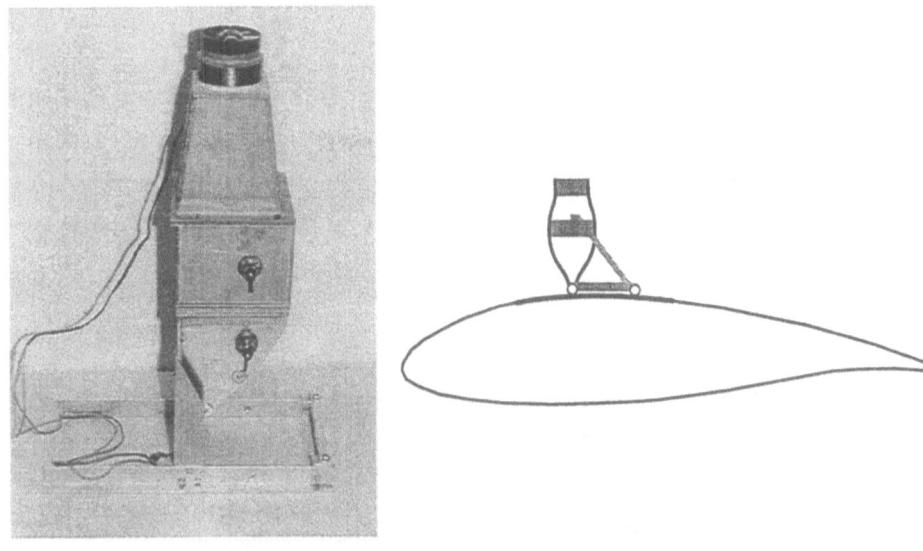

Fig. 4: Plane impinging jet calibration apparatus.

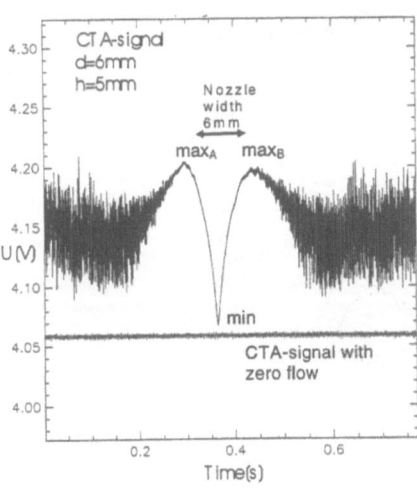

Fig. 5: Hot-film anemometer output signal of the vertically impinging jet.

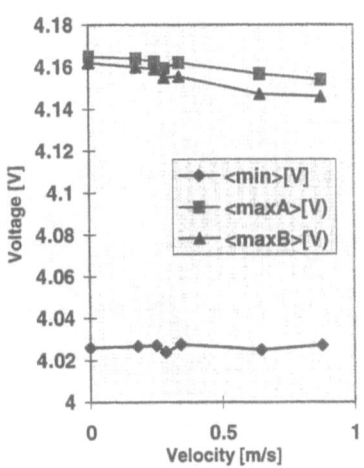

Fig. 6: Amplitude of maxima and minimum when varying traversing speed

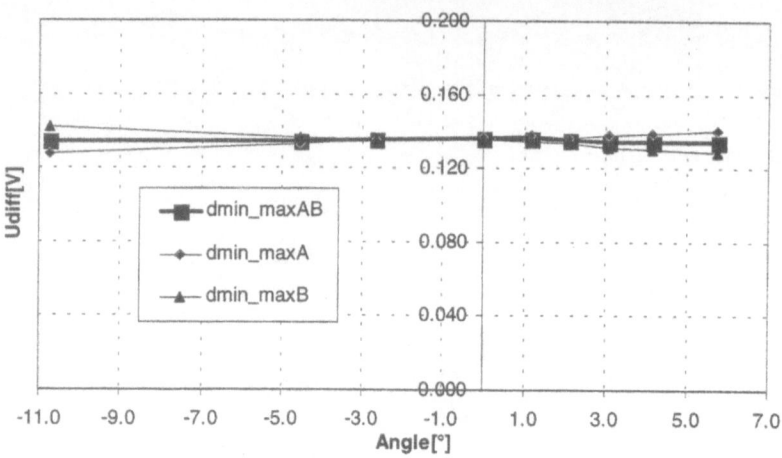

Fig. 7: Amplitude of flow maxima with inclined impinging jet.

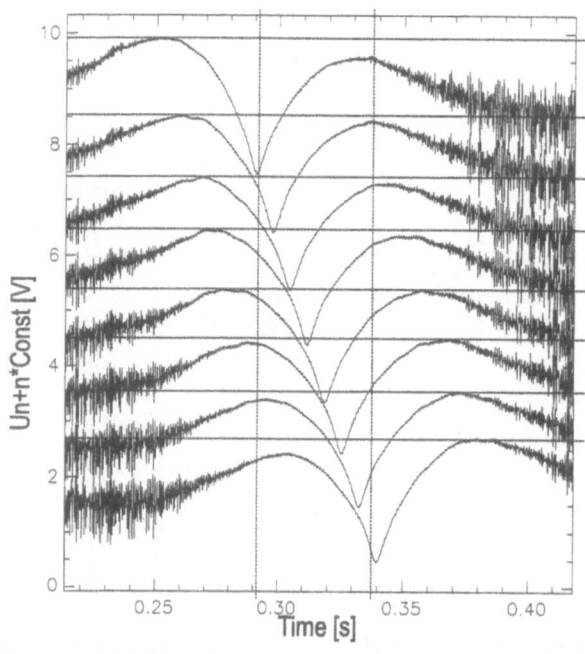

Fig. 8: Eight channel anemometer signal of 8 sensor hotfilm array with 1 mm spacing.

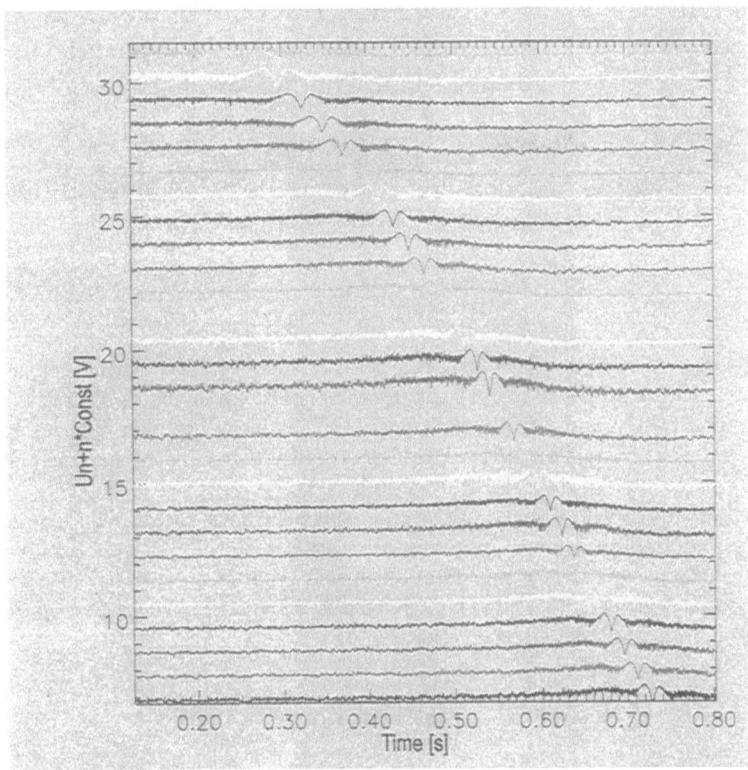

Fig. 9: Recorded voltage signals from 32 sensor array shown in figure 1 for a single jet traverse.

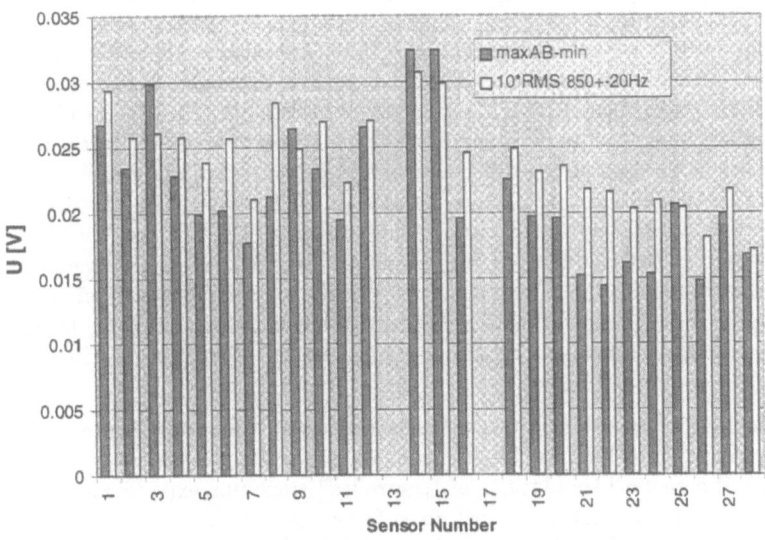

Fig. 10: Maximum-minimum voltage differences and rms values of bandpass filtered voltage signal for every hot-film sensor.

Transition Processes in Mach 6.8 Boundary Layers at Varying Temperature Conditions Investigated by Spatial Direct Numerical Simulation

A. Fezer, M. Kloker

Universität Stuttgart, Institut für Aerodynamik und Gasdynamik (IAG)
Pfaffenwaldring 21, D-70550 Stuttgart, Germany

Summary

Spatial direct numerical simulations (DNS) are carried out to investigate laminar-turbulent transition scenarios under varying temperature conditions of the oncoming flow, distinguishing wind-tunnel and flight conditions. First, Linear Stability Theory is used for both cases to get an overview of the primary stability behaviour of the flat-plate boundary layer where different disturbance modes have to be regarded. Simulations have been performed for fundamental and oblique breakdown in both cases. The investigation of the streamwise amplitude development in connection with the corresponding wall-normal disturbance profiles gives information about which types of modes interact to eventually trigger laminar-turbulent breakdown. The results allow estimations of the probability of the different transition scenarios in practice.

Introduction

Laminar-turbulent transition is of great importance for high-speed aerospace vehicles, because heat flux and shear stresses change significantly during the transition process, leading to strong thermal stresses at turbulence. The knowledge of transition in compressible boundary layers is, compared to the incompressible case, rather incomplete due to the great difficulties in achieving valuable experimental results. Except for some investigations using Linear Stability Theory (LST), almost all results were achieved for "cold" flow, i.e. at flow conditions of wind-tunnel experiments. As Stetson [7] stated after a 10-year experimental research program to investigate the stability of a Mach 6.8 cone boundary layer, results obtained in these experiments are not directly relatable to flight conditions because of the different stability properties of the flows. Therefore direct numerical simulations are a valuable tool to explore transition processes at flight conditions. As a first, important work in this direction, Eißler [3] and Bestek & Eißler [2] investigated at IAG the temperature effects in Mach 4.8 boundary layers. One main conclusion was that the "hot" flow is not necessarily more stable (as widely assumed for simplification) if a realistic wall-cooling by radiation is taken into account. Moreover, the (weakly) nonlinear mechanisms setting the stage for the transition downstream can be different.

The simulations presented here aim at investigating the spatial instability and the laminar-turbulent transition of Mach 6.8 boundary-layer flows along a flat plate. They can be understood as numerical simulations of "controlled" transition experiments. Temporally periodic 2-d and 3-d disturbances with fixed frequency and amplitude are induced, and the reaction of the boundary layer, i.e. the spatial development of the disturbance waves, is simulated by numerical solution of the complete, three-dimensional, unsteady, compressible Navier-Stokes equations.

Numerical method

The flow is considered in a rectangular integration domain (Fig. 1) on the flat plate, not containing any shock wave induced by the leading edge. In a disturbance strip at the wall artificial 2-d and 3-d disturbances are excited by blowing and suction. At the end of the integration domain a buffer domain is appended ($x_3 \leq x \leq x_N$), in which all disturbances of the flow are smoothly damped to zero in order to avoid spurious reflections from the outflow boundary.

The equation of continuity, the Navier-Stokes equations and the energy equation are used in conservative form for the variables ρ, ρu, ρv, ρw and ρe, solved in a disturbance formulation [3]. Lengths are non-dimensionalized with respect to a reference length L. Reference values for velocities, temperature, viscosity, conductivity and density are their freestream values (indicated by subscript ∞). The pressure is normalized with $\rho_\infty^* u_\infty^{*2}$ and the internal energy with u_∞^{*2}, where the superscript * denotes the dimensional quantities. With these definitions, the global and local Reynolds numbers are

$$Re = \frac{u_\infty^* L \rho_\infty^*}{\mu_\infty^*} = 10^5, \quad Re_x = \frac{u_\infty^* x^* \rho_\infty^*}{\mu_\infty^*} = x \cdot Re = R_x^2. \tag{1}$$

The fluid is considered as nonreacting, perfect gas, for which the thermodynamic equation of state is valid. The modelling of the thermodynamic properties of air is implemented for *calorically perfect gas* as well as for *thermally perfect gas* [1, 2, 3].

In streamwise direction the spatial discretization is performed by 4th-order accurate compact finite differences with high modal resolution, which are used in a splitted form. The advantages of split-type compact schemes are discussed in detail by Kloker [5]. Central (viscous terms) and alternating upwind/downwind (convective terms) standard differences [3, 4, 8] are used in normal direction. In spanwise direction the flow is assumed to be periodic, so it is suitable to use a Fourier spectral approach

$$f(x,y,z,t) = \sum_{k=-K}^{K} F_k(x,y,t)e^{ik\beta z} \tag{2}$$

with the basic spanwise wavenumber β. The time integration is performed by a 4-step Runge-Kutta scheme of 4th-order accuracy. A description of the numerical method in more detail is given in [3, 2].

Linear stability analysis of the Ma=6.8 flat-plate boundary layer

First, an overview over the stability behaviour of the $Ma = 6.8$ flat-plate boundary layer shall be worked out with the aid of LST. With this method amplification rates for small disturbances within the relevant frequency parameter range are determined, distinguishing between cold and hot conditions. Following Mack [6] we define wind-tunnel conditions by the following: The stagnation temperature is held constant at $T_0 = 311$ K until, with increasing Mach number, the freestream temperature T_∞ drops below 50 K. For higher mach numbers, including 6.8, T_∞ is held constant at 50 K and T_0 increases. For flight conditions we choose $T_\infty = 220$ K, which corresponds to the atmosphere temperature at an altitude of about 25 km. At cold conditions we assume an adiabatic wall (thus $T_W = 480$ K), and at hot conditions a realistic, cooled wall with $T_W = 975$ K due to the high adiabatic wall temperatures (> 1900 K).

Figs. 2 and 3 show curves of neutral stability as a function of the local Reynolds number R_x and the frequency parameter $F = \omega/Re = 2\pi f^* L/u_\infty^* Re$. Within the area enclosed by the curves small perturbations are amplified, outside they are damped. In the wind-tunnel case (Fig. 2) 2-d waves are amplified as 1st and 2nd mode (Tollmien-Schlichting and acoustic mode respectively), but the instability regions of the modes are melted to one coherent region. 3-d waves are amplified as 1st mode only, analogous to hot conditions. In the latter case 2-d waves are not amplified as 1st mode anymore and the instability region extends to higher dimensionless frequencies. The amplification region of 3-d modes is shifted to higher Reynolds numbers, but the critical Reynolds number (the location where amplification occurs first) nearly keeps constant.

The shown stability diagrams do not contain information about the magnitude of the amplification. Therefore the amplification rates for 6 selected disturbance waves are shown in Fig. 4. For 2-d waves it is obvious that the maximum amplification is relatively large and occurs at relatively high frequency parameters. In these cases, however, growth holds only in small downstream intervals, so the integral amplification is low. Since compressible flat-plate boundary layers are inviscidly unstable, i.e. unstable for $R_x \to \infty$, waves at special (low) frequencies can theoretically have infinite integral amplification rates, in spite of small local rates. For direct numerical simulations a compromise between these two extremes seems to be appropriate. So the dimensionless disturbance frequency for 2-d waves was set to $F = 10 \cdot 10^{-5}$. As can be suspected by comparison of corresponding cases at cold and hot flow (Fig. 4), atmosphere conditions in connection with wall cooling lead to a destabilization of the 2-d 2nd mode (maximum as well as integral), whereas 3-d modes may be stabilized. So it can be expected that 2-d modes will be involved in the transition process at flight conditions.

Results of direct numerical simulations

Exploiting the results of the previous section, parameters have been chosen to simulate fundamental and oblique breakdown at wind-tunnel and atmosphere conditions. These are given in Tab. 1. As for the wall boundary conditions, a radiation-adiabatic wall-temperature boundary condition for the total flow has also been implemented (see [2, 3]), that, however, is not used here to keep correspondence to the LST calculations.

The disturbance strip is in all cases situated roughly at the begin of the linearly unstable region. The following discussion is based on a temporal Fourier decomposition of the unsteady simulation results over the last calculated time period. We use the notation (h, k) for the resulting Fourier modes, describing a wave with frequency $h \cdot F_0$ and wavenumber $k \cdot \beta$. In the fundamental cases the modes $(1,0)$ (2-d, primary wave) and $(1,\pm1)$ (3-d symmetric part wave, in the following referred to as $(1,1)$) are disturbed with the same frequency. In the simulations of oblique breakdown only $(1,1)$ is excited.

Fig. 6 shows the downstream development of the maximum amplitudes (over y) of $(\rho u)'$, considering the most relevant modes for the simulation of fundamental resonance at wind

Tab. 1: Simulation parameters

Simulation	$\mathcal{F}_{WT}$	$\mathcal{F}_{AT}$	$\mathcal{O}_{WT}$	$\mathcal{O}_{AT}$
T_∞ [K]	50	220	50	220
T_W [K]	484 adiabatic	975 cooled	484 adiabatic	975 cooled
$F_{(1,0)} \cdot 10^5$	10.0		—	
$F_{(1,1)} \cdot 10^5$	10.0		5.0	2.0
β	11.0		9.75	4.33
Θ	$\approx 45^0$		$\approx 60^0$	
L [mm]	51.5	17.5	51.5	17.5

tunnel conditions (case $\mathcal{F}_{WT}$). Only a very weak resonant (1,1)-growth, relevant at a 2-d threshold amplitude of almost 30% at $x \approx 15.6$, can be observed, and no initiation of transition is expected. In the corresponding simulation at atmosphere conditions (case $\mathcal{F}_{AT}$) comparable maximum 2-d amplitudes occur; the induced resonant growth (starting at $x \approx 23.8$) however is significantly stronger, so that the fundamental 3-d wave amplitude reaches about 18%. Simultaneously, a considerable damping of the 2-d mode (1,0) takes place, probably caused by a stabilization effect of the strong mean flow distortion (0,0) up to 17%. This seems to be a *self-induced suppression* or *delay* of the transition process, in accordance with the general observation in experiments and simulations that the way to full breakdown in hypersonic boundary layers appears to be the longer the higher the Mach number is. Although the disturbance modes keep a high amplitude level (over 10%), the final breakdown stage is slow. This is verified by looking at characteristic flow structures. A suitable quantity for visualization is the vorticity ω. Fig. 12 shows contour lines of the instantaneous z-component of the vorticity $\omega_z = \frac{\partial u}{\partial y} - \frac{\partial v}{\partial x}$ in the plane $z = \lambda/2$ in case $\mathcal{F}_{AT}$. This representation makes it possible to locate high-shear layers which play a dominant role in the transition process. The region shown extends to approximately four wave lengths. Developed high-shear layers are visible which periodically travel downstream. They remind of the structures connected with the forming of Λ-vortices in incompressible flow just prior to full breakdown. However, the present structures "ride" downstream, but there is no rapid decay that typically exists in incompressible transitional boundary layers starting from the tip of a Λ-vortex.

Nevertheless, it is rather unexpected that, in spite of such high amplitudes of the 2-d disturbance, virtually no resonance appears in the wind-tunnel case. A check of the phase speeds of the disturbance waves (not shown here) reveals no synchronisation which is necessary for the resonance mechanism. To further examine this phenomena we look at the amplitude distributions in wall-normal direction at approximately the x-position where the (1,0)-threshold amplitude is reached ($R_x = 1250$ for wind-tunnel and $R_x = 1450$ for atmosphere conditions). The results from LST (in this terminology called eigenfunctions) are shown in Fig. 5. Although the theory is not valid in the nonlinear regime, it is instructive to recall the different kind the modes look like. As mentioned before, the instability regions of 2-d 1st and 2nd mode merge in the wind-tunnel case, so the eigenfunction constitutes a synthesis of both modes (WT, $\Theta = 0^0$). However, the dominating part is of 2nd mode, visible from the strong near-wall maximum, which is intensified in streamwise direction. The 3-d wave, on the other hand, is a pure 1st mode (WT, $\Theta \approx 45^0$). At hot conditions we have a pure 2nd mode 2-d wave (AT, $\Theta = 0^0$) and a mixed-mode 3-d wave (AT, $\Theta \approx 45^0$).

We now examine the amplitude distributions of mode (1,1) from the numerical simulations (Fig. 8 and 9). In both cases the amplitude distribution at the first plotted x-position looks much like the one from LST. In the wind-tunnel case there is a second maximum far from the wall which does not fit to the LST profile. This is due to a further mode (to make things worse) stimulated by the disturbance strip, a so-called *multiple viscous mode*, which is strongly damped and disappears soon. Further downstream the shape of the distribution stays the same; the wave remains 1st mode. At hot conditions we find a different behaviour. When the 2-d wave reaches its amplitude maximum, the 1st-mode part of the profile disappears and the mode (1,1) looks much like a pure 2nd mode, caused by the resonant interaction with the (1,0) acoustic mode. Hence it is clear that both primary and secondary wave have to be of nearly the same mode type to yield resonance. The question remains whether both modes must have (at least partly) the same identity before the position where the threshold amplitude is reached, or under which circumstances the secondary wave can mutate to the other mode identity.

While in the fundamental case transition is initiated by resonant interaction of a primary 2-d 2nd mode disturbance with a secondary 3-d disturbance, it is caused in the oblique breakdown by nonlinear processes of 3-d 1st mode disturbances alone. This kind of transition was first discovered and investigated by Thumm [8] at IAG. In the simulations presented here the (1,1)-mode has been disturbed at frequencies and spanwise wave lengths for which LST-predicted amplification is strongest (see Tab. 1). In the cold case this disturbance is amplified only in a finite area at comparatively high rates according to LST, whereas in the hot case we have a wide range of instability with nearly constant, but low amplification rates (see Figs 2-4); the Re_x-regions of the simulations have been chosen correspondingly. The maximum amplitudes resulting from the simulations are plotted in Fig. 10 (wind-tunnel cond., case $\mathcal{O}_{WT}$) and Fig. 11 (atmosphere cond., case $\mathcal{O}_{AT}$). During oblique breakdown only modes (h, k) with even sum $h + k$ are nonlinearly generated as well as (0,0). In both cases the directly nonlinearly generated mode (0,2) increases strongly and generates, together with the initial disturbance (1,1), the modes (1,3) and (1,5). The plots show that the oblique-type breakdown is a robust mechanism at wind-tunnel as well as at flight conditions. This is due to the fact that neither a phase speed synchronisation process nor a threshold amplitude are required. The (1,1)-growth is, according to the LST results, stronger in the wind-tunnel case, and amplitude saturation starts early. In the hot case, the growth is slower as expected. As can be seen from the ω_z vorticity iso-contours of case $\mathcal{O}_{AT}$ (Fig 13), the structure deformation in the oblique-type breakdown is much faster (relative to the primary wavelength) and of another type than in the fundamental mechanism.

Conclusions

Linear-Stability-Theory results show a destabilization of the high-frequency acoustic 2-d 2nd mode disturbances at flight conditions if realistic wall cooling is considered. In the simulations of fundamental breakdown for wind-tunnel and atmosphere conditions, comparable disturbances lead to significant fundamental resonance only at atmosphere conditions. As the reason for the resonance deficiency at wind-tunnel conditions the different mode identity of the primary (2nd mode) and secondary (1st mode) disturbance wave has been found. At hot conditions a self-induced stabilization occured at a certain disturbance level, caused by a strong mean-flow distortion, which may delay or even suppress the transition process. Oblique-type transition proves to be a robust mechanism, even at high temperatures, where 3-d modes are, according to linear theory, less amplified.

The financial support of the Deutsche Forschungsgemeinschaft (DFG) within Sonderforschungsbereich 259, Project C4, is gratefully acknowledged.

References

[1] J.D. Anderson. *Hypersonic and high temperature gas dynamics.* McGraw-Hill Book Company, 1989.

[2] H. Bestek and W. Eißler. DNS of Transition in Mach 4.8 Boundary Layers at Flight Conditions. In G. Bergeles W. Rodi, editor, *Engineering Turbulence Modelling and Experiments 3.* Elsevier Science B.V., 1996.

[3] W. Eißler. *Numerische Untersuchungen zum laminar-turbulenten Strömungsumschlag in Überschallgrenzschichten.* Dissertation, Universität Stuttgart, Stuttgart, Germany, 1995.

[4] D. Gottlieb and E. Turkel. Dissipative two-four methods for time-dependent problems. *Math. Comp.,* 30:703–723, 1976.

[5] M. Kloker. A Robust High-Resolution Split-Type Compact FD-Scheme for Spatial Direct Numerical Simulation of Boundary-Layer Transition. In *Applied Scientific Research 59 (4),* pages 353–377. Kluwer Acad. Publishers, NL, 1998.

[6] L.M. Mack. Boundary-layer linear stability theory. In R. Michel, editor, *Special Course on Stability and Transition of Laminar Flow,* AGARD-Report-709, pages 3.1–3.81, 1984.

[7] K.F. Stetson. Hypersonic transition testing in wind tunnels. In M.Y. Hussaini and R.G. Voigt, editors, *Instability and Transition,* volume I, pages 91–100. Springer-Verlag, New York, 1990.

[8] A. Thumm. *Numerische Untersuchungen zum laminar-turbulenten Strömungsumschlag in transsonischen Grenzschichtströmungen.* Dissertation, Universität Stuttgart, Stuttgart, Germany, 1991.

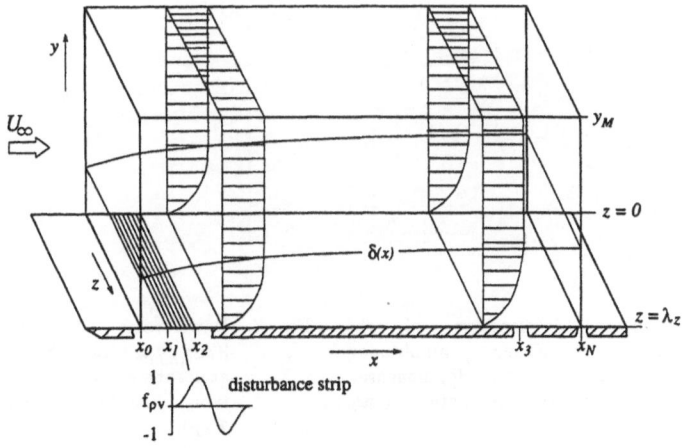

Fig. 1: Integration domain

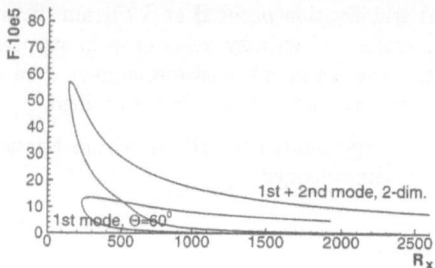

Fig. 2: Neutral stability curves; wind-tunnel conditions, adiabatic wall ($T_w = 480$ K).

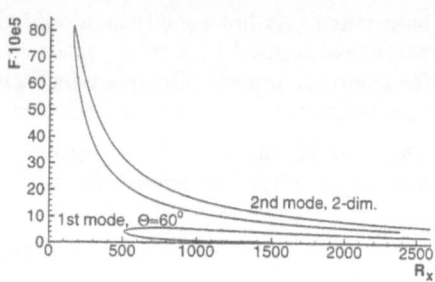

Fig. 3: Neutral stability curves; atmosphere conditions, cooled wall ($T_w = 975$ K).

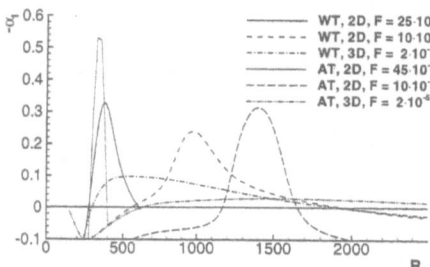

Fig. 4: Amplification rates for selected cases at wind-tunnel (WT) and atmosphere (AT) conditions; 3-d waves with $\Phi \approx 60^0$.

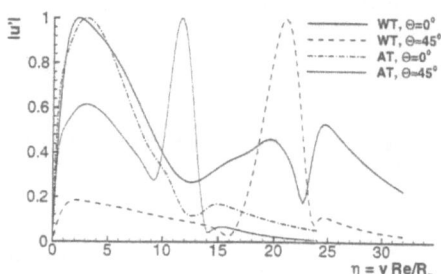

Fig. 5: Normalized eigenfunctions of disturbances at $F = 10 \cdot 10^{-5}$; $R_x = 1250$ (WT) or $R_x = 1450$ (AT).

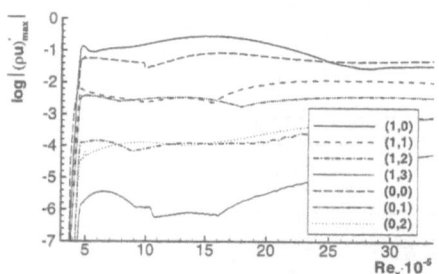

Fig. 6: Downstream development of the $|(\rho u)'_{(h,k)}|$ disturbance amplitudes; wind-tunnel conditions, fundamental-type disturbance combination (case $\mathcal{F}_{WT}$).

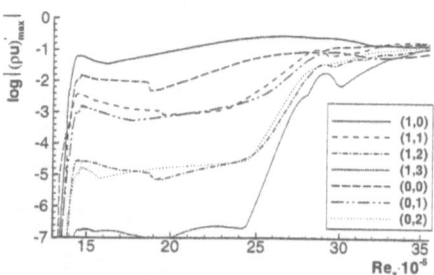

Fig. 7: Downstream development of the $|(\rho u)'_{(h,k)}|$ disturbance amplitudes; atmosphere conditions, fundamental-type disturbance combination (case $\mathcal{F}_{AT}$).

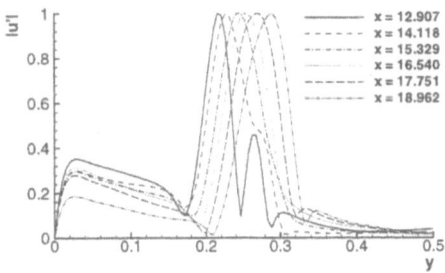

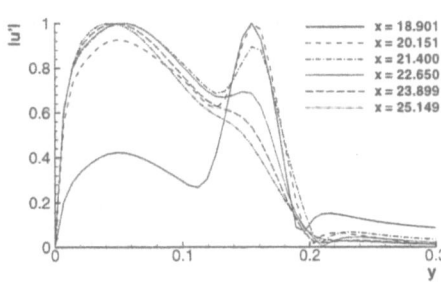

Fig. 8: Normalized amplitude distributions of $u'_{(1,1)}$ at various x-positions; wind-tunnel conditions, fundamental-type disturbance combination (case $\mathcal{F}_{WT}$).

Fig. 9: Normalized amplitude distributions of $u'_{(1,1)}$ at various x-positions; atmosphere conditions, fundamental-type disturbance combination (case $\mathcal{F}_{AT}$).

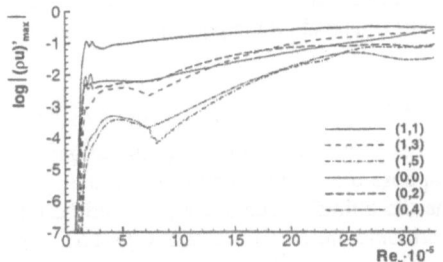

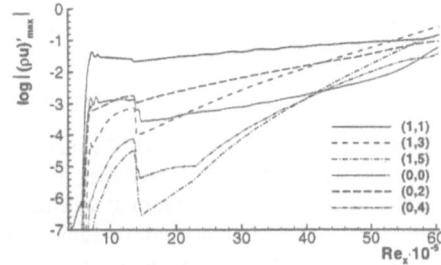

Fig. 10: Downstream development of the $|(\rho u)'_{(h,k)}|$ disturbance amplitudes; wind-tunnel conditions, oblique-type disturbance combination (case $\mathcal{O}_{WT}$).

Fig. 11: Downstream development of the $|(\rho u)'_{(h,k)}|$ disturbance amplitudes; atmosphere conditions, oblique-type disturbance combination (case $\mathcal{O}_{AT}$).

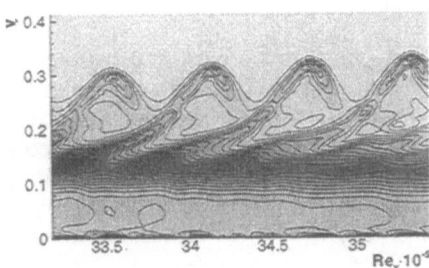

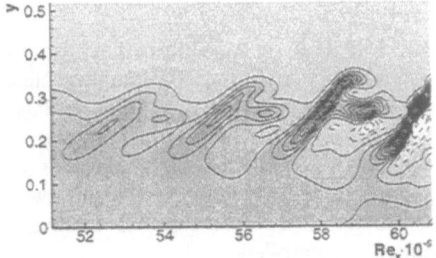

Fig. 12: Iso-contours of the instantaneous vorticity ω_z; atmosphere conditions, fundamental-type disturbance combination (case $\mathcal{F}_{AT}$).

Fig. 13: Iso-contours of the instantaneous vorticity ω_z; atmosphere conditions, oblique-type disturbance combination (case $\mathcal{O}_{AT}$).

145

Investigations of the pressure fields on the oscillating wings by Pressure Sensitive Paint

S.D. Fonov, R.H. Engler, Chr. Klein

Deutsches Zentrum für Luft- und Raumfahrt (DLR) e.V.

Institut für Strömugsmechanik

Bunsenstr. 10, 37073 Göttingen, Germany

S.V.Mihailov, V.E.Mosharov, V.P.Kulesh, V.N. Radchenko.

TsAGI, Zhukowsky, 140160, Russia

E.Schairer,

NASA Ames Research Center, Moffett Field, California, USA

Summary

The paper reviews the main problems encountered during investigations of the oscillating pressure fields by the Pressure Sensitive Paint (PSP) technique. Measurement methodology, theoretical and experimental estimations of the PSP response time are presented. It is shown that currently available pressure sensitive paint formulations can be used for oscillation frequencies up to 20-30Hz. The experimental results obtained on the pitching wing models at Mach numbers from 0.45 to 0.9 are presented.

Introduction

The Pressure Sensitive Paint (PSP) method provides a good opportunity for investigation of unsteady flows. Dynamic parameters of PSP (response time) and measurement system (time, amplitude and spatial resolutions) will determine the ultimate accuracy and should be taken into consideration. To measure periodical non-harmonic pressure fluctuations the PSP has to transfer several harmonics (about ten) of these fluctuations. Measurement system should provide luminescence acquisition in the appropriate time intervals. Standard Image Detector - digital CCD - camera can be used for periodical pressure measurements, bound oscillation of a model, rotating machinery etc., in an integration mode in combination with pulsed illuminator operating in stroboscopic mode synchronized with the main oscillating frequency. Maximal oscillating frequency in this case is restricted by the response time of PSP and excitation light pulse duration.

Background

The PSP layer is two-dimensional array of oxygen sensors, each of them consisting of a polymer binder with the luminophore molecules dissolved in it. The oxygen molecules from the airflow can diffuse into polymer layer. Radiationless energy transfer from the excited luminophore molecules to the oxygen - quenching phenomenon governs the luminescent output. Quenching of luminescence is controlled by the diffusion of the oxygen in the polymer binder of PSP. As a result such PSP characteristics as pressure and temperature sensitivity, time response of PSP to the change of external pressure and spatial resolution are diffusion-controlled.

An analysis of these diffusion-controlled characteristics was presented in earlier papers [1, 2]. To estimate diffusion-controlled characteristics of PSP it was assumed, that PSP has linear calibration characteristics:

$$I/I_O = 1/(a+bp);$$
(1)

where I_0 and I are reference luminescence intensity of PSP at some reference pressure, and at air pressure p, a and b are some coefficients. All the processes would be considered as isothermal to simplify analysis.

Luminescence output of PSP is an integral over the thickness of the PSP layer:

$$I(y,z) = \int_0^h \frac{I_{ex} \Phi_o \varepsilon \cdot c \cdot 10^{-\varepsilon c x} \ln(10)}{a + b \cdot n(t,x,y,z)/0.21\sigma} dx ; \tag{2}$$

where I_{ex} is the excitation light intensity, Φ_o is the luminescence quantum yield at the reference pressure, ε is the extinction coefficient of the luminophore, c is the concentration of luminophore in PSP, σ is the solubility of the oxygen in the PSP binder polymer (0.21 is the fraction of the oxygen in the air), n is the oxygen concentration at a PSP point (x,y,z) at the moment t and h is the thickness of PSP layer ($x=0$ corresponds to the PSP-air boundary, and $x=h$ corresponds to the PSP-model boundary). As the optical density of PSP is $d=\varepsilon ch$ and $I_0=I_{ex}\Phi_o/(1-10^{-d})$, and substituting $X=x/h$, equation (2) may be rewritten as:

$$\frac{I}{I_0} = \frac{d \cdot \ln(10)}{1-10^{-d}} \int_0^1 \frac{10^{-d \cdot X}}{a + b \cdot n(t, X, y, z)/0.21\sigma} dX . \tag{3}$$

Integration of this equation for the constant oxygen concentration with the Henry's Law $n = 0.21\sigma p$ results in equation (1). In the case of variable oxygen concentration in the PSP layer, the luminescence intensity would depend on the oxygen distribution in the PSP binder. A change in the oxygen distribution in the polymer of PSP binder arises in the case of transient or spatial air pressure changes outside the PSP layer and is determined by the mass transfer equation with appropriate boundary conditions:

$$\frac{\partial n}{\partial t} = D \left[\frac{\partial^2 n}{\partial x^2} + \frac{\partial^2 n}{\partial y^2} + \frac{\partial^2 n}{\partial z^2} \right]; \tag{4}$$

where D is the diffusion coefficient of the oxygen in the polymer of the PSP binder. Boundary condition for PSP-model boundary is the condition of the zero normal gradient:

$$\frac{\partial n(t,x,y,z)}{\partial x} \bigg|_{x=h} = 0. \tag{5}$$

The pressure on PSP-air boundary determines the second boundary condition.

Time Response.

If at the moment $t=0$ the pressure changes from the value p_0 to the value p_0+p_1 in all points of the PSP-air boundary $(x=0, y, z)$:

$$n(t,0, y, z) = \begin{cases} 0.21\sigma p_0 & ; t < 0 \\ 0.21\sigma(p_0 + p_1); t > 0 \end{cases}; \tag{6}$$

the oxygen concentration inside the PSP binder begins to change with time. Solution of the equation (4) with the boundary conditions (5) and (6) gives the time dependence of the oxygen concentration distribution in the PSP binder:

$$n(t,x,y,z) = 0.21\sigma \left[p_0 + p_1 \left(1 - \sum_{n=0}^{\tilde{\infty}} \frac{2}{\pi n + \pi/2} \cdot \sin\left(\left(\pi n + \frac{\pi}{2}\right)\frac{x}{h}\right) \cdot \exp\left(-\frac{(\pi n + \frac{\pi}{2})^2 D}{h^2} t \right) \right) \right]. \qquad (7)$$

Consequently, the characteristic time of the oxygen concentration relaxation in the polymer layer is:

$$\tau = \frac{4h^2}{\pi^2 D}. \qquad (8)$$

To get the luminescence intensity relaxation, i.e. PSP time response, equation (3) must be integrated using solution (7). Results of the integration for different optical densities d and pressure step from the vacuum ($p_0 = 0bar$) to one bar ($p_1 = 1bar$) are presented in Fig.1, where $T = t/\tau$ (τ is determined by (8)) and $A = (I - I (p_0 + p_1))/(I (p_0) - I (p_0 + p_1))$- the relative variation of the luminescence intensity. It was assumed that $b \times p_1 = 3a$ - typical for PSP. As the optical density of real PSP is within the range $0.3 \div 1.0$, 99% relaxation of luminescence intensity of real PSP occurs within $(2.5 \div 3)\tau$.

Amplitude-Frequency and Phase-Frequency Characteristics

The periodical pressure component with the frequency ω determines the next condition at the PSP-air boundary:

$$n(t,0,y,z) = 0.21\sigma (p_0 + p_1 \cdot \sin(\omega \cdot t)). \qquad (9)$$

Solution of equation (4) with the boundary conditions (5) and (9) resulted in:

$$n(t,x,y,z) = 0.21\sigma \left[p_0 + p_1 \left(A(\omega,x) \cdot \sin(\omega t) + B(\omega,x) \cdot \cos(\omega t) \right) \right] \qquad (10)$$

with:

$$A(\omega,x) = \frac{ch\left(\sqrt{\frac{2\omega}{D}}(h-\frac{x}{2})\right)\cos\left(\sqrt{\frac{\omega}{2D}}x\right) + \cos\left(\sqrt{\frac{2\omega}{D}}(h-\frac{x}{2})\right)ch\left(\sqrt{\frac{\omega}{2D}}x\right)}{ch\left(\sqrt{\frac{2\omega}{D}}h\right) + \cos\left(\sqrt{\frac{2\omega}{D}}h\right)}$$

$$B(\omega,x) = \frac{sh\left(\sqrt{\frac{2\omega}{D}}(h-\frac{x}{2})\right)\sin\left(\sqrt{\frac{\omega}{2D}}x\right) + \sin\left(\sqrt{\frac{2\omega}{D}}(h-\frac{x}{2})\right)sh\left(\sqrt{\frac{\omega}{2D}}x\right)}{ch\left(\sqrt{\frac{2\omega}{D}}h\right) + \cos\left(\sqrt{\frac{2\omega}{D}}h\right)}.$$

To get the Amplitude-Frequency Characteristics (AFC) and the Phase-Frequency Characteristics (PFC) the pressure value of PSP p_l must be calculated using equation (1). Since the periodical component is usually small ($p_1 << p_0$), this equation may be simplified:

$$p_l \cong p_0 + p_1 \frac{d \cdot \ln(10)}{1 - 10^{-d}} \int_0^1 10^{-dX} \left[A(\omega, Xh) \cdot \sin(\omega t) + B(\omega, Xh) \cdot \cos(\omega t) \right] dX. \qquad (12)$$

This equation allows to determine the Amplitude Frequency Characteristics (AFC) and the Phase Frequency Characteristics (PFC) of PSP. Let

$$\alpha(\omega) = \frac{d \cdot \ln(10)}{1 - 10^{-d}} \int_0^1 10^{-dX} A(\omega, Xh) dX \qquad \beta(\omega) = \frac{d \cdot \ln(10)}{1 - 10^{-d}} \int_0^1 10^{-dX} B(\omega, Xh) dX ; \qquad (13)$$

then AFC may be written as $A = \sqrt{\alpha^2 + \beta^2}$ (Fig.2) and PFC - as $\varphi = arctg(\beta/\alpha)$ (Fig.3) (in these figures $W = 2\omega h^2/D = \omega \tau \pi^2/2$ (τ is determined by (8)).

Measurement Concept

To measure periodical non-harmonic pressure fluctuations the PSP has to transfer several harmonics (about ten) of these fluctuations. Estimations based on Amplitude-Frequency and Phase-Frequency Characteristics for polymer-based PSP show that AFC=0.99 with PFC=-5° occurs at frequency $\omega=1.2/\pi^2\tau_{rel}$ while at frequency 10ω AFC=0.68 and PFC=-35°.

A test setup assembled in a wind tunnel to fulfill PSP measurements on an oscillating model is presented in Fig.4. A pulsed light source was used for PSP excitation at certain phases of the model position. To compensate model movement and non-stability of the excitation light distribution the paint, having additional (reference) non-sensitive to the pressure luminescent output was used. Flash duration of $1\div3ms$ was enough for measurements at the models oscillating frequencies up to $20\div30Hz$. Two digital CCD cameras with appropriate filters acquired the pressure sensitive and reference luminescence intensities.

Investigation of PSP Dynamic Characteristics

Experimental investigation of PSP reaction to periodic pressure change was performed for PSP L2 and PSP R1 that are based on a similar polymer binder [3].

The periodic pressure component was created by harmonically moving piston, driven by an electric motor. The frequency of the periodic pressure component was changed in a range of 0.2-$40Hz$.PSP sample was installed in the vicinity of moving piston in a hermetic cell. The PSP sample was continuously excited through a fused silica window by a stabilized Xe-arc lamp. An optical glass filter selected excitation light of the appropriate spectral range. A photomultiplyer tube through the same quartz window acquired the luminescence light. The luminescence light was separated from excitation light by a carefully chosen optical glass filter. A conventional pressure sensor provided pressure measurements in the cell. Multiplexed analog-digital converter installed in a personal computer measured luminescence intensity and pressure p_s. Pressure values p_s were used to determine the mean pressure value p_{s0}, first harmonic frequency f and amplitude p_{s1}. The PSP luminescence intensity was recalculated to a pressure value p_l using static calibration characteristics and finally PSP's mean value p_{l0}, first harmonic amplitude p_{l1} and phase shift dF were determined. The Amplitude-Frequency Characteristics (AFC) was determined as $p_{l1}(f)/p_{s1}(f)$ and the Phase-Frequency Characteristics (PFC) as $dF(f)$.

Experimentally determined AFC and PFC of PSP L2 with the thickness $h=20\mu m$ and the optical density $d=0.2$ are presented in Fig.5. Experimental results confirm equation (12) and (13) with the characteristic time $\tau=0.2s$ (see equation (8)).

In 1997 a laboratory setup for investigation of PSP response to the step pressure change was assembled in TsAGI. (Fig.6). The topside of the tube cell was closed by a cellulose membrane and an air was evacuated from the cell. A needle destroyed the membrane and movement of a shock wave inside the tube was initiated. The pressure relaxation time in the cell was estimated in a range of $50\div200\mu sec$, while some sonic waves could also appear in this cell with the main frequency about of $2000Hz$ that was determined by the length of the tube.

A continuous luminescent UV lamp performed the excitation of the PSP sample. A photomultiplyer tube acquired the luminescence intensity and its output signal was digitized.

Fig.7 presents the time response of fast PSP F2 paint to the step pressure change. This response corresponds to the theoretical predictions with characteristic time (8) $\tau = 2.6ms$. The Response Time of 99% relaxation of luminescence intensity is equal to $6.5ms$, which correspond well to the theoretical prediction. AFC of this PSP is equal to 0.99 at frequency $7Hz$ and 0.7 at $70Hz$, thus this PSP can be used for unsteady pressure measurements at the frequency up to few dozens Hz. Some samples of very thin PSP F2 was prepared and tested in this cell. Their characteristic times were of the order of $100 \div 300\mu s$.

Pressure field investigation on the pitching models

The first stage of these tests was conducted in the Transonic Wind Tunnel of DLR Göttigen (TWG) and was aimed to explore the measurement technique and PSP potentiality to give acceptable results for the pressure field on the surface of the oscillating model [4]. The wing model with NLR 7301 profile was painted with L4 PSP. The maximum pitching frequency was up to 20Hz. Due to small model displacement (oscillation amplitude was only 0.6^0) and good stability of the light source, standard paint was used. To minimize response time of this paint a very thin layer was applied. The characteristic response time was estimated to about 0.05-0.1sec.

The model was instrumented with the internal Kulite pressure measurement system providing the possibility to determine the static component of the pressure distribution. The plots in Fig.8 show a good agreement between PSP and Kulite measurements.

The second stage of the tests was conducted at the NASA Ames testing facility where it was possible to create oscillations with amplitudes up to 10° and oscillation frequency up to 20Hz. The central section of the wing model having NACA0012 profile was painted with fast binary paint. During the model painting procedure the sample of the paint was prepared and calibrated including measurements of the response time, which was estimated as 8msec. The real response time varies somewhat over the model surface due to different paint thickness. Measurements were conducted at oscillation frequencies 0, 5, 10 and 20Hz and a flow Mach number 0.45. During each image acquisition cycle flash lamp was ignited at a predetermined angle of attack and two images were acquired using separate CCD cameras with appropriate glass filters. Reference images were acquired at the same angle of attack without flow. After image alignment procedure C_p fields were calculated and these fields were transferred to the grid model of the wing section. Special markers were applied to the model surface for image alignment and resection.

Plots in Fig.9 show C_p distributions along a wing chord for 8° angle of attack and different oscillation frequencies. Estimated response time 8msec corresponds to a cutoff frequency of 20Hz, which means that for the model oscillation frequency of 20Hz this paint realization can resolve only one harmonic. This model was not equipped with a conventional pressure measurement system but pressure distributions obtained for zero frequency are in a good agreement with the pressure distributions for this profile type.

Conclusion

Theoretical estimations and experimental research show that currently available PSP can be used for pressure field investigations on the oscillating models at frequencies up 30-40Hz. Binary composition compensates the influence of the model movement and excitation light instability. Further increase of the upper frequency limit can be obtained with a very thin PSP layer, which will require additional efforts for model preparation and dynamic paint calibration.

References

[1] V. Mosharov, A. Orlov, V.Radchenko, M.Kuzmin, N.Sadovskii, "Luminescent Pressure Sensors for Aerospace Research: Diffusion-Controlled Characteristics." 2nd European Conference on Optical Chemical Sensors and Biosensors, Firenze, Italy, April 19-21, 1994.

[2] B.F.Carroll, J.D.Abbitt, E.W.Lukas, M.J.Morris, "Pressure Sensitive Paint Response to a Step Pressure Change." AIAA Paper 95-0483, Jan. 1995.

[3] OPTROD Ltd, Dugin str. Zhukovsky, Moscow reg., 140160 Russia. Fax: 07 095 939 0290, e-mail address: optrod@photo.chem.msu.su Attn. Mosharov.

[4]. S.Fonov, R. Engler, Ch.Klein, V.Mosharov, V.Radchenko, Application of the PSP for Investigation of the Oscillating Pressure Fields, Proc. of AIAA Congress, Albuquerque, Ca, 1998.

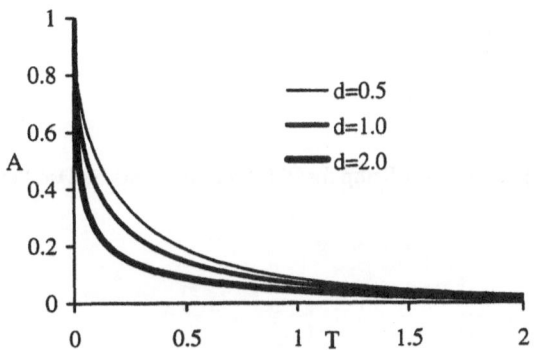

Fig.1. Theoretical relaxation of the luminescence intensity A after the step pressure change. (d - optical density, T – normalized time).

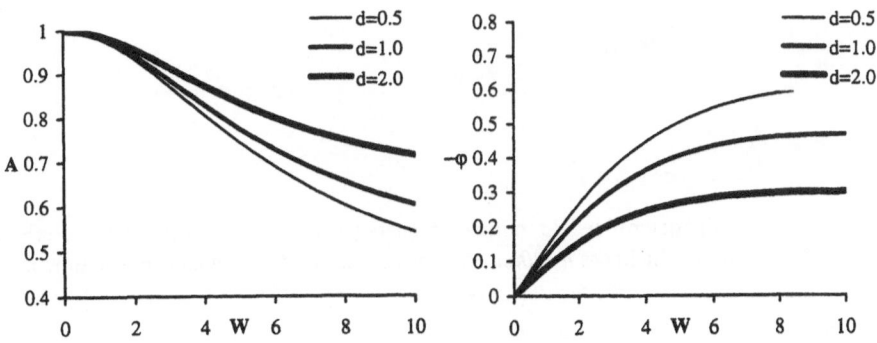

Fig.2. Amplitude-Frequency Characteristic of PSP. (d - optical density, W – normalized frequency)

Fig.3. Phase-Frequency Characteristic of PSP. (d - optical density, W – normalized frequency)

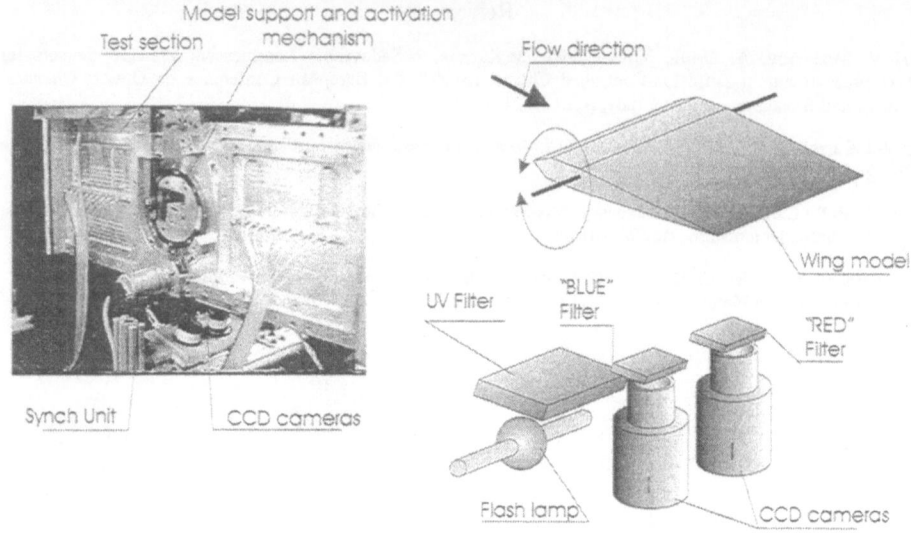

Fig.4. Schematics of Test Setup for PSP Measurements on Oscillating Model

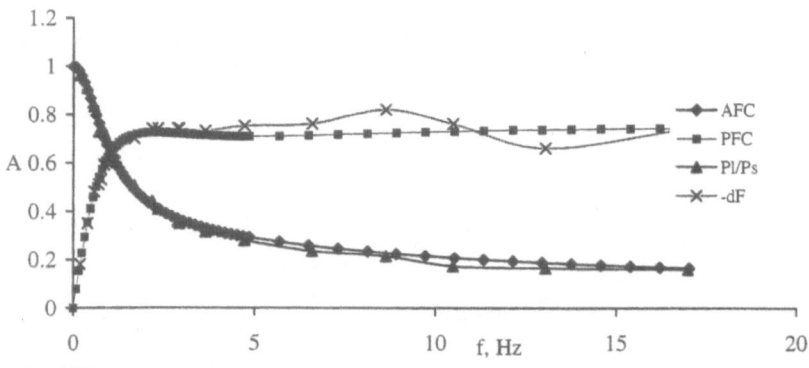

Fig.5. Experimentally determined Amplitude-Frequency and Phase-Frequency Characteristics of PSP L2 with the thickness $h=20\mu m$ and optical density $d=0.2$; characteristic time is $\tau=0.2sec$.

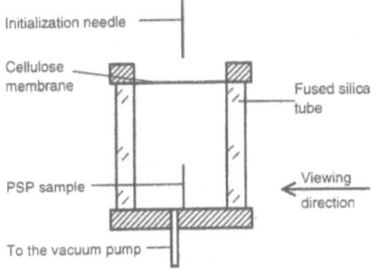

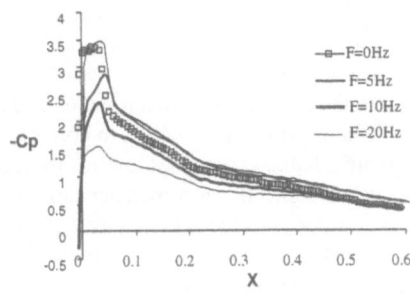

Fig.6. Schematic of a cell for investigation of PSP response to the step pressure change

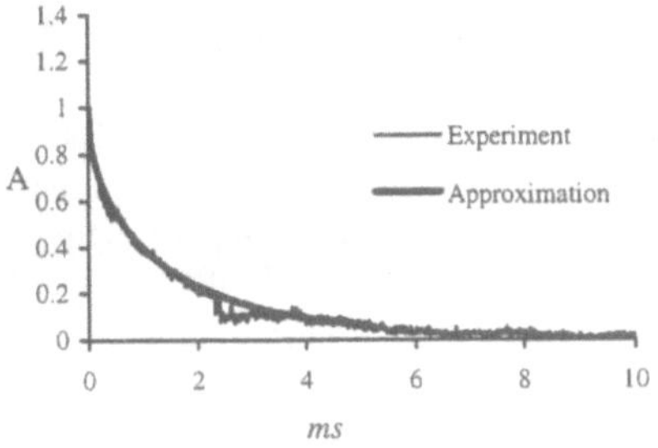

Fig.7. Time response of PSP F2 to the step pressure change, $\tau=2.6msec$.

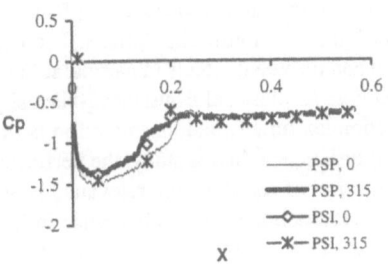

Fig.8. Comparison of PSP (lines) and Kulite (symbols) pressure distributions at the model centerline for 0° 315° phase angles, Ma=0.7, AOA=3°, oscillation amplitude 0.6°, f=5Hz.

Fig.9 Cp distributions (x -normalized coordinates along the wing chord)

Numerical Simulation of Three-Dimensional Transonic Flows Using Advanced Turbulence-Transport Models

M. Franke[1], T. Rung[1], E. Elsholz[2], P. Aumann[2], F. Thiele[1]

[1]TU Berlin, Hermann-Föttinger-Institut für Strömungsmechanik,
Müller-Breslau-Str. 8, D-10623 Berlin, Germany

[2]Daimler-Benz Aerospace Airbus GmbH, Dept. EFV,
Hünefeldstr. 1-5, D-28199 Bremen, Germany

Summary

The accurate numerical simulation of aerodynamic configurations especially at off-design conditions places increasing demands on the quality of numerical flow simulations. Therefore, the accurate representation of visous effects in general and turbulent effects in particular has shifted into the focus of attention in numerical aerodynamics. As classical Boussinesq-viscosity models have proven insufficient to correctly predict complex flow situations, attention is drawn to more reliable approaches towards the modelling of turbulence. In this study, the performance of two advanced linear turbulence-closure models in flows of aerodynamic relevance is analyzed with respect to the accuracy of flow physics representation and the applicability in industrial computing. Three exemplary three-dimensional test cases are investigated to assess these issues. The results demonstrate the superior predictive accuracy of more sophisticated modelling practices in complex flow situations as well as the feasibility of such computations employing state-of-the-art supercomputing systems.

Introduction

Until recently, the numerical simulation of aerodynamic flows of industrial relevance has generally been restricted to design conditions, where major portions of the flow field are governed by inviscid phenomena whereas viscous effects are confined to attached near-wall regions. Therefore, the computational modelling of turbulence-transport effects has usually remained rather simple and has quite often been restricted to algebraic mixing-length models. However, the models used in such calculations offer only a limited range of validity and are often unable to accurately predict intricate flow situations such as flows exposed to adverse pressure gradients. As the industrial need increases for simulations at off-design conditions, more accurate approaches towards the modelling of turbulence are required. Such models are not widely used in industry as yet, the main reason for this being the increased computational costs.

The principal aim of this study is the comparison of two application-oriented advanced linear two-parameter turbulence models with a conventional one with respect to the capability of representing intricate flow situations in complex industrial applications and the evaluation of the computational surplus in supercomputing environments, demonstrating the feasibility of such approaches.

Computational Approach

The computations are performed with the Finite-Volume Code FLOWer, which is being devised by DLR, DASA Airbus and several universities under the aegis of the MEGAFLOW project [1], where the development, integration and validation of advanced turbulence models is being performed at the Technical University of Berlin. FLOWer solves the compressible, three-dimensional Reynolds-averaged Navier-Stokes equations using a cell-vertex scheme on block-structured grids. Spatial discretization is based on central differences with added artificial dissipation, an explicit five-stage Runge-Kutta scheme is employed to integrate in time. Convergence acceleration is achieved by local time stepping, implicit residual smoothing and a multigrid algorithm. Turbulence models available besides algebraic mixing-length models include several linear one- and two-equation eddy-viscosity models. The code is highly portable and runs on a variety of parallel, vector and scalar computers. For the computations presented in this work, a CRAY T3E-900 parallel system and a NEC SX-4/36 H2 vector system have been employed.

Turbulence Modelling

Various studies have demonstrated the inability of the most prominent standard Boussinesq-viscosity turbulence-transport models to render the fundamental physics of turbulence, especially in flows exposed to non-equilibrium high-load conditions. Although a general approach would, arguably, be based on a second-moment closure, present efforts try to adopt a computationally cheaper solution. To remedy the shortcomings of the classical approach, two advanced linear two-equation models have been developed which are investigated here. They are introduced in brief, details can be found in the respective references.

The first model stems from an explicit solution to the second-moment closure in the limit of equilibrium turbulence. This explicit algebraic Reynolds-stress model can be regarded as a generalized (non-linear) two–parameter model, which retains the predictive benefits of the second–moment closure methodology, while numerical advantages of the Boussinesq-viscosity concept are conserved. Additional key features of the modelling practice are topography-independent low-Re formulations, obediance of the realizability principle, consistency to the hydrodynamic stability theory and an approximately self-consistent representation of non-equilibrium turbulence. Based on the resulting formulation, a family of low-Re number eddy-viscosity models (EVM) is devised [2], the simplest member being a linear truncation of the non-linear constitutive relation, cast here in terms of a Wilcox $k - \omega$ model (LEA $k - \omega$).

The second advanced model investigated here is a local linear two-parameter model (LLR $k-\omega$) derived from realizability and non-equilibrium turbulence constraints [3]. The coefficients of the stress–strain relation and the turbulence-transport equations are all functions of the non-dimensional invariants of the mean strain and vorticity rates. The approach tries to accomplish consistent stress–strain distributions not only in plane shear flow, but also in more general flow situations.

155

The models mentioned above are compared to their conventional counterpart, viz. the Wilcox $k - \omega$ model [4].

Results and Discussion

ONERA M6 Wing

The basic validation has been performed on the ONERA M6 wing at $M = 0.84$, $Re = 11.7 \cdot 10^6$ and two different angles-of-attack, viz. $\alpha = 3.06°$ and $\alpha = 5.06°$. A grid of approximately 1.5 million grid nodes and 44 blocks with 16705 surface points has been employed. As is the case with all meshes employed in this study, a distance of the first grid point to the wall of $y^+ = 1$ to 3 and a presence of 5 to 8 grid nodes in the region $y^+ < 12$ was aspired. Computations were performed on a CRAY T3E-900. Pressure measurements are available in selected wing sections [5]. It should be noted that the resolution of the grid is comparable to two-dimensional testcases in streamwise (353 points) but not in spanwise direction (65 points), the streamwise discretization therefore being superior to the one normally found in 3D meshes. For $\alpha = 3.06°$, the flow is characterized by a λ-structure of two shocks merging at 85% half span. Downstream of the second shock, the flow is almost two-dimensional without separation occurring. In contrast, for $\alpha = 5.06°$, the shock merging of the λ-structure takes place at 62% half span, with shock-induced separation ocurring behind the unified shock in the outboard region.

Fig. 1 compares the computed pressure coefficient with the experimental data in two selected wing sections for both angles-of-attack. It can be observed that for the attached case, all models lead to satisfactory predictions concerning the suction peak and the predicted shock locations. At $\eta = 0.80$, the solution deviates somewhat from the experiments. This can be attributed to the fact that this section is very close to the shock merging position, which is obviously not met precisely, a likely reason being the suboptimal mesh resolution in spanwise direction. In contrast, in the case with shock-induced separation, the models yield different results in the wing section where separation is present, with the advanced models' prediction of the shock location being more accurate. This leads to the important conclusion that in cases where the flow structure remains rather simple, the use of advanced models is not advantageous due to their computational overhead. However, with complex phenomena present, the use of such models leads to an increase in predictive accuracy, justifying the surplus effort quantified later in this chapter.

DLR F4 Wing-Body Configuration

As a second well-known test case, the DLR F4 wing-body has been computed at $M = 0.75$, $Re = 3.0 \cdot 10^6$ and $\alpha = 0.93°$. For this case, a rather fine grid of approximately 6.8 million grid nodes and 128 blocks with 63778 surface points has been employed for parallel computations on a CRAY T3E-900. Due to symmetry reasons, only half of the configuration was calculated. In certain regions of the domain, however, this mesh is still suffering from deficiencies, for example an insufficient boundary layer resolution on the fuselage affecting the front region of the wing-body junction. Nonetheless, the evaluations performed here are justifiable as they are concerned with regions in which the grid is well-defined. Comparisons are made for both LLR and LEA $k - \omega$ models with Wilcox $k - \omega$, the experimental reference data are taken from [6].

Fig. 2 depicts the computed pressure coefficient and streamlines on the configuration surface for a computation using the LEA model. The stagnation zones on the fuselage and the cockpit, the separation line on the rear part of the fuselage as well as the shock on the wing surface are clearly distinguishable. A comparison of computed and measured pressure coefficients in two representative wing sections is given in Fig. 3. It can be seen that the suction peak is resolved by all models, while the shock location is predicted too far downstream in the outer section. Nevertheless, the predictive accuracy of the advanced models proves to be superior as compared to the standard Wilcox model; in the inner section the shock location is met almost exactly.

A remarkable feature of the flow is a vortex-shedding area at the rear of the wing-body-junction, see Fig. 4. The LEA and Wilcox $k - \omega$ models basically yield the same flow topology featuring a distinct focus-type separation, the LEA prediction being more pronounced than the Wilcox solution. This is not surprising since the LEA modifications, as already mentioned before, are plugged onto a Wilcox $k - \omega$ model. The LLR result is distinctively different. A small focus is still visible, however, the major part of the recirculating flow remains attached and leaves the wing not via the focus separation but via the trailing edge. Unfortunately, as oil flow patterns are not available for this case, it is impossible at this stage to decide which is the physically correct solution. This clearly points out the pressing need for additional test cases including oil flow patterns and velocity profiles on three-dimensional aerodynamic configurations for validation purposes.

Generic Wing-Body-Tail Configuration

Moving towards more realistic configurations, a generic wing-body-tail geometry has been investigated at $M = 0.82$, $Re = 2.5 \cdot 10^6$ and $\alpha = 2.50°$. Again, due to the symmetry of the set-up, only half of the configuration was computed; calculations were performed on a grid of approximately 1.2 million grid nodes and 8 blocks with 22137 surface points on a NEC SX-4. These preliminary simulations on a rather coarse mesh are a necessary step towards the determination of the general flow features; this information is needed for generating a satisfactory high-resolution grid. Measurements are available for a very similar configuration [7], which differs from the present one in the outer sections of the wing. Therefore, the measurements on the fuselage and the inboard part of the wing can be used as a database for these simulations. Comparisons are made for the LLR $k - \omega$ with the Wilcox $k - \omega$ model.

Fig. 5 shows the computed pressure coefficient and surface streamlines obtained with the LLR $k - \omega$ model in a side view from below. Remarkable features of the flow include a comparatively strong deviation of the surface streamlines on the fuselage due to the wing, a merging of streamlines aft of the wing-body-junction attributable to a "squeeze-off" separation and a dislocation of the separation line away from the lower symmetry line onto the horizontal stabilizer. These flow features are observed in the Wilcox $k - \omega$ simulation as well, however, they are less pronounced. A comparison of computed and measured pressure coefficients in two inboard wing sections and the fuselage symmetry lines is given in Fig. 6. It can be observed that the shock location on the upper side of the wing is predicted more accurately by the LLR model as compared to the Wilcox computation. However, the suction peak is not met by neither, owing to an insufficient mesh resolution of the wing leading edge. The pressure distribution on the fuselage is predicted well by both models, except for the region under the horizontal stabilizer in the aft part of the fuselage, where the wing wake interacts with the local boundary layer. Again, this is due to grid deficiencies.

Computational Performance

The computations presented in this paper have mainly been obtained on a massively parallel supercomputer, viz. a CRAY T3E-900. In terms of additional costs for the application of the models presented here, the LLR $k - \omega$ model requires approximately 25% more time per iteration than Wilcox $k - \omega$, LEA $k - \omega$ about 17%, while basically retaining the same memory requirements despite significantly more elaborate algorithms. This is due to the fact that the computational surplus of the advanced models is at least in part counterbalanced by the higher performance rates achievable. Table 1 displays the performance rates obtained with LLR, LEA and Wilcox $k - \omega$ models for the ONERA M6 and the DLR F4 cases, respectively.

The performance figures also demonstrate the superiority of massively parallel architectures. On state-of-the-art vector computers, such as the NEC SX-4, FLOWer reaches a maximum performance of about 1.4 GFLOPS. In contrast, for the ONERA M6 case with 1.5 million grid nodes, at least 2.3 GFLOPS could be achieved on the MPP system. The difference becomes even more pronounced when the DLR F4 case with 6.8 million nodes is considered, where at least 6.9 GFLOPS were obtained, which is far beyond the performance of current vector computers.

Conclusion

In this study, investigations concerning the application of two advanced linear two-equation turbulence-closure models to three-dimensional, transonic configurations were performed. It is demonstrated that, while yielding the same results for rather simple flows, the LEA and LLR $k - \omega$ models show an improved predictive accuracy as compared to the standard Wilcox model in comparatively complex flow situations, the LLR model being slightly more accurate in predicting shock locations. The computational overhead resulting from their employment remains limited, therefore their use in the industrial design process is recommendable. This is especially vaild as both models are local and therefore well suited for the use on MPP computers. It should be mentioned, however, that in order to accurately resolve the flow physics, the models presented here are even more dependent on high-quality meshes to develop their predictive potential than the standard models generally used in aerodynamic computations.

It is to be expected that the role of advanced turbulence-closure models will increase in future aerodynamic design processes, as they promise a superior flow physics representation at a tolerable computational surplus. Future work should include the application of non-linear models to aerodynamic flows in order to further advance the flow physics representation in flows featuring pronounced 3D features. The eventual aim is to supply the designer with a whole range of models of varying complexity from which he can choose one which is appropriate to the complexity of the flow problem under consideration, thereby minimizing the overall computational simulation time in the design process.

Acknowledgement

This work was sponsored by the German Ministry of Education and Research (BMBF) under the umbrella of the MEGAFLOW project (Grants No. 20A9501F and 20A9505H). The parallel computations were performed on the CRAY T3E-900 of ZIB Berlin, the vector computations utilized the NEC SX-4/36 H2 of HLRS Stuttgart.

References

[1] Becker, K.; Kroll, N.; Rossow, C.C.; Thiele F.: *Numerical Flow Calculation for Complete Aircraft - The MEGAFLOW Project.* DGLR-JT98-043, DGLR-Jahrestagung, Bremen, 1998.

[2] Rung, T.; Lübcke, H.; Franke, M.; Xue, L.; Thiele, F.; Fu, S.: *Assessment of Explicit Algebraic Stress Models in Transonic Flows.* To be presented at: 4th International Symposium on Engineering Turbulence - Modelling and Measurements, Corsica, France, 1999.

[3] Rung, T.; Thiele, F.: *Computational Modelling of Complex Boundary-Layer Flows.* Proc. 9th Intl. Symposium on Transport Phenomena in Thermal-Fluid Engineering. Singapore, 1996.

[4] Wilcox, D.C.: *Turbulence Modeling for CFD.* DCW Industries, Inc., La Cañada, CA, USA, 1993.

[5] Schmitt, V.; Charpin, F.: *Pressure Distributions on the ONERA M6 Wing at Transonic Mach Numbers.* AGARD AR-138, 1979.

[6] Redeker, G.; Müller, R.; Ashill, P.R.; Elsenaar, A.; Schmitt, V.: *Experiments on the DFVLR F4 Wing-Body Configuration in Several European Windtunnels.* AGARD-FDP Symposium, Naples, Italy, September 1987.

[7] Coustols, E.; Séraudie, A.; Mignosi, A.: *Rear Fuselage Transonic Flow Characteristics for a Complete Wing–Body Configuration.* Journal of Aircraft, Vol. 34, No.3, 1997, pp. 337–345.

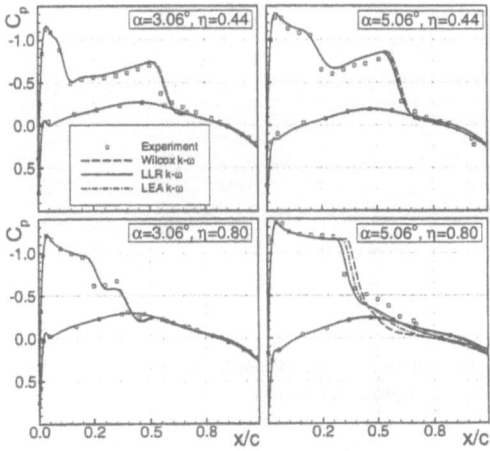

Figure 1 ONERA M6: Comparison of measured and computed pressure coefficient in two wing sections for $\alpha = 3.06°$ (left) and $\alpha = 5.06°$ (right) using LLR, LEA and Wilcox $k - \omega$ models.

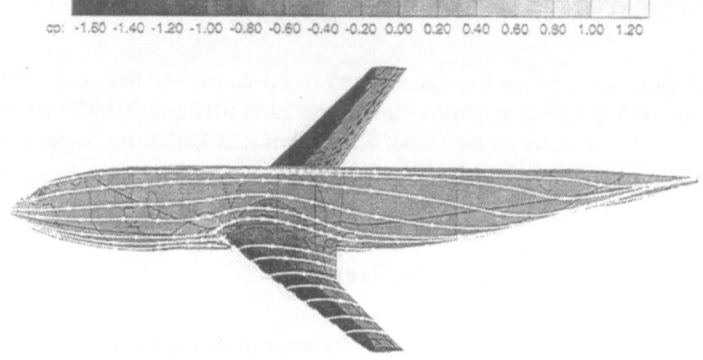

Figure 2 DLR F4: Computed surface pressure coefficient and surface streamlines at $Ma = 0.75$, $Re = 3.0 \cdot 10^6$ and $\alpha = 0.93°$ using the LEA $k - \omega$ model.

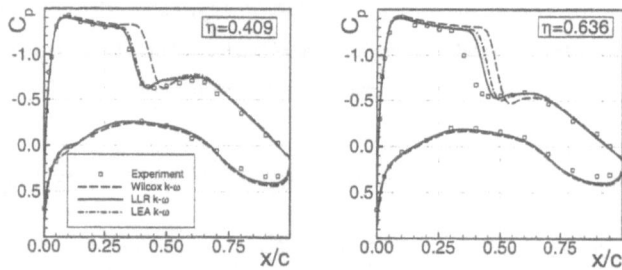

Figure 3 DLR F4: Comparison of measured and computed pressure coefficient in two wing sections using LLR, LEA and Wilcox $k - \omega$ models.

Figure 4 DLR F4: Streamlines on the wing near the wing-body-junction outlining the local vortex shedding area, the black lines depicting block boundaries.

160

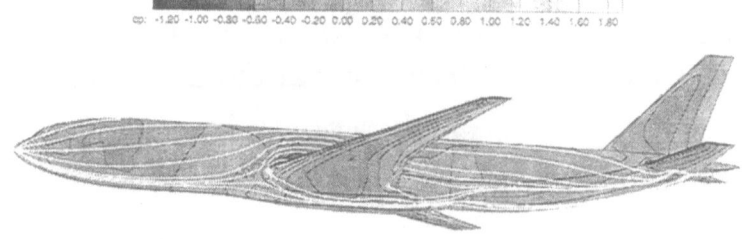

cp: -1.20 -1.00 -0.80 -0.60 -0.40 -0.20 0.00 0.20 0.40 0.60 0.80 1.00 1.20 1.40 1.60 1.80

Figure 5 Generic Wing–Body–Tail: Computed surface pressure coefficient and surface streamlines on the fuselage at $M = 0.82$, $Re = 2.5 \cdot 10^6$ and $\alpha = 2.50°$ using the LLR $k - \omega$ model.

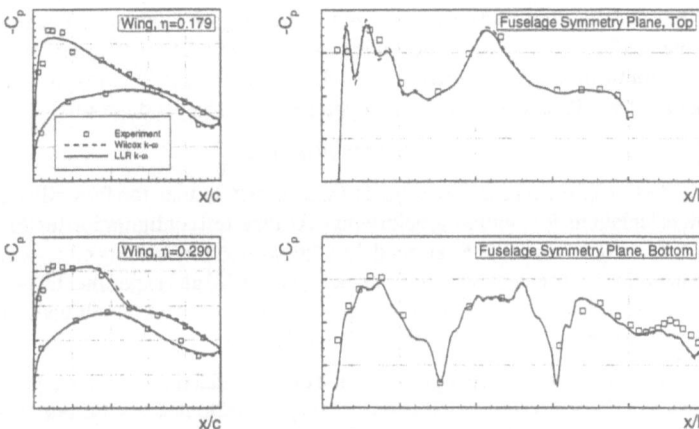

Figure 6 Generic Wing–Body–Tail: Comparison of measured and computed pressure coefficient in two inboard wing sections (left) and the fuselage symmetry plane (right) using LLR and Wilcox $k - \omega$ models.

Table 1 Comparison of performance rates on the CRAY T3E-900.

Case	ONERA M6			DLR F4		
Number and Size of Blocks	44 - 36465 Pts.			128 - 53625 Pts.		
Model	Wilcox	LEA	LLR	Wilcox	LEA	LLR
Performance per PE [MFLOPS]	51.6	53.5	56.3	53.9	56.2	58.7
Total performance [GFLOPS]	2.32	2.41	2.53	6.95	7.24	7.57

161

Numerical Simulation of Unsteady Vortical Flow about Delta Wings Oscillating at High Incidence

W. Fritz

Daimler–Benz Aerospace AG, Military Aircraft Division

D–81663 München, Germany

Summary

The unsteady three–dimensional flow field over an oscillating, $65°$ swept delta wing is investigated by the numerical solution of the compressible, time–dependent, Reynolds–averaged Navier–Stokes equations. The spatial discretization is performed with a Jameson–based finite volume scheme using structured grids. The time integration is done by a modified version of the dual time stepping scheme, which is stable for arbitrarily large and small time steps. All unsteady calculations have been carried out in a hyperbolic C–O type mesh consisting of 2,016,657 grid points and using the Baldwin–Lomax turbulence model with the Degani–Schiff modification. The resolution in time was 24 time steps per cycle, giving an increment of $15°$ in the phase angle. Results are presented for the low speed testcase (M=0.4, Re=$3*10^6$) at $\alpha_0=9°$ and $\alpha_0=15°$ for the reduced frequency $\omega^*=0.56$ and oscillation amplitudes of $\Delta\alpha=3°$ and $\Delta\alpha=6°$.

1. Introduction

The main objective of the present investigation was to demonstrate the feasibility of 3–D unsteady flow calculations for oscillating delta wings. An ideal test configuration for this study was the TA 15 delta wing, a cropped, $65°$ swept delta wing with a sharp leading edge, which was investigated in the past very extensively by numerical [1], [2], [3] and experimental [4], [11] studies. Dynamic measurements over a wide range of mean angles of attack, amplitudes and frequencies for this delta wing at low speeds are given in [11].

Implicit methods which use a linear approximation of the spatial flux operator are not very suitable for unsteady Navier–Stokes calculations, as they are restricted to CFL numbers of about 5000, which is far to low for an effective numerical scheme. Some recent 3–D unsteady Navier–Stokes calculations [6], [7], [8] confirm this experience and in those calculations between 7500 and 10000 time steps were necessary for one cycle of oscillation, giving a nondimensional time step size of about 0.001 for the resolution in time, whereas a time step size between of 0.5–1.0 would be fine enough for physical requirements. In numerous 2–D test calculations it was found that at the moment the dual time stepping scheme [5] is the only scheme, which is stable for arbitrarily large physical time steps. Using a slightly modified formulation, it is also stable for arbitrarily small time steps.

2. Description of the Method

For the time accurate solution of the unsteady Navier–Stokes equations a modified version of the original dual time stepping scheme is used (second order in time):

$$w_{(l+1)}\left(1 + \frac{3\Delta t^*}{2\Delta t}\right) = w_l - \Delta t^*\left(-\frac{2}{\Delta t}w^n + \frac{1}{2\Delta t}w^{n-1} + \frac{1}{V}R(w)\right). \tag{1}$$

Here w is the conservation vector consisting of the components $\{\varrho, \varrho u, \varrho v, \varrho w, \varrho e\}$, V the volume of the cell i, j, k and R the spatial flux operator approximating the net fluxes of the above components across the cell faces. The superscript $(n+1)$ stands for the new, unknown time–level, n and

$(n-1)$ mean the known solutions at the time–levels n and $(n-1)$ and the subscript l stands for the iteration level. Δt is the real, physical time step and Δt^* is the second, fictitious time step. The above formulation is not very suitable for the application of the multi–grid strategy, as the residual doesn't vanish in the converged solution. Therefore equation (1) is rearranged as:

$$w_{(l+1)} = w_l - \Delta\tau^*\left(\frac{3}{2\Delta t}w_l - \frac{2}{\Delta t}w^n + \frac{1}{2\Delta t}w^{n-1} + \frac{1}{V}R(w)\right), \qquad (2)$$

with the modified time step:

$$\Delta\tau^* = \frac{\Delta t^*}{1 + \frac{3\Delta t^*}{2\Delta t}} \ .$$

The scheme (2) is stable for arbitrarily large and small time steps. It is also stable within the multi–grid strategy, as within each iteration step only the spatial operator has to be approximated. For a sequence of $l=1,2,3,...m$ iterations the iterative solution of (2) will converge towards the value w^{n+1}, which is the solution for the new, unknown time–level. For the numerical solution, the scheme according to (2) was implemented into the multi–block, steady Euler/RANS solver FLOWer 112.8 [9], which was developed by the DLR. The Baldwin/Lomax turbulence model in combination with the Degani–Schiff modification for vortical flow was used in all calculations. More details about the implementation are given in [10].

3. Results and Discussion

The geometry of the delta wing can be seen in Figures 1 and 2. It is a cropped delta wing with $65°$ leading edge sweep. The geometry is normalized by the root chord, giving a root chord $c_r=1.0$ and a half span $b_h=0.3985$ in the computations. The under wing mounted fuselage, which is present in the experimental model, was not modelled in the computations. The physical coordinate system has its origin in the apex of the wing. The x–axis direction is in streamwise direction, the direction of the y–axis is normal to the surface and the z–axis is aligned with the spanwise direction. As it can be seen in Figures 1 and 2, a C–O type mesh is used. This mesh type enables a very fine resolution of the sharp leading edge but has a singular line from the wing apex towards the upstream far field. This grid, which was finally used for the unsteady calculation consists of 193 *129*81 grid points (streamwise, circumferential and normal direction). The spacing of the first grid points in normal direction is 0.00001, resulting in y^+ values between 0.2 and 0.8 at the wing surface. All calculations have been performed with a Reynolds Number $Re_\infty=3.1*10^6$ and with a Mach Number of $M_\infty=0.4$. (In the experiments of Löser [11] the Mach Number is $M_\infty=0.12$, the higher value in the calculations has been chosen to avoid convergence problems).

The influence of the grid resolution and the influence of the turbulence model have been studied in numerous steady RANS calculations. The essential results are given in Figure 3, where the surface pressure distribution in the cross section x/c is represented. As it can be seen from the left hand side of the figure, a very fine grid resolution is absolutely necessary for a satisfactory Navier–Stokes solution. The also presented Euler results show the typical effect of an Euler solution for this flow type: because of the missing secondary vortex, the primary vortex is too strong and its axis is more outboard. Only at very high angles of attack, where no secondary vortex is present, the Euler solution gets closer to the reality. The right hand part of Figure 3 shows a comparison of the surface pressure distribution for the two different turbulence models, both in the 193*129*81 grid. (The calculations with the k–ω model have been done with a pilot version of FLOWer 113, which has been provided by the DLR). Although the eddy viscosity distribution for these turbulence models are completely different, the surface pressure distributions are nearly identical. Therefore it seams, that the eddy viscosity distribution within the primary vortex is of only weak influence on the global flow field and the differences due to the turbulence models concentrates on the secondary vortex. Here the algebraic turbulence model predicts high values for the eddy viscosity at the interaction region of the vortices, the k–ω model (it is assumed, that the k–ω results are more realistic) predicts moderate values of the eddy viscosity within the region of the secondary vortex.

With the knowledge of the steady precalculations, the unsteady calculations were carried out using the 193*129*81 grid and the Baldwin/Lomax turbulence model with the Degani/Schiff modification. Harmonic pitch oscillations about the axis x/c=0.5625, described by the sinusoidal motion

$$\alpha(t) = \alpha_0 + \Delta\alpha \cdot \sin(\omega t) \tag{3}$$

were simulated in the calculations. Here ω is the natural angular frequency of the motion, and is given by the nondimensional reduced angular frequency $\omega^* = (\omega c_{ref})/u_\infty$. For all unsteady calculations a time resolution of 24 time steps per oscillation cycle was used. This corresponds to an increment in the phase angle of 15°. and a nondimensional stepsize of 0.987. Additional investigations have shown, that there was no significant influence on the results by using a finer time resolution. (In unsteady transonic flow with moving shock waves, a much finer time resolution would be required). The Reynolds Number and the reduced frequency were taken as Re=3.1*10^6 and ω^*=0.56, the same values as in the experiment [11].

The Figures 4 and 5 show the hysteresis effect on c_l at the mean angle of attack $\alpha_0 = 9°$ and the amplitudes $\Delta\alpha = 3°$ and $\Delta\alpha = 6°$ together with the results of an unsteady Euler calculation. The Navier Stokes results agree much better with the experiment than the Euler results. The differences, which are still present between the experiment and the Navier–Stokes results may be caused by the underwing body, which was not simulated in the present calculations.

Figures 6, 7 and 8 show the unsteady wall pressure distributions in terms of magnitude and phase angles of the first 3 Fourier harmonics of the sinusoidal oscillation for different mean angles of attack for the cross–sections at x/c=0.6 and at x/c=0.8. The agreement of the magnitudes of all three Fourier harmonics is excellent in both sections. The plots of the phase of the first harmonic p_1 show also a very good agreement between calculation and experiment. In the phases of the second harmonic p_2 there is a constant shift of 135° between the numerical and experimental results (the jumps in the phase of the higher harmonics are caused by the fact that these modes are defined modulo 360°/n). The reason for this constant shift is not quite clear. It is not an effect of the turbulence model as it was also observed in unsteady Euler calculations. It is also not an influence of a lack in convergence, as fully converged solutions and solutions with a finer time resolution have shown the same behavior. In the phase of the third harmonics, the uncertainty in the data is large on a wide range due to the very low levels of the amplitudes. But at the tip section of the upper surface, where a certain magnitude of the amplitude is present, there is again a constant shift of 240° in the phase of the third harmonics between the numerical and experimental results.

In Figures 9 and 10 the variation of the total pressure loss contours in the cross section x/c=0.8 during the oscillation are presented. This cross section is situated behind the oscillation axis, at increasing angles of attack it moves downward and at decreasing angle of attack upward. Figure 9 shows the relations for $\alpha_0 = 9°$ and the amplitude $\Delta\alpha = 6°$. During the upward motion (left side) both, primary and secondary vortex are growing and moving towards inboard. At decreasing angle of attack both vortices weaken and at the lowest angle of attack they vanish completely.

At $\alpha_0=15°$ and the amplitude $\Delta\alpha=6°$ the relations are more complex as it can be seen in Figure 10. At the maximum angle of attack, the primary vortex is isolated and has no connection to the feeding sharp leading edge. During pitching down, the primary vortex breaks down which can be seen at the suddenly blow up of the vortex core. At smaller angles of attack a regular structure of the primary vortex finally develops again.

4. Conclusions

It has been shown, that modified dual time stepping scheme is a very effective method for the solution of the unsteady Navier–Stokes equations in high resolution grids. It allows arbitrarily large time steps, so that the resolution in time can really be adapted to the physical requirements of the problem. The unsteady vortical flow past an oscillating delta wing has been successfully computed up to a medium angle of attack $\alpha_0=15°$ and an oscillation amplitude $\Delta\alpha=6°$. The agreement between the computed and the experimental pressure distributions is very good. Due to the availability of all field data, the numerical simulation can show very interesting details of unsteady vortical flow fields. A prerequisite for any realistic numerical simulation is an accurate geometric discretization of all relevant details, i. e. the use of high quality grids.

5. References

[1] Wagner B., Hitzel S. M., Schmatz M. A., Schwarz W., Hilgenstock. A., Scherr S.: Status of CFD validation on the VORTEX FLOW Experiment. AGARD CP–437 (1988) Vol. 1., pp. 10–1 to 10–10.

[2] Longo J. M. A.: Verification des Rechenverfahrens CEVCATS für die Berechnung von Wirbelströmungen um Deltaflügel. In: ZfW 20/1996, Springer–Verlag, pp 213–226, 1996.

[3] van den Berg J. I., Brandsma F. J.: Navier–Stokes Results for a Sharp Edged Delta Wing. Contribution of NLR to Phase VI of IEPG Panel III, SG6–TA 15. NLR Report NLR CR 94592 C, 1994.

[4] Hummel, D.: Documentation of Separated Flows for Computational Fluid Dynamics Validation. In: AGARD Conference Proceedings No. 437, 1988.

[5] Jameson A.: Time–Dependent Calculations Using Multigrid, with Applications to Unsteady Flows past   Airfoils and Wings. AIAA Paper 91–1596, 1991.

[6] Menzies M. A., Kandil O.A.: Natural Rolling Responses of a Delta Wing in Transonic and Subsonic Flow. AIAA Paper 96–3391, 1996.

[7] Obayashi S., Chiu I. T., Guruswamy P.: Navier–Stokes Computations on Full Wing–Body Configurations with Oscillating Control Surfaces. AIAA Paper 93–3687, 1993.

[8] Ekaterinaris J. A.: Numerical Investigation of Dynamic Stall of an Oscillating Wing. AIAA Journal, Vol. 33, No. 10, Oct. 1995.

[9] Kroll N., Radespiel R., Rossow C–C.: Accurate andEfficient Flow Solvers for 3D Applications on Structured Meshes. VKI Lecture Series 1994–05, 1994.

[10] Fritz, W.: Unsteady Navier–Stokes Calculations for a Delta Wing Oscillating in Pitch. ICAS–Paper ICAS–98, 21st ICAS Congress, September 1998, Melbourne.

[11] Löser, Th.: Dynamic Force and Pressure Measurements on an Oscillating Delta Wing at Low Speeds. DLR Report IB 129–96/9, 1996.

6. Figures

Figure 1: 3–D Grid arrangement around the TA 15 delta wing

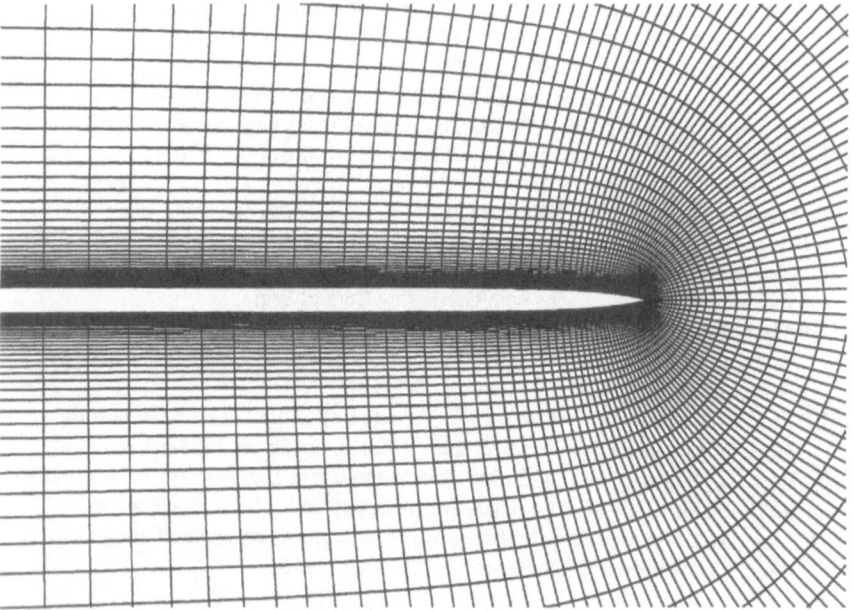

Figure 2: Grid in the cross section x/c=0.8

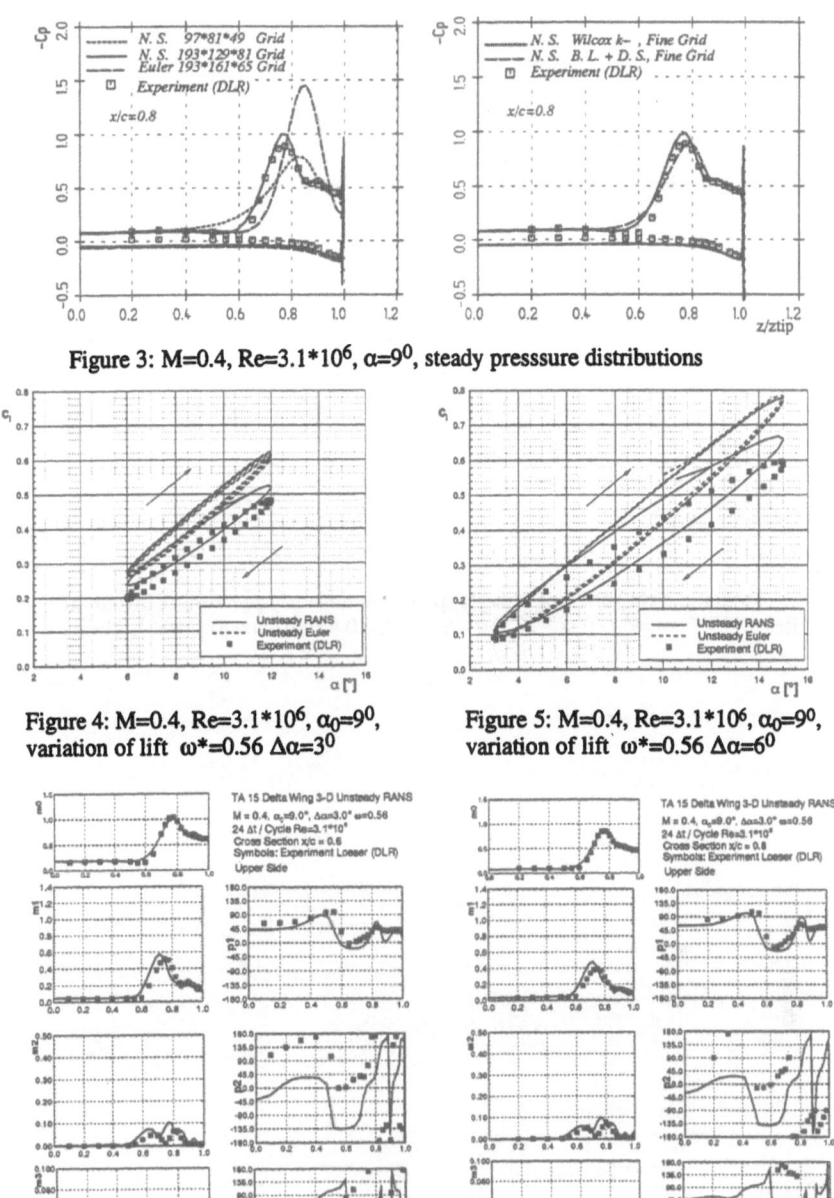

Figure 3: M=0.4, Re=3.1*10⁶, α=9⁰, steady presssure distributions

Figure 4: M=0.4, Re=3.1*10⁶, α_0=9⁰, variation of lift ω*=0.56 Δα=3⁰

Figure 5: M=0.4, Re=3.1*10⁶, α_0=9⁰, variation of lift ω*=0.56 Δα=6⁰

z/ztip   z/ztip

Figure 6:   Magnitude (m_i) and phase angle (p_i) of the unsteady pressure distribution. M=0.4, Re=3.1*10⁶, α_0=9⁰, ω*=0.56, Δα=3⁰, x/c = 0.6 (left) and x/c=0.8 (right).

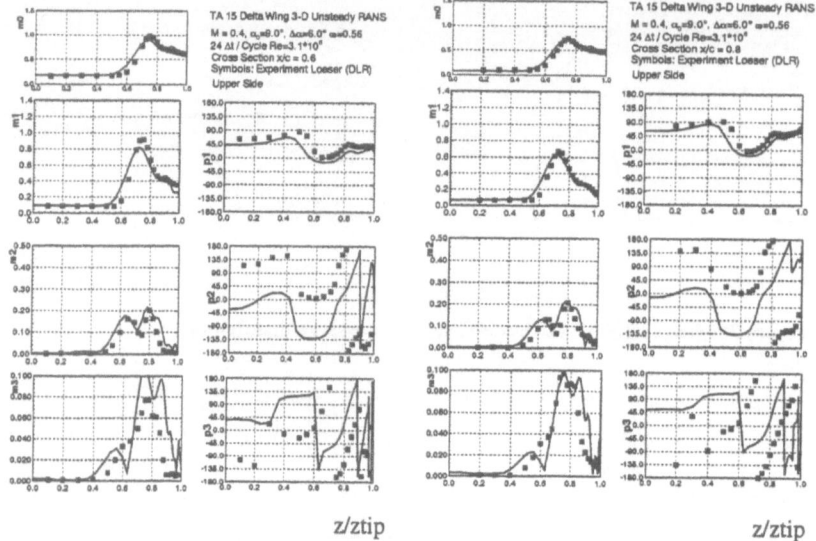

z/ztip   z/ztip

Figure 7:   Magnitude (m_i) and phase angle (p_i) of the unsteady pressure distribution. M=0.4, Re=3.1*10^6, α_0=9°, ω*=0.56, $\Delta\alpha$=6°, x/c = 0.6 (left) and x/c=0.8 (right).

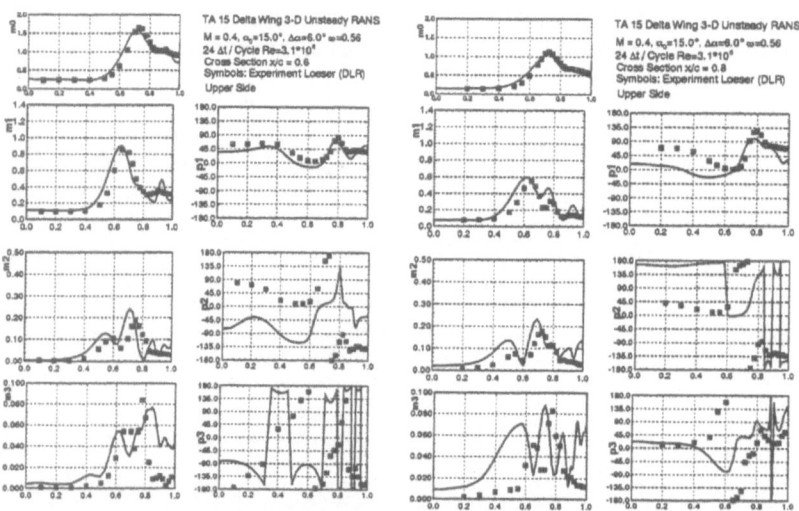

Figure 8:   Magnitude (m_i) and phase angle (p_i) of the unsteady pressure distribution. M=0.4, Re=3.1*10^6, α_0=15°, ω*=0.56, $\Delta\alpha$=6°, x/c = 0.6 (left) and x/c=0.8 (right).

168

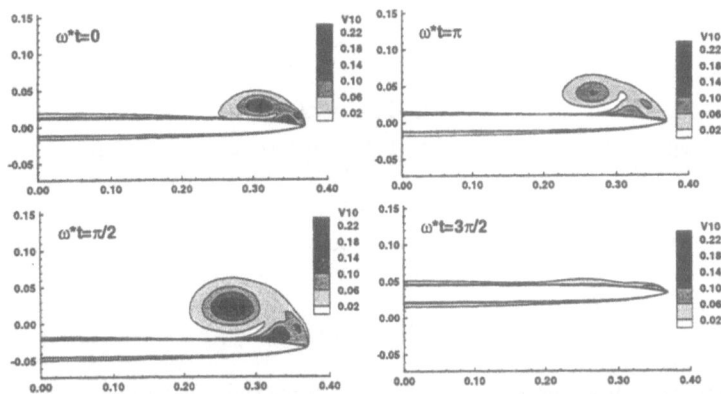

Figure 9:   Unsteady motion of the delta wing: Total pressure loss contours at cross
            section x/c=0.8. M=0.4, Re=3.1*10^6, a$_0$=9^0, w*=0.56, $\Delta\alpha$=6^0 .

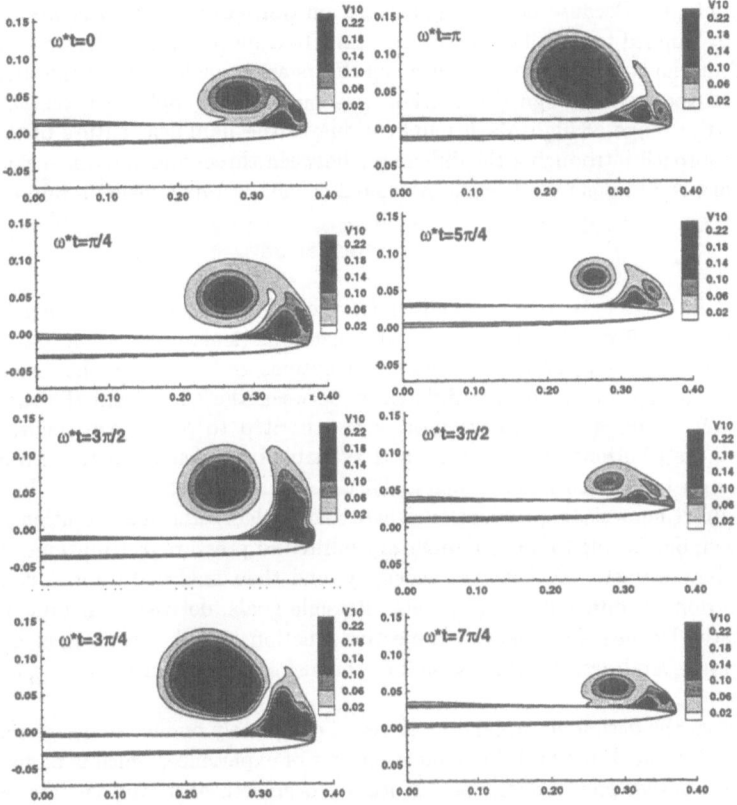

Figure 10:   Unsteady motion of the delta wing: Total pressure loss contours at cross
            section x/c=0.8. M=0.4, Re=3.1*10^6, a$_0$=15^0, w*=0.56, $\Delta\alpha$=6^0 .

169

Aerodynamic Optimization of Airplane Wings using Analogy Methods

O. Frommann

Synaps Ingenieur-Gesellschaft mbH, Fahrenheitstr. 1, D-28359 Bremen, Germany

D. Forbrich

Daimler Benz Aerospace Airbus, Hünefeldstr. 1-5, D-28183 Bremen, Germany

Summary

Wing design based upon inverse methods has become the standard procedure in the past. However, the design is limited to one single design point and the quality of the result depends upon the experience of the designer. Direct optimization allows to avoid these disadvantages. Because of the long calculation times of three-dimensional flow solvers it was only applied to airfoil design in the past. To achieve a benefit for the complete wing, hybrid methods, where three-dimensional effects are considered by geometrical distortion, are used today. Although this method was successfully applied to some cases, it often fails in the range of high Mach numbers due to the nonlinear nature of this flow. Another approach introducing the differences between three- and two-dimensional results as boundary conditions into a two-dimensional calculation may yield an improvement.

Introduction

In general, there exist two possibilities for the direct design of aerodynamic structures, which are the inverse design method and direct numerical optimization. Using the inverse design method one prescribes pressure distributions and calculates the corresponding geometries in dependence on the differences between the actual and the target pressure, e.g. [1, 2]. This is an iterative process and limited to one design point, because the pressure distributions are only valid and available in this single point. Inverse design is possible for two-dimensional as well as for three-dimensional geometries. In contrast, the use of direct numerical optimization, where search algorithms like evolution strategies are employed, one is able to find solutions not limited to pressure distributions. The problem has to be parametrized, e.g. the geoemtry, and then analyzed using a flow solver and may be done in different design points. Possible goals, derived from these analyses and formulated by means of a so called ojective function, may be the drag or a sum of drag values for several design points as well as the airfoil or spar thickness, lift, etc.

In the past the design of airplane wings relied completely on inverse design methods combined with some direct airfoil modifications out of experience, which is performed only at one single design point. Here, starting from proven solutions, wing sections and pressure distributions were modified and the results afterwards analysed using a 2-D flow solver. The quality of the design was proven by windtunnel measurements and 3-D Navier-Stokes calculations. The resulting knowledge could lead to an improved design in the next design step. One of the disadvantages of this concept is the long time one cycle takes, especially for a windtunnel test. Therefore, in this case a high quality 3-D flow solver is essential to improve the design before the windtunnel test. However, the higher the demands for the design targets are, it is getting more difficult to choose the right airfoil modifications and to predict which pressure distribution is the most suitable for the design target. Also, off-design points cannot be considered directly or geometrical constraints like thicknesses fulfilled.

Another possibility consists of the use of the three-dimensional inverse method using a special inverse flow solver like the inverse FLOWer-code. Here, the whole 3-D geometry of the wing will automatically be modified during the computation in order to match the sectionwise prescribed pressure distributions. Although this is advantageous with respect to the consideration of three dimensional effects, it still does only allow to create multipoint designs through the intuition and experience of highly skilled designers.

In contrast, direct optimization enables the designer to produce multipoint designs without much experience and enhances the flexibility for the description of other characteristics for the design, e.g. thicknesses. The only crucial thing is the construction of the objective function to obtain an appropriate measure of merit for the overall design. Previously used weighting of design points and penalties on constraints requires a thorough knowledge about the interdependencies and sensitivities. This proved to be rather uncertain and often leads to a trial and error process to develop a satisfying objective function. To overcome this disadvantage an approach based upon Fuzzy Logic, exploiting the knowledge of the designer, was proposed recently [3, 4]. This method yields a simple and standardized procedure to reliably transform usually only vaguely and verbally formulated objectives into a mathematical form.

However, because of the relatively high number of necessary flow calculations, the application was restricted to two-dimensional cases in the past. The calculation using three-dimensional flow solvers is too time consuming for the present day computer generation using a realistic number of design parameters and design points. In order to derive a benefit from optimization for three-dimensional designs, the effect of the three-dimensional flow has to be considered properly while at the same time only the faster two-dimensional analysis codes should be used. Therefore, the use of an analogy method, using the knowledge about three-dimensional effects in two-dimensional codes, is indispensable.

Analogy method

The optimization tool PointerPro™ [5, 6, 7, 8] recently used at Daimler Benz Aerospace Airbus [9] for airfoil design is complemented by an additional design step. Starting from the result of a three-dimensional Navier-Stokes calculation of a clean wing, new airfoil geometries are determined by an inverse design step coupled with a two-dimensional analysis code. These geometries, the analog airfoils, should yield the true three-dimensional pressure distributions from the Navier-Stokes calculation when analysed with the two-dimensional code. The geometric differences between these analog airfoils and the starting basic loft airfoils are considered in the following direct optimization step with PointerPro™. This way the three-dimensional effects are transformed into geometric distortions. Optimizations are carried out for each of the basic loft airfoils, allowing the adherence of local lift coefficients and multipoint demands. The result of the optimization is than converted into a new lofted wing which is calculated under the previous flow conditions by the three-dimensional Navier-Stokes code. With these results new analog airfoils are obtained and so on. The procedure should converge after some cycles to an improved solution. For the three-dimensional Navier-Stokes calculation the FLOWer code [11, 12] is used, while the two-dimensional calculations are carried out using XLS6 [13], a code based upon full potential theory coupled with a boundary layer calculation.

The geometric distortion for an inboard wing section is shown in Fig. 1, where the airfoil differences between the real three-dimensional geometry (actual wing section) and the analog geometry are obvious. The analog two-dimensional pressure distribution is almost identical to the three-dimensional one with the exception of the trailing edge pressure. It is neither possible nor desirable to match this pressure in the two-dimensional result. This would make the inverse design step more difficult respectively the distortions excessive. Therefore, the remaining pressure differences are compensated directly.

Applications

One example that proves the convergence of this method is a fixed camber A3XX wing design test case. The design has been carried out with 23 design parameters for each airfoil at four design points, Fig. 2. In Fig. 3 the improvement of the total drag coefficient calculated by the three-dimensional Navier Stokes code is plotted in dependence on the number of iterations for this analog design loop, which supplies evidence of the convergence. The pressure distributions at six airfoil sections along the loft are shown for the start solution and the last iteration in Fig. 4 and Fig. 5. One notices that the shock at the outboard sections was reduced significantly resulting in lower wave drag. Another improvement was attained at the crank section, where the minimum pressure on the lower side was increased. This means an advantage with respect to the integration of engines, which cause an additional acceleration of the flow and therefore pressure reduction that would result in a strong shock on the lower side with increasing wave drag. The development of the pressure distribution for the section at 65% wing span during the iteration is shown in Fig. 6. It is visible that the pressure level was reduced significantly on both airfoil sides, resulting in a decreased wave drag.

A comparable design performed with the old manual method can be found in Fig. 7. This reveals an additional drag of fourteen counts. One can estimate the improvement potential at the outer wing pressures whereas the optimized solution is worse in the crank area where for this high Mach number strong three dimensional effects are present. The other two regions with significant three-dimensional influences, the root and tip regions, were omitted for this test case and adopted from an existing design.

Although this looks rather promising, the present method is not yet suitable as an every day production tool. The more sophisticated the design is the more difficulties arise. The transformation of three-dimensional effects into geometry distortions is limited and thus of diminishing usability in the range of highly nonlinear flow effects. As an example the result from a design of a variable camber wing is shown in Fig. 9 and the start solution in Fig 8. Although an improvement was obtained the result is not very satisfying. The convergence of the method was not as good as in the fixed camber case and therefore the drag reduction much smaller. This makes clear that the present analogy method is not sufficient enough.

Improvement potential

Based upon the experience with geometric analogies an alternative approach is conceivable. In contrast to putting the information about the three-dimensionality into the geometry it may be possible to consider them with respect to the flow properties. The differences of the flow variables between three- and two-dimensional results or spanwise gradients could be introduced as some kind of boundary conditions into a two-dimensional calculation. For example, these differences may be used in every grid cell. This seems to be a more physically based approach since flow effects are considered as such. Although this is a more consistent approach and may yield better results, it is not guaranteed to lead to a converging method. It will be necessary to try this by the adaption of an existing code. Even though this approach is of numerical nature and mathematically possibly not satisfying, it constitutes the engineering solution to a present problem.

Conclusion

Alltogether, it is possible to perform a whole wing design with this method in two to three weeks, accelerating the design cycle by a large amount. Soon it will also be possible to use the parallel computing properties of workstations [10] to speed up the design even more. Therefore, in the future more time can be spend on analysing the effects of other parameter variations on the design and a multidisciplinary design will be the ultimate target. An improvement of this method may be achieved in the future by considering three-dimensional effects with respect to the flow properties instead of using geometry distortions.

References

[1] Matsushima, K., Takanashi, S.: *An inverse design method for transonic multiple wing systems on integral equations*, AIAA Paper 96–2465, 14th Applied Aerodynamics Conference, New Orleans, LA, June 17-20, 1996.

[2] Matsushima, K., Takanashi, S.: *Non-planar wing design by Navier-Stokes inverse computation*, AIAA Paper 92-0285, 30th Aerospace Sciences Meeting, Reno, NV, Jan. 6-9, 1992.

[3] Frommann, O.: *Objective Function Construction for Multipoint Optimization Using Fuzzy Logic*, EUROGEN97, Nov. 28–Dec. 5, Trieste, Italy, 1997.

[4] Frommann, O.: *Conflicting Criteria Handling in Multiobjective Optimization Using the Principles of Fuzzy Logic*, AIAA Paper 98–2730, 16th AIAA Applied Aerodynamics Conference, Albuquerque, NM, June 15-18, 1998.

[5] Van der Velden, A. und Forbrich, D.: *Use of Aerodynamic Shape Optimization for a Large Transonic Aircraft*, 15th AIAA Applied Aerodynamics Conference, Atlanta, June 23-25, 1997.

[6] Van der Velden, A., Frommann, O.: *Use of Aerodynamic Shape optimization for the A3XX*, DASA Airbus interner Bericht, 1998.

[7] Van der Velden, A.: *Aerodynamic Shape Optimization*, AGARD-R-803 AGARD-FDP-VKI Special Course, April 1994.

[8] Synaps, Inc.: *"PointerPro^{TM} 4.2"*, www.synaps-ing.de

[9] Forbrich, D.: *Aerodynamische Flügeloptimierung für ein großes transsonisches Verkehrsflugzeug*, Jahresbericht STAB-Projektgruppe „Flügel großer Streckung", 1997.

[10] Axmann, J.K., Hadenfeld, M., Frommann, O.: *Parallel Numerical Airplane Wing Design*, New Results in Numerical and Experimental Fluid Mechanics, Contributions to the 10th AG STAB/DGLR Symposium, Braunschweig, Germany, Vieweg-Verlag, 1997.

[11] Becker, K., Kroll, N., Rossow, C.C., Thiele, F.: *Numerical Flow Calculation for Complete Aircraft - The MEGAFLOW Project*, Proceedings of the DGLR Symposium, Bremen, Sept. 1998.

[12] Becker, K., Kroll, N., Rossow, C.C., Thiele, F.: *MEGAFLOW: Collaborative Development of a Numerical Flow Simulation Package*, 1st CME Congress, Bremen, June 1998.

[13] Dargel, G., Thiede, P.: *Viscous Transonic Airfoil Flow Simulation by an Efficient Viscous-Inviscid Interaction Method*, AIAA Paper 87-0412, Jan. 1987.

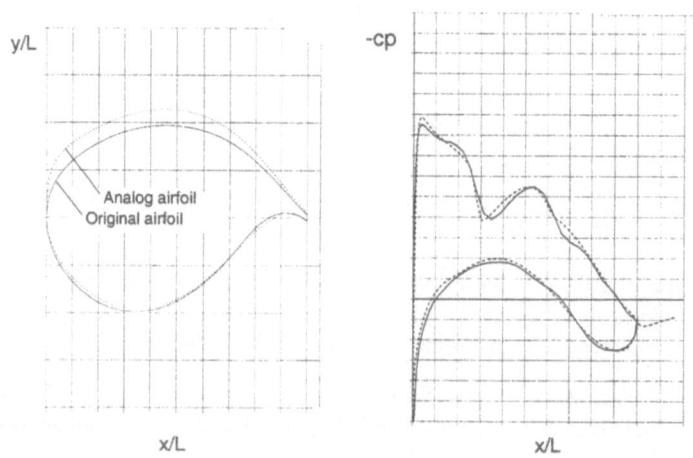

Fig. 1: Original and distorted airfoil geometry, three-dimensional and analog pressure distributions

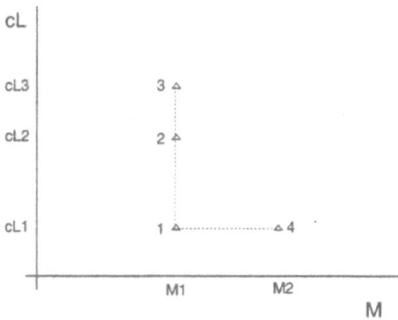

Fig. 2: Design points

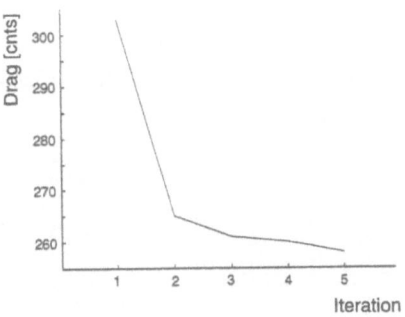

Fig. 3: Drag reduction in dependence on iteration step

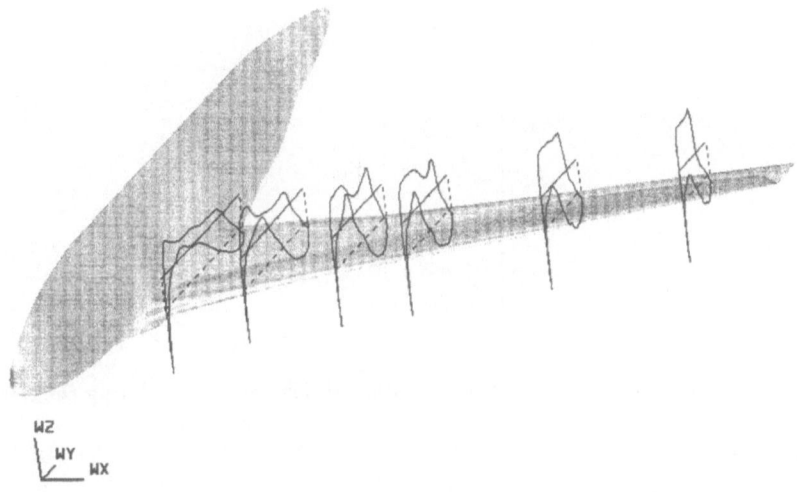

Fig. 4: Pressure distributions of start solution (Fixed camber A3XX-wing)

175

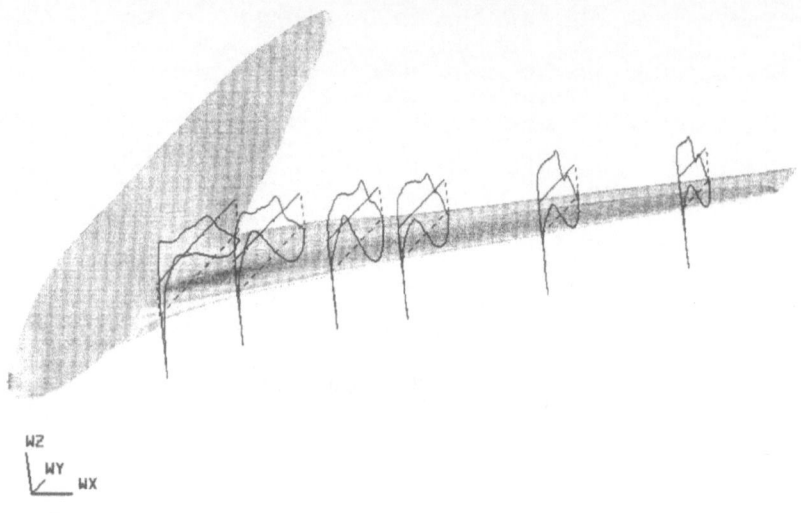

WZ
WY
WX

Fig. 5: Pressure distributions of last solution (Fixed camber A3XX-wing)

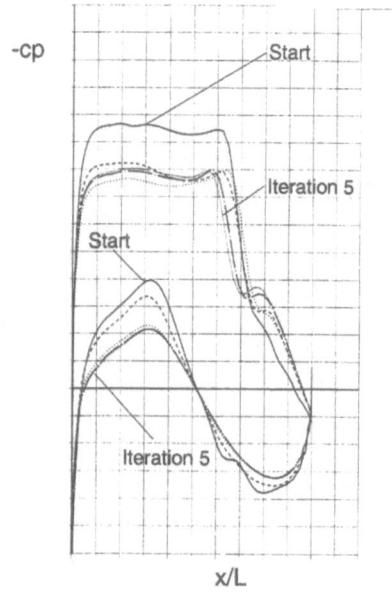

Fig. 6: Development of pressure distribution at 65% wing span during the iteration

Fig. 7: Pressure distributions of manually designed wing

Fig. 8: Pressure distributions of start solution (Variable camber A3XX-wing)

Fig. 9: Pressure distributions of last solution (Variable camber A3XX-wing)

Efficient Computation of 3D-Flows for Complex Configurations with the DLR-TAU Code Using Automatic Adaptation

T. GERHOLD, J. EVANS

Deutsches Zentrum für Luft- und Raumfahrt e.V.
Institut für Strömungsmechanik
Bunsenstraße 10, D-37073 Göttingen, Germany
Fax: +49 551 709 2416
email: Thomas.Gerhold@dlr.de, John.Evans@dlr.de

Summary

This paper deals with local grid adaptation in an unstructured method and its capability of increasing the accuracy of the predictions. The paper briefly presents an extension of the standard h-refinement cases that reduces the number of inserted grid points needed for the conservation of the grid-conformity, which is of importance especially for large scale grids. With the DLR-TAU code and the use of fast grid generators for the hybrid grid technique it is possible to obtain CFD-results in only days after starting from CAD-data. This has been achieved for the complete aircraft configuration DLR-F6 of which the inviscid and turbulent results are presented and compared with experimental and structured grid results.

Introduction

In industrial environments the prediction of flows becomes more and more of importance. At the same time the applications become more complex, because often complete configurations are of interest. Thus, also with increasing computer power the overall turnaround times remain large, especially when a lot of interactive work is needed. This is true when using e.g. established multi-block methods, because most of the time needed is spent on the grid generation process. Unstructured/hybrid-grid methods appear to be a remedy, because they allow a much higher degree of automation. However, the more requirements are put on the initial grids the more work has to be spent to fullfill them by paying attention to e.g. local spacings, parameters or source terms needed.

Thus, our strategy is to accept any given grid, that is sufficiently refined in regard of surface curvature (which can be automated) and which has a certain number of structured sublayers, which cover the boundary layer over surfaces where viscous flow is expected. In order to get a sufficient accuracy the grid can be adapted during the calculation depending on the local flow phenomena, like shocks, wakes, shear layers or boundary-layer spacings. This also allows to use one initial grid for different flow conditions.

Even if this strategy may be accepted for Euler flows, nowadays, it still has to be proven that the final results for complex 3D turbulent computations are comparable to structured grid results, especially in terms of accuracy. In this respect, the current status of the DLR-TAU code is reflected. We focus on the adaptation and the demonstration of the system applied to a complete aircraft configuration and the comparison to structured grid results.

The DLR-TAU Code

The flow solver is based on a three-dimensional finite volume scheme for solving the Reynolds-averaged Navier-Stokes equations. The dual-grid technique is employed, which, together with the edge-based data structure, allows to run the code on any type of cells. The edge-based data is provided by a preprocessing module, which currently supports hexahedral, prismatic, pyramidal and tetrahedral elements. The flow variables are stored in the centers of the dual grid, i.e. the vertices of the primary grid. The temporal gradients are discretized using a multi-step Runge-Kutta scheme. The inviscid fluxes are calculated either by a Roe- or AUSM-type 2nd-order upwind scheme, or by employing a central method with scalar dissipation. The gradients of the flow variables are determined by employing a Green-Gauß formula. The viscous fluxes are discretized using central differences. For time-accurate calculations the wellknown dual-time stepping approach for 1st-, 2nd- or 3rd-order discretization in physical time is implemented.

The turbulence models available in the code are the one-equation Spallart-Allmaras model, the Wilcox $k\omega$-model and the SST-model according to Menter. Transition can be fixed by assigning a flag (turbulent or laminar) to each surface-grid triangle or quadrilateral. On grid points closer to a laminar than to a turbulent surface, the production of turbulence is limited to be smaller or equal the local destruction rate.

In order to accelerate the convergence to steady state, local-time stepping, residual smoothing and a multigrid technique based on agglomeration of the dual-grid volumes are employed. The agglomeration of the dual cells is done with special attention to cells in structured parts of the grid (e.g. prismatic layers) in order to retain the structure also on the coarse-grid levels. The grids on different levels have identical properties, which allows to employ the same discretization algorithms. However, on coarse grids modified dissipation or 1st-order operators are used for central discretization or upwinding, respectively. Optimization for different architectures is achieved by vector- or cache-type coloring of the edges, on which most of the work is done. For parallel computations a domain-decomposition is employed providing a subset of dual-grids. MPI is used as message passing interface.

The Adaptation Method

The adaptation module is based on the bisection of edges in the primary grid. That means a new point can only occur as the midpoint of two connected points (p_i, p_j). Let E be the set of all Edges of the primary grid. The set of edges to be bisected is defined by $E_{bis} = \{(p_i, p_j) | (p_i, p_j) \in E, \phi(x_{p_i}, x_{p_j}) < c\}$, where $\phi : I\!R^3 \times I\!R^3 \to I\!R_+$ is an indicator function and $c \in I\!R_+$ a threshold value depending on the wanted number of new points. The choice of the function ϕ is of course essential for the efficiency of the adaptation itself. Several indicator functions are implemented. The one used for the calculations shown is based on differences between certain flow values on the corresponding points:

$$\phi(x, y) := \max_{k=1,\ldots,N} \{C_k \frac{|f_k(x) - f_k(y)|}{f_{k,\infty}}\} \cdot |x - y|^\alpha,$$

where $f(x) = (\rho(x), |\vec{v}(x)|, P_t(x), H_t(x))^T$ is the vector of different values, C_k, α are user defined values and $f_{k,\infty} := \max_{(p_i, p_j) \in E}\{|f_k(x_{p_i}) - f_k(x_{p_j})|\}$ are the overall componentwise maxima.

Since we require a conforming mesh (i.e. no hanging nodes), more work than simply inserting new points has to be done. We have to connect new points, such that the edges build elements the preprocessing module can handle. If this is not possible, due to a limited number of permitted element subdivisions, additional edges have to be bisected until the resulting mesh is conforming. On one hand this implies a certain amount of overhead, on the other hand the more additional points are used the smoother the resulting grid will be. Moreover, since usually more than one adaptation step is desirable, the further adaptation steps have to be done with care in order to avoid highly stretched elements. One way to do so is to store the performed changes of elements and to allow only certain combinations on one particular element (e.g. if a tetrahedron is split into two ones, in a next adaptation step none of these is permitted to be split in the same way again). In the adaptation module several loops over all elements are performed until no additional point is generated. This guarantees a conforming mesh.

The memory and time requirements of the whole process depend on the used data-structures. We store the edges of the grid in an array, which contains the two point numbers, a flag which indicates if the edge is allowed to be bisected (which it might not be due to a former refinement on an adjacent element) and a flag in which the number of the new point is stored. Moreover, for each element the number of the edges, that build the element, and a flag with the type of refinement are stored. The possible refinement cases depend on the element type (tetrahedra, prisms, hexahedra, pyramids).

In literature (e.g. [3]) four different refinement cases on tetrahedra are in use. Any occuring case, that does not match with one of these, implies additional points. Therefore we extended the number of cases, such that for any combination of marked edges no additional points are necessary when dealing with initial or isotropically refined elements. This leads to ten different cases on tetrahedra (figure 1). On prismatic elements, which are used to cover the boundary layer because of their semi-structured nature, we do not consider the wall-normal edges. Otherwise a new point on a wall-normal edge would result in a whole new layer, if we conserve the element type, which is desirable in the boundary layer, because we want to maintain the wall-normal rays. That means prisms are basically treated like triangles. Nevertheless, if we think of piles of prisms, a new point on one edge of a prism does mean additional points distributed along the whole corresponding pile (figure 1). Another aspect of prism layers is the adaptation of $y+$-values. This is achieved by redistribution of points on wall-normal rays.

Since we usually deal with curved surfaces, the curvature has to be taken into account when adding new points onto such a surface. This is done by computing the surface-normal for each surfacepoint thereby considering discontinuities in the first derivative and then calculating the corresponding Bézier-Spline on which the new point is positioned. After the calculation of the coordinates of the new points the solution values are interpolated linearly onto these points, which saves a lot of computational time for the restart process.

Test Case: DLR-F6 Configuration

For the DLR-F6 wing-body-pylon-nacelle configuration pressure distributions from wind-tunnel measurements in 8 sections and reference data from structured grid computations with the MEGAFLOW system are available. For more detailed information see [2], [4]. The flow conditions are $M_\infty = 0.75$, $Re = 3 \cdot 10^6$ at an angle of attack of $\alpha = 0.98°$.

First, the inviscid computation with the TAU code is presented, which has been performed on a tetrahedral grid. The initial grid consists of 402804 points, the three times adapted grid consists of 852089 points (30% new points inserted by adaptation each 200 multigrid cycles). The left part of figure 3 indicates the efficiency of the adaptation defined by the ratio $\#E_{bis}/(\#E_{bis} + \#E_{add})$, where $\#E_{add}$ is the number of additional points needed in order to get a conforming grid. The increase of efficiency due to our extended method becomes obvious. The right part of figure 3 depicts the convergence history in terms of the residual and the lift-coefficient during three adaptation steps. Figure 4 shows the increase of accuracy in section 4 due to the adaptation in terms of reduction of pressure loss and sharpening of the shock-resolution. The Euler solution does of course not converge to the viscous experimental data. In figure 2 both surface grids (initial and 3 times refined) are shown.

The simulation of the turbulent flow is presented next. The structured grid results were obtained with the FLOWer code (see [4]) employing a central discretization and the Baldwin-Lomax as well as the Wilcox $k\omega$-model using grids with 3.3M and 3.5M grid points, respectively. The TAU code used the central scheme and the Spalart-Allmaras turbulence model. The initial grid is composed of a prismatic sublayer, tetrahedra in the outer region and partly pyramidal elements between both. It consists of 2M grid points. The computation was started fully turbulent for the first 1000 multigrid cycles (4 level-W). The following 500 cycles were computed with fixed transition. The comparison (not shown here) indicates that nearly no effect on the pressure distributions can be observed, only the Cf-distribution is influenced by the different transition management. The defined transition lines are similar to the ones in the reference computations, except close to the pylon, where we chose different settings. Because the true transition line is not known, anyway, we defined the pylon surface to be turbulent, whereas the wing surface on the lower side near the leading edge is defined to be laminar. Consequently, in the junction of wing and pylon laminar and turbulent surfaces are connected. In the reference computation the lower wing surface near the pylon has also been set turbulent. This computation was continued by two adaptation cycles in step 1500 and 2081, thereby increasing the size of the grid by 40 and 30 percent, respectively. The final grid consists of 3.6M points. These steps become visible in figure 5 indicating the convergence history in terms of the density residual and the aerodynamic coefficients. The latter match well with the reference data, which is, as clearly visible, due to the help of the adaptation.

Figures 6 and 7 show a cut through the initial grid and the adapted grid, respectively, together with the isolines of the x-component of velocity (first line at 0., increment 0.05). It can bee seen that not only the shock region above the upper side of the wing is refined, but also the wakes of the wing and the nacelle are adapted. Figure 8 shows the distribution of pressure and skin friction in section 2, 3, 4, and 6. The Cp-distributions indicate the improved accuracy of the adapted results. Shocks are resolved sufficiently and e.g. the whiggle, which can clearly be seen in the initial solution near the trailing edge is reduced. To our experience it is mainly caused by the O-type topology of the prismatic layer. Using different types of grids this disturbance disappears (compare e.g. with results of [5]). The Cf-plots also show that the adapted solution is in quite good agreement with the reference data, except near the transition location and on the lower part of the wing near the junction of the pylon and the wing (section 3 and 4). The latter observation is caused by the different transition managements, which is indicated by comparison to the fully turbulent computation (not shown here). The slower increase of Cf at the transition line, than observed with the other turbulence models, is a wellknown behaviour of the Spalart-Allmaras model (see also [5]). The prediction of the SA-model is closer to the BL-solution of the structured grid computation, which seems to be more realistic than the $k\omega$-result in this case.

Finally, we want to mention the resources needed for the turbulent simulation. The hybrid grid generation needed less than two days starting from CAD data. The flow solver needed for the whole run on the 3 different grids 16 hours wallclock time on 4 processors NEC-SX4 (memory: 625MB per 1M points). The adaptation and the preprocessing were run on a workstation in less than 4 hours altogether.

Conclusion

The results show that the adaptation used in the DLR-TAU code is efficient and important for the accuracy of a CFD simulation. Starting from a given grid, the solution is driven against the reference data in a few adaptation cycles. This automation prevents the user to put a lot of knowledge, i.e. a lot of effort and time, in the set up of the initial grid. The final hybrid-grid results obtained are in terms of Cp and Cf in quite a good agreement with experimental and structured-grid data. The turbulent results for the complete aircraft configuration have been produced in less than 7 working days including the grid generation.

Acknowledgements

The DLR-TAU code is partly developed in the frame of the MEGAFLOW project [2].

To some extend the adaptation module is developed within the frame of AVTAC. The AVTAC project (Advanced Viscous Flow Tools for Complete Civil Transport Aircraft Design) is a collaboration between British Aerospace, DASA, CASA, Dassault Aviation, SAAB, Alenia, DLR, ONERA, CIRA, FFA and NLR. The project is managed by British Aerospace and is funded by the CEC under the IMT initiative (Project Ref: BRPR CT97-0555).

The initial hybrid grid for the F6-configuration was made available by CentaurSoft for the demonstration of the gridgeneration system. We would like to acknowledge the immediate commitment and the rapid success (2 days from CAD data-ftp to a ftp retransfer of the grid).

References

[1] Gerhold, T.; Galle, M.; Friedrich, O.; Evans, J.; 1997: "Calculation of Complex Three-Dimensional Configurations Employing the DLR-TAU-Code", AIAA paper 97-0167.

[2] Kroll, N.; Rossow, C.C.; Becker, K.; Thiele, F.: "MEGAFLOW - A Numerical Flow Simulation System", ICAS 1998, Melbourne, Australia.

[3] Mavriplis, D.J.: "Adaptative Meshing Techniques for Viscous Flow Calculations on Mixed Element Unstructured Meshes", ICASE Report No. 97-20.

[4] Ronzheimer, A.; Brodersen, O.; Monsen, E.; Rudnik, R.; Rossow, C.C.: "Numerical Calculations of Aerodynamic Coefficients for the DLR-F6 Configuration with MEGAFLOW", To be presented at the DGLR-Fach-Symposium, Strömungen mit Ablösung, Berlin, Nov, 10-12, 1998.

[5] Schwamborn, D.; Gerhold, T.; Hannemann, V.: "On the Validation of the DLR-TAU Code", To be presented at the DGLR-Fach-Symposium, Strömungen mit Ablösung, Berlin, Nov, 10-12, 1998.

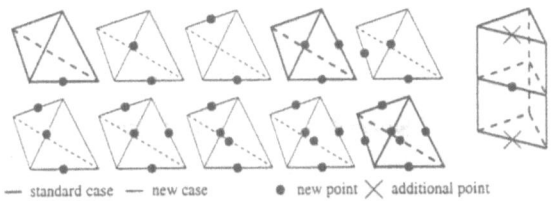

— standard case — new case ● new point ╳ additional point

Figure 1: Permitted refinement cases on tetrahedra (left) marking of one prism pile (right)

Figure 2: Initial (left half) and final (right half) surface grid

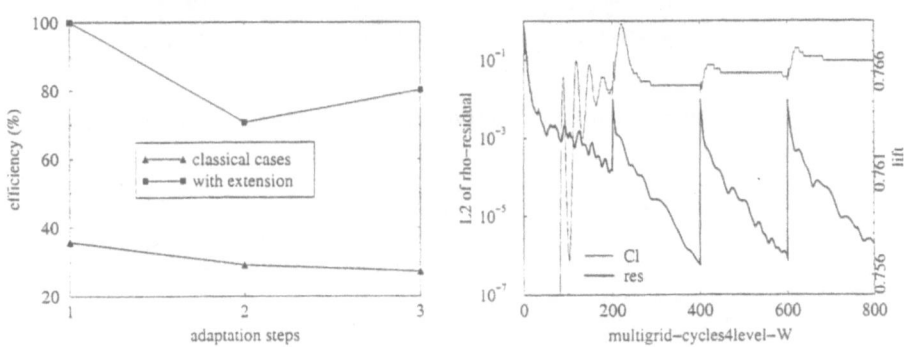

Figure 3: Efficiency of the adaptation method (left) and convergence history (right)

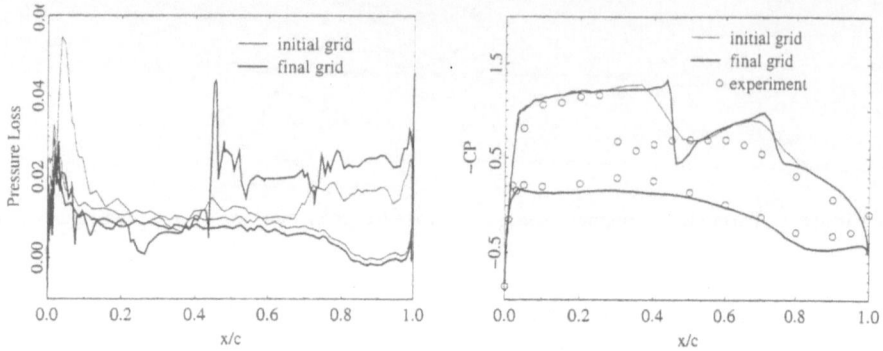

Figure 4: Pressure loss (left) and pressure distribution (right) in section 4 ($\eta = 0.377$) (Euler)

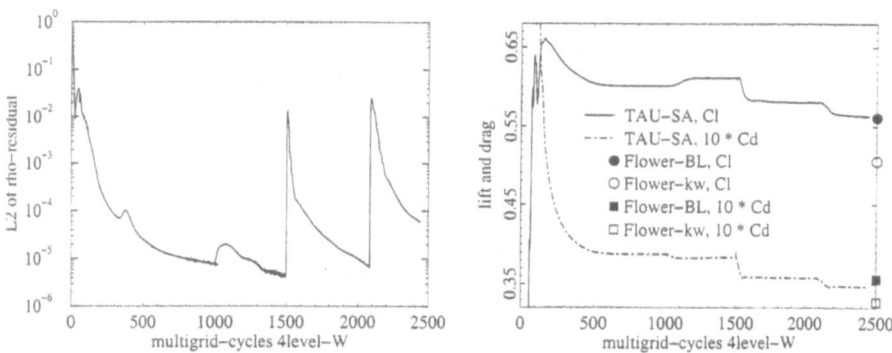

Figure 5: Convergence history of the complete adaptation cycle (turbulent)

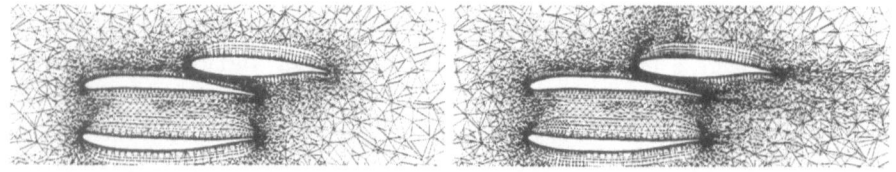

Figure 6: Cut through grid in section 3, initial (left) and final grid (right)

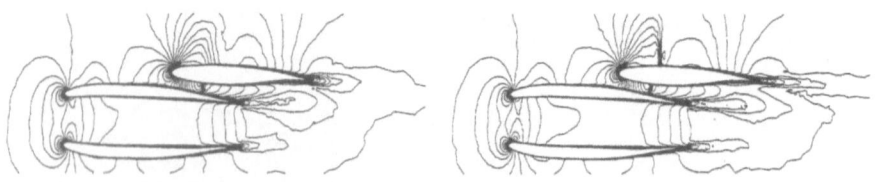

Figure 7: Isolines of x-component of velocity in section 3, initial (left) final grid (right)

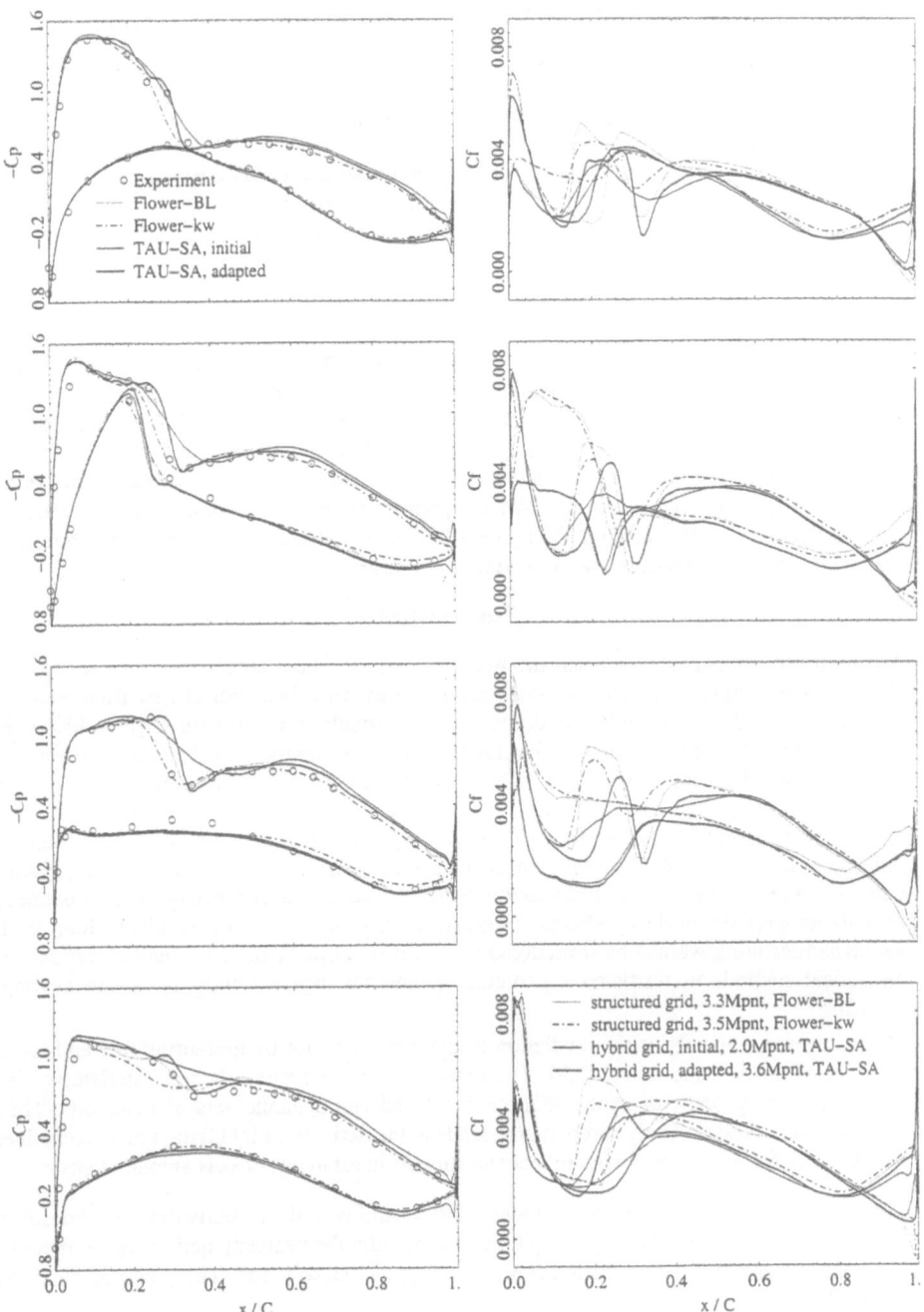

Figure 8: Cp- (left) and Cf-coefficient (right) chordwise in section 2, 3, 4 and 6 (1st, 2nd, 3rd and 4th row corresponding to $\eta = 0.238, 0.331, 0.377, 0.512$, the nacelle axis is at $\eta = 0.35$)

Numerical Prediction of Dynamic Derivatives for Lifting Bodies with a Navier-Stokes Solver

P. Giese, R. Heinrich, R. Radespiel

Deutsches Zentrum für Luft- und Raumfahrt (DLR) e.V.
Institut für Entwurfsaerodynamik
Lilienthalplatz 7, D-38108 Braunschweig, Germany

Summary

Standard technique for the prediction of dynamic derivatives is the use of wind tunnel simulations with expensive technical equipment and complex data processing. Numerical flow computations, on the other hand, can simulate arbitrary model motion and hence, all desired derivatives are directly accessible. The prediction of dynamic derivatives with the help of an efficient Navier-Stokes solver is investigated in the present work. The assessment is performed for a lifting body that represents the X-24A experimental vehicle. Comparisons with wind tunnel and flight test data indicate good agreement. We can also demonstrate that the numerical approach requires reasonably low computation resources.

Introduction

The high cost of flight testing and the need to guarantee high safety measures for manned flight calls for rather small uncertainty levels on the dynamic behavior of new flight vehicles. An accurate prediction of unsteady derivatives is particularly needed for flight vehicles that exhibit an unstable behavior in their longitudinal or lateral motion. The challenge is even more severe for vehicles with a large flight trajectory that includes strongly varying flight conditions [1]. Some existing prediction methods are based on semiempirical formulas for the individual aircraft components. This is the DATCOM-type approach. The answers of these methods are rather rough and they often fail to display the critical behavior of a particular configuration. Other methods are based upon linearized potential equation and they use oscillating boundary formulations for the unsteady effects. These methods provide accurate results as long as the aerodynamics are governed by irrotational flow and small perturbation. Many of the existing theoretical methods are restricted to particular geometries such as slender bodies, delta wings, etc. [2].

Most common, however, is the prediction of dynamic behavior by measuring free and forced oscillations in the wind tunnel [2,3]. The extraction of the desired dynamic derivatives is a rather complex process of applying mechanical and aerodynamic sets of equations. These relate the measured unsteady aerodynamic loads to the derivatives by taking into account inertia, elastic deformations and wind tunnel and support interference effects among others.

The objective of the present work is to assess the capability of the recently developed unsteady Navier-Stokes solver FLOWer [4] to predict numerically the unsteady derivatives in presence of strongly nonlinear flow effects such as shocks and separations. The advantages of this approach are:

- simulation of arbitrary vehicle motion
- direct determination of dynamic derivatives
- simulation of flight conditions in terms of Mach and Reynolds numbers.

As a model we choose the shape of the experimental hypersonic flight vehicle X-24A for which wind tunnel data and flight data are available. Dynamic derivatives are of interest here as this vehicle is famous for its special "riding qualities" due to the unusual distribution of lateral stability parameters and control efficiency. At transonic flow conditions shocks appear on the lower and upper surface of this vehicle. Flow separations take place along the upper surface and on the winglet. The present work contributes to verification and validation of the numerical method for this complex flow situation.

Numerical Approach

Following Jameson [8], the DLR multi-block code FLOWer [4] has been extended to the solution of unsteady viscous flows. Because the FLOWER code employs a cell-vertex scheme with fluxes obtained by central averaging, artificial dissipation is added in order to avoid spurious oscillations in the iterative solution. The dissipation is scaled by the adaptive coefficients $k^{(2)}$ and $k^{(4)}$ for second and fourth order dissipative terms. Additionally rigid body motions with three translational and rotational degrees of freedom have been implemented. The so called Reynolds-averaged Navier-Stokes equations are transformed to the moving coordinate system with time-averaged turbulent quantities. The integral form of the three-dimensional Reynolds-averaged Navier-Stokes equations in a moving Cartesian coordinate system can be written as

$$\frac{d}{dt}\int_V \vec{W}dV + \int_{\partial V} \bar{\bar{F}} \cdot \vec{n}dS + \int_V \vec{G}dV = 0,$$

with $\vec{W}$ the vector of conserved quantities in the moving coordinate system . V denotes an arbitrary control volume with boundary ∂V and the outer normal $\vec{n}$. The source term $\vec{G}$ contains the time derivatives of the cartesian base vectors of the moving coordinate system. In the convective flux tensor $\bar{\bar{F}}^c$ the velocities of the control volume surfaces generated through moving coordinate system are taken into account. The spatial discretization leads to an ordinary differential equation for the rate of change of the conservative flow variables in each grid point

$$\frac{d}{dt}(\vec{W}_{ijk}V_{ijk}) + \vec{R}(\vec{W}_{ijk}) = 0$$

with the total residual $\vec{R}_{ijk}$ for the approximation of the sum of inviscid and viscous net fluxes and the source term for a particular control volume arrangement with volume V_{ijk} surrounding the grid node (i,j,k). The time derivative is discretized using a backward difference operator of second order accuracy which represents a numerical time-step-solution. The implicit scheme

$$\frac{3W^{n+1}V^{n+1}}{2\Delta t} - \frac{2W^nV^n}{\Delta t} + \frac{W^{n-1}V^{n-1}}{2\Delta t} + R(W^{n+1}) = 0$$

is solved iteratively using the Dual Time Stepping Method of Jameson [8].

The grid generation package MEGACADS [4] is used to create the coordinate grids around the X-24A lifting body. The grid system employed consists of 4 blocks. The main part of the body is covered by an O-type topology in the circumferential direction. Additional 3 blocks are used to cover the nose cap and the rear base region. Figure 1 shows that surface grid points are clustered along the nose and the leading edge of the winglet. The 3-D field mesh contains grid clustering near the surface. The distance of the first grid point away from the wall is 10^{-5} times the body length in order to resolve viscous boundary layer at Reynolds numbers of about 10×10^6. The finest mesh under consideration contains 835584 cells for the half model. Additionally, a medium mesh with 104448 cells and a coarse mesh with 13056 cells are generated by omitting every second grid point in each coordinate direction.

Our numerical simulation approach for the dynamic derivatives resembles existing wind tunnel procedures in that we use forced harmonic motion of the model. However, the present longitudinal and lateral motions distinguish between pure rotation that induces additional asymmetric velocities and pure change of the angle of incidence with symmetric velocities.

1. Determination of the damping derivatives $c_{mq} + c_{m\dot\alpha}$

Pure harmonic rotation is applied. The reduced frequency, k, is estimated from the undamped natural longitudinal frequency,

$$k = \frac{\omega \cdot l_{ref}}{2V_\infty} \quad , \text{ with } \omega = \sqrt{-(\partial M/\partial\alpha)/\Theta}$$

where Θ is the inertia moment of the original vehicle and the pitching moment is taken from flight data. The flow solver yields the unsteady distribution of aerodynamic coefficients by using the dual time stepping scheme [4,8]. A typical residual convergence of the inner loop used to converge each physical time step is displayed in Figure 2. It is seen that about 20-40 multigrid iterations are needed to converge each time step during the harmonic oscillation.

Figure 3 indicates that the initialization phase of the flow takes about one quarter of a complete oscillation period. By using the time history, $c_m(t)$, one can define a linearized approach [2,6]

$$c_m(t) = c_m(\alpha_o) + c_{m_\alpha}\Delta\alpha(t) + (c_{m_q}q(t) + c_{m\dot\alpha}\Delta\dot\alpha(t)) \cdot l_{ref}/2V_\infty \ .$$

The pitching motion is

$$\Delta\alpha(t) = \Delta\alpha_A \sin\omega t \ ,$$

that is equal to $\Delta\theta(t)$ if the rotation axis passes throughth the moment reference point. Because of $q(t) = \Delta\dot\theta(t) = \Delta\dot\alpha(t)$ one can display and analyse time history

$$c_m(t) = a_o + a_1\sin(\omega t) + a_2\cos(\omega t)$$

such that a_2 determines the sum of pitch and incidence damping coefficients.

$$c_{mq} + c_{m\dot\alpha} = a_2/(k\Delta\alpha_A) \ ,$$

where a_2 is evaluated by a numerical curve fitting process.

2. Pure harmonic incidence oscillation

The incidence damping, $c_{m\dot\alpha}$, is determined in an analogous manner. Here, the pure oscillation of incidence is kinematically simulated with an oscillating vertical hub velocity that is normal to the average angle of attack, α_o:

$$\Delta\dot y_{Hub}(t) = -\Delta\alpha_A\sin(\omega t)V_\infty\cos\alpha_o \ .$$

The hub motion is then

$$\Delta y_{Hub}(t) = (\Delta\alpha_A l_{ref}/2k)\cos\alpha_o\cos(\omega t) \ .$$

In general the hub motion yields a non-constant freestream velocity. A good alternative is to capture the unsymmetric x-velocity differences for $\pm\Delta\alpha$ wih a corrective hub velocity in x-direction $\Delta x_{Hub}(t) = -\Delta\dot y_{Hub}(t)\sin\alpha_o$. The small differences to the constant freestream velocity are corrected for the time dependent force and moment coefficients c(t) by

$$c_{corr}(t) = c(t)/(V_{eff}^2/V_\infty^2) \ ,$$

where V_{eff} is the time dependent effective freestream velocity.

The determination of the lateral dynamic derivatives is performed in an analogous way. The numerical effort needed for the lateral flow simulation doubles because the full model must be represented in lateral motion. Normaly the direct determination of c_{mq} inquires the simulation of combined incidence- and vertical hub oscillations (see Figure 4). For subsonic speeds pure pitch damping, c_{mq} can also be determined from the difference of the sum of pitch and incidence damping and the pure incidence damping coefficient.

Results

The computations for the X-24A configuration, which is an earlier configuration of the future X-38 vehicle, are presented in several steps. At first code verification results are presented for symmetric longitudinal motion. Code validation is then possible by using comparisons with wind tunnel and flight data [7]. Then the derivatives due to yaw and roll motion are analaysed for the Mach range 0.5 - 1.1 with a mean Reynold number of 6×10^7. Fully turbulent flow was assumed because measurements in wind tunnels are made with tripping strips around the nose of the fuselage and along the leding edges of the winglets. The results are presented by using the body-fixed coordinates.

Symmetric derivatives

For forced incidence oscillations the effects of varying grid density and numerical dissipation coefficients have been carefully investigated. As a test case we choose transonic flow conditions at 15° angle of attack that matches one trajectory point of the future X-38 vehicle. Figure 5 confirms that the effect of doubling the grid density on the resulting symmetric derivative $c_{mq} + c_{m\dot\alpha}$ is reduced for the finer mesh. Also the effect of changing the adaptive coefficients $k^{(2)}$ and $k^{(4)}$ is reasonably small. Additionally, the effect of changing the reduced frequency, k, from 0.05 to 0.1 on the damping coefficient is rather small and this justifies the approach used here to determine the derivatives. Computations with the k-ω turbulence model instead of Baldwin-Lomax on the medium mesh do not show any major effect on the damping coefficient.

Figure 6 displays the comparison of the damping derivatives with the wind tunnel and flight data [7]. The medium mesh has been used to determine the Mach number effect. The typical damping characteristics that are experimentally observed for subsonic and supersonic Mach numbers are well reproduced by the numerical computations. We find that the fine mesh result for $c_{mq} + c_{m\dot\alpha}$ is about 10 per cent below the wind tunnel data. Our numerical computations show that c_{mq} alone is about 60 per cent of the sum $c_{mq} + c_{m\dot\alpha}$ (at least for $M_\infty = 0.95$).

Asymmetrical derivatives

Figure 7 displays the yaw damping coefficients $c_{nr} - c_{n\dot\beta} \cos\alpha_0$ as function of the Mach number. The computation of pure $c_{n\dot\beta} \cos\alpha_0$ due to pure yaw angle is simulated by oscillation of horizontal hub. Good agreement can be demonstrated for pure derivative c_{nr}, which is simulated with combined yaw angle and horizontal hub oscillation.

Some comparisons for the pure roll damping c_{lp} are presented in Figure 9. The pure c_{lp}-values are determined from the results $c_{lp} + c_{l\dot\beta} \sin\alpha_0$ of roll motion and corrections from the pure $c_{l\dot\beta}$-values of the produced yaw motion. There is sufficient agreement for the transonic range. Note the large scatter of experimental data. Only for the Mach number 0.5 we predict a

189

larger roll damping than experimentally measured. The first cross derivatives c_{np} and c_{lr} are shown in Figure 11. The damping hysteresis of the time history is so small (see Figure 8 and 10) that numerical scattering may be one of the reasons for the unsufficient agreement for the c_{lr}-damping derivative. Note that all asymmetric damping derivatives are referenced on $(p, r)l_{ref}/V$ based on coefficients with reference lenght l_{ref}.

Concluding remarks

Successful verification and validation for predictions of unsteady derivatives with the NS-solver FLOWer has been demonstrated. Arbitrary (combined) motions have been performed which are not feasible in wind tunnels. Reasonable computing times on a IBM Workstation RS/6000 have been achieved. These are about 5 cpu-h for symmetrical and 15 cpu-h for asymmetrical calculations on medium mesh. Therefore the code and the related knowhow can be used to create aerodynamic data bases. The grid generation takes about 3-4 weeks on the base of a complete and gapless iges data file of the model surface.

References

[1] Kirsten, P. W.:
A Comparison and Evaluation of Two Methods of Extracting StabilityDerivatives from Flight Test Data. AGARD CP 172, 1992, pp. 1-18 bis 18-2.

[2] Fuchs, H.:
Prediction of Dynamic Derivatives. AGARD CP 451, 1979, pp. 6-1 bis 6-15.

[3] Determann, O.; Heege, K.:
Erweiterung der Ausrüstung und Windkanalerprobung der Mobilen Oszillierenden Derivativwaage MOD.
BMFT LFW 7701 I/3 Teil I und II 1079, Institut für Flugmechanik, TH- Darmstadt.

[4] Heinrich, R.; Pahlke, K.; Bleecke, H.:
A Three-Dimensional Dual-Time Stepping Method For The Solution Of The Unsteady Navier-Stokes Equations.
Two Day Conference 17-18 July 1996, London ; UK The Royal Aeronautical Society DLR-Institut for Design Aerodynamics, D-38108 Braunschweig, Germany.

[5] Brodersen, O.; Hepperle, M., Ronzheimer, A.:
The Parametric Grid Generation System MegaCads.
Proceedings of the 5th International Conference on Numerical Grid Generation in Computational Field Simulations, Ed B. K. Soni, J. F. Thomson, P. Eiseman. NSF Missisippi State, 1996, pp. 353-362.

[6] V. d. Decken, J.; Schmidt, E.; Schulze, B.:
On The Test Procedure Of The Derivatives Balance Used in West-Germany.
AGARD CP 235, 1978, pp. 5-1 bis 6-17.

[7] Kirsten, P.W.:
Wind Tunnel and Flight Test Stability and Control Derivatives for the X-24 Lifting Body.
Air Force Flight Test Center Technology Document No. 71-7.

[8] Jameson, A. J.:
Time dependent calculation using multigrid, with applications to unsteady flows past airfoils and wings. AIAA-Paper 91-1596.1991.

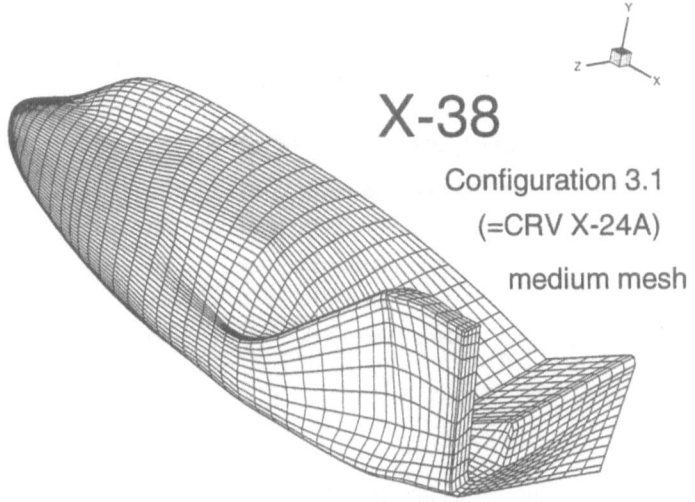

X-38

Configuration 3.1

(=CRV X-24A)

medium mesh

Fig. 1 : surfgrid of the investigated Configuration

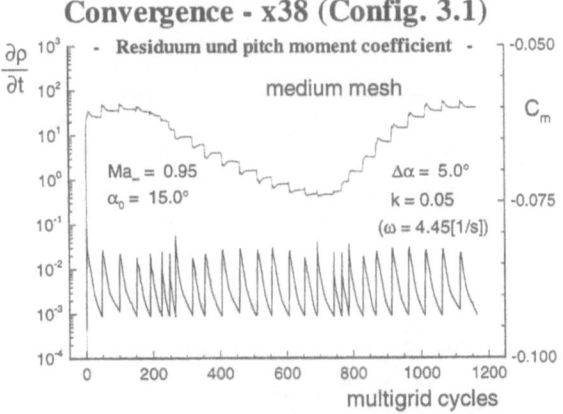

Convergence - x38 (Config. 3.1)

Fig 2 : Typical convergence hystory of a complete oscillation period

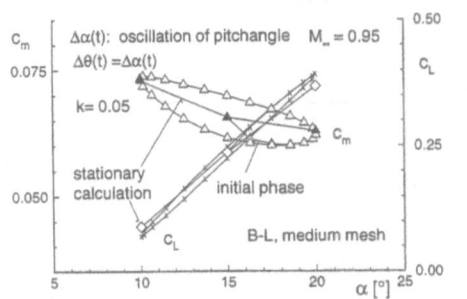

Fig . 3 : Lift- and pitching moment characteristics for harmonic α–oscillation

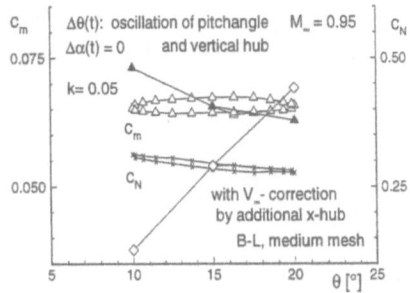

Fig. 4 :Normal force- and pitching moment
characteristics for pure θ–oscillation

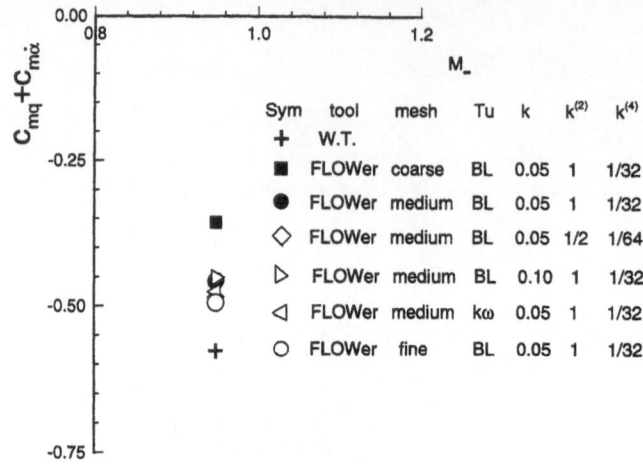

Fig. 5 : Numerical influences on longitutinal motion
(Example $c_{mq} + c_{m\dot{\alpha}}$)

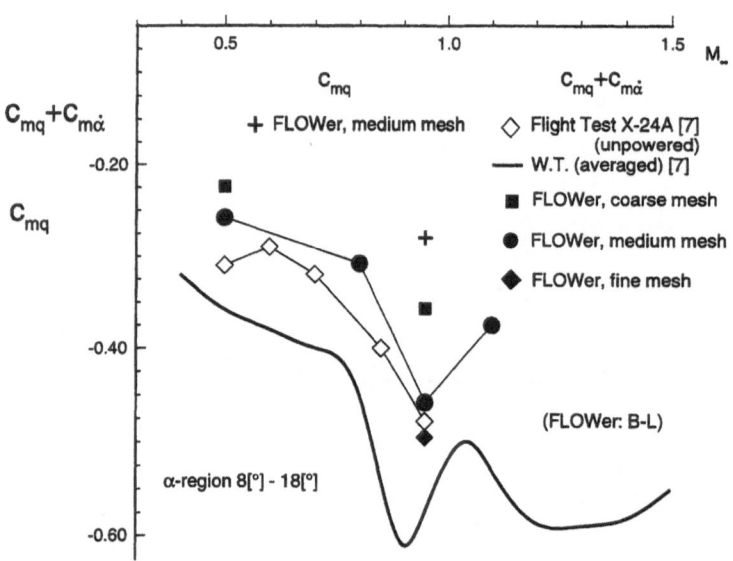

Fig. 6 : Comparisons of the longitudinal motion derivatives

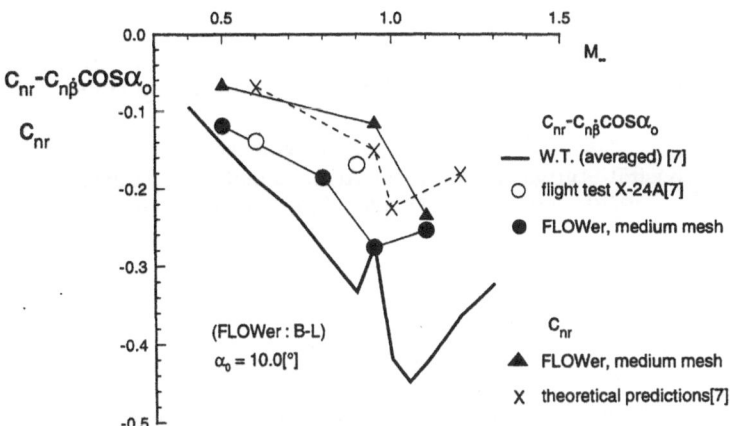

Fig 7 : Comparisons of the yaw damping derivatives

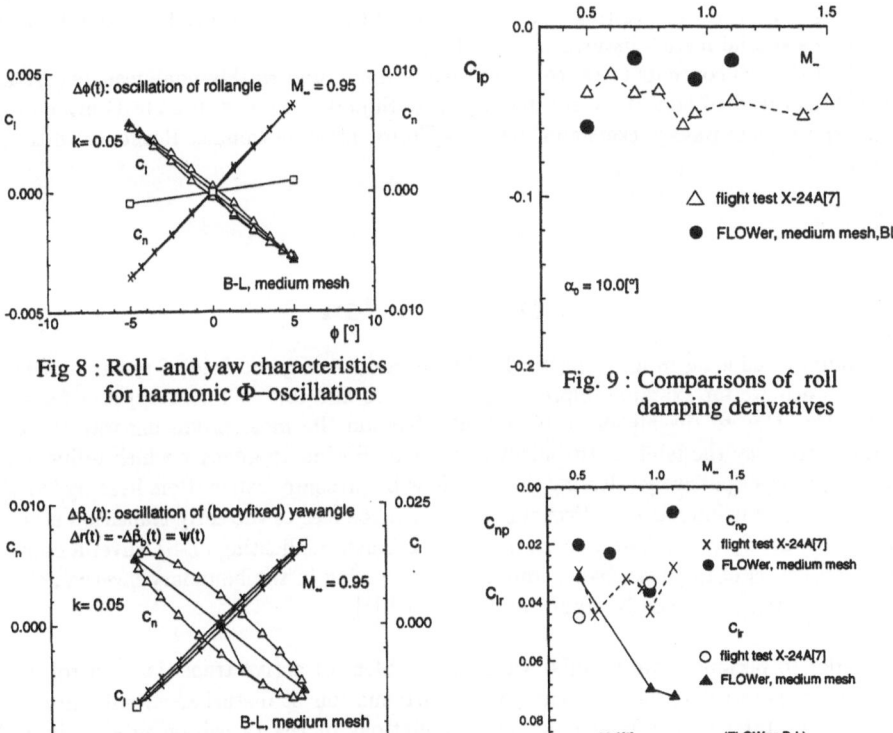

Fig 8 : Roll -and yaw characteristics for harmonic Φ–oscillations

Fig. 9 : Comparisons of roll damping derivatives

Fig 10 : Roll -and yaw characteristics for harmonic ψ–oscillations

Fig. 11 : Comparisons of pure cross damping derivatives

Investigations of Active Control of Wave Packets and Comparable Disturbances in a Blasius Boundary Layer by DNS

C. Gmelin, U. Rist, S. Wagner
Universität Stuttgart, Institut für Aerodynamik und Gasdynamik
Pfaffenwaldring 21, D-70550 Stuttgart, Germany

Summary

Under natural conditions the route to transition is somewhat different to the one observed in controlled experiments or numerical investigations. The incoming disturbances can rather be considered as a sequence of wave packets than as waves of a certain frequency and propagation angle. This leads to investigations of the propagation process of wave packets and similar disturbances in a Blasius boundary layer as a test case to active transition control. For testing and comparing different approaches for active transition control a special model-disturbance was designed.

Several (linear) concepts to cancel or attenuate linear and weakly nonlinear disturbances which can lead to transition are investigated: Superposition of discrete Fourier modes, superposition of wave packets and the FIR-(Finite duration Impulse Response)-filter concept.

Introduction

Nowadays, reducing drag, respectively skin friction is a great field of research with the aim to achieve for example improved performance and less fuel consumption for future aircrafts. Due to the higher turbulent skin friction the most promising way to reduce drag is to delay the laminar-turbulent transition. Beside approaches which influence the boundary layer passively (altering the base-flow by pressure distribution, heating/cooling) active control is intended to affect unsteady disturbances of the early transition stages in a direct way. By superposition of anti-phase Tollmien-Schlichting (T-S) waves a considerable reduction of the disturbance amplitudes occurring in the boundary layer is achieved and hence transition can be significantly delayed [5].

In order to become more familiar with the problem of active transition control and in a later stage, to develop strategies for the attenuation of disturbances with non-linear amplitude, three-dimensional numerical simulations of the transition process in a 2D-Blasius boundary layer were performed. The integration domain which was used (Fig.1) contains two independent disturbance-strips where 2D- and 3D- disturbances with any frequency and spanwise wave number can be introduced. The first strip (*ST1*) generates initial T-S waves which evolve downstream whereas the second strip should generate disturbances to cancel the initial waves.

Numerical method

All simulations were performed with the spatial DNS-code developed by Konzelmann, Rist and Kloker ([7], [6], [2], [3]) in a rectangular integration domain. The flow is split into a steady 2D-part (Blasius base flow) and an unsteady 3D-part. The x-(streamwise) and y-(wall-normal) directions are discretized with finite differences of fourth-order accuracy and in the spanwise dimension z a spectral Fourier approach is applied. Time integration is performed by a fourth order Runge-Kutta scheme. The utilized variables are nondimensionalized with: $\tilde{U}_\infty = 30\frac{m}{s}$, $\tilde{\nu} = 1.5 \cdot 10^{-5}\frac{m}{s^2}$ and $\tilde{L} = 0.05m$ which here is the distance of the integration domain from the leading edge of the flat plate ($Re_{\delta 1} = 543$):

$$x = \frac{\tilde{x}}{\tilde{L}} \quad , \qquad y = \frac{\tilde{x}}{\tilde{L}} \cdot \sqrt{Re} \quad , \qquad z = \frac{\tilde{z}}{\tilde{L}} \quad , \qquad t = \tilde{t} \cdot \frac{\tilde{U}_\infty}{\tilde{L}} \quad ,$$

$$u = \frac{\tilde{u}}{\tilde{U}_\infty} \quad , \qquad v = \frac{\tilde{v}}{\tilde{U}_\infty} \cdot \sqrt{Re} \quad , \qquad w = \frac{\tilde{w}}{\tilde{U}_\infty} \quad , \qquad Re = \frac{\tilde{U}_\infty \tilde{L}}{\nu} = 10^5 \quad ,$$

where u, v and w are the components of the unsteady velocity disturbances.
This leads to the dimensionless Frequency $\beta = \frac{2\pi \tilde{f} \tilde{L}}{\tilde{U}_\infty}$, where $\tilde{f}$ is the Frequency in $[Hz]$.

Control concepts

Three (linear) control concepts are compared here. All are based on the idea to cancel a disturbance by the superposition of the initial disturbance in antiphase. Fig.1 illustrates the general control procedure: an initial disturbance is forced by $ST1$ and a control disturbance emanates from $ST2$ propagating downstream. Without interaction both disturbances should produce the same time signal with opposite sign at a control point (CP) far downstream. Superposition of both signals leads to a decrease or cancellation of their original amplitude.

Two of the discussed concepts (superposition of Fourier-modes and wave packets) are off-line concepts, thought to investigate the principles of active wave cancellation, but they could easily be extended to in-line control with knowledge of the predominant receptivity properties. The last concept described below shows on-line control with off-line "training" of the control unit which is in this case a FIR-filter. Using on-line implementations the sensor point (SP) drives the second disturbance generator.

Superposition of Fourier modes: In a first step an 'uncontrolled' simulation with linear TS-waves generated by the first disturbance strip ($ST1$) alone is performed. The time signal of the spanwise wall-vorticity ω_z at the control point (CP, Fig.1) is recorded and a Fourier-transform in time is performed to obtain amplitude and phase of every single frequency. The second step is to calculate a further (reference-) run with harmonic TS-waves emanating from the second strip ($ST2$). Again the time signal at CP is Fourier-analyzed. Amplitude and phase of the forcing input for each frequency of the second run is tuned to obtain the same amplitude but opposite phase at the control point. As a third step a simulation is performed, where both disturbances (the first as before and the second, tuned one) are superposed. Fig.2 shows the analytic superposition of two sine waves with the same frequency but of different amplitude and phase. It is obvious that the result of the superposition is less sensitive to amplitude deviations than to phase errors around the optimal phase $\Theta = \pi$.

In cases with non-harmonic disturbances, like wave packets for instance, the signal must be divided in time windows, where each window is treated separately. This causes some problems occurring at the passage from one time window to the next. To overcome these problems we thought it would be better to cancel non-harmonic disturbances by super-position of wave packets.

Superposition of wave packets: Basically the same procedure as for the control with Fourier-decomposition is applied but due to the need of a time-frequency decomposition a continuous wavelet-transform of the uncontrolled run with a Morlet-wavelet [1] is performed. Similar to the Fourier-control the time signal of some short control-pulse emanating from $ST2$ is wavelet transformed now. The problem here is to find an appropriate combination of control wavepackets to cancel the initial disturbance. This leads to an iterative process, where the initial disturbance and the control packet are adjusted at the point in the β/t-plane where the initial disturbance has its maximum. Off-line superposition of both yields the new 'initial' disturbance for the iteration until a certain cancellation is reached.

FIR-filter: Often used in experimental setups, the FIR filter is an on-line concept which has to be trained for each flow condition or has to be adapted continuously. To avoid time-consuming calculations in our case the filter was trained once to obtain the filter coefficients for the subsequent runs.

For the temporal behaviour of the filter the following relationship is applied:

$$f(t) = \sum_{i=0}^{N-1} h_i \cdot s(t - i\Delta t) \tag{1}$$

where:

$f(t)$	:	forcing amplitudes at $ST2$,
h_i , $\quad i = 0....N-1$	:	N filter coefficients,
$s(t - i\Delta t)$	:	wall vorticity amplitudes at the sensor point (SP) delayed by i timesteps,
N	:	number of filter coefficients,

After a Fourier-transform, this relation reads in the frequency-domain:

$$F(\beta) = H(\beta) \cdot S(\beta) \tag{2}$$

$$H(\beta) = \frac{F(\beta)}{S(\beta)} \tag{3}$$

where:

$S(\beta)$	:	Fourier-transformed signal obtained at the sensor point SP (complex),
$F(\beta)$	:	Fourier-transformed input for $ST2$ (complex),
$H(\beta)$	:	complex transfer function.

The transfer function H respectively the filter coefficients h contain all informations about the receptivity properties of the sensor as well as of the second disturbance strip and the linear development between both. To train a FIR-filter the time signals of a simulation with 'successful' cancellation can be used. To obtain the complex transfer function H only the time signals s and f are needed. The back transform of H into the physical space yields the filter coefficients h_i in eqn.(1). Due to the Fourier transform the resolution of the transfer function depends on the length of the coefficient vector h_i.

Test cases

To evaluate the different approaches several test cases were selected and performed:

- Single linear 2D and 3D wave packets.
- A sequence of 2D wave packets with random amplitude and length. Comparisons with time-traces obtained in free-flight tests by A. Seitz [8] showed the same frequency range and similar wavelet-patterns (Fig.3).
- Harmonic 2D and 3D waves.

Application of active control

To compare the different approaches simulations of the control of the linear wave packet sequence were performed (Fig.4). Fig.5 shows the appropriate filter coefficients respectively the transfer function of the FIR filter with a frequency resolution of $\Delta\beta = 1$. Application of the different control strategies (Fig.6) shows a reduction in amplitude of about 1.5 to 2.5 orders of magnitude at a fixed position downstream. Best reduction is achieved using FIR filters with high resolution of the transfer function.

The results using other concepts show their specific disadvantages: Especially the wave packet control generates additional disturbances in a frequency range above $\beta \approx 20$ because of the non-orthogonality of the control wave packets. This means that every pulse, cancelling the initial disturbance at a specific point of the (β, t)-plane produces new disturbances in its vicinity.

After obtaining good results using the FIR-filter control, this concept was applied to nonlinear harmonic waves (a single 2D-wave or a pair of oblique waves). The mode $(1,0)$ (the first index gives multiples of the frequency β, the second multiples of the fundamental spanwise wave number γ) and, in the oblique case $(1,\pm1)$ were forced at $ST1$. In both cases due to the nonlinear amplitude numerous higher harmonics are generated which travel downstream with the phase-speed of the fundamental wave. Reduction of the fundamental amplitude causes the higher-harmonics to attenuate with a damping rate according to the linear stability theory (Fig.7 and Fig.8). In the oblique case in addition to the direct higher harmonic modes $(2,2)$ and $(3,3)$ a strong stationary mode $(0,2)$ arises (Fig.8 $\cdot - \cdot - \cdot$) which is very stable and hardly influenced by the attenuation of the fundamental mode. The amplitude of the fundamental wave is removed not as successfully as in the linear case because of the nonlinear development downstream. The reason for being able to control the waves in this case is the phase speed which remains almost unchanged, compared to the linear case. This enables us to catch the right phase, leading to only small deviations from the amplitude necessary if optimal control were applied (Fig.2).

Active control via FIR-filters at different streamwise positions (i.e. varying $ST2$) was also used in a K-breakdown scenario (Fig.9). Here the modes $(1,0)$ and $(0,1)$ were forced at $ST1$ while the $(1,\pm1)$ modes were generated as direct higher-harmonic of the two fundamental modes. Fundamental resonance of $(1,0)$ and $(1,\pm1)$ occurs at $x \approx 2$. Due to the onset of resonance the phase of the interacting modes is synchronized, respectively their phase speed become equal (Fig.10 a)). With application of control to the 2D-mode the phase-coupling is broken up and the waves evolve independently with a large steady mode $(0,1)$ remaining (Fig.10 b)).

As a result the control is more effective when applied at an early stage of transition as in [4] where the phase which is needed for the success of control is more predictable and hence a better reduction of the amplitude is achieved.

Conclusions

Linear concepts are suitable to reduce linear as well as weakly nonlinear disturbances in boundary layers, when the phase speed of the influenced disturbance is close to the one predicted by linear stability theory. It is also possible to suppress unsteady modes in an oblique or K-breakdown scenario whereas steady modes that arise due to resonant interactions remain very stable and resist a successful control of the disturbances, because of their different nature with respect to time.

Acknowledgments

The support of this research by the Deutsche Forschungsgemeinschaft DFG under contract Be 1192/7-1 is gratefully acknowledged.

References

[1] M. Farge. Wavelet transforms and their application to turbulence. In *Ann. Rev. of Fluid Mech.*, volume 24, pages 395–457, 1992.

[2] M. Kloker. *Direkte Numerische Simulation des laminar-turbulenten Strömungsumschlages in einer stark verzögerten Grenzschicht.* Dissertation, Universität Stuttgart, 1993.

[3] U. Konzelmann. *Numerische Untersuchungen zur räumlichen Entwicklung dreidimensionaler Wellenpakete in einer Plattengrenzschicht.* Dissertation, Universität Stuttgart, 1990.

[4] E. Laurien and L. Kleiser. Numerical simulation of boundary-layer transition and transition control. *J. Fluid Mech.*, 199: pages 403–440, 1989.

[5] H. W. Liepmann and D. M. Nosenchuck. Active control of laminar-turbulent transition. *J. Fluid Mech.*, 118: pages 201–204, 1982.

[6] U. Rist. *Numerische Untersuchung der räumlichen, dreidimensionalen Störungsentwicklung beim Grenzschichtumschlag.* Dissertation, Universität Stuttgart, 1990.

[7] U. Rist and H. Fasel. Direct numerical simulation of controlled transition in a flat-plate boundary layer. *J. Fluid Mech.*, 298: pages 211–248, 1995.

[8] A. Seitz. Private communication.

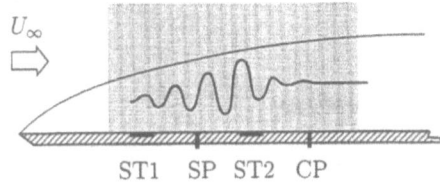

Fig.1: Integration domain and basic setup of the simulation.
(ST1 and ST2: disturbance strips,
SP: sensor point, CP: control point)

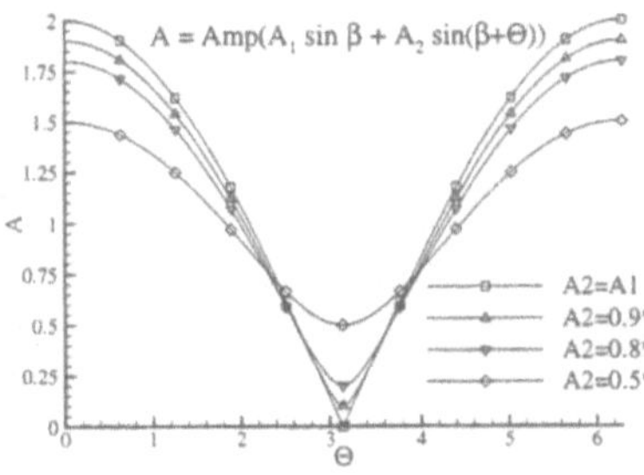

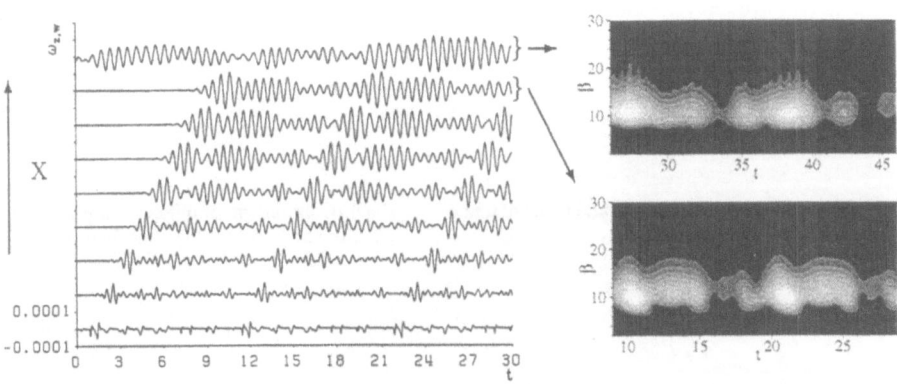

Fig.2: Superposition of two sine waves of the same frequency β

$$A = \text{Amp}(A_1 \sin \beta + A_2 \sin(\beta + \Theta))$$

A2=A1
A2=0.9*A1
A2=0.8*A1
A2=0.5*A1

Fig.3: Time signals of spanwise vorticity at the wall at different streamwise positions and wavelet spectra (contours of constant wavelet amplitude versus dimensionless time and frequency) of a random sequence of wave packets compared to results obtained in free-flight measurements (uppermost signal) [8].

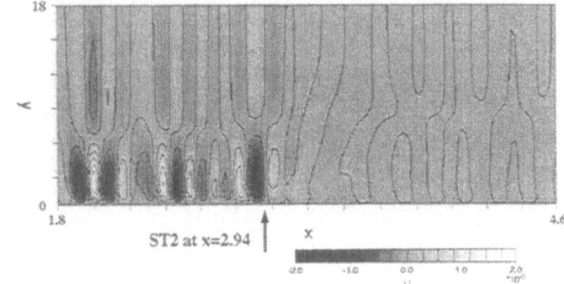

Fig.4: Instantaneous u-velocities of a random sequence of wave packets attenuated via FIR filter.

ST2 at x=2.94

199

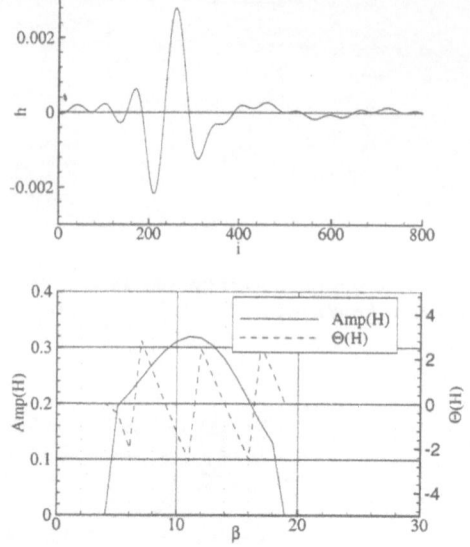

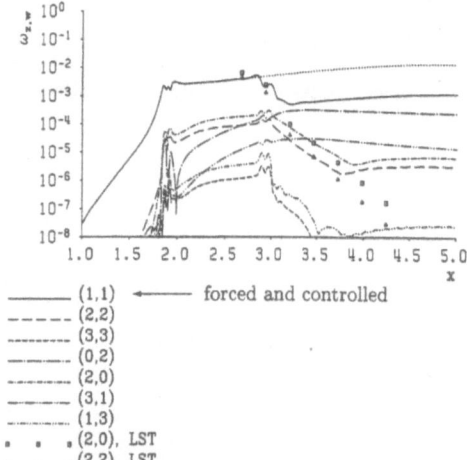

Fig.5: FIR-filter coefficients (a) and transfer function for the linear case (b).
$x_S = 2.41, x_{ST2} = 2.94$

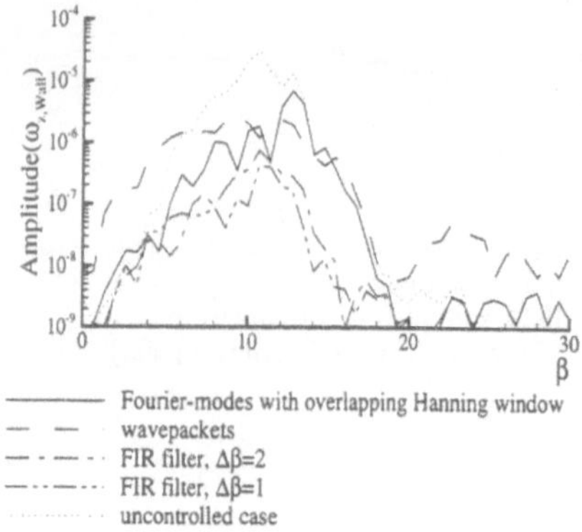

Fig.6: Comparison of different control strategies applied to the 'natural' 2D-disturbance (x=4.5)

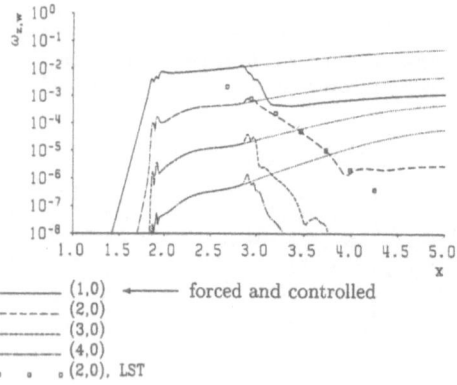

Fig.7: 2D nonlinear case, controlled applied at $x = 2.94$.
Fundamental frequency $\beta = 10$.

Fig.8: 3D nonlinear oblique case, control applied at $x = 2.94$.
Fundamental frequency $\beta = 10$,
fundamental spanwise wavenumber $\gamma = 10$.

200

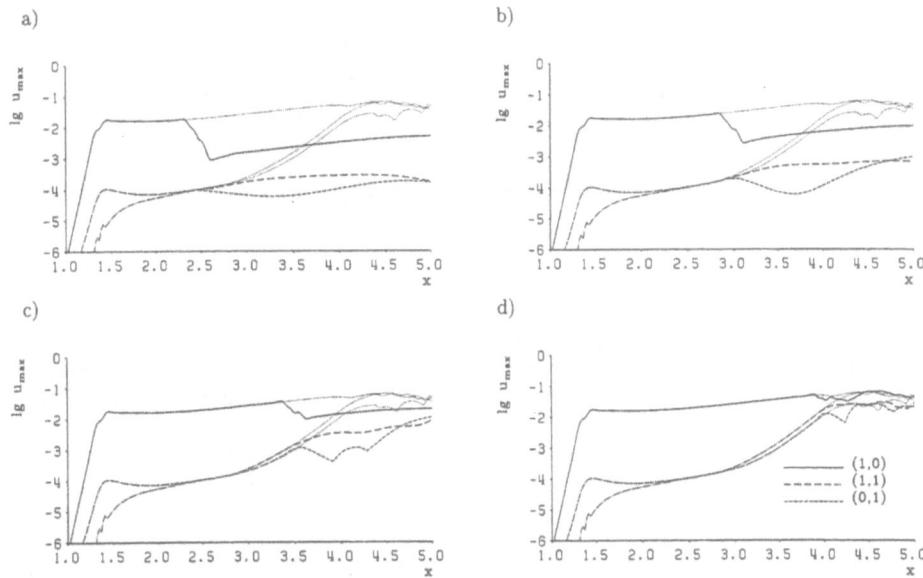

Fig.9, a)-d): K-breakdown with active control of the (1,0) mode applied at $x_{ST2} = 2.41, 2.94, 3.46$ and 3.99. (dotted lines: uncontrolled case)

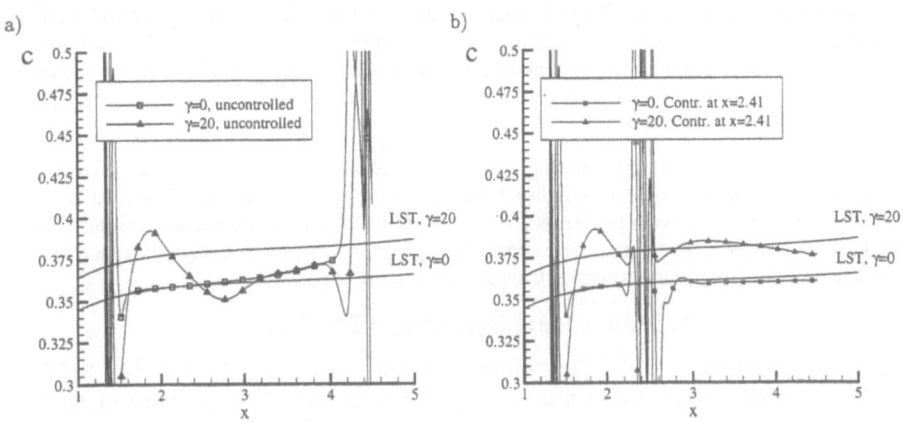

Fig.10: Phase speed c in comparison to the linear stability theory (LST), fundamental resonance. a): uncontrolled case, b): control applied at $x = 2.41$

Multidisciplinary Aerodynamics/Structure Dynamics High-Performance Simulation and Optimization

W. Haase, E.H. Hirschel, J. Krammer, H. Rieger

Daimler-Benz Aerospace AG, Military Aircraft Division

D-81663 München, Germany

Summary

Multidisciplinary computation methods are developed at Dasa Military Aircraft. The paper gives an overview of the design problems related to aerodynamics/structure dynamics couplings in aircraft design. The development approach of the elements of a future multidisciplinary simulation and optimization system is sketched. Selected results from the development of these elements are discussed. They are steady and unsteady flow computation, structural computation and optimization, multidisciplinary and multiobjective optimization and finally the time-domain aeroelastic coupling in the transonic regime.

1. Introduction

Aircraft airframes are flow-dominated extreme light-weight structures. Flow/structure interactions of many kinds pose very important design problems. High-performance computation has the potential to open new ways for the treatment of coupled aerodynamic/structure problems in the development and definition processes of an aircraft [1]. Also increasing performance demands of aircraft can be met in future with of help of disciplinary and interdisciplinary numerical simulation and optimization methods on adequate physical and mathematical/numerical modelling level. Industry will use such methods in order to reduce development (and operational) risks and definition and development cost, and also to improve product quality.

The paper gives an overview of the topics and approaches in interdisciplinary aerodynamics/structure dynamics/aeroelastics high-performance computation at Dasa Military Aircraft. In Chapter 2 the major design problems are discussed, where high-performance computation will have its impact. The development approach is given in Chapter 3. Chapter 4 reports on current activities, concluding remarks, Chapter 5, close the paper.

2. Motivation: Major Design Problems

During the years 1995 and 1996 Dasa Military Aircraft together with Dasa Airbus, MTU, BMW Rolls-Royce, DLR, GMD and several universities studied the potential of high-performance computation in aircraft industry and research [2]. This study was initiated because it became evident at that time, that the development in computer power will permit soon the use of cost-effective, design-problem adequate disciplinary and multidisciplinary simulation and optimization methods in the industrial design processes, Fig. 1. For the airframe industry the following application areas were identified, which are dominated by multidisciplinary interrelations of aerodynamics, structure dynamics et cetera, Fig. 2:

- Fidelity of aerodynamics surfaces, especially of wings, high-lift systems and control surfaces, in view of static and dynamic structural loads and the related deformations and structural integrity.

- Dynamic aerodynamic derivatives and control surface transients, also in view of structural loads.

- Dynamic answer of the structure to aerodynamic and external forces, and aeroelastic stability (flutter, buffeting).

- Layout of notch filters for the flight-control system (structural frequencies versus actuation-chain frequencies) (for Dasa Military Aircraft this belongs to the special, highly important topic "layout, optimization and proof of the controlled elastic aircraft", because unstable aircraft make necessary an integrated aerodynamics/structure dynamics/flight dynamics/flight control system design).

- Dynamic structural loads due to the granularity of turbulent boundary layers and separation/vortices.

- Antenna integration and correction of aeroelastic airframe deformations.

These application areas show that multidisciplinary high-performance simulation and optimization in aircraft design reaches far beyond the classical aeroelastic design (and its computation topics) towards the holistic design and optimization of the controlled elastic aircraft.

3. The Development Approach

The development approach of the multidisciplinary numerical simulation and optimization methods at Dasa Military Aircraft has the following objectives:

- Provision of multidisciplinary simulation and optimization methods, which use established disciplinary codes in a "weak" coupling mode.

- Use of central and/or distributed (workstation cluster) high-performance parallel computer architectures.

- Provision of graphical user interfaces with process control and intervention possibilities, and of multidisciplinary post-processing and visualization tools.

In the present treatment of aeroelastic and structure-dynamics problems predominantly linear aerodynamic methods are employed, dynamic structure problems are treated in the frequency domain, the flow/structure coupling is made, for instance, in the skeletal surface of a wing. The couplings with flight-dynamics and flight control are made sequentially. The new treatment basically must evolve out of the present capabilities in order to reduce risks and to ensure acceptance of users in the aircraft definition and development processes.

The major work elements are:

- General adaptation of disciplinary codes.

- Extensions of aerodynamic codes including grid generation to describe moving surfaces (general deformations of airframe, control-surface deflections). The FLOWer code [3] was adopted as basic code.

- Development of weak coupling procedures, extension to transonic flow problems, i.e. coupling of Euler/RANS codes and structure-mechanics codes in the time domain.

The work is performed in co-operation with other Dasa companies, DLR, GMD, universities in national and international (EU) projects. The core work is done in the frame of the project AMANDA ("A Multidisciplinary High Performance Numerical Simulation and Development System for Aircraft"), which was defined [4] jointly by industry and research on the basis of the memorandum [2].

4. Current Activities

In this chapter an overview is given over the present development status of the main elements of the future multidisciplinary simulation and optimization system at Dasa Military Aircraft, which will become fully operational after the year 2000. These elements are steady and unsteady flow computation, aerodynamic shape optimization, structural optimization, multidisciplinary and multiobjective optimization, and time-domain aeroelastic coupling in the transonic regime.

- Steady and unsteady flow computation

Basic requisites for numerical multidisciplinary aerodynamics/structure dynamics simulations are aerodynamic codes for steady and unsteady flow computations together with efficient pre- and postprocessing tools. Figs. 3 and 4 demonstrate the present capabilities. Fig. 3 [5] shows for the Eurofighter/Typhoon configuration at M = 0.9 and H = 10.8 km that the full configurational complexity of an aircraft with external load and propulsion system simulation can be treated today. The employed Reynolds-averaged Navier-Stokes (RANS) code is FLOWer [3], applied to a block-structured grid.

Results of an unsteady flow simulation for a delta wing in harmonic pitch oscillation, M = 0.4, Re = $3 \cdot 10^6$, $\alpha = 9° \pm 6°$ [6], are given in Fig. 4. The code employed is the FLOWer code [3], both in Euler and RANS mode. Fig. 4a shows a fair agreement of lift as function of angle of attack with experimental data already after two oscillations for both the Euler and the RANS mode. The averaged pressure distribution on the leeward side of the wing at x/L = 0.8 is given in Fig. 4b. Amplitude and phase of the first and second harmonic, also at that location, Fig. 4c, are in good agreement with the experiment, too, except for the phase of the second harmonic, which is due to evaluation problems of the experimental data.

- Structural computation and optimization

This topic is well developed in industry. Structural optimization for whole aircraft configurations can ensure minimum weight, while strength, buckling, flutter et cetera demands are fulfilled. The computer demands are smaller than those of aerodynamic simulations, because the latter have a much higher granularity of the discretized field. Fig. 5 gives the optimization history of the structural weight optimization of a wing [7]. The initial weight was reduced after ten iterations by about 25 per cent, while the percentage of violated constraints went down from 95 per cent to 1.5 per cent, which is a very acceptable result in structural design optimization. The used code is the LAGRANGE code [8], which is the standard structural optimization code at Dasa Military Aircraft.

- Multidisciplinary and multiobjective optimization

This topic presently advances with large strides, [9]. Here an example of an application of a genetic algorithm optimization scheme is shown [10]. The problem is the optimization of the aeroelastic flap effectiveness for a composite delta wing with respect to spanwise divided and deflected flaps and various relevant structural load cases (problem parametrization in Fig. 6a). The overall objective function is minimum weight and maximum roll rates, at M = 1.2, sea level and $\alpha = 5.73°$. Fig. 6b gives the resulting objective space with the Pareto frontier (weight function at the ordinate and the roll rate function at the abscissa). All points represent (local) optima. The Pareto frontier gives the designer many "optimum pairs" of wing weight and roll rate, which are then employed in the overall aircraft trade-off and optimization processes. Note that the Pareto frontier not necessarily has a monotonic behaviour.

- Time-domain aeroelastic coupling in the transonic regime

The aeroelastic coupling in the time domain with non-linear flow codes is the largest challenge for multidisciplinary aerodynamic/structure dynamics/flight dynamics/control system simulation and optimization problems. A result from the exploratory study [11] is given in Fig. 7. The planform of the wing, Fig. 7a is that of the AGARD I-Wing, with NACA 65004 wing sections. The dominant structural modes with their frequencies are shown in Fig. 7b and a Euler result in Fig. 7c. Fig. 7d finally gives the (damped) oscillations of the first generalized coordinates.

5. Concluding Remarks

Flow/structure couplings are a major concern in aircraft design. They are seen today, at least for military aircraft, in the larger frame of the design and optimization of the control system "aircraft". Multidisciplinary high-performance simulation and optimization has a large potential to improve aircraft design. The aircraft industry develops together with research new approaches and methods. At Dasa Military Aircraft presently elements of an multidisciplinary computation system are developed and adapted. Emphasis is put on aerodynamic codes for moving and deforming surfaces (grid generation is a major bottle neck) and coupling approaches and systems. Flow-physics (transition, turbulence) and structure-physics models are also severe bottle necks. Because the new multidisciplinary simulation and optimization tools are very complex, and demand equally complex application definitions and results interpretations in the actual aircraft design processes, large demands on knowledge and skill of the development staff make changes in engineering and scientific education necessary.

6. References

[1] E.H. Hirschel: Perspektiven des Höchstleistungsrechnens in der Luftfahrtindustrie". Statustagung HPSC 98, München 1997, Dasa-LMLE3-AERO-MT-985, 1997.

[2] E.H. Hirschel et al.: "Höchstleistungsrechnen in der Luftfahrtindustrie und -forschung". Denkschrift, Dasa München, DLR Göttingen, 1996.

[3] N. Kroll: "National CFD Project MEGAFLOW – Status Report". NNFM 60, Vieweg, Braun-schweig/Wiesbaden, 1997, pp. 15-23.

[4] E.H. Hirschel, M. Faden, A. Brenneis, A. Beckert: "Höchstleistungsrechnen in der Luftfahrtindustrie und –forschung, Phase 1, Teil 1, Vorschlag des Verbundprojektes AMANDA". Dasa/MT63/AMANDA/R/0001/A, 1998.

[5] S. Hitzel, S. Leicher, W. Schwarz, personal communication, 1998.

[6] W. Fritz: "Numerical Simulation of Unsteady Vortical Flow About Delta Wings Oscillating at High Incidence". This Volume.

[7] S. Schweiger, S. Krammer, H. Hörnlein: "Development and Application of the Integrated Structural Design Tool LAGRANGE". 6th AIAA/NASA/ISSMO Symposium on Multidisciplinary Analysis and Optimization". Sept. 4-6, 1996/Bellevue, WA.

[8] J. Krammer: "Practical Architecture of Design Optimization Software for Aircraft Structures taking the MBB-LAGANGE Code as an Example". AGARD-Lecture Series No. 186, 1992.

[9] E.H. Hirschel, K. Becker, W. Haase, L. Fornasier: "Dasa Aerodynamic Design Optimization-Topics and Approaches". In: Computational Fluid Dynamics '98, Volume 2 (K.D. Papailiou, D. Tsahalis, J. Périaux, D. Knörzer, eds.), Wiley, Chichester, 198, pp. 429-434.

[10] W. Haase, personal communication, 1998, ESPRIT Project 20082, FRONTIER. Open System for Collaborative Design Optimization Using Pareto Frontier.

[11] H. Ringelschwendner: "Aeroelastische Analyse eines frei schwingenden Flügels mit einem unstrukturierten Euler-Verfahren", Diplomarbeit TU München, Dasa-S-PUB-591, 1997.

7. Figures

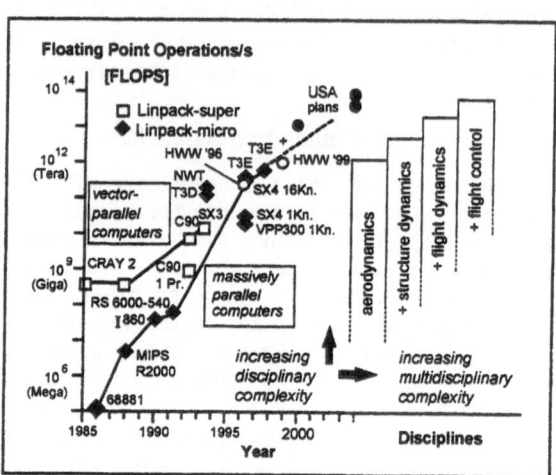

Fig. 1   The key for the development : rise of computer power and its future positive benefit/cost relation for industry [1]

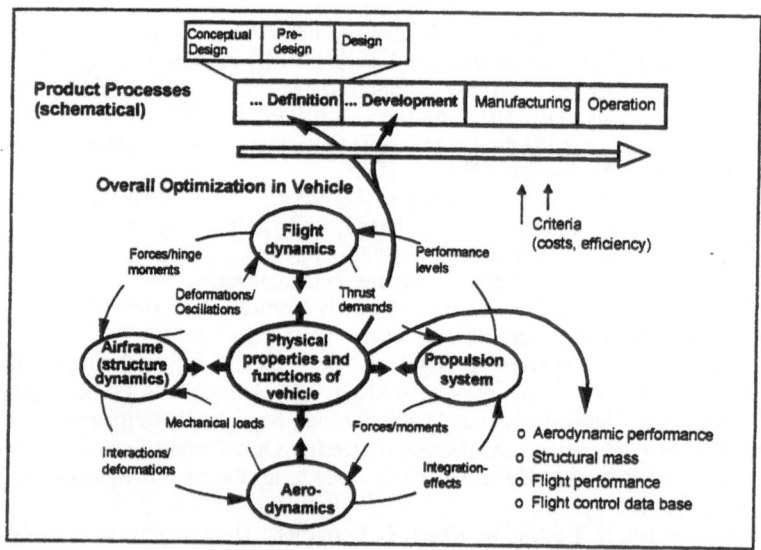

Fig. 2   Application areas of multidisciplinary high-performance simulation and optimization [1]

Fig. 3   Computed pressure distribution on the Eurofighter/Typhoon
configuration at M = 0.9, H = 10.8 km [5]

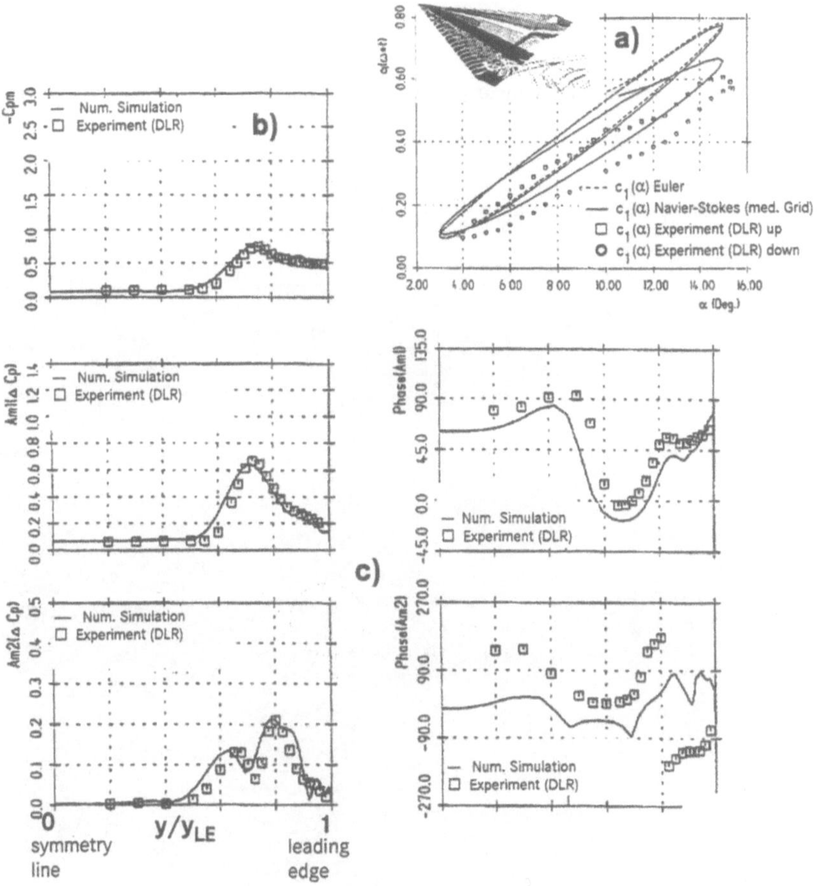

Fig. 4   Selected results of unsteady flow simulation for delta wing in harmonic pitch
oscillation, M = 0.4, Re = 3 · 10⁶, α = 9° ± 6°, [6], a) lift as function of angle of
attack, b) averaged pressure and c) amplitude and phase of first and second harmonic on the leeward side of the wing at x/L = 0.8.

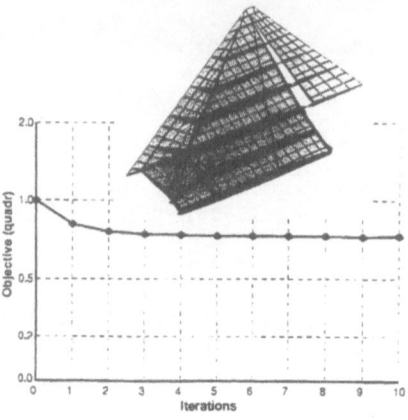

Fig. 5   Optimization history of a wing structure, 25 per cent weight reduction [7]

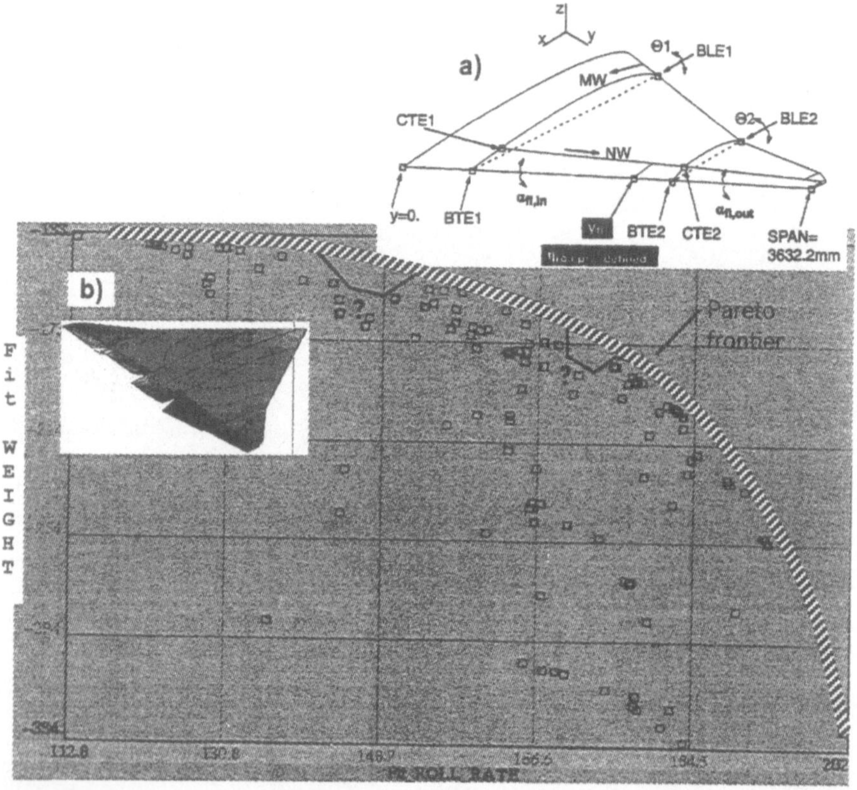

Fig. 6   Application of the FRONTIER genetic algorithm to multidisciplinary and multiobjective
          optimization of the flap effectiveness of a delta wing at M = 1.3, sea level, α = 5.73° [10], a)
          problem parametrization, b) objective space with Pareto frontier (wing weight vs. roll rate)

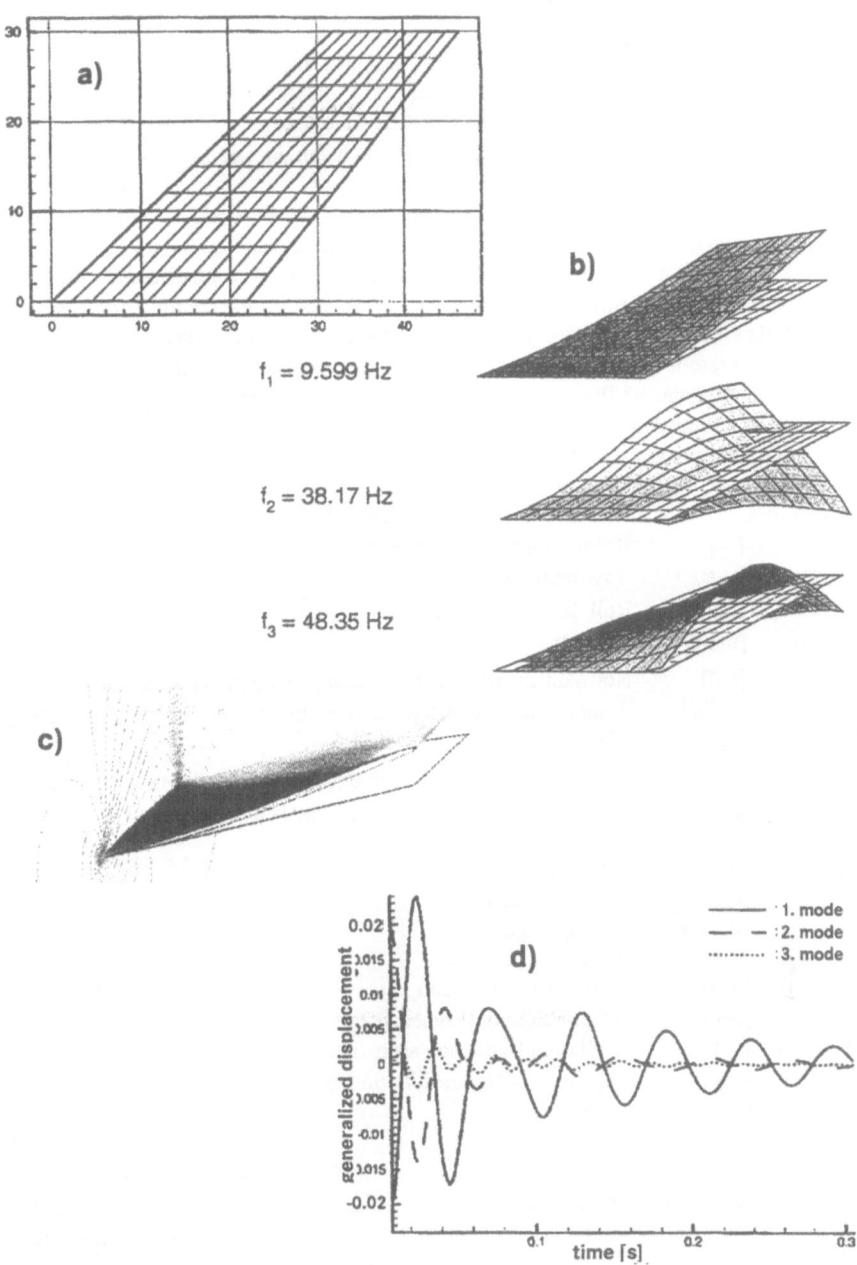

$f_1 = 9.599$ Hz

$f_2 = 38.17$ Hz

$f_3 = 48.35$ Hz

——— : 1. mode
— — : 2. mode
·········· : 3. mode

Fig. 7   Aeroelastic coupling in the time domain, AGARD I-Wing, M = 0.901, inviscid flow [11], a) wing planform, b) dominant structural modes, c) flow computation result, d) oscillation of the first generalized coordinates

Yaw Control Using Asymmetric Forebody Strakes at High Angles of Attack

Peter R. Hakenesch

Daimler-Chrysler Aerospace - Military Aircraft
81663 München, Germany
e-mail: peter.hakenesch@m.dasa.de

Summary

Experimental data of side forces, yawing and rolling moments induced on a 1/7.5 scale model of the X31 aircraft due to asymmetric deployed forebody strakes are discussed for large angles of attack. Emphasis of the work is placed on demonstrating the control power of deployable nose strakes at angles of high incidence where conventional rudder control is no longer efficient, if present at all. The potential benefits of yaw control using forebody nose strakes include enlargement of the flight envelope, reduction of the size of the vertical tail with subsequent reduction in structure weight, total drag, radar cross section and wave drag while flying in the supersonic speed regime.

List of Symbols

C_l	[-]	Rolling moment coefficient, body axis system
C_n	[-]	Yawing moment coefficient, body axis system
C_{lp}	[rad^{-1}]	Roll damping derivative $\quad C_{lp} = \dfrac{\partial C_l}{\partial p}$
v	[m/s]	Velocity
$\dfrac{\Omega \cdot b}{2 \cdot v_\infty}$	[rad]	Normalized speed of revolution, aerodynamic axis system
Ω	[rad·s^{-1}]	Rotation rate around velocity vector, aerodynamic axis system
b	[m]	Wing span
$\dfrac{p \cdot b}{2 \cdot v_\infty}$	[rad]	Normalized speed of revolution, body axis system
p, q, r	[rad·s^{-1}]	Rotation rates around x_f, y_f, z_f, body axis system
S_{ref}	[m^2]	Reference area
l_{ref}	[m]	Reference length
α	[deg]	Angle of attack
β	[deg]	Sideslip angle
η	[deg]	Flap deflection
φ	[deg]	Radial Strake position
θ	[deg]	Radial Position of fuselage tip
ω	[s^{-1}]	Oszillation frequency around x_f
A	[deg]	Amplitude of the oscillation frequency ω
x, y, z	[m]	Coordinates

Subscripts

a	Aerodynamic axis system
f	Body axis system
ref	Reference
∞	Free stream conditions

Abbreviations

LAMP	Large Amplitude Test Facility
BAR	Bihrle Applied Research
C	Canard
LEF	Leading Edge Flap
MAC	Mean Aerodynamic Chord
R	Rudder
TEF	Trailing Edge Flap
W	Winglet

1. Introduction

Aircraft incorporating a highly swept delta wing geometry are prone to instabilities around the roll-axis while flying at high angles of attack[1-3]. Those instabilities are caused by the asymmetric breakdown of vortices contributing largely to the nonlinear lift of delta wings. While performing roll maneuvers at high angles of attack, the asymmetric breakdown of the wing leading edge vortices is mainly due to the variation of the effective angle of attack at the wing leading edge as a function of wing span and roll rate. As a consequence, self propellant rolling moments may be induced by creating asymmetric lift distributions from one wing to the other. This typical roll damping characteristic was also observed at the X31 experimental aircraft at higher angles of attack during flight testing. The focus of this work is to provide an understanding of the flow phenomena leading to this unstable roll characteristic and how to actively control the vortices that are generated at the forebody, using deployable strakes in order to improve roll stability and to enhance yaw control at high angles of attack. The idea to make use of forebody vortices to generate control forces and moments has been studied extensively in recent years[4-8].

2. Experimental Setup

Two 1/7.5 scale wind tunnel models of the X31, Fig. 1, were under investigation, one equipped with a six-component balance for the measurement of forces and moments and one equipped with pressure instrumentation to record the surface pressure distribution. Tests were performed in the LAMP[9] wind tunnel of BAR Neuburg, Germany, a low speed facility for static as well as dynamic testing. The nominal test conditions as well as the model references are listed in table 1. In order to validate the existing dynamic flight test data[10], forced oscillation tests were performed with the wind tunnel model. Rolling moment derivatives were measured by applying a forced oscillation ω around the longitudinal axis of the aircraft. The amplitude A as well as the variation of the different oscillation frequencies are listed in table 1. The derivation of damping derivatives from rotary balance testing is described in detail in[11]. Results of the dynamic wind tunnel tests compared very well with the data obtained by parameter identification from previous flight tests of the X31, Fig. 2, and proved the suitability of the LAMP facility for further investigations of the flow characteristics. The departure of the roll damping from natural stability to instability and back to a naturally damped state in the angle of attack range from $27° < \alpha < 55°$ can be clarified by the results of surface pressure measurements. Those tests were carried out under quasi-stationary test conditions recording the surface pressure of 505 pressure gauges. The model was rotated around the wind vector axis at constant rotation rates Ω as listed in table 1. In order to investigate the influence of forebody strakes on the lateral stability, different sets of strakes where examined. They were inclined 30° off the plane of symmetry and started right at the nose tip, their length ranging from 1.5" to 5",

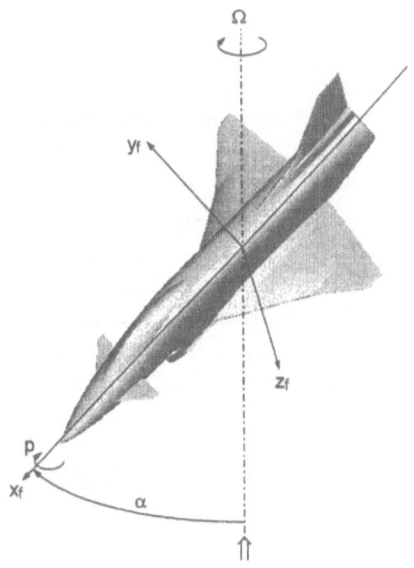

Fig. 1: X31 Model in LAMP facility

Fig. 3. Symmetric, i.e. left and right side mounted, as well as asymmetric configurations, i.e. left side only, were tested. Besides static force coefficients, dynamic derivatives were also determined while spinning the model around the wind vector axis at different rotation rates. Plotting the force and moment coefficients obtained during dynamic testing versus the normalized speed of revolution indicates the damping characteristics for the relevant coefficient. A stable behavior is characterized by a curve running from the upper left to the lower right quadrant and undamped behavior is marked by a curve running from the lower left to the upper right quadrant respectively, Fig. 8 and Fig. 9.

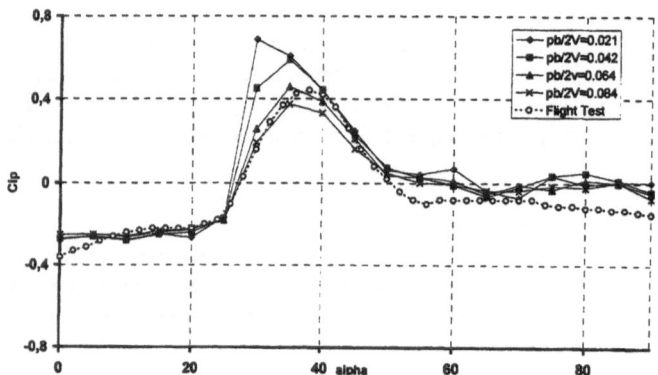

Fig. 2:   **X31 roll damping at different rotation rates**
LEF=-40°/-32°, TEF=0°/0°, C=-40°, R=0°

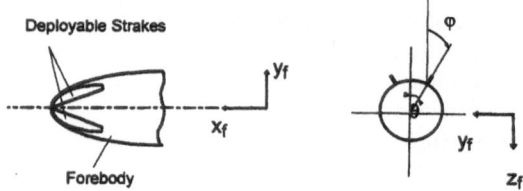

Fig. 3:   **Deployable forebody strakes**

Tab. 1:   **X31 Model References and Nominal Test Conditions**

Model scale		1:7.5				
Reference area S_{ref}	[m²]	0.3738				
Reference length l_{ref} = MAC	[m]	0.502				
Moment reference point cg_{ref} = 0.25 MAC	[m]	0.908				
Wing span b	[m]	0.968				
Free stream velocity V_∞	[m/s]	10.86				
Revolution around velocity vector $\Omega \cdot b/2 \cdot V_\infty$	[rad/s]	±0.05, ±0.1, ±0.2, ±0.3				
Revolution around x_f $p \cdot b/2 \cdot V_\infty$	[rad/s]	0.020	0.041	0.062	0.082	0.100
Oscillation frequency ω	[s⁻¹]	0.5529	1.1121	1.6713	2.2242	2.7835
Oscillation amplitude A	[deg]	40	40	40	40	40

3. Results

3.1 Pressure Measurements

At moderate angles of attack i.e. $0° > \alpha > 27°$, the effective angle of attack at the down going wing is increased by the additional angle of attack due to the rotation rate of the aircraft and creates higher vorticity than in the static case. On the opposite side, the up moving wing experiences a reduced effective angle of attack and therefore decreased vorticity on this wing. This causes asymmetric lift distributions on the two wings and creates an additional stabilizing rolling moment, Fig. 4. Further increase of the angle of attack, i.e. $\alpha > 27°$, leads to vortex breakdown on the down going wing with subsequent loss of lift on this wing. The flowfield on the up going wing is still dominated by the intact vortex , thus still creating considerable vorticity on this wing. The combination of the asymmetric lift distribution on the two wings leads to a propellant rolling moment, Fig. 5. Increasing the angle of attack beyond $\alpha > 55°$, causes also vortex breakdown on the up going wing with subsequent loss of lift also on the second wing. The propellant rolling moment therefore disappears and the rolling moment returns back to a naturally damped state as shown in Fig. 2.

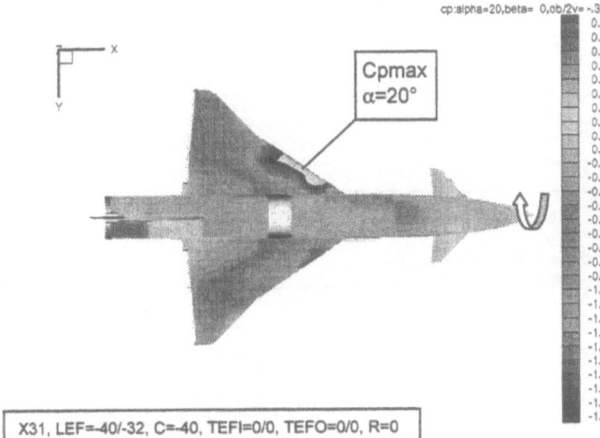

X31, LEF=-40/-32, C=-40, TEFI=0/0, TEFO=0/0, R=0

Fig. 4: X31 Top view

$\Omega \cdot b/2 \cdot V_\infty = -0.3$

$\alpha = 20°$

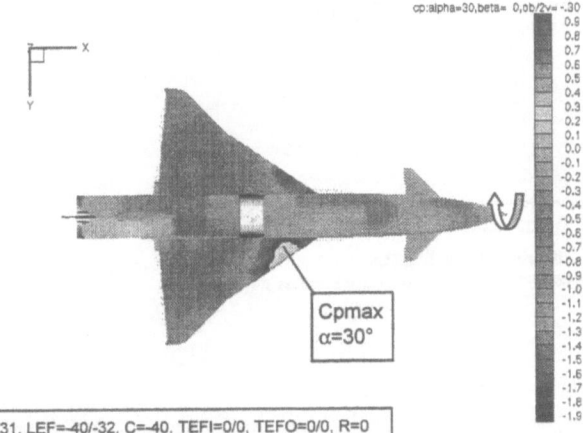

X31, LEF=-40/-32, C=-40, TEFI=0/0, TEFO=0/0, R=0

Fig. 5: X31 Top view

$\Omega \cdot b/2 \cdot V_\infty = -0.3$

$\alpha = 30°$

3.2 Force Measurements

3.2.1 Influence of strake length

In a first step, the length of the forebody strakes was continuously diminished from 5" to 1.5". With regard to yawing moment only moderate changes in effectiveness were measured at angles of attack of $\alpha < 40°$, Fig. 6. The influence on rolling moment was even less pronounced over the entire angle of incidence range investigated, i.e. $10° < \alpha < 60°$, Fig. 7. Therefore, all subsequent tests were carried out with the 1.5"-strakes.

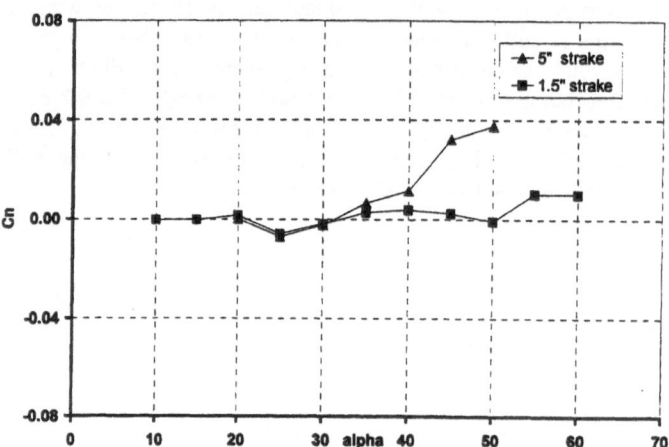

Fig. 6: Yawing moment for different strake lengths,
LEF=-40°/-32°, TEF=0°/0°, C=-40°, R=0°, both strakes deployed

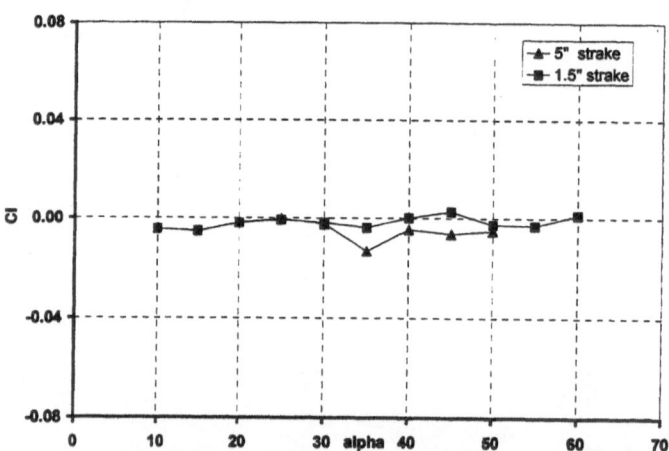

Fig. 7: Rolling moment for different strake lengths,
LEF=-40°/-32°, TEF=0°/0°, C=-40°, R=0°, both strakes deployed

3.2.2 Roll control

The second objective of this test series was to reduce the self propelling rolling moments that caused the departure of the roll damping from stable to unstable at the angle of attack range from 27°<α<55° . The rolling moments at different speeds of revolution of the configuration LEF=-40°/-32°, TEF=0°/0°, C=-40°, R=0°, no strakes deployed, designated baseline configuration, were measured while spinning the model around the wind vector axis at different revolution rates and compared to the same configuration with the left forebody strake deployed, Fig. 8 and Fig. 9. The results obtained so far indicate very little effect on roll stability by asymmetric strake deployment and suggest further investigations for an optimized positioning of the forebody strakes.

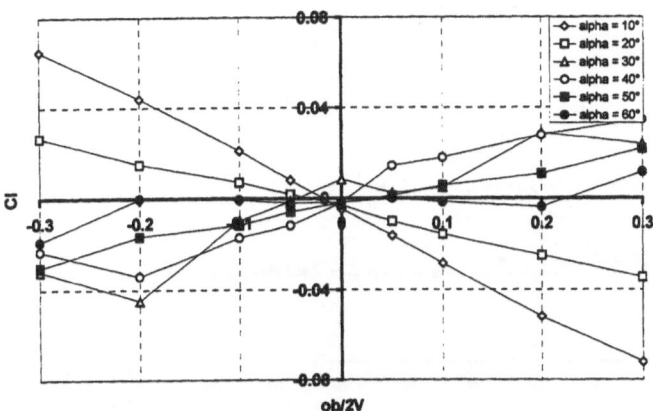

Fig. 8:   Rolling moment for symmetric baseline configuration vs. speed of revolution

LEF=-40°/-32°, TEF=0°/0°, C=-40°, R=0°, β=0°, no strakes

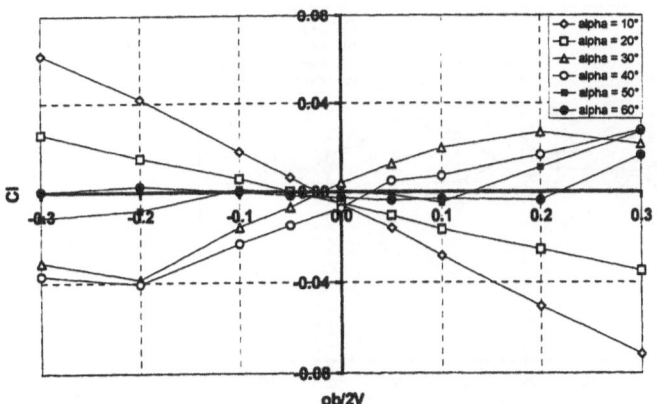

Fig. 9:   Rolling moment for asymmetric strake deployment vs. speed of revolution

LEF=-40°/-32°, TEF=0°/0°, C=-40°, R=0°, β=0°, 1.5"-left strake deployed

3.2.3 Yaw control

So far, no significant control forces, either around roll- or yaw-axis are generated at moderate angles of attack, i.e. $\alpha < 30°$. At higher angles of incidence however, the available control power is drastically increased using asymmetrically deployed forebody strakes. By deploying only one strake, a turn is initiated to the opposite direction. This applies also for sideslip conditions, i.e. the turn rate can be further increased by deploying the strake on the leeward side of the nose. The yawing moments that can be achieved at sideslip angles of up to $\beta = 20°$ are even larger than the yawing moments due to a 30°-rudder deflection, Fig. 10. If the aircraft is already flying at a sideslip condition, yawing moments can be generated to turn it back by deploying the forebody strake on the windward side, Fig. 11. The efficiency of the strakes with regard to yaw control does however diminish with increasing sideslip angle.

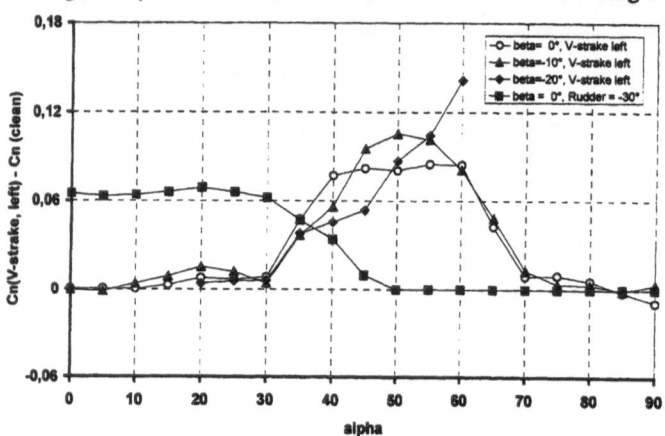

Fig. 10:   Effectiveness of left 1.5" forebody strake on yawing moment at negative sideslip
LEF=-40°/-32°, TEF=0°/0°, C=-40°, R=0°

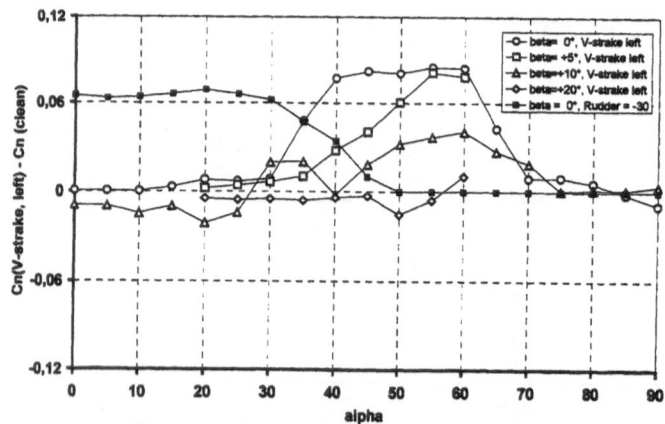

Fig. 11:   Effectiveness of left forebody strake on yawing moment at positive sideslip
LEF=-40°/-32°, TEF=0°/0°, C=-40°, R=0°, 1.5"-left strake deployed

At angles of attack beyond $\alpha > 45°$, the conventional rudder loses almost completely its effectiveness. In order to maintain lateral stability, artificial stability and/or thrust vectoring is

216

required. The use of forebody strakes for yaw control can reduce the deflection margins of ailerons and/or tailerons that are required for artificial stability and therefore enhance the agility and redundancy in case of thrust vectoring. As hardly any increment in pitching and rolling moment is generated by asymmetric strake deployment, the suitability of the forebody strakes for directional control is underlined.

4. Conclusions

In order to overcome the limitation of rudder effectiveness and to reduce the undamped roll characteristics of swept delta wing aircraft at high angles of attack, the efficiency of asymmetric deployed forebody strakes was investigated. It was shown, that the results were more or less independent of strake length. This should account for the fact, that the creation of vorticity at the aircraft forebody is mainly influenced by small disturbances at the nose of the fuselage within the laminar onset of the boundary layer. Up to angles of attack of $\alpha=30°$, no considerable control forces or moments are generated. At higher angles of incidence however, considerable yawing moments are created whose magnitude is in the order or even above the efficiency of a rudder deflection of $\eta_R=30°$. The use of asymmetric deployed forebody strakes not only enables the initiation of a turn, but also provides sufficient control power to stop a yawing movement and turn the airplane back to a no sideslip attitude. Though the forebody strakes did not provide any improvement for the unstable roll characteristics, they showed that very little rolling moment is induced due to strake deployment and therefore decoupled yawing moments are generated. The potential of forebody strakes for directional control becomes especially important at an angle of attack range where the conventional rudder loses its effectiveness completely. Further investigations should focus on the optimization of radial strake positioning to enable also roll control and enhance roll stability as well as an in depth analysis of flap efficiency of forebody strakes for yaw control.

References

[1]   Jobe C.E., Hsia A.H.: *Critical States and Flow Topology on a 65° Delta Wing*, AIAA 94-3479.

[2]   Ericsson L.E.: *Fluid/Motion Coupling in Conceptional Supermaneuvers*, AIAA 96-0787.

[3]   Klute S.M., Telionis D.P.: *The Unsteady Characteristics of the Flow Over an F/A 18 at High Alpha*, AIAA 96-824.

[4]   Phillips E.H.: *F/A-18 HARV Exploits Forebody Controls*, Aviation Week & Space Technology, Nov. 20, 1995.

[5]   Malcolm G.N., Ng T.T., Lewis L.C.: *Development of non-conventional control methods for high angle of attack using vortex manipulation*, AGARD-CP-465.

[6]   Malcolm G.N.: *Forebody Vortex Control*, AGARD-R-776: Special Course on Aircraft Dynamics at High Angles of Attack: Experiments and Modeling p.6-1 - p.6-40,   03/1991.

[7]   Ng T.T., Malcolm G.N.: *Aerodynamic Control Using Forebody Strakes*, AIAA-91-0618.

[8]   Fisher D.F., Cobleigh B.R.: *Controlling Forbody Asymmetries in Flight - Experience With Boundary Layer Transition Strips*, NASA-TM-4595, July 1994.

[9]   Bihrle W.: Applied Research: *Description of the LAMP facility Neuburg, FRG*, Personal communications.

[10]   Plaetschke E., Weiss S., Rohlf D.: *Comparison of Parameter Identification Results from Pilot and Flutter Test Box Input Maneuvers*, DLR Braunschweig Institut für Flugmechanik, IB 111-95/15.

[11]   Bihrle W.: *Use of Rotary Balance Data in the Prediction of Aircraft Dynamics*, AGARD-AR-265, pp.188-208, Dec. 1990.

Investigation of Stall Characteristics of an A3XX Relevant Airfoil up to High Reynolds Numbers in the Technology Program HAK 2

H. Hansen, I. Szabo

DaimlerChrysler Aerospace Airbus GmbH
Hünefeldstr. 1-5, D-28199 Bremen, Germany

Summary

The A3XX wing is designed for high Mach numbers and has thin, sharp airfoils (Fig. 1) which have a critical stall characteristic in high-lift configuration at Reynolds numbers obtainable in conventional wind tunnels. Lift is abruptly lost due to the sudden burst of a short laminar bubble. The inability of computational methods to either calculate or sufficiently approximate leading-edge (LE) stalls made an intense study of these phenomena in the BMBF (German ministry of education and research) sponsored technology program HAK 2 (High-Lift concepts 2) necessary. The objective of the following study was the experimental investigation of stall characteristics for this airfoil type for a range of Re numbers, including high Re numbers, in order to be able to make first predictions about the development of stall characteristics for the A3XX at flight Re numbers and to obtain a database for the validation and development of computational methods.

Situation

Low-speed studies at the DNW (German-Dutch wind tunnel) of an early wing design for the A3XX showed a critical stall characteristic for both the clean wing and the wing with deployed high-lift devices despite the relatively high Re numbers obtainable. A leading-edge stall was observed in both cases which dramatically reduced maximum lift. It has been observed that this stall type is defined by the sudden burst of a short laminar bubble on the leading-edge. Since this stall type is a strong function of the Re number, it can be assumed that at flight Re numbers of 40 million this stall type either disappears altogether or is very weakened. In the DNW Re numbers of 3 million are obtainable with this model. In order to make relevant observations about the flight stall characteristics of the model in the DNW a leading-edge modification was developed which, to a large degree, simulates stall characteristics and c_{Amax} under DNW conditions (Fig. 2). Computational methods under use at DA (DaimlerChrysler Aerospace Airbus GmbH) had considerable difficulties with these stall characteristics. Further development of these methods was started under the technology program HAK 2 and an improvement was achieved in the Re number range covered by conventional wind tunnels. Using these computational methods, a leading-edge modification was implemented that experimentally simulates the calculated flight stall characteristics in the DNW. This configuration was then successfully tested in the DNW. Further high Reynolds number experiments with the original airfoil, i.e. without LE modification, were deemed necessary to validate the concept of the LE modification.

2D airfoil model for the cryogenic wind tunnel Cologne (KKK)

There is a lack of experimental data in the range above Re=10 million and different computational methods show large variations of maximum lift and stall characteristics for these Re numbers. A high Re number experiment was deemed necessary to validate both the design concept of the leading-edge modification and the computational methods used, as well as to obtain data to assess the flight characteristics of the A3XX. An airfoil model with a chord length of 600mm was constructed based on the outboard wing of the DNW model without the LE modification. This model was tested in the KKK under cryogenic conditions at Re numbers of up to 14 million. That is on the same order of magnitude as the flight Re number on the outboard wing in the high lift flight range (ca. 18 million). The model is shown in the KKK in Fig. 3, the main dimensions and the position of the pressure taps are shown in Fig. 4. Lift was integrated over 54 pressure taps on the middle airfoil section. Drag was not measured. The two-dimensionality was checked by two additional rows of chordwise pressure taps and by three rows of spanwise pressure taps at the leading-edge, at 30% chord and at the trailing edge. Wall interference effects were reduced through blowing out through slots in front of the airfoil. For the entire Re number range flow was observed with the help of mini-tufts on the upper side of the airfoil. Oil flow images were made at room temperature for better observation of stall characteristics.

Detailed measurements of the clean configuration were made, further measurements of the high lift configuration will follow.

Experimental results and comparison with computational methods

It was possible to measure a large range of Ma and Re numbers with this airfoil model (Re = 2.8 - 14.5 million, Ma = 0.1 - 0.2, Fig. 5). A few important results of these experiments will be presented.

In Fig. 6 examples of lift curves as functions of Re and Ma numbers are shown. The strong effect of the Re number on the lift curve is clearly visible. At low Re numbers lift is abruptly lost in the linear part of the curve, while at higher Re numbers the lift curve first shows a gradual decrease of the lift gradient followed by an abrupt loss of lift. A study of these lift curves would classify the stall characteristics at the lower Re number as a typical leading-edge stall that changes to a combined leading- and trailing-edge stall for higher Re numbers. An analysis of the pressure distributions (Fig. 7) show that at high Re numbers (Re=14 million) a leading edge stall still occurs, although it is not as pronounced as at low Re numbers, as the change in c_{pmin} shows. The KKK experiment shows a weaker tendency towards a trailing-edge stall compared to the computationally expected stall characteristic at flight conditions (Fig. 2), although the rise in maximum lift due to Re number effects is predicted quite well. Therefore it can be assumed that the leading edge modification in the DNW simulates the flight stall and maximum lift characteristics under wind tunnel conditions. The stall characteristics with the leading-edge modification are somewhat too favorable (Trailing-edge stall with LE modification at low Re numbers compared with a combined stall without LE modification at high Re numbers).

Fig. 8 shows the experimental maximum lift as a function of the Re number compared with the maximum lift as calculated by the three 2D computational methods in use at DA: EPPLER [1], HILI [2] and MSES [3]. The experimental results are shown for Ma=0.2, with and without blowing on both sides of the airfoil. At low Re numbers, blowing has a large effect on maximum lift, which becomes markedly less at high Re numbers. The amount of blowing was chosen so that the three spanwise and the three upper chordwise pressure measurements show a high two-dimensionality. The 2D flow was also controlled through observation of the mini-tufts. Fig. 9 shows an example of the spanwise distribution of cp and the effects of blowing at c_{Amax}: there are only weak spanwise effects at the tunnel walls before c_{Amax} that can be minimized through blowing. In the post-stall region the example in Fig. 9 shows strong spanwise effects with separation all along the right side of the airfoil. These effects, as can be seen on all three spanwise stations, can be reduced by blowing. The leading-edge pressure taps, as is to be expected, are affected the most.

Therefore, when comparing experimental results with computational methods it is to be assumed that the experimental results with blowing yield the correct results. As is visible in Fig. 10, both computational methods EPPLER and HILI show a good qualitative agreement with the observed dependence of maximum lift on Re number although the computational methods show higher maximum lift than is actually observed at lower Re numbers. The highest code quality MSES method by M. Drela (Euler isentropic multi-element code with a fully coupled boundary layer formulation, able to calculate separation bubbles [3]) calculates a maximum lift that is clearly too high, although the Re number characteristics are, for the most part, correctly described. The comparison of all three calculated lift curves with the experimental results show that both HILI and EPPLER in most cases describe the leading-edge stall correctly while MSES indicates a trailing-edge stall. It has to be noted that HILI would also calculate the same stall characteristic with a too high maximum lift if it were not adapted for an empirical pressure difference rule which limits the maximum increase in pressure obtainable without separation. After this point the calculated lift is set as constant. It is astonishing to note how well EPPLER (Panel method without boundary layer coupling, with very effective separated region modeling [1]) calculates both maximum lift and lift curve. This confirms observations made at DA that this method, which can only be used for clean airfoils, works very well for a fast analysis of stall and maximum lift characteristics as a function of Re number. The leading-edge modification for the DNW model was also calculated using this method.

Fig. 11 shows that MSES is capable of reproducing both the size and position of the short laminar bubble that is observed at low Re numbers. At higher Re numbers (Re=14 million) the calculations still show a very short laminar bubble which can not be observed in the experimental results. Further investigations have to be made as to whether an early transition is responsible for these experimental effects. It could also explain the loss of maximum lift at high Re numbers which can not be observed in any of the computational methods described above. This loss of maximum lift was also observed at other Ma numbers and is unusual for 2D flows that are not influenced by attachment lines and cross flow instabilities. Other observed flow phenomena, e.g. the strong lift enhancing effect of an oil film on the leading-edge can not be gone into in this article.

Conclusions

Due to the observed critical stall characteristics of an A3XX relevant wing design a series of experiments with a large 2D airfoil model for the cryogenic wind tunnel in Cologne were conceived and carried out. With this airfoil model and the use of cryogenic technology the flight Re numbers of the A3XX outboard wing were achieved. The experiment was used to examine the new concept of a leading-edge modification of a large DNW-A3XX model by computational methods. The objective of this leading-edge modification was not to improve the flight characteristics of the aircraft, but specifically to modify the existing wing in such a manner that the calculated expected maximum lift and, more importantly, the stall characteristics of the aircraft at flight Re numbers was obtained at DNW Re numbers. This concept was, for the most part, validated through the experiments in the KKK.

This experiment provided a lot of data for the validation of computational methods in spite of the expected difficulties of measuring under cryogenic conditions. It can be seen that in spite of the simple single element configuration high quality computational methods, which do without empirical stall models, have great difficulty calculating the very complex process surrounding the burst of a short bubble on the leading edge. This data is also being and will continue to be used to validate Navier-Stokes methods. Preliminary Navier-Stokes calculations show that considerable development of these methods is still required. An exchange of experimental data with developers of computational methods is therefore very much desired.

This 2D model is a very interesting basis for further investigations under flight conditions. It should be used, both in clean and high lift configuration, and supplemented by extensive measurement technology to create a national, 2D, high Re number validation database for computational methods. This database will be added to by validation experiments within the HAK 2 technology program at the DNW with the large 3D model H8Y, on which the 2D model is based, as well as the H8Y half-model which will soon be measured at high Re numbers (Re=11 million) at Onera's F1 wind tunnel. The necessity of these fundamental experiments, which also make an important contribution to A3XX high lift development, could clearly be proved.

References

[1] R. Eppler, D. M. Somer: *A Computer Program for the Design and Analysis of Low-Speed Airfoils.* NACA TM 80210 1980.

[2] G. Dargel, H. Jakob: *Berechnungsverfahren für viskose Klappenprofilströmungen - Rechenprogramm HILI.* DA-Bericht TE2-1632 1988.

[3] M. Drela, M. Giles: *ISES: A Two-Dimensional Viscous Aerodynamic Design and Analysis Code.* AIAA-86-0424, January 1987.

[4] W. Valarezo, V. Chin: *Method for the Prediction of Wing Maximum Lift.* Journal of Aircraft, Vol. 31, No. 1, Jan.-Feb. 1994 pp. 103-109.

[5] D. Reckzeh: *Ermittlung des Ablöseverhaltens am H8Y-Flügel mit dem Druckdifferenzenkriterium.* DA-Bericht EFP-279/97 1997.

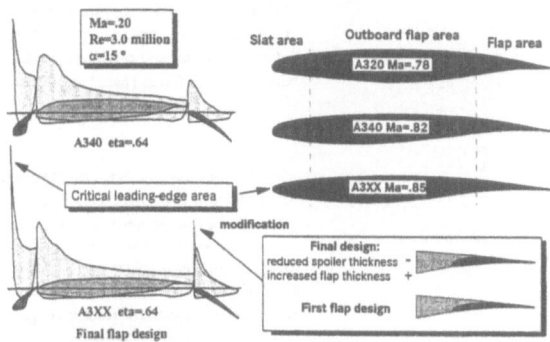

Fig. 1: Difficulties with the thin A3XX wings

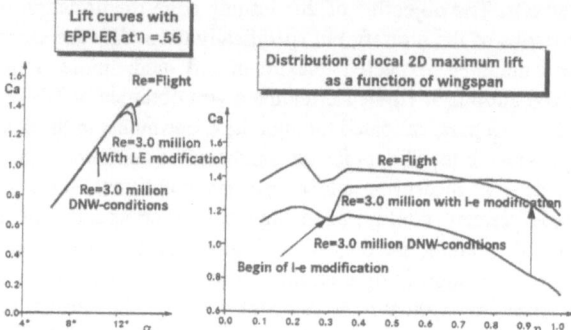

Fig. 2: Leading-edge modification as a means of computational simulation of flight conditions

Fig. 3: H8Y Megaliner 2D-airfoil in the cryogenic wind tunnel Cologne

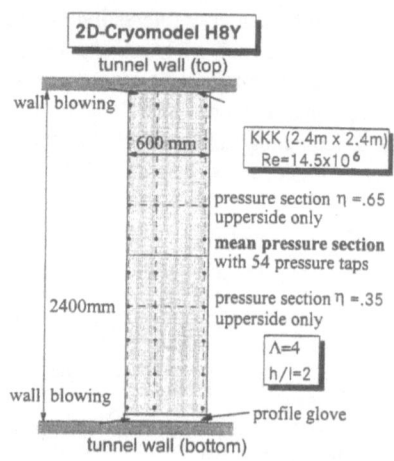

Fig. 4: Measuring equipment and dimensions of the H8Y Megaliner 2D-Airfoil in the KKK

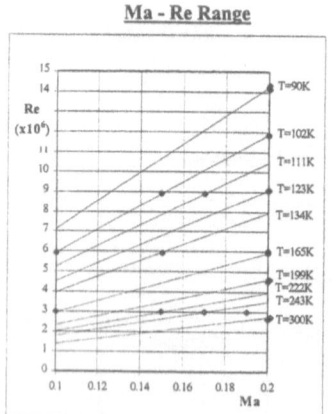

Differences Phases 1/2

Phase 1:
- clean configuration
- 1 chordwise row of pressure taps at η = 50% (54 taps)
- 1 spanwise row of pressure taps at x/l=30% (10 taps us)
- motor on only one side of airfoil, no brake
- Gap between airfoil and tunnel roof of about 5 mm
- blowing only on one side of airfoil (preliminary experiment)

Phase 2:
- clean configuration und first high lift experiments
- additional rows of pressure taps:
  2 chordwise rows at η = 35%, 65% (11 taps each us)
  2 spanwise rows at x/l = 1%, 97,5% (10 taps each us)
- motor on one side of airfoil, with brake and position measurement on opposite side
- new airfoil end pieces and reduction of gaps
- blowing on both sides of the airfoil

Fig. 5: Ma and Re range and description of both experimental phases

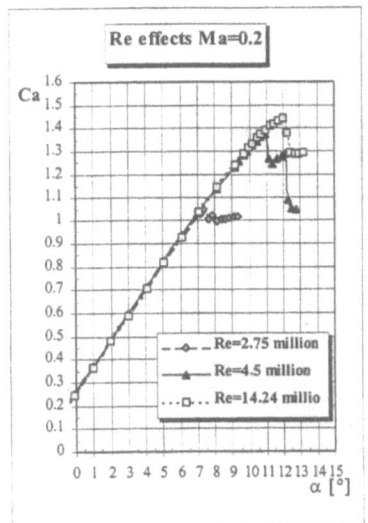
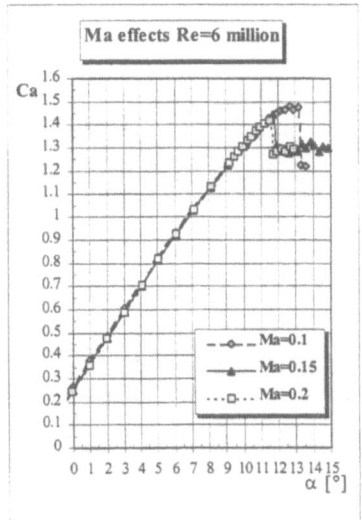

Fig. 6: Re and Mach effects on the 2D-Clean airfoil

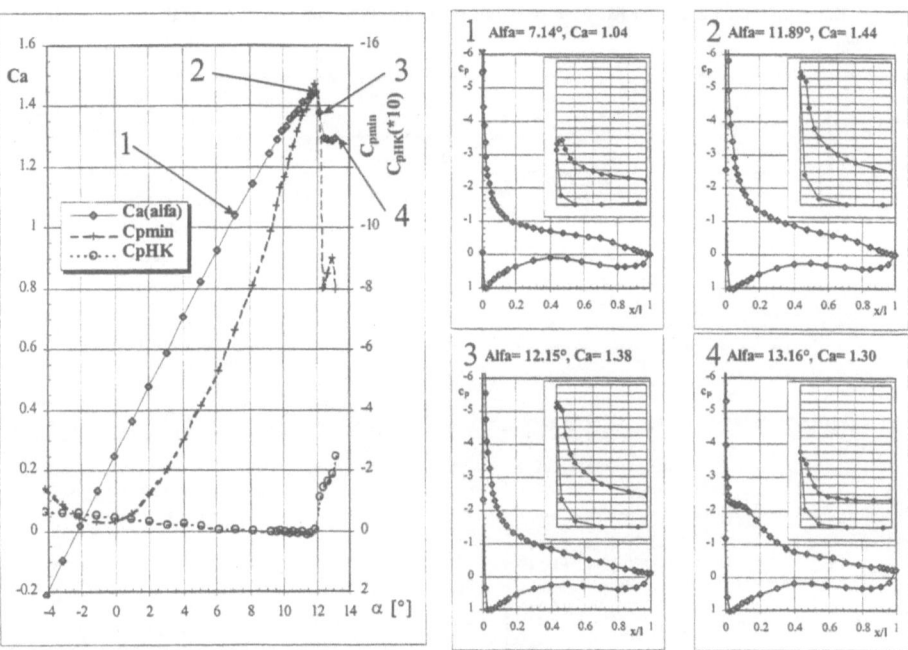

Fig. 7: Lift curves and pressure distributions for Re=14.24 Mio. and Ma=.20

223

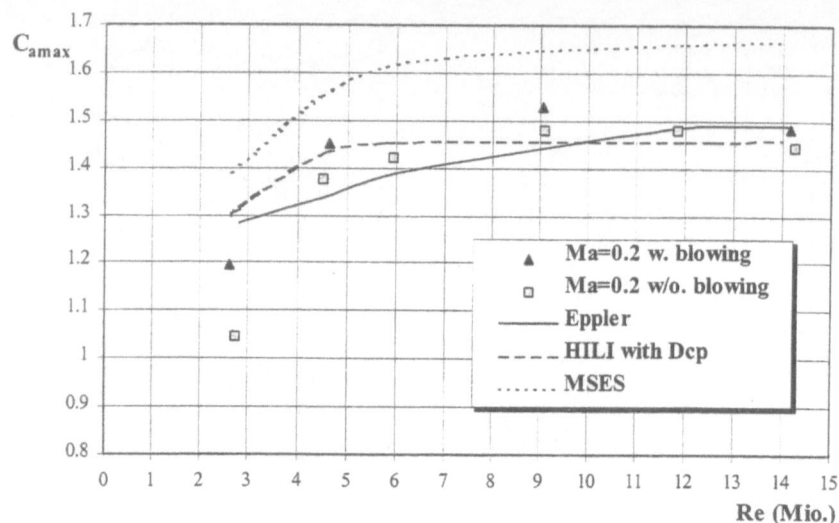

Fig. 8: Experimental maximum lift as a function of Re number and compared with 2D computational methods

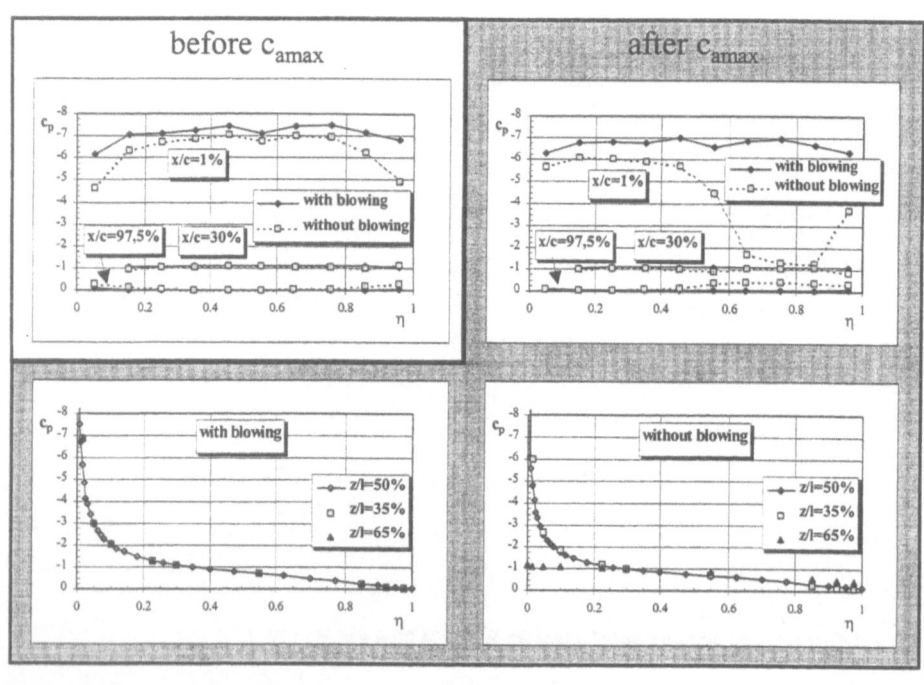

Fig. 9: Effects of wall blowing on spanwise pressure distribution

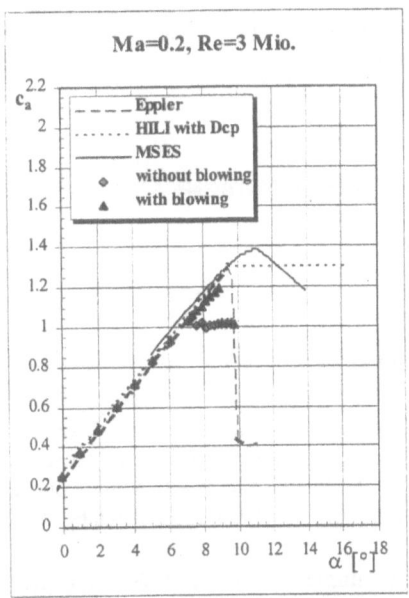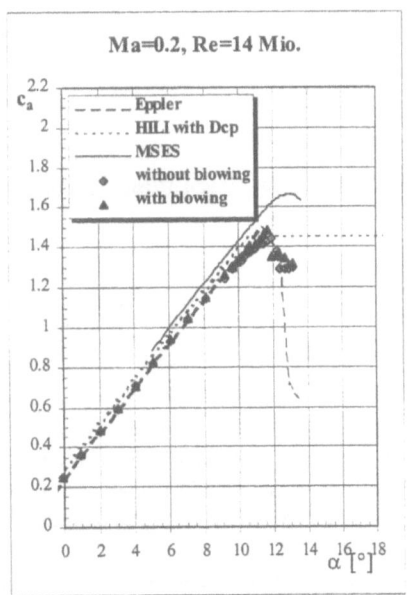

Fig. 10: Comparison of experimental and computational lift curves as a function of Re number

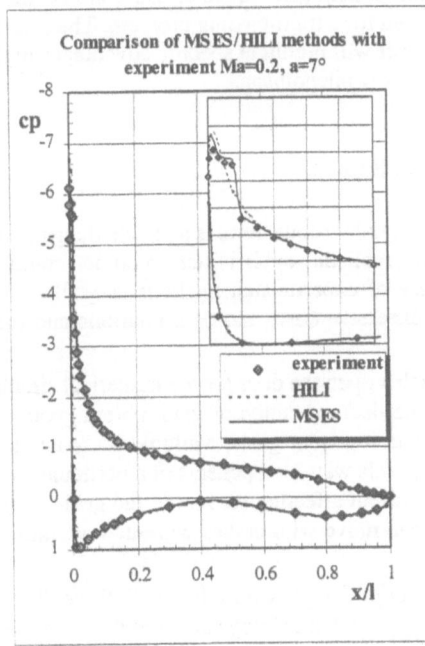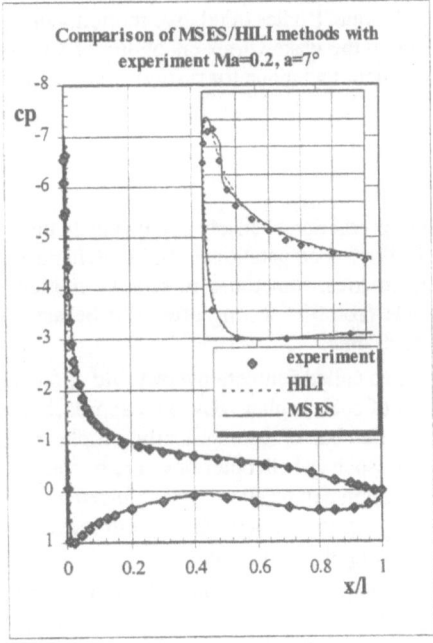

Fig. 11: Comparison of size and position of experimental and computational short bubble as a function of Re number

Numerical Simulation of Three-Dimensional Flows Using the Chimera-Technique

Ralf Heinrich[*], Nikolai Kalitzin[+]

[*]DLR Institute of Design Aerodynamics, Lilienthalplatz 7
D-38108 Braunschweig, Germany

[+]Institut für Akustik und Sprachkommunikation
D-01062 Dresden, Germany

Summary

The Chimera-technique has been implemented in the three-dimensional, block-structured Navier-Stokes solver FLOWer to enhance the flexibility of the code. Overlapping meshes can be used, which remove the constraints induced by the inter block connectivity. On the one hand the grid generation effort can be reduced, on the other bodies in relative motion can be handled easily.

After a brief introduction into the Chimera-technique implemented in FLOWer, a variety of applications will be shown, starting with validation test cases using the viscous and inviscid flow around airfoil-flap configurations. The integration of a flap-track-fairing on a transport aircraft wing will underline the idea of merging component grids together using the Chimera-technique. Bodies in relative motion will be presented for a train-passing problem. The simulation of the wake vortex encounter of a generic aircraft will highlight specific advantages of the Chimera-technique for resolving complex physical flow phenomena.

Introduction

The prediction of steady and unsteady three-dimensional viscous flows is a big challenge in the field of computational fluid dynamics. Many realistic applications include complex geometries, sometimes in relative motion. The use of experimental methods as well as full scale prototype testing often fail because of the excessive costs, model limitations and time constraints.

The so called Chimera overset grid approach [1] helps open the door for the numerical simulation of such applications. The approach involves the decomposition of the problem geometry in a number of geometrically simple overlapping component grids. Multiple-body applications, such as helicopter flow, are treated naturally in this way. Components of a particular configuration can be altered, or changed completely, without affecting the rest of the grid system. Grid components associated with moving bodies can move with bodies without stretching or distorting the grid system.

The computational incentive for employing an overset grid approach for unsteady three-dimensional flow is multiple. The flow solution process is applied to topologically simple component grids. Body-fitted component grids are ideally suited to regions of thin shear flows such as boundary-layers, wakes, etc. All the advantages associated with structured data are realizable in the approach, included highly efficient implicit flow solvers, vectorization, and parallelism.

As a logical consequence the Chimera-technique has been implemented in the existing block-structured Navier-Stokes solver FLOWer within the national CFD-project MEGAFLOW [4], to enhance the flexibility of the code. Therefore the experiences made with the cell-centred Euler-Solver ROTCATS, specially designed for helicopter applications, were very helpful [2][3]. Some of the basic algorithms could be transferred to FLOWer only with slightly modifications.

Description of the Algorithm

Basic features of FLOWer

FLOWer solves the three-dimensional Reynolds-averaged full Navier-Stokes equations [4]. The second order spatial discretization is based on a finite volume cell-vertex method. The fluxes are approximated using a central discretization operator. To avoid spurious oscillations, in FLOWer the well known scalar dissipation model of Jameson is implemented [6]. A 5 stage Runge-Kutta scheme is used for the time-integration. The convergence can be accelerated using local time-stepping, implicit residual smoothing and a multigrid method. Unsteady solutions can be obtained by using the basic explicit scheme, or the implicit dual-time stepping method can be chosen [7][5]. All blocks are now allowed to move independently, having three translational and rotational degrees of freedom.

Chimera aspects

In a Chimera-style overset grid approach, domain connectivity is achieved through interpolation of necessary intergrid boundary information from solutions in the overlap region of neighbouring grid systems. Consider, for example, the simple two grid discretization of the NACA0012 airfoil shown in Fig. 3. The physical domain is decomposed into a body-fitted grid near the airfoil surface and a background Cartesian grid which extend out to the farfield boundaries. The Cartesian mesh completely overlaps the airfoil grid. The information transport from the off-body Cartesian grid is realized by interpolating solutions from the Cartesian mesh at the near-field boundary of the body-fitted grid (quadratic symbols). Two layers of points are interpolated, to guarantee a second order spacial discretization inside the field. A similar transfer of information from the near-body solution back to the off-body solution is required. The off-body Cartesian grid has no natural boundaries that overlap the near-body grid. The Chimera-style of overset gridding makes it possible, to create an artificial boundary (hole-boundary) within the off-body grid system, and thereby establish the required near-body to off-body connectivity.

A hole boundary for this example is created by excluding the region of the off-body Cartesian grid that is overlapped by the airfoil. The resulting hole is excluded from the remaining off-body solution. Conditions for the hole boundary points (circular symbols) are interpolated from the solution in the near-body airfoil grid.

To cut holes inside grids, several techniques have been implemented in FLOWer. For example a hole can be defined by a simple geometry, like a box or a cylinder. More complex holes can be defined by an auxiliary mesh, or simply by grouping two or more hole-definition geometries. In Fig. 1 the principle hole definition algorithm is graphically summarized. In a first step, iblank is set to 0 inside the hole-definition geometry. Afterwards the hole-boundary points are identified (black bullets). In the third step one layer of points is added. Thus the second order spacial discretization is guaranteed for all other points (iblank = 1).

227

For the interpolation of the hole-boundary and nearfield-boundary points, donor cells have to be located in the corresponding partner grid. In a first step generally always the nearest point of the corresponding grid is located. Several techniques are used to speed up the searching process. If the donor grid is of equidistant Cartesian type, the indices of the donor cells can be computed directly. For non equidistant Cartesian meshes, the algorithm only needs three sweeps, one in each coordinate direction, to locate the nearest point. For curvilinear grids, the algorithm starts searching on a coarser grid level. When the nearest point on this level is found, the algorithm switches to the next finer level to search in the cell layers around the located point, until the finest level is reached (Fig. 4).

After locating the nearest point in the corresponding mesh, the donor cell, including the point, has to be located in the surrounding layers. Therefore all cells in the neighbouring layers are divided into 6 tetrahedra (Fig. 2). For the donor tetrahedra, surrounding the point, trilinear interpolation coefficients a1, a2 and a3 are calculated. The interpolation used is non-conservative [2].

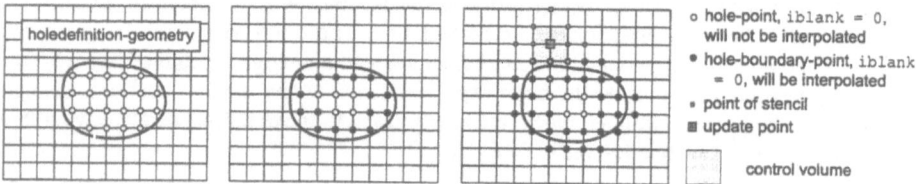

Fig. 1   Hole definition

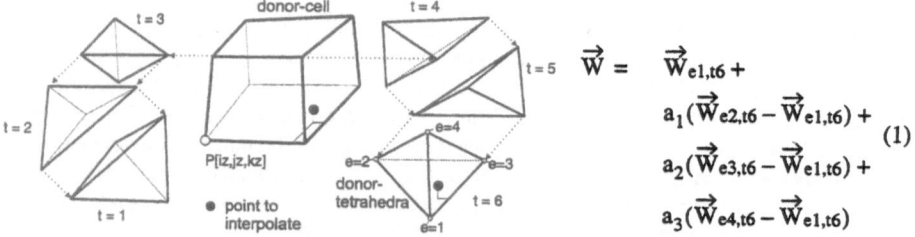

$$\vec{W} = \vec{W}_{e1,t6} + a_1(\vec{W}_{e2,t6} - \vec{W}_{e1,t6}) + a_2(\vec{W}_{e3,t6} - \vec{W}_{e1,t6}) + a_3(\vec{W}_{e4,t6} - \vec{W}_{e1,t6}) \qquad (1)$$

Fig. 2   Decomposition of a grid cell into 6 tetrahedra

The array iblank can directly be used within the Runge-Kutta time stepping, to exclude hole points from the update procedure. So the Runge-Kutta update is changed from

$$\vec{W}_{i,j,k}^{(s)} = \vec{W}_{i,j,k}^{(o)} - \alpha_s \frac{\Delta t}{V} \vec{R}_{i,j,k}^{(s-1)} \quad \text{to} \quad \vec{W}_{i,j,k}^{(s)} = \vec{W}_{i,j,k}^{(o)} - iblank_{i,j,k} \alpha_s \frac{\Delta t}{V} \vec{R}_{i,j,k}^{(s-1)} . \quad (2)$$

$\vec{W}$ denotes the conserved quantities, s the Runge-Kutta stage number, α the Runge-Kutta coefficients, $\vec{R}$ the Residual, Δt the timestep and V the control volume. By using the iblank array, it is not necessary to provide special branching logic to avoid hole points, and all vector and parallel properties of the basic algorithm remain unchanged. So it is in fact very easy, to adapt an existing multiblock flow solvers of documented accuracy and efficiency for overset grids, which is a major advantage of the Chimera-method.

The hole-cutting-procedure as well as the needed searching-algorithms for location of the donor cells can be handled outside the flow-solver. This makes sense especially for steady calculations, because the additional Chimera-information has to be generated only once. For unsteady applications including grids in relative motion, for each timestep this information has to be supplied. Then the integration of the Chimera-package inside the flow-solver is useful. This idea is followed in the FLOWer code. In the present state of the implementation of the method, a user needs special expertise to set up a Chimera-mesh and computation. To open the door for non-specialists using the code and mesh-generating system, additional work is required in the direction of automation.

Results

For validation purposes results obtained on standard block-structured meshes (that means multiblock meshes without overlapping domains) were compared to results of Chimera-computations. Fig. 5 shows results of the viscous flow around a modified NACA 63-215 profile with a flap. The Chimera-mesh set-up uses two component meshes around profile and flap and one Cartesian background mesh. To facilitate the comparison the point distributions on profile and flap are the same for both computations, as well as the wall normal distance in the boundary-layer. The agreement of both cp-distributions is excellent. The difference of the global forces is 0.2% for the lift and 0.5% for the total drag.

An exact solution of the plane potential flow about a Kármán-Trefftz profile with flap can be obtained [8]. Fig. 6 shows the exact calculated cp-distribution on profile and flap compared to a Chimera-computation. The overall agreement is very good. Only the peak on the leading edge of the flap is a little bit underestimated. The Chimera-mesh contains two overlapping component c-meshes. The flap mesh is embedded inside the profile mesh. In the figure only every second grid line is shown.

Fig. 7 shows an example for the integration of a flap-track-fairing (FTF) on a wing of a transport aircraft. A C-H mesh has been generated around the wing. This component mesh is embedded inside a simple H-mesh, to transport numerical and physical disturbances to the far-field boundary. The FTF component mesh has been generated independently of the wing mesh. The input needed of course is the surface geometry of the FTF and the wing. To guarantee a good communication between the FTF component mesh and the wing mesh, points have been concentrated in spanwise direction on the wing in the region of the location of the FTF. In general, the grid resolution of communicating blocks should be comparable in the overlapping area if high gradients have to be resolved. In the present case the density of the background mesh is much coarser, compared to the wing mesh in the region of the hole boundary. Because of the relative small gradients along the hole boundary, no negative influence on the overall accuracy of the solution can be noted. In the presence of shocks, the resolution of the background mesh would be insufficient. Along the outer-boundary of the FTF-mesh almost no discontinuity of the pressure lines can be seen, which is an indicator of the correct implementation of the Chimera-technique (and of course of the correct setup of the Chimera-logic).

Fig. 8 shows an unsteady example including bodies in relative motion. Two ICE-similar high-speed trains were passing with a speed of 250 km/h. The component meshes around the trains consist of 3 standard blocks and are embedded in a Cartesian background mesh. The time integration is performed explicitly. The cp-distribution shows the imprint of the pressure wave induced by train I on the surface of train II.

Several actual research projects work on problems related to the wake vortex fields generated by large aircrafts. Especially small aircraft can run into difficulties encountering a wake vortex disturbed flow field. Within the EU-project WAVENC (wake vortex evolution and wake vortex encounter) experimental studies have been performed in the DNW wind tunnel [9]. A wake generating model has been placed 17 m in front of a smaller (half wing span) generic aircraft (SWIM geometry). The vortex disturbed flowfield has been measured in a plane about 1 m in front of the model. The SWIM-geometry has been placed at several positions in the vortex disturbed flowfield. Force and moment coefficients have been measured as well as pressure distributions.

Euler simulation shall be performed by prescribing the vortex disturbed flowfield as an inflow boundary conditions. Standard methods will run into difficulties transporting the vortices from the inflow boundary to the generic aircraft without reducing the vortex strength unacceptably. Of course well suited block-structured meshes for special positions of the vortices can be generated, but the man power needed is unacceptable high, especially if you think of studying the influence of the aircraft position in the vortex disturbed flowfield. So the idea is to use special vortex transport meshes, which can be placed at several positions of the flowfield, using the Chimera-technique. Additionally to the technique of vortex transport meshes, the component grid idea has consequently been followed for the mesh setup of the SWIM-geometry (Fig. 9, every second mesh line is shown). Grids have been generated around wings, vertical tail, horizontal tails and body. The mesh generating procedure only takes one day for a non-specialist using MEGACADS, which is the grid generating system within MEGAFLOW.

Fig. 10 shows a first promising result using the idea of vortex transport meshes. For test purposes only one vortex is presribed at the inflow boundary. In the lower right corner gives an impression of the vortex-transport mesh (every forth grid line is displayed). The influence of the clockwise rotating vortex on the pressure distribution can be seen clearly, which results in an anti-clockwise rolling moment. Unfortunately the computation of the global force and moment coefficients is not possible if overlapping surface meshes are used in the actual version of FLOWer. This will be fixed in the nearer future. A detailed comparison of numerical simulations with the experimental data base will be performed in the first half of 1999.

Conclusions

The flexibility of the DLR block-structured Navier-Stokes solver FLOWer has been significantly enhanced by implementing the Chimera-technique. All advantages associated with structured data are realizable in the approach. Decomposition of complex domains into a number of overlapping components simplifies the mesh-generating process. The wake vortex encounter problem shows the capabilities of the Chimera-method for resolving physical phenomena. Additionally, applications with bodies in relative motion can now be handled easily. In the present state of the implementation of the method, a user needs special expertise to set up a Chimera-mesh and computation. To open the door for non-specialists using the flow solver and mesh-generating system, additional work has to be done in the direction of automation.

References

[1]   Steger, J.; Dougherty, F. C., Benek, J. *A Chimera Grid Scheme*. Advances in Grid Generation, K.N. Ghia and U. Ghia, eds., ASME FED-vol 5. pp. 59-69.

[2]   Pahlke, K. *Berechnung von Strömungsfeldern um Hubschrauber im Vorwärstsflug durch die Lösung der Euler-Gleichungen* PHD-thesis, to be published as DLR-FB.

[3]   Schwarz, T. *Berechnung der Umströmung einer Hubschrauber-Rotor-Rumpf-Konfiguration auf Basis der Euler-Gleichungen mit der Chimären Technik* DLR-IB 129 - 97/23, 1997.

[4]   Kroll, N. *National CFD-Poject MEGAFLOW - Status-Report* - In: Notes on Numerical Fluid Mechanics, Volume 60, eds.Körner, H. and Hilbig, R., Vieweg 1997.

[5]   Heinrich, R.; Pahlke, K.;Bleecke, H. *A Three Dimensional Dual-Time Stepping Method for the Solution of the Unsteady Navier-Stokes Equations* In: Proc. Unsteady Aerodynamics, London, 17.-18.7.1996, ISBN 1857680820 (1996).

[6]   Jameson, A.; Schmidt, W.; Turkel, E. *Numerical Solutions of the Euler Equations by Finite Volume Methods Using Runge-Kutta Time Stepping Schemes* AIAA-Paper 81-1259, 1981.

[7]   Jameson, A. J. *Time dependent calculation using multigrid, with applications to unsteady flows past airfoils and wings* AIAA-Paper 91-1596, 1991.

[8]   Williams, B. R. *An exact Test Case for the Plane Potential Flow About Two Adjacent Lifting Aerofoils* Ministry of Defence, Reports and Memoranda No. 3717, 1971.

[9]   Hegen, G. H. *Wake encounter test in DNW wind tunnel - test number: 98-1116*, NLR-report NLR-CR-98291, 1998.

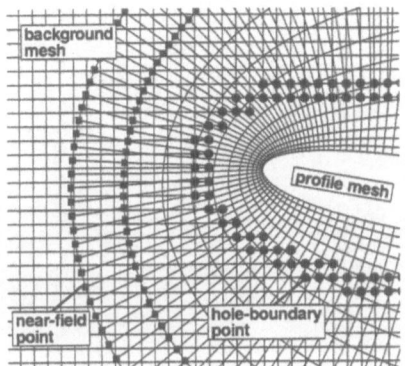

Fig. 3   Near field boundary and hole boundary
points

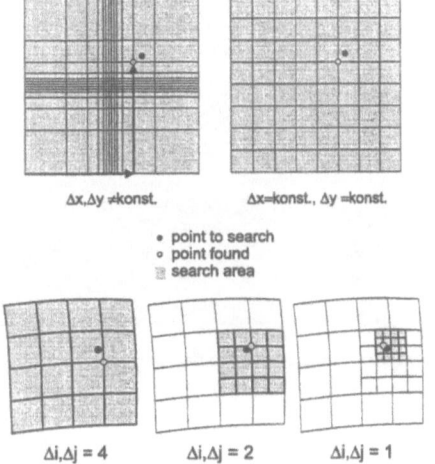

Fig. 4   Searching algorithms

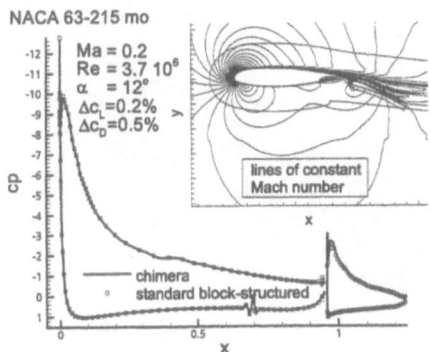

Fig. 5   Comparison of standard block-structured
and Chimera-method for the viscous flow
around a profile flap configuration

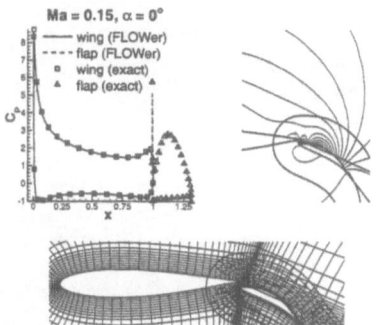

Fig. 6   Comparison of numerical and exact solution
for the flow around a Kármán-Trefftz pro-
file-flap configuration

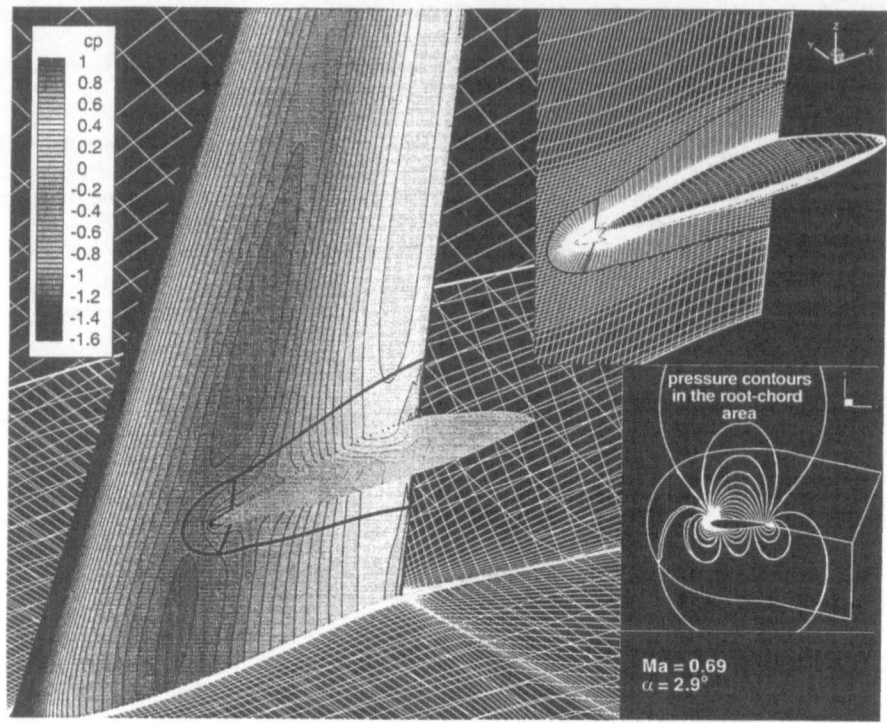

Fig. 7   Integration of a flap-track-fairing

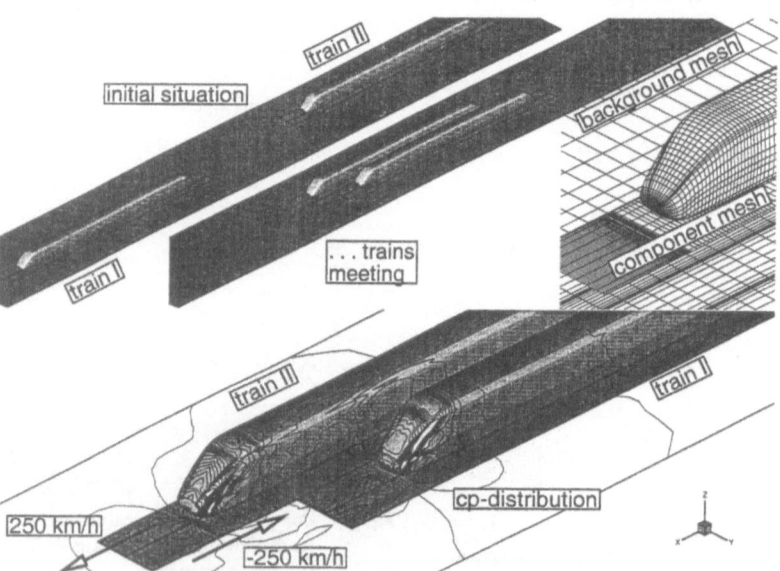

Fig. 8   Unsteady simulation of the meeting of high speed trains

232

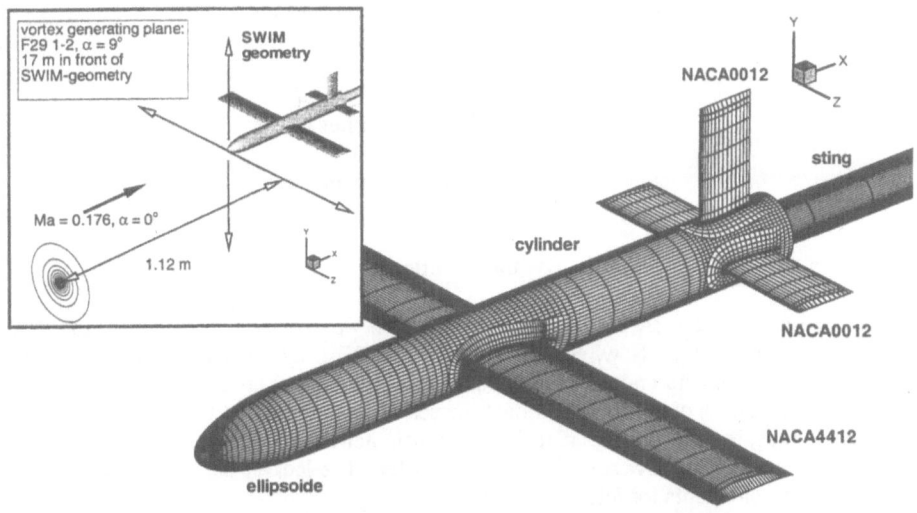

Fig. 9   Chimera-surface meshes of components for a generic aircraft (SWIM geometry)

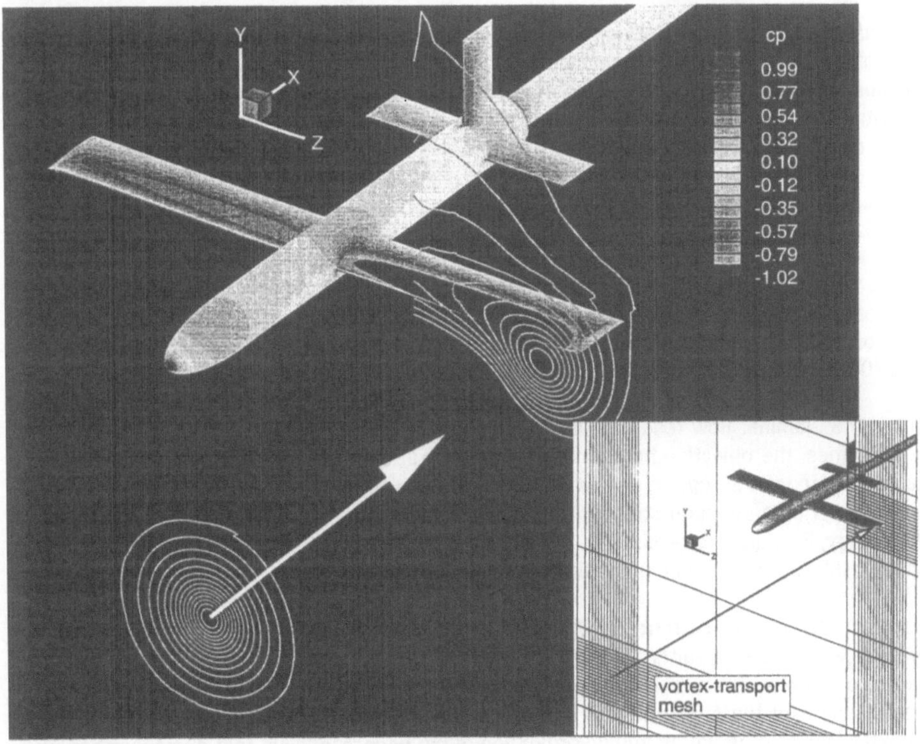

![Numerical simulation figure]

Fig. 10   Numerical simulation of a wake vortex encounter

First Results from the „A 320 HLF Fin" Programme

Rolf Henke

DaimlerChrysler Aerospace Airbus GmbH
Hünefeldstraße 1 - 5
D-28199 Bremen, Germany

Introduction

An initial report on the „A 320 HLF Fin" programme was given in Erlangen at the STAB symposium in 1994, [1]. Now, concurrrent with this STAB Symposium (Berlin, 1998), the programme is nearing its end with the completion of the final flight test campaign. Because we do not yet have all of the flight test data, no scientific evaluation of these can be made at this point. Thus, the present report reviews the main activities of the programme within the larger European laminar flow research context. A tentative look at our results will indicate some of the requirements for future activities.

Technical Background and Programme History

Laminar flow research in Germany and Europe during the past 15 years has focused on the understanding of flow physics, including transition mechanisms such as Tollmien-Schlichting Instability (TSI), Cross Flow Instability (CFI), and Attachment Line Transition (ALT). At first the activities concentrated on Natural Laminar Flow (NLF), i.e. laminarisation by profile shaping. Later on laminarisation by suction in the nose region together with NLF, so-called „Hybrid Laminar Flow „ or „HLF," has been investigated. Naturally, the main tasks of laminar research programmes were experiments at high Reynolds numbers. Flight tests with an NLF glove on a Fokker 100 [2], and wind tunnel experiments with large HLF half models [3] were among the events that took place within Europe. To a certain extent it has been possible to extrapolate data from NLF experiments in the wind tunnel to flight data. However, because of the possible influence of noise and turbulence in the tunnel, this data has been questioned in case of HLF. The influence of noise and turbulence remains one of the unknowns within the „A 320 HLF Fin" programme.

In the U.S., laminar flow research has received massive funding in civil as well as in military programmes, the objective being more on the operational side. Meanwhile, basic research has dominated the European approach. Corresponding to this approach, the European Airbus partners set up a common strategy for „Laminar Technology (LaTec)" as part of their overall 3E (Energy, Economy, Environment) Technology Plan, with flight tests as major highlights. The „A 320 HLF Fin" programme is the first step within that LaTec strategy [4].

As precursor of the current „A 320 HLF Fin" programme, ONERA, France, worked on theoretical aspects for aircraft selection. Based on the results, they and DLR, Germany carried out two wind tunnel test campaigns with a 1:2 scale HLF fin model. Because of DA 's responsibility for the A 320 fin and interest in laminar technology, leadership of the present „A 320 HLF Fin" programme and co-ordination of the necessary flight test demonstration was delegated to DA. In the partner countries, the „A 320 HLF Fin" programme has been taken up in the national research programmes, in Germany as the „ELAS" programme [5]. Furthermore, a connection to the EC programme HYLDA has evolved through the circumstance of HYLDA renting the A 320 HLF Fin for its own flight tests.

Both in light its technical content as well as its organisational structures, which, within Germany, for example, included an extraordinary cooperation of industry, research establishments, universities and suppliers, the A 320 HLF Fin programme must be considered as an essential part of the overall laminar flow research activities within Europe. The partnership, in the preliminary programmes as well as the „A 320 HLF Fin" programme itself, has proven successful both technically as well as from the management side.

Programme Description

Due to its connection with Airbus as well as its nature as a flight test demonstration, the programme was multinational as well as multidisciplinary. The partners agreed on objectives, and, based on these objectives, the content of the programme was established. Because this largest of Airbus technology programmes of the 1990's was a pilot programme, both technical and managerial objectives were of concern.

Technical objectives:
- Modification of the A 320 vertical tailplane to establish laminar flow well over the fin box, without modifying the box itself. Because the A 320 fin already has a suitable profile for laminarity—one of the results of the theoretical study—only the fin nose had to be modified to include a suction system.
- Flexibility of the HLF system in order to learn about the HLF physical boundaries. Thus, the suction specification was not only focused on the cruise case but also on a variation of Mach number and altitude resulting in a large Reynolds number range.
- Rating of the HLF technology for transport aircraft application. This objective had already influenced the structural and system concept phase, as well as the final definition of the flight test programme. Though the next step in the LaTec programme is an operational fin flight test, the basis for this will remain the current A 320 HLF Fin programme. Thus, e.g. de- or anti-contamination has already been addressed in this programme.

Management objectives:
- Transcend the earlier monodisciplinary orientation by implementing an international, interdisciplinary programme management.
- Agree on work shares and cost shares within a technology programme by means of special prototype contracts between all programme participants.
- Use an Airbus flying test bed for technology demonstration, because the requirements on the A 320 test aircraft as laminar test bed are far different from its normal daily use.

The programme included the typical disciplines: Aerodynamics, systems, structure and manufacture, flight test preparation, flight test, and integration (i.e. project office). The interdisciplinary nature of this programme can be explained using the example of development progress on the suction nose. The suction distribution will be specified in aerodynamics, the system group as well as the structure group would then convert this into hardware, and the flight test preparation group then has to integrate the hardware into the aircraft. Any change in any discipline leads to a reaction in the whole programme: If a chosen compressor does not fulfil the suction requirements, for example, either the specification has to be adapted or the system concept has to be changed. Or, for another example, if the suction ducts cannot be integrated in the aircraft, again, either the specification must be adapted, or the system, structure of integration concept has to be changed. These interrelations cannot be approached serially, rather, they must be dealt with as a complex system, interdisciplinarily, from the beginning. Because the partners had to be active in each discipline, appropriate national and international steering groups were established.

Aerodynamics including transition measurement techniques:

The main aerodynamic task was the specification of the suction system, first, to establish system and structural concepts, and later, within a close interdisciplinary co-operation of all areas, to reach a common „specification freeze". Nevertheless, the specifications had to be adapted at various times during the manufacture phase. Check-out measurements have been supervised in aerodynamics, also.

Furthermore, the so-called „Gaster Bump," a passive device against attachment line transition ALT, Fig. 1., was designed here. Though this was designed based on wind tunnel experience, the „bump" had to be adapted for the flight test programme which required far more than the wind tunnel in terms of Mach and Reynolds number as well as yaw angles.

The measurement techniques concept could also be based on experience gained in several wind tunnel test campaigns in the earlier programmes. Many sensors had been tested before, making the main task within the „A 320 HLF Fin" programme to adapt these techniques and the control and monitoring concept to the test aircraft.

Fig. 1: Gaster Bump

The most striking measurement technique was the use of infrared cameras in the horizontal tailplane for transition detection, and use of a traversable wake rake, see Fig. 7. Other measurement techniques were: hot films in the suction nose and on the fin box, a boundary layer rake, a CPM probe (Computational Preston tube Method), very dense pressure distributions in three spanwise positions, an array of piezo electric sensors integrated in a foil, sensors, such as temperature probes, pressure tubes and flow meters, in the suction ducts, and sensors in the suction fan.

Apart from the piezo electric foil and the suction system sensors, the measurement techniques, i.e. the complete chain from the sensor up to the monitoring station, were tested in a first flight test, the „Turbulent Reference Flight Test TFT." Thus, problems with the measurement techniques were solved and some primary procedures were established concerning the use of the A 320 test bed, as well, in advance of the actual laminar flight test campaign.

Fig. 2: Altitude Chamber Test

Suction System

Two objectives were pursued in the systems area:

- The suction system itself had to be designed and manufactured, and then integrated into the aircraft based on the aerodynamic specification, and

- the control, adjustment and monitoring of the system had to be integrated into the aircraft data management system, so that a suction distribution could be set, based on system internal plus aircraft external data.

The first objective is a typical systems task. Together with the DA systems department this task was handled by two German supplier companies after a worldwide call for tender. For the suction device, a three-stage axial compressor was chosen. This maintained its design point by introducing a bypass duct to the system. Due to the required flexibility in different altitudes it was not possible to run that fan on ground.

The second objective, i.e. the system integration into the aircraft data system was handled by a newly developed DA central data processor, which worked as an interface between the suction system and the aircraft data management. This processor was the base for the whole HLF control and monitoring concept, including the measurement techniques.

The complete setup including flight crew training was tested in an altitude chamber test before the HLF flight test, Fig. 2, so that many problems were solved before the flight tests.

Suction Nose Boxes

The concept, design and manufacture of the suction nose boxes was divided into two structural parts:
- The load carrying structure, consisting of the outer suction panel, machined parts in the nose region and a titanium honeycomb further downstream was brazed (diffusion bonding) by DA.
- The structure behind that, forming the suction chambers to which the suction tubes were connected. These chambers were Laser beam welded (LBW) to the load carrying structure under a British Aerospace subcontract.

The complete manufacture chain was tested over a series of test pieces, some of which were for pressure and static tests. In the course of each step in the manufacturing process, the aerodynamics department measured the changes in the suction characteristics due to that step.

Nevertheless, the only major problems within the whole programme arose in that area. Though these will not be described in detail here, solutions to the problems of structure and manufacture will play a key role in future programmes.

Fig. 3: Suction Nose Box Manufacture

After completion, the nose boxes as shown in Fig. 3 had to pass a final check out test and were then handed over to the flight test installation (FTI) group for installation into the A 320 fin at the DA plant in Stade.

Flight Test Installation

The FTI group's task was to integrate the suction system, suction nose boxes, sensors and monitor stations into the aircraft, using design drawings that reflected the integration concept. This effort culminated in the the so-called „working party," during which these elements were physically integrated. In order not to ground the test aircraft for too long a time, a second fin in which the suction nose boxes, the heating mat and sensors were installed before it was delivered to Toulouse for integration, was used. Thus, this integration was just an exchange of that fin by the original one, while parallel work was done on the horizontal tail plane (HTP) and in the cabin.

The main work steps during the working party in Toulouse were:
- Installation of the equipped suction fin including the measurement techniques, Fig. 4

- Installation of the suction system including by-pass, noise protection, cables etc., <u>Fig. 5</u>
- Installation of all other measurement techniques including the infrared cameras in the HTP
- Installation of the monitor stations

Fig. 4: HLF Fin Installation

Fig. 5: Suction System Installation

All work steps took place in the usual Airbus work share, with CASA being responsible for the HTP, Aerospatiale for data acquisition, DA for the fin, etc., with some adaptation to the special character of the HLF equipment.

Flight Tests
Turbulent Reference Flight Test TFT
The TFT took place in March, 1995; the main objectives were:
- Qualification of the measurement techniques for the HLF flight test.
- Qualification of the Gaster Bump under all Mach number, altitude and yaw conditions.
- Acquisition of the turbulent reference data, which were also used for a final check of the suction specification

All three goals were met, moreover, valuable experience was gained within the „management objectives" described previously. However, some systems were subsequently changed by different partners, e.g. the hot film data storage system, so that single elements were adapted but not the complete chain from the sensors to the data storage system. Because of this, more integrative work needed to be done during the next working party. Concerning the Gaster bump, it looked as if ALT would not be a problem during the laminar flights, although it

Fig. 6: HLF Fin Preparation for Flight Test

was not clear how the perforated suction nose surface would influence that effect.

Hybrid Laminar Flow Flight Test (HLFT)
The HLFT, which took place in autumn, 1998, was the major event in the „A 320 HLF Fin" programme. The actual flight tests, including the tasks contracted by HYLDA, were performed after the working party. These tests included many ground tests, an engine test for EMC etc., a first flight to open the flight envelope, and shake down flights to check all new systems (e.g. sensors, suction system, monitors).

Fig. 7: A 320 HLF Fin

Maximum use was made of all flights by adapting the test procedures to consider all constraints, even, for example, the bad weather conditions at that time of the year. The optimised order of events of a typical flight test was:

Pre-flight checks as in Fig. 6, briefing, flight test as in Fig. 7, post flight checks directly after the flight and later on in the hangar, data storage, quick look analysis for preparation of the next flight, and restoring the aircraft for the next flight, e.g., doing small repairs, cleaning, etc. During the use of the aircraft for some special flight tests in-between the HLFT, the suction fin was protected by a special self-adhesive foil.

After completion of the flight tests, the original fin was put back on the aircraft, and the aircraft itself was refurbished in two phases. The suction fin is presently being stored at DA in Bremen where some ground tests, such as geometrical measurements of sensor positions, and local pressure loss measurements on the suction surface, are planned.

Results

At the time of this report, not all qualified data, i.e., checked, corrected and officially released by the responsible partner, were available. Therefore, no quantitative statements can be made for the time being. Instead of that, a qualitative discussion of the data using two examples, and an outline of the steps to follow will be given.

Fig. 8: IR Image

Fig. 9: Transition Line from IR Image

Infrared Image Technique: The infrared image was composed of three images in spanwise position on both sides of the fin. These composite images were produced by newly developed software onboard during each flight test data point, Fig. 8.

Now each of these images has to be evaluated by taking into consideration the results of the pre- and post flight checks (e.g. concerning contamination), the actual configuration (e.g. position of artificial disturbances), and the relevant boundary conditions such as clouds etc. Furthermore, they must be geometrically equalised, as sketched in Fig. 9.

Because DLR is responsible for the IR technique, it will release the IR images to the partners. Afterwards, special teams will work on the

239

evaluation, first taking a quick look to rate the HLFT, and afterwards, in a future programme, doing a detailed analysis. This analysis will then be used to validate theoretical models, to adapt HLF design methods, and to prepare future HLF experiments and/or applications.

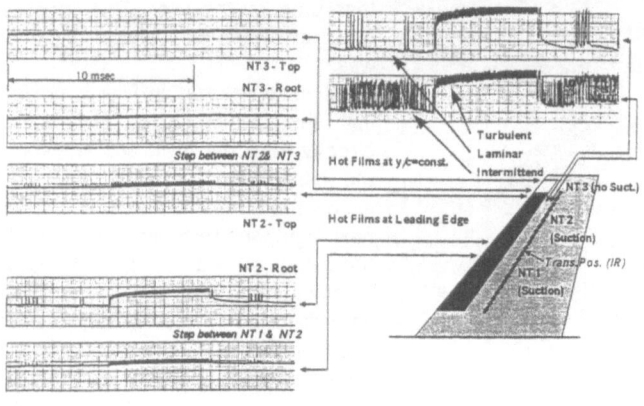

Fig. 10: Typical Hot Film Signals

Hot Film Sensors: In Fig. 10, some exemplary hot film signals in yaw condition at M=0.82 are shown. Disturbances can be detected and followed over the different hot film positions so that the source of a disturbance can be evaluated. All hot film signals have to be evaluated as a single sensor in order to check them using the IR technique and to provide frequency information for boundary layer stability calculation methods, as well as for the design tools. Finally, they have to be correlated with eachother in order to detect and follow disturbances as shown in the example.

These are just two examples of how the evaluation process of this huge data collection must take place. Furthermore, the pressure distributions together with the geometrical information must be evaluated as input for the stability codes, the suction data must be processed, static and total pressure data of the wake rake must be taken for drag calculations, the piezo electric sensors must be evaluated singularly as well as being compared with other methods.

But even now, some major qualitative statements can be derived from the data:
- The effect of system internal noise was less critical than expected after the altitude chamber tests.
- If a passive device such as the Gaster bump works properly, the suction distribution, i.e. the number, position and size of suction chambers can be reduced.
- The surface quality is less critical than expected after the wind tunnel tests, this counts for surface imperfections as well as suction surface inhomogeneity.

Compared to the outcomes of the various wind tunnel tests, it seems that the sensitivity of HLF laminar flow in flight is less critical than expected. This means that application of the HLF technology is not only possible in principle, but that its realisation is possible with today's state of the art technology, based on the results gained from the HLFT.

Summary and Outlook

All goals of this programme have been reached or even exceeded, despite severe problems that occurred during the manufacturing process. The feasibility of HLFC for transport aircraft has been proven. A direct measurement of the fuel consumption reduction in flight has also been accomplished. The existing physical models have been validated, though some of the methods will have to be adapted to the HLFT results. Meanwhile, a large and valuable basic research data base has been created for future use, including extrapolation to HLF applications. This data base, we believe, is unique worldwide and gives the European effort a certain advantage in the race to produce a more fuel-efficient plane. In addition, because the European approach to achieving laminar flow is based solidly on knowledge rather than on trial-and-error, we have a better chance of reaching our goals efficiently. Indeed, some major steps toward the application of HLF on series aircraft have already been achieved.

Three main needs must be addressed for the future:
- Concepts for an industrial system from the nose box to the suction device.
- Concepts for in-service use e.g. de-/anti-icing or de-/anti-contamination.
- Operational proof of concept, including the validation in the next flight test campaign.

All three will again be based on a preliminary aerodynamic specification, which will undergo an interdisciplinary adaptation during the validation process. As we have seen, beyond the „A 320 HLF Fin" programme, which had clear aerodynamic experimental objectives, there are still some questions that need to be answered in other areas. Though laminar flow remains primarily an aerodynamic task, in order to use it in aircraft design, engineering spirit will be called for in all disciplines to answer these questions.

Acknowledgements

First, the author wishes to thank the institutions involved in the programme: The German Ministry of Education, Science, Research and Technology, the European Commission, the partner companies: Airbus Industrie, Aerospatiale, British Aerospace Airbus, Construccione Aeronauticas SA, DaimlerChrysler Aerospace Airbus, the research establishments DLR and ONERA and the Technical University of Berlin, the supplier companies AOA Gauting, Nord-Micro and AS&T (UK) and other subcontractors. However, along with these institutions, it was primarily individuals who made this programme a success. The HLFT crew, who performed a highly professional job under difficult conditions, should be mentioned. Together with colleagues in all areas from technology to design to manufacture, the personal commitment and effort of all of these individuals have contributed to the success of this technology validation programme. We hope that this success will support the production sites in Germany and Europe in the future, which should be the main aim of industrial research.

References

[1] R. Henke: Das Programm „A 320 HLF Fin": Eine multidisziplinäre Herausforderung neuer Qualität.
9. DGLR Fach-Symposium STAB, Erlangen (D), October 1994.

[2] N. Voogt: Flight Testing of a Fokker 100 Test Aircraft with Laminar Flow Glove.
2. CEAS European Forum on Laminar Flow Technology, Bordeaux (F), June 1996.

[3] R. Henke, F. Garcon: The European HLF wind tunnel experiment.
1. European Forum on Laminar Flow Technology, Hamburg (D), March 1992.

[4] R. Henke et al.: The „A 320 HLF Fin - Programme": Objectives and Challenges.
2. CEAS European Forum on Laminar Flow Technology, Bordeaux (F), June 1996.

[5] R. Henke: HLF-Technologie auf einem Airbus Erprobungsträger.
Jahrbuch 1996 II der DGLR, Deutscher Luft- und Raumfahrtkongress 1996, Dresden (D), September 1996.

Fully Developed Turbulent Flow
in Conduits with Circular Cross Section

T. J. Hüttl and R. Friedrich
Lehrstuhl für Fluidmechanik, Technische Universität München,
Boltzmannstr. 15, 85748 Garching, Germany

Summary

Several direct numerical simulations have been performed in order to study the fully-developed, statistically steady turbulent flow in straight, curved and helically coiled pipes with circular cross section. The incompressible Navier-Stokes equations, explicitly derived in an orthogonal helical coordinate system, are integrated numerically by using a second order accurate finite volume method. Pipe curvature induces a secondary flow, which has a strong effect on the flow quantities. Turbulence is significantly inhibited by streamline curvature and the flow almost relaminarizes for high values of the curvature parameter κ. The effect of torsion τ on the mean axial velocity is much weaker, compared to the curvature effect, but it can strongly influence the secondary flow. The Reynolds stresses, that are of relevance for turbulence modelling, change with increasing torsion in a non-negligible way. The evaluation and evolution of time series show remarkable differences between toroidal and helical pipes.

Introduction

The direct numerical simulation of fully developed turbulent flow of an incompressible, Newtonian fluid in straight, curved and helically coiled conduits with circular cross section, constant curvature κ and torsion τ is the subject of the present investigation. The turbulent flow in straight pipes has, due to its simple geometry, already been investigated in detail by means of direct numerical simulation, see [5]. Curved or helically coiled pipes are frequently used in practice in order to save space, to satisfy geometric requirements or to take benefit of the characteristics of the induced secondary flow (influence on heat and mass transfer). Curved and coiled pipes are often used in heat exchangers, chemical reactors, exhaust gas ducts of engines, or any kind of pipelines, tubes and conduits transporting gases and liquids. Although the systematic theoretical and experimental investigation of flow in pipes with curvature and torsion is just of recent origin, this flow configuration has always been classified as more complex than the flow through straight ducts. Until now only few investigations have been made to predict the turbulent flow in curved or coiled pipes. Boersma and Nieuwstadt performed a DNS of fully developed turbulent flow in a toroidal pipe for $Re_\tau = Ru_\tau/\nu = 230$, $\kappa = 0.1$, see [3, 4], and large-eddy simulations (LES) for higher Reynolds numbers, [1, 2]. Hüttl and Friedrich reported some results of turbulent flow in helically coiled pipes [12, 11], but several new aspects will be presented here.

Computational Method

The geometry of a helical pipe can be viewed as a pipe of radius R wound around a cylinder of constant radius $(r_a - R)$, [12]. With the pitch p_s, defined by the increase in elevation per revolution of coils $2\pi p_s$, the curvature κ and the torsion τ of the helical pipe axis can be calculated from

$$\kappa = \frac{r_a}{(r_a^2 + p_s^2)} \quad \text{and} \quad \tau = \frac{p_s}{(r_a^2 + p_s^2)}. \tag{1}$$

As introduced by Germano [7, 8], a helical coordinate system can be established with reference to the master Cartesian coordinate system. By using the helical coordinates s for axial direction, r for radial direction and θ for circumferential direction, the position of any given point X inside the helical pipe can be described by the vector $\vec{x}$

$$\vec{x} = \vec{P}(s) - r\sin(\theta - \tau s)\,\vec{N}(s) + r\cos(\theta - \tau s)\,\vec{B}(s). \tag{2}$$

Here $\vec{T}$, $\vec{N}$ and $\vec{B}$ are the tangential, normal and binormal directions to the generic curve of the pipe axis at the point of consideration, [12].

The Navier-Stokes equations, derived in the orthogonal helical coordinate system, are used to describe the flow through curved or coiled pipes. A finite volume method on staggered grids is used to discretize the spatial derivatives and source terms in the governing equations. It leads to central differences of second order accuracy for the mass and momentum fluxes across the cell faces. Boundary conditions are required for all boundaries of the computational domain. At the walls impermeability and no-slip boundary conditions are realized. Velocity components which are needed on the pipe axis, are obtained by interpolation across the axis. In the circumferential direction all variables are periodic by definition. In axial direction periodic boundary conditions are used, too. For helically coiled pipes the rotation of the coordinate system along the pipe axis must be taken into account and the perfect matching of the cells at the in- and out-flow boundaries must be ensured by choosing a suitable combination of axial length and number of grid points in θ-direction. A semi-implicit time-integration scheme treats all those convection and diffusion terms implicitly and with second order accuracy which contain derivatives in θ-direction. The remaining convection terms are integrated in time with a second order accurate leapfrog-step. An averaging step all the 50 time steps avoids possible $2\Delta t$-oscillations. Diffusive terms with derivatives in s- and r-directions are treated with a first-order Euler backward step. The size of the time step is selected according to a linear stability argument. The use of a projection step leads to a 3D Poisson problem for the pressure correction, which is solved by a Conjugate Gradient method for unsymmetric matrices.

The pipe radius R, the mean friction velocity u_τ and the time $t_{ref} = R/u_\tau$ are used as scaling variables. The dimensionless mass density is set to 1. The mean friction velocity is defined as the square root of the mean wall shear stress, [10]. The dimensionless curvature $\kappa = R\kappa'$ and torsion $\tau = R\tau'$ are nondimensionalized by the pipe radius R. The Reynolds, Dean and Germano numbers based on these scaling quantities are:

$$\text{Re}_\tau = \frac{Ru_\tau}{\nu}, \qquad \text{De}_\tau = \sqrt{\kappa}\text{Re}_\tau, \qquad \text{Gn}_\tau = \tau\text{Re}_\tau. \tag{3}$$

The computational domains are pipes of $15.23R$ in length and diameter $D = 2R$. They are resolved by $256 \times 70 \times 180$ cells in (s, r, θ)-directions. The grid spacing is nonequidistant only in r-direction. Several pipe configurations are computed for the same Reynolds numbers: one straight pipe DP, two toroidal pipes DTSC, DT and several helically coiled pipes, but only results of one case DXXH will be shown here. The toroidal case DT has also been computed by Boersma and Nieuwstadt [3, 4]. The geometrical and flow parameters are listed in Table 1.

Table 1: Geometrical and flow parameters.

case	κR	τR	r_a/R	p_s/R	Re_τ	Re_b	De_τ	Gn_τ	u_b/u_τ	$\lambda \cdot 10^2$
DP	0.0	0.0	∞	0.0	230	6812	0.0	0.0	14.80	3.64
DTSC	0.01	0.0	100.0	0.0	230	6926	23.0	0.0	15.06	3.53
DT	0.1	0.0	10.0	0.0	230	5624	72.7	0.0	12.23	5.34
DXXH	0.1	0.165	2.68	4.43	230	5576	72.7	37.9	12.12	5.44

The turbulence Reynolds number Re_τ based on friction velocity and pipe radius has been taken as 230 in all cases. Therefore the Reynolds number $Re_b = 2u_b R/\nu$, based on bulk velocity u_b and pipe diameter $2R$, varies between 5624 and 6926. Some of the results are compared with case DPU which is equivalent to the simulations of Unger, see [5].

Computational parameters are usually discussed in wall units. The pipe length is 3503 wall units, i.e roughly three times the length of a streaky structure. The length along the inner side is 90 per cent of that and therefore still sufficiently large. The axial grid spacing Δs^+ is 15 along the outer side and 12.3 along the inner side (case DT). The radial grid spacing varies between 1.03 and 5.38 (in the core region). Close to the wall the circumferential grid spacing is 8.01. If one takes the variation of the friction velocity along the pipe circumference into account (Figure 3), Δs^+, based on local values, would have values of 18 along the outer side of the curved pipe and 6.6 along the inner side. It will, therefore, be interesting to see the effect of axial grid refinement in succeeding computations. Statistical results presented below have been obtained by time-averaging and spatial averaging in streamwise direction over those points which have the same (r, θ)-position as the point of the inflow plane (The s-coordinate in case DXXH does not connect these points). On the average 200 statistically independent time samples of the flow field have been used to obtain stable statistics.

Numerical Results

A secondary flow is induced by centrifugal forces when flow passes through a curved pipe. Starting from the pipe core region the fluid is driven outward towards the wall where it bifurcates feeding two recirculation zones with a flow along the pipe walls. Between the inner wall and the outer wall a pressure difference is built up and the maximum axial velocity moves from the pipe's center to the outer wall. This behaviour is well known for laminar flow, [9], and it is also shown by the mean quantities of turbulent flow through a curved pipe in Figure 1 and 2. The maximum axial velocity $\langle u_s \rangle$ is near the outer wall (right) at the radial position $r = 0.56R$ for case DTSC and at $r = 0.86R$ for the case DT. Near the inner wall (left) the axial velocity $\langle u_s \rangle$ is very low. The contour lines are symmetric with respect to the horizontal plane for toroidal pipe flow. The mean velocity components in radial, $\langle u_r \rangle$, and circumferential direction, $\langle u_\theta \rangle$, are not zero, like for straight pipe flow, because a secondary flow is induced by centrifugal forces. Two recirculation zones are established leading to high values of the circumferential velocity component near the upper and lower walls. Due to this secondary flow, fluid with high axial velocity is transported to the left near the upper and lower wall. The shape of contour lines reflect this transport. Two secondary flow cells rotate the fluid and the slow fluid is forced to the center. The vector plot of the mean secondary flow (Figure 2) shows the two secondary flow cells. In a helically coiled pipe, (case DXXH), the two secondary flow cells are slightly turned clockwise and the lower one is appreciably damped. The mean pressure $\langle p \rangle$ rises almost linearly from the inner to the outer wall, see [11]. No local pressure minima are shown at the centers of the recirculation cells and therefore we call them recirculation zones rather than vortices. Figure 2 already indicates that the main reason for the secondary flow is the centrifugal effect due to pipe curvature, but strong torsion modulates this effect noticeably. We recall that the elevation of the pipe after one revolution is about 10 times the radius r_a of the helix. The mean axial velocity $\langle u_s \rangle$ in Figure 1 is only marginally modified by torsion. Figures 3 and 4 show the axial and circumferential components of the wall shear stress normalized with u_τ and plotted as functions of θ. The highest axial shear stress is on the outer part of the bend. The secondary flow leads to wall shear stress components $\tau_{w,\theta}$ which reach more than 60 per cent of the mean axial wall stress $\tau_{w,m}$. The effect of strong torsion increases the amplitude of $\tau_{w,\theta}$ near $\theta = \pi$ by nearly 15 per cent and decreases it even more near $\theta = 0$. While turbulent flow through straight pipes is characterized by four axisymmetric components of the Reynolds stress tensor, there are six non-zero components in the case of curvature and torsion. Their amplitudes reach values of the order of u_τ^2 locally, as demonstrated in Figures 5 - 7, see also [12]. The strong spatial variations of these quantities are due to the complex structure of the mean flow. While the overall level of the turbulence fluctuations is reduced due to curvature, especially in the near-wall regions, torsion enhances the turbulence activity in the core region. These results show that reliable statistical predictions of flow through curved or coiled pipes are a great challenge to turbulence modellers since already the production of turbulent kinetic energy depends on all components of the Reynolds stress tensor. Friedrich [6] points out that two-equation turbulence models must fail for helically coiled pipes.

An interesting insight into the complexity of turbulent flow is given in Figures 8 - 13. Time series, probability density distributions (PDD), temporal correlation functions and spectral functions are shown there for one position near the upper wall ($\theta = 0$, $y^+ = 14$). In contrast to the other simulations, the curve of $PDD(u_s)$ has two maxima for case DT. Almost no small-scale turbulence can be seen in the time series of case DT. The case DXXH produces long periods with small fluctuations between periods with strong fluctuations. The spectral functions of case DT show a peak at $\omega = 12.5$, which corresponds to the oszillations with a period length of 0.5 dimensionless times. The simulation with smaller curvature (case DTSC) and the helically coiled pipe flow (case DXXH) do not show such peaks in the spectral function.

Conclusions

Direct numerical simulation data have been used to study fully developed turbulent flow in conduits with circular cross section. Due to pipe curvature a secondary flow is induced and the structure of the mean flow is very complex. This leads to strong spatial variations of the six non-zero components of the Reynolds stress tensor and to a variation of the wall shear stress along the circumference. While the turbulence activities are damped in strongly curved toroidal pipes, they can be enhanced again in helically coiled pipes. This makes the reliable turbulence modelling of effects of curvature and torsion a big task.

Acknowledgement. We gratefully acknowledge the support of the HLRS in Stuttgart and the LRZ in Munich.

References

[1] Boersma, B.J. and Nieuwstadt, F.T.M. - Large Eddy simulation of turbulent flow in a curved pipe, In: Tenth symposium on turbulent shear flows, The Pennsylvania State University, 1, Poster Session 1, P1-19 - P1-24 (1995).

[2] Boersma, B.J. and Nieuwstadt, F.T.M. - Large-Eddy Simulation of Turbulent Flow in a Curved Pipe, Transaction of the ASME, Journal of Fluids Engineering, vol. 118, pp. 248 - 254 (1996).

[3] Boersma, B.J. and Nieuwstadt, F.T.M. - Non-Unique Solutions in Turbulent Curved Pipe Flow, In: J.-P. Chollet et al. (eds.), Direct and Large-Eddy Simulation II, Kluwer Academic Publishers, pp. 257-266, (1997).

[4] Boersma, B.J. - Electromagnetic effects in cylindrical pipe flow, Ph.D. Thesis, Delft University Press, (1997).

[5] Eggels, J.G.M., Unger, F., Weiss, M.H., Westerweel, J., Adrian, R.J., Friedrich, R. and Nieuwstadt, F.T.M. - Fully developed turbulent pipe flow: a comparison between direct numerical simulation and experiment. J. Fluid Mech. 268, 175-209 (1994).

[6] Friedrich, R. - Direct Numerical Simulation of Incompressible Turbulent Flows. Invited paper. - In: Proc. ACFD3, Dec. 7-11, 1998, Bangalore India, 1998.

[7] Germano, M. - On the effect of torsion on a helical pipe flow., J. Fluid Mech., vol. 125, pp. 1-8 (1982).

[8] Germano, M. - The Dean equations extended to a helical pipe flow., J. Fluid Mech., vol. 203, pp. 289-305 (1989).

[9] Hüttl, T.J. and Friedrich, R. - Fully Developed Laminar Flow in Curved or Helically Coiled Pipes -In: Jahrbuch 1997 der Deutschen Gesellschaft für Luft- und Raumfahrt - Lilientahl - Oberth e.V. (DGLR), Tagungsband "Deutscher Luft- u. Raumfahrtkongress 1997, DGLR-Jahrestagung, 14. - 17. Okt. in München", Band 2, DGLR-JT97-181, pp. 1203-1210 (1997).

[10] Hüttl, T.J., Friedrich, R. - High Performance Computing of Turbulent Flow in Complex Pipe Geometries. – In: High Performance Computing in Science and Engineering '98, Transactions of the High Performance Computing Center Stuttgart (HLRS) 1998, E. Krause, W. Jäger (Eds.), Springer Verlag, pp. 236-251, 1999.

[11] Hüttl, T.J., Friedrich, R. - Direct numerical simulation of turbulent flows in curved and helically coiled pipes. – In: Proc. ACFD3, Dec. 7-11, 1998, Bangalore India, 1998.

[12] Hüttl, T.J., Friedrich, R. - Influence of curvature and torsion on turbulent flows in helically coiled pipes. – To appear in: Proc. 4th International Symposium Engineering Turbulence Modelling and Measurements, May 24-26, 1999, Corsica, France, 1998.

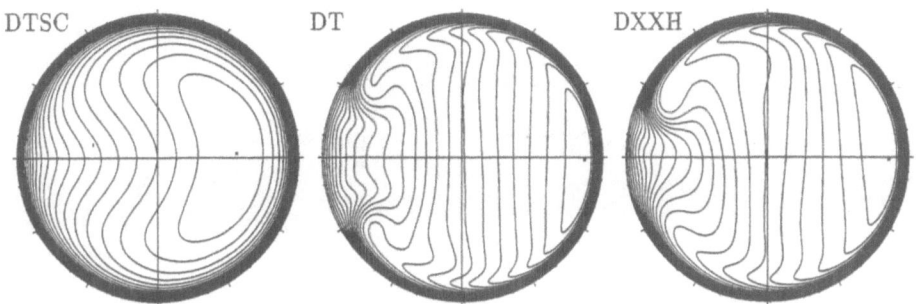

Figure 1: Contour lines of the mean axial velocity component $\langle u_s \rangle$ in two curved pipes and one helically coiled pipe ($s = 3$).

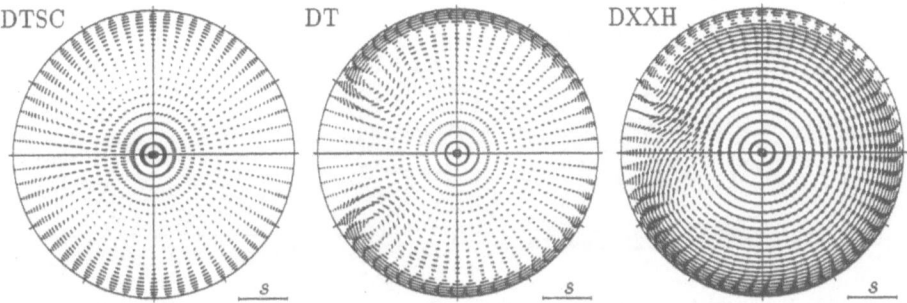

Figure 2: Vector plot of the mean secondary flow in two curved pipes and one helically coiled pipe ($s = 3$).

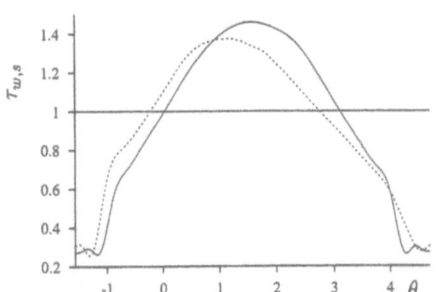

Figure 3: Profiles of the mean wall friction in axial direction $\tau_{w,s}$: —— DT, - - - - DXXH.

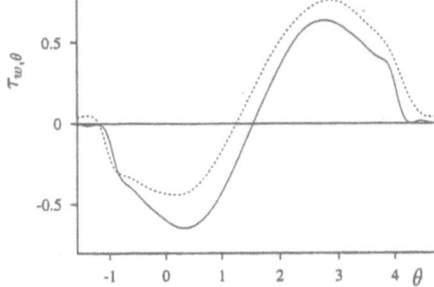

Figure 4: Profiles of the mean wall friction in circumferential direction $\tau_{w,\theta}$: —— DT, - - - - DXXH.

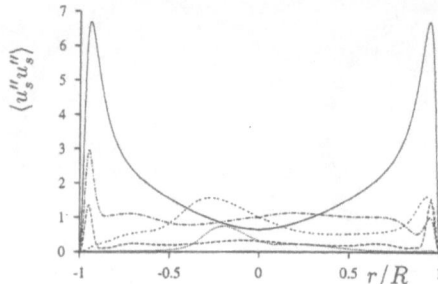

Figure 5: Profiles of the Reynolds stress $\langle u_s'' u_s'' \rangle$: ——— DP, - - - - DT ($\theta = 0$), $\cdots\cdots$ DT ($\theta = \pi/2$), —·—· DXXH ($\theta = 0$), - - -- -- DXXH ($\theta = \pi/2$).

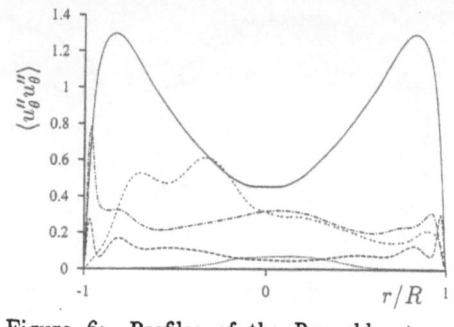

Figure 6: Profiles of the Reynolds stress $\langle u_\theta'' u_\theta'' \rangle$: ——— DP, - - - - DT ($\theta = 0$), $\cdots\cdots$ DT ($\theta = \pi/2$), —·—· DXXH ($\theta = 0$), - - -- -- DXXH ($\theta = \pi/2$).

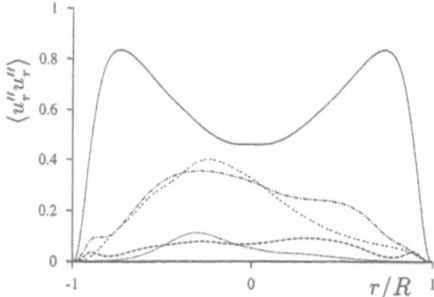

Figure 7: Profiles of the Reynolds stress $\langle u_r'' u_r'' \rangle$: ——— DP, - - - - DT ($\theta = 0$), $\cdots\cdots$ DT ($\theta = \pi/2$), —·—· DXXH ($\theta = 0$), - - -- -- DXXH ($\theta = \pi/2$).

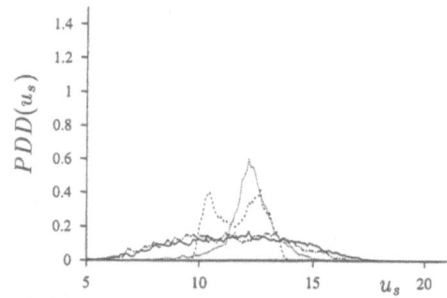

Figure 8: Probability density distributions of the axial velocity component u_s at the position $\theta = 0$, $y^+ = 14$, $\Delta u_s = 0.1$: ——— DTSC, - - - - - DT, $\cdots\cdots$ DXXH, —·—· DPU.

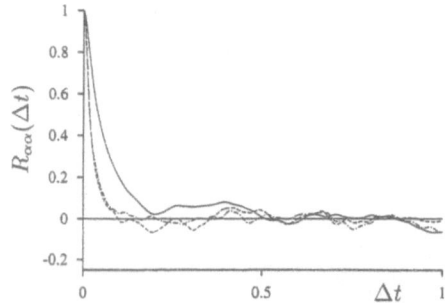

Figure 9: Temporal correlation function $R_{\alpha\alpha}$ of the three velocity components of simulation DTSC at the position $y^+ = 14$, $\theta = 0$: ——— $R_{u_s u_s}$, - - - - $R_{u_\theta u_\theta}$, —·—· $R_{u_r u_r}$.

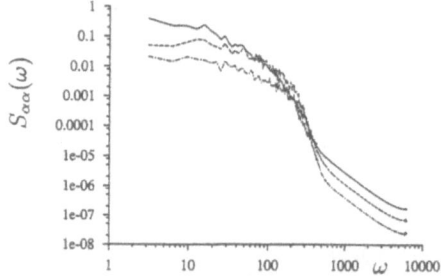

Figure 10: Spectral function $S_{\alpha\alpha}$ of the three velocity components of simulation DTSC at the position $y^+ = 14$, $\theta = 0$: ——— $S_{u_s u_s}$, - - - - $S_{u_\theta u_\theta}$, —·—· $S_{u_r u_r}$.

248

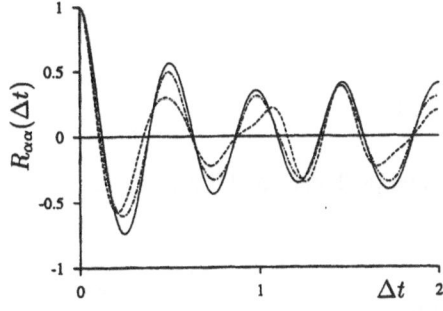

Figure 11: Temporal correlation function $R_{\alpha\alpha}$ of the three velocity components of simulation DT at the position $y^+ = 14$, $\theta = 0$: ——— $R_{u_s u_s}$, - - - - $R_{u_\theta u_\theta}$, —·—· $R_{u_r u_r}$.

Figure 12: Spectral function $S_{\alpha\alpha}$ of the three velocity components of simulation DT at the position $y^+ = 14$, $\theta = 0$: ——— $S_{u_s u_s}$, - - - - $S_{u_\theta u_\theta}$, —·—· $S_{u_r u_r}$.

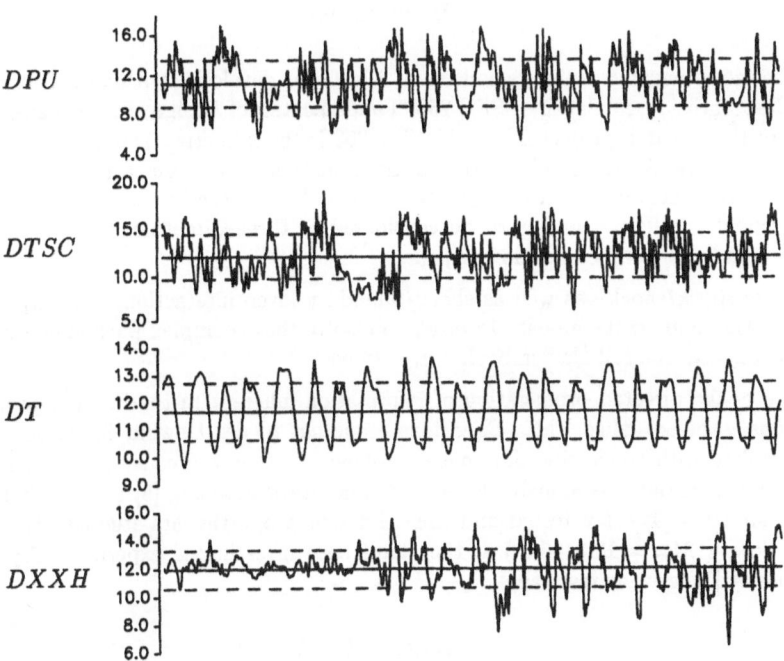

Figure 13: Time series of the axial velocity component at the Position $y^+ = 14$, $\theta = 0$ (near upper wall). Ten dimensionless times $t_p = R/u_\tau$ are shown for each case.

Numerical Simulation of Laminar Symmetric Corner Flows in the Hypersonic Regime

J. van Keuk, J. Ballmann

Lehr- und Forschungsgebiet für Mechanik der RWTH Aachen,
Templergraben 64, 52062 Aachen, Germany

Summary

This work focuses on the numerical simulation of laminar, three–dimensional, symmetric corner flows in the hypersonic regime. For this purpose an algorithm based on a Finite–Volume–Method with cell vertex approach and explicit time integration is used. Different Upwind–Splitting methods have been implemented with emphasis on differences among these methods regarding the overall quality of the corresponding results. A description of the schemes is presented and the results are compared with experimental data as well as with results from other authors for two different corner flow problems.

Introduction

The generic hypersonic configuration ELAC 1 has been developed within the framework of the collaborative research center SFB 253 "Fundamentals of Design of Aerospace Planes". Focus of the research project C3 in the SFB 253 is the numerical simulation of the flow field in the air intake of ELAC 1. Previous investigations have been carried out for a two–dimensional model and the results can be found in [11]. If the intake sidewalls are taken into account the flow field has to be considered as three–dimensional and phenomena similar to a hypersonic corner flow appear.

There are shock/shock– as well as shock/boundary layer–interactions and large regions of separation and reattachment. In order to resolve these complex phenomena correctly one has to solve the full Navier–Stokes Equations.

Because of the directed propagation of information inherent in the inviscid part of the equations – the so–called Euler-Equations – one has to use Upwind–Splitting methods when dealing with convection dominated problems as they are considered in this work. Two different variants – namely the AUSM flux vector splitting [8] and Roe's flux difference splitting [9] – are tested and their different properties are studied. Hypersonic corner flows ($M_\infty = 12.3, 16.0$) as they have been investigated experimentally [7] and numerically [1] serve as test cases.

Physical Model

For a simply connected, three–dimensional domain V with boundary ∂V the Navier–Stokes Equations can be written as

$$\int_V \frac{\partial \mathbf{U}}{\partial t} dV = \oint_{\partial V} \left\{ (\mathbf{F}_c - \mathbf{F}_v) \, e_{n_x} + (\mathbf{G}_c - \mathbf{G}_v) \, e_{n_y} + (\mathbf{H}_c - \mathbf{H}_v) \, e_{n_z} \right\} dA = 0 \qquad , \qquad (1)$$

where $\mathbf{U} = (\rho, \rho u, \rho v, \rho w, \rho E)^T$ is the vector of conservative variables. ρ, u, v, w, E denote the density, the cartesian components of velocity and the total energy. $\mathbf{F}_c$, $\mathbf{G}_c$ and $\mathbf{H}_c$ represent the convective flux functions

$$
\mathbf{F}_c = \begin{bmatrix} \rho u \\ \rho u^2 + p \\ \rho u v \\ \rho u w \\ u(\rho E + p) \end{bmatrix} \quad , \quad \mathbf{G}_c = \begin{bmatrix} \rho v \\ \rho v u \\ \rho v^2 + p \\ \rho v w \\ v(\rho E + p) \end{bmatrix} \quad , \quad \mathbf{H}_c = \begin{bmatrix} \rho w \\ \rho w u \\ \rho w v \\ \rho w^2 + p \\ w(\rho E + p) \end{bmatrix} \quad (2)
$$

in $x-$, $y-$ and $z-$direction, respectively. The corresponding expressions for the diffusive flux functions $\mathbf{F}_v$, $\mathbf{G}_v$ and $\mathbf{H}_v$ and the formulation of the stress tensor components (symmetrical) with Stokes' hypothesis of vanishing pressure viscosity ($\mu_v = \frac{2}{3}\mu$) are the standard ones [11]. The vector of heat conduction is modeled by Fourier's law ($\mathbf{q} = -k\,grad T$). Sutherland's formula is adapted for the laminar viscosity as a function of the static temperature.

$$
\mu(T) = \frac{\sqrt{\gamma} M_\infty}{Re_\infty} \left(\frac{T}{T_\infty} \right)^{\frac{3}{2}} \frac{T_\infty + 110K}{T + 110K} \quad . \quad (3)
$$

For the heat conductivity the Prandtl number is assumed to be constant, which leads to the formula

$$
k = \mu c_p \frac{1}{Pr} \quad \text{with} \quad Pr = 0.72 \quad , \quad (4)
$$

where c_p denotes the specific heat at constant pressure. Referring to the experiment [7], where the flow was observed as laminar and the free–stream temperature was very low, neither turbulence nor real gas effects are taken into account here. Finally, the following boundary conditions are assumed at solid walls: $u = v = w = 0$, $T_w = const.$, and in addition according to the boundary-layer theory $(grad p)\mathbf{e}_n = 0$.

Basic Flow Solver

The DLR FLOWer–Code (Version 114 extended to full Navier–Stokes formulation) forms the basis for the numerical algorithm used here to calculate numerical solutions for the three–dimensional hypersonic corner flow. For the numerical solution the computational domain is divided into a set of non–overlapping hexahedral cells. Curvilinear coordinates ξ, η, ζ are introduced and the conservation laws in integral form are applied to each cell. Approximating the volume and surface integrals by the mean value theorem and the midpoint rule, one obtains the following Finite–Volume expression in semi–discrete form

$$
\frac{d\mathbf{U}_{i,j,k}}{dt} = -\frac{1}{V_{i,j,k}} \left[\Delta_\xi \left(\hat{\mathbf{F}}^c - \hat{\mathbf{F}}^v \right) + \Delta_\eta \left(\hat{\mathbf{G}}^c - \hat{\mathbf{G}}^v \right) + \Delta_\zeta \left(\hat{\mathbf{H}}^c - \hat{\mathbf{H}}^v \right) \right] = \mathbf{Res}_{i,j,k} \quad , \quad (5)
$$

where

$$
\Delta_\xi(\bullet) = (\bullet_{i+\frac{1}{2},j,k} - \bullet_{i-\frac{1}{2},j,k}), \Delta_\eta(\bullet) = (\bullet_{i,j+\frac{1}{2},k} - \bullet_{i,j-\frac{1}{2},k}), \Delta_\zeta(\bullet) = (\bullet_{i,j,k+\frac{1}{2}} - \bullet_{i,j,k-\frac{1}{2}}). \quad (6)
$$

$\hat{\mathbf{F}}$, $\hat{\mathbf{G}}$ and $\hat{\mathbf{H}}$ are the transformed flux vectors in $\xi-$, $\eta-$ and $\zeta-$direction, respectively. For example:

$$
\hat{\mathbf{F}}_{i+\frac{1}{2},j,k} = \left(\mathbf{F}_{i+\frac{1}{2},j,k} e_{n_x} + \mathbf{G}_{i+\frac{1}{2},j,k} e_{n_y} + \mathbf{H}_{i+\frac{1}{2},j,k} e_{n_z} \right) dA_{i+\frac{1}{2},j,k} \quad . \quad (7)
$$

The propagation of information being inherent in the inviscid part of the equations takes place along the characteristics of the Euler–Equations and must be considered when dealing with hypersonic flow problems. Therefore, several Upwind–Splitting methods (e.g. van Leer, AUSM, HLLE [2], Roe) have been implemented into the original FLOWer code. Two of them will be described in the next section. Without the advective part, Eq.(1) would become parabolic so that a numerical formulation with a central discretization operator would be appropriate. That is applied to the diffusive part.

Time integration of Eq.(5) is performed by an explicit 5–Stage Runge Kutta scheme. For the simulation of flow problems with asymptotically steady state solutions, acceleration of convergence techniques like local time–stepping and the well known multigrid method in FMG–FAS formulation are available in the FLOWer–Code.

Upwind–Splitting Methods

A large variety of different approaches for capturing the directed disturbance propagation by splitting of the inviscid flux vector have been presented in the past [6]. Basically, one distinguishes between flux vector splitting and flux difference splitting. The first one is to consider the eigenvalues of the flux' Jacobian of the inviscid flux vector and to split the flux terms according to the sign of the associated propagation speeds. Those methods are known as flux vector splitting schemes, e.g. [10]. They are simpler than the methods based on the second approach of Godunov [3], where again more physical properties are introduced into the numerical scheme. Godunov considered a cell–wise constant distribution of the conservative variables in the computational domain. This idea resulted in local Riemann problems at the cell interfaces, which can be solved either exactly [3] or approximately, e.g. [9]. Hence, properties derived from the complete local solution of the Euler–Equations are introduced in the numerical method.

For simplicity reasons the following considerations are restricted to the one–dimensional case. The methods are extended to multi–dimensional flow problems by the Finite–Volume–Method described in the previous section.

AUSM Flux Vector Splitting

The $\underline{A}$dvection $\underline{U}$pstream $\underline{S}$plitting $\underline{M}$ethod was proposed by Liou and Steffen in 1993 [8]. The main idea in this scheme is to recognize that the inviscid flux vector consists of two physically distinct parts, namely *advective* and *pressure* terms. These two kinds of terms are then split separately, leading to the following expression for the flux at the cell interface.

$$
\mathbf{F}_{i+\frac{1}{2}} = \frac{1}{2} M_{i+\frac{1}{2}} \left[\begin{pmatrix} \rho a \\ \rho a u \\ \rho a H \end{pmatrix}_i + \begin{pmatrix} \rho a \\ \rho a u \\ \rho a H \end{pmatrix}_{i+1} \right] - \frac{1}{2} |M_{i+\frac{1}{2}}| \Delta_{\frac{1}{2}} \begin{pmatrix} \rho a \\ \rho a u \\ \rho a H \end{pmatrix} + \begin{pmatrix} 0 \\ p_i^+ + p_{i+1}^- \\ 0 \end{pmatrix},
$$

$$(8)$$

where $\Delta_{\frac{1}{2}}\{\bullet\} = \{\bullet\}_{i+1} - \{\bullet\}_i$. and $M_{1/2} = M_L^+ + M_R^-$. According to the original van Leer splitting [10] and to Liou / Steffen [8], the split Mach numbers $M^\pm$ and pressure terms $p^\pm$ can be expressed as follows

$$
M^\pm = \begin{cases} \pm\frac{1}{4}(M \pm 1)^2 & \text{if } |M| \le 1 \\ \frac{1}{2}(M \pm |M|) & \text{otherwise} \end{cases} \quad, \quad p^\pm = \begin{cases} \frac{p}{4}(M \pm 1)^2(2 \mp M) & \text{if } |M| \le 1 \\ \frac{p}{2}\frac{(M \pm |M|)}{M} & \text{otherwise} \end{cases}.
$$

$$(9)$$

The method described so far is only of first order accuracy in space. This is not sufficient for the numerical solution of the Navier–Stokes Equations for reasons of consistency, because of the second order derivatives appearing in the diffusive terms. It is therefore necessary to increase the formal order of accuracy in space in the discretization of the convective terms. For the AUSM method this is done by means of the so–called MUSCL–Extrapolation (Monotonic Upstream Scheme for Conservation Laws). The idea is to consider a piecewise linear reconstruction of the flow variables within each control volume leading to left and right extrapolated values [6]. To guarantee the TVD–property of the scheme one has to use limiter functions avoiding overshoots of the reconstructed variables in the vicinity of strong gradients (e.g. shocks). The one used in this work is the so–called "minmod"–limiter [6].

Roe Flux Difference Splitting

In Roe's method the approximation of the solution of the nonlinear Riemann problem is replaced by the exact solution of the linearized problem, that has to be extended for the approximation of discontinuous solutions. Writing the flux difference between two neighbouring states as a linear wave decomposition and summing up the contribution of the single waves gives the following expression for the inviscid flux vector at the cell interface.

$$\mathbf{F}_{i+\frac{1}{2}} = \frac{1}{2}\left[\mathbf{F}_{i+1} + \mathbf{F}_i - \sum_k \alpha_k |\lambda_k| \mathbf{r}_k\right] \quad \text{with} \quad \alpha_k = \mathbf{l}_k(\mathbf{U}_{i+1} - \mathbf{U}_i) \quad . \quad (10)$$

λ_k is the propagation speed of the k–th wave of the linearized Riemann problem and $\mathbf{l}_k$, $\mathbf{r}_k$ are the corresponding left– and right–eigenvectors [6]. The so constructed flux formulation consists of a central part supplemented by an upwind term, which has to be computed using average values. For averaging approaches see [4,6]. A problem appears when one of the eigenvalues changes sign. For centered expansion fans with sonic point the scheme then leads to an expansion shock and generates non–physical solutions such as the so–called "carbuncle phenomenon" when calculating hypersonic blunt body flows. To circumvent this difficulty Harten [5] proposed a modification of the modulus function in Eq.(10)

$$|\lambda_k| = \begin{cases} \frac{1}{2}(\frac{\lambda_k}{\delta} + \delta) & \text{for} \quad |\lambda_k| < \delta \\ |\lambda_k| & \text{else} \end{cases} \quad , \quad (11)$$

where δ is a small number often referred to as "entropy fix".

For Roe's method an alternative variant to formally reach second order in space accuracy is used. It was proposed by Harten and Yee and is known as "modified flux approach" [12,4]. The authors of this paper prefer this proposal instead of the MUSCL–Extrapolation, because of its superior stability properties in connection with Roe's method.

Results

In order to verify the capacity of the modified FLOWer–Code to solve complex hypersonic flow fields, the three–dimensional symmetric corner flow problem in the laminar regime is investigated. This problem has been studied in the literature in detail, so that the validation will be performed using both numerical [1] and experimental results. According to the experiments of Hummel [7] the following geometrical and free stream conditions have been chosen:

	δ	M_∞	Re_∞/m	Pr	$T_\infty [K]$	$T_W [K]$
Experiment a)	$8°$	12.3	$5.0 \cdot 10^6$	0.72	45.3	300.0
Experiment b)	$8°$	16.0	$1.7 \cdot 10^6$	0.72	25.9	300.0

In Fig.1 the different physical phenomena characterizing a hypersonic corner flow problem are sketched in the y, z–plane. The x–direction is the oncoming flow direction. Altogether, there is a system of five shocks. Two of them are the primary shocks, which are generated by the wedged sidewalls on the left and the lower side of the computational domain. The other three are caused by the Mach disk (irregular reflection) in the corner, where the primary shocks interact. Two contact discontinuities, produced by the different states on either side of the interactions, result from the interaction of the shocks (triple points) towards the symmetry plane. In addition, two reflected shock waves interact with the boundary layers and are reflected as expansion fans, which are transformed in compression waves when interacting with the contact discontinuities. The impinging shocks cause the boundary layers to separate producing two vortex systems.

In the present work, numerical simulations using both AUSM and Roe FDS have been performed corresponding to Experiment a). Fig.2 shows the computed density contours for Roe's scheme, where a grid consisting of $100 \times 100 \times 100$ grid points has been used. According to [7] the contours for cross section $x = 0.09m$ are plotted. All the flow phenomena mentioned above, even the secondary separation, are qualitatively well resolved. Next, in Fig.5 the results for the same cross–section, the Mach number and the pressure distribution, are compared with those of [1], showing a good agreement. This is satisfactory considering the fact that two differently expanded solution domains, different grids and in particular totally different numerical codes have been used. In contrast to the present work, [1] solves the parabolized Navier–Stokes Equations with a space marching method on computational grids composed of 120×120 cells.

Fig.3 displays for both numerical results a comparison with the experimental data of Hummel [7] for the wall pressure ($x = 0.09m$). The overall agreement between the measured and computed values is satisfying. The present work shows slightly better results in the near–corner region, whereas [1] predicts the distribution in the reattachment zone more precisely. But this fact may be due to the smaller solution domain in the present work that prevents the capturing of the effect of reattachment. In addition, the region of the secondary separation is enlarged and the agreement of the present results using AUSM with [1] for the location and extension of the secondary separation is satisfactory. One negative feature of the Roe method became also visible. For stability reasons a value of 0.1 had to be chosen for the parameter δ and hence, due to the additional numerical viscosity, the secondary separation is nearly vanishing. In Fig.4 the pressure distribution obtained with AUSM is displayed and the spurious pressure oscillations are visible in friction dominated regions and behind shocks. These effects are known negative properties of the original AUSM scheme [11].

Finally, Fig.6,7 show a comparison between measured and computed values for the surface pressure and the computed Mach number contours corresponding to Experiment b) ($x = 0.09m$). The physical consequences of increasing the Mach number and simultaneously decreasing the Reynolds number are similar in experiment and computation. The boundary layer is much thicker and the reflected shock is much more fanned out when impinging on it. Unfortunately the agreement for the surface pressure is not as good as in Experiment a), especially in the separation region. This may be due to the fact that the same grid has been used for both computations and, considering the thickness of the boundary layer in Experiment b), perhaps the number of grid points was not sufficient to resolve the separation behaviour correctly. Therefore, subject of future work in this project are grid-sensitivity studies in order to verify this suspect.

Conclusions

A study has been performed to investigate the prediction quality of different Upwind–Splitting methods (AUSM, Roe FDS), when simulating laminar, three–dimensional, symmetric corner flow problems in the hypersonic regime. The results have been compared with numerical results of other authors as well as with experimental results for two different test cases. Although not perfect, the overall agreement for the complex flow features characterizing this problem and the wall pressure distribution appears quite satisfactory for both AUSM and Roe FDS. Unfortunately, Roe's method requires sometimes too much artificial viscosity for stability, whereas AUSM in the original formulation shows spurious pressure oscillations in friction dominated regions as well as behind shocks.

It has been shown that a perfect upwind method is not yet available, but both AUSM and Roe FDS are able to resolve the major flow phenomena of even complex hypersonic flow problems as well as the wall pressure distributions in an overall acceptable quality.

Subject of future work in this project will be the extension of the gas model to real gas.

Acknowledgements

This paper has been prepared within the frame of a PhD project in the Collaborative Research Center (SFB 253) "Fundamentals of Design of Aerospace Planes" at the RWTH Aachen, Germany.

References

[1] D' Ambrosio D., Marsilio R., Pandolfi M. (1995): *Shock-Induced Separated Structures in Symmetric Corner Flows.* AIAA Paper 95-2270.

[2] Einfeldt B., Munz C.D., Roe P.L. and Sjögreen B. (1991): *On Godunov-Type Methods near Low Densities.* J. Comp. Phys. **92**, pp. 273-295.

[3] Godunov S.K. (1959): *A Difference Scheme for Numerical Computation of Discontinuous Solutions of Hydrodynamic Equations.* Math. Sbornik **47**, pp. 271-306.

[4] Grotowsky I.M.G., Ballmann J. (1996): *A Numerical Algorithm for Calculating Flows in Hypersonic Inlets.* ZFW **20**, pp. 95-104.

[5] Harten A. (1983): *High Resolution Schemes for Hyperbolic Conservation Laws.* J. Comp. Phys. **49**, pp. 357-393.

[6] Hirsch C. (1990): *Numerical Computation of Internal and External Flows.* **2**, Computational Methods for Inviscid and Viscous Flows, J. Wiley & Sons.

[7] Kipke K., Hummel D. (1975): *Untersuchungen an längsangeströmten Eckenkonfigurationen im Hyperschallbereich.* ZFW **23**, Heft 12, pp. 417-429.

[8] Liou M.S., Steffen C.J. (1993): *A New Flux Splitting Scheme.* J. Comp. Phys. **107**, pp. 23-39.

[9] Roe P.L. (1981): *Approximate Riemann Solvers, Parameter Vectors and Difference Schemes.* J. Comp. Phys. **43**, pp. 357-372.

[10] van Leer B. (1982): *Flux Vector Splitting for the Euler Equations.* Lecture Notes in Physics **170**, pp. 507-512.

[11] van Keuk J., Ballmann J., Schneider A., Koschel W. (1998): *Numerical Simulation of Hypersonic Inlet Flows.* AIAA Paper 98-1526.

[12] Yee H.C. (1987): *Upwind and Symmetric Shock-Capturing Schemes.* NASA TM 89464.

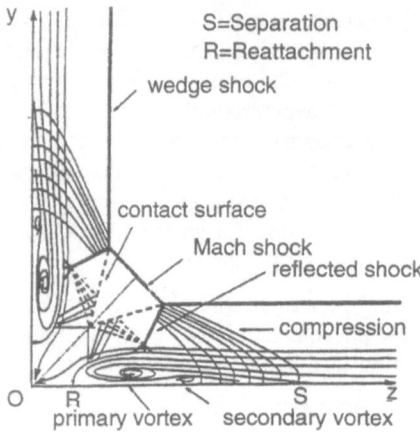

Figure 1. Physical phenomena in the flow field.

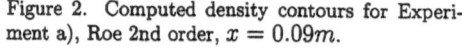

Figure 2. Computed density contours for Experiment a), Roe 2nd order, $x = 0.09m$.

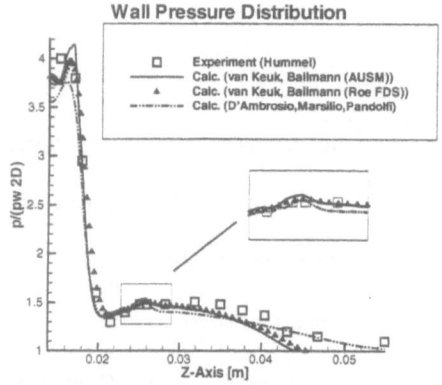

Figure 3. Surface pressure distribution (Calculations (van Keuk, Ballmann / D'Ambrosio, Marsilio, Pandolfi), Experiment a) (Hummel)).

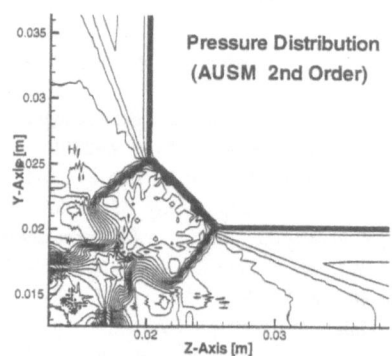

Figure 4. Computed pressure contours for Experiment a), AUSM 2nd order, $x = 0.09m$.

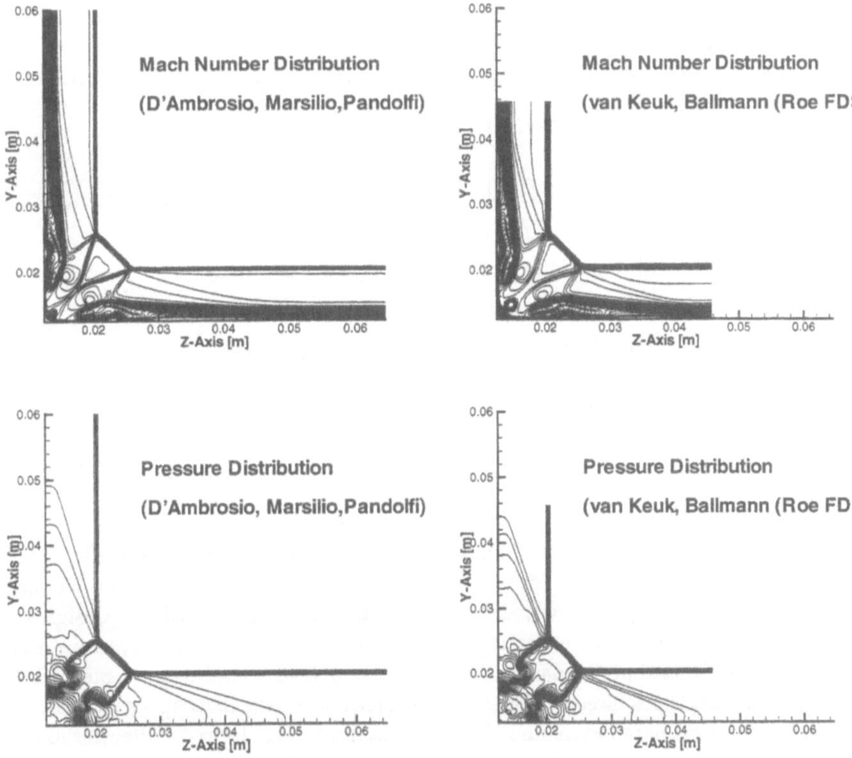

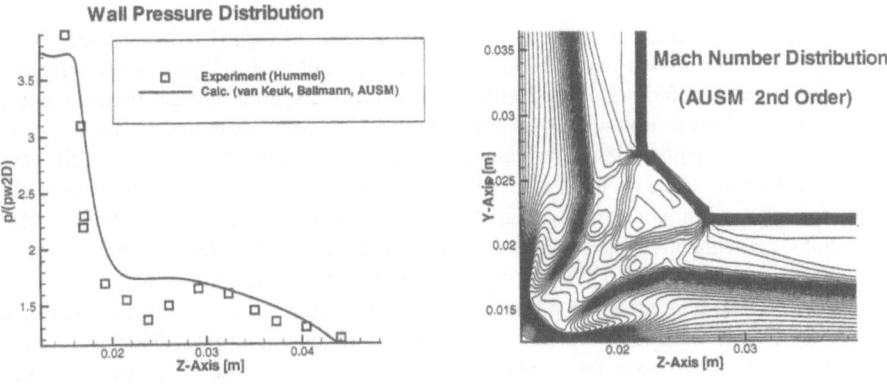

Figure 5. Comparison between computed results [1] and present results for Mach number (on top) and pressure distribution, case of Experiment a), $x = 0.09m$.

Figure 6. Surface pressure distribution (calculation (van Keuk, Ballmann) for Experiment b) (Hummel)).

Figure 7. Computed Mach number contours for Experiment b), AUSM 2nd order, $x = 0.09m$.

A Conceptual Design Methodology for the Estimation of the Local Shock Locations on Optimized Transonic Wings

T. Krißler

Synaps Ingenieur-Gesellschaft mbH
Fahrenheitstr. 1, D-28359 Bremen, Germany
Daimler-Benz Aerospace Airbus GmbH
Hünefeldstr. 1–5, D-28183 Bremen, Germany

Summary

A conceptual design methodology shall be developed for use within multidisciplinary aircraft design. The methodology estimates the local shock locations on the upper surface of wings being optimized with respect to total drag as a function of the design parameters Ma, t/c, c_L and Re. In order to achieve this a database of drag optimized cross sections, called optimized performance designs, was analyzed accordingly. A methodology to estimate the local shock sweep angle using the method for the local shock locations as a function of the design parameters and the wing planform was derived. The methodology to estimate the local shock sweep angle will be used within another conceptual design method to predict the minimum wave drag of a transonic wing.

Introduction

Within the scope of multidisciplinary conceptual aircraft design [1] [2], a transonic aerodynamic module to predict transonic effects of a wing shall be developed. Typically, about 300 parameters and 50 design variables and constraints are used to describe the complete configuration, whereas about 150 parameters and 15 design variables of these are used for aerodynamic purposes [3]. The objective of such a methodology is to describe transonic aerodynamic effects with a limited number of parameters and predict the correct trends for use within conceptual design. Processing time becomes critical for use of the methodology within multidisciplinary design automization. For this reason, the use of time intensive codes (e.g. 3-D Navier Stokes, 3-D Euler) is not appropriate.

Aerodynamic characteristics are a result of the wing planform, the spanwise load distribution, the distribution of twist, camber, thickness and local flow data. These parameters shall be used as input data for a method to predict the wavedrag of a transonic wing. It was the idea to design cross sections using an optimizer with total drag as the objective. The resulting designs are single point designs with lowest possible total drag, called optimized performance designs.

A database of optimized performance designs was produced for the design parameters Ma, c_L, t/c, $Re = 2 \cdot 10^6$. The idea for this database originates from A. Van der Velden. In earlier studies a database of about 80 optimized performance designs was produced. A methodology to predict the wavedrag as a function of the design parameters Ma, t/c, c_L, Re and the shock sweep angle ϑ_S was established by A. Van der Velden based on this database. In order to use this methodology to predict the minimum wave drag of a transonic wing, the local shock sweep angle must be known along with the local design parameters. The local shock sweep angle is a function of the local chordwise shock location on the upper surface of the wing and the wing planform. It is the objective of this study to find a formulation for the local shock sweep angle ϑ_S as a function of the design parameters Ma, t/c, c_L and Re in order to obtain a methodology to predict the minimum wave drag of a transonic wing.

Methodology

The relationship between cross section data and design parameters can be derived using appropriate methods. A function of the relationship must be found, ensuring physically plausible trends towards the borders of design parameter ranges. Daimler-Benz Aerospace Airbus (DA) inhouse codes were linked to the Synaps Inc. optimization tool $PointerPro^{TM}$ [4] to set up the database of optimized performance designs. The original database contained about 80 cross section designs and the code used was $VICWA$ [5] (Full potential code with finite-difference boundary layer calculation). During the work on this project, the database was enlarged to about 120 cross section designs. This was necessary for a better estimation of the relationship of the shock sweep angle to the design parameters near the borders of design parameter ranges. Since the calculation limit of the $VICWA$ code is reached at high Mach numbers ($Ma > 0.76$) for $t/c < 0.07$ and $t/c > 0.13$, another code $XLS6$ had to be used (Full potential code with integral boundary layer calculation). To ensure the consistency of the database, the original database was recalculated and enlarged appropriately. All calculations are 2-D, without sweep angle. The design parameters have been varied within the following ranges:

$$0.7 < Ma < 0.8$$
$$0.5 < c_L < 0.9$$
$$0.05 < t/c < 0.16.$$

Typical optimized performance cross section designs of the database are shown in Fig. 1.

Determination of the shock locations

The upper surface pressure distributions ($c_{P_{US}}$) of the optimized performance designs in the database were analyzed to determine the chordwise location of the shock. The location of $c_{P_{US}}$ passing the critical pressure c_P^* with a positive pressure gradient was used as a criterion for the shock location. In the case of multiple shocks, the shock farthest downstream was used. Possible shocks on the lower surface of the cross section are not of interest and therefore not considered.

$$(x/c)_{S_{dat}} = MAX \left((x/c) \mid_{\left(c_{P_{US}} = c_P^* , \frac{dc_P}{dx} > 0 \right)} \right) \tag{1}$$

with

$$c_P^* = c_P \mid_{(Ma=1)}. \tag{2}$$

An appropriate function was formulated to find the shock location as a function of the design parameters Ma, c_L and t/c for constant Reynolds numbers. A cubic function with mixed terms in the design parameters Ma, c_L and t/c was used as a first estimate.

$$(x/c)_S = \sum_{\substack{i,j,k=1 \\ i+j+k \leq 3}}^{3} A_{ijk} \, x_1^i \, x_2^j \, x_3^k \tag{3}$$

with

$$
\begin{aligned}
x_1 &= (Ma - Ma_0) \\
x_2 &= (c_L - c_{L0}) \\
x_3 &= ((t/c) - (t/c)_0).
\end{aligned}
$$

Using the shock location derived from a cross section in the database $(x/c)_{S_n}$ and the shock location using equation (3) the local deviation of the model E_{locn} for the specific cross section n may be obtained as follows (e.g. $e = 2$ for the quadratic deviation).

$$E_{locn} = (|(x/c)_S - (x/c)_{S_n}|)^e. \tag{4}$$

The global deviation E_{glob} of the function for all N datapoints of the database may then be derived using equation (5).

$$E_{glob} = \frac{1}{N} \sum_{n=1}^{N} E_{locn}. \tag{5}$$

Due to the resulting nonlinear equation system, an optimizer was used to calculate the coefficients A_{ijk} of the function. The coefficients A_{ijk} were varied until the minimum of the global deviation E_{glob} was reached to achieve this. While calculating the coefficients, a physically correct description of the shock location using the function was of major importance. Primarily a good reproduction of trends in shock location development was intended. An exact agreement of the shock locations determined from the database with shock locations calculated using the model was not of interest, because mainly trends in shock location development are required in multidisciplinary design, Figs. 3, 4, 5. The derived formula principally consists of an absolute term, linear terms in Ma, c_L and t/c, functions of Ma^2 and $(t/c)^2$ and the mixed terms $Ma \cdot c_L$, $Ma \cdot t/c$ and $c_L \cdot t/c$. As a consequence of a physically plausible formulation of the shock location, third order terms were omitted. This results in a quadratic function of the shock location with Ma, c_L and t/c as parameters.

260

$$(x/c)_S = Q_1(Ma,\ c_L,\ t/c) \tag{6}$$

with

$$Q_1 \quad \text{quadratic function}.$$

Appropriate asymptotes and limits had to be introduced to correctly describe the shock location near the boundaries of the design parameter ranges. Near higher cross section thickness ($(t/c) \rightarrow 1$), linear asymptotes were used. A quadratic function in c_L for the coefficients of the asymptotes proved to be appropriate. Hence, the following relationship for the aymptotes was derived:

$$(x/c)_{S_{asy}} = L(t/c) = Q_2(c_L) \cdot (t/c) + Q_3(c_L) \tag{7}$$

with

$$L \quad \text{linear function}$$
$$Q_2,\ Q_3 \quad \text{quadratic functions}.$$

A steady transition of the quadratic function of the shock locations and the linear asymptotes was achieved using suitable weighting functions. The following formulation for the shock location is obtained, Figs. 6, 7, 8.

$$(x/c)_{S_{proj}} = W_1 \cdot (x/c)_S + W_2 \cdot (x/c)_{S_{asy}} \tag{8}$$

with

$$W_1,\ W_2 \quad \text{weighting functions}.$$

The mean deviation of the mathematical formulation for the shock location is about 5%, calculated for all designs within the database. Thus, a formula for the shock location $(x/c)_S$ as a function of the parameters Ma, c_L and t/c is provided which can be used as a conceptual design methodology. The formula for the shock location is only defined for parameter combinations resulting in shock locations between the cross section leading- and trailing-edges ($0 < (x/c)_S < 1$). Parameter combinations resulting in shock locations outside the cross section are not relevant for practical design purposes.

With the shock location following equation (8) and the wing planform the shock sweep angle can be calculated as described in the following paragraph.

Estimation of the local shock sweep angle

The local shock sweep angle is a function of the chordwise shock location and the wing planform. For an isobaric concept, the following relationship can be used (Fig. 2(a)):

$$\vartheta_S = \arctan((1 - x_S)\tan(\vartheta_{LE}) + x_S \tan(\vartheta_{TE})). \tag{9}$$

261

For a non-isobaric concept, a mean shock sweep angle between neighbouring cross sections can be calculated using equation (10). In this case local shock locations must be transformed into a global coordinate system (Fig. 2(b)). However, this is only applicable for regions with mainly two-dimensional flow (not at the wing root or tip) and may be used as a first estimate for the shock sweep angle in conceptual design.

$$\overline{\vartheta_S} = \arctan\left(\frac{d\overline{x}_S}{dy}\right). \tag{10}$$

The model predicts the correct trends in shock location development as a function of the design parameters. The behaviour near the borders of the parameter ranges is physically plausible.

Conclusion

A conceptual design methodology was developed to predict the local shock sweep angle of the upper surface shock of an optimized transonic wing. The method is a mathematical function of the local design parameters Ma, c_L, t/c and the wing planform. A database of optimized performance designs (cross sections, optimized with respect to lowest possible total drag) was analyzed and a formulation of the shock location as a function of the design parameters was derived from it. The methodology for the shock sweep angle is used with a method for the wave drag to predict the minimum total drag of a transonic wing. Within multidisciplinary design automization at DA and Synaps the conceptual design methods have already been used for A3XX orientated wing design studies.

References

[1] Van der Velden, A., Aerodynamic Shape Optimization, AGARD-R-803 AGARD-FDP-VK1 Special Course, April 1994.

[2] Van der Velden, A., Tools for Applied Engineering Optimization, AGARD-R-803 AGARD-FDP-VK1 Special Course, April 1994.

[3] Sobieczky, H., Seebass, R., Mertens, J., Dulikravich, G., Van der Velden, A., New design concepts for high speed air transport, CISM courses and lectures no. 366, International Center for Mechanical Sciences, Springer, 1997.

[4] Synaps Inc., "PointerPro^{TM}4.2", www.synaps-ing.de

[5] Dargel, G., Ein Programmsystem für die Berechnung transsonischer Profil- und konischer Flügelströmungen auf der Basis gekoppelter Potential- und Grenzschichtlösungen, DGLR Bericht 92-07, Köln, 1992.

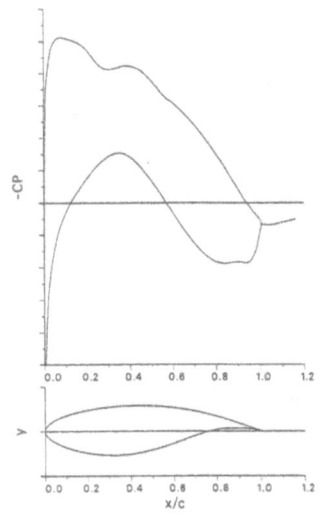
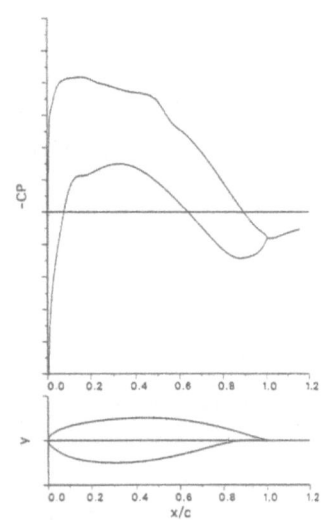

(a) Optimized performance design,
$Ma = 0.72$, $c_L = 0.7$, $(t/c) = 0.11$

(b) Optimized performance design,
$Ma = 0.76$, $c_L = 0.5$, $(t/c) = 0.10$

Fig. 1: Typical cross section designs of the database

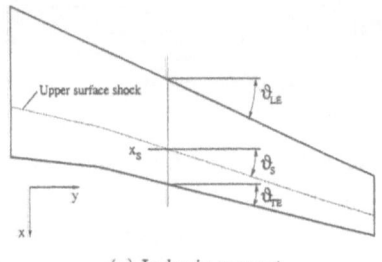

(a) Isobaric concept

(b) Non isobaric concept, for regions
with mainly 2-D-flow

Fig. 2: Local shock sweep angle

263

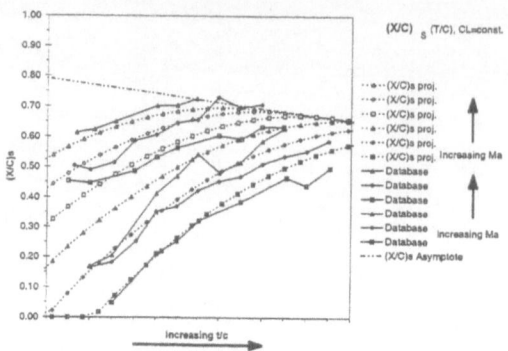

Fig. 3: Database shock location $(x/c)_{S_{db}}$ versus formulation $(x/c)_{S_{proj}}(Ma, t/c)$ for $c_L = 0.5$

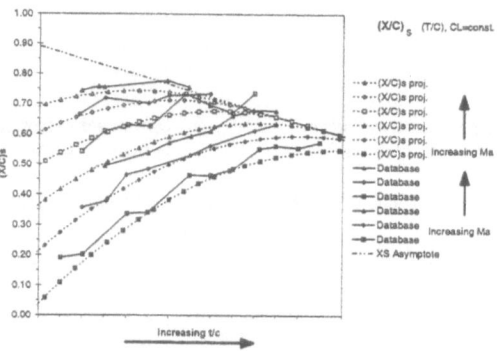

Fig. 4: Database shock location $(x/c)_{S_{db}}$ versus formulation $(x/c)_{S_{proj}}(Ma, t/c)$ for $c_L = 0.7$

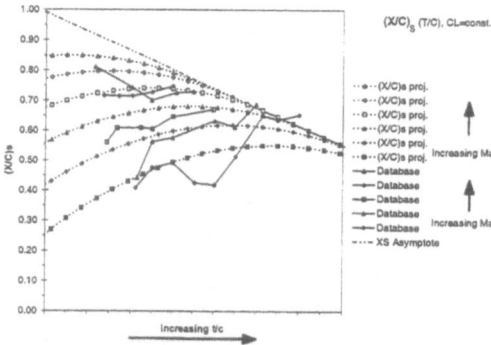

Fig. 5: Database shock location $(x/c)_{S_{db}}$ versus formulation $(x/c)_{S_{proj}}(Ma, t/c)$ for $c_L = 0.9$

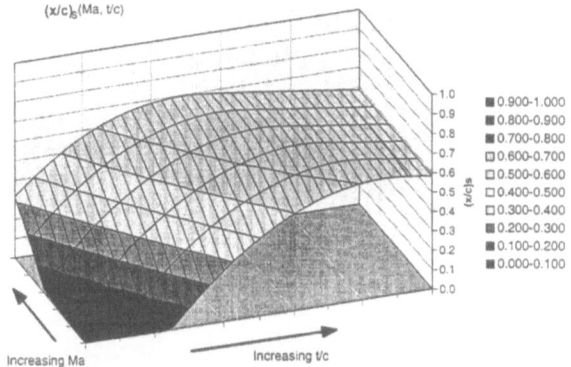

Fig. 6: Shock location $(x/c)_{S_{proj}}(Ma,\ t/c)$ for $c_L = 0.5$

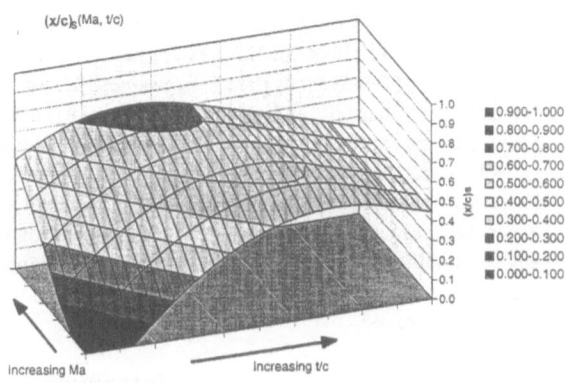

Fig. 7: Shock location $(x/c)_{S_{proj}}(Ma,\ t/c)$ for $c_L = 0.7$

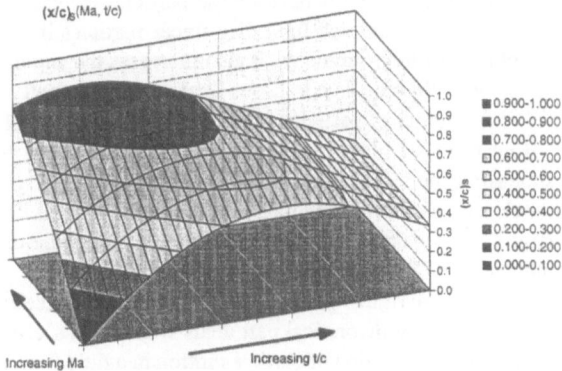

Fig. 8: Shock location $(x/c)_{S_{proj}}(Ma,\ t/c)$ for $c_L = 0.9$

Investigation of the flow tracking capabilities of tracer particles for the application of PIV to supersonic flow fields

Neven Lang

Aerodynamisches Institut

Rheinisch-Westfälische Technische Hochschule (RWTH) Aachen

Wüllnerstraße zw. 5 u. 7

D-52062 Aachen, Germany

Summary

The transport of particles is analysed for two different types of supersonic flow fields. The vicinity of a shock front and a flow field similar to a solid body rotation as an approximation of vortex centers were investigated. The influence of particle size and density are theoretically investigated by an analysis of Basset-Boussinesq-Oseen's equation which is the governing equation describing particle motion in a fluid flow. Results show that even small particles need a distance of approximately 10 mm downstream of a shock to adjust to the ambient fluid velocity. Velocity measurements with Particle-Image Velocimetry confirm the theoretical results. A theoretical investigation of a vortical flow field shows a promising low error of particle motion versus fluid flow. However the examination of the long time ejection shows an accurate motion only for small and light particles. Therefore measurements were taken at the vortical flow over a delta wing. A comparison with numerical results shows that even areas of high vorticity were accurately resolved.

Introduction

Modern experimental techniques for the measurement of fluid velocities in gaseous and liquid flows are based on the estimation of the velocity of tracer particles which are added to the flow. Contradictory requirements on these particles are caused by the flow tracking capabilities and the amount of scattered light. Big tracer particles increase the amount of scattered light that simplifies the measurements. It also increases the slip of particles versus fluid flow. For a measurement technique as e.g. Laser-Doppler Anemometry (LDA) or Particle-Image Velocimetry (PIV) the ability of particles to follow the fluid flow is of fundamental importance.

Regarding the flow tracking capabilities two major problems arise when PIV is applied. If supersonic flow fields are investigated large velocity gradients are likely to occur in form of shock waves or expansion fans. Moreover in vortical flows areas of high vorticity exist where large centrifugal forces act on the particles. Therefore the accuracy of measurements is mainly influenced by the particle size and density.

The purpose of this paper is to investigate theoretically and experimentally the flow tracking capabilities. The theoretical part deals with the Basset-Boussinesq-Oseen's equation which is the governing equation of particle motion in a fluid flow. Different solutions of this equation are presented for a flow field with an oblique shock and a fluid flow similar to a solid body rotation.

The Basset-Boussinesq-Oseen's equation is simplified for a flow field with an oblique shock. The results are compared with PIV experiments over a flat plate with a sharp leading edge and to theoretical and experimental investigations of Melling [8] and Thomas and Bütefisch [11].

Subsequently the behaviour of particles is investigated in a vortical flow field by another solution of the BBO equation. The flow field of a vortex center was approximated by a solid body rotation. The results of the theoretical investigation predict a promising low error of the particle versus the fluid motion. Finally the flow over a delta wing in supersonic flow is examined. Results from numerical simulation are compared to the experiments.

Dimensional analysis of Basset-Boussinesq-Oseen's equation

If particle–particle and particle–fluid interactions are neglected the motion of spherical particles in a fluid flow is described by the Basset-Boussinesq-Oseen's (BBO) equation [7].

$$
m_p \frac{du_p}{dt} = \frac{1}{8}\pi\eta d_p Re_p c_D(u_f - u_p) + \frac{1}{6}\pi d_p^3(\rho_p - \rho_f)g + \frac{1}{6}\pi d_p^3 \rho_f \frac{Du_f}{Dt}
$$

$$
+ \frac{1}{12}\pi d_p^3 \rho_f \frac{d}{dt}(u_f - u_p) - \frac{3}{2}\pi d_p^2 \eta \int_0^t \left(\frac{d/d\tau(u_p - u_f)}{\sqrt{\pi\eta(t-\tau)/\rho_f}} \right) d\tau. \tag{1}
$$

The subscripts p and f represent particle and fluid relevant variables, 0 indicates the reference quantities. Velocity and density are denoted by u and ρ, the viscosity of the fluid by η. Usually the drag coefficient c_D according to Stokes formulation is used. The particle Reynolds number is $Re_p = \rho_f(u_p - u_f)d_p/\eta$. The total derivative of the particle and fluid motion is represented by d/dt and D/Dt respectively. According to Newton's second law the left hand side of Eqn. (1) is the product of particle mass and acceleration that equals the external forces acting on the particle. Those are the drag of a sphere with a diameter d_p, the gravity, the pressure gradient force, the virtual mass and the Basset history term. This term defines the resistance caused by the unsteadiness of the flow field.

An estimate of the order of magnitude of the different terms of the BBO equation is obtained by dimensional analysis. Therefore dimensionless parameters are introduced. They are $\overline{t - \tau} = (t - \tau)/T$ and $\bar{t} = t\Omega$ for the time variables, $\bar{u}_p = u_p/u_{p,0}$ and $\bar{u}_f = u_f/u_{f,0}$ for the particle and fluid velocities and $\bar{v} = v/u_{p,0}$ with the slip velocity $v = u_p - u_f$. Furthermore, for the density $\bar{\rho}_p = (\rho_p - \rho_f)/\rho_p$ is introduced.

$$
\frac{d\bar{u}_p}{d\bar{t}} = \frac{1}{St}\bar{v} + \delta\pi_1\pi_2\,\bar{\rho}_p + \delta\pi_1 \frac{D\bar{u}_f}{D\bar{t}} +
$$

$$
\frac{1}{2}\delta\frac{d\bar{v}}{d\bar{t}} + \frac{\pi_3}{St}\int_{\bar{t}_0}^{\bar{t}} \frac{\partial\bar{v}/\partial\bar{\tau}}{\sqrt{\bar{t}-\bar{\tau}}}d\bar{\tau}. \tag{2}
$$

The different dimensionless parameters are the Stokes number $St = (\rho_p d_p^2\Omega)/(18\eta)$, the ratio of particle and fluid densities $\delta = \rho_f/\rho_p$, the ratio of velocities $\pi_1 = u_{f,0}/u_{p,0}$, the influence of gravity $\pi_2 = g/(u_{f,0}\Omega)$ and a further parameter $\pi_3 = \sqrt{(\rho_f d_p^2)/(4\pi\eta T)}$. For the presented PIV measurements the influence of gravity forces and the Basset integral is neglected because $\pi_2 \ll 1$, $\delta\pi_1 \ll 1$ and $\pi_3 \ll 1$ while the order of magnitude of the Stokes number is $St = \mathcal{O}(1)$. Next the particle trajectories are calculated with the help of further simplifications for a flow field with an oblique shock.

Flow field with an oblique shock

In case of the particle motion downstream of an oblique shock, the pressure force is neglected due to a uniform flow field which is equivalent to $D\bar{u}_f/D\bar{t} = 0$. Furthermore, the influence of the virtual mass is neglected because $\delta \ll 1/St$ and as such the particle deceleration is described by

267

$$\frac{\partial u_p}{\partial t} = -\frac{3}{4} Re_p \, c_D \, \frac{\eta}{\rho_p d_p^2} \, v \,. \tag{3}$$

There are serveral formulations for the drag coefficient c_D when the particle shape is assumed to be spherical.

$$c_{D,Melling} = \frac{24}{Re_p(1+Kn_p)}, \quad c_{D,Oseen} = \frac{24}{Re_p}(1 + \frac{3}{16}Re_p),$$

$$c_{D,Abr.} = 0.292(1 + \frac{9.06}{\sqrt{Re_p}})^2, \quad c_{D,Gold.} = \frac{24}{Re_p}(1 + \frac{3}{16}Re_p - \frac{19}{1280}Re_p^2 + ...) \,. \tag{4}$$

Melling [8] modified Stokes' drag formulation for compressible flows. It is applicable to $Re_p < 1$. The Knudsen number is the ratio of mean free path of the gas molecules l to the particle diameter $Kn = l/d_p$. For the present conditions of the experiments in the vicinity of a shock it was estimated to be of order $\mathcal{O}(0.1)$ [12]. The formulations of Oseen, Abraham and Goldstein are applicable for higher Re_p [1, 2, 10].

Equation (3) can be exactly integrated when the drag coefficient c_D is calculated by the formulations of Melling, Oseen or Abraham. Using the drag coefficient according to the study of Goldstein Eqn. (3) can be numerically integrated using e.g. a fourth-order Runge-Kutta method [9].

Using Eqn. (3) the relaxation distance $\Delta x'_{relax}$ was investigated. It is defined as the required length of the particles where the velocity lag $v = u_p - u_f$ is reduced by a factor $1/e$.

Solid body rotation

The particle motion in a vortical flow field is investigated by approximating a vortex center through a solid body rotation. Therefore the fluid velocity components are $u_{f,x} = -\Omega/(2y)$ and $u_{f,y} = \Omega/(2x)$ where Ω is the vorticity. Furthermore, without gravity forces and the Basset history term the complex coordinate $z = x + iy$ is used to simplify Eqn. (2) to a second order differential equation

$$\left(1 + \frac{\delta}{2}\right)\ddot{z} + \left(\frac{\Omega}{St} - \frac{1}{4}\delta\Omega\,i\right)\dot{z} + \left(\frac{\Omega^2\delta}{4} - \frac{1}{2}\frac{\Omega^2}{St}\,i\right)z = 0 \tag{5}$$

the exact solution of which has the form

$$z = c_1 e^{\lambda_1 t} + c_2 e^{\lambda_2 t} \tag{6}$$

with the complex eigenvalues

$$\lambda_{1,2} = \frac{(-\alpha \pm \sqrt{\alpha^2 - 4\beta\gamma})}{2\beta} \tag{7}$$

and

$$\alpha = \frac{\Omega}{St} - i\,\frac{\Omega\delta}{4}, \beta = 1 + \frac{\delta}{2}, \gamma = \frac{\Omega^2\delta}{4} - i\,\frac{\Omega^2}{2St} \,.$$

At $t = 0$ the particle location is assumed to be $z(t = 0) = r_0$ and its velocity $\dot{z}(t = 0) = i\,v_\theta = i\,\frac{\Omega}{2}r_0$. Then c_1 and c_2 can be calculated by

$$c_1 = \frac{-r\lambda_2 + i\,v_\theta}{\lambda_1 - \lambda_2} \quad \text{and} \quad c_2 = r - c_1 \,. \tag{8}$$

The largest real part of the eigenvalues λ_1 describes the ejection or entrapment behaviour of particles in a solid body rotation.

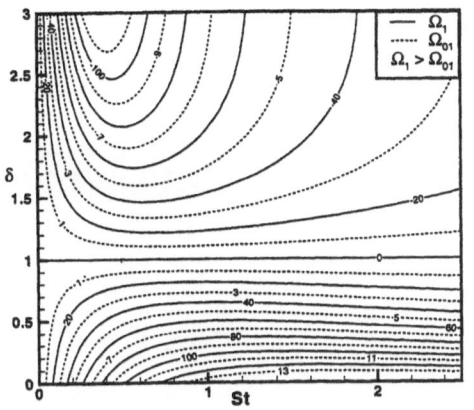

Figure 1 Isolines of the largest eigenvalue λ_1 for the solution of the BBO equation for a flow field similar to a solid body rotation.

Figure 1 shows isolines of the real part of λ_1 for two different values of Ω depending on the Stokes number St and the density ratio δ. For heavy particles ($\delta < 1$) a positive real part of λ_1 is obtained. This means that particles are ejected from the vortex core. A decreasing δ causes a larger real part of λ_1 which corresponds to a faster ejection of particles. A faster ejection of particles is also caused by constant δ and increasing small Stokes numbers which is equivalent to a larger particle diameter at a constant vorticity Ω. This effect changes at larger values of St which is due to a rising influence of the drag force. If St and δ are constant and the vorticity Ω is decreased a slower ejection will take place. If light particles are used ($\delta > 1$) they behave just in the opposite way. For example increasing δ at constant St causes a faster entrapment.

Apparatus

Velocity measurements were carried out using PIV in the AIA trisonic wind tunnel. It is a suction type wind tunnel enabling Mach numbers from $Ma = 0.2$ to 4. The testing time is 3 s to 10 s depending on the Mach number.

A flow field with an oblique shock was generated by a flat plate with a sharp leading edge at angle of attack. Velocity measurements were taken in the vicinity of the shock front in order to compare experiments and theory. The test was carried out at a Mach number $Ma = 2.02$ and an angle of attack $\alpha = 6.25°$. The corresponding Reynolds number was $Re_L = 2.4 \cdot 10^6$ based on the length of the plate in streamwise direction $L = 0.2\,m$.

The first stage of a model of the hypersonic configuration ELAC 1 was used for the investigation of the particle motion in a supersonic vortex system [3, 4]. The scale of the model is 1:240 and the corresponding Reynolds number is $Re_L = 3.6 \cdot 10^6$ with $L = 0.3\,m$ at $Ma = 2$. The angle of attack was $\alpha = 10°$, which causes flow separation at the leeward side of the model near to the rounded leading edges. The model was mounted on a y,z–traverse which enabled measurements at different positions of the model in spanwise direction and perpendicular to the model surface.

The velocity measurements were realized by the Particle-Image Velocimetry. Figure 2a shows a sketch of the experimental set up. A Nd:YAG–Laser with an output energy of 170 mJ at a wavelength of $\lambda = 532\,nm$ and a pulse duration of 8 ns was used. The double cavity system allows a delay of the laser pulses down to 200 ns. The first telescope forms approximately a parallel beam. With the help of a second telescope and a cylindrical lens the light sheet is formed. This sheet is cast through the housing of the wind tunnel and the wall of the laval nozzle into the test section. The length of the beam supply is approximately 15 m. Due to vibrations caused by the wind tunnel the whole set up was mounted on shock absorbers. Photographs of the pulsed light sheet were taken during a run of the wind tunnel with a 55 mm Leitz lens at an aperture number $F^\# = 2.8$. The magnification was $M = 0.13$. A high resolution Kodak Technical Pan film with a maximum resolution of 320 $lines/mm$ was used to resolve even small

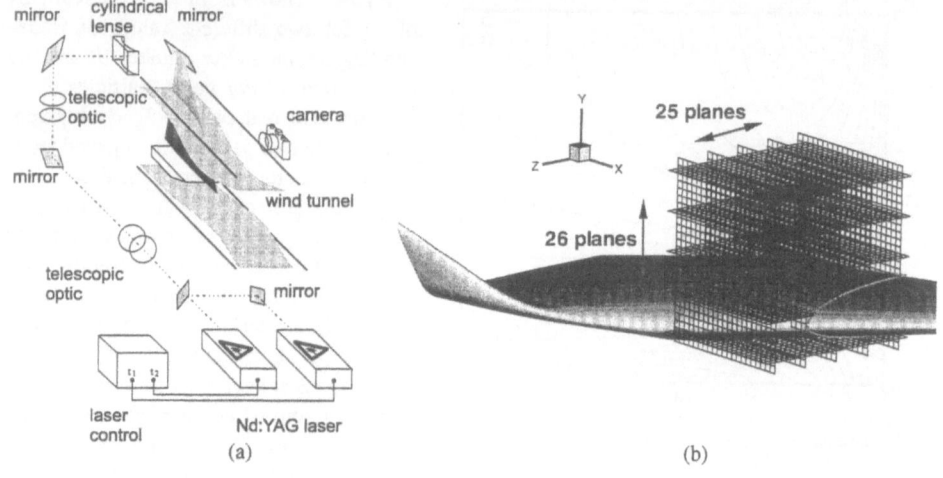

Figure 2 Set up for application of PIV to the AIA trisonic wind tunnel (a). Orientation of the different planes of measurement for application of PIV to ELAC 1 (b)

structures in the flow. The photographs were evaluated by the use of a semi optical system with autocorrelation as e.g. explained in Liu et al. [6]. In the present experiments autocorrelation is appropriate because no recirculation occurs.

An olive oil aerosol was used as tracer particles for the PIV measurements. The density of the used oil was $\rho_p = 910 kg/m^3$. Due to the necessary amount of scattered light particles with a mean diameter $\bar{d}_p = 1.95 \mu m$ were used [4]. When measurements were taken at the flat plate the pulse delay between first and second laser pulse was $\Delta t = 2.1 \mu s$. It was gradually shortened to $\Delta t = 1.5 \mu s$ when the primary vortex area of the delta wing configuration was investigated.

Theoretical and experimental results

Flow field in the vicinity of a shock

The particle motion downstream of the shock was calculated for the wind tunnel test conditions, that is at $Ma = 2.02$, a shock angle $\sigma = 35°$, a stagnation temperature $T_0 = 293 \, K$ and a total pressure $p_0 = 980 \, hPa$. The relaxation distance $\Delta x'_{relax}$ was calculated with different formulations for the drag coefficient. The formulation of Melling yields $\Delta x'_{relax,Mel.} = 2.5 \, mm$. Because of the larger drag coefficient the relaxation distance according to Oseen, Abraham and Goldstein decreases to $\Delta x'_{relax,Os.} = 1.4 \, mm$ and $\Delta x'_{relax,Abr.} = \Delta x'_{relax,Gold.} = 1.5 \, mm$, respectively. LDA measurements in the vicinity of a shock confirm these results [11]. Using TiO_2 or lycopodium increases the relaxation length to $\Delta x'_{relax,TiO_2} = 6.8 \, mm$ and $\Delta x'_{relax,lyc.} = 0.15 \, m$ if $c_{D,Abr.}$ is used.

The result from one PIV measurement is described in Figs. 3a,b. Figure 3a shows the evaluated velocity distribution over the flat plate at $Ma = 2.02$ and $\alpha = 6.25°$. The dashed line indicates the position of the oblique shock. The change of the flow direction can be seen downstream of the shock, indicating the tracking capabilities of the particles. The flow velocity changes from $516 \, m/s$ to $481 \, m/s$ and the direction of the flow is deflected by $6.25°$. The fluid velocity normal to the shock front is $294 \, m/s$ upstream and $231 \, m/s$ downstream of the shock.

In Fig. 3b experimental and theoretical results are compared. The experimental results show

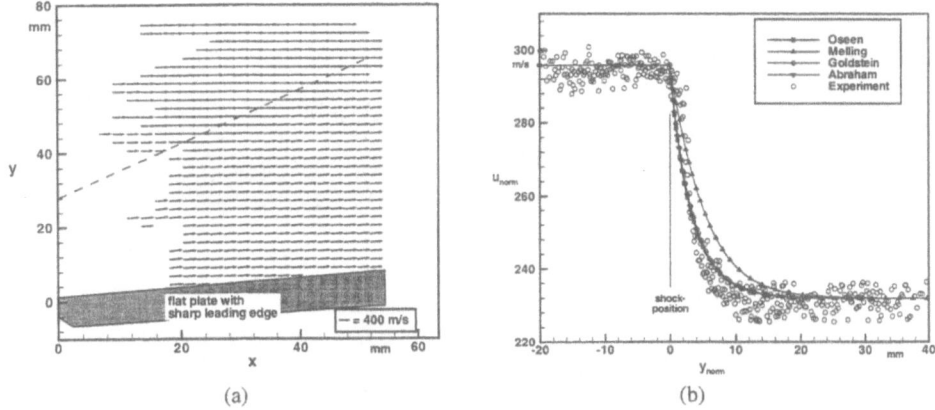

Figure 3 Velocity distribution at a flat plate with sharp leading edge at $Ma = 2.02$ and $\alpha = 6.25°$ obtained by PIV measurements. The dashed line indicates the shock front (a). Measured velocity component normal to the shock front and theoretically expected particle response (b).

that the particles were decelerated over a distance of approximately $10\ mm$ to the ambient fluid velocity. The formulation of Melling estimates a slower deceleration of the particles. It is caused by the limitation of $c_{D,Melling}$ to $Re_p < 1$. The particle Reynolds number is $Re_p = 3.34$ when the particles enter the decelerated fluid flow at the shock front. The formulations of Oseen, Abraham and Goldstein give a good approximation of the particle response, since those formulations are applicable to higher Re_p. 5% of $\bar{v}' = (u_p - u_f)/(u_p - u_f)_0$ are reached at 10.6 mm, 10.4 mm to 10.7 mm according to the formulations of Oseen, Abraham and Goldstein.

The drag coefficient formulations are still applicable downstream of the shock because the droplets remain spherical. This can be verified by the Weber number of $We = 0.081$ describing the ratio of aerodynamic forces to surface tension.

Vortical flow field

Based on the results of a numerical simulation of the flow field around the delta wing configuration ELAC 1 the vorticity was determined to $\Omega = 30\ 000\ 1/s$ at a radius of $r_0 = 0.02\ m$ [3]. The density of the air flow is $\rho_f = 0.33\ kg/m^3$ leading to a Stokes number $St = 0.48$ and a density ratio $\delta = 3.6 \cdot 10^{-4}$. The difference of the motion of particles versus the fluid flow is calculated using Eqn. (6). The particle motion is integrated for the duration of a PIV measurement which is equivalent to the delay between first and second pulse. By this method the error of particle motion versus fluid motion in the radial direction $\epsilon_r = |(r_p - r_f)/r_f|$ and the error concerning the angular velocity $\epsilon_{\dot{\varphi}} = |(\dot{\varphi}_p - \dot{\varphi}_f)/\dot{\varphi}_f|$ were investigated. They were calculated to be $\epsilon_r = 0.043\%$ and $\epsilon_{\dot{\varphi}} = 0.028\%$.

Due to the described initial conditions at $t = 0$ the calculated errors for typical tracer particles as Titaniumdioxid (TiO_2) or lycopodium are nearly alike. Therefore, the long time ejection of heavy particles ($\delta < 1$) was calculated for different values of vorticity to describe the influence of particle density and size. Results were achieved for the olive oil droplets TiO_2 with $d_p = 1.95\ \mu m$ and $\rho_p = 4000\ kg/m^3$ and lycopodium with $d_p = 30\ \mu m$ and $\rho_p = 800\ kg/m^3$. The choice of particles for low speed wind tunnel experiments is of less importance in comparison to tests in supersonic flow. Fig. 4a describes the ejection of the different particle types in a solid body rotation with $\Omega = 1000\ 1/s$ according to PIV measurements at the hypersonic

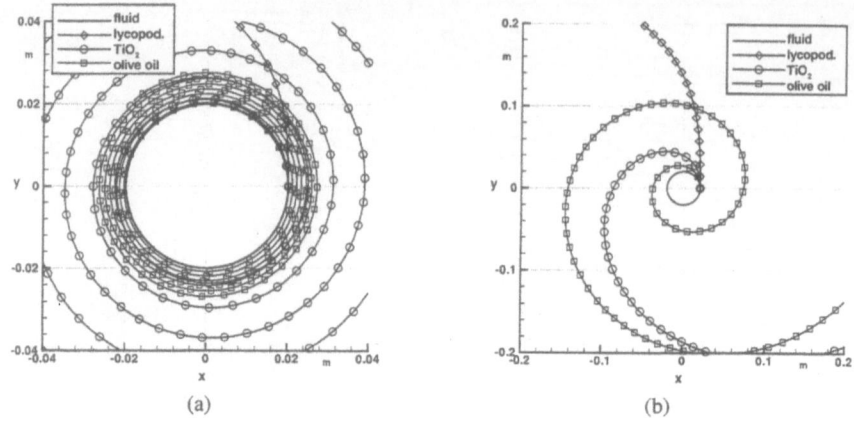

Figure 4 Long time ejection of different particles in a solid body rotation for $\Omega = 1000\frac{1}{s}$ (a) and $\Omega = 30000\frac{1}{s}$ (b)

configuration ELAC 1 in a low speed wind tunnel [5]. Because of the low density and the small particle size, the oil aerosol shows the best flow tracking capabilities. The lycopodium achieves the worst particle response caused by the large particle diameter, although the difference between TiO_2 and the olive oil aerosol is small. Figure 4b depicts the particle motion for $\Omega = 30\,000$ $1/s$. Just the oil aerosol shows an accurate particle response while TiO_2 or lycopodium do not give accurate results for PIV measurements in supersonic flows.

Thus, the investigation of the vortex over the delta wing configuration ELAC 1 was carried out with the oil aerosol. Every plane described in Fig. 2b was averaged over 12 measurements. After the reconstruction of the different sets in spanwise direction and perpendicular to the model surface all velocity components are known in a three-dimensional volume of the flow field. The velocity distribution of a cross section at a relative chord length of $x/l = 61\%$ is shown in Fig. 5a. The corresponding vorticity distribution is gray shaded.

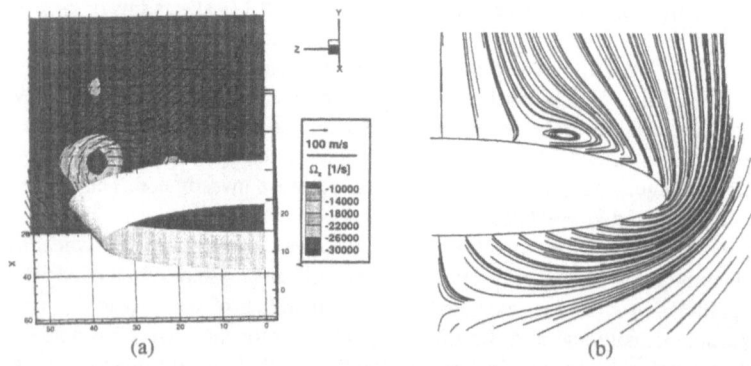

Figure 5 a) Velocity distribution in a cross section of the delta wing configuration ELAC at $x/l = 61\%$ at $Ma = 2.02$ and $\alpha = 10°$. The vorticity distribution is colour coded. b) Streamlines from numerical simulation at $x/l = 61\%$ at $Ma = 2$ and $\alpha = 10°$ [3].

Figure 5a depicts the roll up in the primary vortex of the separated flow. Vorticity increases to a value of $\Omega = 30\,000\ 1/s$ in the vortex center which was expected from the results of the numerical simulation [3]. Juxtaposing experimental and numerical results a good agreement concerning the vortex topology was obtained.

Conclusion

Theoretical and experimental investigations of the flow tracking capabilities show that shock waves cannot be accurately resolved by the use of Particle-Image Velocimetry. This is caused by the relaxation length needed by particles to decelerate to the ambient fluid flow. If the center of a vortex core is approximated as a solid body rotation the solution of the Basset-Boussinesq-Oseen's equation estimates an accurate response for an olive oil aerosol with a diameter of $1.95\ \mu m$. This means the PIV concept can be used to the accurate analysis of supersonic flow as long as it is applied only to those areas of the flow where no shock discontinuities occur.

Acknowledgement

These investigations were supported by the Graduiertenkolleg "Transportvorgänge in Hyperschallströmungen" of the German Research Association DFG.

References

[1] F.F. Abraham. Functional dependence of drag coefficient of a sphere on Reynolds number. *Phys. Fluids*, 13(8):2194 – 2195, 1970.

[2] S. Goldstein. The forces on a solid body moving through viscous fluid. *Proc. Roy. Soc., London*, 123:216 – 235, 1929.

[3] A. Henze, E.M. Houtmann, M. Jacobs, and V.N. Vetlutsky. Comparison between experimental and numerical heat flux data for supersonic flow around ELAC 1. *Z. Flugwiss. Weltraumforsch.*, 20:61–70, 1996.

[4] N. Lang. PIV measurements in sub- and supersonic flow over the delta wing configuration ELAC. In *8. Int. Symp. on Flow Vis., Sorrento, Italy*, Paper No. 205, Sept. 1998.

[5] N. Lang. Reconstruction of 3-D steady incompressible flow field out of 2-D PIV measurements in wind tunnels. In *9. Int. Symp. on Appl. Laser Techn. to Fluid Mech., Lisbon, Portugal*, pages 19.4.1–19.4.8, July 1998.

[6] Z.-C. Liu, C.C. Landreth, R.J. Adrian, and T.J. Hanratty. High resolution measurement of turbulent structure in a channel with Particle Image Velocimetry. *Exp. in Fluids*, 10:301 – 312, 1991.

[7] M. R. Maxey and J.J. Riley. Equation of motion for a small rigid sphere in a nonuniform flow. *Phys. Fluids*, 26(4):883 – 889, 1983.

[8] A. Melling. Tracer particles and seeding for Particle Image Velocimetry. *Meas. Sci. Technol.*, 8:1406–1416, 1997.

[9] W.H. Press, S.A. Teukolsky, W.T. Vetterling, and B.P. Flannery. Numerical recipes in FORTRAN. Cambridge Univ. Press, 1992.

[10] L. Schiller and H. Schmiedel. Widerstandsmessungen an Kugel und Scheibe bei kleinen Reynoldsschen Zahlen. *Zeitschr. für Flugtechnik und Motorluftschiffahrt*, 21:497–501, 1928.

[11] P.J. Thomas and K.-A. Bütefisch. An investigation of the influence of the size distribution of seeding particles on LDA velocity data in the vicinity of a large velocity gradient. *Phys. Fluids A*, 5(11):2807–2814, 1993.

[12] W.G. Vincenti and C.H. Kruger Jr. *Physical Gas Dynamics*. Krieger Publishing Company, 1965.

Turbulence Behaviour of Swirled Dual-Stream Nozzle Flows Without and With Combustion

B. Lehmann, J. Mante

Deutsches Zentrum für Luft- und Raumfahrt (DLR)
Institut für Antriebstechnik
Abt. Turbulenzforschung, Berlin, Germany
Müller-Breslau-Str. 8, 10623 Berlin
e-mail: bernhard.lehmann@dlr.de

Summary

Experimental results are reported which were measured as flow-field data for numerical modelling and with the aim to explore the flow dynamics in a model combustion chamber with and without the presence combustion. The flow was driven by a dual-flow swirl nozzle. With the help of laser-Doppler and flow visualisation techniques the isothermal and the combusting flow fields were mapped with all coefficients of the Reynolds-stress tensor. Frequency spectrum analysis was executed of the velocity-components' fluctuations and of the acoustic emission. The results show a strong and stable anisotropy of the turbulence and an absolutely unstable behaviour that forms a highly coherent helical structure at least in the cold and non-reacting zone upstream of the elevated flame. Most of the discussed results compare the non-reacting with the reacting flow-fields where methane gas was applied as a fuel.

Introduction

The necessity to reduce gaseous and acoustical pollution of the atmosphere and of the living area of people has emphasised the research on any kind of combustion systems in technical devices with a high power consumption. This holds, especially, for the combustion processes of stationary and of aero gas turbines.

Actual developments aim at the increase of the combustion effectiveness as well as at the increase of power density under the above-mentioned criteria. An extended analysis of an isothermal post-nozzle model flow was made by Hoffmeister et al. [1] with the means of hot-wire anemometry and also by means of an LDA technique.

Similar research is going to be done at the Institute for Propulsion Technique of DLR and covers the different functional units and the physical fundamentals of the combustion process of gas turbines (Blümke et al. [2], Brandt et al. [3], Hassa et al. [4], [5], Hassa [6], Lehmann et al. [7], [9], Michel and Lehmann [8]). This paper reports the results of an aerodynamic analysis of the isothermal flow and of the combusting flow in a model combustion chamber downstream of an individual atomisation nozzle. For the case of burning flow we got rid of the influence of liquid fuel effects by using methane gas (CH_4) as a fuel gas.

From the large variety of nozzle layouts as they are in use for combustion systems, we used the model layout of a two-flow swirl nozzle with a gas injection slit at the inner contour of the inner air nozzle. But the discussion of results includes also the experiences obtained from experiments with other model nozzles that apply the same two-flow technique.

The aim of the experiments is to build up data bases to provide input data for model computations and to compare them with the results of modelling. For actual Reynolds-stress modelling the measurements try, as far as possible, to acquire the complete Reynolds-stress tensor of the flow field all over the model combustion chamber.

The Measurement Object

The measurement object is a dual air-flow atomisation nozzle (Fig. 1) based on the concept of a research nozzle as used in the national KEROMIX-project of German industrial partners, universities and DLR. For the experiments it was provided with a circular slit-like outlet for methane (CH_4) as a fuel gas instead of a liquid-fuel spray. Two separated concentric air flows enter from a common plenum chamber and interact downstream of the nozzle exit plane in order to enhance the mixing process with the fuel. Both air flows swirl into unique senses of direction. The central flow picks up the methane gas from the 1.2 mm wide circular slit at the inner nozzle contour 10 mm upstream of the exit. Thus the mixing process starts as an interaction of the fuel-gas flow with the wall shear-layer of the inner air flow. Then mixing continues downstream of the nozzle exit in the free shear layer between both partial air flows. In the course of the isothermal measurements the unswirled fuel-gas flow was replaced by a third stream with the same amount of volume flow.

The model combustion chamber was a cylinder with an irregular octagonal cross-section formed of four 40 mm wide quarz windows and in between of four nearly 30 mm wide iron wall stripes. The free axial length of the chamber was 130 mm and ended in a nozzle-like contraction to half the chambers' cross-section. Exchangeable windows of quarz or glass plates were inserted alternatively for the thermal or isothermal optical measurements.

Measurement Techniques

Velocity measurements. Velocities were measured by means of a three-component laser-Doppler measurement device. It was composed of a two-component system and a single-component one with the different probe volumes intersecting by an angle of 90 degrees (Fig. 2). The six measuring light beams of this orthogonally working system were taken from an individual argon laser. The two-component system used each the blue and the green laser light for both components and a second portion of the green light served to measure the third component with the single-component system. Signal separation was achieved by means of optical filters to separate blue and green signals. The two component's green signals were split by the application of intensity discrimination under the use of forward-scattering modes.

Signal analysis was executed by means of three Dantec BSAs. In most cases they were driven under a hardware triple-coincidence mode. This enables the experimenter to acquire the signals of the three velocity components coincidentally within a time window of two microseconds and to reject the not coincident signals. Coincident signal acquisition was the excluding demand for measuring the correlation coefficients of the Reynolds-stress tensor.

The optical scatter particles consisted of solid SiO_2. They were nearly ball-shaped and fairly monodisperse providing diameters of 0.8 micrometers with an uncertainty of 20 % maximum. These particles proved to be very suitable for the forward-scatter measurements.

Laboratory equipment. The nozzle-air was taken from a pressure supply system and the fuel gas from pressure-gas bottles. A permanently exhausting device kept the laboratory air clean from the optical scatter particles and the atmosphere was prevented from being heated up due to combustion.

A computer-controlled three-axis traversing system served to scan the flow field. A vertical rod enabled the complete optical system together with the laser head to be vertically traversed in parallel with the flow axis. The measurement object was fixed on a horizontal dual-axis traversing frame in order to control its field coordinates at constant axial position.

Temperature measurements. The temperature field of the combustion flow was measured by means of a coated thermocouple with a diameter of 1 mm. The sensor material consisted of Pt-PtRh. The thermocouple was introduced and traversed perpendicularly in reference to the main flow direction. This technique includes a systematic error source due to the thermal conduction of the thermocouple along its axis.

Light-sheet visualisation. Laser light was chopped by means of a bragg cell and fed through a multimode light fibre to the light-sheet optics. The light could be directed into the flow field under different orientations. Frequency and Phase of the light flashes were controlled by the acoustic signal of a microphone which was positioned nearby the nozzle-exit plane and about 300 mm away from the nozzle axis. Suitable electronic apparatus served to filter and amplify the acoustic signal and to trigger it in order to shift the phase of the light flashes.

Acoustic measurements. As far as the measurements of acoustic spectra are mentioned, they have been made by means of an ¼ inch microphone positioned half a meter away from the nozzle axis and in the nozzles' exit plane.

Experimental Conditions

In gas-turbine systems a pressure drop of 3% of the system's total pressure is typically applied to drive the atomising and the mixing processes for combustion. Therefore, due to the atmospheric environment for the reported measurements a constant pressure drop of 3000 Pa was used to supply the experimental system for the isothermal and for the thermal measurements. The pressure drop was measured and controlled as the pressure difference between the plenum chamber upstream of the flow twisters against the environmental pressure.

In the case of combustion a constant methane gas-flow without any swirl was applied delivering a constant combustion power of 42 kW. It was controlled by use of a volume-flow rotameter. The air-fuel mass-flow ratio was AFR=23.6 and the air number $\lambda=1.37$. The flame always burnt in the so-called elevated state.

The main state of the discussed flows is governed by the identical swirl senses of the two partial air flows. Further measurements are planned to be done with a related counter-swirl nozzle configuration.

Results

Mean velocity and turbulence profiles. In Fig.3 the mean velocities and the turbulence levels of the velocity components are plotted along the axis of the model combustion chamber comparing the cases of isothermal and thermal flow. Normalisation of the experimental data has been done for these and for all further plots with the exit diameter D=15 mm of the inner nozzle and with the velocity U_{tot} =68.6 m/s calculated from the total pressure as applied to the nozzle's plenum chamber.

An axial back-flow region extends for the isothermal flow in the chamber's centre all over its length. In the thermal case the axial flow reverses to forward direction at x/D>3.33. The maximum of the thermal backward velocity at x/D=1.1 seems to be introduced purely due to the combustion process. Along the first short length up to x/D=0.5 downstream the results of both the isothermal and the thermal cases coincide.

Considerable difference of the turbulence values exists already at the nozzle exit for the iso-thermal and thermal conditions as well as for the axial and radial turbulence degrees u'/U_{tot} and v'/U_{tot}.

This gives a hint to the later on discussed anisotropic features of turbulent activity in the flow field. The maximum at x/D=0.7 of axial turbulence together with the mean back-flow maxi-mum at x/D=1.1 mark the region where the main reaction zone of the elevated flame should exert maximum thermal influence upstream.

Fig.4 is a selection of velocity profiles across the combustion chamber of the three velocity components. The selected axial coordinates relate to positions of the main reaction zone on the axis being situated in front of (x/D=0.33), behind (x/D=1.67) and considerably down-stream (x/D=3.33) of it. An only slight influence of the combustion is exerted to the mean flow at x/D=0.33 with the tendency of an increase of the profiles' extreme values and with a slight broadening of the profiles. Considerably more deviations of isothermal and thermal U/U_{tot} profiles are found at x/D=1.67. The thermally increased central back-flow compensates with an increased forward flow in an off-axis ring-shaped area. Far more downstream (x/D=3.33) the axial component begins to reverse at the chambers axis whilst a weak ring-shaped minimum with still existing back-flow remains to exist and the main forward flow volume being shifted nearer to the wall region.

The radial velocity v acts mainly in the near-nozzle region and reduces quickly along the way downstream. Due to a restricted optical access into the flow chamber the radial component could be measured only along a reduced radial area compared with the other coordinates.

The combustion exerts a remarkable effect to the circumferential velocity w. Fig 5 shows im-pressively that in the near-axis region the velocity profiles deviate from the isothermal case in the sense of an accelerated solid-body rotation. The enhanced swirl-flow region develops in the reaction-zone and keeps existing all over the chambers length with an almost constant ro-tation speed. An estimation gives nearly 700 rps for this rotating fluid. This number does not compare with the later-on discussed frequency of absolute flow instability.

According to the positions of Fig.4 the turbulence profiles are plotted in Fig.6. At x/D=0.33 the profiles of isothermal and thermal flows show slight differences of u and w turbulence which start to arise from the chamber axis. The radial v turbulence shows larger deviations with partly reversed tendencies for the different cases. At x/D=1.67 and x/D=3.33 the differ-ence of turbulence levels continue to increase and the axial region seems to be a zone of pre-ferred thermally induced turbulence..

Fig.7 compares the turbulence degrees of the three velocity components for the cases without and with combustion. The profiles show considerable differences of the individual compo-nents' turbulence levels for the isothermal flow and with the tendency of an increase also for the thermal flow. This demonstrates the existence of a pronounced anisotropy of turbulence that exists already in the highly swirled cold flow and continues to exist also in the reacting flow. At the larger distances downstream the anisotropy seems to be additionally amplified due to combustion effects and not smeared by the thermally produced stochastic turbulence as it could be expected.

Flow-field instability. Another important feature of the highly swirled flow is the formation of an absolute instability the activity of which, in many cases, can be already heard with the human ear as a clear tone. The laser-Doppler measurements enabled us to measure the related fluctuations and their frequency spectra of the three velocity components. Fig. 8 shows an ex-ample of LD-measured spectra at the position x/D=0.33 and y/D=0.53 with the presence of combustion. These kinds of spectra develop already in the cold and non-reacting zone and show mutual variation of the spectral power magnitudes of the different components that depends on the measuring position in the flow. The spectra were measured with the same frequencies also in the isothermal flow.

Michalke [9] picked up this problem and made an analytical instability analysis. With the assumption of the axial and the azimuthal velocity components' profiles adapted to our measurements he deduced the existence of a self-induced absolute instability in a wide range of the swirl parameter. He calculated its frequency a which was almost the same as we found experimentally. The frequency is of the order of 1.7 kHz and depends for the same nozzle only on the driving pressure difference.

A phase-related light-sheet visualisation technique in the cold flow demonstrated the existence of a strong helical vortex hose the separation point of which revolves along the nozzle edge with the measured instability frequency.

More downstream of the nozzle and in the reacting flow field we failed to acquire the related spectra due to the additional combustion turbulence which superimposed the weaker coherent motions.

Fig. 9 shows acoustic spectra measured by means of a microphone. In the cold flow the combustion chamber damps the emitted acoustical power maximum to a tenth of the original magnitude. With the presence of combustion the acoustical spectral power irradiated through the chamber walls obtains an order of magnitude which is nearly hundred times stronger than that one of the original flow instability.

Nevertheless it must be assumed that the swirl-induced instability and the related helical flow structure govern the large-scale mixing process all over the flow chamber and contribute to the dynamics and to the quality of the combustion process. It cannot be said up to now whether this coherent fluctuation exerts positive or negative influence to the formation of combustion products of interest and whether, at all, it can be globally detected.

Conclusions

The reported experimental results relate to the special nozzle as shown in Fig. 1 but, at least for cold flows, they were found in a similar manner also in other flow systems which imply swirl nozzles of different layouts. The presence of combustion changes the central back-flow zone by introducing a reversion point of the axial flow at a well-defined axial position downstream of the nozzle.

The combustion concentrates vortical strength along the chamber's axis forming a vortex hose with a reduced diameter and increased rotation speed compared with the cold-flow case.

A pronounced anisotropy of turbulence is produced by the swirling flow mechanisms together with the shear-layer activity between the different partial flows. It keeps existing also under the presence of combustion and seems to be amplified in certain flow regions, especially nearby the chamber's axis.

Flow-field instability is introduced already into the cold flow and forms a helical large structure which revolves with the detected frequency which Michalke [9] analytically identified as a mechanism of absolute instability. In the presence of combustion this instability can be clearly detected upstream of the main combustion zone where the flow is still cold. But the related coherent structure is supposed also to be active in and downstream of the main combustion region though it is not detectable there with the applied measurement techniques due to the high level of combustion-induced turbulence.

Up to now, we cannot say whether this instability mechanism exerts positive or negative effects to the combustion process. An answer to this question might be found by means of future experiments considering global features like combustion efficiency, chemical and acoustical pollution together with the instability parameters.

REFERENCES

[1] Hoffmeister, M., Hertwig, K., Kreul. K.-J., Kretschmar, H., Erler, K.: Modellierung von turbulenten Drall-strahlen für Brennkammern von Flugtriebwerken. VDI-Fortschritt_Bericht, Reihe 7, Strömungstechnik, Nr. 291, 1996, VDI-Verlag GmbH, Düsseldorf

[2] Blümcke, E., Eickhoff, H., Hassa, Ch. 1989, Untersuchungen zur turbulenten Partikeldispersion an einer Luftstromzerstäuberdüse, VDI-Berichte Nr 765, pp. 635-644.

[3] Brandt, M., Hassa, C., Eickhoff, H. 1992, An Experimental Study of Spray-Gasphase Interaction for a Co-Swirling Airblast Atomizer, Proc. Eighth Annual European Conference on Liquid Atomization and Spray Systems, Sept./Oct. 1992, Amsterdam, The Netherlands.

[4] Hassa, C., Blümcke, E., Brandt, M., Eickhoff, H. 1992, Experimental and Theoretical Investigation of a Research Atomizer/Combustion Chamber Configuration, Proc. International Gas Turbine and Aeroengine Congress and Exposition, June 1992, Cologne, Germany.

[5] Hassa, C., Deick, A., Eickhoff, H. 1993, Investigation of the Two-Phase Flow in a Research Combustor under Reacting and Non-Reacting Conditions, AGARD Conference Proc. 536 on 'Fuels and Combustion Technology for Advanced Aircraft Engines, May 1993, pp. 41-1 to 41-12, Fiuggy, Italy.

[6] Hassa, C, 1994, Experimentelle Untersuchung der turbulenten Partikeldispersion in Drallströmungen, thesis, research report no. 94-20, Deutsche Forschungsanstalt für Luft- und Raumfahrt, Institut für Antriebstech-nik, Cologne, Germany.

[7] Lehmann, B., Hassa, C., Helbig, J.: Three-Component Laser-Doppler-Measurements of the Confined Model Flow behind a Swirl Nozzle. Eighth International Symposium on Applications of Laser Techniques to Fluid Mechanics, 8.-11. Juli, 1996, Lissabon, Portugal.

[8] Michel, U., Lehmann, B.: Experimentelle Untersuchung der Strömung hinter Dralldüsen für Gasturbinen-Brennkammern. DGLR-Fachausschußsitzung „Schadstoffarme Verbrennung in Fluggasturbinen", 6.-7. Nov. 1997, Dresden.

[9] Michalke, A.: Absolute inviscid instability of a ring jet with back-flow and swirl. European Journal of Mechanics - B/Fluids, vol. 18 , n° 2, 1999.

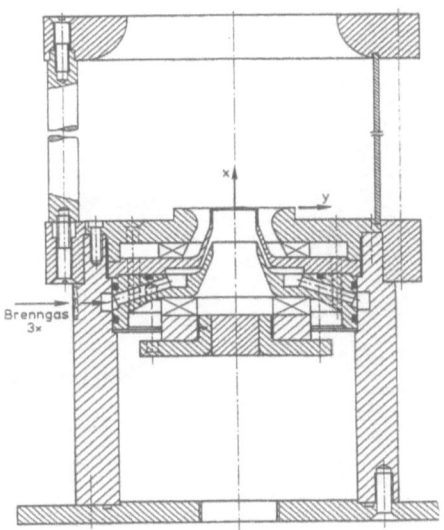

Fig.1 Schematic of nozzle and combustion chamber layout

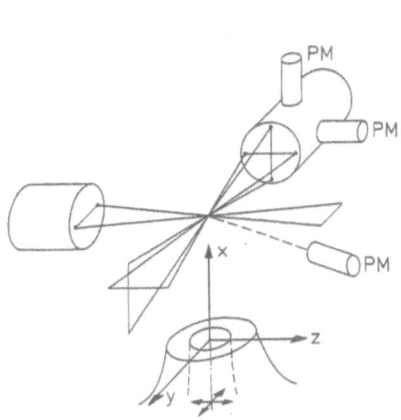

Fig.2 Optical arrangement

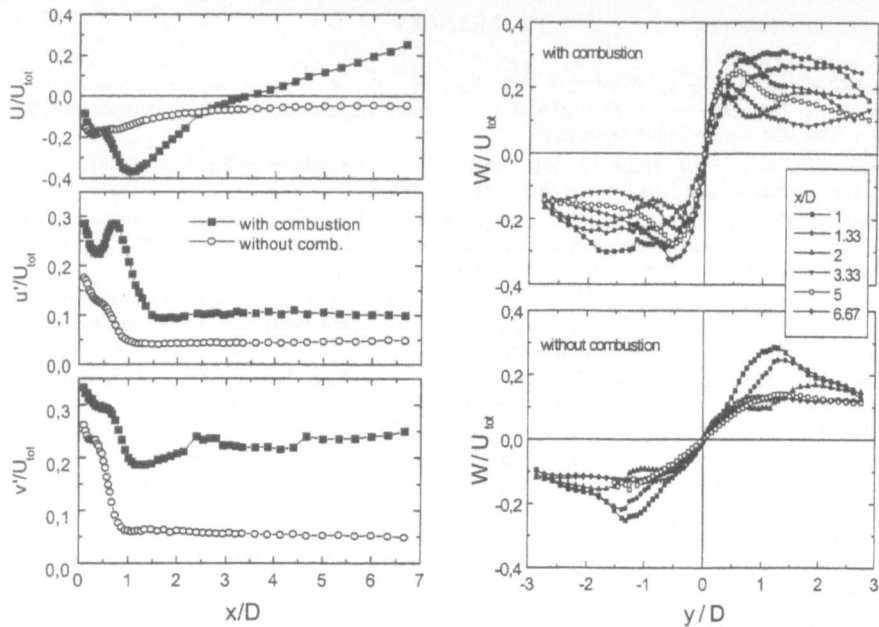

Fig 3 Mean velocity and turbulence data
along the axis of combustion chamber

Fig.5 Azimuthal velocity distributions with
and without the presence of combustion

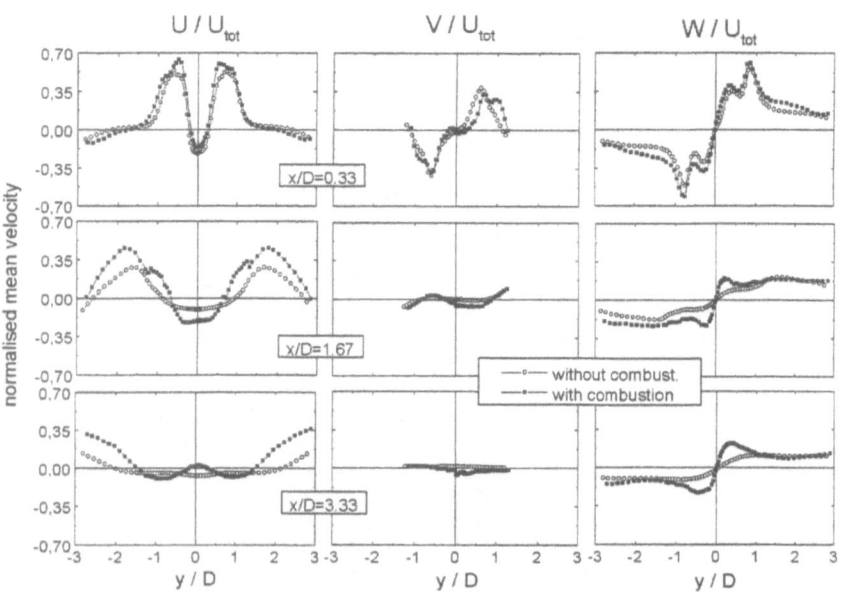

Fig.4 Mean velocity profiles with and without combustion at different axial positions

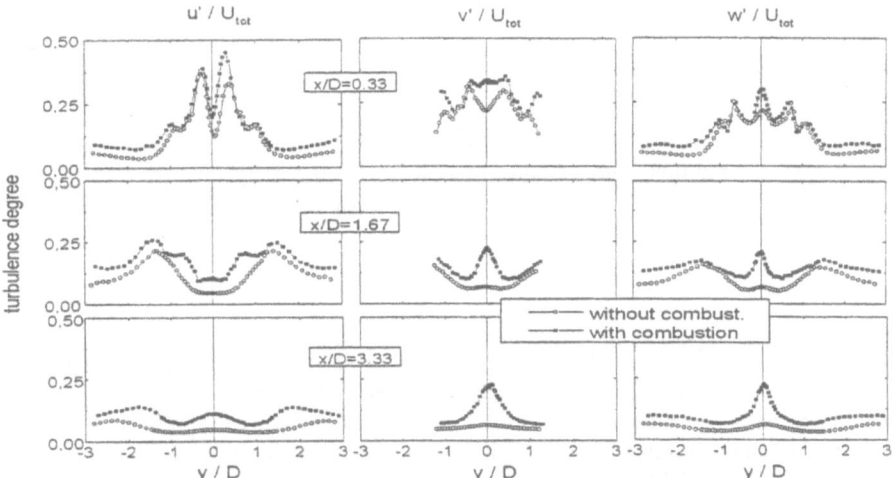

Fig.6 Turbulence level profiles with and without combustion at different axial positions

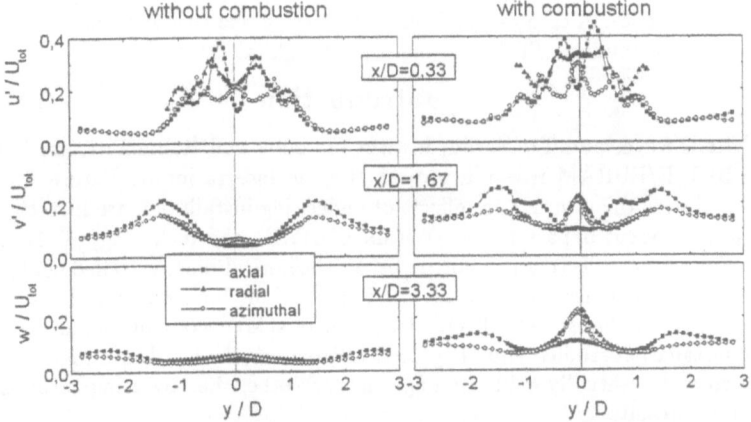

Fig. 7 Turbulence levels of the different velocity components compared

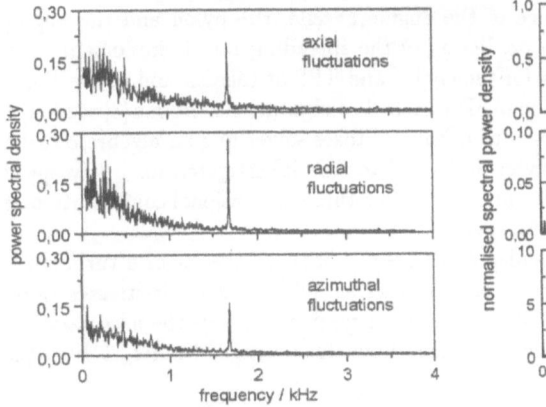

Fig. 8 Velocity fluctuation spectra with combustion at x/D= 0,33mm and y/D=0,53

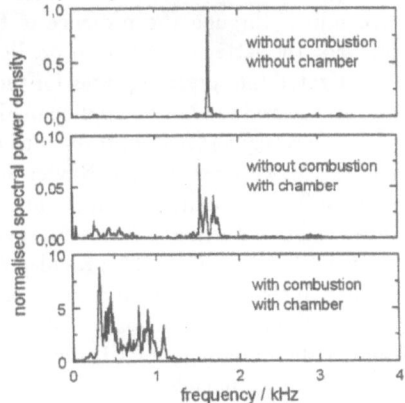

Fig.9 Acoustic frequency spectra under different conditions

CFD Calculation of Isolated Jet Engines with Emphasis on Jet Mixing

J.A. Lieser*, I. Wallbruch[†]

* BMW Rolls-Royce GmbH, Eschenweg 11, 15827 Dahlewitz, Germany

† TU-Berlin, ILR, Marchstr. 12, 10587 Berlin, Germany

Summary

Computations of confluent heated jets are presented. Four CFD codes using different turbulence models and mesh topologies are used. The calculated spreading rates are compared with experimental data for a typical axisymmetric exhaust system with a hot and a cold internal mixing jet leaving the nozzle and mixing with an external flight stream. The applicability of different turbulence models for confluent heated jets is assessed. A three-dimensional isolated engine configuration with inlet and exhaust system is calculated with selected codes. A comparison is made between a real turbofan engine and the related turbo powered simulator (TPS) for high speed flight at cruise.

Introduction

BMW Rolls-Royce and TU-Berlin, Institut für Luft- und Raumfahrt, are both involved in the BRITE/EURAM research project "Engine Integration on Future Aircraft (ENIFAIR)" [1],were the interference effects of underwing installation are investigated for engines with different bypass ratios. Within ENIFAIR, the technology of Turbo Powered engine Simulators (TPS) is promoted. In the theoretical part, installed configurations as well as isolated engines and high lift configurations without engines are computed with Navier-Stokes methods. The ENIFAIR partners shared computational effort for parametric installation studies and they used different methods. These methods had to be compared and eventually calibrated against each other, i.e. by using turbulence models with similar results.

Changes of the pressure distribution on the wing occur due to the disturbance of the streamlines through the presence of the engine nacelle, the pylon and the edge of the jet. For the latter an accurate prediction of the spreading rate is important. Pflug [2] investigated the spreading rates for the engine and TPS at take off and cruise conditions, but his method can not be applied in 3D installed configurations. Bolms [3] studied the jet flow at low flight Mach-numbers with a Navier-Stokes solver and an algebraic turbulence model. In the present work Navier-Stokes solvers are investigated for a low speed case with available experimental data and applied to a three-dimensional configuration at high speed cruise condition.

Within ENIFAIR, ONERA provided experimental data for the jet of a turbofan engine with an internal annular mixer. Mean velocities as well as Reynolds-stresses were measured in the wake of the nozzle. Thus, these measurements include the necessary information for the validation of CFD methods using Reynolds-averaged Navier-Stokes equations and turbulence models.

The turbulent jet has been extensively investigated [4, 5, 6] and analytical solutions based on similarity assumptions and mixing length theory are presented in [7, 8]. Pflug used parabolic jet equations in combination with a mixing length model and two equation k-ε models. This jet model was applied to configurations similar to those presented in the present paper and showed good agreement with experiments [2]. The model is limited to axisymmetric jet flow with axial pressure gradients.

For complex geometry such as integrated engines, several flow phenomena, such as boundary layer, free shear layer (jets) separation and free vortices influence the integral values like lift and drag in which we are interested. To fulfil this need of universality, the full 3D Navier-Stokes equations have to be used. We assume that with today's second order accurate numerical schemes the Reynolds averaged mean flowfield of a jet is solved accurately. When using the Reynolds-averaged Navier-Stokes equations the turbulent mixing has to be modelled by turbulence models. Unfortunately a universal turbulence model for industrial applications is not existent.

As pointed out by Wilcox [8], experimental results of jets can be reproduced by fitting the mixing length to the problem. Thus, for an ideal jet configuration the modelling of the turbulence is not a difficult task if the model can be adjusted to the application. Using representatives of $k - \epsilon$ and $k - \omega$ models, Wilcox showed that the calculated spreading rates of jets differ from the experimental values for all models. A big disadvantage of the $k - \omega$ model for free shear flows is its dependency on free stream values of ω. The $k - \epsilon$ model gives better results in case of free shear layers. Therefore, Menter [9] proposed a $k-\omega$ model which combines the advantages of the $k-\omega$ model for wall bounded separating flows with the advantages of the $k - \epsilon$ model for free shear layers. Unfortunately, this model is not available to us up to now.

A special problem exists for the round jet. While the standard sets of constants are reasonable for most of free shear layer flows, they fail to predict the round jet. The reason is that the turbulent structure in the round jet is different from that in plane jets and shear layers. The spreading rate is consequently lower than that for a plane jet. The circular like shaped vortices in the mixing layer are stretched due to the spreading of the jet. An adaptation of the constants such as proposed by Rodi [4] or an implementation of a "measure of vortex stretching" as proposed by Pope and extended by Rubel [5, 6] is necessary to predict the spreading rate of the round jet. Unfortunately those modifications still lack universality, since they fail in other applications such as the radial jet. Recent models, often called realisable models, use a variable constant in the eddy viscosity equation and a modified production term in the ϵ equation [10]. It is proposed, that they overcome the plane/round jet problem. They will be investigated when the next versions of FLUENT and FLOWer are available

In case of the jet engine exhaust system, the Mach-numbers are high and the geometry is complex. Therefore, besides the turbulent mixing the pressure field, the boundary layer development, compressibility effects and the boundary conditions may also be important. The experiments show large standard deviation values behind the nozzle (y/D=0.5, D is the nozzle exit diameter) and at the axis behind the bullet base (y/D=0). We therefore assume that vortex shedding and longitudinal vortices behind the bullet may occur which cannot be computed by our time-averaged steady approach.

Morris et al. [11] modelled large scale structures as linear instability waves in axisymmetric compressible shear layers. Although this kind of modelling is not a full turbulence closure, it showed some fundamental effects such as round jet/plane jet anomaly and effects of convective Mach-number. The convective Mach-number, defined as the ratio of the velocity difference to the sum of the speeds of sound $M_c = (U_1 - U_2)/(a_1 + a_2)$, goes up to values of about 0.37 in our case. It is therefore expected that the spreading rates may be about 20 percent lower than in incompressible flows. The turbulent Mach-number $M_t = 2k/a^2$, were k is the turbulent kinetic energy and a the speed of sound, is low in the jet behind a common exhaust system. Compressibility is therefore assumed to be unimportant for the turbulence modelling in our applications.

Numerical Methods

All methods used solve the Reynolds averaged Navier-Stokes equations in combination with two/one-equation, eddy viscosity turbulence models. The codes are more or less commercially available and were not modified within this work. The codes are used by the partners in research and commercial projects.

The calculations were performed with CFX TASCflow, FLOWer, FLUENT UNS and CFD-ACE. They are all based on finite volume methods. FLOWer uses central differencing with artificial dissipation, FLUENT-UNS uses upwind flux vector splitting and in CFD-ACE a smart scheme was chosen which switches between central differences to first or second order upwind schemes depending on local flow variables. In all codes, a second order accurate scheme was selected. FLOWer is the only one using an explicit time stepping scheme. All other schemes use pressure based implicit solution algorithms. In TASCflow and FLUENT UNS, we used a $k - \epsilon$ two layer zonal model. In the free shear layers the model reduces to the model proposed by Launder and Spalding [12]. In TASCflow, the production terms in the turbulence equations are modified to avoid unrealistic increases in regions of strong streamwise gradients. In CFD-ACE, the $k - \epsilon$ model of Chien [13] was used, where the constants in the turbulence equations are slightly different from the two layer zonal model. The different near wall formulation may lead to different velocity profiles at the trailing edges of nozzle, mixer and bullet. FLOWer uses the Wilcox [8] $k - \omega$ model without extensions, which is easy to apply for complex geometries since no wall distance is needed. A Spalart Allmaras one-equation model is also available [14]. FLUENT UNS can use unstructured meshes with several types of cell geometry. Since the effort for mesh generation can be reduced significantly by using unstructured meshes we included a triangular mesh in our studies. When using triangular meshes, it is hard to resolve the boundary layer down to the inviscid sublayer without significantly increasing the number of cells. Therefore, we used a high Reynolds-number turbulence model instead of a low Reynolds-number model for the triangular mesh.

Axisymmetric Turbofan Nozzle

SNECMA provided test data to the ENIFAIR partners to validate the codes against experimental data. A scaled long duct confluent flow geometry with blunt bases at nozzle exit, mixer and centerbody or bullet is given (see figure 1). Mean velocity components as well as averaged fluctuating velocities and standard deviations were measured in the confluent jet behind the nozzle up to a distance of 15.99 diameters. The free stream Mach-number is low (M=0.24 or 80 m/s). The total pressures and temperatures for the fan and core flow, respectively, are given. In the CFD computations, they are assumed to be constant over the radius. The fan pressure and temperature ratio is $p_{t,Fan}/p_\infty = 1.55$, $T_{t,Fan}/T_\infty = 1.184$ and for the core $p_{t,Core}/p_\infty = 1.49$, $T_{t,Core}/T_\infty = 3.118$. Turbulence data were not given or measured inside the nozzle.

3D Turbofan

Within ENIFAIR a Turbofan, a very high and an ultra high bypass ratio engine are studied. The aim is the comparison of the different engine types for cruise and take off as well as a comparison of real engines (named engines) and TPS. Since the massflow in the real engine is higher than for the TPS, were the core massflow is feeded in separately, the intake of the TPS is narrowed in such a way, that for a certain flight stream Mach-number the maximum Mach-number on the nacelle is similar (same spillage drag). The computational effort is distributed over all partners and a mesh was made only for the engine geometries. The presented Turbofan in cruise was chosen as a cross check test case, which is calculated by all partners including the authors. The engine is calculated in cruise condition, were the jet should behave very similar to the axisymmetric test case and for take off condition were the angle of attack is 12 degree (not presented here). The cross check of codes will be done for cruise. The take off condition is calculated only by BMW Rolls-Royce. The VHBR calculations of TU-Berlin are not completed yet.

The influence of engine geometries on the jet development was studied by Pflug [2] and other authors in [15]. In our study we use the Navier-Stokes codes, which will later on be used for installed calculations of interference effects and apply them for high speed cases.

Results

Axisymmetric Turbofan Nozzle

Because some of the partners in ENIFAIR suffered from convergence problems we agreed to do some simplifications for the common mesh study. The mesh was provided by ONERA with triangular mesh blocks behind the blunt nozzle and mixer bases. We agreed to skip this blocks which means that the mixer and nozzle geometry is a little bit longer compared to the experiment. A mesh block was implemented into the bullet and the exit boundary condition from the bullet was set to freestream flow values. For the additional FLOWer and UNS calculations BMW Rolls-Royce created two extra meshes with blunt bases at nozzle, mixer and bullet and found that the simplifications are not important for the overall results. The additional meshes are shown in figure 2.

Figure 1 shows the Mach-number distribution for the exhaust system. At the outer edge of the nozzle a small supersonic region occurs. After the shock the flow accelerates again. At the beginning the outer shear layer has a wave-like form. Behind the bullet one can see a long region with low velocity belonging to the 80 m/s exit condition. After about 10 nozzle diameters the outer shear layer reaches the axis.

Because the spreading rate is difficult to define in the multi-jet system we plotted the axial velocity at the axis in figure 4. The progression of the velocity along the axis provides the information about spreading rates. Behind the bullet a recirculation zone appears. All codes (except UNS with triangular mesh) predict the recirculation zone too long, regardless which boundary condition is used at the bullet base. We assume that either the core velocity profile at the core jet entry, which is rectangular in the numerical simulation, is responsible or instabilities occurred in the experiments which are in fact indicated by measured high standard deviations on the axis behind the bullet.

After one nozzle diameter the recirculation ends and the core velocity meets the axis. The acceleration depends on the spreading rate and the core profile and is similar for all codes (except FLUENT UNS using a triangular mesh). The velocity on the axis decreases again, when the outer jet boundary reaches the axis at about ten nozzle diameters behind the nozzle. From figure 4 the advantage of the $k - \epsilon$ models over the $k - \omega$ model can be seen. The spreading rate is too low for the $k - \omega$ model. Therefore the Spalart Allmaras model was also applied and seems to perform much better.

In figure 5 the results from the additional generated meshes are compared with the common mesh for UNS and FLOWer. UNS on a triangular mesh is the only one to show the velocity increase behind the bullet correctly. We assume that the recirculation is shorter due to a higher numerical diffusion in the triangular mesh. The result in the new FLOWer mesh is slightly better compared with the results on the common mesh. Additionally the solution converged five times faster due to the lower aspect ratios in the farfield.

In figure 6 the velocity profiles behind the nozzle are shown. Due to numerical problems, the TASCflow calculation was done with velocity inlet conditions for the jets. The prescribed velocities were too high, as can be seen at the velocity profile directly behind the nozzle. At the outer edge of the jet and in the wake of the internal mixer significant discrepancies between experiment and calculations are observed. We suppose, that this is due to the fact, that we are using constant flow values at the fan and core boundary conditions whereas in the experiment or in real engines a parabolic shaped duct flow profile can be expected. Therefore gradients, especially behind the mixer, are to sharp. Pflug [2] started his calculation with the first measured velocity profile and got good results. In most cases the details of the blunt bullet and the shape of the flow quantities at the boundary are not known. More investigations are necessary to clarify the influence of these details on the development of the jet.

The prediction of the Reynolds-stresses shown in figure 7 is satisfactory in the two outer shear layers. But near the axis the deviation between calculation and experiment is very high. The levels of Reynolds-stresses are lower from the beginning in TASCflow. Modifications implemented in TASCflow for the production term in regions of high axial gradients, which appear at the nozzle exit, are assumed to cause the better results.

Turbofan Engine

The aim of the isolated engine calculations are to study differences in the flow field of a real engine and a TPS. The TPS is calibrated for thrust over rotary speed. Thus the thrust drag balance can be simulated in the wind-tunnel experiment with integrated engines. The main differences in the flow field is the temperature in the core jet and the massflow through the intake. While the core jet in a real engine is hot, the core jet of a TPS is quite cold. The air used for driving the turbine has about ambient temperature and is cooled down while expanding in the turbine. The velocity in the engine core jet is higher than in the TPS core jet at similar Mach-numbers. The velocity gradient is the relevant factor for the mixing process. Figures 8 and 9 show the Mach-number distribution for a three-dimensional isolated turbofan real engine and TPS in the vertical symmetry plane. At the nozzle exits, the Mach-numbers are similar, but the mixing of the jets develop different. As the temperature in the engine core decreases, the speed of sound decreases too and the Mach-number increases in axial direction. The situation is different in the core jet of the TPS. The spreading rate of the common jet is higher for the real engine, which means that the influence on the wing will be underpredicted with TPS simulations. As pointed out, the intake highlight area of a TPS is normally reduced to account for the lower massflow. We used an identical mesh (geometry of the real engine) for real engine and TPS and thus the differences can be seen in figures 8 to 10. The pressure distribution presented in figure 10 exhibit the differences in the intake and around the intake lip. The incoming streamtube is thicker in case of the engine, and thus the stagnation point is more outside. Consequently lower pressures occur at the outer side of the intake lip in

case of the TPS.
The viscous calculations on 1.6 million mesh points took about 20 hours in parallel mode on 4 SGI R10000 processors using 3 level multigrid.

Conclusions

Although the jet flow is a multiple investigated flowfield, the calculation of a realistic exhaust system with CFD is still a challenging task. Since gradients of interest are not limited to the region closed to the walls certain requirements to mesh generation and farfield definition have to be fulfilled for jet calculations. The development of the jet is of parabolic type and depends on the starting conditions. Thus details of boundary conditions and geometry in the exhaust system of the engine become very important. Turbulence models approved for wall bounded flows and extended for effects of vortex stretching effects in jets have to be used and improved for complex flowfields including jets. Further investigations of the effects of boundary conditions on turbulent mixing and model modification for vortex stretching in jets are necessary.

CFD can point out the differences between the real engine and a TPS. For high flight Mach-numbers the spreading rate of the jet is higher for the real engine. Thus the interference may be underpredicted in windtunnel experiments with TPS. Calculations of isolated engines, which are less costly than installed calculations, will be used to investigate unexpected jet interference effects found in the windtunnel experiments.

Acknowledgements

The presented work was performed within the BRITE/EURAM project ENIFAIR. The permission of publication is gratefully acknowledged. We thank our ENIFAIR partners SNECMA from France and CIRA from Italy for providing meshes and experimental data.

References

[1] Burgsmüller, W., Rollin, C., and Rossow, C.: Engine Integration on Future Transport Aircraft - The European Research Programs DUPRIN/ENIFAIR -. In *ICAS-5.6.1*, 1998.

[2] Pflug, M. and Haberland, C.: On Numerical Jet Flow Simulation of Current and Future High By-Pass Engines. In Hoheisel, H., (editor), *Aspects of Engine-Airframe Integration for Transport Aircraft. Proceedings of the DLR Workshop*, DLR Mitteilung 96-01, pp. 25–1,18, 1996.

[3] Bolms, H.-T.: Calculation of Viscous Engine Jet Flow. In Hoheisel, H., (editor), *Aspects of Engine-Airframe Integration for Transport Aircraft. Proceedings of the DLR Workshop*, DLR Mitteilung 96-01, pp. 25–1,18, 1996.

[4] Rodi, W. and Spalding, D.: A Two-Paramter Model of Turbulence, and its Application to Free Jets. *Wärme und Stoffübertragung*, Vol. 3, pp. 85–95, 1970.

[5] Pope, S.: An Explanation of the Turbulent Round-Jet/Plane-Jet Anomaly. *AIAA Journal*, Vol. 16 (No. 3), pp. 279–281, 1978.

[6] Rubel, A.: On the Vortex Stretching Modification of the $k - \epsilon$ Turbulence Model: Radial Jets. *AIAA Journal*, Vol. 23 (No. 7), pp. 1129–1130, 1985.

[7] Schlichting, H.: *Grenzschicht-Theorie*. Verlag G.Braun, Karlsruhe, 1982.

[8] Wilcox, D. C.: *Turbulence Modeling for CFD*. DCW Industries, Inc., 1993.

[9] Menter, F.: Improved Two-Equation $k - \omega$ Turbulence Models for Aerodynamic Flows. TM-103975, NASA, 1992.

[10] Shih, T.-H., Liou, W. W., Shabbir, A., and Zhu, J.: A New $k - \epsilon$ Eddy-Viscosity Model for High Reynolds Number Turbulent Flows - Model Development and Validation. *Computers Fluids*, Vol. 24 (No. 3), pp. 227–238, 1995.

[11] Viswanathan, K. and Morris, P.: Predictions of Turbulent Mixing in Axissymmetric Compressible Shear Layers. *AIAA Journal*, Vol. 30 (No. 6), pp. 1529–1536, 1992.

[12] Launder, B. and Spalding, D.: The Numerical Computation of Turbulent Flows. *Computer Methods in Applied Mechanics and Engineering*, Vol. 3, pp. 269–289, 1974.

[13] Chien, K.-Y.: Predictions of Channel and Boundary-Layer Flows with a Low-Reynolds-Number Turbulence Model. *AIAA-Journal*, Vol. 20, p. 33, 1982.

[14] Spalart, P. and Allmaras, S.: A One-Equation Turbulence Model for Aerodynamic Flows. *AIAA 92-439*, 1992.

[15] Hoheisel, H., (editor): *Aspects of Engine-Airframe Integration for Transport Aircraft. Proceedings of the DLR Workshop*, DLR Mitteilung 96-01, 1996.

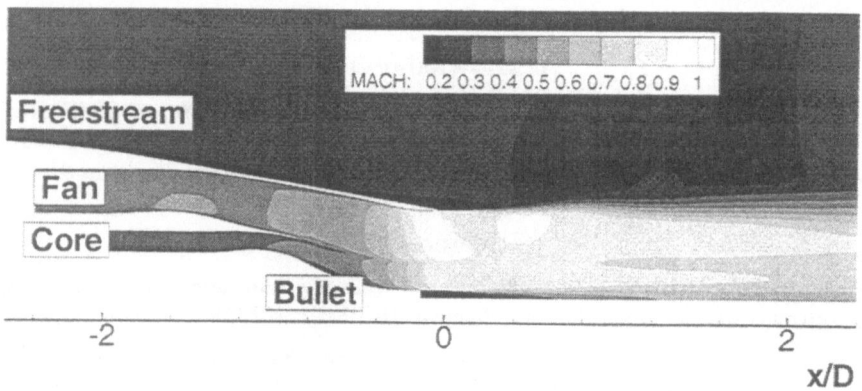

Fig. 1: Axisymmetric Exhaust system: Mach-number distribution. Calculated with TASCflow

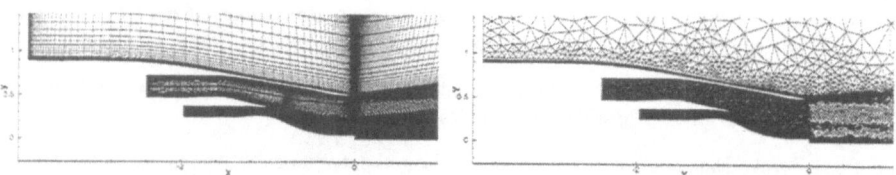

Fig. 2: New FLOWer mesh with reduced aspect ratios. Bullet closed, blunt nozzle and mixer.

Fig. 3: Unstructured triangular mesh. Bullet closed, blunt nozzle and mixer.

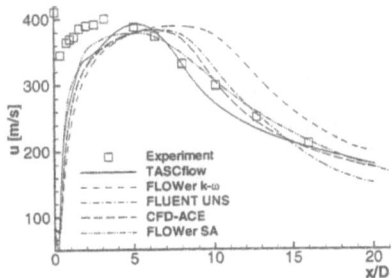

Fig. 4: Axisymmetric exhaust system. Velocity along the jet axis calculated with different CFD methods and turbulence models on identical meshes.

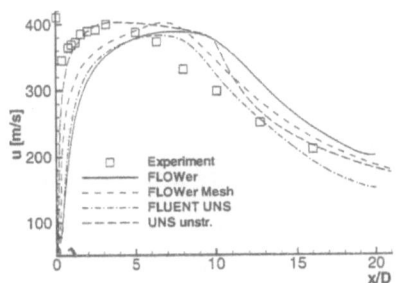

Fig. 5: Axisymmetric exhaust system. Velocity along the jet axis calculated with FLOWer. Comparison of new structured and unstructured mesh.

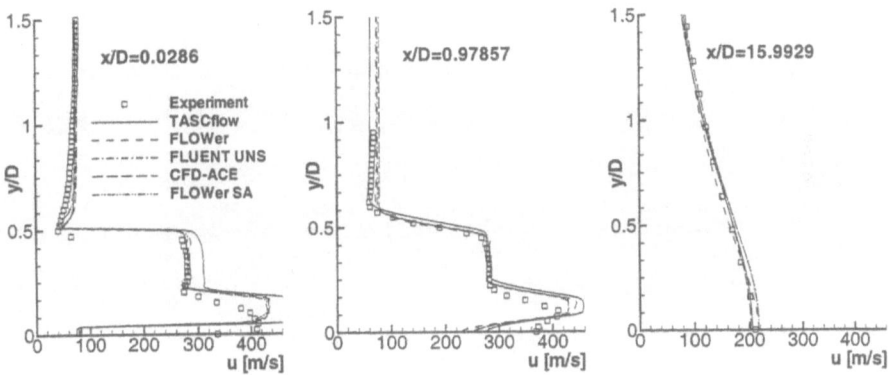

Fig. 6: Axisymmetric exhaust system. Comparison of velocity profiles.

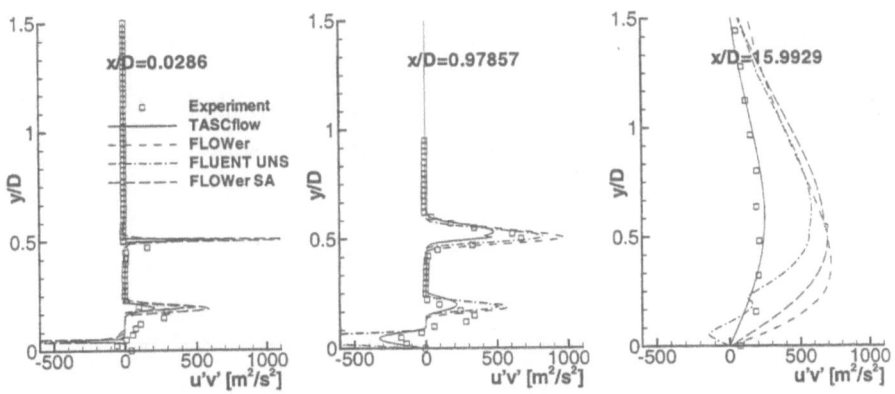

Fig. 7: Axisymmetric exhaust system. Comparison of Reynold-stress profiles.

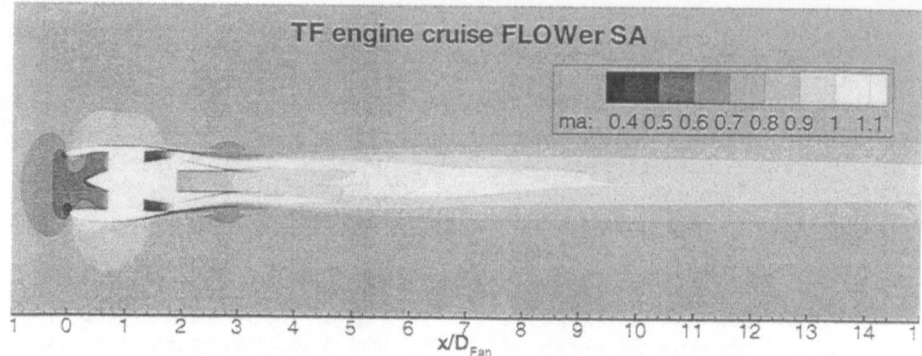

Fig. 8: Mach-number Distribution Turbofan Engine in cruise calculated with FLOWer and Spalart Allmaras model

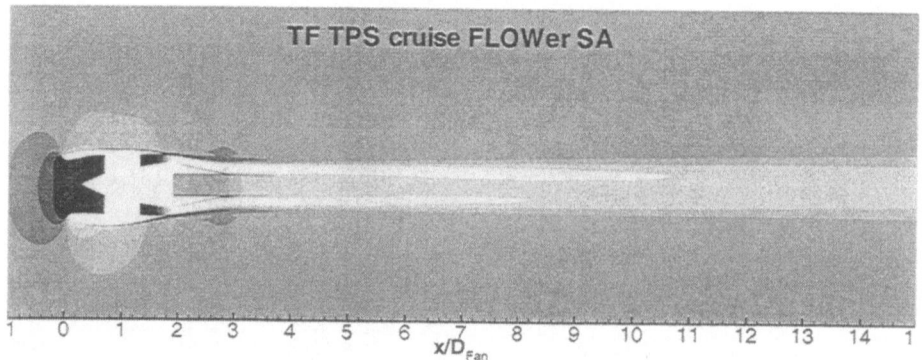

Fig. 9: Mach-number Distribution Turbofan TPS in cruise calculated with FLOWer and Spalart Allmaras model

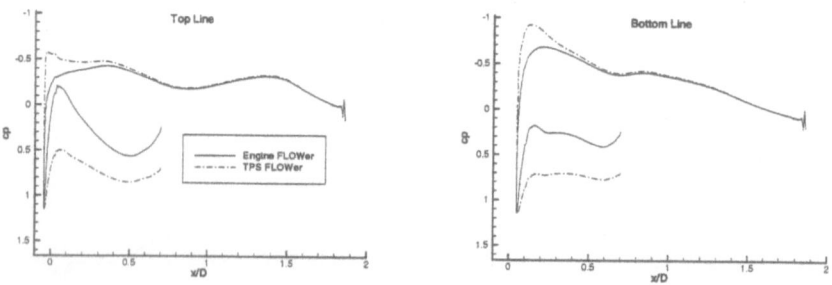

Fig. 10: Turbofan engine in cruise condition. Comparison of pressure distribution for two different codes and engine versus TPS for FLOWer

Design of an integral probe for temperature and flow vector measurement

T. Loeser, K. de Groot, K.H. Horstmann

DLR, Institut für Entwurfsaerodynamik

Lilienthalplatz 7, 38108 Braunschweig, Germany

Summary

This paper describes the design of a probe for inflight flow measurements at low Mach numbers. The probe contains sensors for flow vector and temperature measurement within a single housing. The flow vector will be measured by means of a five hole probe with additional taps for static pressure measurement. Two sensors with different dynamic behaviour are used for the temperature measurement. The influence of geometric layout on the pressure distribution as well as the flowfield around the probe have been investigated numerically with potential flow and Navier-Stokes methods. The pneumatic layout of a pressure measurement system of high accuracy and good dynamic behaviour is described.

Introduction

The objective of the newly founded *Sonderforschungsbereich 420* (SFB 420) entitled ,Flight Measurement Techniques' is the improvement with respect to accuracy and time resolution of measurement in ,unfriendly' environment.

One of the basic goals is the substantial improvement of the inflight wind vector measurement concerning accuracy and resolution in time. The wind vector is measured as a difference of the aircraft movement with respect to the ground measured by inertial and satellite supported systems and the aerodynamic flow vector measurement. The determination of the aircraft movement has been improved rapidly with the advantages of the Global Positioning System GPS. Differential systems allow an accuracy of position determination better than 0.1 m. Objective of the SFB 420 is a corresponding improvement of the inflight flow measurement.

Within that scope a flow temperature probe and a flow vector probe had to be developed, which had to fulfill very high requirements in regard of accuracy as well as resolution in time. In cooperation with the Institute for Electric Measurement Techniques (EMT) of the Technical University of Braunschweig (TU BS) a probe geometry integrating the measurement of both values of interest into a single case has been developed.

This probe will be mounted on a noseboom of the Dornier Do-128 twin-engine aircraft of the Institute of Flight Guidance and Control of the TU BS.

Aerodynamic design

The probe was designed to fulfill the following accuracy requirements: the resolution with respect to time has to be better than 20 ms, corresponding to a frequency resolution of 50 Hz. The accuracy of the flow temperature has to lie within 0.1 K, the accuracy of the magnitude of the flow vector has to be better than 0.1 m/s at an airspeed of 50 m/s while the angular accuracy of the flow vector has to lie within $0.1°$.

For the measurement of the flow vector components a five hole probe (non-nulling mode) has been chosen. Because of the expected range of angle of attack and angle of sideslip it was not necessary to use a seven hole probe with its greater range of flow angles. To achieve the best sensitivity of the probe, the difference of the static pressures indicated by the opposing holes at a given deflection angle has to be as high as possible. Among other geometric details, the influence of the shape of the probe head on the probe sensitivity has been investigated by means of the higher order panel method HISSS developed by L. Fornasier [1]. Head shapes commonly found at multi hole probes include hemispherical, pyramid-shaped and conical types, see Bryer, Pankhurst [2]. In addition to the above mentioned shapes a head based on the Cassini function has been investigated. With this shape the head blends into the shaft without a jump in curvature, which reduces the tendency of the flow to separation.

The influence of the head shape on the difference of static pressure of two opposing holes for a given angle of attack is shown in **Fig. 1**. It can be seen that the conical head shapes with 45° and 60° angle have a much lower sensitivity than the hemispherical and the Cassini-type shapes, which are of roughly the same sensitivity. The contours of the different heads are shown in the lower part of **Fig. 1**.

Having chosen the Cassini-type head as being best suited, the ideal length to diameter ratio has to be determined. **Fig. 2** shows that increasing the length of the head yields at symmetrical onflow (hollow symbols) to a significant decrease of the pressure gradient, which leads from the pressure minimum to roughly the freestream value. This reduces the risk of laminar separation bubbles, which may lead to Reynolds number dependency as well as to hysteresis effects of the probe. The probe sensitivity, characterized by the difference in pressure ΔC_p of opposed holes is much less affected by the length to diameter ratio, as can be seen from the filled symbols in Fig. 2. The location of the ΔC_p peak on the probe axis dictates the position of the four circumferential holes. HISSS calculations at different angles of attack indicated very little dependency of the ΔC_p peak location with respect to the angle of attack.

The static pressure also has to be measured at the probe. This is accomplished via 6 circumferential, interconnected holes in the cylindrical shaft behind the probe head. To minimize errors the position of these holes should be where the static pressure has exactly the onflow value. This position depends on the length of the probe shaft as well as on the probe contour behind the shaft and the noseboom. The influence of the probe shaft length on the static pressure can be seen in **Fig. 3**.

The length of the probe shaft is important because it determines the length of the tubes connecting the holes in the probe head with the pressure sensor, which is housed in the rearward cylindrical section of the probe. It is well known that the dynamic response of tube systems becomes worse with increasing length [3], [4]. The dynamic response for a tube system to be used in this probe has been investigated with the program described in [3]. Amplitude ratio and phase delay in the interesting frequency range are shown in **Fig. 4** for 3 different tube inner diameters. It can be seen that a reduction of the diameter generally increases the phase delay, whereas for the amplitude ratio in the interesting frequency range an optimum diameter (1.5 mm in this case) exists.

The effects of viscosity cannot be reproduced by the panel method, which was used for the above investigtions. For that reason the flowfield around the final probe geometry has been calculated by the Reynolds-averaged Navier-Stokes (RANS) code FLOWer, descibed in [5]. A 2-block structured grid with roughly 300000 cells and a Baldwin-Lomax turbulence model with Degani-Schiff extension have been used for the calculations shown. The Mach number has been increased from 0.18 to 0.3 in order to reduce calculation time. Moving the point of laminar to turbulent transition to the end of the probe head lead to a laminar separation bubble at higher angles of attack, as can be seen on the flow vector plot in **Fig. 5**. Moving the transition point forward up to the point of minimum static pressure eliminated the separation bubble in the calculations.

Flow visualization and stethoskopic test in a wind tunnel have revealed that at Reynolds numbers slightly below the typical inflight Reynolds numbers small separation bubbles could be prevented by addition of a circular turbulator. From the RANS calculations it can also be stated that the vortices originating from the sides of the probe shaft at higher angles of attack do not reach into the volume needed for the temperature measurement at the rearward part of the probe.

Pneumatic design

A 8-channel pressure transducer without pneumatic multiplexer (ZOC from Scanivalve Corp.) is used for the measurement of the pressures. This transducer can be operated in different modes (zero point measurement, operation and calibration), so that this transducer can be calibrated immediately before and after the measurements. By that, the measurement error can be kept below 0.2 %. A further error reduction is achieved by a thermal insulation and a controlled heating device, leading to a nearly constant temperature of the sensors. Because the transducer is switched between the different operation modes my means of two different control pressures (+3.5 bar and -0.5 bar), a pneumatic system as shown in **Fig. 6** is required. **Fig. 6** depicts the measuring mode; the five holes in the probe head are connected directly to the sensor inputs Px, the static pressure measured at the probe shaft is connected to the reference connection Ref.

The measurement of the static pressure is accomplished with a slow but highly accurate absolute pressure transducer (Setra Systems Inc.) together with the ZOC. Calibration pressures will be supplied in flight by two external probes. Because the calibration pressures supplied correspond to dimensionless coefficients of $Cp = +1$ and $Cp = -1$, they are perfectly adapted to the probe pressures at any airspeed. The calibration sensor will be a high accuracy pressure sensor (Digiquartz, Paroscientific Inc.).

Because the inflight calibration has to be accurate also in non-steady condition, e.g. with the aircraft climbing or descending, the dynamic response of all tubes involved has to be optimized. The dynamic response of the tubes connecting the calibration sensor with the calibration probes has be be the same as the response of tubes connecting the calibration sensor with the ZOC transducer. Because all tubes also should be as short as possible, some of the valves have to be installed in the vicinity of the noseboom's root, as can be seen in **Fig. 6**.

Data acquisition

For control of the pneumatic devices and data acquisition a VME bus system using a Power PC processor and the real time operating system OS 9000 will be used. The electric signals from the pressure transducer are AD-converted without multiplexers with a 16 bit resolution and a maximum sampling rate of 50 kHz simultanously. The synchronization with the airplane's onboard instrumentation and other installed measurement devices is done by analysis of the GPS PPS (Pulse per second) signals.

Temperature measurement

In order to fulfill the requirements with respect to accuracy and time resolution the measurement of the air temperature is done by an array of slow but highly accurate sensors in the conical part behind the probe shaft and a fast but less accurate sensor outside of the probe casing for measurement of the higher frequency components of the temperature. Within the scope of the SFB 420, the Institute for Electric Measurement Techniques (EMG) of the TU BS is responsible for the development of the temperatur sensors.

The ‚slow sensor‘ array is fed with air at low velocity by means of orifices in the probe. The exchange of air in the interior is driven by the difference in static pressure at the inlet and outlet openings. For the ‚fast‘ sensor the EMG is currently developing several alternative designs, the first prototype of the probe incorporated a sensor based on the analysis of the transit time of ultrasonic pulses in the reflection ring enclosing the probe shaft, using the temperature dependancy of ultrasound velocity.

Calibration

A first calibration of the probe has been performed in the DNW-NWB wind tunnel. The flow qualities and the accuracy of the wind tunnel‘s calibration will be checked by means of LDA measurements performed by the PTB (Physikalisch-Technische Bundesanstalt, national metrology institute). Measurements with increasing and decreasing angle of attack α, as shown in **Fig. 7**, indicate the absence of hysteresis effects. Another preliminary result of this tunnel entry is shown in **Fig. 8**. The dimesionless pressures of the five holes as well as the measured static pressure at a tunnel freestream velocity $U_\infty = 60$ m/s are shown as functions of angle of attack α and angle of sideslip β. The pressures measured at all holes show a very high linearity with respect to the flow angles. Further evaluation of the measured data is in progress and will be presented in **Fig. 8**.

For the correlation of the measured pressures with the flow vector (α, β, U_∞) different approaches are being evaluated. One of them is based on locally modeled meshes, using a program developed by the DLR Institute for Flight Mechanics, further described in [6]. Whether this or the classic polynomial approach as described in [7] will be used has still to be evaluated.

Conclusion

The probe described in this paper was developed with the aim of improving the accuracy and at the same time increasing the dynamic response with respect to common aerial probes. The accuracy is improved by careful design of the probe with special attention to the head, which leads to a high sensitivity without the risk of laminar flow separation.

Accuracy is also improved by using a system capable of inflight calibration. The temperature stabilized pressure transducers presently in use will be replaced by more accurate sensors currently under development at the EMG. Attention is also being paid to the pneumatic system and calibration techniques.

References

[1]   Fornasier, L.: *HISSS - A Higher-Order Panel Method for Subsonic and Supersonic Attached Flow About Arbitrary Configurations.* In: Panel Methods and Fluid Mechanics, Notes on Numerical Fluid Mechanics, Vol. 21, Vieweg Verlag (1987).

[2]   Bryer, D.W., Pankhurst, R.C.: *Pressure-probe methods for determining wind speed and flow direction.* National Physical Laboratory, HMSO, London (1971).

[3]   Nyland, T.W., Englund, D.R., Anderson, R.C.: *On the dynamics of short pressure probes: some design factors affecting frequency response.* NASA TN D-6151 (1971).

[4]   Bergh, H., Tijdeman, H.: *Theoretical and experimental results for the dynamic response*

of pressure measuring systems. NLR-TR F.238 (1965).

[5] Kroll, N., Radespiel, R., Rossow, C.-C.: *Accurate and Efficient Flow Solvers for 3D Applications on Structured Meshes.* VKI Lecture Series 1994-05 (1994).

[6] Giesemann, P., Thielecke, F.: *Kalibrierung einer Siebenloch-Sonde mit Lokalmodell-Netzen.* DLR-IB 111 - 98/46 (1998).

[7] Bohn, D., Simon, H.: *Mehrparametrige Approximation der Eichräume und Eichflächen von Unterschall- bzw. Überschall-5-Loch-Sonden.* Archiv Technisches Messen, Lieferung 470, pp R31 - R37, März 1975.

[8] Homeier, L.: *Windkanaluntersuchungen an einer Fünflochsonde und Vergleich verschiedener mathematischer Methoden zur Modellierung der Kalibrierdaten.* Diplomarbeit 267, Inst. für Strömungsmechanik, TU Braunschweig, (1999).

Figures

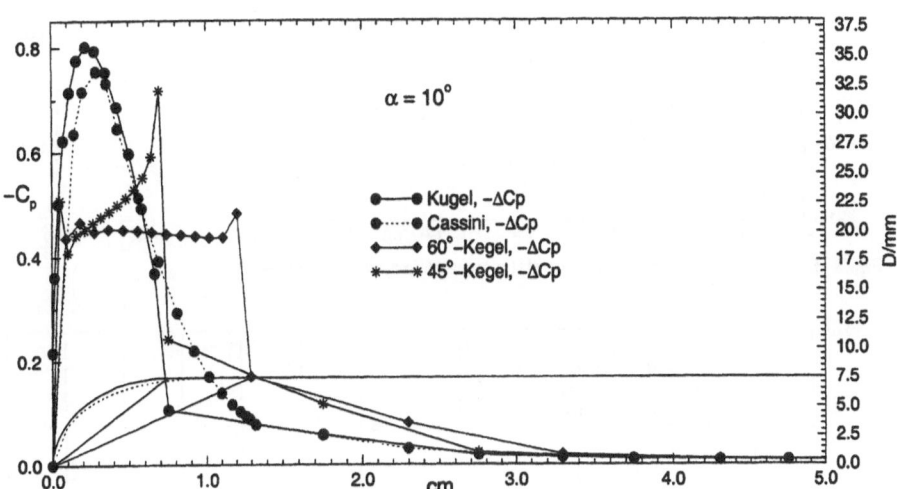

Fig. 1 Probe sensitivity as a function of head shape.

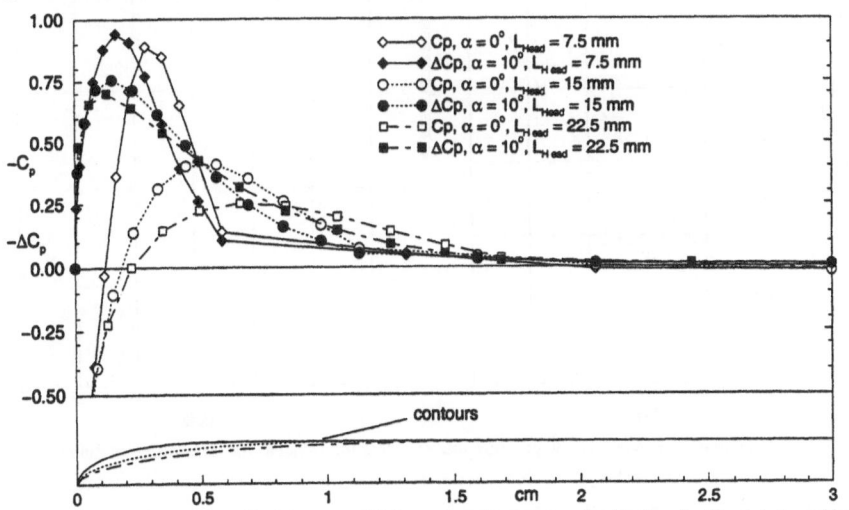

Fig. 2 Influence of probe head length on sensitivity and static pressure distribution for Cassini-shaped head.

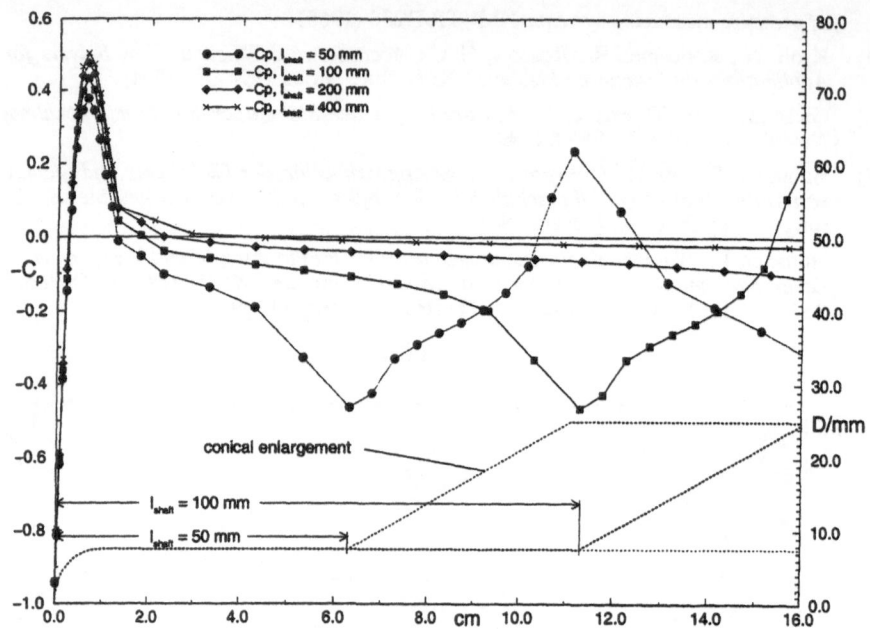

Fig. 3 Influence of probe shaft length on static pressure in symmetrical flow

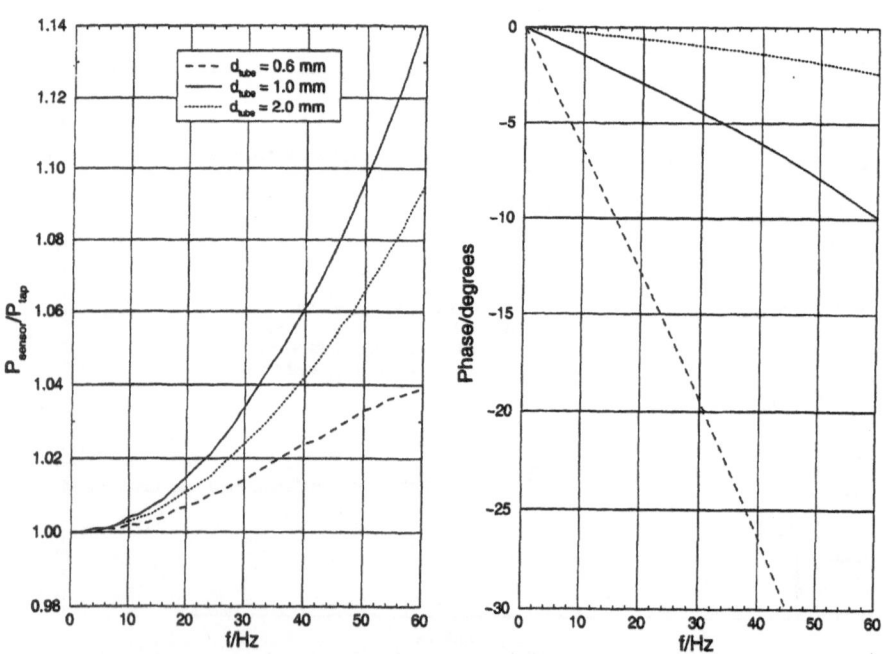

Fig. 4 Calculated dynamic response of the pressure measurement tube system for three different diameters

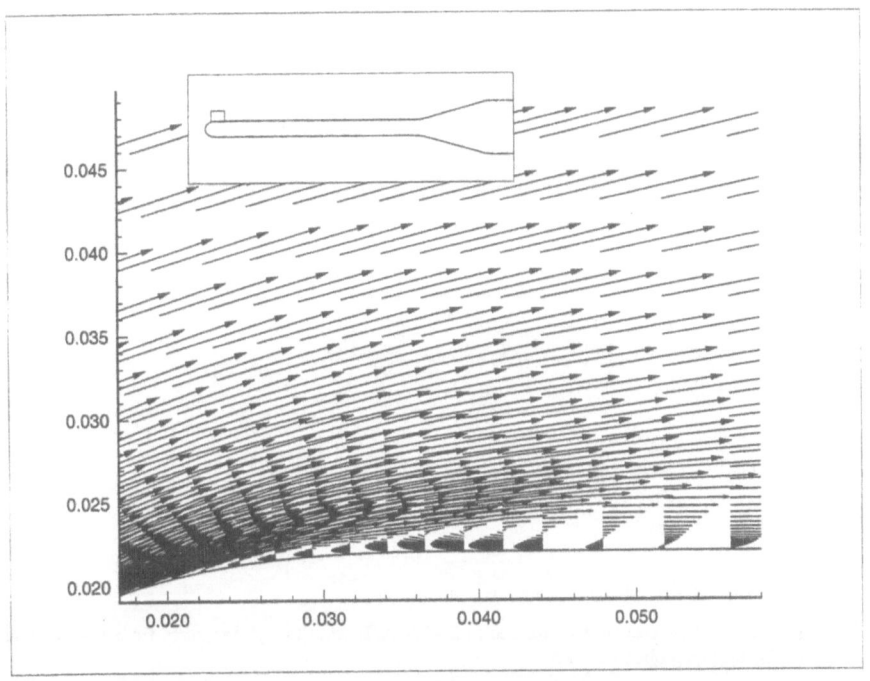

Fig. 5 Result of RANS calculation at $\alpha = 15°$ with transition prescribed at dimensionless x coordinate 0.05 showing laminar separation bubble

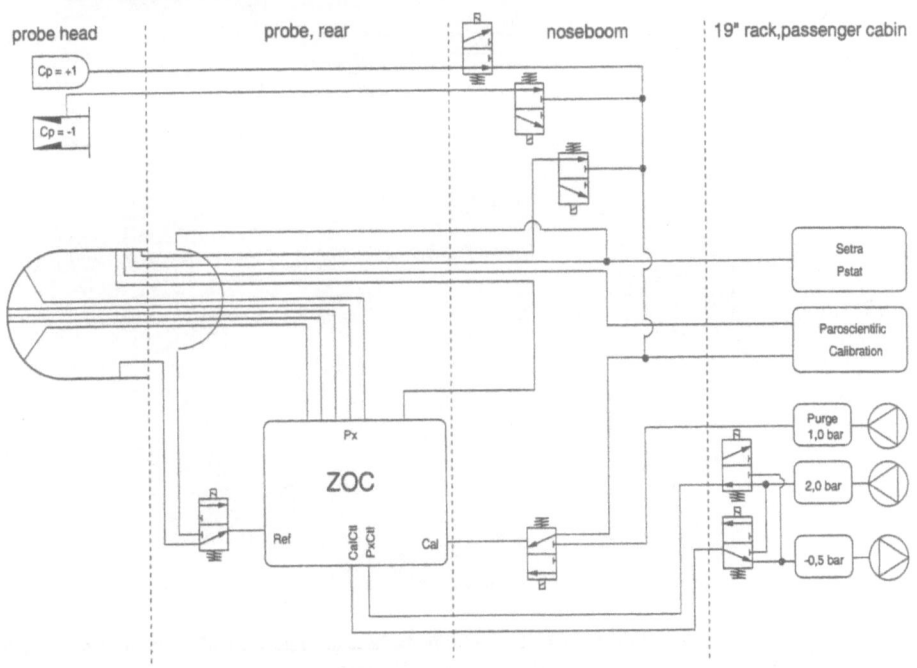

Fig. 6 Schematic lay.out of pneumatic arrangement (measuring mode shown)

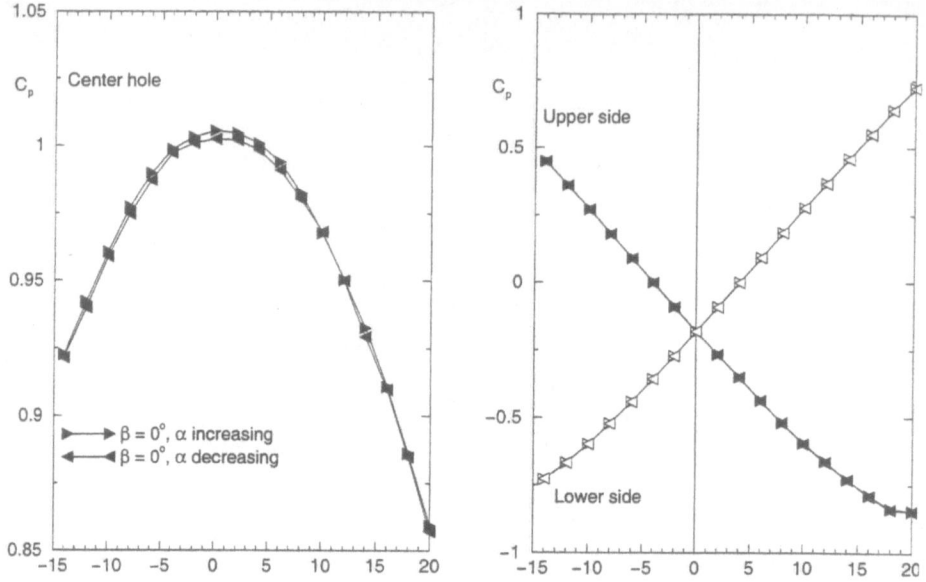

Fig. 7 Result from hysteresis test performed in DNW-NWB wind tunnel. Pressures from taps on upper side, lower side and probe tip shown.

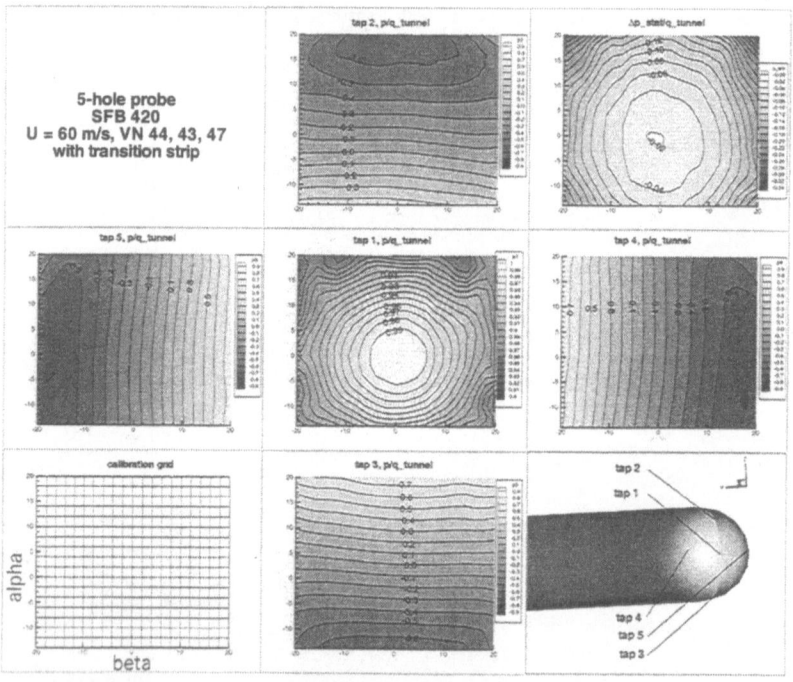

Fig. 8 Pressure coefficients of the holes located at the probe head and measured difference in static pressure as functions of angle of attack and angle of sideslip at U = 60 m/s.

Using Zonal Grids for Direct Numerical Simulation of Turbulent Boundary Layers with Pressure Gradients

Michael Manhart

Lehrstuhl für Fluidmechanik

Technische Universität, München

Boltzmannstr. 15, 85747 Garching

Germany

Summary

A method for direct numerical simulation (DNS) with zonal grids is evalutated for the case of spatially developing boundary layers. In the method presented here, the grid is refined locally in all three directions in the near wall region. It will be shown for the cases of zero and adverse pressure gradient boundary layers that the use of local grid refinement leads to considerable savings of computational resources without loss of accuracy compared to a full grid simulation.

Introduction

Separated turbulent flows are still a challenging task for any simulation method. Fully reliable results to date can only be achieved by Direct Numerical Simulation (DNS). In a DNS all relevant turbulent length and time scales have to be resolved. Because of limited computational power, up to now only low or moderate Reynolds numbers and simple geometries can be investigated with DNS. Although computational power has been continuously increasing during the last years, a DNS of a moderately complex flow, like a separated turbulent boundary layer, at a moderate Reynolds number requires the use of the fastest available computers and the most efficient algorithms currently available. Using a locally refined computational grid in critical regions and a coarse grid in less critical regions of the flow is a promising way to save computer resources in a DNS.

The use of locally refined (zonal) grids in LES and DNS is a relatively new approach. Kravchenko et al. [3] performed a zonal grid DNS of turbulent plane channel flow using a combined B-spline/spectral method. Sullivan et al. [9] used zonal grids to perform a LES of a planetary boundary layer. Like Kravchenko et al. [3], they used spectral interpolation at the boundary between coarse and fine grid. Spectral interpolation delivers highly accurate results but has the disadvantage of being restricted to relatively simple geometries with periodic boundaries. A first step to overcome this drawback has been done by Manhart [5] in a zonal grid DNS of turbulent plane channel flow using a finite volume code.

The purpose of the current research is to extend the zonal grid approach to the DNS of turbulent boundary layers with pressure gradients and separation. The approach has been implemented in a well-tested finite volume code for LES and DNS of turbulent flows in complex geometries (Werner and Wengle [11] and Manhart and Wengle [4]). It has already been tested for a turbulent plane channel flow, which is a simple geometry (Manhart [5]). In the present paper the application of the approach to a spatially developing flow, viz. a turbulent boundary layer with adverse pressure gradient, is documented.

Numerical Methodology

The basic scheme. Our approach is based on a finite volume formulation of the Navier-Stokes equations on a staggered Cartesian non-equidistant grid. The spatial discretization is of second order (central) for the convective and diffusive terms. For the time advancement of the momentum equation, an explicit second-order time step (leapfrog with time-lagged diffusion term) is used.

The Poisson equation for the pressure is solved by an iterative point-wise velocity-pressure iteration like that described in Hirt et al. [2]. It can be used as a single-grid iteration or as a smoother in a multigrid cycle. The maximum divergence is chosen in order to keep the maximal velocity error below $\Delta u_{max} \leq 10^{-5} U_\infty$ (according to the relation $div_{max} = \Delta u_{max} / \Delta x_{min} \cdot l / U_\infty$). Here, U_∞ and l are the characteristic velocity and length scales, respectively.

Zonal grid algorithm. The refinement for the local grids is done by dividing one coarse grid cell into 8 fine grid cells. The coarse and the fine grid are arranged in an overlapping way, so that the coarse grid is defined globally (global grid) and the fine grid is defined only locally (zonal grid). Each second cell-face of the local fine grid lies exactly on a coarse grid cell-face.

The coarse-grid and the fine-grid solutions are fully coupled. The coupling is achieved by transferring the fine-grid solution in the overlap region to the coarse grid. This so-called restriction is done at certain steps within the solution algorithm. We use averaging over four cell faces for the velocities and averaging over 8 grid cells for the pressure restriction. The solution on the coarse-grid level in the non-overlapping region serves as a boundary condition for the fine grid. For solving the Poisson equation on both levels, we use the pressure correction on the coarse grid as a new pressure estimate for the fine grid in a multigrid cycle. On the local grid, a Neumann boundary condition is used for the pressure at the fine grid/coarse grid interface. It has been found that this treatment is superior to a Dirichlet boundary condition for the fine grid pressure correction at the grid interface (Manhart [5]). For a more detailed description of the algorithm see Manhart [6].

Zero pressure gradient boundary layer

Configuration and boundary conditions. In order to validate the zonal grid approach for the case of a developing turbulent boundary layer, three direct numerical simulations of a zero pressure gradient boundary layer have been performed at different grid resolutions. The geometry of the flow is sketched in Figure 1.

The streamwise, spanwise and wall-normal directions are denoted by x, y and z, respectively. The Reynolds number of the performed simulations based on the inlet momentum thickness is $Re_\theta = 670$. The dimensions of the computational box have been chosen as $L_X/\delta_0 = 40.96$, $L_Y/\delta_0 = 3.2$ and $L_Z/\delta_0 = 3.8$. We have run three different grid resolutions: starting from a coarse grid ("apg2") two refined runs have been done, one with a locally refined grid near the wall up to $z^+ = 60$ ("apg6") and one with a fully refined grid over the total domain ("apg4", see table 1). The use of the zonal grid reduced the CPU-time spent during one time step significantly from 8.0 seconds ("apg4") to 2.5 seconds ("apg6"). The coarse grid run took 1.5 seconds per time step on the machine used

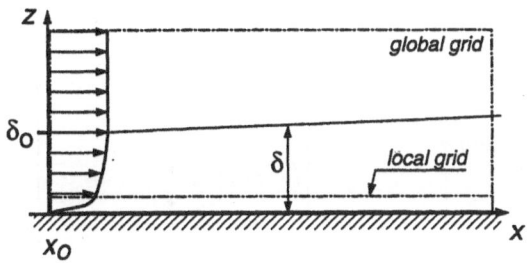

Figure 1: Geometry of the boundary layer simulations (not to scale).

Table 1: Parameters of the grids used for evaluation of the local grid approach.

Case	"apg2"	"apg6"		"apg4"
		global	local	
N_X	512	512	1024	1024
N_Y	80	80	160	160
N_Z	96	96	32	192
N_{TOT}	$3.9 \cdot 10^6$	$3.9 \cdot 10^6$	$5.2 \cdot 10^6$	$31.4 \cdot 10^6$
Δx^+	26	26	13	13
Δy^+	13	13	6.5	6.5
Δz^+_{min}	3.2	3.2	1.6	1.6

(a Fujitsu VPP700, 8 PE's).

At the inlet plane, a time dependent boundary condition is used to trigger turbulent fluctuations in the boundary layer. It is generated by taking fluctuations from a position 10 inlet boundary layer thicknesses δ_0 downstream and superposing them onto a time mean velocity profile corresponding to the Reynolds number considered. As time mean velocity profiles we have taken the profiles from the simulations of Spalart [8]. The fluctuations are exponentially damped at a wall distance higher than δ_0 in order to prevent the boundary layer from growing in time.

On the exit plane and the top surface, the velocity derivatives normal to the boundary are set to zero and for the pressure we use a Dirichlet boundary condition $p = 0$. At the lower boundary (the wall), a no-slip boundary condition is applied, and the spanwise direction is assumed to be periodic.

Results. The quality of the simulations has first been checked by an inspection of the global quantities like the skin friction coefficient c_f. The evolution of the skin friction coefficient over the momentum thickness Reynolds number is plotted in Figure 2 and compared with Coles' correlation [1]. The skin friction coefficient in the coarse grid simulation is too high, which is an indication of an insufficient grid resolution. The results of the global and the zonal fine-grid simulations approach the empirical values to within

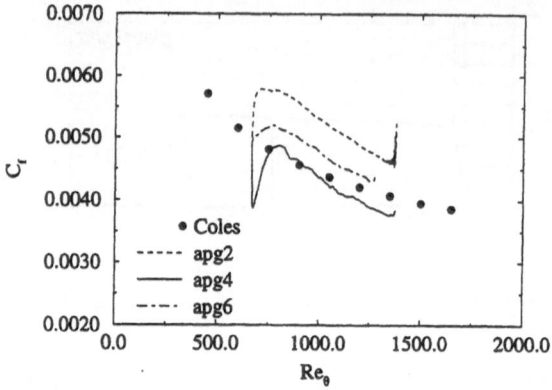

Figure 2: Evolution of skin friction coefficient with momentum thickness Reynolds number. Symbols: Coles [1] .

10% accuracy. In the global fine grid simulation, a minimum occurs immediately after the inflow. This collapse of the skin friction is due to a side effect of the parallelisation (in this run, a domain decomposition was used in spanwise direction). This "bug" has been fixed in the zonal-grid run leading to a smoother development of the skin friction coefficient.

The achieved accuracy improvement of the locally refined grid over the coarse grid run can be seen in the time averaged velocity profile shown in Figure 3. Up to a wall distance of about $z^+ = 100$ a remarkable accordance of the locally refined and the fully refined grid runs can be observed. Only with the higher near-wall resolution the "law of the wall" $U^+ = log(z^+)/0.41 + 5.0$ can be reproduced.

In Figure 4, the RMS-values of the streamwise velocity fluctuations are compared with the results of a spectral code (Spalart [8]). In the nearest wall region ($z^+ \leq 20$) the coarse grid values are too high, whereas in the logarithmic layer they are too low. The locally refined grid improves the situation in the near wall region only – the RMS-values stay too low in the logarithmic region, where the grid interface is located. An effect of that interface is hardly visible.

It can be stated that by using a locally refined grid in the near wall region a significant improvement of the solution can be achieved without the computational cost of a fully refined grid.

Adverse Pressure gradient boundary layer

For the simulation of the adverse pressure gradient boundary layer we used the grid "apg6" with a local refinement at the wall. (The adverse pressure gradient run is denoted hereafter with "apg6dp"). The only alteration compared to the zero pressure gradient boundary layer is the description of a variable pressure at the upper boundary of the computational domain. The pressure distribution has been derived from Bernoulli's equation using a u_∞ which meets the experiment of Watmuff [10]. This experiment has been selected because it is well documented and there is already a DNS available for comparison (Na and Moin

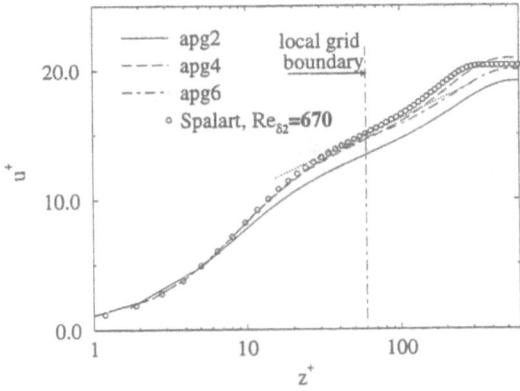

Figure 3: Time averaged velocity profile in inner coordinates. Symbols: Spalart [8]. Dotted line: log law $U^+ = log(z^+)/0.41 + 5.0$.

[7]). In the experiment the upper wall of the wind tunnel has been adjusted to accelerate the flow before the adverse pressure gradient (deceleration) has been applied. In our simulation, we let a zero pressure gradient boundary layer develop for 5 inlet boundary layer thicknesses δ_0 before we apply the favourable pressure gradient at the top surface at $x = 0.0$. This leads to a smooth onset of the pressure gradient at the wall (Figure 5). The falling pressure is connected to a rising skin friction and the rising pressure to a falling skin friction (Figure 6). In the simulation, this effect seems somehow delayed as compared to the experiment. For the falling pressure region ($x/\delta_0 \leq 15.0$) this is not surprising because in the experiment the adverse pressure gradient has been applied for a longer streamwise distance than in case of the simulation. For the adverse pressure gradient region this is not so obvious and for an explanation of the higher c_f in this region probably a longer streamwise extent would be necessary in the simulation. Nevertheless, the skin friction difference between simulation and experiment is less than 10%.

Because of the limited space, we restrict ourselves in the following to the comparison of two averaged streamwise velocity profiles. In Figure 7, a profile in the favourable pressure gradient and in Figure 8 one in the adverse pressure gradient region have been selected. At $x/\delta_0 = 7.8$, the coincidence between simulation and experiment is fully satisfying. At $x/\delta_0 = 23.4$, there are slight differences between simulation and experiment in a small section not affecting the overall agreement of the profiles.

Conclusions

We have presented a method for DNS of spatially developing flows with the use of locally refined grids. Considerable savings of computational resources can be achieved by refining the grid in a near-wall region only. The evaluation of the method in a zero pressure gradient boundary layer indicates the following observations. The time averaged velocity profile (and the global values connected with it) is improved considerably by the use of a zonal grid refinement near the wall. In the RMS-values the effect is less pronounced,

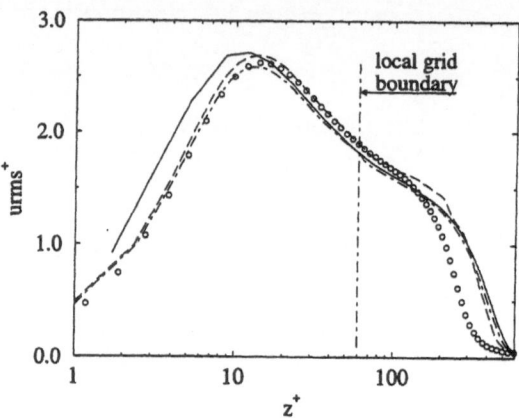

Figure 4: RMS-values of streamwise velocity fluctuations in inner coordinates. Symbols: Spalart [8].

i.e. visible in the refined region only. Here and also in the case of an adverse pressure gradient boundary layer, a negative effect of the grid interface on the statistical values can hardly be detected.

Notes and Comments. We gratefully acknowledge the support of the HLRS in Stuttgart and the LRZ in Munich. The work has been supported by the DFG (FR 478/15).

References

[1] D. Coles. The turbulent boundary layer in a compressible fluid. In *Report R-403-PR*. The Rand Corporation, Santa Monica, CA, 1962.

[2] C.W. Hirt, B.D. Nichols, and N.C. Romero. Sola – a numerical solution algorithm for transient fluid flows. In *Los Alamos Sci. Lab.*, Los Alamos, 1975.

[3] A.G. Kravchenko, P. Moin, and R. Moser. Zonal embedded grids for numerical simulations of wall-bounded turbulent flows. *J. Comp. Phys.*, 127:412–423, 1996.

[4] M. Manhart. Vortex shedding from a hemisphere in a turbulent boundary layer. *Theoretical and Computational Fluid Dynamics*, 12(1):1–28, 1998.

[5] M. Manhart. Zonal direct numerical simulation of turbulent plane channel flow. In R. Friedrich and P. Bontoux, editors, *Computation and visualization of three-dimensional vortical and turbulent flows. Proceedings of the Fifth CNRS/DFG Workshop on Numerical Flow Simulation*, volume 64 of *Notes on Numerical Fluid Mechanics*. Vieweg Verlag, 1998.

[6] M. Manhart. Direct numerical simulation of turbulent boundary layers on high performance computers. In *High performance Computing in Science and Engineering 1998*. Springer Verlag, to appear.

[7] Y. Na and P. Moin. Direct numerical simulation of turbulent boundary layers with adverse pressure gradient and separation. Report No. TF-68, Thermosciences Division, Department of mechanical engineering, Stanford University, 1996.

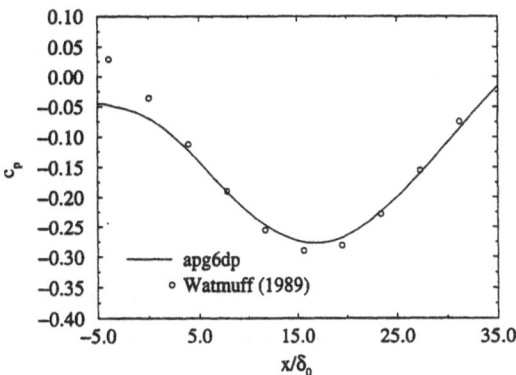

Figure 5: Streamwise development of c_p in the experiment of Watmuff [10] and in the simulation.

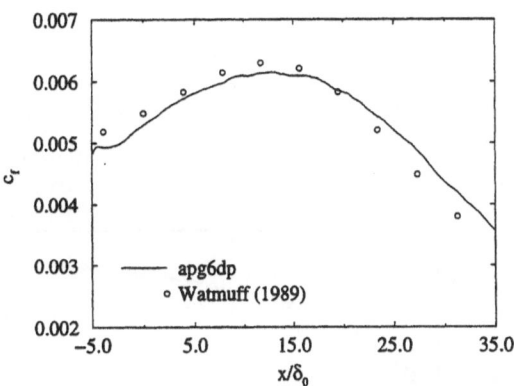

Figure 6: Streamwise development of c_f in the experiment of Watmuff [10] and in the simulation.

[8] P.R. Spalart. Direct simulation of a turbulent boundary layer up to $r_\theta = 1410$. *J. Fluid Mech.*, 187:61–98, 1988.

[9] P.P. Sullivan, J.C. McWilliams, and C.-H. Moeng. A grid nesting method for large-eddy simulation of planetary boundary-layer flows. *Boundary-Layer Meteorology*, 80:167–202, 1996.

[10] J.H. Watmuff. An experimental investigation of a low reynolds number turbulent boundary layer subject to an adverse pressure gradient. In *Ann. Res. Briefs*, pages 37–49. Center for Turbulent Research, 1989.

[11] H. Werner and H. Wengle. Large-eddy simulation of turbulent flow over and around a cube in a plate channel. In F. et al. Durst, editor, *Turbulent Shear Flows 8*, Berlin, 1993. Springer.

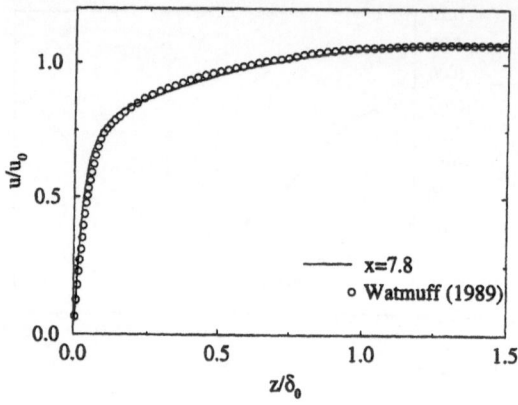

Figure 7: Averaged streamwise velocity at $x/\delta_0 = 7.8$. Symbols: Watmuff [10].

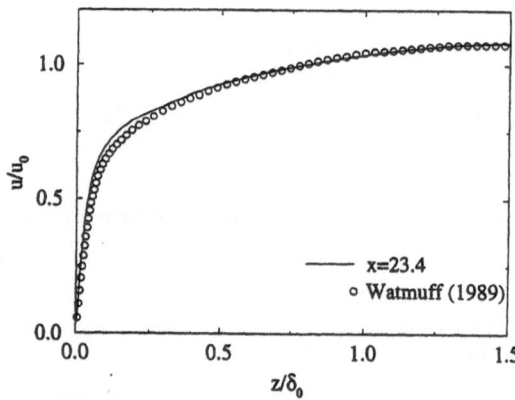

Figure 8: Streamwise development of c_f in the experiment of Watmuff [10] and in the simulation.

Transitional Structures in a Laminar Separation Bubble

U. Maucher, U. Rist, S. Wagner
Universität Stuttgart, Institut für Aerodynamik und Gasdynamik
Pfaffenwaldring 21, D-70550 Stuttgart, Germany

Summary

Transition in laminar separation bubbles (LSB) is investigated by direct numerical simulation (DNS). A 2D Tollmien-Schlichting (TS) wave is forced upstream of the LSB. Its amplitude saturates in the re-attachment region. Additional pulse-like 3D modes are excited in the LSB with very low ($\equiv$ 'linear') amplitude. They start to grow in time. Recently, it was observed that a mechanism of secondary, temporal amplification exists if the Reynolds number is sufficiently high [2]. It is shown, that a 3D instability of 2D high-shear layers close to the wall, which are present periodically in the LSB supports this mechanism. The temporal growth finally results in non-linear 3D amplitudes and the 3D modes saturate. Strong non-linear interactions take place. However, there are significant similarities compared to the state of linear 3D amplitudes. It is concluded, that the mechanism of temporal secondary amplification is likely to have major impact on the onset of transition and on the self-sustaining character of the 3D modes in the re-attachment region at sufficiently high Reynolds numbers.

Introduction

A boundary layer subject to a strong adverse pressure gradient is susceptible to separation. In the separated region, disturbance waves, so-called Tollmien-Schlichting waves, are strongly amplified and transition to turbulence takes place. The increased mixing causes momentum transfer towards the wall and finally forces the boundary layer to re-attach. Since a LSB may influence the flow around an entire airfoil and possibly has strong impact on its aerodynamic properties, a more detailed understanding of the physical processes in LSB is crucial for their prediction and control.

First three-dimensional DNS of LSB have been performed by Rist [4] and Rist et al. [5]. They investigated the interaction of different combinations of 2D waves and 3D waves at low Reynolds number ($Re_{\delta_{1,s}} = 1250$, where $\delta_{1,s}$ is the displacement thickness at laminar separation). The region upstream of the LSB and the front part of the LSB are dominated by primary convective instability of the 2D TS-wave. As the 2D amplitude becomes large, secondary instability sets in. The strong secondary amplification breaks down in the region downstream of the LSB, although the 2D amplitude continues to be very large. Transition to fine-scale turbulence takes place only if the amplitude of the 3D-waves gains large values *inside* the separated region. Otherwise the domain downstream of the LSB is characterized by large-amplitude large-scale 2D-vortices (saturated TS-waves).

If the Reynolds number is increased to approximately $Re_{\delta_{1,s}} = 2500$, a new mechanism of secondary instability becomes to be effective. In the presence of a large-amplitude 2D-wave, 3D-modes are temporally amplified. This mechanism has been firstly reported by Maucher et al. [2]. The investigations presented here focus on this mechanism and its impact on the transition in a LSB.

Numerical method

The numerical method was developed in the research-group "Transition and Turbulence" of the *Institut für Aerodynamik und Gasdynamik, IAG*. It is based on the complete, incompressible Navier-Stokes equations in vorticity-velocity formulation. A flat-plate boundary-layer is subjected to an adverse pressure gradient by prescribing the edge velocity distribution u_e at the upper boundary of a rectangular integration domain (Figure 1a). The streamwise (x) and wall-normal (y) direction are discretized by 4th-order accurate finite differences. In spanwise (z) direction a spectral ansatz is applied and the time integration is performed with a 4th-order accurate Runge-Kutta scheme. At the inflow boundary steady flow is assumed. In disturbance strips at the wall, artificial 2-D and 3-D disturbances can be excited by periodic or pulse-wise, wall-normal suction and blowing. Upstream of the outflow boundary a buffer domain is applied, which damps the unsteady component of the flow smoothly to zero. A refined, very accurate boundary-layer interaction model (Maucher et al. [3]) captures the displacement effect (index v) of the boundary layer (in particular of the LSB) and models its influence on the initially prescribed potential velocity distribution u_p yielding an instantaneous edge-velocity distribution: $u_e(x,t) = u_p(x) + u_v(x,t)$.

All variables are non-dimensionalized by a reference length $\hat{L}$, by the velocity $\hat{U}_\infty$, and the variables in wall-normal direction y and v are streched with the square-root of the Reynolds number $Re = \hat{U}_\infty \hat{L}/\hat{\nu}$, where ˆ denotes dimensional variables and $\hat{\nu}$ is the kinematic viscosity:

$$x = \hat{x}/\hat{L}; \ y = \sqrt{Re}\,\hat{y}/\hat{L}; \ z = \hat{z}/\hat{L}; \ u = \hat{u}/\hat{U}_\infty; \ v = \sqrt{Re}\,\hat{v}/\hat{U}_\infty; \ w = \hat{w}/\hat{U}_\infty.$$

This leads to the definition of the non-dimensionalized vorticity components:

$$\omega_x = \frac{1}{Re}\frac{\partial v}{\partial z} - \frac{\partial w}{\partial y}, \ \ \omega_y = -\frac{\partial u}{\partial z} + \frac{\partial w}{\partial x}, \ \ \omega_z = \frac{\partial u}{\partial y} - \frac{1}{Re}\frac{\partial v}{\partial x}. \tag{1}$$

Test case

In experiments in the Laminar Wind Tunnel of the institute, a LSB on a wing section with a chord-length of $\hat{c} = 0.615m$ was investigated (Würz et al. [6]). The chord Reynolds number was $Re_c = 1.2 * 10^6$. The free-stream velocity $\hat{U}_\infty$ is $29.3\frac{m}{s}$. In DNS the reference length $\hat{L}$ is chosen to be $6.15cm$. Hence, the Reynolds number is 120000, the chord length $c = \hat{c}/\hat{L} = 10$. The computational grid has 197 and 2754 points in wall-normal and streamwise direction, respectively. A TS wave length is discretized with approximately 160 grid points.

In the experiments, two boundary-layer edge-velocity distributions $u_p(x)$ have been measured (Figure 1b). The first one (squares) refers to a flow with separation bubble. For the second, turbulent one (triangles), separation has been suppressed by fixing a turbulator upstream of the separation bubble. For the computations, the turbulent distribution has been prescribed as initial condition u_p at the free-stream boundary. Maucher et al. [3] show that the displacement effects are captured by the interaction model and finally the edge velocity distribution u_e almost resembles the experimental measurements with separation bubble.

The shape parameter H_{12} characterizes the stability properties of a boundary-layer profile in a coarse manner. Thus, by means of the shape parameter general properties of the DNS can be compared with the experiment. Figure 2a shows the shape parameter for two computations. In a 2D computation (dashed line) a TS-wave was forced by periodic suction and blowing. A second, transitional computation has a spanwise resolution of 44 spectral modes and a rapid breakdown of the 2D-wave into small 3D structures begins in the re-attachment region characterized by the drop of the shape parameter. Generally, the agreement with the experiment (squares) is good in both cases. However, downstream of the LSB the shape parameter in the 2D case remains well above the value typical for a turbulent boundary layer, since transition to turbulence is suppressed due to the lack of three-dimensionality.

Transient development: secondary temporal amplification

At sufficiently high Reynolds numbers 3D-disturbances are temporally amplified in the re-attachment region of the LSB once they are present. Similar to the mechanisms of secondary instability theory according to Herbert [1], Maucher et al. [2] found 3D amplification with subharmonic and fundamental frequency with respect to the forced 2D-wave for different spanwise wave numbers. In recent investigations the secondary temporal growth of 3D-modes in a large range of spanwise wave numbers was investigated. A 2D DNS with a periodically forced TS-wave was superposed with 3D-waves with fixed spanwise wavenumber γ by short pulse-like 3D excitation with very low amplitude ($\gamma = 2\pi/\lambda_z$, where λ_z is the spanwise wave length). They vanish for low γ. For higher γ up to very high values they grow exponentially with the temporal growth rate $\beta_i = \frac{\partial}{\partial t}[\ln A(t_0 + t) - \ln A(t_0)]$. Figure 2b shows the temporal amplification rates β_i obtained from DNS. The open symbols denote subharmonic amplification, the filled symbols show fundamental amplification.

Temporal secondary disturbance-amplification was found for $\gamma \gtrsim 10$, respective $\gamma/\alpha_{TS} \gtrsim 0.5$, where α_{TS} is the streamwise TS wave-number $\alpha_{TS} = 2\pi/\lambda_{TS}$, $\lambda_{TS} \approx 0.3$. For low spanwise wave-number γ the subharmonic mechanism is observed. If γ is increased above $\gamma \approx 42$ the fundamental mechanism dominates. At $\gamma \approx 160$ a strong increase of the amplification rate sets on, while the dominance of either the fundamental or the subharmonic mechanism alternates. For validation additional computations with a refined grid were performed (240 grid points per TS wave length, circles). Only for large $\gamma > 300$ differences appear.

In Figure 3a the streamwise mean-velocity component $\bar{u}$ and streamlines $\Psi = const$ in the vicinity of the LSB are shown. The reverse flow region is marked by a strong change from dark to light grey as the velocity becomes negative. Since the y-axis is scaled by a factor of 10 the actual wall normal extend of the separation bubble from $x \approx 7.2$ to $x \approx 8.0$ is very small. The mean-flow profiles in the separation bubble (sub-figure b, $x = 7.35...8.01$, $\Delta x = 0.06$) show characteristics typically observed in transitional separation bubbles. The profiles are shifted by 0.2 from one x-station to the next for better clearness. In the front part of the bubble ($x < 7.72$) the reverse flow velocity is small. At $x \approx 7.77$ a counter-rotating vortex is present in the mean flow underneath the mean recirculation region (solid streamline $\Psi = 0$). Downstream of this vortex the height of the bubble decreases while the reverse-flow velocity in the bubble becomes large.

To investigate the secondary amplification mechanism, the 3D flow field for $\gamma = 160$ is compared with the 2D flow. Upstream of $x = 7.84$, the 3D rms-amplitude exhibits a complex pattern (Figure 4a), which cannot be explained by secondary (convective) amplification due to the 2D TS-wave. Instead, the TS-amplitude (Figure 4b) grows continuously without any maxima upstream of $x = 7.84$. In contrast, the *instantaneous*

3D flow-field in Figure 4c shows local strong amplitude maxima in this region. They remain almost at a constant place during approximately one half of each TS-period T_{TS}, i.e. as long as reverse 2D flow is present in the re-attachment region, while their amplitude is temporally growing. Obviously, local maxima of the rms-pattern (marked by boxes) are due to the temporal growth in this phase. Two of the instantaneous 3D maxima are directly related to the formation of 2D high-shear layers (Figure 4d, boxes). The roll-up of the free shear-layer is highlighted by contours of large vorticity (d). In the second half of the TS-period the 2D velocity becomes positive and the 3D perturbations are convected downstream. Smearing of the rms-pattern and the increased 3D rms-amplitude downstream of $x = 7.84$ is caused by convection.

It is remarkable that the 3D rms-amplitude as well as the instantaneous 3D values are low in the free shear-layer above the recirculation area. The onset of three-dimensionality happens inside the separation bubble, where 3D perturbed fluid is present from the previous TS-period and where instantaneous high-shear layers are strongly unstable with respect to temporally growing 3D modes.

Final state, transition in LSB

At the end of the transient phase, the temporal 3D growth finally leads to large 3D amplitudes. Complex interactions between the 2D wave and the secondary 3D modes take place. The re-attachment region develops towards turbulence and the 2D wave deviates from the transient case. Finally, the amplitude of the secondary 3D modes saturates and an equilibrium state between 2D and 3D modes ends the transient phase. The 2D wave attains its amplitude maximum in the re-attachment zone at $x = 8.0$ (Figure 5a - solid line). The flow field is dominated even far downstream of the separation bubble by the TS frequency, which exceeds the 2D higher harmonics (dash-dotted) and the 3D modes (dashed and symbols), by far. A variety of 3D modes with different spanwise wave number but the same frequency $\beta = \beta_{TS}$ is plotted. The modes with lower γ are denoted by dashed lines, modes with large γ are marked with symbols. The secondary amplification rates are maximum for $\gamma \approx 500$ (Figure 2b). In contrast, in the final state modes with wave numbers lower than $\gamma = 160$ (dashed lines) have a larger amplitude downstream of the LSB than the modes with larger γ (symbols). Obviously, the amplification characteristics are modified in the state with saturated 3D amplitude in favour of lower spanwise wave numbers. To get a more detailed insight, amplitude spectra for several streamwise positions in the separated region are plotted versus the spanwise wave number γ in Figure 5b. For increasing x, the single curves are shifted upwards by $\Delta u = 0.003$, each. Generally, the amplitude is decreasing for $\gamma > 300$. Upstream of $x \approx 7.88$ the maximal amplitudes are at $\gamma \approx 150$–300. Further downstream lower spanwise wave numbers dominate.

In comparison with the transient case, the reverse mean flow velocity is weaker in the final state and the counter-rotating vortex at $x \approx 7.8$ near the wall is smaller due to non-linearity (Figure 6). As in the transient case the instantaneous flow in the re-attachment zone shows the same change between a phase with strong reverse flow and the formation of a counter-rotating vortex related to the roll-up of the free shear-layer and a subsequent phase with large streamwise velocity when a 2D vortex is shedding (convection phase). Similar to the transient case temporal growth of 3D disturbances is observed in the phase with instantaneous strong reverse flow near localized high-shear layers. At the beginning of the phase with strong reverse flow a elongated shear-layer exists at $0.015 < y < 0.02$ (Figure 7b). As its amplitude strengthens also 3D disturbances are growing and the instantaneous streamwise velocity-component of the 3D mode with the spanwise wave number $\gamma = 160$, $u_{\gamma=160}$ is concentrated directly underneath the high shear-layer (Figure 7a). For $7.8 < x < 7.88$ this mode has the largest amplitude of all 3D modes.

In contrast to the transient case, however, the high shear-layer and the related growing 3D modes travel upstream during the progressive roll-up of the free shear-layer simultaneously lifting from the wall. At the end of the temporal growth phase dynamic interactions between the 3D perturbations concentrated at $x \approx 7.86$ with the free shear-layer take place (7c and the respective 2D vorticity 7d).

The comparison of contour surfaces of the spanwise vorticity in the transient case (Figure 8a and b, $\gamma = 160$) with the final state (Figure 8c – f) indicates a quite similar behaviour in both cases. The left column shows the flow field at the end of the temporal 3D growth phase ($t = t_0$). In the transient case (Figure 8a) as well as in the 3D part of the final state (Figure 8c, $\omega_{z,3D} = \pm 0.05$) three-dimensional perturbations left over from the previous TS period are present near the wall ($x > 7.84$ in the transient case respective $x > 7.92$ in the final state). These perturbations have been amplified in the temporal growth phase and are lifted away from the wall upstream of the respective x-stations just mentioned. The total spanwise vorticity (2D plus 3D, $\omega_{z,total} = \pm 0.1$) in the final state (Figure 8e) proves, that the free shear layer is already strongly affected in the region where the 3D modes are raised ($x \approx 7.88$), whereas, further downstream the free shear-layer is still almost two-dimensional. When the streamwise convection sets on (right column, 1/4 TS-period later), the 3D structures are stretched in streamwise direction during their streamwise convection. Although the spanwise distribution in the final state is irregular due to the self-excited character of the 3D modes (d), the general structure is quite similar to the transient case (b). The total vorticity in the final state (e) highlights the rapid breakdown into fine three-dimensionality. Just one fourth TS-period after the state in Figure 8e the free shear-layer in the whole domain downstream of $x = 7.88$ broke down into 3D structures and is propagating downstream.

Although the process of 2D vortex shedding is time periodic (in a coarse manner), the small 2D and 3D structures change significantly from one TS-period to the next. Thus, it is impossible to investigate 3D structures during more than one period to determine whether a 3D structure with a certain spanwise wave-number has either fundamental or subharmonic frequency. A quantitative comparison of amplification mechanisms with the transient case therefore fails.

The rapid breakdown of the 2D vortex seems to be in contradiction to the observation in Figure 5a, that a 2D mode with the TS-frequency dominates the boundary layer even considerably downstream of the LSB. However, this is to be interpreted as a periodic change of 3D perturbed shedding vortices and intermediate calm phases in the wake of the LSB. It takes a remarkably long way downstream, until fully developed turbulence occurs and the 2D TS-amplitude looses its dominance.

Conclusions

A periodically occuring self-sustaining mechanism is observed in the re-attachment zone of the LSB. Instantaneously occurring small-scale high-shear layers related to strong reverse flow are locally amplified and get convected downstream together with the large-scale 2D motion of the oscillating free shear-layer, thereby destroying it. The temporal amplification of 3D modes, respectively the presence of self-sustaining non-linear saturated 3D modes, might be the most important mechanism forcing the onset of turbulence in the re-attachment region of a LSB at large Reynolds number.

Acknowledgments: The financial support of this research by the Deutsche Forschungsgemeinschaft DFG under grant Ri 680/1-1,2,3 is greatfully acknowledged.

References

[1] T. Herbert. Secondary instability of boundary layers. In *Ann. Rev. of Fluid Mech.*, volume 20, pages 487–526, 1988.

[2] U. Maucher, U. Rist, and S. Wagner. Secondary instabilities in a laminar separation bubble. In H. Körner and R. Hilbig, editors, *New Results in Numerical and Experimental Fluid Mechanics*, volume 60 of *NNFM*, pages 229–236. Vieweg, 1997.

[3] U. Maucher, U. Rist, and S. Wagner. A refined method for DNS of transition in interacting boundary layers. AIAA 98-2435, 1998.

[4] U. Rist. Nonlinear effects of 2D and 3D disturbances on laminar separation bubbles. In S.P. Lin, editor, *Proc. IUTAM-Symposium on Nonlinear Instability of Nonparallel flows*, pages 324–333, New York, 1994. Springer.

[5] U. Rist, U. Maucher, and S. Wagner. Direct numerical simulation of some fundamental problems related to transition in laminar separation bubbles. In Dèsidèri *et. al.*, editors, *Computational Methods in Applied Sciences '96*, pages 319–325. John Wiley & Sons Ltd, 1996.

[6] W. Würz and S. Wagner. Experimental investigations of transition development in attached boundary layers and laminar separation bubbles. In H. Körner and R. Hilbig, editors, *New Results in Numerical and Experimental Fluid Mechanics*, volume 60 of *NNFM*, pages 413–420. Vieweg, 1997.

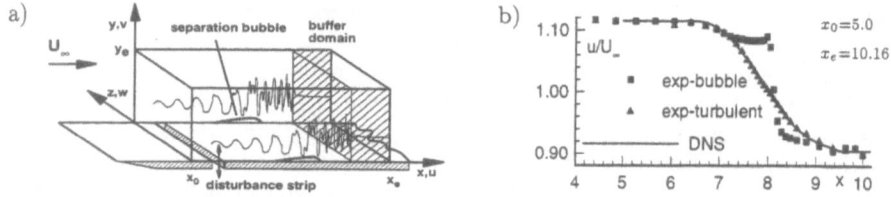

Fig. 1: a) Integration Domain and b) the edge-velocity distribution in the experiment (symbols) and in DNS.

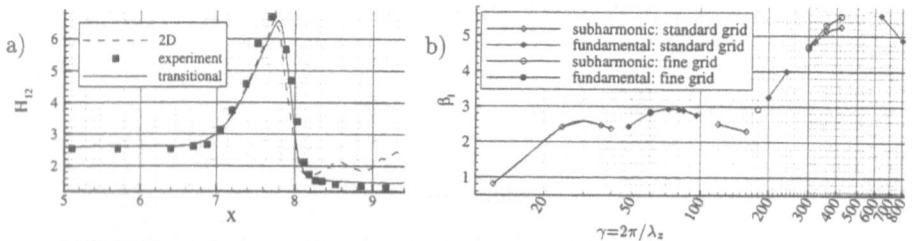

Fig. 2: a) Comparison of the shape parameter in the experiment (symbols) and in DNS.
b) Dependency of the secondary temporal amplification rate β_i from the spanwise wave number γ with the standard grid (rhombi) and a refined grid (circles).

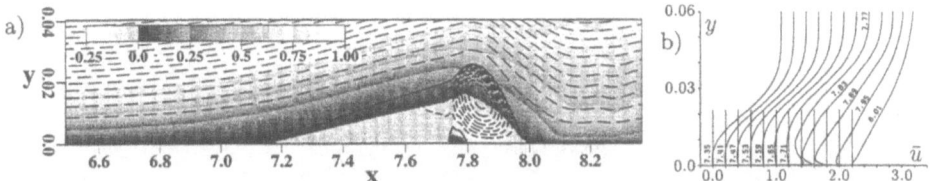

Fig. 3: Small 3D amplitude: a) streamwise mean-velocity component $\bar{u}$ and stream-lines $\Psi = const$: negative Ψ, short dashes, $\Delta\Psi = 0.0002$; positive Ψ, long dashes, $\Delta\Psi = 0.002$. b) respective mean-flow profiles at $x = 7.35, 7.41 \ldots 8.01$ (shifted by 0.2 from one x-station to the next).

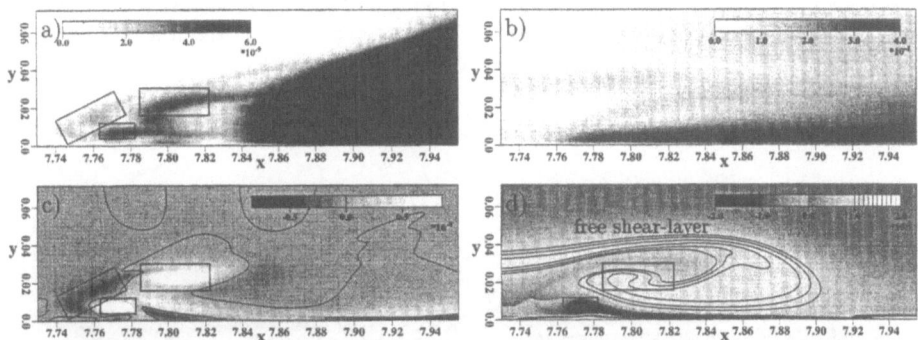

Fig. 4: Secondary temporal disturbance amplification: a) $u_{\gamma=160}$-rms amplitude, b) u-amplitude of TS-wave, c) instantaneous $u_{\gamma=160}$-distribution, d) respective instantaneous 2D vorticity.

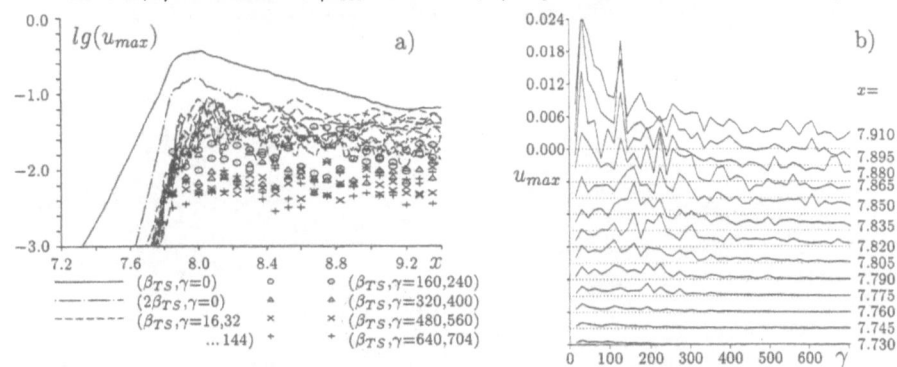

Fig. 5: Final state: a) amplification curve and b) 3D amplitude versus spanwise wave number γ: amplitude scale for uppermost curve ($x = 7.91$), other curves shifted by $\Delta u_{max} = -0.003$, respectively.

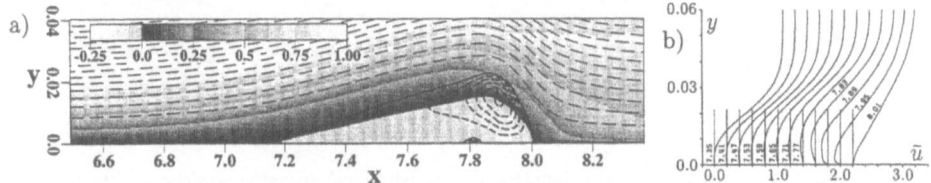

Fig. 6: Final state: a) streamwise mean-velocity component $\bar{u}$ and stream-lines $\Psi = const$: negative Ψ, short dashes, $\Delta\Psi = 0.0002$; positive Ψ, long dashes, $\Delta\Psi = 0.002$. b) respective mean-flow profiles at $x = 7.35, 7.41 \ldots 8.01$ (shifted by 0.2 from one x-station to the next).

313

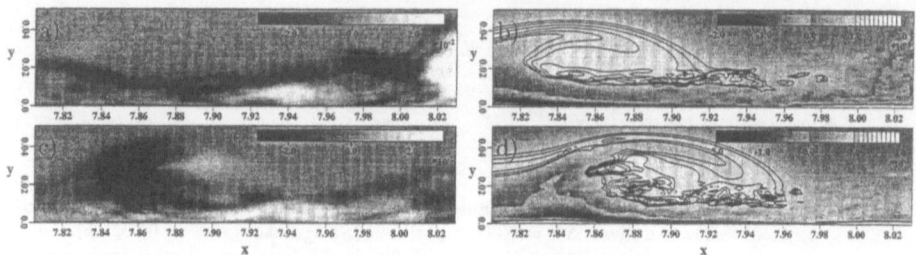

Fig. 7: Final state: instantaneous distribution of a) and c) streamwise velocity $u_{\gamma=160}$, b) and d) respective 2D vorticity ω_z at two different time steps during (top) and at the end of temporal growth (bottom).

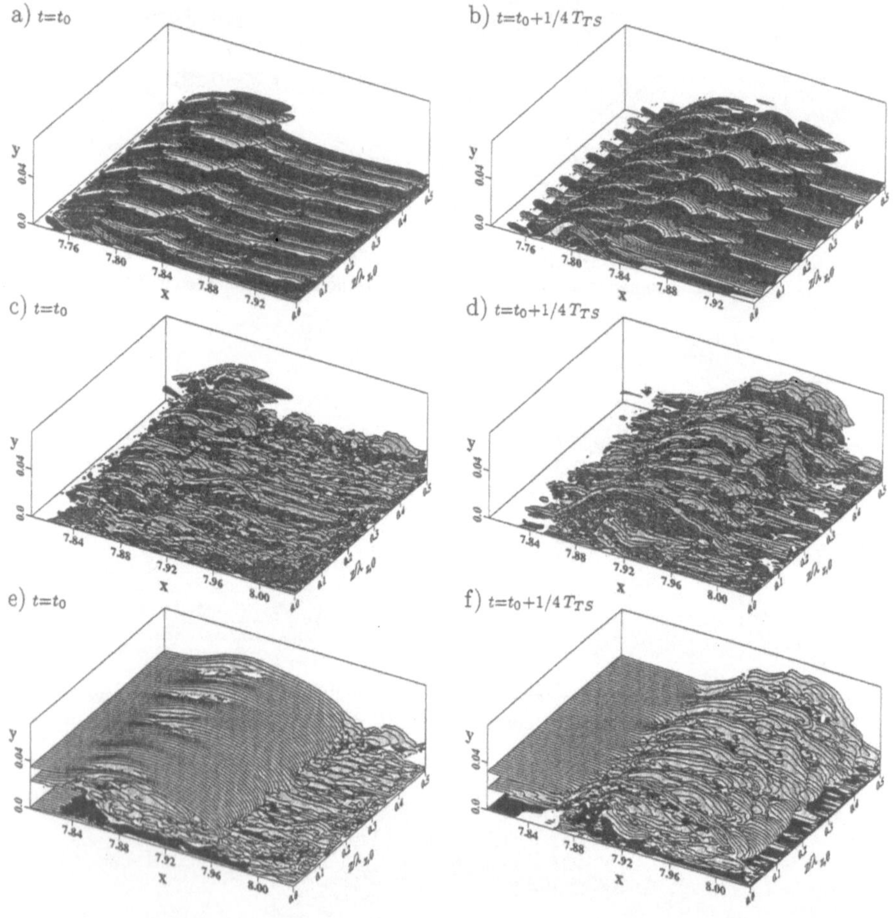

a) $t=t_0$

b) $t=t_0+1/4\,T_{TS}$

c) $t=t_0$

d) $t=t_0+1/4\,T_{TS}$

e) $t=t_0$

f) $t=t_0+1/4\,T_{TS}$

Fig. 8: Surfaces of constant positive (light) and negative (dark) spanwise vorticity ω_z: (a) and (b) transient case ($\gamma = 160$), (c) – (f) final state, (c) and (d) 3D part only, (e) and (f) total spanwise vorticity (2D plus 3D). Left figures: at the end of temporal 3D growth ($t = t_0$). Right figures: $1/4\,T_{TS}$ later. $\lambda_{z,0}$: spanwise extend of the integration domain.

Some Important Results of the Technology Programme RaWid

Josef Mertens
DaimlerChrysler Aerospace Airbus GmbH
Technology Programmes, Flight Physics
28183 Bremen, Germany

Summary

The programme RaWid aims at technologies for improvement of aerodynamic performance at transonic cruise speed. The first chapter covers theoretical design tools. The next deals with application of the tools in aerodynamic designs oriented towards a MEGALINER. A third chapter covers interdisciplinary investigations for special aerodynamic technology realisations like Variable Camber (VC), Hybrid Laminar Flow (HLF), Shock-Boundary Layer interference Control (SBLC) and separation control. Experimental technologies and the verification experiments are included in the experimental chapter. RaWid is interconnected to other technology and research programmes dealing with aerodynamic design and further steps of structure and system design for more efficient aircrafts. In RaWid, close cooperation with DLR and universities is realized.

Introduction

The German national technology programme RaWid (cruise drag reduction) startet in 1994, at first funded by company resources only; since mid 1995 it is also sponsored by the German ministry for research and technology (BMBF). As an industrial programme RaWid is oriented towards two objectives:
- technologies to improve the product and
- technologies to improve production.

Aerodynamics is design technology to generate the geometry; it has ist main influence in preliminary design. Here the objectives translate to:
- Improve the product mainly by reduced fuel consumption due to lower drag and structure weight; this will result in further lowered gross weight. Eventually, it enables longer range and reduces emissions. But besides mission flexibility by improved payload/range performance, increased speed flexibility is another item to be considered.
- Improve geometry development by the generation of faster and more reliable tools for the aerodynamic design. This enables higher quality designs, shortening of the design cycles, better interdisciplinary balance and earlier provision of reliable aerodynamic data to generate the performances with reduced margins to be offered to the customers.

Main technology areas to achieve those goals are
- optimisation of the whole aircraft configuration,
- variable camber (VC),
- reduction of friction drag, firstly by laminarisation.

Theoretical Aerodynamic Design Tools

Several presentations on this symposium deal with theoretical investigations of the RaWid framework, e.g. concerning transition prediction or the MEGAFLOW development.

Here I select the new design tool generated for aerodynamic wing optimisation, fig. 1. In RaWid a method was developed to optimize transonic airfoils using the established DA-analysis tools [1]. It was the first multipoint airfoil design tool covering several transonic and subsonic design and constraint points. Then, it was included in an engineering environment for rapid design work [2]. To reduce design cycle time, the MEPO parallel computing environment was developed [3] and connected to the design method. Next step was to generate a complete wing surface out of several generating airfoils; in the fully automatic procedure the constraints of wing lofting had to be respected [4]. Aerodynamically, the individual wing sections were coupled by a rather simple lift distribution procedure which respects for the induced effects [2]. Last step was the combination of those tools with an automatic mesh generation for a FLOWer analysis to validate the previous design. The design iteration cycle was closed using the FLOWer analysis to update 3D-corrections -so called analog modifications- in the 2D design steps [5].

Aerodynamic Design Work

At DA, a design methodology was developed for aircraft design [6], fig. 2. To use this methodology for transonic aircraft -previously it was only used for supersonic transports-, a new aerodynamic module was generated to estimate performance in the nonlinear transonic flow regime; elements of this module are presented in [7]. Additionally, the accuracy of the estimations was improved, to respect for the mature starting point in transonic aircraft design. To cover the most important interdisciplinary repercussions on aerodynamic design, a new structure module was introduced including besides the load, stress and weight estimation also the static aeroelastic deformation [8]. Now, the main parameters for a MEGALINER wing specification were optimized (planform, load and thickness distribution). It resulted in a reduction of wing structure plus fuel weight of between 2% and 4% of MTOW, depending on the accuracy assumptions for the aerodynamic and structure predictions. Next step under work is the aerodynamic verification of this rough estimation: the complete 3D-wing geometry is designed applying the above mentioned design iteration process [5].

The last RaWid aerodynamic wing design with experimental verification was the TVC2-wing, a MEGALINER wing using VC-technology. Fig. 3 shows performance achieved at typical conditions:
In the left diagram performance M*L/D over CL is given at constant Mach. At lower cruise CL an L/D-improvement of about 3.5% is achieved with respect to the fixed camber polar (3°), at high cruise CL about 4%. Both values are achieved with the flap settings indicated allowing still some minor improvements for higher or lower flap settings.
The right diagram marks the buffet onset limits over Mach. Using only the simple collective camber variation, 5° camber increase buffet onset at Mach 0.85 about 20% with respect to the uncambered wing or about 9% with respect to the optimum L/D camber. This improvement increases with Mach number. Using differential camber, buffet limits can be increased further; additional 10% should be possible when using limit flap deflections at the huge inner wing.

One achievement of RaWid was the design and wind tunnel verification of a laminar glove for an A340 HLF-flight test [9]. Fig. 4 shows the glove planform and the modified airfoil. The pressure distribution design was tested in the ARA transonic wind tunnel. Additionally, a Krüger flap design was tested in the DNW wind tunnel which proved the low speed performance and handling qualities with a wing glove only on the right hand side[10].

System Engineering

The flap system design for VC-wings was investigated in cooperation with the departments for structure, loads, systems, weight, cost estimation and the project office. This is required to evaluate VC benefits, because some improvements produce indirect effects, e.g. via load reduction for the wing and other aircraft parts like the tail. Especially for load reduction on large aircraft, VC enables load redistribution via additional loading of the inner wing by flap deflection at the inner wing; in contrast, the „classical" load redistribution via deloading of the outer wing (and additional aircraft pitch) becomes ineffective, because the flexibility of large wings strongly reduces efficiency of the outer flaps or ailerons. To overcome this problem, an all speed flap concept was developed by the two technology programmes RaWid and HAK (High Lift Configurations), fig. 5. At the downstream end of the (Fowler) flap an actuated tab is applied. It allows for
- deflections to improve performance at cruise or take-off,
- reduction of loads, mainly via inboard deflections,
- extension of buffet onset via local load limitation when buffet starts,
- increased drag for steep descents,
- adoption to reduced wake vortex generation and
- improvement of roll control at all speeds via combined usage of tabs and spoilers.

Hybrid laminar flow (HLF) system design still requires many new solutions to be developed. Some were investigated in RaWid. Several tests were installed to investigate in service deterioration of the perforated plates. One prominent example is shown in fig. 6: The Airbus Super Transporter „Beluga" carried for over a year a perforated plate instead of a similar shaped simple metal sheet. It was placed in the stagnation region. Each month the plate was exchanged against an identical second one and investigated in the lab. On plate was cleaned after each time in the lab, one accumulated the contaminations. But both were treated by the usual „Beluga" procedures, including aircraft cleaning. Results of all investigations were, that no significant long term deterioration was found.
Fig. 7 shows an installation example for a laminar flow system with a Krüger flap in a wing leading edge. For all wing sections, solutions were found.
In fig. 8, a 1,5m long demonstrator is shown for a new structure concept for a HLF suction nose. This concept has so far proven superior to all other concepts investigated by DA. It can be adapted to wing and stabilizer requirements.

Experiments

Several experimental techniques were developed in RaWid, so e.g. a contribution to the PSP development (Pressure Sensitive Paint) in cooperation with DLR, see e.g. the presentation by R.H. Engler, Chr. Klein, P. May. Here I mention a system to improve quality and decrease time for wind tunnel model manufacture, called WinGS (Wind tunnel model Geometry control System). Usually wind tunnel models are macined by numerically controlled milling and finally finished by hand. Control of this hand finishing is on measuring machines along generating lines. To do this, the model has to be removed from the working table to the measuring machine. So twice accuracy is lost: measurement only at lines and each time two recalibration steps. Fig. 9 shows an example of a fin model. Now, the whole surface control is plotted by the new system WinGS. Along the control lines, accuracy is ok. But in between, tolerances are violated. After finish work using the new system WinGS, no model removal was required and surface was within the tolerances everywhere.

References

[1] A. Van der Velden, D. Forbrich: *Use of Aerodynamic Shape Optimization for a Large Transonic Aircraft*, AIAA-Paper 97-2274, 15th Applied Aerodynamics Conference, June 23-25, 1997, Atlanta, GA, USA.

[2] A. Van der Velden: *AEROPT 4.3 & Pointer Pro 1.3, User Manual, Developer Manual*, User's Guide, Synaps, Inc., Atlanta, GA, USA, May 1998.

[3] J.K. Axmann, M. Hadenfeld, O. Frommann: *Parallel Numerical Airplane Wing Design*, New Results in Numerical and Experimental Fluid Mechanics, Contributions to the 10th AG STAB/DGLR Symposium, Braunschweig, Germany 1966, Notes on Numerical Fluid Mechanics, Vol. 60, Editors H. Körner & R. Hilbig, Vieweg, 1997, pp. 48-55.

[4] O. Frommann: *Tragflügelstrak und -parametrisierung für die aerodynamische Optimierung*, MEGAFLOW-Ber. DA-EFV-008/97, 15.01.1997, 4 S., 1 Tab., 2 Abb..

[5] O. Frommann, D. Forbrich: *Aerodynamic Optimization of Airplane Wings using Analogy Methods*, Presentation 11th DGLR-Fach-Symposium AG STAB, 10.-12. Nov. 1998, TU Berlin.

[6] A. Van der Velden, D. von Reith: *Multi-disciplinary SCT design at DASA*, Proceedings of the 7th European Aerospace Conference EAC '94 „The Supersonic Transport of Second Generation", 25-27 October 1994, p. 449-464.

[7] T. Krißler: *A project method to predict the local shock sweep angle on a transonic wing*, Presentation 11th DGLR-Fach-Symposium AG STAB, 10.-12. Nov. 1998, TU Berlin.

[8] R. Kelm, M. Grabietz: *Multidisciplinary Aspects of Aeroelasticity in the Pre-design Phase for a new Aircraft*, CEAS International Forum on Aeroelasticity and Structural Dynamics, 17-20 June, 1997, Rome, Italy, 16 p., 30 fig..

[9] S. Schmid-Göller, H. Hansen: *Design of a Laminar Glove for the A340 and First Results of a High Speed Wind Tunnel Test*, New Results in Numerical and Experimental Fluid Mechanics, Contributions to the 10th AG STAB/DGLR Symposium, Braunschweig, Germany 1966, Notes on Numerical Fluid Mechanics, Vol. 60, Editors H. Körner & R. Hilbig, Vieweg, 1997, pp. 283-287.

[10] R. Gramlow: *Design of a Krüger for the A340 Laminar Glove and First Low Speed Wind Tunnel Results*, Presentation on the 10. DGLR-Fach-Symposium der AG STAB, 11.-13. Nov. 1996, DLR Braunschweig, DA-Report EFP-023/97, 17.01.97, 2 pages, 3 fig..

Figures

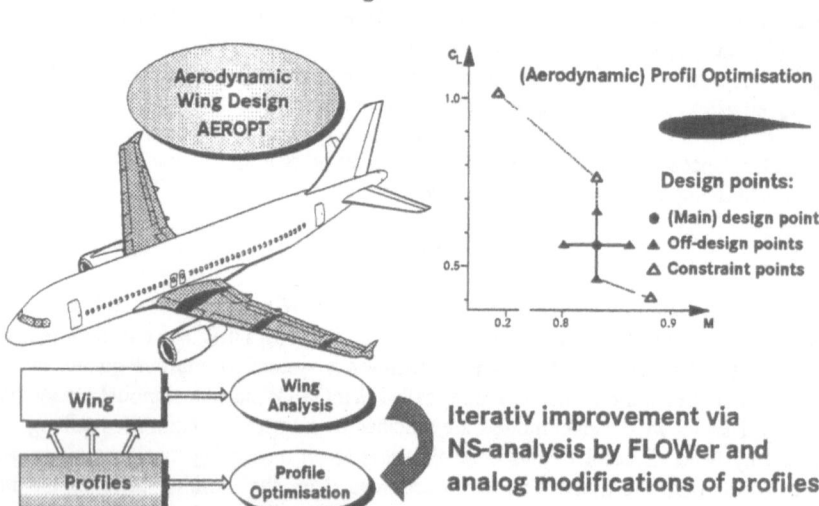

using suitable parameterisation for automated lofting

<u>Figure 1:</u> Wing Design and Wing Optimisation

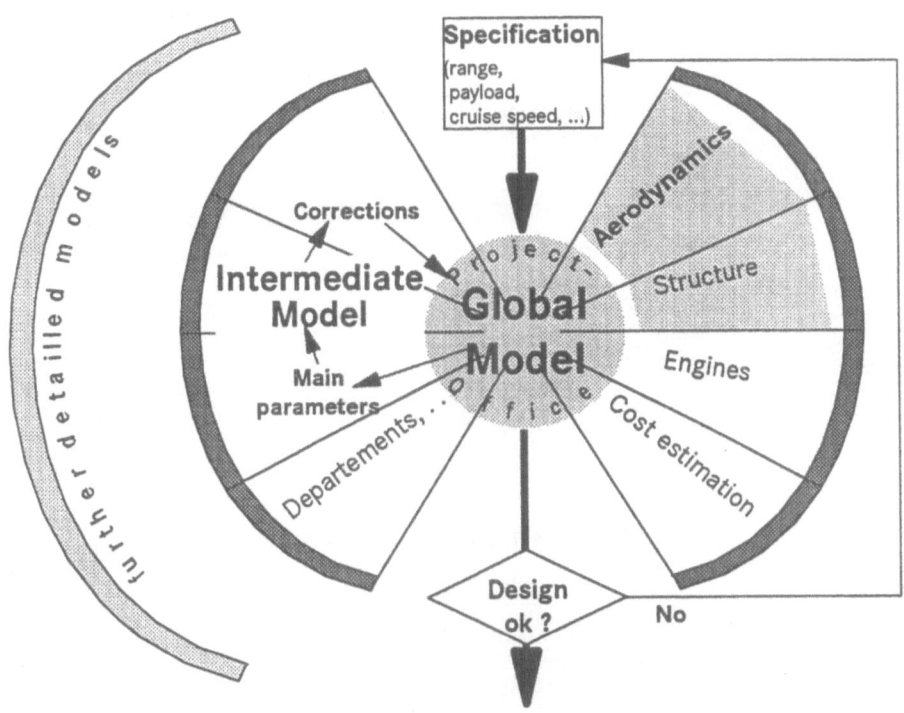

Figur 2: DA Design Methodology

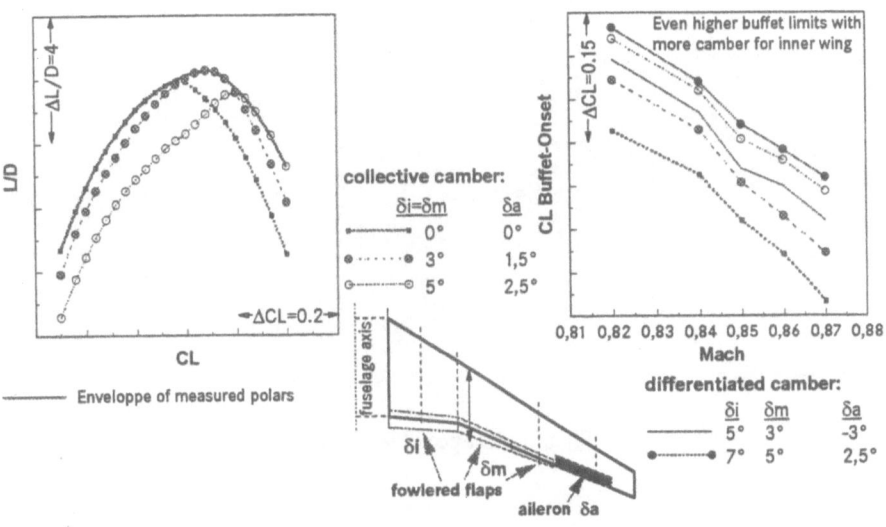

Figur 3: Performance Improvement Using Flaps (RaWid Test TVC2)

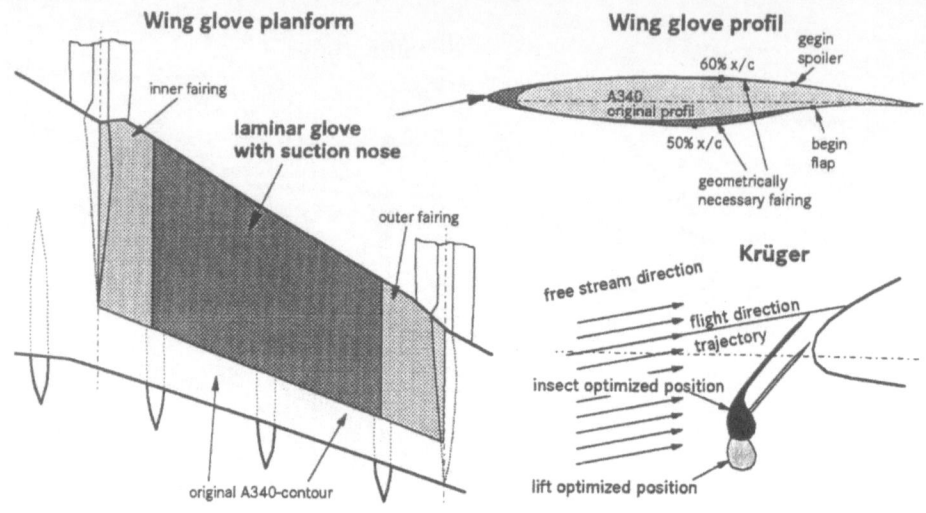

Figur 4: Laminar Glove for A340

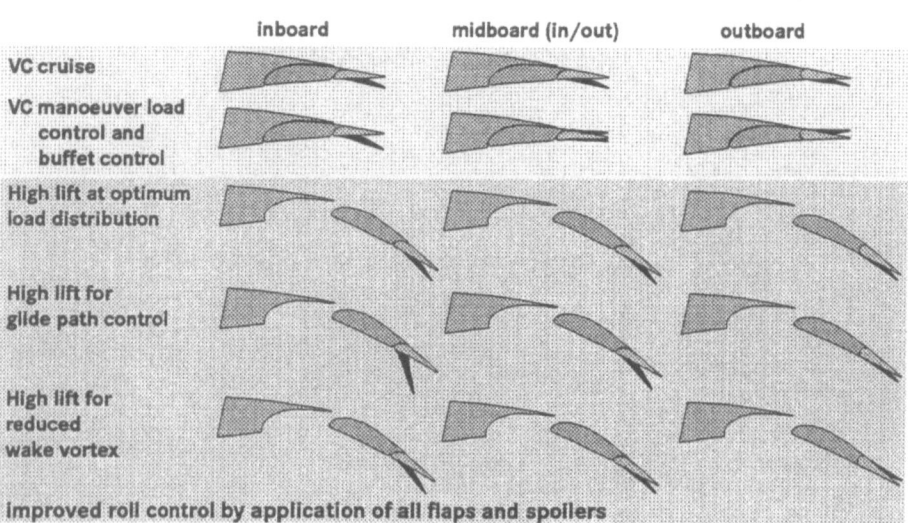

Figur 5: All Speed Flap Concept

Ti-sheet with ca. 200.000 suction holes Ø ca. 50 µm is investigated for contamination

For a long time laminar technology will provide the largest potential for energy saving. Important component tests are ongoing.

In **ELAS** (A320 Laminar Fin) the first European HLF flight test with a transport aircraft was performed.

Figur 6: „Beluga" tests with perforated suction sheet

Integration of Krüger flap and Suction System

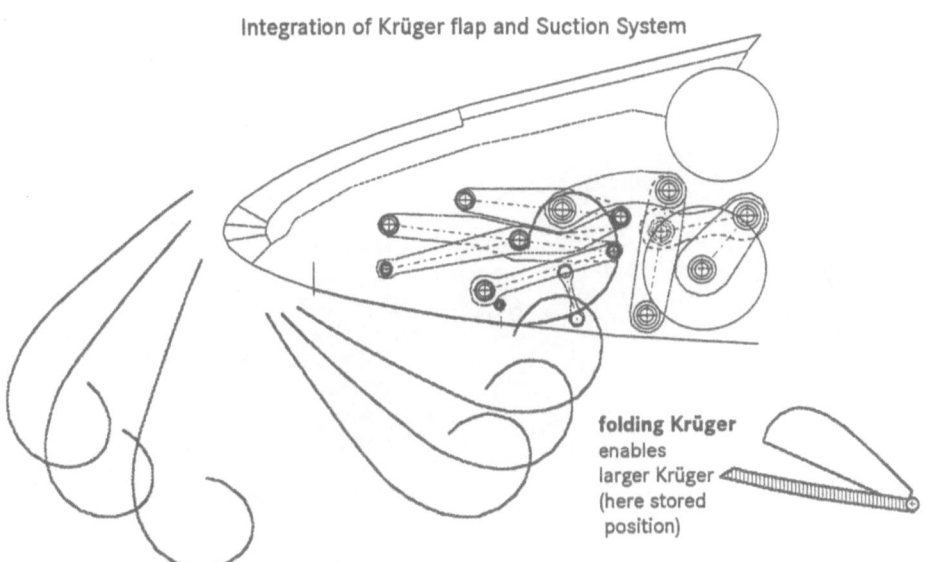

folding Krüger enables larger Krüger (here stored position)

Figur 7: Study for Wing Laminarisation

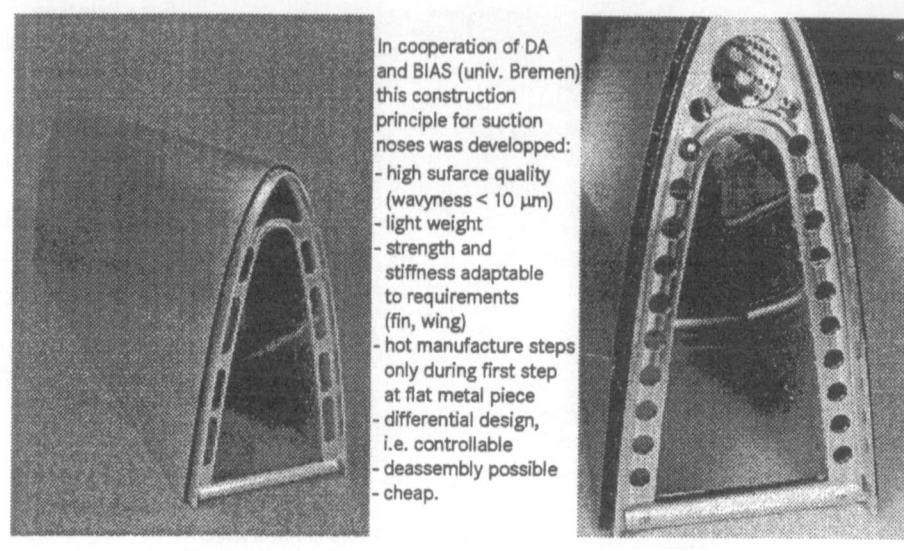

In cooperation of DA and BIAS (univ. Bremen) this construction principle for suction noses was developped:
- high sufarce quality (wavyness < 10 µm)
- light weight
- strength and stiffness adaptable to requirements (fin, wing)
- hot manufacture steps only during first step at flat metal piece
- differential design, i.e. controllable
- deassembly possible
- cheap.

Figur 8: RaWid Construction Principle for Suction Noses

Continuous optical control of the model surface during finish work now for regular use in the model shop in Varel

conventional measuring lines

conventionally manufactured and controlled modell fin, vizualisation using the geometry system WinGS.

0.26
0.22
0.17
0.12
0.08
0.03
-0.02
-0.06
-0.11
-0.16
-0.20

Antons, TVF3, 14.10.97, 7:40

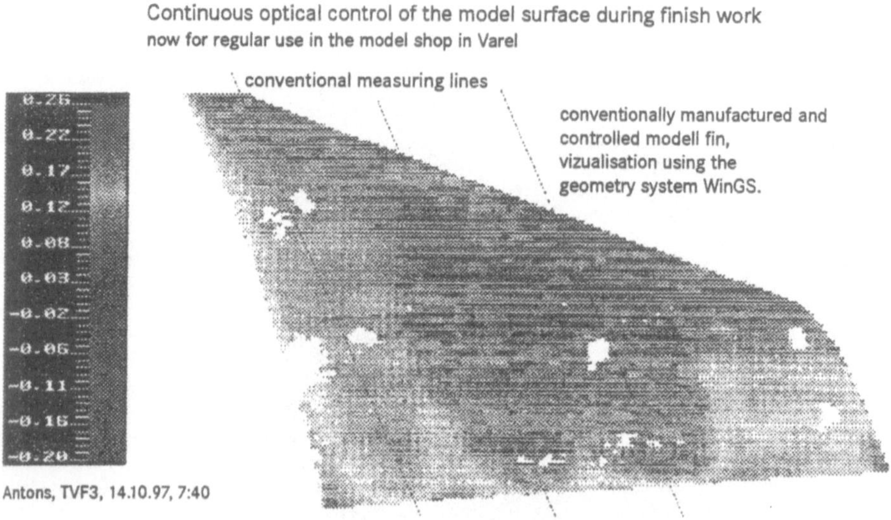

Figur 9: WinGS System for Model Geometry Monitoring

Direct Numerical Simulation of Suction through Discrete Holes in a Three-Dimensional Boundary Layer

R. Messing, M. Kloker, S. Wagner

Institut für Aerodynamik und Gasdynamik
Universität Stuttgart
Pfaffenwaldring 21, 70550 Stuttgart
Germany

Summary

Spatial direct numerical simulations using the full incompressible unsteady Navier-Stokes equations are employed to explore the effects of suction through a row of holes on an accelerated 3-D laminar boundary-layer flow. By variation of the hole diameter, spanwise hole spacing and volume flow rate per hole, the influence of different hole configurations on the generation and downstream development of hole-induced disturbances is investigated. It turns out that the induced crossflow vortices are distinctly larger in amplitude than the desired stabilizing mean-flow distortion. For high suction rates, the flow becomes instationary downstream of the holes in spite of the stationary boundary conditions.

Introduction

The application of hybrid laminar flow control (HLFC) technology to large aircraft is an attractive way to reduce drag. HLFC for wings consists of a combination of surface suction, applied in the leading edge region, and of an extended region of favourable pressure gradients, attained by profile shaping to reduce the instability with respect to disturbances triggering turbulence. Due to structural reasons, suction at the wall is in practice applied through laser-drilled micro-holes, typically 50 μm in diameter, rather than through spanwise slits. Physical details of the 3-D boundary-layer flow over swept wings with perforated suction panels are yet almost unknown. There is hardly any quantitative experimental data for a flow field in the vicinity of such suction holes available because of the microscopic scale of the flow dynamics. Indeed there is still a lack of knowledge about the effects of localised suction flow on the generation of unstable crossflow vortices in the 3-D boundary layer. These modes can compromise the stabilizing influence of the spanwise constant mean-flow distortion. Spatial direct numerical simulation (DNS) represents an ideal tool to investigate the modification of stability and receptivity of transitional 3-D boundary layers by suction through discrete holes.

The scope of the present work is to examine the excitation and streamwise development of the hole-induced disturbances introduced by different hole configurations. For this purpose several simulations have been performed varying the hole diameter d, the spanwise hole spacing s and the volume flow rate per hole $\dot{V}$.

Governing equations and Numerical Method

The governing equations are the full 3-D incompressible unsteady Navier-Stokes equations in vorticity-velocity formulation

$$\frac{\partial \vec{\omega}}{\partial t} - (\vec{\omega} \cdot \nabla)\, \vec{u} + (\vec{u} \cdot \nabla)\, \vec{\omega} = \nu \nabla^2 \vec{\omega} \qquad (1)$$

323

with
$$\vec{u} = (u, v, w) \quad \text{and} \quad \vec{\omega} = (\omega_x, \omega_y, \omega_z).$$

The solution of the Navier-Stokes equations is obtained in two substeps. First, a 3-D steady base flow (denoted by the index B) is calculated which satisfies the assumption of an infinite swept plate ($w_B \neq 0$, $\partial f_B / \partial z = 0$). In a second step the propagation of disturbances (denoted by a prime), generated by a suction hole for example, is simulated within the integration domain. Thus, the total flow variables are decomposed in the following way:

$$u = u_B + u', \quad v = v_B + v', \quad w = w_B + w'$$
$$\omega_x = \omega_{xB} + \omega'_x, \quad \omega_y = \omega_{yB} + \omega'_y, \quad \omega_z = \omega_{zB} + \omega'_z. \tag{2}$$

In the numerical method, velocities are normalized by the freestream velocity $\overline{U}_\infty$, and spatial variables by a reference length $\overline{L}$, and wall-normal quantities stretched by $\sqrt{Re}$, yielding

$$x = \frac{\overline{x}}{\overline{L}}, \ y = \frac{\overline{y}}{\overline{L}} \cdot \sqrt{Re}, \ z = \frac{\overline{z}}{\overline{L}}, \ u = \frac{\overline{u}}{\overline{U}_\infty}, \ v = \frac{\overline{v}}{\overline{U}_\infty} \cdot \sqrt{Re}, \ w = \frac{\overline{w}}{\overline{U}_\infty}, \ t = \overline{t}\frac{\overline{U}_\infty}{\overline{L}}. \tag{3}$$

The reference Reynolds number is defined as $Re = \overline{U}_\infty \overline{L}/\overline{\nu}$ ($\overline{\nu}$ is the kinematic viscosity). The simulations are carried out in a rectangular body-fixed integration box on the plate (Fig. 1). For the calculation of the steady base flow a velocity distribution $U_e(x)$ is imposed at the freestream boundary $y = y_e$ according to the approach of Falkner and Skan ($U_e = U_0 x^m, m = \beta_H/(2 - \beta_H)$, β_H-Hartree Parameter). The advantage of using a Falkner-Skan-Cooke (FSC)-type boundary layer is to cover all essential properties of a 3-D flow in the leading edge region of a wing at a low number of well-defined parameters (β_H, local angle of the external streamline φ_e). For the calculation of the disturbance flow three coupled vorticity transport equations and three Poisson equations have to be solved [4]. At the freestream boundary at $y = y_e$, potential flow with vanishing disturbance vorticity is assumed. For the wall-normal velocity an exponential decay is imposed at the freestream boundary. The no-slip condition is satisfied for u' and w' at the wall. Within the hole with diameter $d = 2 \cdot r$ steady suction can be modeled by prescribing a wall-normal velocity distribution

$$v'(r) = -v_c cos^3 \left(\frac{\pi r}{d}\right), \ r = \sqrt{(x - x_L)^2 + (z - z_L)^2}, \ 0 \leq r \leq \frac{d}{2}. \tag{4}$$

The hole centre is located at x_L in streamwise and z_L in spanwise direction. Outside the hole v' is zero at the wall. At the inflow boundary the flow is assumed to be laminar and all disturbances are set to zero. Upstream of the outflow boundary the vorticity disturbances are artificially suppressed in a buffer domain in order to avoid undue reflexions (Fig.1). The numerical treatment of the outflow zone is described in detail by Bonfigli & Kloker [1]. If we assume that disturbances are not amplified over the whole span, periodicity can be enforced:

$$f'(x, y, z, t) = f'(x, y, z + \lambda_z, t), \ \left.\frac{\partial^n f'}{\partial z^n}\right|_{x,y,z,t} = \left.\frac{\partial^n f'}{\partial z^n}\right|_{x,y,z+\lambda_z,t} \tag{5}$$

For a simulation of a row of equally spaced holes the use of periodic boundary conditions in z permits limiting the extension of the integration domain in spanwise direction to one hole spacing s. Thus, the disturbance-flow equations can be effectively discretized in spanwise direction by a fully complex Fourier spectral representation

$$f'(x, y, z, t) = \sum_{k=-K}^{K} F_k(x, y, t) e^{ik\gamma_0 z} \qquad (6)$$

where γ_0 denotes the basic spanwise wave number, related to the spanwise width λ_z of the computational domain through $\gamma_0 = 2\pi/\lambda_z$. The width λ_z is identical with the hole spacing s in this work. Applying the ansatz (6) to the disturbace flow equations yields the governing equations in spectral space where all variables are fully complex [1].

The spatial derivatives are basically discretized with 6th-order compact Finite Differences in x-and y-direction. The step size Δy can be halved in a wall zone. The nonlinear terms of the vorticity transport equations are evaluated pseudospectrally using the 2/3-rule to transform data from physical space to Fourier space and vice versa with no aliasing. Time integration is done with a fourth-order Runge-Kutta method. For all other numerical details not reported here see [1, 4].

DNS of boundary layer flows with suction through micro-holes cause long computing times even on up-to-date supercomputers. In view of the goal to investigate effects of several suction-device parameters on the boundary layer flow it is desirable to minimize the computing time requirements. The solution of the v'-Poisson equation [1, 4]

$$\frac{1}{Re}\frac{\partial^2 V_k(x,y)}{\partial x^2} + \frac{\partial^2 V_k(x,y)}{\partial y^2} - \frac{k^2\gamma_0^2}{Re} = ik\gamma_0\Omega_{z\,k}(x,y) - \frac{\partial\Omega_{z\,k}(x,y)}{\partial x} \qquad (7)$$

is principally computationally expensive and thus offers potential for computing time reductions.

As a standard this linear equation system is solved with a LSOR-method, and a multi-grid algorithm (with 2~4 V-cycles) to accelerate convergence. During validation of the numerical method for cases considered in this work it turned out that six V-cycles were necessary not to unduly disturb the marching in time to steady state. In this case 55%-60% of the total computing time is needed to solve the v'-Poisson equation. These long computing times can be reduced by replacing the LSOR multigrid method by a direct method using a Fourier approach for the x-discretisation terms. The x-derivatives in equation (7) are discretized with 4-th order compact differences. Expanding the wall-normal velocity $V_k(x, y)$ and the right hand side in sin-series in streamwise direction, the partial differential equation (7) changes into a set of independent ordinary differential equations in y-direction. The ordinary differential equations are then discretized with 6-th order compact differences. The solution $V_{mk}(y)$ in double Fourier space (streamwise and spanwise) is finally transformed back to spanwise Fourier space. Due to the fact that FFT-algorithms can be used to perform the sin-transforms, the number of floating-point operations for a whole simulation decreases by 50%-60%. Finally the implementation of the direct solution of the v'-Poisson equation has halved the total computing time. We point out here that the Fourier ansatz in x-direction is used as an effective means to solve the algebraic equations as obtained by Finite-Difference discretisation (exploiting the properties of matrix eigenvectors, see [3]), and not for basic discretisation.

Numerical Results

The parameters of the simulations are chosen to match the experiments with a 1:2 model of the ATTAS wing with LFC glove and are summarized in the following table.

$\overline{U}_\infty = 159 m/s$	freestream velocity	$Re = 100000$	reference Reynolds number
$\overline{L} = 0.00943 m$	reference length	$\beta_H = 0.4$	Hartree-Parameter
$Re_{\delta 1} = 450$	Re at x_0 based on displacement thickness	$\varphi_e = 35^0$	local sweep angle at the inflow boundary

To resolve the suction holes ($d = 50\mu m$) adequately, a small step size in streamwise direction is necessary leading to an integration domain length extending about 3 mm downstream of the hole. Within this region only a slight increase in the edge velocity U_e can be observed. According to the weak acceleration the local angle of the external streamline φ_e decreases by about 1.5°.

Steady suction through a row of holes represents a disturbance source for a laminar 3-D boundary layer. A modal analysis in spanwise direction of the physical suction distribution provides the range of discrete spanwise wave numbers excited at the wall. In this geometrical consideration two aspects have to be taken into account. The spanwise distance between two holes s determines the fundamental spanwise wave number $\gamma_0 = 2\pi/s$. For the given FSC base flow, crossflow vortices with $\gamma \geq \gamma_{crit} \approx 150$ are damped within the whole integration domain (Fig. 2). Thus, hole spacings $s \leq 395\mu m$ do not excite amplified crossflow-vortex disturbances according to linear stability theory (LST, numerical solution of the Orr-Sommerfeld equation, spatial approach) applied to the unperturbed base flow.

The generation of spanwise higher harmonics depends on the hole-diameter/hole-spacing ratio. To clarify this aspect a modal analysis was performed for two different hole diameters ($d = 50\mu m$ and $d = 100\mu m$) at fixed hole spacing $s = 500\mu m$. In Fig. 3a the spanwise normal-velocity distribution along the hole centerline is plotted over one fundamental wave length λ_z ($\lambda_z = s$). The corresponding spectral amplitude distribution can be seen in Fig. 3b. All phases are zero since the hole is located at $z = 0$. Fig. 3b illustrates that small holes excite higher wave number modes. The larger the hole at fixed hole spacing the less is the number of relevant modes excited by the row of holes. If the hole degenerates to a strip, i.e. the diameter tends to infinity, only the mode $K = 0$ has non-zero amplitude. This mode represents spanwise constant suction through a strip sucking indeed fluid off the boundary layer as desired. The modes $K > 0$ introduce no net mass flow. They modulate the strip suction into suction through holes. Thus the modes $K > 0$ mainly are undesired disturbances induced by the suction holes.

To start, a simulation with very low suction rate was carried out (negligible mean flow distortion). Fig. 4 shows the growth rate for a crossflow-vortex mode with $\gamma = 115$ (DNS: $\alpha_i = -\frac{d}{dx} ln\frac{A}{A_0} = -\frac{1}{A}\frac{dA}{dx}$). Up to about 60 hole diameters downstream of the hole the crossflow mode is damped ($\alpha_i > 0$), and up to 70 hole diameters less amplified than predicted by LST (hole location at $(x_L - x_0)/\Delta x = 200$). After this transient region the amplification rates evolve as expected (LST predicts slightly less amplification). Such a transient region with decreasing amplitudes of the stationary crossflow modes downstream of their excitation has also been observed in other DNS studies of 3-D boundary layers without suction at our research group.

To examine the effects of hole diameter d, hole spacing s and suction rate per hole $\dot{V}$ on the 3-D boundary layer, a parametric study has been conducted. Table 2 summarizes the relevant parameters.

Table 2: Parameters of hole configurations

Case 1	Case 2	Case 3	Case 4	Case 5	Case 6
$\dot{V} = 2.5 * 10^{-9} m^3/s$			$\dot{V} = 12.5 * 10^{-9} m^3/s$		
$v_c = 0.0089$	$v_c = 0.0356$	$v_c = 0.0356$	$v_c = 0.0445$	$v_c = 0.178$	$v_c = 0.178$
$s = 500\mu m$	$s = 500\mu m$	$s = 250\mu m$	$s = 500\mu m$	$s = 500\mu m$	$s = 250\mu m$
$d = 100\mu m$	$d = 50\mu m$	$d = 50\mu m$	$d = 100\mu m$	$d = 50\mu m$	$d = 50\mu m$

The results will be presented in the frequency-spanwise wavenumber spectrum (β, γ) where the first and the second number denote multiples of the fundamental frequency β_0 and the fundamental spanwise wave number γ_0, respectively. The Fourier analysis in time was performed over 200 time steps which corresponds to a non-dimensional time intervall $T_0 = 0.032$ with the fundamental frequency $\beta_0 = 2\pi/T_0$. The downstream disturbance amplitude development (u'_{max} over y) is plotted in logarithmic scale over the downstream direction x. The hole centre is located at $(x_L - x_0)/\Delta x = 200$, except for cases 5 and 6 where the hole centres are situated at $(x_L - x_0)/\Delta x = 300$.

Case 1 confirms the observation at low suction rates that downstream of the hole the amplitudes of crossflow modes decrease, which is not in agreement with LST. The integration domain extends, depending on the chosen hole diameter, to $20d$ to $40d$ downstream of the hole and lies well inside the above-mentioned transient region. The flow attains a steady state with negligible instationary modes. Crossflow modes with small wave numbers have a larger receptivity than the desired mode (0,0). All modes attenuate, but the mode (0,1) keeps a higher amplitude than the mean flow distortion even far downstream of the suction hole (Fig. 5). This receptivity behaviour of the 3-D boundary layer is unfavourable since the mode (0,1) can grow downstream if (0,0) is small. The (0,0) decay is expected since any locally exerted 2-D distortion must fade away far downstream in laminar flow. A smaller hole with constant volume flow rate per hole $\dot{V}$ causes higher amplitudes for modes with higher wave numbers as can be seen in Fig. 6. The (0,0) curve is identical to that of case 1 since the volume flow rate per spanwise unit length is identical and no relevant nonlinear contribution of the (0,k) modes exists yet.

Reducing the spanwise hole spacing s results in the highest initial amplitudes for all cases considered so far. Notice that modes (0,k) in Fig. 7 correspond to modes (0,2k) in the previous Figures 5 and 6, due to the doubled fundamental spanwise wave number. Comparing modes with equal spanwise wave numbers in case 2 and 3 one finds slightly weaker damping (despite doubled favourable (0,0)) and higher initial amplitudes in case 3. However, no mode should grow further downstream since $k\gamma_0 > \gamma_{crit}$.

For cases 1,2 we have to keep in mind that the observed damping of mode (0,1) is not attributed to suction, but to the transient region downstream of the disturbance excitation. Cases 4,5,6 correspond to cases 1,2,3 but with fivefold volume flow rate $\dot{V}$, respectively. In case 4 the amplitude decay of all modes (0,k) is weaker compared to case 1, in spite of the fivefold (0,0) amplitude. This is an indication for the onset of relevant nonlinear phenomena. Again the flow field attains a steady state.

Cases 5 and 6 are qualitatively different from the other cases. They do not attain a steady state. This can be seen from the the large amplitudes of the instationary modes (1,0),(2,0) and (1,1) in Fig. 9. The crossflow-vortex modes are no longer damped but strongly amplified downstream of the hole until they saturate (Fig. 10). At this stage of the investigation it is not yet clear how an instationary flow field can occur which leads to transition in spite of stationary boundary conditions. Similiar results were obtained by Meitz & Fasel [2] who carried out DNS of a Blasius boundary layer with discrete suction. In the case of strong suction rates they observed vortices that emerged from the hole, became unstable and finally broke up leading to an instationary flow field. They attributed the instability to vortex instability or to an instability of the recirculation region that formed in between the vortices. Up to now the global instability mechanism is completely unknown and is the subject of further investigations.

Conclusions

The effect of three different suction configurations within a single row of holes on a 3-D FSC boundary-layer flow was investigated by spatial DNS. Clearly, suction through holes induces crossflow modes. The receptivity of the flow to these undesired modes is, unfavourably, by a factor of 3 ~ 4 larger than to the „slit-suction" mode (0,0). Thus, suction through an array of hole rows with possible partial cancellation of the induced crossflow modes is necessary to raise (0,0) without undue (0,k) induction. For weak suction with small (0,0) a transient region extending to about 70 hole diameters downstream of the hole exists wherein primary unstable crossflow modes are damped. For strong (but still practised) suction, self-induced unsteadiness is observed. The understanding of the underlying instability mechanism could provide valuable knowledge for the successful design and application of HLFC.

Acknowledgement

This research has been financially supported by the German Ministry of Education, Science, Research, and Technology (BMBF) under contract number 20 A 9505 G.

References

[1] Bonfigli, G.; Kloker, M.: *Spatial Navier-Stokes Simulation of Crossflow-Induced Transition in a Three-Dimensional Boundary Layer*. In Nitsche, W.; Hilbig, R. (Eds.): New results in numerical and experimental fluid dynamics. NNFM, 11. STAB/DGLR Symposium, Berlin, November 1998, Vieweg Verlag.

[2] Meitz, H.L., Fasel, H.F.: *Navier-Stokes simulations of the effects of suction holes on flat plate boundary layer*. In Kobayashi, R. (ed.): Laminar-Turbulent Transition. Proc. IUTAM-Symposium, Sendai, Japan, September 1994, Springer Verlag, 1995.

[3] Swarztrauber, P. N.: *The Methods of Cyclic Reduction, Fourier Analysis and the FACR Algorithm for the Discrete Solution of Poisson's Equation on a Rectangle*. SIAM Review, Vol. 19, No.3, pp.490-501,1977.

[4] Wassermann, P., Kloker, M.: *Direct numerical simulation of the development and control of boundary-layer crossflow vortices*. In Nitsche, W.; Hilbig, R. (Eds.): New results in numerical and experimental fluid dynamics. NNFM, 11.STAB/DGLR Symposium, Berlin, November 1998, Vieweg Verlag.

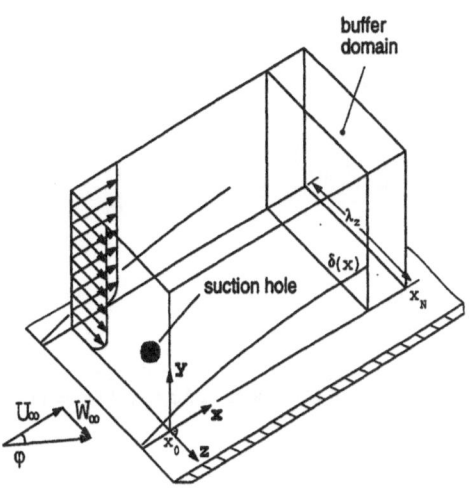

Fig. 1 Integration domain

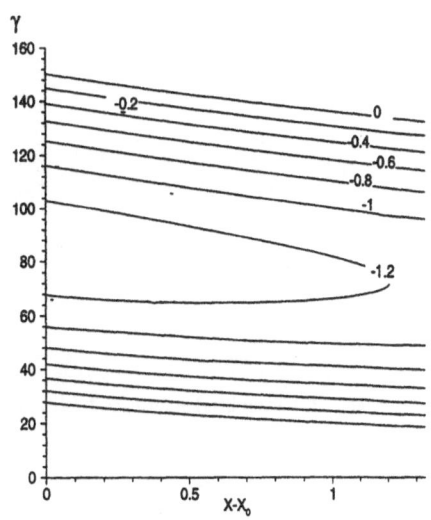

Fig. 2 Contour lines of constant amplification rates $-\alpha_i$ for steady crossflow modes as obtained by Linear Stability Theory

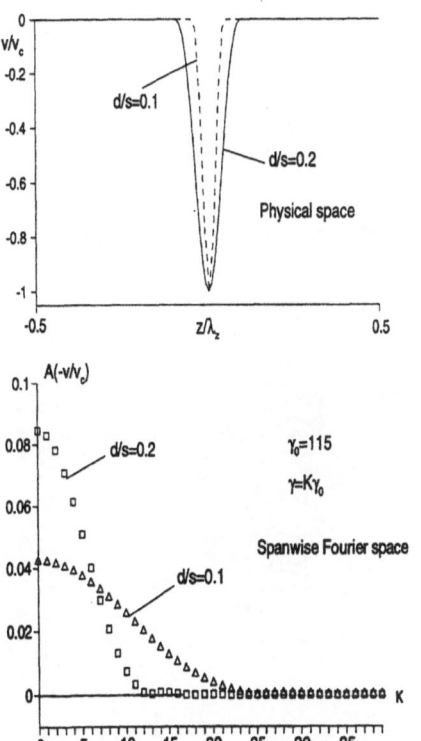

Fig. 3a Spanwise v'-distribution at the wall along the hole centerline

Fig. 3b Corresponding discrete spectral amplitude distribution

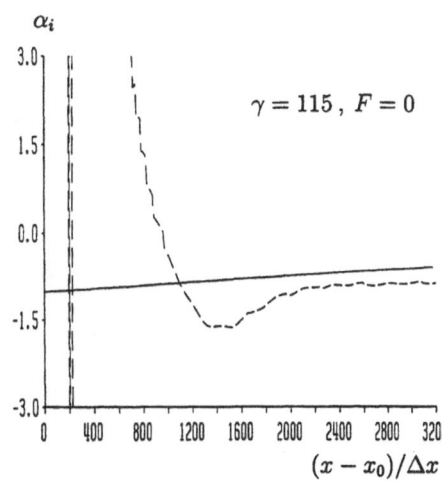

-------- Direct Numerical Simulation

———— Linear Stability Theory

Fig. 4 Comparison of growth rates at very weak suction, $\Delta x = 0.42 * 10^{-3}$, hole located at $(x - x_0)/\Delta x = 200$

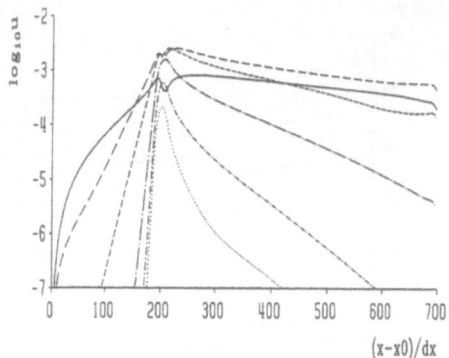

Fig. 5 u'-amplitudes for case 1 ($\gamma_0 = 115$)

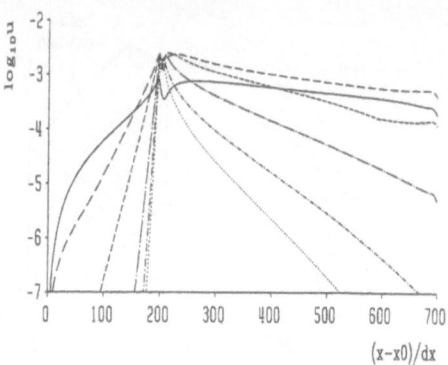

Fig. 6 u'-amplitudes for case 2 ($\gamma_0 = 115$)

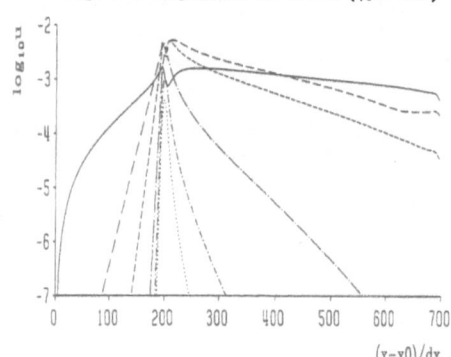

Fig. 7 u'-amplitudes for case 3 ($\gamma_0 = 230$)

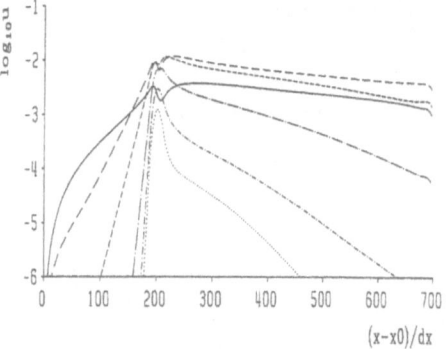

Fig. 8 u'-amplitudes for case 4 ($\gamma_0 = 115$)

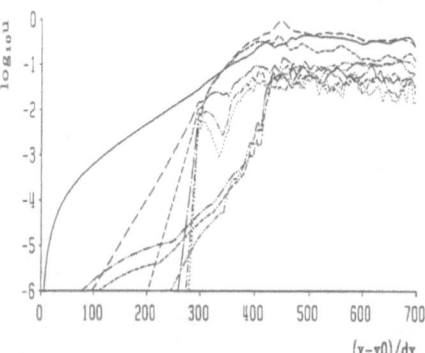

Fig. 9 u'-amplitudes for case 5 ($\gamma_0 = 115$)

———— (0,0)	—·—·— (0,5)	—··—·· (1,0)
— — — (0,1)	·········· (0,8)	———— (1,1)
———— (0,2)	············ (0,10)	—·—·— (2,0)

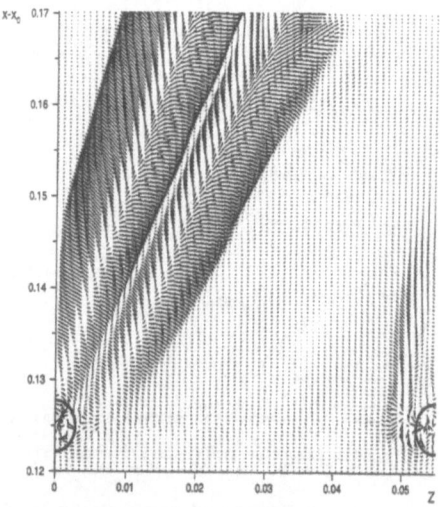

Fig. 10 Vectors of the instantaneous disturbance velocities u', w' in the plane $y = 0.5\Delta y$ above the wall for case 5

An analysis of vortex breakdown predicted by a time-accurate Euler code

J. Müller, D. Hummel

Institut für Strömungsmechanik, TU Braunschweig
Bienroder Weg 3, 38106 Braunschweig
Germany

Summary

Time-accurate Euler-calculations have been carried out for the non-moving delta wing of the VFE-configuration at $M_\infty = 0.2$ and $\alpha = 20$ deg. The calculations predict periodic oscillations of the force and moment coefficients caused by a spiral-type vortex breakdown of the leading-edge vortex in the rear part of the wing. A detailed analysis of the breakdown process indicates that the flow becomes unsteady: The instantaneous vortex axis forms a spiral in space which rotates with respect to time at a reduced frequency of $\omega^* = 15.3$. The numerical influence of the time discretization methods (global time-stepping and dual time-stepping) is negligible if the time resolution is sufficiently high. The influence of the artificial numerical damping coefficient $k^{(4)}$ is considerable; it affects all unsteady characteristics as well as the breakdown position. The latter has been used for calibration which leads to the value $k^{(4)} = 1/40$ in the present calculations.

1. Introduction

The complex vortical flow over a sharp edged delta wing, especially the unsteady phenomenon of vortex breakdown at high angles of attack, has been the subject of numerous experimental and computational research activities in the last 50 years (see e.g. [1]-[3]). In 1986 a joint program known as the "International Vortex Flow Experiment (VFE)" has been started in order to provide comparable experimental data for a normalized delta wing [4]. The VFE-configuration consists of a 65 deg swept and cropped delta wing with an aspect ratio of 1.38 and a symmetrical airfoil. During the last decade comprehensive experimental and numerical investigations have been carried out for this configuration (e.g. [5]-[7]). The simulation of vortex breakdown using a *non time-accurate* Euler or Navier-Stokes code yields quite good results according to experimental observations although especially adapted grids must be used [6]. However, these non time-accurate methods are not able to predict the reduced frequency of the unsteady motion. The present paper focuses on the analysis of vortex breakdown for the non-moving delta wing of the VFE-configuration using a *time-accurate* Euler code. The calculations have been carried out for a Mach number of $M_\infty = 0.2$ and an angle of attack of $\alpha = 20$ deg.

2. Numerical method

The numerical simulation is based on the solution of the three-dimensional, unsteady, compressible Euler equations in integral form using the DLR code *FLOWer* [8]. The spatial discretization of the solution algorithm is characterized by a cell vertex scheme and central differences. For the damping of numerical oscillations first and third order dissipative terms are added to the governing equations. The artificial viscosity is mainly controlled by the user-defined damping coefficient $k^{(4)}$. In the present paper two different time discretization methods were applied: A modified "global time-stepping" scheme [9] and a "dual time-stepping" scheme based on the method of A. J. Jameson [10]. For the calculations a structured elliptically smoothed O-O-grid with 943 245 gridnodes was generated. The farfield boundary extension is about eight wing chord lengths c in all directions and the first spacing normal to the wing is about 0.01 wing chord lengths or less.

3. Results

3.1 Influence of the numerical discretization

The influence of the **time discretization** was analyzed at $M_\infty = 0.2$ and $\alpha = 20$ deg. The artificial damping coefficient was choosen as $k^{(4)} = 1/40$. A non time-accurate local time-stepping scheme with 1000 multigrid cycles has been applied to produce a starting solution for the subsequent global and dual time-stepping calculations. The dual time-stepping calculations have been carried out with 20 inner iterations n_i per time step Δt. Fig. 1 shows the dependence of the lift coefficient C_L on the non-dimensional time parameter $t^* = t \cdot U_\infty/c$ for the global time-stepping calculations and for different amounts of the non-dimensional time step $\Delta t^* = \Delta t \cdot U_\infty/c$. For all values of Δt^* a periodic oscillation of the lift coefficient occurs and for decreasing time step Δt^* a converged solution turns out. The same behaviour is observed for the drag and the pitching moment coefficient.

Fig. 2 shows the resulting mean value $\overline{C}_L = \frac{1}{n}\sum_{t_1^*}^{t_2^*} C_L$, standard deviation $C_L' = \sqrt{\frac{1}{n-1}\sum_{t_1^*}^{t_2^*}\left(C_L - \overline{C}_L\right)^2}$ and oscillation period $T^* = T \cdot U_\infty/c$ for the different time-stepping schemes and for varying time step Δt^*. $\overline{C}_L$, C_L' and T^* are evaluated for four cycles of the C_L-oscillation. With decreasing time step Δt^* all unsteady quantities converge asymptotically to one value for both time-stepping schemes. Therefore, the influence of the time discretization is negligible if the time step Δt^* is small.

In the following the global time-stepping scheme with a time step of $\Delta t^* = 0.0012$ was choosen since the dual time-stepping scheme needs more than twice the calculation time to achieve the same accuracy. Therefore, the global time-stepping scheme is recommended for flow problems with high frequency oscillations like vortex breakdown.

The effects of the **numerical dissipation** are governed by the grid resolution and the artificial damping coefficient $k^{(4)}$. For a given grid the influence of $k^{(4)}$ on the unsteady characteristics $\overline{C}_L$, C_L' and T^* is remarkable, see Fig. 3. In addition, the vortex breakdown position x_B/c varies also within the range $0.8 < x_B/c < 1.0$. (The criterion for the determination of the breakdown position will be presented subsequently.) In agreement with the experimental breakdown position $x_B/c \approx 0.9$ according to H.-C. Oelker [5] a value of $k^{(4)} = 1/40$ was choosen.

3.2 Structure of the spiral-type vortex breakdown

For $M_\infty = 0.2$ and $\alpha = 20$ deg the global time-stepping method with a time step of $\Delta t^* = 0.0012$ and an artificial damping coefficient of $k^{(4)} = 1/40$ predicts a force and moment oscillation period of $T^* \approx 0.41$. This corresponds to a reduced frequency of $\omega^* = 2\pi/T^* = 15.3$. Therefore, the time resolution of this oscillation is about 340 iterations per period. For a detailed analysis an oscillation period between $3.7 \leq t^* \leq 4.1$ has been chosen:

Fig. 4 shows lines of constant total pressure losses for a crossflow plane at the trailing-edge $(x/c = 1.0)$ and for different times $t^* = 3.7$; 3.8; 3.9 and 4.0. Their position on the lift oscillation is marked on the upper right corner of the first four sub-figures. In addition, the mean value and the standard deviation of total pressure losses calculated for this oscillation period are shown. For all times a distinct region with total pressure losses can be observed. The point of maximum total pressure loss is always excentered and it rotates with respect to time in the same sense as the leading-edge vortex is turning around. This leads to a very flat region of maximum total pressure losses for the mean value and to a ring-shaped region of maximum fluctuations for the standard deviation.

The location of the maximum total pressure losses for all crossflow planes at a given t^* leads to the instantaneous leading-edge vortex axes as shown in Fig. 5. In the same manner a mean vortex axis can be evaluated. Downstream of $x/c \approx 0.9$ the instantaneous vortex axes deviate from the mean axis and spiral in space against the sense of rotation of the leading-edge vortex. In addition, the instantaneous vortex axes turn around with respect to time in the sense of the leading-edge vortex with the same frequency as the force and moment oscillations. An evaluation of the axial velocity with regard to the mean vortex axis $\bar{u}_{ax}$ indicates a reverse flow region ($\bar{u}_{ax} < 0$) which is surrounded by the spiraling instantaneous vortex axes (see the boundary surface of zero axial velocity in Fig. 5). Therefore, the time-accurate Euler calculations clearly predict a spiral-type vortex breakdown in the rear part of the wing which causes the oscillations of the force and moment coefficients.

3.3 Phases of the breakdown process

For the location of the breakdown position a further detailed analysis was carried out. Several phases of the breakdown process can be distinguished:

$x/c \approx$ 0.80:	Fig. 6 shows the progress of several quantities along the mean vortex axis. At first, a significant increase of the mean value of the pressure coefficient $\bar{c}_p$ and a decrease of the axial velocity $\bar{u}_{ax}/U_\infty$ take place downstream of $x/c \approx 0.8$.
0.85:	Fig. 7 shows the axial velocity $\bar{u}_{ax}/U_\infty$ for a longitudinal section along the mean vortex axis. A considerable change of the velocity profiles perpendicular to the mean vortex axis can be observed. At $x/c \approx 0.85$ the axial velocity profiles change their character from a "jet-like" to a "wake-like" shape with a local minimum of axial velocity at the mean vortex axis.
0.90:	Further downstream (see again Fig. 6), a remarkable decrease of the mean value and an increase of the standard deviation of the total pressure losses can be observed along the mean vortex axis at $x/c \approx 0.9$. This is the same position at which the instantaneous vortex axes deviate from the mean vortex axis (Fig. 5). In the present paper this criterion has been used for the determination of the breakdown position, because this is the point of "beginning unsteadiness". Fig. 8 shows lines of constant total pressure losses (mean value and standard deviation) for the same longitudinal section along the mean vortex axis. It turns out, that the decrease of the mean value of total pressure losses on the mean vortex axis is accompanied by a considerable enlargement of the region of total pressure losses downstream of $x/c \approx 0.9$.

Again the ring-shaped region with maximum values of the standard deviation above and below the mean vortex axis can be determined. Upstream of $x/c \approx$ 0.9 the fluctuations are approximately zero.

0.95 - 1.05: Within the spiraling unsteady flowfield a reverse flow region is embedded and two stagnation points ($\overline{u}_{ax} = 0$) turn out along the mean vortex axis at $x/c \approx$ 0.95 and $x/c \approx 1.05$ (see Fig. 6 and 7).

1.10: Finally, the spiral-type of the flow ends at $x/c \approx 1.1$ (see Fig. 5).

4. Final remarks

The present work has shown, that a time-accurate Euler code with a proper amount of artificial numerical dissipation is able to predict the spiral-type vortex breakdown very well according to experimental results. Further calculations on the VFE-configuration are planned for different angles of attack between 15 deg and 45 deg, i.e. for the whole region of vortex breakdown dominated flow. In addition, time-accurate Navier-Stokes calculations will be carried out in order to replace the artificial numerical dissipation by the real viscous damping and to reduce the artificial dissipative effects.

It is gratefully acknowledged, that the Institute of Design Aerodynamics of the DLR Braunschweig provided the Euler and Navier-Stokes code *FLOWer* as well as the grid generation program *MegaCADs*. Special thanks are expressed to R. Heinrich for his assistance.

5. References

[1] Lambourne, N. C.; Bryer, D. W.: The bursting of leading-edge vortices. Aeronautical Research Council, R. & M. No. 3282, London (1962).

[2] Hummel, D.: On the vortex formation over a slender delta wing at large angles of attack. AGARD CP-247, 1978, pp. 15-1 to 15-17.

[3] Nelson, R. C.; Visser, K. D.: Breaking down the delta wing vortex – The role of vorticity in the breakdown process. AGARD CP-494, 1990, pp. 21-1 to 21-15.

[4] Drougge, G.: The international vortex flow experiment for computer code validation. ICAS-Proceedings 1988, Vol. 1, pp. XXXV-XLI.

[5] Oelker, H.-C.: Aerodynamische Untersuchungen an kurzgekoppelten Entenkonfigurationen bei symmetrischer Anströmung. Dissertation TU Braunschweig 1990, ZLR-Forschungsbericht 90-01 (1990).

[6] Strohmeyer, D.; Orlowski, M.; Longo, J. M. A.; Hummel, D.; Bergmann, A.: An analysis of vortex breakdown predicted by the Euler equations. 20th ICAS-Proceedings 1996, Vol. 1, pp. 1189-1200.

[7] Fritz, W.: Unsteady Navier-Stokes calculations for a delta wing oscillating in pitch. Paper at the 21st ICAS Congress, Melbourne (1998).

[8] Kroll, N.; Radespiel, R.; Rossow, C. C.: Accurate and efficient flow solvers for 3D applications on structured meshes. VKI Lecture Series 1994-04 "Computational Fluid Dynamics", Brussels, March 21-25 (1994).

[9] Pahlke, K.; Blazek, J.; Kirchner, A.: Time-accurate Euler computations for rotor flows. Royal Aeronautical Society, 1993 European Forum of "Recent developments and applications in aeronautical CFD", Bristol, UK, September 1-3 (1993).

[10] Jameson, A. J.: Time dependent calculations using multigrid with applications to unsteady flows past airfoils and wings. AIAA Paper 91-1596 (1991).

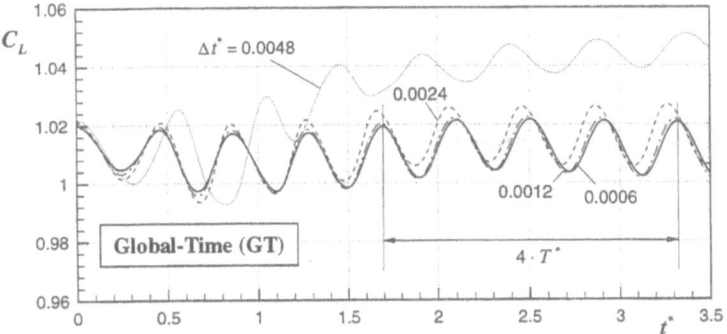

Fig. 1: Influence of the time step Δt^* on C_L ($M_\infty = 0.2$; $\alpha = 20$ deg; $k^{(4)} = 1/40$)

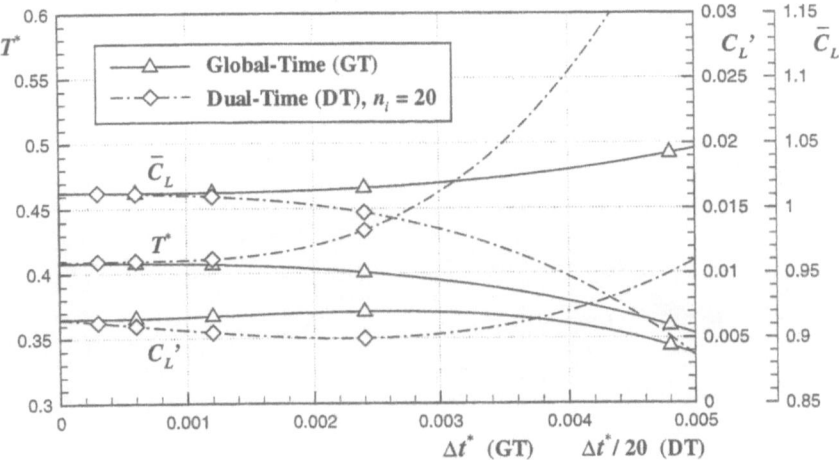

Fig. 2: Influence of the time step Δt^* on the unsteady characteristics $\overline{C}_L$, C_L' and T^* ($M_\infty = 0.2$; $\alpha = 20$ deg; $k^{(4)} = 1/40$)

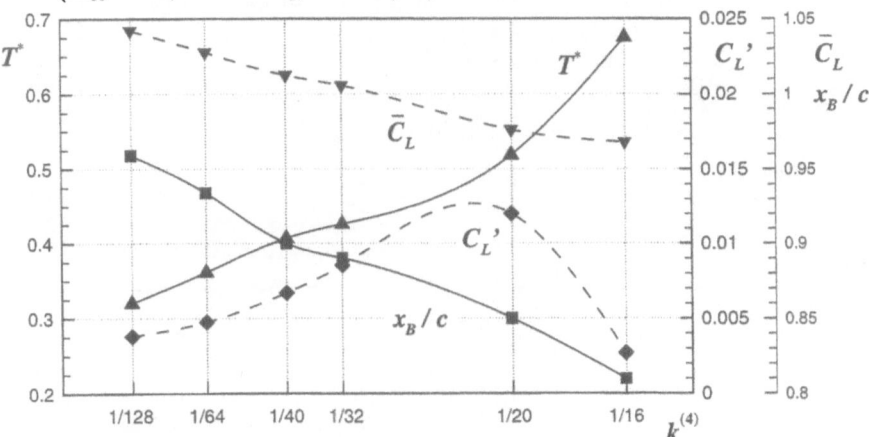

Fig. 3: Influence of $k^{(4)}$ on the unsteady characteristics $\overline{C}_L$, C_L', T^* and the breakdown position x_B/c ($M_\infty = 0.2$; $\alpha = 20$ deg; GT: $\Delta t^* = 0.0012$)

335

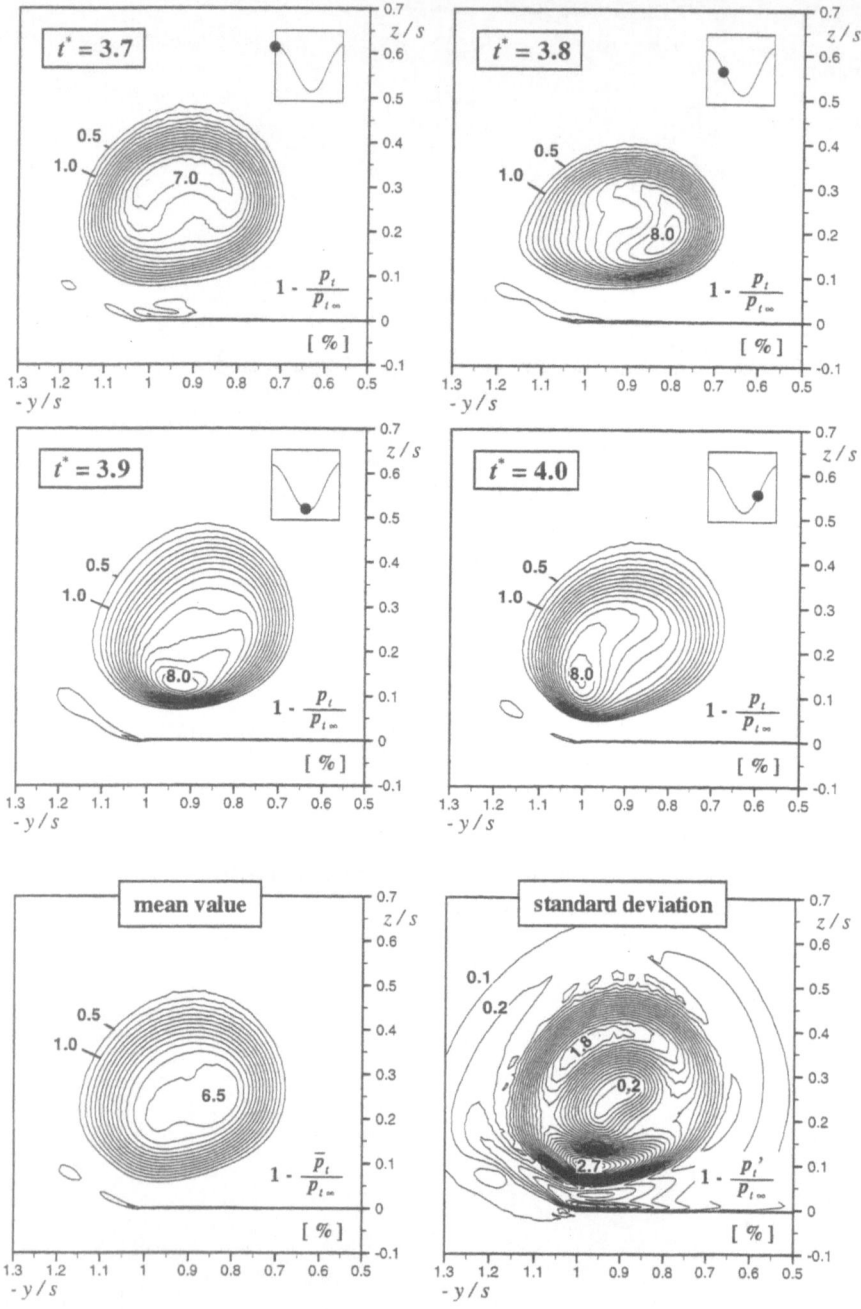

Fig. 4: Lines of constant total pressure losses $1 - p_t/p_{t\infty}$ at $x/c = 1.0$
($M_\infty = 0.2$; $\alpha = 20$ deg; $k^{(4)} = 1/40$; GT: $\Delta t^* = 0.0012$)

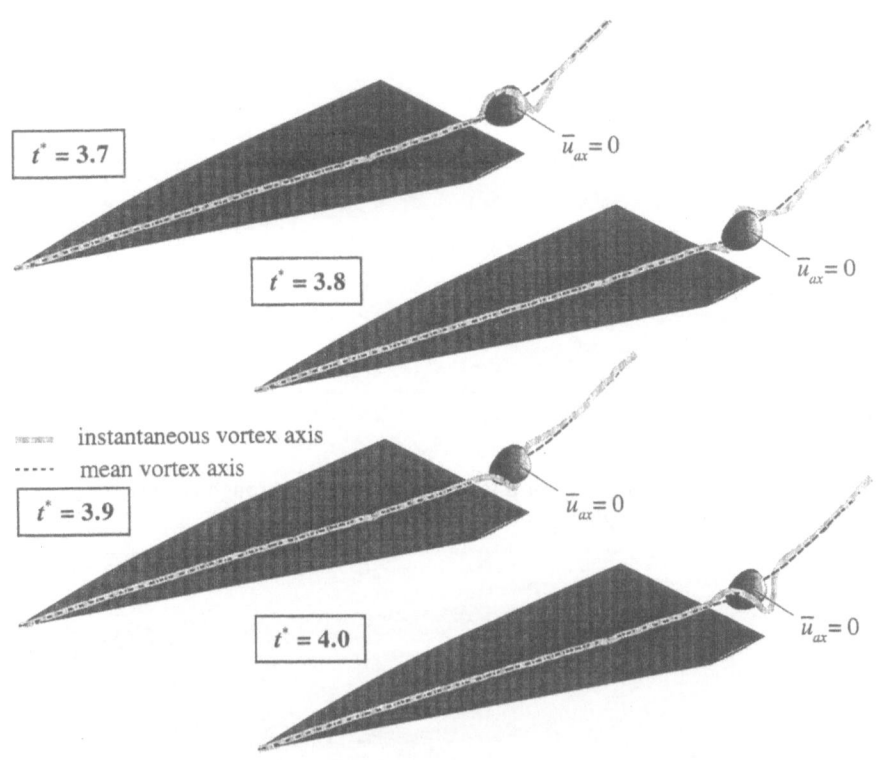

Fig. 5: Instantaneous and mean vortex axes for different times t^*
($M_\infty = 0.2$; $\alpha = 20$ deg; $k^{(4)} = 1/40$; GT: $\Delta t^* = 0.0012$)

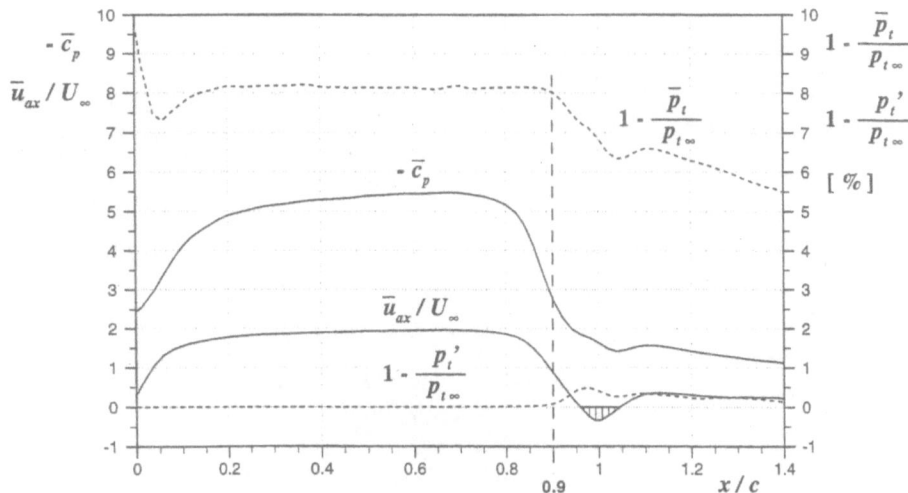

Fig. 6: Pressure and velocity distributions along the mean vortex axis
($M_\infty = 0.2$; $\alpha = 20$ deg; $k^{(4)} = 1/40$; GT: $\Delta t^* = 0.0012$)

337

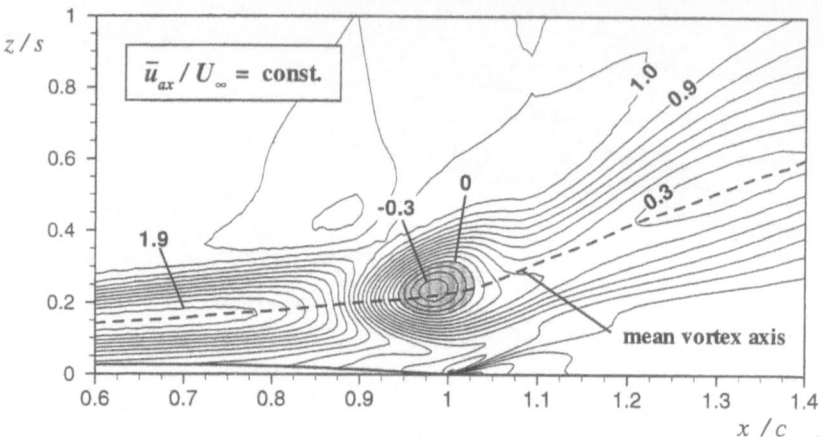

Fig. 7: Axial velocity $\bar{u}_{ax}/U_\infty$ for a longitudinal section through the mean vortex axis ($M_\infty = 0.2$; $\alpha = 20$ deg; $k^{(4)} = 1/40$; GT: $\Delta t^* = 0.0012$)

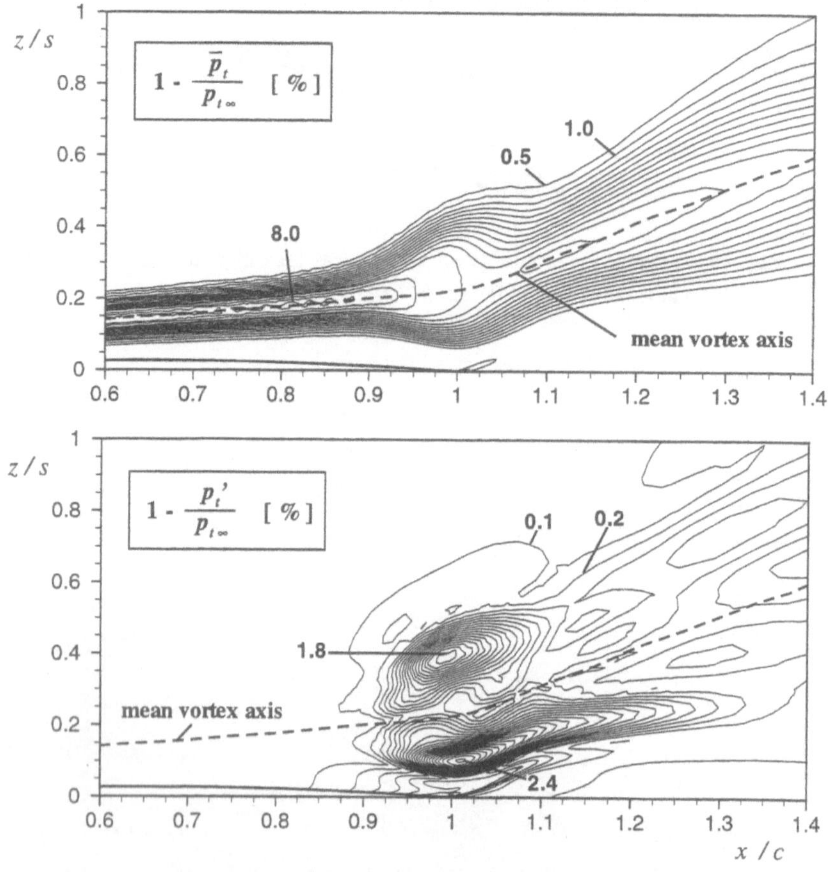

Fig. 8: Total pressure losses $1 - \frac{p_t}{p_{t\infty}}$ for a longitudinal section through the mean vortex axis ($M_\infty = 0.2$; $\alpha = 20$ deg; $k^{(4)} = 1/40$; GT: $\Delta t^* = 0.0012$)

Controlled Elliptic Grid Generation

P. Niederdrenk[1], O. Brodersen[2]

[1]DLR, Institut für Strömungsmechank, Bunsenstraße 10, D-37073 Göttingen, Germany
[2]DLR, Institut für Entwurfsaerodynamik, Lilienthalplatz 7, D-38108 Braunschweig, Germany

Summary

The numerical solution to the quasi-linear elliptic grid generation equation usually proceeds in a staggered iterative manner. In an outer loop the control functions and the coefficients containing first order derivatives are updated. Then the linearized equation with known control functions is solved in an inner loop [1].
In a recent paper [2] the control functions were analytically defined from a comparison with the metric identity and numerically specified line-wise via so-called target points.

In this paper we show that one may equally well disregard the control functions altogether and obtain the solution to the equivalent grid generation equation directly from a suitable average of the line-wise specified target central points [2], thus saving one of the afore-mentioned iterative loops.

Introduction

In the generation of three-dimensional structured grids from a solution to the Poisson equation for the position vector $\underline{r}$ (see e.g. [1])

$$\underline{S}_1^2(\underline{r}_{\xi\xi} + P\underline{r}_\xi) + \underline{S}_2^2(\underline{r}_{\eta\eta} + Q\underline{r}_\eta) + \underline{S}_3^2(\underline{r}_{\zeta\zeta} + R\underline{r}_\zeta)$$
$$+ 2\underline{S}_1\underline{S}_2\underline{r}_{\xi\eta} + 2\underline{S}_2\underline{S}_3\underline{r}_{\eta\zeta} + 2\underline{S}_3\underline{S}_1\underline{r}_{\zeta\xi} = 0 \tag{1}$$

with surface normal vectors defined as

$$\underline{S}_1 = \underline{r}_\eta \times \underline{r}_\zeta, \quad \underline{S}_2 = \underline{r}_\zeta \times \underline{r}_\xi, \quad \underline{S}_3 = \underline{r}_\xi \times \underline{r}_\eta$$

. the so-called control functions P, Q, R serve to adjust to our objectives the spacing and the orientation of the grid lines in the field. For a converged solution of these equations there is a one-to-one correspondence between the derivatives of the position vector and the control functions. Thus, for an already known grid these functions are easily found by solving the Poisson equations for them. In most numerical codes the control functions are evaluated either from an initial algebraic grid with subsequent directional, often Laplace-like smoothing in the interior field, or from a local, iterative update along the grid boundaries only and subsequent suitable interpolation into the field.

In a previous paper [2] the full control functions were split into a matrix-like system

$$P = P_i + P_j + P_k$$
$$Q = Q_i + Q_j + Q_k \qquad (2)$$
$$R = R_i + R_j + R_k$$

where with the cell volume V the partial line-wise control functions are analytically defined as

$$P_i = \frac{V}{\underline{S}_1^2}\left(\frac{\underline{S}_1^2}{V}\right)_\xi, \quad P_j = \frac{V}{\underline{S}_1^2}\left(\frac{\underline{S}_1\underline{S}_2}{V}\right)_\eta, \quad P_k = \frac{V}{\underline{S}_1^2}\left(\frac{\underline{S}_1\underline{S}_3}{V}\right)_\zeta,$$

$$Q_i = \frac{V}{\underline{S}_2^2}\left(\frac{\underline{S}_2\underline{S}_1}{V}\right)_\xi, \quad Q_j = \frac{V}{\underline{S}_2^2}\left(\frac{\underline{S}_2^2}{V}\right)_\eta, \quad Q_k = \frac{V}{\underline{S}_2^2}\left(\frac{\underline{S}_2\underline{S}_3}{V}\right)_\zeta, \qquad (3)$$

$$R_i = \frac{V}{\underline{S}_3^2}\left(\frac{\underline{S}_3\underline{S}_1}{V}\right)_\xi, \quad R_j = \frac{V}{\underline{S}_3^2}\left(\frac{\underline{S}_3\underline{S}_2}{V}\right)_\eta, \quad R_k = \frac{V}{\underline{S}_3^2}\left(\frac{\underline{S}_3^2}{V}\right)_\zeta.$$

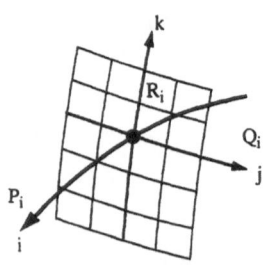

Fig. 1: Forces acting on a point when proceeding in i-direction

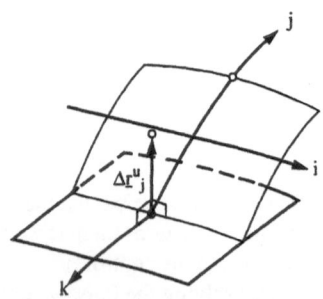

Fig.2: Specification of central point by distance and angle relative to face at j-1

Their physical meaning may be understood by considering a single coordinate line, for example one, on which i varies (see fig.1). The components of the left column vector in (2) control via P_i changes of the spacing along that line, via Q_i and R_i changes of the line's orientation relative to faces i=const. The analogue holds for the lines in j- and k-directions.

For each interior grid point the control functions are locally defined by the interior point itself and the 26 points surrounding it. Since the grid generation equation is elliptic, we may at each iteration step of its solution assume the points on the local cell boundaries to be momentarily known and specify only the position of the unknown central point relative to the cell boundaries.

Doing so line-wise as sketched in figure 2 for the j-direction we prescribe the position vector of the central point relative to the upstream boundary at j-1 by a vector $\Delta\underline{r}^u_j$ specified via its angles with the i- and k-directions and the distance to the upstream cell boundary. Proceeding in the same manner from the downstream boundary at j+1 the target central point $\underline{r}_{cj}$ in j-direction follows from its weighted positions defined from up- and downstream cell faces:

$$\underline{r}_{cj} = w_j[\underline{r}_{j-1} + \Delta\underline{r}^u_j] + (1 - w_j)[\underline{r}_{j+1} + \Delta\underline{r}^d_j] .$$

Directional control of spacing and angle, here relative to faces j=const, is exerted via the difference in up- and downstream weights. Analogous specifications of the line-wise target central points in k- and i-directions, $\mathfrak{r}_{ck}$ and $\mathfrak{r}_{ci}$, complete the procedure. Desired spacings may be measured from an initial algebraic grid or interpolated from the boundaries into the field. Besides a local prescription desired angles may be specified from normals to opposing bounding cell faces or from projections of these normals onto appertaining faces (for example the normal to j=const in fig.2 may be projected onto a face k=const or i=const) or according to a straight line connecting points on opposing boundaries.

Solution to the Grid Generation Equation

With the target central points specified the nine partial control functions, $P_i(\mathfrak{r}_{ci})$, $Q_i(\mathfrak{r}_{ci})$, $R_i(\mathfrak{r}_{ci})$, $P_j(\mathfrak{r}_{cj})$, $Q_j(\mathfrak{r}_{cj})$, $R_j(\mathfrak{r}_{cj})$, $P_k(\mathfrak{r}_{ck})$, $Q_k(\mathfrak{r}_{ck})$, $R_k(\mathfrak{r}_{ck})$, are evaluated line-wise and superimposed according to equation (2) to yield the three full control functions needed to solve the grid generation equation (1).

On the other hand, if it were possible to directly define a unique target central point - say for instance from a suitable superposition of line-wise target central points -, then there would be no more need to evaluate the control functions and subsequently solve the grid generation equation. Because of the local duality between the position vector and the control functions the local solution to the grid generation equation would just retrieve the target central point. Now we will show that such a relation between the central point and partial control functions does also exist along a single grid line and that a linear superposition of partial control functions corresponds to a weighted superposition of line-wise target central points solving the elliptic grid generation equation.

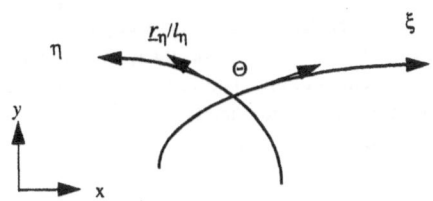

Fig.3: Rotation of unit tangent vectors

For simplicity and brevity we confine our consideration to the two-dimensional case. In an earlier paper [3] line-wise identities had been derived from a rotation of unit vectors tangent to the grid lines (see fig. 3) and differentiation with respect to ξ and η. Here they are rewritten as equations (4) with s_ξ and l_η as the spacings in the respective directions.

$$l_\eta^2 \left[\mathfrak{r}_{\xi\xi} - \left(\frac{s_{\xi\xi}}{s_\xi} + \cot\theta \ \theta_\xi \right) \mathfrak{r}_\xi \right] - \mathfrak{r}_\xi \mathfrak{r}_\eta \mathfrak{r}_{\xi\eta} + s_\xi^2 \left(\frac{s_{\xi\eta}}{s_\xi} + \frac{l_\eta}{s_\xi} \frac{\theta_\xi}{\sin\theta} \right) \mathfrak{r}_\eta = 0,$$

$$s_\xi^2 \left[\mathfrak{r}_{\eta\eta} - \left(\frac{l_{\eta\eta}}{l_\eta} + \cot\theta \ \theta_\eta \right) \mathfrak{r}_\eta \right] - \mathfrak{r}_\xi \mathfrak{r}_\eta \mathfrak{r}_{\xi\eta} + l_\eta^2 \left(\frac{l_{\eta\xi}}{l_\eta} + \frac{s_\xi}{l_\eta} \frac{\theta_\eta}{\sin\theta} \right) \mathfrak{r}_\xi = 0. \tag{4}$$

We cast these relations into the form

$$l_\eta^2 [\mathfrak{r}_{\xi\xi} + P_i \mathfrak{r}_\xi] - \mathfrak{r}_\xi \mathfrak{r}_\eta \mathfrak{r}_{\xi\eta} + s_\xi^2 Q_i \mathfrak{r}_\eta - V\underline{R} = 0,$$

$$s_\xi^2 [\mathfrak{r}_{\eta\eta} + Q_j \mathfrak{r}_\eta] - \mathfrak{r}_\xi \mathfrak{r}_\eta \mathfrak{r}_{\xi\eta} + l_\eta^2 P_j \mathfrak{r}_\xi + V\underline{R} = 0, \tag{5}$$

which become grid generation equations, once we prescribe the control functions

$$P_i = \frac{V}{l_\eta^2} \left(\frac{l_\eta^2}{V}\right)_\xi, \quad P_j = \frac{V}{r_\xi r_\eta} \left(\frac{r_\xi r_\eta}{V}\right)_\eta \quad ; \quad Q_i = \frac{-V}{r_\xi r_\eta} \left(\frac{r_\xi r_\eta}{V}\right)_\xi, \quad Q_j = \frac{V}{s_\xi^2} \left(\frac{s_\xi^2}{V}\right)_\eta, \quad (6)$$

where $V = |r_\xi \times r_\eta| = s_\xi l_\eta \, \sin\theta,$ $\qquad\qquad VR = l_\eta l_{\eta\xi} r_\xi - s_\xi s_{\xi\eta} r_\eta.$

For a central discretization the unknown central point occurs only in the second order line-wise differences. Locally, i.e. assuming the eight points on the cell boundary to be momentarily known, there is a direct correspondence between r_{ci} and the control functions P_i and Q_i as well as between r_{cj} and the control functions P_j and Q_j. The addition of equations (5) corresponds to a linear superposition of the control functions and just yields the two-dimensional elliptic grid generation equation. Solving equations (5) for the central unknown points,

$$l_\eta^2 \, r_{ci} = \frac{1}{2} l_\eta^2 [r_{i-1} + r_{i+1} + P_i r_\xi] + \ldots - \frac{1}{2} \, VR, \qquad (5a)$$

$$s_\xi^2 \, r_{cj} = \frac{1}{2} s_\xi^2 [r_{j-1} - r_{j+1} + Q_j r_\eta] + \ldots + \frac{1}{2} \, VR, \qquad (5b)$$

and adding these relations tells us, how to weigh our line-wise specified target central points by comparing them with the central difference solution to the elliptic grid generation equation:

$$(l_\eta^2 + s_\xi^2) \, r_{i,j} = (l_\eta^2 r_{ci} + s_\xi^2 r_{cj}) = l_\eta^2 [r_{li} + g(r_{ti} - r_{li})] + s_\xi^2 [r_{lj} + g(r_{tj} - r_{lj})]. \qquad (7)$$

Equation (7) is a direct solution to the grid generation equation with correspondingly specified control functions. On the right hand side of (7) we have defined the target central points (index t) line-wise relative to their Laplace positions (index l) for vanishing weight g and vanishing control functions in order to control the grid, where g tends to one; and to smooth it, where g tends to zero. Introducing this definition into equations (5a) and (5b) clearly shows the line-wise correspondence between objective central points and control functions:

$$r_{ti} - r_{li} = 0.5(P_i r_\xi + \frac{s_\xi^2}{l_\eta^2} Q_i r_\eta) , \qquad\qquad r_{tj} - r_{lj} = 0.5 \left(\frac{l_\eta^2}{s_\xi^2} P_j r_\xi + Q_j r_\eta\right).$$

Relation (7) is readily extended to three dimensions. At present the line-wise target central points are updated for each iteration in a Jacobi-like fashion. In contrast to the earlier procedure [2], where the solution to the grid generation equation was updated in an inner iteration loop embedded in an outer one to update the control functions, here we have only one iterative loop. The resulting position vector is underrelaxed by a factor in the order of 0.8.

Examples

The new procedure was first applied to retrieve almost the same local improvements done to an algebraic C-O-grid about a single wing as they are shown in Reference [1]. After integration into MegaCads [4], the structured grid generation system of the MEGAFLOW project [5], the method has been used to improve block-wise algebraic grids in two and three dimensions. Figure 4a shows an algebraic H-type grid for the upper half of an airfoil geometry. The result of the elliptically controlled grid with Dirichlet boundary conditions is visible in Figure 4b. An orthogonalisation of the grid lines towards the boundaries is obtained as desired. This can be observed in detail for the nose region of the airfoil in Figure 4c and 4d. The control slowly vanishes after a certain number of grid lines off the boundaries, specified by the user.

A three-dimensional application of the method is presented in Figure 5a-c. A block of the multiblock Navier-Stokes grid for the DLR-F6 configuration [5] has been chosen to show the possibility of controlling the grid not only off fixed boundaries but also relative to specified grid planes in the interior floating with the resulting solution. The block contains two clusterings of grid cells in spanwise direction, as a result of the resolution of the pylon/nacelle boundary layers, and the wing tip. In streamwise direction a higher density of grid lines can be found near the nacelle. In Figure 5b, a grid plane close to the nacelle is presented. The algebraic grid has non-orthogonal grid lines at the boundaries and in the interior (Figure 5b, left). Applying grid control not only off the boundaries but also relative to the mid-face in the clustered region the distribution of grid lines can be improved significantly (Figure 5b, right). This can also be observed for a grid plane close to the inflow boundary of the block (Figure 5c).

The algorithm is implemented into MegaCads in a way that it can also be applied to grid blocks with boundary faces that are mapped onto themselves. An example is a grid block with C-O-topology for a wing tip. Figure 6a explains the grid topology. In the i- and k-direction the faces collapse onto themselves. In this case the objective was to orthogonalize the first 15 grid lines of the wing tip boundary-layer grid and only a few cells at the other boundaries, including the imax- and kmax-face, to achieve smooth interfaces to neighbouring blocks. Figure 6b displays resulting grid faces. In figure 6c the kmin outflow face is presented and the normal grid lines through the boundary-layer are visible for this O-type grid.

Conclusions

Freed from the control functions and based on an automatic specification of target points only a rather simple and flexible method has been developed to allow for a local control of elliptic grid generation anywhere in the field though not everywhere simultaneously. Joining regions of grid control to those of grid smoothness we did not encounter any problems with grid overlap, though there is no guarantee for a mapping free of folding in the controlled areas. Integrated into MegaCads the method can easily be used for a wide range of applications.

References

[1] Thompson J.F.: *A general Three-Dimensional Elliptic Grid Generation System on a Composite Grid Structure.* Computer Methods in Applied Mechanics and Engineering 64, 1987, pp. 377.

[2] Niederdrenk P. : *On the Control of Elliptic Grid Generation.* Proc. of the 6th Intern. Conf. on Numerical Grid Generation in Comp. Field Simulations, eds. M. Cross et al., NSF Eng. Center, Mississippi, 1998, pp. 257.

[3] Niederdrenk P. : *Solution Adaptive Grid Generation by Hyperbolic / Parabolized Hyperbolic P.D.E.s.* in Numerical Grid Generation in Comp. Fluid Dynamics and Related Fields, eds. A.S. Arcilla et al., North-Holland, Amsterdam, 1991, pp. 173.

[4] Brodersen O., Ronzheimer A., Ziegler R., Kunert T., Wild J., Hepperle M.: *Aerodynamic Applications Using Megacads.* Proc. of the 6th Intern. Conf. On Numerical Grid Generation.
in Comp. Field Simulations, eds. M. Cross et al., NSF Eng. Center, Mississippi, 1998, pp. 793

[5] Kroll N., Rossow C.-C., Becker K., Thiele F. : *MEGAFLOW - A Numerical Flow Simulation System.* 21st ICAS Congress, Melbourne, 1998.

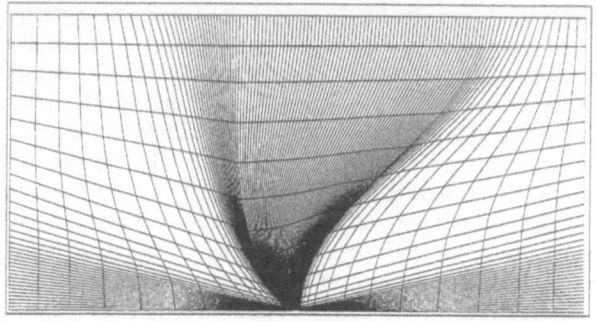

Fig. 4a: Algebraic grid for the upper side of an airfoil; every 2nd grid line

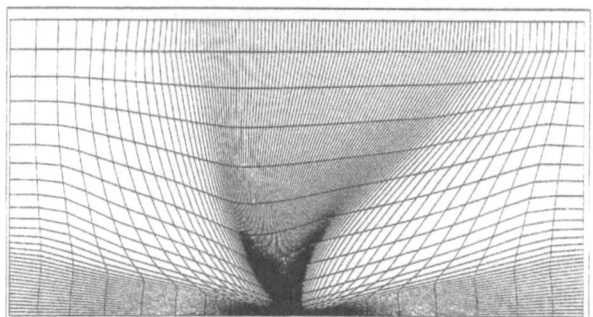

Fig. 4b: Elliptically controlled grid for the upper side of an airfoil; every 2nd grid line

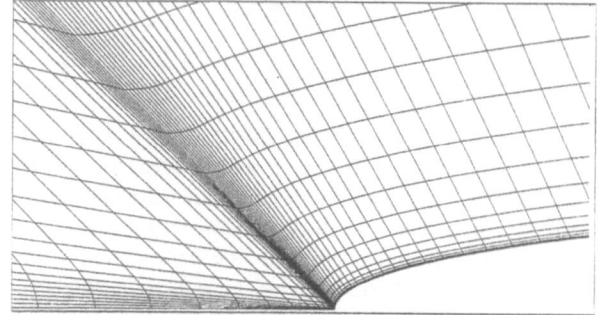

Fig. 4c: Algebraic grid for the upper side of an airfoil nose region; every 2nd grid line

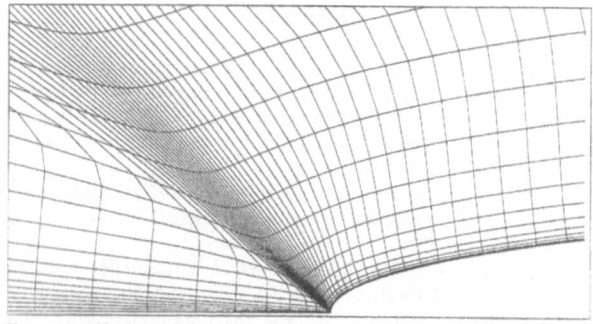

Fig. 4d: Elliptically controlled grid for the airfoil nose region of fig.4c; every 2nd grid line

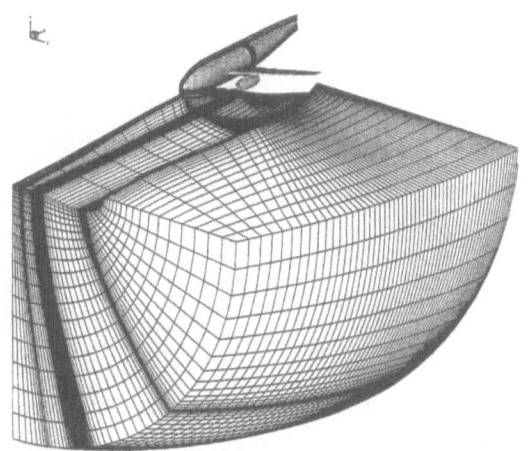

Fig. 5a: Algebraic grid block for DLR-F6, grid clustering in two directions

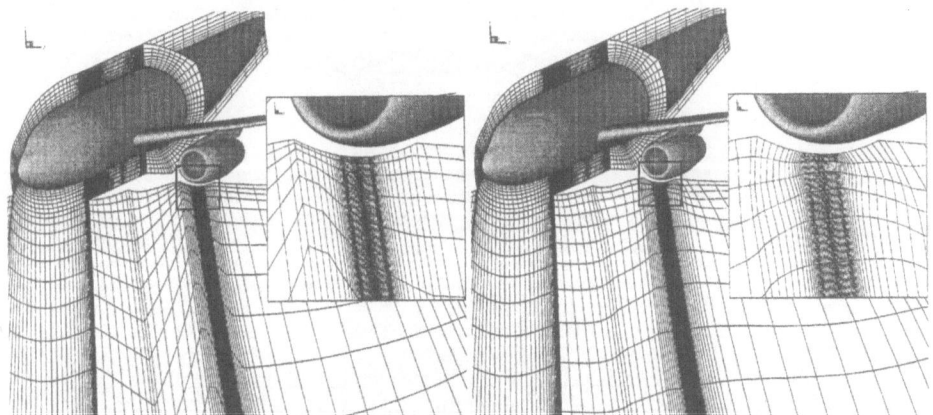

Fig. 5b: Grid planes at the nacelle region, algebraic (left), elliptic (right)

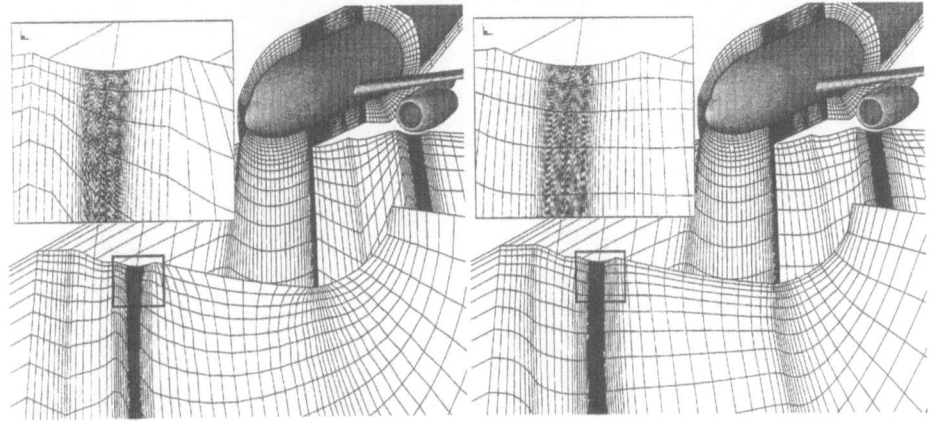

Fig. 5c: Grid planes at the far field inflow region, algebraic (left), elliptic (right)

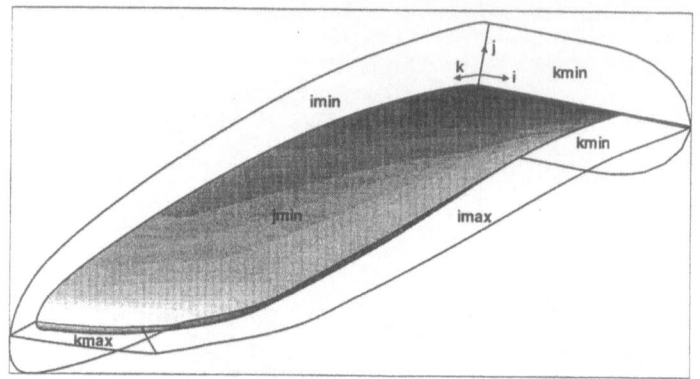

Fig. 6a: C-O grid topology for a wing tip, imax and kmax faces folded onto themselves

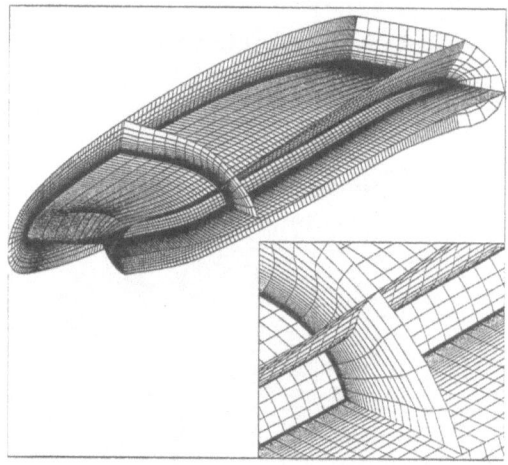

Fig. 6b: C-O grid topology for a wing tip, elliptically controlled grid

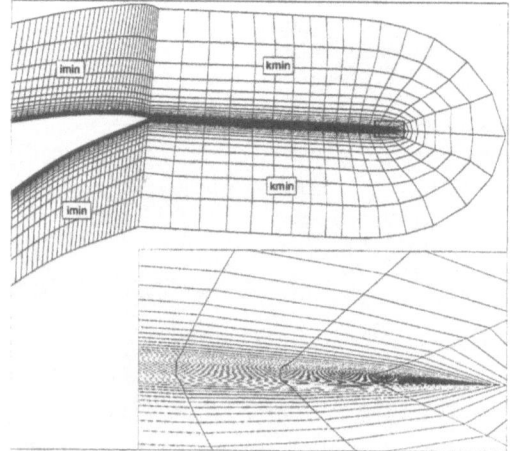

Fig. 6c: C-O grid topology for a wing tip, k-outflow-face, elliptically controlled grid

346

CFD-Methods for the Design Process of High-Lift-Configurations

D. Reckzeh

DaimlerChrysler Aerospace Airbus GmbH
EFP
Hünefeldstr. 1-5, 28199 Bremen, Germany

Deutsches Zentrum für Luft- und Raumfahrt e.V.
Institut für Entwurfsaerodynamik
Lilienthalplatz 7, 38108 Braunschweig, Germany

Summary

The design process of high-lift systems for transport aircraft configurations is not only based on requirements of field performance and the shape of the cruise-wing, but also on structure and system-technology [1]. Due to the complexity of the flow around a 3D high-lift wing its aerodynamic design depends mainly on wind-tunnel experiments, although CFD-methods are starting to play an important role. Different CFD-methods are used at DaimlerChrysler Aerospace Airbus in the high-lift-system design according to the complexity of the stage of development. The shaping of the multielement profiles is based on 2D-methods. For comparison of the performance of different designs, the characteristics of the high-lift wings are predicted by using a quasi-3D-method. For chosen designs calculations with a full 3D-method are performed to evaluate the effects on the complex high-lift-configuration with fuselage, engine/pylon, etc.

Introduction

For the aerodynamic development of the high-lift-systems of the recent Airbus types up to A340 (FIG. 1) CFD played only a minor role. Due to the complexity of the flow around a high-lift-wing it was only possible to check the achieved performance in wind tunnel experiment. CFD was used for finding the right shape of the spanwise stations by 2D calculations. It was not possible to include calculations on 3D high-lift wings in the timetable of the design process.

In recent design work on the Megaliner the necessity to evaluate a large number of different designs in a short time leads to an increased importance of CFD methods for high-lift development. New methods were introduced and now they are beginning to establish their place next to the wind-tunnel-experiments.

In the German research program HAK [2] („high-lift-concepts") new technologies are developed. A goal is to introduce high-lift systems for a Megaliner-configuration (FIG. 2) which show better performance and less complexity. But also the development of adequate CFD-methods is a field of activity.

Requirements for effective CFD in aerodynamic design

For aerodynamic design of high-lift systems CFD-methods which can be included in the working process and especially in the given time frame are necessary. On one hand it is important that validated methods produce reliable results. On the other hand they must distinguish themselves through acceptable expense, which means moderate calculation time and geometry modelling which can be executed by the design engineer and not only by a CFD-specialist. For this reason today's time intensive Navier-Stokes and Euler methods are suitable only for long-term studies. For design work 2D/3D panel methods and lifting surface methods are used. The work with these tools consists of developing a shape of the high-lift-wing which respects the given requirements and calculating its performance, first in 2D for the spanwise profile cuts, then in 3D for the total wing. Inverse methods are not used, mainly due to the large number of constraints by the general aircraft layout which have to be observed.

CFD-methods used in high-lift-design

At DaimlerChrysler Aerospace Airbus during the work done on Airbus high-lift-systems several CFD-methods were developed or introduced.

Currently the different geometry tools in use are brought together in the program system *AEROTOOL*. With this environment it is possible to select the planform and type of the high-lift devices on the wing, to design the shape of multielement profiles with regard to the given constraints and to calculate the performance of the high-lift profiles, in order to redesign as necessary.

The method used for calculation of the 2D-flow over a multielement airfoil (named *HILI* [3]) can cope with up to three elements or an aileron at the trailing edge and a slat or a Krüger-Flap at the leading edge. It is a panel method with coupling of an integral boundary layer method. Reynolds-number dependent viscous effects and separation on leading- or trailing-edge can be predicted. In the input data sheet the setting of the elements (angle, gap and overlap) can be easily varied. The calculated pressure distribution and the force coefficients are used to evaluate the shape and the setting of the elements on spanwise stations of the high-lift-wing.

When a 3D-planform is chosen the prediction of the performance of the wing is possible with the quasi-3D-method *Q3D* [4]. In this tool the Truckenbrodt lifting surface method is coupled with the calculated 2D-performance of the multielement profiles at discrete spanwise stations. The 2D-calculations are performed with the method *HILI*. Lift slope, spanwise lift distribution, maximum lift, pitching moment and drag can be compared for different configurations. With these calculations for example the effect of the spanwise extension of the high-lift system or the performance of a slat-configuration in comparison to a Krüger-configuration can be evaluated.

For the examination of flow characteristics around a complete aircraft in high-lift-configuration the method *VSAERO* [5] is used. *VSAERO* is a 3D-panel-method with the option to calculate viscous effects. In principal there is no limitation to the complexity of the geometry. An example for the grid-model of a high-lift-configuration is given in FIG. 3. Further it is possible to perform boundary-layer calculations along on-body-streamlines to predict viscous effects. Not only separation on the undisturbed wing, but also effects of nacelle and pylon or fairings on local separation are predicted in good agreement with experiments. Reynolds-number effects are also well predicted. Additionally a post-processing method for predicting separation based on analysis of the shape of local pressure distribution was developed [6].

Up until now modelling high-lift-geometry with setting-variation of the elements was so work-intensive that *VSAERO* was not used in the design process. As an additional problem, the quality of the solution of a panel method depends on careful modelling especially in the regions of the gaps between the elements, discontinuous leading- and trailing edges, slat- and flap ends and the wing root. This means that the available CAD-data normally is not suitable for a panel-model without redistributing the gridpoints.

Due to these requirements a new procedure for preparing the input data was developed [6]. The input data routine is based on the automated geometry routines from the quasi-3D-method *Q3D*. This makes it possible to avoid the time-intensive use of a CAD-system for every new position of the elements. Further the quasi-3D-calculation can generate a starting solution for the wakes of the elements. Various tests showed that a convergent solution of calculations with wake iterations can only be obtained with a realistic starting solution. The quasi-3D starting solution of the wake has already such high quality that, if the wake-vortex field behind the configuration is of no interest, the wake iterations can be omitted. This leads to a large decrease of calculation time. Additionally a generic separation bubble model can be included in the cove on the lower side of the slat and the main-element. This leads to a more realistic smooth pressure distribution in these regions.

Design work on the H8Y research-configuration

The Megaliner-configuration H8Y was developed as research model in the technology program HAK. This configuration was built as full-model for testing in the DNW (FIG. 2). A large variety of advanced high-lift systems can be tested, such as slat or Krüger-Flap as leading edge devices and different flaps, for example extended to the wing tip with unslotted and slotted tab devices on the flap end. Additionally a simplified configuration was tested to provide a data basis for validation of the CFD-tools in use. This data led to improvements in the quality of the theoretical predictions.

CFD was also used for design and prediction of performance. In FIG. 4 examples for 2D-calculations on spanwise cuts of the H8Y wing are shown. In the pressure distributions the effect of an increased slat angle is visible. While the suction peak on the slat decreases, the loading of the main element increases. Results of varying angle, gap and overlap helped to choose an initial setting of the elements for the wind-tunnel-test. But also the effects of such setting-variations on the performance of the 3D-wing can be calculated. FIG. 5 shows the influences of varying slat angles on the maximum lift on the wing. The tests in DNW generally confirmed these results.

It became clear that the effects of engine/pylon installation on the wing cause local effects which can not be neglected and may be limiting for maximum lift. The strongly increased local angle of attack on the inboard side of the nacelles, the influence of the pylon/wing junction and the break in the leading-edge device often lead to a beginning of separation in these regions. Such 3D-effects can be predicted with *VSAERO*. FIG. 6 shows *VSAERO*-results for the H8Y-configuration with the „ideal" high-lift-system. In FIG. 7 the influence of the flow around the nacelles on the pressure distribution of the leading-edge is displayed. The free streamlines emphasise the acceleration of the flow around the nacelle and the high local angle of attack on the wing leading-edge.

Evaluation of a canard-wing in high-lift conditions

The use of a canard wing combined with the horizontal tailplane can probably be an alternative for a Megaliner-configuration to keep the required control effectiveness without a heavy tail unit. A canard-design for a Megaliner-configuration has been developed by DLR [7].

This canard design was transferred to the H8Y-configuration to check the performance under high-lift-conditions [8]. A *VSAERO*-model for H8Y with canard and high-lift wing was built and calculations were performed for various flow conditions. The results include pressure and velocity distributions on the body and over the canard surface, streamline calculations and viscous calculations on canard, wing and fuselage.

In FIG. 8 the calculated pressure distributions for the configuration with and without canard are presented. The downwash effect of the canard on the inboard-region of the main-wing is visible. Loading and suction peaks are reduced compared to the configuration without canard. In FIG. 9 off-body streamlines are plotted. FIG. 10 shows that a stronger velocity component in spanwise direction only occurs in direct proximity to the fuselage. Most of the canard-span is in a flow nearly parallel to the undisturbed onset flow.

Important results come from the calculation of the viscous effects on the canard. In FIG. 11 the distribution of the friction coefficient is shown for the angle of attack when heavy, lift limiting separation on the canard is indicated. In the figures the iterated wake of the canard is also shown. For a typical wind-tunnel Reynolds-number (based on reference chord of the wing) of Re=3 Mio. the flow separates with a laminar separation on the leading-edge already at $\alpha=4°$. For assumed flight-conditions at Re=40 Mio. the separation does not occur before $\alpha=13°$. A turbulent separation is indicated, but it is still located very near the leading-edge.

The results predict that wind-tunnel testing of the canard, in order to check the performance in acceptable agreement to flight conditions, due to the strong Reynolds number effects is only possible at sufficiently high Reynolds numbers.

The presented calculations for the H8Y-canard-configuration give a good example for the benefits of the use of CFD in high-lift design. With a sufficiently validated method it is already possible in the stadium of modelling the wing to check the expected performance. For testing at wind-tunnel conditions possible Reynolds-number effects can be predicted. Further, the measured performance in wind-tunnel facilities at low Reynolds-numbers can be scaled to flight-conditions.

Conclusions

The use of CFD in high-lift design can bring a great benefit in quality, cost and time. But it is clear that the advanced Navier-Stokes-methods of today are not usable for this task because of their complexity in handling and the problems with adequate grid generation for a 3D-wing in high-lift configuration. Therefore robust and well known tools such as 2D panel methods are intensively used, and with smart tools like the quasi-3D-method fast statements on wing-performance can be made early in the design process, so that it is possible to compare several different designs against each other. With more work-intensive tools this would not be possible for the design engineer who normally is only in second order a CFD specialist. The examination of the 3D high-lift configuration is possible with a 3D panel method like *VSAERO* and becomes more and more important since modelling is possible in an acceptable time. To obtain reliable results for a large variety of configurations for different flow conditions still intensive validation work has to be done. Further development will lead to more exact prediction of the separating flow on complex configurations and its effect on the performance of the aircraft.

References

[1] Hilbig R., Flaig A., *High-Lift Design for Large Civil Aircraft*, AGARD-CP-515, 1992.

[2] Hansen H., *Überblick über das Technologieprogramm Hochauftriebskonzepte (HAK)*, DGLR Jahrestagung 1998, DGLR-JT98-042.

[3] Dargel G., Jakob. H., *Berechnungsverfahren für viskose Klappenprofilströmungen-Rechenprogramm HILI*, DASA Airbus-Bericht TE 2-1683.

[4] Jakob H., *Zur rechnerischen Vorhersage des Maximalauftriebs an Flügel-Rumpf-Konfigurationen*, DASA Airbus-Bericht EF-19 93.

[5] *VSAERO User Manual*, Analytical Methods Inc., Redmont, Washington.

[6] Reckzeh D., *Abschlußbericht HAK II-Unterauftrag: Rechenverfahren in der Hochauftriebsaerodynamik*, DASA Airbus-Bericht EF-08/98.

[7] Wichmann G., Rohardt C.-H., *Numerical Investigation of the Flow around a Fuselage-Canard Configuration for a Long Range High Capacity Aircraft* (10th AG-STAB-DGLR Symposium), New Results in Numerical and Experimental Fluid Mechanics, pp.389, Vieweg Verlag, NNFM60, 1996.

[8] Repmann C., *Einsatz von Rechenverfahren und Analyse von Windkanalmessungen für den Entwurf von Hochauftriebssystemen für Verkehrsflugzeuge*, Diplomarbeit TU Braunschweig 1998.

FIG. 1: Airbus A340 in High-Lift Configuration

FIG. 2: Megaliner High-Lift Model in DNW

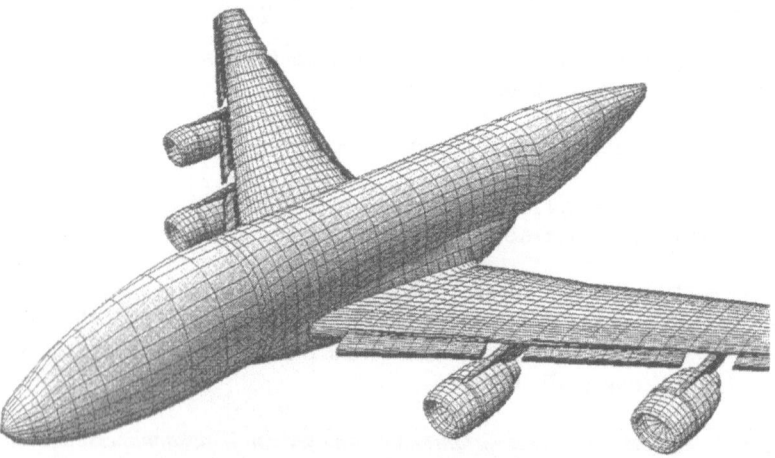

FIG. 3: Megaliner High-Lift CFD-Model

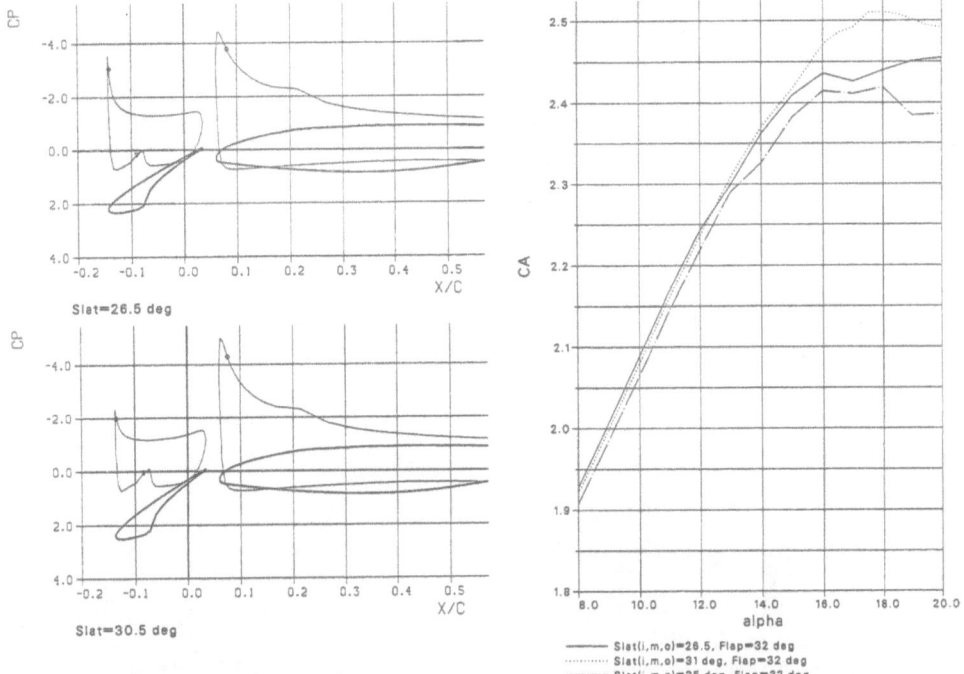

FIG. 4: 2D calculations (*HILI4*) for optimization of H8Y slat angle

FIG. 5: Quasi-3D calculation (*Q3D*) of H8Y wing performance with slat angle variation

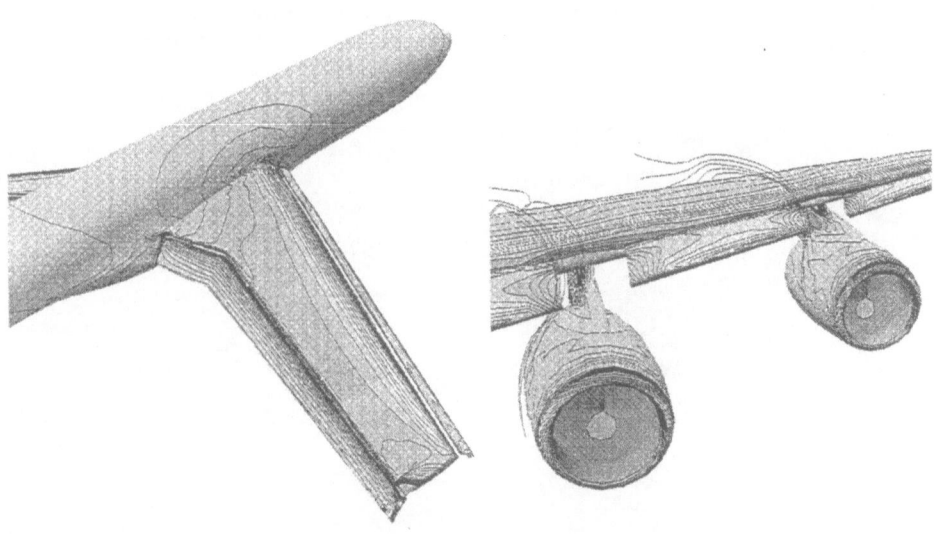

FIG. 6: 3D calculation *(VSAERO)* of pressure distribution on the H8Y high-lift configuration

FIG. 7: 3D calculation *(VSAERO)* of pressure distribution with engine installation effects

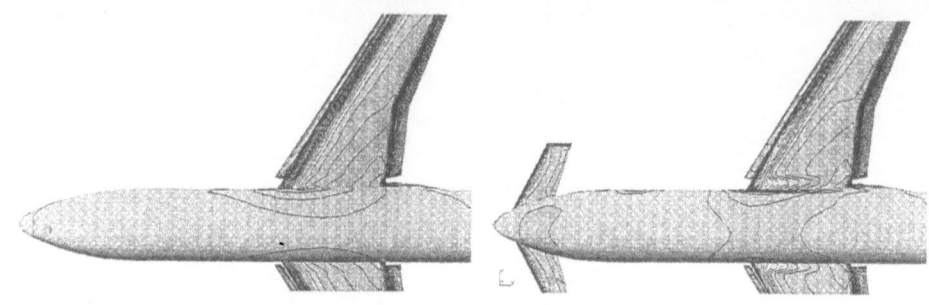

FIG. 8: Pressure distribution on H8Y wing without and with canard

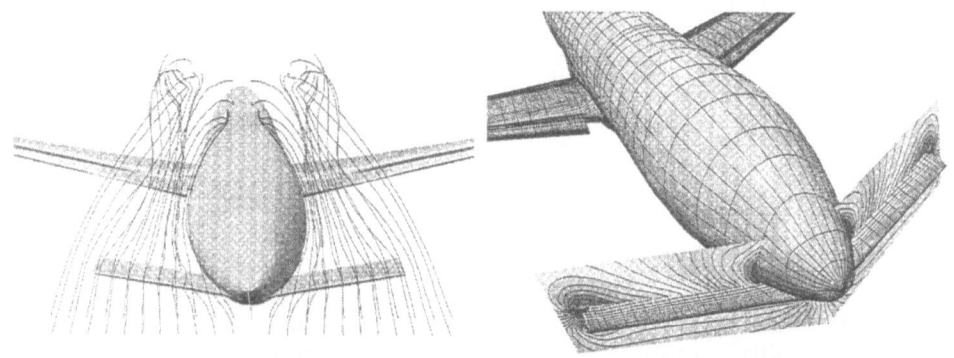

FIG. 9: Off-body streamlines around H8Y canard

FIG. 10: Velocity distribution in y-direction around H8Y canard

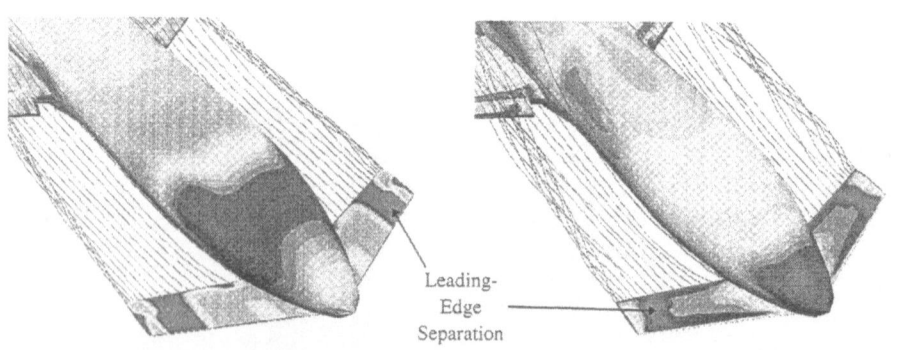

Leading-Edge Separation

Re=3 Mio., α=4°   Re=40 Mio., α=13°

FIG. 11: Distribution of friction coefficient on H8Y Canard for different Reynolds-Numbers
(darkest shading ⟹ C_f=0)

A Simple Flux Splitting Scheme for Compressible Flows

C.-C. Rossow

DLR, Institut für Entwurfsaerodynamik, Lilienthalplatz 7, D-38022 Braunschweig, Germany

Summary

A flux splitting scheme suitable for the computation of compressible flows is developed. In order to establish a robust and accurate, however simple method for the discretization of the convective fluxes, elements of the well known LDFSS- and HCUSP-flux splitting schemes are used. The resulting formulation is based on terms with regard to the Mach number, it almost completely avoids the computation of an intermediate state at a cell face, and it shows no violation of the entropy condition. Application to 2D transonic and hypersonic, inviscid and viscous flows shows that despite its simplicity, the scheme rivals other most advanced schemes commonly in use, with respect to both accuracy and robustness.

1. Introduction

The requirements on discretization methods suitable for the simulation of technically relevant flows cover a very broad spectrum, and are sometimes even contradicting: In order to obtain accurate solutions, only a minimum amount of numerical viscosity may be permitted, especially for the resolution of boundary layers in viscous flows. For flows with strong shock waves, however, robustness becomes of primary importance, especially near vacuum conditions, where the prediction of negative values for positive quantities like pressure and density is likely to occur. In such situations, the proper amount of numerical viscosity needs to be supplied to achieve a converged solution. Accuracy is said to be best established by flux-differencing schemes according to Roe [1], however flux-differencing lacks robustness at strong shock waves. Flux-vector splitting schemes following van Leer [2] on the other hand are known to be extremly robust, but are inadequate to resolve shear layers in viscous flow. Therefore, the development of appropriate discretization schemes recently focusses on the construction of hybrid flux-splitting formulations, where the accuracy of flux-difference splitting is to be combined with the robustness of flux-vector splitting. The most popular of the current upwind techniques is the Advection Upstream Splitting Method (AUSM) of Liou and Steffen [3]. In this approach, the inviscid flux at a cell interface is split into a advective contribution, which is upwinded in the direction of the flow, and a pressure contribution, which is upwinded based on acoustic considerations. Edwards [4] extended this formulation to the LDFSS discretization. With similar considerations as for AUSM, Jameson [5] introduced a Convective Upwind Split Pressure (CUSP) scheme. The ability of the CUSP-scheme to accurately and efficiently compute inviscid and viscous flows has been demonstrated by Swanson et al. in [6].

In the present work, an alternative formulation for an upwind scheme is developed. In the scheme proposed, elements of the CUSP- and of the LDFSS-scheme are used to construct a simple but robust and accurate method. A key element of the new scheme is the almost complete avoidance of the computation of an intermediate state at a cell interface. Instead, the left and right Mach numbers are used for the treatment of the interface flux.

2. Governing Equations

We consider the two-dimensional Navier-Stokes Equations for compressible flow. The system of partial differential equations in strong conservation form is given by

$$\frac{\partial \vec{W}}{\partial t} + \frac{\partial \vec{F}}{\partial x} + \frac{\partial \vec{G}}{\partial y} = 0 \quad , \tag{1}$$

where

$$\vec{W} = [\rho, \rho u, \rho v, \rho E]^T$$

represents the vector of conservative variables. Here ρ, u, v, and E denote density, Cartesian velocity components, and total energy per unit mass, respectively. The flux-density vectors for the x- and y-direction, $\vec{F}$ and $\vec{G}$, may be split in convective and viscous parts according to

$$\vec{F} = \vec{F}^c - \vec{F}^v; \quad \vec{G} = \vec{G}^c - \vec{G}^v. \tag{2}$$

Setting the viscous parts of the flux-density tensor, $\vec{F}^v$ and $\vec{G}^v$, to zero, the Euler equations governing inviscid flow are obtained. For the viscous terms, central discretization leads to stable numerical approximations. As it is well known, this is not the case for the convective terms of the equations. Therefore, these are the terms on which attention is focussed when discretising the governing equations.

In the construction of hybrid flux-splitting upwind schemes, it was recognized that the convective parts of the flux-density vectors may be further subdivided into contributions related to advection and pressure [3, 5]:

$$\vec{F}^c = \vec{F}^\phi + \vec{F}^P \quad ; \quad \vec{G}^c = \vec{G}^\phi + \vec{G}^P, \tag{3}$$

where the contributions due to pressure are defined as

$$\vec{F}^P = [0, p, 0, 0]^T \quad ; \quad \vec{G}^P = [0, 0, p, 0]^T, \tag{4}$$

and the contributions related to advection are given by

$$\vec{F}^\phi = u\vec{\phi} \quad ; \quad \vec{G}^\phi = v\vec{\phi}, \tag{5}$$

with the vector of advected quantities defined by

$$\vec{\phi} = [\rho, \rho u, \rho v, \rho H]^T. \tag{6}$$

3. Hybrid Flux-Splitting Scheme for Compressible Flow

In the following, the discretization of the convective fluxes with the hybrid flux-splitting scheme will be outlined. The discretization scheme uses the splitting of the convective fluxes in advective and pressure terms as given by eqs. (3) – (6). The derivation of the new scheme will be given purely from an engineering point of view.

3.1 Flux-Splitting, advective terms

The derivation starts with the treatment of the advective part of the convective fluxes given in eqs. (5) and (6). Consider the central discretization of any component of the advective part of the flux-density vector normal to a cell interface:

$$F^\phi_{cen} = \frac{1}{2}\left(q_n^L + q_n^R\right) \cdot \frac{1}{2}\left(\phi^L + \phi^R\right), \tag{7}$$

where L and R denote the states on the left and right side of a cell interface, respectively, q_n denotes the normal component of velocity through the interface, and ϕ is an arbitrary component of the vector of advected quantities $\vec{\phi}$. Eq. (7) may be converted to an upwind formulation with regard to the term $\left(\phi^L + \phi^R\right)$ by

$$F^\phi_{up} = \frac{1}{2}\left(q_n^L + q_n^R\right) \cdot \frac{1}{2}\left(\phi^L + \phi^R\right) - \frac{1}{2}\left(\left|q_n^L + q_n^R\right|\right) \cdot \frac{1}{2}\left(\phi^R - \phi^L\right). \tag{8}$$

The remaining problem is to establish an upwind formulation for the normal interface velocity q_n to substitute the arithmetic average in eq. (8). In the present scheme, this is established by defining:

$$q_n = \frac{1}{2}\left(q_n^L + q_n^R\right) - \frac{1}{2}\beta^M \cdot c^{av} \cdot \left[M^R \cdot sign\left(M^R\right) - M^L \cdot sign\left(M^L\right)\right] \tag{9}$$

where

$$\beta^M = \max\left(0, \ 2 \cdot M^{max1} - 1\right), \quad M^{max1} = \min\left[\max\left(\left|M^L\right|, \left|M^R\right|\right), 1\right]. \tag{10}$$

356

The average speed of sound c^{av} is simply computed by the arithmetic average of the left and right state:

$$c^{av} = \frac{1}{2}\left(c^L + c^R\right),$$ (11)

and the left and right Mach numbers M^L and M^R are computed using q_n^L and q_n^R, normalized by c^{av}. Note that this is analogous to the LDFSS discretization [4].

The blending function β^M corresponds directly to the choice of Jameson in the CUSP scheme [5], where the blending function was based on the acoustic eigenvalues. The difference to the formulation in [5] is the use of the maximum Mach number M^{max1} defined by eq. (10). The upwinding of q_n is regarded as a purely advection-governed process. Therefore, whenever the normal velocity exceeds the speed of sound, this should be reflected in the discrete scheme by fully upwinding the advection speed.

Using q_n as defined by eq. (9) and substituting this into eq. (8), the following upwind formulation for the advective part of the convective flux density tensor may be obtained:

$$
\begin{aligned}
F^\phi = &\frac{1}{2}\left(q_n^L + q_n^R\right) \cdot \frac{1}{2}\left(\phi^L + \phi^R\right) && \text{(central flux)}\\
&-\frac{1}{2}(\phi^L + \phi^R) \cdot \frac{1}{2}\beta^M \cdot c^{av} \cdot \left[M^R \cdot sign(M^R) - M^L \cdot sign(M^L)\right] && \text{(upwinding normal velocity)}\\
&-\frac{1}{2} \cdot c^{av} \cdot max\left(\left|M^L\right|, \left|M^R\right|\right) \cdot \left(\phi^R - \phi^L\right) && \text{(dissipative flux)}.
\end{aligned}
$$ (12)

Note that in eq. (12) the average $0.5\left|q_n^L + q_n^R\right|$ is replaced by the maximum of M^L and M^R, multiplied by c^{av}. Therefore, the dissipative flux $\left(\phi^R - \phi^L\right)$ may be scaled by the "downwind" Mach number in case the flow is accelerated. However, viewing the scheme as a central plus dissipative flux, using the maximum parameter for scaling the dissipation should always provide more than sufficient numerical viscosity to establish a stable scheme. For decelerating flow, where shocks may occur, the formulation of eq. (12) is leading to a perfect upwinding.

3.2 Flux splitting, pressure terms

Following the procedure for upwinding the normal velocity at a cell interface eq. (9), the upwinding of pressure is achieved by

$$F^P = \frac{1}{2}\left(p^L + p^R\right) - \frac{1}{2} \cdot \beta^P \cdot \left[p^R sign(M^R) - p^L \cdot sign(M^L)\right],$$ (13)

where

$$\beta^P = max\left(0, \ 2 \cdot M^{min1} - 1\right), \quad M^{min1} = min\left[min\left(\left|M^L\right|, \left|M^R\right|\right), 1\right].$$ (14)

However, in contrast to eq. (9), here the minimum Mach number at a cell interface is used. The reason for this is that adding pressure differences to the momentum equations leads to a sharpening of the shock-profile. Increasing the scaling provokes over- and undershoots in pressure. Omitting these terms on the other hand leads to a broad smearing of the shock. A reasonable scaling is thus provided by the minimum Mach number.

3.3 Flux-splitting, complete operator

Assembling all terms for the advective and pressure contributions, the complete normal convective flux-density vector $\vec{F}_n^c$ at a cell interface reads:

$$\vec{F}_n^c = \vec{F}^\phi + \vec{F}^P,$$ (15)

where

$$\vec{F}^{\bullet} = \frac{1}{2}\left(q_n^L + q_n^R\right)\cdot\frac{1}{2}\left(\vec{\phi}^L + \vec{\phi}^R\right)$$

$$- \frac{1}{2}\left(\vec{\phi}^L + \vec{\phi}^R\right)\cdot\frac{1}{2}\beta^M \cdot c^{av} \cdot\left[M^R \cdot sign\left(M^R\right) - M^L \cdot sign\left(M^L\right)\right]$$

$$- \frac{1}{2}\cdot\max\left(\left|M^L\right|,\left|M^R\right|\right)\cdot c^{av}\cdot\left(\vec{\phi}^R - \vec{\phi}^L\right); \tag{16}$$

$$\vec{F}^P = \frac{1}{2}\left(\vec{p}^L + \vec{p}^R\right) - \frac{1}{2}\cdot\beta^P\cdot\left[\vec{p}^R\cdot sign\left(M^R\right) - \vec{p}^L\cdot sign\left(M^L\right)\right];$$

$$\vec{\phi} = \begin{bmatrix} \rho \\ \rho u \\ \rho v \\ \rho H \end{bmatrix} \quad ; \quad \vec{p} = \begin{bmatrix} 0 \\ p\cdot n_x \\ p\cdot n_y \\ 0 \end{bmatrix} \quad ; \quad \vec{n} = \begin{bmatrix} n_x \\ n_y \end{bmatrix},$$

and $\vec{n}$ denotes the normal vector at a cell interface with components n_x and n_y. Since the basic scheme splits the advective and pressure terms based on the Mach numbers of the left and right state at a cell interface, in the following the scheme will be referred to as Mach number-based Advection Pressure Splitting (MAPS) scheme.

4. Results

First, the MAP-Splitting scheme is investigated for one-dimensional inviscid flow. In order to clearly identify the capabilities and limits of the scheme, only a first order implementation is investigated. For comparison, the AUSM-scheme of [3], the flux-difference splitting following [1], and the HCUSP-scheme of [5] are evaluated. The first test case is a supersonic flow with an inflow Mach number of $M_\infty = 32$ in order to simulate a strong discontinuity. As can be seen in Fig. 1, the MAPS-scheme captures the shock with one interior point, with a slight overshoot. AUSM exhibits the well know pressure oscillations, whereas the Roe- and the HCUSP scheme capture the shock without over- or undershoots with only one interior point. Note that no entropy-fix is used in the Roe-scheme.

The second test case is the flow in a Laval nozzle. The results computed for this case are displayed in Fig. 2. For the MAPS-scheme the transition from subsonic to supersonic flow in the nozzle is smooth, and the shock is captured with one interior point. AUSM needs one point more over the shock-profile, and shows a slight glitch at the sonic transition. The Roe- and the HCUSP-schemes however are not able to handle the sonic transition. As noted earlier, in the Roe-scheme no entropy-fix is employed, so this behaviour might be expected. For the HCUSP-scheme, the necessity for an entropy-fix has yet not been indicated in the literature. However, in [5] it was mentioned that the eigenvalues employed for the HCUSP scheme have the same sign as the acoustic eigenvalues of the Roe-scheme, and change sign at the sonic line similar to the Roe-scheme. That implies that also for the HCUSP-scheme one eigenvalue vanishes at the sonic line, and a violation of the entropy condition as shown in Fig. 2 is likely to occur. Note that the difficulties of flux-vector and flux-difference splitting schemes at the sonic point were already observed in [7] by Moschetta and Gressier. In [7] the standard schemes are modified based on kinetic theory. The present scheme does not to suffer from these deficiencies.

For two-dimensional flows, the basic solution scheme is a cell centered, finite volume scheme, where the time integration is performed using a 5-stage Runge-Kutta scheme. The MAPS-scheme is implemented in the pattern of a central discretization plus artificial dissipation, and the artificial dissipative terms are evaluated at every odd stage of the time-marching scheme. For second order accuracy, the components of the advection vector $\vec{\phi}$ are reconstructed. In order to control the reconstruction, the SLIP limiter of [5] is used following the implementation given in [6]. In order to accelerate convergence towards the steady state, local time-stepping, implicit residual averaging, and Multigrid are used. The influence of turbulence is modelled according to Baldwin and Lomax [8].

The first two-dimensional test case computed is the inviscid transonic flow around the NACA 0012 airfoil. The onflow conditions are chosen to the standard values of $M_\infty = 0.8, \alpha = 1.25°$, and an O-mesh with 160 cells around the airfoil and 32 cells in normal direction is employed. Fig. 3 shows the convergence history and the computed pressure distribution. The pressure distribution shows that both the strong upper shock and the weak lower shock are well captured. Note that no smearing of the lower shock occurs, but it is well captured with one interior point.

The next case is the viscous flow around the RAE2822 airfoil at freestream conditions corresponding to Case 9 of [9]: $M_\infty = 0.73, \alpha = 2.79°, Re = 6,500,000$. The computational mesh is a C mesh where 256 cells are on the airfoil surface, 32 cells are in the wake region, and 64 cells are emanating normally from the airfoil surface to the outer farfield boundary. Fig. 4 displays the convergence history and the computed pressure distribution. In order to assess the accuracy of the MAPS-scheme, Fig. 5 shows a comparison of lift and drag values obtained with MAPS and a standard scheme with scalar dissipation [10]. Regarding the computed lift and drag values as functions of the inverse of the number of the mesh cells, as shown in Fig. 5, the MAPS scheme clearly demonstrates second order accuracy. The improved accuracy of the MAPS scheme can be estimated by the fact that it provides the values of the scheme with scalar dissipation already on a corresponding coarser mesh.

The last test case computed is the hypersonic 2-D viscous flow over a blunt wedge. For this computation, the node centered code of Radespiel et al. [11] was used, where the HCUSP formulation presented in [6] was substituted by the MAPS-scheme. The onflow conditions are $M_\infty = 10, \alpha = 0°, Re = 10,000$ based on the diameter of the wedge, and adiabatic wall conditions were used. The computational mesh is constructed with 128 cells on the wedge surface and 64 cells in normal direction. Fig. 6 displays the pressure distribution by contour plots, where $\Delta c_p = 0.1$. The bow-shock is captured without oscillations, and as can be seen in the enlarged view, in almost one cell.

5. Conclusion

A hybrid flux splitting scheme was derived for the solution of the compressible Euler and Navier-Stokes equations. The scheme uses a Mach number-based splitting of the advective and pressure terms of the convective flux-density tensor. The derivation of the scheme was solely based on engineering principles, using elements of established schemes like LDFSS and HCUSP. The resulting scheme almost completely avoids the calculation of an intermediate state at a cell interface, and it is very easy to implement since only left and right state Mach numbers are necessary. Accuracy and robustness of the scheme rival that of other presently in use, most advanced high-resolution/high-robustness schemes, as was demonstrated by the computation of inviscid and viscous, transonic and hypersonic, 1D and 2D flows. It was observed that the proposed scheme does not violate the entropy condition, as encountered in other formulations.

6. References

[1]   Roe, P.L. *Approximate Riemann Solvers, Parameter Vectors and Difference Schemes.* Journal of Computational Physics, Vol. 43, pp 357-372, 1981.

[2]   Van Leer, B. *Flux Vector Splitting for the Euler Equations.* Proceedings of 8th Int. Conference on Num. Meth. For Hyperbolic Conservation Laws, Springer Verlag, Berlin, 1982.

[3]   Liou, M.-S.; Steffen, C.J. *A New Flux Splitting Scheme.* Journal of Computational Physics, Vol. 107, pp 23-29, 1993.

[4]   Edwards, J.R. *A Low-Diffusion Flux-Splitting Scheme for Navier-Stokes Calculations.* Computers & Fluids, Vol. 26, No. 6, pp 653-659, 1997.

[5]   Jameson, A. *Artificial Diffusion, Upwind Biasing, Limiters and their Effect on Accuracy and Multigrid Convergence in Transonic and Hypersonic Flow.* AIAA Paper 93-3559, 1993.

[6]   Swanson, R.C.; Radespiel, R.; Turkel, E. *On Some Numerical Dissipation Schemes.* Submitted to Journal of Computational Physics.

[7]   Moschetta, J.-M.; Gressier, J. *The Sonic Point Glitch Problem: A Numerical Solution.* Proc. Of 16th International Conference on Numerical Methods in Fluid Dynamics, Arcachon, France, 1998.

[8]   Baldwin, B.S.; Lomax, H. *Thin Layer Approximation and Algebraic Turbulence Model for Separated Turbulent Flows.* AIAA paper 78-257, 1978.

[9]   Cook, P.H.; Mc Donald, M.A.; Firmin, M.C.P. *Aerofoil RAE2822 Pressure Distributions and Boundary Layer and Wake Measurements.* AGARD-AR-138, 1979.

[10]   Jameson, A.; Schmidt, W.; Turkel, E. *Numerical Solutions of the Euler Equations by Finite Volume Methods Using Runge-Kutta Time Stepping Schemes.* AIAA Paper 81-1259, 1981.

[11]   Radespiel, R.; Kroll, N. *Accurate Flux Vector Splitting for Shocks and Shear Layers.* Journal of Computational Physics, Vol. 105, pp 207-223, 1993.

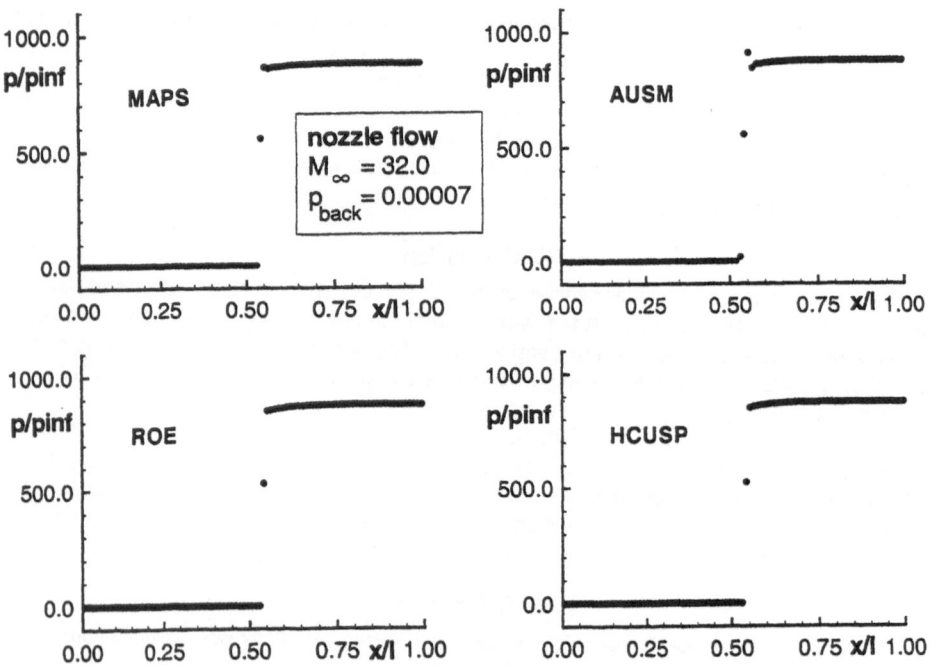

Fig. 1: Comparison of upwind schemes for supersonic nozzle flow

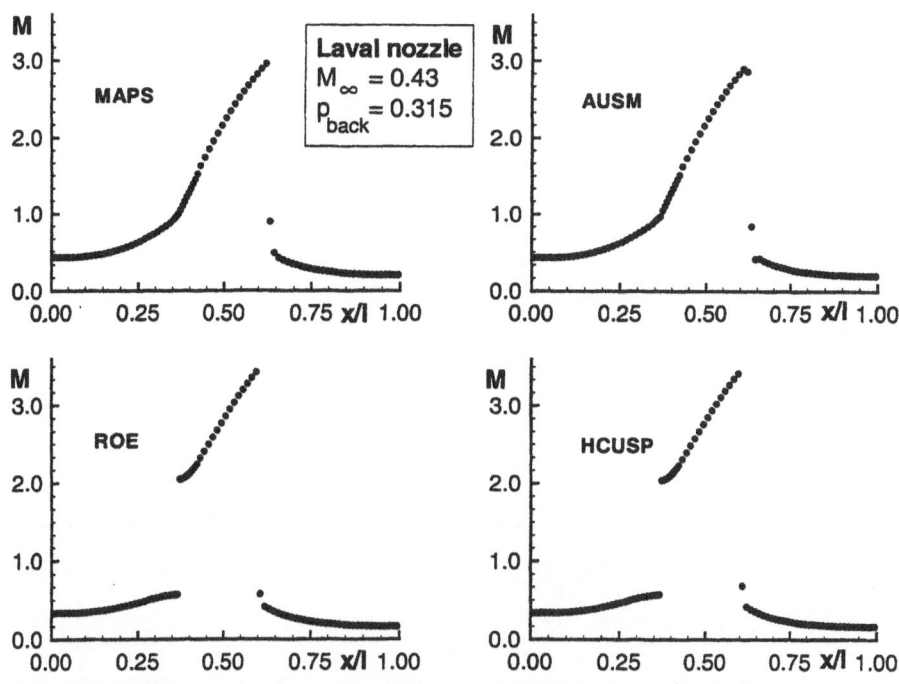

Fig. 2: Comparison of upwind schemes for flow in Laval nozzle

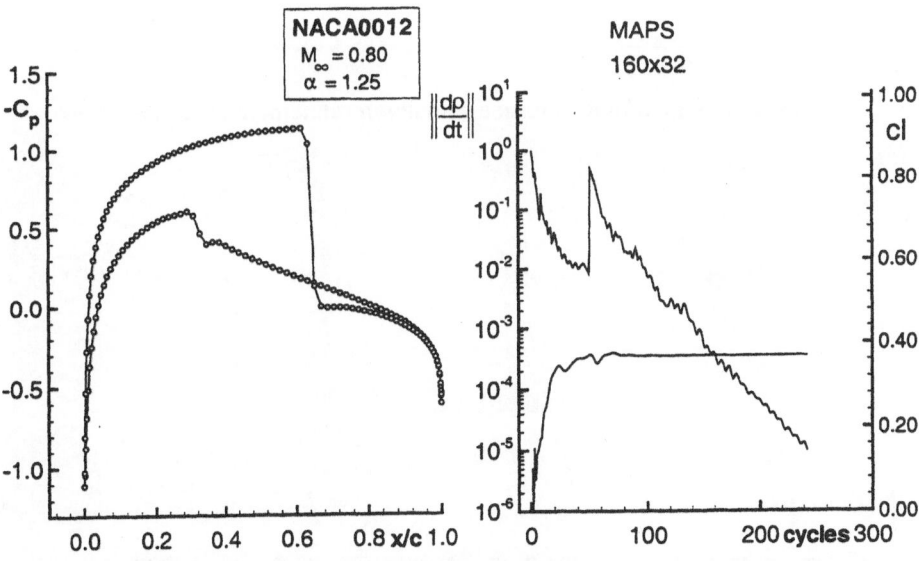

Fig. 3: Airfoil pressure distribution and convergence history for inviscid flow

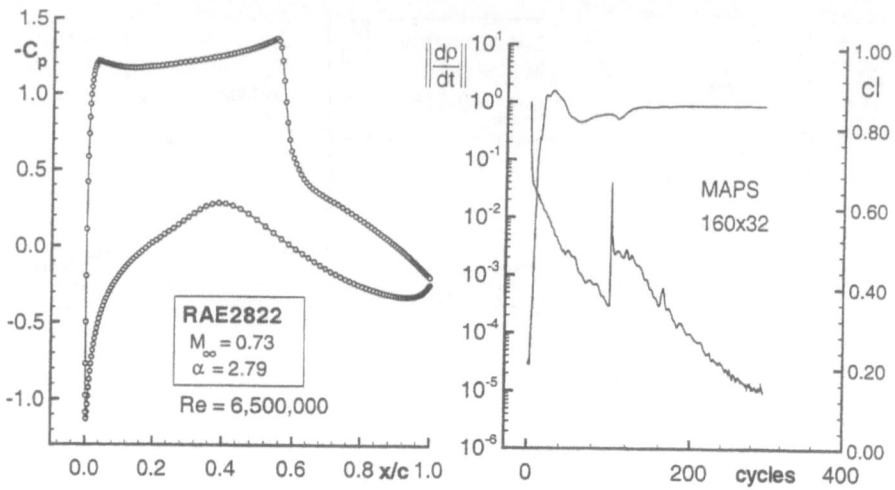

Fig. 4: Airfoil pressure distribution and convergence history for viscous flow

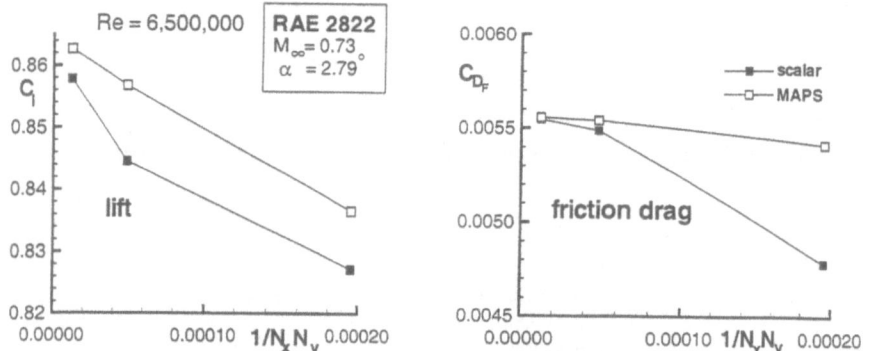

Fig. 5: Evolution of total forces with mesh refinement for viscous flow

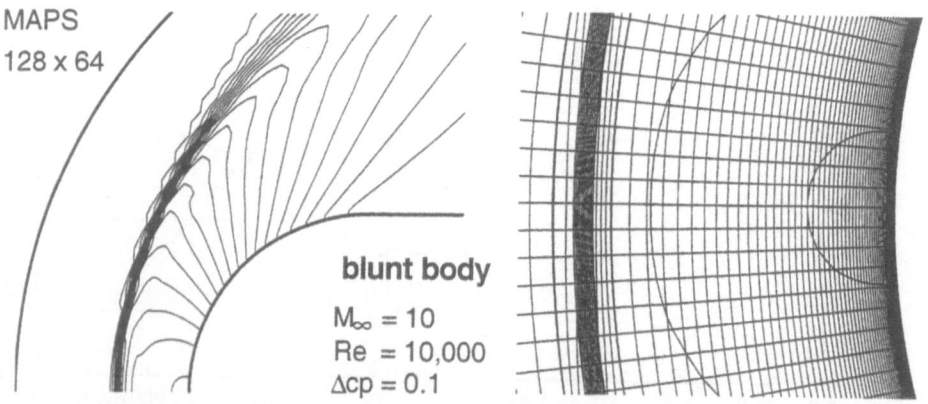

Fig. 6: Field pressure distribution for hypersonic flow around blunt body

Numerical Flow Simulation for a Wing/Fuselage Transport Aircraft Configuration with Deployed High-Lift System

R. Rudnik, A. Ronzheimer, J. Raddatz

German Aerospace Center (DLR)
Institute of Design Aerodynamics
Lilienthalplatz 7, D-38108 Braunschweig, Germany

Summary

The flow simulation system MEGAFLOW is applied to the Navier-Stokes investigation of the ALVAST wing/fuselage configuration with deployed high-lift system. The comparison of the numerical results to experimental data reveals good agreement and satisfying robustness of the convergence process. The Reynolds-number influence is estimated by increasing the Reynolds-number one order of magnitude, causing essentially an increase of lift on all three components. Within the framework of wake vortex investigations Euler computations have been conducted. For the same angle of attack as well as for the same lift coefficient as used in experiments, the neglect of viscosity yields a deviation from the loading of the single components of the high-lift wing. First evaluations of the wake vortex flowfield behind the configuration yield plausible results.

Introduction

At present, theoretical methods for the computation of high lift flows in an industrial environment are essentially drawn from the class of coupled viscous/inviscid interactions (VII) methods, [1], [2]. Due to extensive experience and well established knowledge about their shortcomings, these methods are successfully applied to 2D problems, yielding satisfying results in the linear range of the C_l-α curve. State of the art for more realistic 3D high-lift configurations in industrial applications is the use of panel methods with viscous code modules, allowing to take most of the geometric complexity of the complete configuration into account, though the prediction quality is often only moderate.

Assessing the simulation capabilities of the tools, that are currently in industrial use in terms of simulation quality, efficiency and ability to treat complex geometries, it becomes obvious, that the major reasoning for the application of Navier-Stokes methods with respect to high-lift flows is basically twofold. On the one hand there is a demand for improved simulation of flow phenomena, that are of relevance already in 2D flows, such as wake confluence, direct consideration of compressibility effects, geometry-induced separation in coves of deployed high-lift devices, which represent the major drawbacks of VII methods. Additionally, pressure induced separations have to be treated properly, especially for typical multiple element airfoils in landing configurations, where this type of separation dominates the nonlinear C_l-α regime and finally causes $C_{l,max}$. Although not yet completely captured, recent developments of Navier-Stokes methods, especially with regard to turbulence modeling and accuracy, show promising improvements for this class of problems [3].

On the other hand it will be necessary to incorporate Navier-Stokes methods for detailed 3-dimensional investigations. With respect to high-lift configurations 3-dimensional effects are caused primarily by geometric features, as wing sweep, engine mounting, wing-fuselage and wing-tip area, flap and slat cutouts and flap-track fairings. As VII-methods are already in use for such type of configurations, Navier-Stokes investigations will have to prove that the considerable computational effort is justified by an improved accuracy, and the capability to deal with flow situations, in which the coupled VII-methods reach their limits. Such situations are stronger related to effects of the flowfield itself, than to the pure three-dimensionality of the configuration. Examples are three-dimensional separation, jet/flap interference, or flows, which depend strongly on the development and course of the different element wakes.

Although the objective of computing complex 3-dimensional high-lift configurations with Navier-Stokes methods is in general undisputed, only very limited experience has been gathered yet in this area, mainly due to the difficulty of grid generation and deficiencies in robustness and efficiency of the flow solvers. Therefore, the investigation of the present high-lift configuration has been selected as a major validation milestones within the context of the MEGAFLOW Navier-Stokes simulation system [3]. Representing a collaboration of industry, DLR and several universities, the MEGAFLOW project intends to provide an accurate and efficient CFD simulation system for the computation of complete aircraft in cruise and high-lift configuration. In order to prepare for the comparatively complex wing/fuselage configuration with deployed high-lift system, the Navier-Stokes method, as used in the present investigation, has been successfully applied to various multiple element airfoils and a generic rectangular wing with a deployed half span flap [3], [4]. Flowfield results of the Navier-Stokes computation are presented, as well as an approximation of Reynolds-number influences and some selected results of a preliminary Euler investigation of the same configuration.

Aircraft Configuration

The counterpart of the numerical model of the present study is the DLR ALVAST wind tunnel model [5]. It is representing an AIRBUS type twin-engine transport aircraft. The high-lift configuration considered here consists of a fuselage without empennage and a low wing equipped with a full span slat and two Fowler flaps, departed by an engine cut-out. The slat extends up to the tip, whereas the outboard flap ends at about 85%half-span. The blunt trailing edges of the wing components are included in the numerical model. Aside from the slat and flaps fastening, the major geometric simplification is the extension of slat and inboard flap to the fuselage, so that the spanwise slat and flap gaps are neglected. The slat length is 15% local chordlength, the slat deflection angle is 19.5°. The fowler-flaps have a chordlength of 35%, the corresponding deflection angle amounts to 20°. This setting represents a take-off configuration.

Numerical Method

Grid Generation

The generation of the computational grid for the high-lift configuration is carried out using the interactive grid generation tool MEGACADS [6], which allows the parametric construction of structured multiblock grids. Once a basic grid has been generated, the parametric concept and the restart capability of MEGACADS minimizes the required user-input for moderate geometric changes like variations of flap or slat setting. The computational domain around the ALVAST high-lift configuration is subdivided into 48 blocks, containing a total of about 8 million grid points. The surface grid on the configuration and a wing section grid is depicted in **Fig. 1**. An H-topology is used in the spanwise direction with 112 cells placed on the wing. In the wing sections embedded C-type topology grids are wrapped around the single elements. On the slat surface 128 cells are placed in the streamwise direction, 378 cells on the main wing, and 168 cells on the flap surface, respectively. Normal to the surface the wing boundary layer is resolved with about 25 cells. The first wall spacing amounts to 10^{-5} local chordlength, corresponding to an average y^+ of 1. The complete computational domain extends to about 8 half-spans in the streamwise direction and about 2.4 half-spans in the spanwise direction.

Flow Solution

The flowfield has been computed using the flow solver of the MEGAFLOW code system, which is the FLOWer code, version 115. A detailed description of the capabilities of the code is given in [3] and [7]. The present computations focus on the solution of the three-dimensional Reynolds-averaged Navier-

Stokes equations. The basic features of the code for subsonic and transonic steady applications are a central differencing space discretization together with second and forth order artificial dissipation and explicit 5-stage Runge-Kutta time stepping with implicit residual smoothing. The eddy viscosity is evaluated using a slightly modified version of the Wilcox k-ω 2-equation turbulence model [8]. For the turbulence equations a point-implicit treatment of the source terms is used, the convective fluxes are discretized using a first order upwind scheme. Being the first attempt on a configuration of such degree of complexity, multigrid acceleration is not used here.

Results

Navier-Stokes results for windtunnel conditions

To assess the simulation quality of the FLOWer code a comparison with windtunnel experiments of the full-span model test in DNW [9] is presented. The flow conditions are Mach number M = 0.22, Reynolds Number based on reference chord length Re = $2.\text{x}10^6$ and an angle of attack of α = 12.03°. Fig. 2 depicts computed isobars on the upper surface of the wing and the central part of the fuselage. Major three-dimensional effects, indicated by spanwise isobar gradients, are visible at the wing tip and at the slat fuselage intersection. In contrast to this the interference of the fuselage with the main wing and flap intersection is comparatively moderate. The isobars on the flaps indicate weak gradients in the spanwise and in the flow direction. To compare the numerical results with windtunnel data the pressure distributions at two sections at 26.5% (D2) and at 78% (D7) half span are evaluated for the three components, Fig. 3. For the present angle of attack only a moderate suction peak appears on the slat. The pressure rise due to the wing fuselage interaction at section D2 on the slat produces a deformed type of pressure distribution. As observed in the isobars, the spanwise deviations in the pressure distributions of the main wing are small. Also the pressure distributions for both flaps have the same characteristics and levels. The overall agreement between experimental and numerical results is good. Differences occur at the slat trailing edges and at the flap suction peaks. A probable source for the deviations at the slat, section D2, is the missing modeling of the spanwise gap at the fuselage intersection, which will lower the circulation around the element and consequently cause a pressure rise. The reasons for the other deviations in the pressure distributions are not yet clear and might be attributed to numerical or experimental sources like grid resolution or aerodynamic twist of the model during the tests. The aerodynamic force coefficients are presented in Fig. 4. The Navier-Stokes computation overestimates lift by 4.5% and drag by 6.8%. Despite these differences the numerical data meet the experimental drag polar, because the lift and drag deviations compensate each other. The convergence history is shown in Fig. 5. The efficiency and robustness is comparable to two-dimensional cases. Due to the low freestream Mach-number about 10.000 iterations are necessary to achieve sufficiently time-independent solutions, corresponding to a CPU time of about 300h on a NEC SX4 in sequential mode with a code performance of about 750 MFLOPS,

Variation of Reynolds-Number

To estimate Reynolds-number effects on the flow solution the Reynolds-number has been increased by one order of magnitude from 2 to 20 million, which is close to flight conditions. Fig. 6 shows pressure distributions in section D2 for windtunnel condition and increased Reynolds-number. The general effect of the Reynolds-number variation is an increase of circulation on all three elements due to the decreased boundary layer thickness, leading to a lower pressure level on the complete upper surfaces. The lower surfaces of the elements are only slightly affected. Concerning the force coefficients, Fig. 4, the lift increases about 6.7%, whereas the drag variation amounts only 1.6%. In this context, it has to be considered, that the grid resolution is not adapted to the decreased boundary layer thicknesses. Consequently the grid resolution is different for both Reynolds-numbers, so that these results can only serve as an estimation of the effects. Concerning the non-dimensional first wall spacing, the increased Reynolds-number causes the averaged y^+ to increase to about 1.5.

Within the framework of the European Wake Vortex programme EUROWAKE, DLR conducts Euler computations for the ALVAST high-lift configuration. The aim of this investigation is to assess the prediction capabilities of the Euler simulation in terms of wake vortex characteristics. Nevertheless, neglecting viscosity raises the problem of defining adequate freestream parameters corresponding to windtunnel or free flight measurements. A first conceivable approach is to compute for the same angle of attack as measured, a second one is to compute for the same lift coefficient. The resulting pressure distributions of both approaches are compared at section D2 with the Navier-Stokes and experimental results in **Fig. 7**. The experimental lift coefficient is obtained by lowering the angle of attack from the initial value of 12.03° to 7.5°. The pure neglect of viscosity leads to an overprediction of circulation on all three elements, see dotted lines. The decrease of the angle of attack has a different impact on the single elements. As the dashed lines indicate, the pressure level is considerably higher on the slat upper surface, whereas it is nearly unaffected on the flap. At the main wing upper surface the suction peak pressure level is higher, while the trailing edge pressure level is lower than the experimental values. Obviously the adjustment to the experimental total lift coefficient causes a complete different loading of the single components of the high-lift system. The corresponding force coefficients, which are also depicted in **Fig. 4**, clearly prove the overprediction of lift and the decrease in drag due to the neglect of viscosity. Although the adjustment to the experimental spanwise lift distribution, which is an important parameter for the development of the aircraft wake, is not straight forward, the evaluation of the wake vortices in a crossflow plane shows qualitatively feasible results, as the crossflow velocity distribution in **Fig. 8** proves. A quantitative comparison to experimental data will follow, as soon as the experimental data are available.

Conclusion and Future Work

The Navier-Stokes investigation of the ALVAST wing/fuselage high-lift configuration using the MEGAFLOW simulation package demonstrates the capability of the code to compute such type of configuration in good agreement with experimental results in terms of pressure distribution and global coefficients. The convergence history reveals satisfying robustness. An estimation of the Reynolds-number influence is given by increasing the Reynolds number one order of magnitude. The major effect is an increase in lift of about 7%. In order to evaluate wake vortex properties of the ALVAST configuration, Euler computations are conducted for the same angle of attack as well as for the same lift coefficient as obtained from experimental investigations. Neither of both parameters settings yields the same component loading as observed for the experiments. Although this parameter may strongly influences the development of wake vortex flow behind the configuration, first evaluations of the crossflow velocity and streamlines reveal plausible results.

Future investigations will be devoted to the increase of numerical efficiency by using multigrid acceleration. Also preconditioning of the eigenvalues [3] in the time step approximation and for the artificial dissipation will be incorporated to improve both, efficiency and accuracy. With respect to the latter aspect also the grid resolution has to be assessed. Moreover, the computation of more points on the polars is planned. As the current k - ω turbulence model has shown deficiencies in the nonlinear range of the C_L - α curve, improved k - ω turbulence models will be applied especially for higher angles of attack.

With special regard to the further investigations of the wake vortex flowfield, the incorporation of the spanwise gaps and the wing/fuselage junction appears to be necessary.

References

[1]   Arnold, F.: *Numerical Flow Simulation on High-Lift Configurations at Daimler-Benz Aerospace Airbus*. Proceedings of the High-Lift and Separation Control Conference, Univers. of Bath, U.K., pp. 8.1-8.9, 29.03-31.03.1995.

[2] Nield, B. N.: *An overview of the Boeing 777 high lift aerodynamic design*. Aeronautical Journal, pp.361-371, Nov. 1995.

[3] Kroll, N., Rossow, C.-C., Becker, K., Thiele, F: *MEGAFLOW-A Numerical Flow Simulation System*. 21st ICAS congress, 1998, Melbourne, 13.09-18.09.1998, ICAS-98-2.7.4, 1998.

[4] Rudnik, R., Ronzheimer, A., Schenk, M.: *Berechnung von zwei- und dreidimensionalen Hochauftriebskonfigurationen durch Lösung der Navier-Stokes Gleichungen*. DGLR-Jahrestagung 1996, Dresden, 24.09-27.09.1996, DGLR-Jahrbuch 1996, Bd. II, JT 96-104, S. 717-726, 1996.

[5] Kiock, R.: *The ALVAST Model of DLR*. DLR IB 129-96/22, 1996.

[6] Brodersen, O., Hepperle, M., Ronzheimer, A., Rossow, C.-C., Schöning, B.: *The Parametric Grid Generation system MEGACADS*. Proc. of the 5th Intern. Conference on Numerical Grid Generation in Computational field Simulations 1996, Mississippi, Ed.: Soni, B.K., Thompson, J.F., Häuser, J., Eisemann, P., pp. 353-362, 1996.

[7] Kroll, N.: *National CFD Project MEGAFLOW-Status Report*. Notes on Numerical fluid Mechanics, Vol. 60, pp. 15-23, Vieweg-Verlag, Braunschweig, 1997.

[8] Rudnik, R.: *Untersuchung der Leistungsfähigkeit von Zweigleichungs-Turbulenzmodellen bei Profilumströmungen*. DLR FB 97-49, 1997.

[9] Puffert-Meißner, W.: *ALVAST Half-Model Wind Tunnel Investigations and Comparison with Full-Span Model Results*. DLR IB 129-96/20, 1996.

Figures

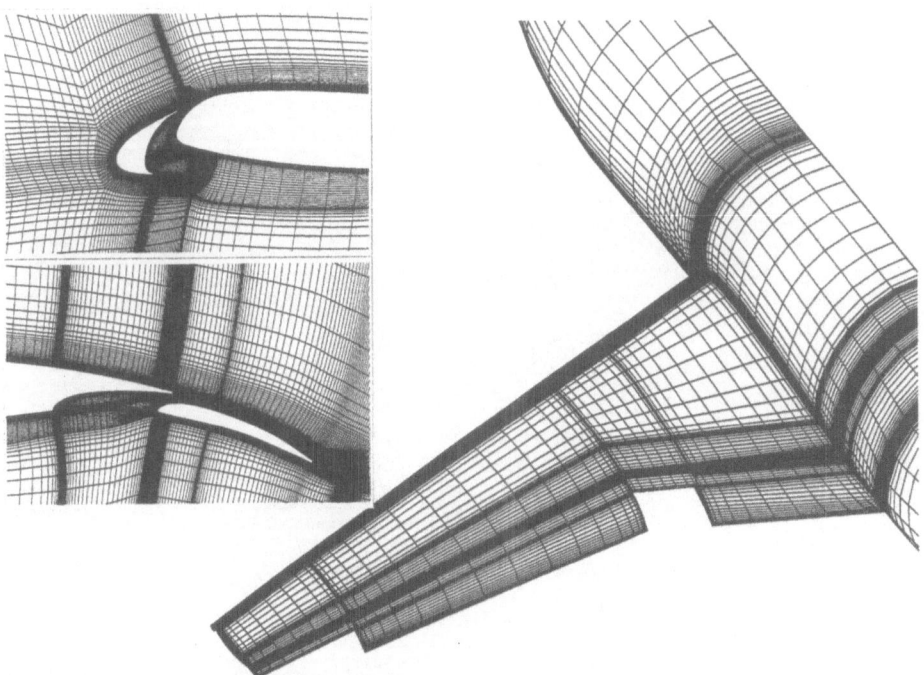

Fig. 1 Surface grid (every other point omitted) and grid in wing section

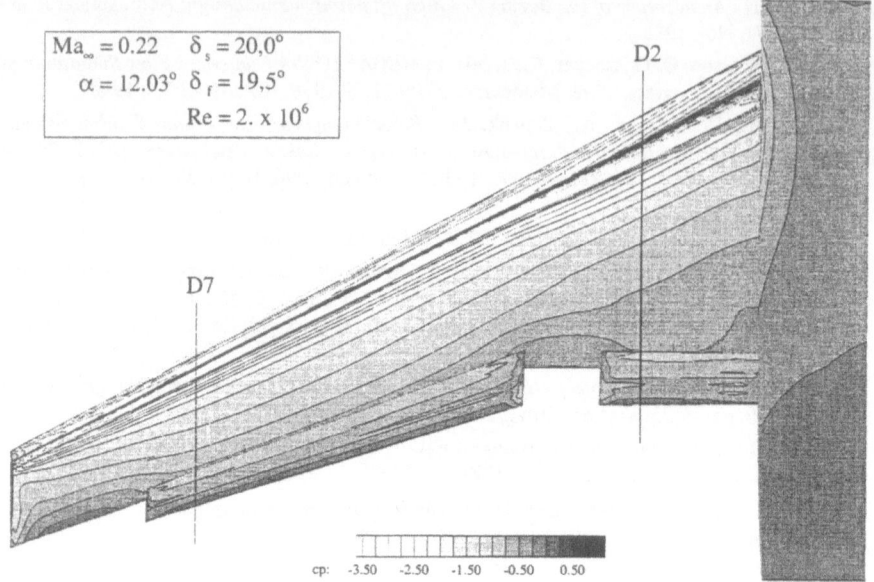

Fig. 2 Isobars on upper surface of ALVAST high-lift configuration

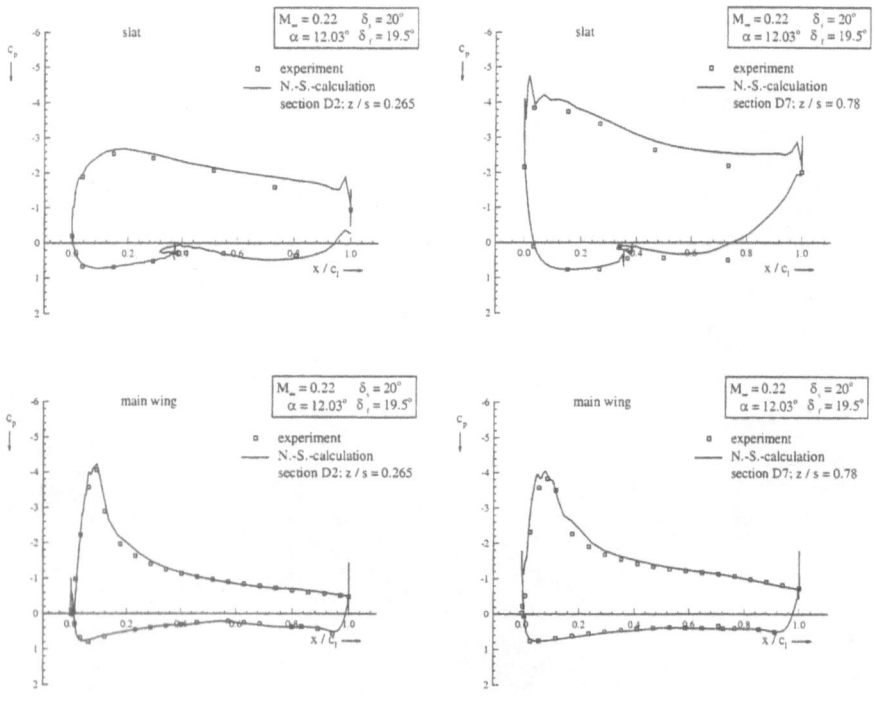

Fig. 3 Slat and main wing c_p distributions in 2 wing sections

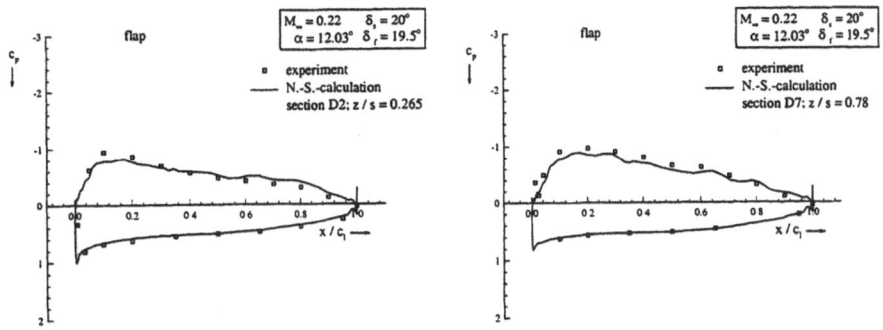

Fig. 3 (cont.) flap c_p distributions in 2 wing sections

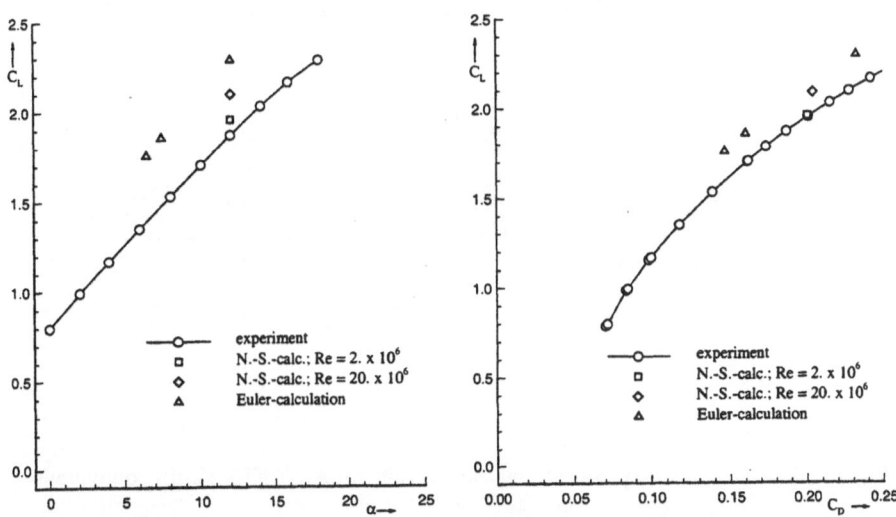

Fig. 4 Aerodynamic coefficients for ALVAST high-lift configuration

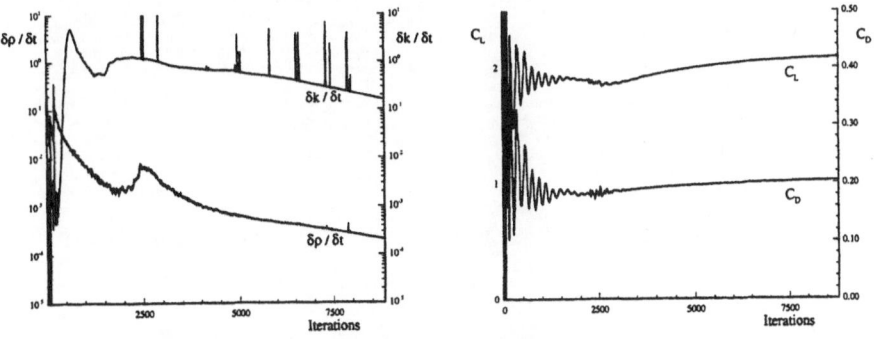

Fig. 5 Convergence history for ALVAST Navier-Stokes computation; $Re = 20 \cdot 10^6$

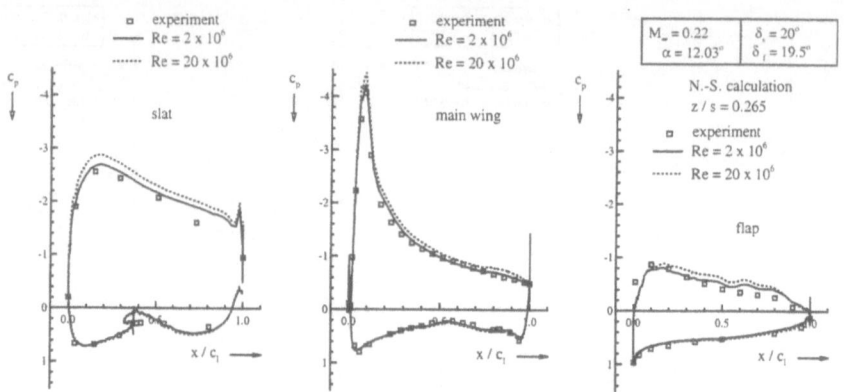

Fig. 6 Reynolds-number influence on c_p distribution in section D2

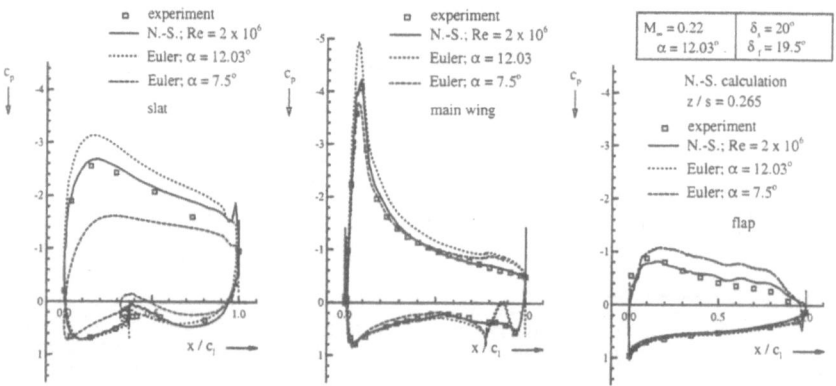

Fig. 7 Comparison of Nav.-Stokes and Euler results for ALVAST high-lift configuration

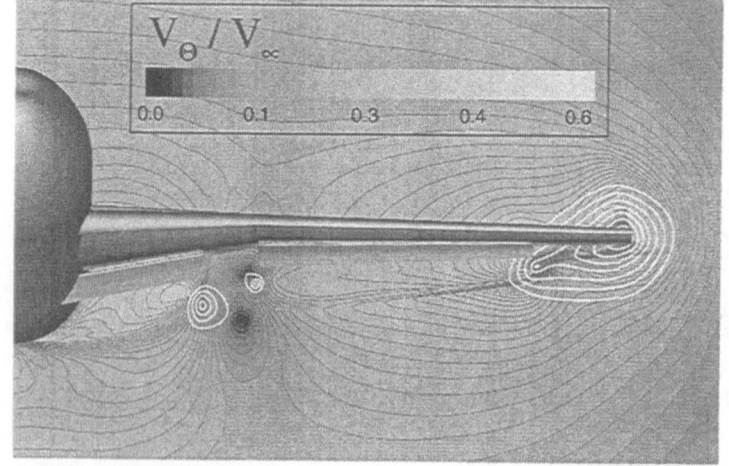

Fig. 8 Crossflow velocity behind ALVAST high-lift configuration; Euler computation

A NUMERICAL SIMULATION OF A 2D- VISCOUS HYPERSONIC FLOW USING THE TIME-MARCHING METHOD OF CHARACTERISTICS

SANAKNAKI H. and BALLMANN J.

Lehr- und Forschungsgebiet für Mechanik

RWTH-Aachen, 64 Templergraben, D-52062 Aachen, Germany

Summary

The method of characteristics in space-time is used to solve the Navier-Stokes equations describing a hypersonic flow in the leading edge region of a circular cylinder. Because emphasis is put on the method, we assume laminar flow of a perfect gas in this paper. We proceed similarly as for unsteady and inviscid flows. First we transform the gas dynamic balance equations for viscous and heat-conducting gases from their non-conservative form into their characteristic form for establishing the compatibility conditions whose integral's approximation leads to an explicit time-marching scheme for unsteady flows. Using a grid fitted between the bow shock and the surface of the body and starting from assumed approximate initial conditions and an assumed approximate initial position of the shock, a steady solution is then searched. This solution is compared with a solution of the method of characteristics for the inviscid case and with the solution for the viscous flow obtained by an implicit upwind finite volume method.

Introduction

Usually, for simulating steady hypersonic flows with subsonic regions, finite difference, finite volume or finite element methods are formulated as time-marching techniques, using more or less sophisticated upwind and reconstruction techniques in space. We aim also to solve the system of gas dynamic conservation laws with a time-marching technique but using the method of characteristics in two dimensions space and time.

The multi-dimensional method of characteristics has been successfully used to simulate inviscid nonequilibrium hypersonic flows (e.g. [1] and [2]). It is possible to extend its concept to the viscous and heat-conducting flow case although this is unusual. In fact, instead of having a set of purely hyperbolic equations in the Euler case where the use of characteristics is mathematically adequate, the Navier-Stokes equations are of hyperbolic-parabolic nature. Nevertheless, physically, the characteristic directions in space-time represent the trajectories of propagation of waves. Thus, their existence is evident for viscous flows, too. The upwind techniques in the Navier-Stokes solvers are based on their use. Therefore, we proceed for the viscous case the same way as in the inviscid case to establish a numerical scheme. The dissipative terms appearing in the so-called compatibility conditions are considered primarily as source terms. But the numerical stability condition must take into account the parabolic nature of the dissipative terms as well as the hyperbolic nature of the inviscid Euler part of the equations.

The Governing Equations

Neglecting the body forces and the body energy supply, the governing partial differential equations (PDE's) are deduced from the gas dynamic conservation laws for a viscous and heat-conducting perfect gas in the following non-conservative form :

$$\text{Momentum :} \qquad \rho \frac{D\underline{v}}{Dt} = div\underline{\underline{\sigma}} \,, \qquad (1)$$

Internal Energy :
$$\rho \frac{De}{Dt} = \underset{\sim}{\sigma} : \underset{\sim}{D} - div\underset{\sim}{q} , \qquad (2)$$

Continuity :
$$\frac{1}{\rho} \frac{D\rho}{Dt} + div\underset{\sim}{v} = 0 . \qquad (3)$$

where :

ρ density , $\quad \underset{\sim}{v}$ velocity vector , $\quad e$ internal energy , $\quad \underset{\sim}{\sigma}$ stress tensor ,

$\underset{\sim}{D}$ strain rate , $\quad \underset{\sim}{q}$ heat flux vector , $\quad \frac{D}{Dt}$ material derivative with respect of time.

The Characteristic Formulation of the Governing Equations

The compatibility conditions are the result of a transformation of the system of PDE's into the so-called *characteristic form*. The equations are written in a space-time contravariant system of coordinates x^0, x^1, x^2, x^3 (see [1], [3] and [6]) where x^0 is a coordinate related to the time t by means of the linear relation $dx^0 = c\,dt$ where c is an arbitrary constant. All space-time entities will be identified in the rest of this paper by the sign *. It appears convenient to transform the thermodynamic variables as follows :
$\mathcal{R} = Log\ \rho;\ \ \mathcal{P} = Log\ p;\ \ \mathcal{T} = Log\ T \quad$ (p and T are the pressure and the temperature).
The law of perfect gas reads then : $\mathcal{P} = \mathcal{R} + \mathcal{T} + Log\ R$,
where $R = 287\ J/(kgK)$ is the specific gas constant for the considered gas (air).
We deal with a two-dimensional flow in a plane normal to the axis of a circular cylinder. The plane is spanned by the covariant base vectors g_1, g_2. Taking the aforementioned transformations into account, we get :

$$\begin{pmatrix} v^* & 0 & \frac{a^2}{\gamma}\underset{\sim}{g}^1 & 0 \\ 0 & v^* & \frac{a^2}{\gamma}\underset{\sim}{g}^2 & 0 \\ \gamma g_1 & \gamma g_2 & v^* & 0 \\ \tilde{0} & \tilde{0} & v^* & -\gamma v^* \end{pmatrix} \begin{pmatrix} \nabla^* v^1 \\ \nabla^* v^2 \\ \nabla^* \mathcal{P} \\ \nabla^* \mathcal{R} \end{pmatrix} = \begin{pmatrix} -v^{*\eta}v^\lambda \Gamma^1_{\lambda\eta} \\ -v^{*\eta}v^\lambda \Gamma^2_{\lambda\eta} \\ -\gamma v^\lambda \Gamma^\nu_{\lambda\nu} \\ 0 \end{pmatrix} + \begin{pmatrix} e^{-\mathcal{R}} div \underset{\sim}{\mathcal{T}_v} \underset{\sim}{g}^1 \\ e^{-\mathcal{R}} div \underset{\sim}{\mathcal{T}_v} \underset{\sim}{g}^2 \\ (\gamma-1)e^{-\mathcal{P}}\left(\underset{\sim}{\mathcal{T}_v} : \underset{\sim}{D} - div\underset{\sim}{q}\right) \\ (\gamma-1)e^{-\mathcal{P}}\left(\underset{\sim}{\mathcal{T}_v} : \underset{\sim}{D} - div\underset{\sim}{q}\right) \end{pmatrix} . \quad (4)$$

Or in abreviated form : $\quad \underset{\approx}{A} \nabla^* \underline{U} = \underline{R} = \underline{m} + \underline{d}$, $\qquad (5)$

where $\nabla^* \underline{U}$ is the gradient of the vector of unknowns $\underline{U} = {}^t(v^1, v^2, \mathcal{P}, \mathcal{R})$, v^1 and v^2 are the contravariant components of the velocity vector, γ is the ratio of specific heats, $\underset{\sim}{\mathcal{T}_v}$ is the tensor of viscous stresses, g^i $(i = 1, 2)$ are the contravariant base vectors, Γ^i_{jk} are the symbols of Christoffel and summation convention is used for double greek indices.
The model of Navier-Stokes (4) differs from the Euler model only by the presence of the dissipative terms $\underline{d}$ on its right-hand side. The metric terms $\underline{m}$ appear in both models.

The Source Terms

The metric terms $\underline{m}$ depend only on the used type of spatial system of coordinates [5]. We use here only the Cartesian system and thus they do not appear furthermore.
For the dissipative terms $\underline{d}$, the usual approach [8] is chosen and we get :

$$\underline{d} = \begin{pmatrix} d_1 \\ d_2 \\ d_3 \\ d_4 \end{pmatrix} = \begin{pmatrix} e^{-\mathcal{R}}\left(\frac{4\mu}{3} v^1_{,11} + \frac{\mu}{3} v^2_{,12} + \mu\, v^1_{,22}\right) \\ e^{-\mathcal{R}}\left(\frac{4\mu}{3} v^2_{,22} + \frac{\mu}{3} v^1_{,12} + \mu\, v^2_{,11}\right) \\ (\gamma-1)e^{-\mathcal{P}}\left\{\mu\left[\frac{4}{3}\left((v^1_{,1})^2 + (v^2_{,2})^2 - v^1_{,1}v^2_{,2}\right) + (v^1_{,2} + v^2_{,1})^2\right]\right. \\ \left. + k\,(T_{,11} + T_{,22})\right\} \end{pmatrix} , \quad (6)$$

$$d_4 = d_3$$

where the coefficients b_i are calculated for each variable using the solutions at the nine neighbouring grid points which belong to the convex envelope of the intersecting area of the Monge cone and the LIVP, by means of a least-square procedure. The discretized compatibility conditions form a system of six equations for four unknowns. It is solved by means of the least-square method for over-determined linear systems [9].

The averaging operators in equations (8) involve the averages of the dissipative terms. These are functions of the partial differentials of the solution in the point P^* (Fig. 1) and thus unknowns in an explicit scheme like ours. Since a steady solution is searched, we adopt the less expensive solution to this problem by taking them the same as those in the point (i, j) in the LIVP. This assumption is *exact* when the steady solution is reached.

Our numerical scheme is of second-order. The partial differentials of the dissipative terms are established as the derivatives of the polynomial eq. (9). Establishing them this way has the advantage of little effort without loss of the order of accuracy of the scheme.

The Stability Criteria : We use an explicit scheme, hence the time step influences the stability of the solution. Because of our emphasis on the method, we considered a perfect gas. Thus, regarding the Euler part of the conservation laws, the time step is first constrained by the CFL criterion which is fulfilled by the procedure of interpolation in eq. (9). We used a time step of $0.5 \ 10^{-9}$ s corresponding to a CFL number ≤ 0.75.

On the other hand, the consideration of the viscosity and heat conduction requires the use of a time step restriction corresponding to the heat transport equation [12]. However, for the smallest cells near the wall where the temperature reaches its highest value, the observation during the calculation has shown that the time step admitted by this criterion has been of the order of 10^{-8} s which is larger than the time step admitted by the CFL criterion. This unusual observation is caused by the high speed of sound due to the high temperature and may be different if real gas effects are taken into account.

The Boundary Conditions : We consider a no-slip condition on the wall, which is supposed adiabatic. Theoretically, at the outflow boundary all variables should be specified because the transport terms with parabolic nature are kept in the governing equations. Nevertheless, because of the purely supersonic inflow and the mainly supersonic outflow, we proceeded here the same way as for the Euler equations. Of course, there will appear some defect when a part of the outflow region is subsonic, like in part of the boundary layer. To alleviate the defect, we used therefore for all variables a second order extrapolation from the interior points which is certainley not perfect.

On the region of the symmetry line of the flow, the solution of the grid line j=1 is established from that of the grid line j=3 using the mirror principle.

At the bow shock, we use the same boundary treatment like in the case of an inviscid flow [1]. Briefly, the boundary conditions on the bow shock are determined by resolving the Rankine-Hugoniot relations and one compatibility condition of type eq. (7). The shock velocity distribution occuring during the computational process is found iteratively by a Newton-Raphson scheme [9] for every shock point. It is considered as the velocity of the anterior points of the grid and, together with the value 0 at the wall, it is used to move the points on the grid line j=constant with a linearly distributed velocity. This velocity vanishes identically when the steady state of the solution is reached.

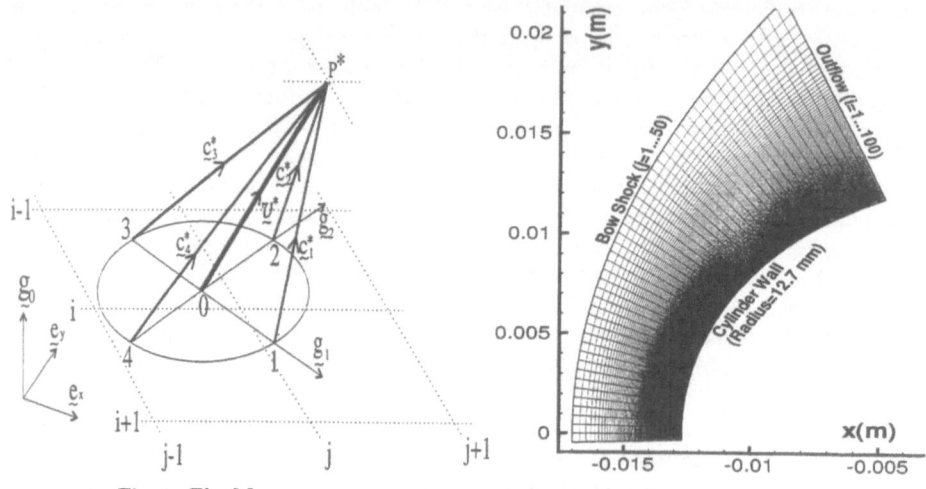

Fig. 1: The Monge cone   Fig. 2: The grid of discretization

The equations (8) form an overdetermined linear system of six equations for the four unknowns : $v_{P^*}^1, v_{P^*}^2, \mathcal{P}_{P^*}$ and $\mathcal{R}_{P^*}$.

The Parameters of Computation

The Grid of Discretisation : We use the structured grid shown in (Fig. 2) fitted between the bow shock and the wall. It has 50 points in the stream direction and 100 points in the radial direction. It is clustered near the cylinder wall to better resolve the boundary layer. The sizes of the cells grow exponentially in the radial direction from the wall toward the shock. The smallest cell which is positioned at the stagnation point has a radial extension of 0.02 mm corresponding to $\frac{1}{700}$ of the cylinder diameter d or $\simeq 10\, d/Re$ where $Re = \frac{\varrho_\infty U_\infty d}{\mu_\infty}$ is the Reynolds number equal to 12000 in the here treated example.

The Initial Conditions : The computation is started from an assumed approximate solution of the flow field and an assumed approximate position of the bow shock. The Rankine-Hugoniot jump relations for oblique shocks are used to establish the variables behind the shock. The pressure and the density are put constant along the radial grid lines. The velocity components along these lines are taken as varying exponentially from their zero value at the wall (the no slip condition) to their values at the shock.

The Procedure of Computation : For the interior points of the grid, the base vectors g_1 and g_2 are taken as the cartesian ones $\underline{e}_x$ and $\underline{e}_y$. The solution in each new point P^* at time level $t + \Delta t$ (Fig. 1) is deduced from the solution at time level t in the local initial value plane (LIVP) which is purely space-like. First, the position of the point 0 relative to the point P^* is determined going back along the space-time particle path to the LIVP : $-[v]_0^* \Delta t$. The positions of the points 1, 2, 3, 4 in the LIVP are then determined using the directions $\underline{c}_i^*$, $i = 1, \ldots, 4$, respectively. The variables in these points are approximated in the LIVP by the biquadratic approximation polynomial :

$$\phi = b_1 + b_2\, x + b_3\, y + b_4\, xy + b_5\, x^2 + b_6 y^2 \qquad (9)$$

where the coefficients b_i are calculated for each variable using the solutions at the nine neighbouring grid points which belong to the convex envelope of the intersecting area of the Monge cone and the LIVP, by means of a least-square procedure. The discretized compatibility conditions form a system of six equations for four unknowns. It is solved by means of the least-square method for over-determined linear systems [9].

The averaging operators in equations (8) involve the averages of the dissipative terms. These are functions of the partial differentials of the solution in the point P^* (Fig. 1) and thus unknowns in an explicit scheme like ours. Since a steady solution is searched, we adopt the less expensive solution to this problem by taking them the same as those in the point (i, j) in the LIVP. This assumption is *exact* when the steady solution is reached. Our numerical scheme is of second-order. The partial differentials of the dissipative terms are established as the derivatives of the polynomial eq. (9). Establishing them this way has the advantage of little effort without loss of the order of accuracy of the scheme.

The Stability Criteria : We use an explicit scheme, hence the time step influences the stability of the solution. Because of our emphasis on the method, we considered a perfect gas. Thus, regarding the Euler part of the conservation laws, the time step is first constrained by the CFL criterion which is fulfilled by the procedure of interpolation in eq. (9). We used a time step of $0.5 \ 10^{-9} \ s$ corresponding to a CFL number ≤ 0.75.

On the other hand, the consideration of the viscosity and heat conduction requires the use of a time step restriction corresponding to the heat transport equation [12]. However, for the smallest cells near the wall where the temperature reaches its highest value, the observation during the calculation has shown that the time step admitted by this criterion has been of the order of $10^{-8} \ s$ which is larger than the time step admitted by the CFL criterion. This unusual observation is caused by the high speed of sound due to the high temperature and may be different if real gas effects are taken into account.

The Boundary Conditions : We consider a no-slip condition on the wall, which is supposed adiabatic. Theoretically, at the outflow boundary all variables should be specified because the transport terms with parabolic nature are kept in the governing equations. Nevertheless, because of the purely supersonic inflow and the mainly supersonic outflow, we proceeded here the same way as for the Euler equations. Of course, there will appear some defect when a part of the outflow region is subsonic, like in part of the boundary layer. To alleviate the defect, we used therefore for all variables a second order extrapolation from the interior points which is certainley not perfect.

On the region of the symmetry line of the flow, the solution of the grid line j=1 is established from that of the grid line j=3 using the mirror principle.

At the bow shock, we use the same boundary treatement like in the case of an inviscid flow [1]. Breefly, the boundary conditions on the bow shock are determined by resolving the Rankine-Hugoniot relations and one compatibility condition of type eq. (7). The shock velocity distribution occuring during the computational process is found iteratively by a Newton-Raphson scheme [9] for every shock point. It is considered as the velocity of the anterior points of the grid and, together with the value 0 at the wall, it is used to move the points on the grid line j=constant with a linearly distributed velocity. This velocity vanishes identically when the steady state of the solution is reached.

The Results

The flow conditions are taken from an experiment executed by Hornung [15]. Convergence has been achieved after about 10000 time steps in a computing time of about 6 hours on a HP C200 workstation. The results are presented in (Fig. 3) and (Fig. 4) together with other solutions, for comparison.

Comparing with the inviscid solution (Fig. 3) which was also obtained by the time-

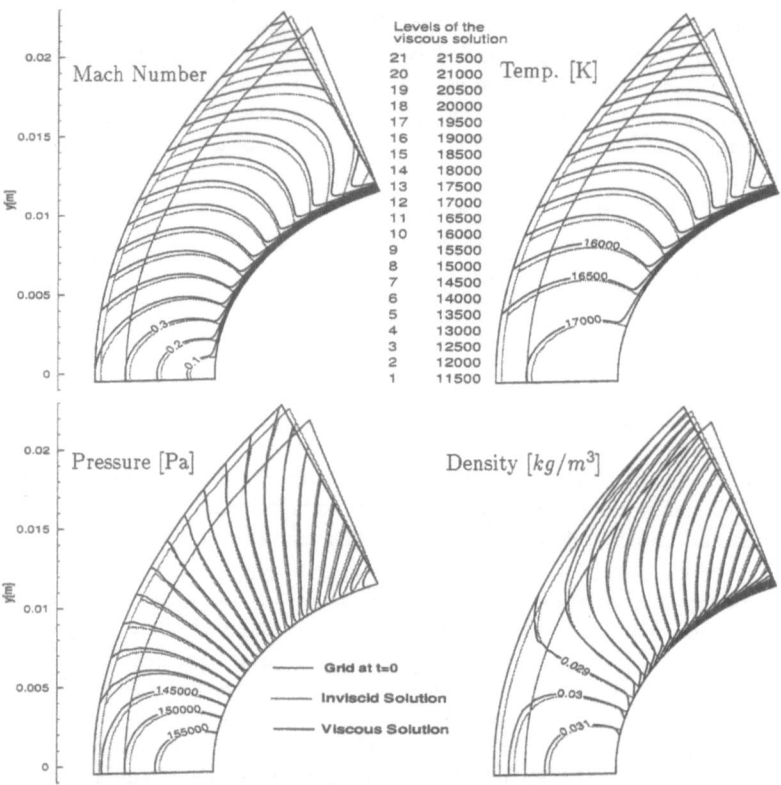

Fig. 3: Comparison of viscous and inviscid solutions of the method of characteristics. Cylinder diameter d = 25.4 mm, Re= 12000, M_∞ = 6.3, P_∞ = 2908 Pa, T_∞ = 1833 K.

marching method of characteristics [1], we observe that the position of the bow shock in the viscous case is about 0.5 millimeter (4 per cent of the cylinder radius) farther away from the cylinder wall. This is due to the so-called *viscous interactions* [10]. Consequently, the contour lines of all variables undergo a slight displacement with respect to the inviscid solution. Furthermore, the comparison of the solutions outside the boundary layer shows a high level of agreement and throughout a reasonable physical structure, without regard to the real gas effects not taken into account here. For this reason, the temperature in the shock layer is too high [10] because no kinetic energy of the flow is absorbed by its endothermic chemical decomposition or the excitation of higher degrees of freedom. The temperature reaches near the wall an unrealistic value of 21500 K in the outflow region, even higher if the computational domain is extended downstream. This defect is removed in an upcoming paper whose purpose is different from that of the actual paper.

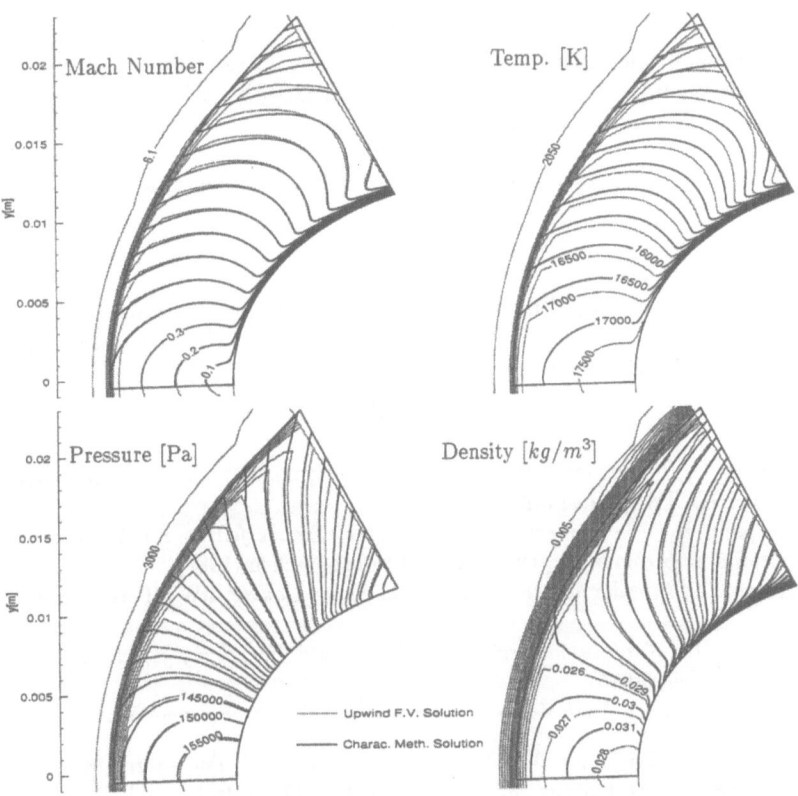

Fig. 4: Comparison with a viscous solution of an upwind-finite-volume scheme. Cylinder diameter d = 25.4 mm, Re= 12000, $M_\infty = 6.3$, $P_\infty = 2908\ Pa$, $T_\infty = 1833\ K$.

By comparison with the viscous solution (Fig. 4) obtained for the same physical data with an upwind finite volume scheme [13] [14], we observe that the bow shock computed with the method of characteristics is located in the middle of the bow shock region of the shock-capturing finite volume solution. The Mach number fields of both methods are also in good agreement. The structures of the solutions of the thermodynamic variables concur very well between both methods, but on the other hand, their numerical values differ slightly. The pressure fields agree well in the subsonic flow region but they deviate progressively in the downstream direction. The temperature and the density fields exhibit a substantial deviation in the symmetry line region which attains about 5 per cent near the stagnation point. This difference decreases in the downstream direction. We believe that these aspects should be less important, when real gas effects or non-equilibrium are taken into account because then the temperature in the shock layer drops drastically.

Conclusion

It has been shown in this paper that the time-marching method of characteristics which is usually applied only to deal with inviscid flows, can be extended to the viscous and heat-conducting flows in a similar sence like the upwind methods used for solutions to the Navier-Stokes equations. It supposes simply to deal with the transport terms as source terms. We used this method to simulate a stationary hypersonic flow of a perfect gas around a circular cylinder. Outside the boundary layer, the results are in good agreement with those for the inviscid case. The overall agreement with a solution obtained with an upwind finite volume scheme is also reasonable. In spite of the unrealistic perfect gas assumption for hypersonic flow, the results are encouraging for further progress like the extension to the case of real gas flow models or gas with chemical and thermodynamic non-equilibrium.

Acknowledgment

The authors express their gratitude to the *Deutsche Forschungsgemeinschaft* for awarding a PostDoc fellowship in the GK *"Transport Phenomena in Hypersonic Flows"* to Mr. Hussein SANAKNAKI to achieve this work.

References

[1] Klomfass A., *Hyperschallströmungen im thermodynamischen Nichtgleichgewicht*, Dissertation RWTH-Aachen (1995).

[2] Grünspahn K., *Ein numerisches Bicharacteristikenverfahren zur Berechnung stark instationärer Strömungen*, Dissertation RWTH-Aachen (1987).

[3] Ballmann J., *Introduction to Gasdynamic, Space Course*, RWTH-Aachen, Paper 10 (1991).

[4] Ballmann J., *Kontinuumsmechanik - Lehr- und Forschungsgebiet für Mechanik*, RWTH-Aachen (1995).

[5] Flügge W., *Tensor Analysis and Continuum Mechanics*, Springer Verlag (1972).

[6] Sanaknaki H., Ballmann J., *On a Multi-Dimensional Time-Marching Method of Characteristics for Viscous and Heat-Conducting Flows*, to be published.

[7] Butler D. S., *The Numerical Solution of Hyperbolic Systems of Differential Equations in Three Independent Variables*, Proceedings of the Royal Society, Vol. 255a, 232-252 (1966).

[8] Bird R. B., Stewart W. E., Lightfoot E., *Transport Phenomena*, John Wiley and Sons (1960).

[9] Dahlquist G., Björck A., *Numerical Methods*, Prentice Hall, (1974).

[10] Anderson J. D., *Hypersonic and High Temperature Gasdynamics*, McGraw-Hill (1989).

[11] Vincenti W. G, Kruger Ch. H., *Introduction to Physical Gas Dynamics*, Wiley (1965).

[12] Richtmyer R. D., Morton K. W., *Difference Methods for Initial-Value Problems*, Interscience Publishers (1967).

[13] Grotowsky I.M.G., *A numerical algorithm for calculating flows in hypersonic inlets*, ZFW 20, pp. 95-104 (1996).

[14] Schubert A., *Grundlagen für die numerische Simulation von laminaren und turbulenten Hyperschallströmungen*, Dissertation RWTH-Aachen (1998).

[15] Hornung H. G., *Non-Equilibrium Dissociating Nitrogene Flow over Spheres and Cylinders*, Journal of Fluid Mechanics, Vol. 53, 49-176 (1976).

Hot Wire Measurements of the Suction Velocity and its Distribution on the ELFIN II-Wing Model with Hybrid Laminar Flow Technology

S. Schaber, H.E. Fiedler

Technische Universität Berlin
Herrmann-Föttinger-Institut für Thermo- und Fluiddynamik
Müller-Breslau-Str. 8, D-10623 Berlin
Germany

Summary

Successful experiments with the ELFIN II wing model were carried out in the transonic wind tunnel S1 of ONERA with a boundary layer suction system. However questions regarding the local distribution of the suction velocity remained. Following the completion of these experiments, the local suction velocities were measured utilizing hot wire anemometry in the absence of a free stream. A mesh of over 13000 data were collected to visualize the velocity distribution over the 18 suction compartments and their 51 chambers. Low suction velocities (0.5 m/s maximum) make it necessary to consider the effects of free convection on the hot wire probe, and the position of the probe relative to the chambers. A map of the velocity field is expected to help determine origins of turbulent wedges in the boundary layer flow.

Introduction

The ELFIN-II wing model was designed for experiments with laminar boundary layers to investigate hybrid laminar flow under transonic conditions. The work was done in the European Laminar Flow INvestigations Programme (ELFIN II) with several partners: Daimler-Benz Aerospace Airbus, Aerospatiale, British Aerospace Airbus, Dassault, ONERA, DLR, TU-Delft and FFA.

The Wing Model ELFIN II

The wing has a span of 4.5 m and a cord length at the root of 2.25 m. The weight is over 3 t, and the sweep angle is 32° which can be changed to 36° (Fig. 1). Investigations were made up to Ma = 0.84. The suction system is confined to the leading edge area, up to 20% of chord length. The perforated surface consists of holes of about 50 micron diameter. Underneath is a system of 51 chambers which are grouped into 18 suction compartments (Fig. 2). The spanwise length of the perforated area is about 2600 mm. Decelerated fluid particles are sucked through the micro perforation to stabilize the boundary layer flow, allowing laminar flow to last for up to 60% of chord length. By avoiding transition near to the leading edge the reduction of friction drag and boundary layer displacement thickness decreases the overall drag.
There are several influences on the flow that can increase Tollmien-Schlichting-instabilities and crossflow-instabilities. The effects of single and distributed roughness elements, noise,

vibration, temperature effects and separation are well known, but the influence of inhomogeneous suction distribution is less clear. Fasel, Joslin and Meitz [1, 2] investigated the influence of the single hole suction, and Deyhle [3] and Reimann [4] the inhomogeneous suction distribution. Lay, Fletcher and Poll [5] and MacManum and Eaton [6] investigated the influence of strong suction through single holes and rows of holes on the transition.

The influence of a spanwise gradient of the suction velocity in a local area increases the growth of instabilities and leads to transition. Strong discontinuities in the velocity distribution induce lateral components of the velocity parallel to the wall which may lead to longitudinal vortices and turbulent wedges. Weak discontinuities may be tolerable. The tolerance level is still unknown. During the wind tunnel testing of the ELFIN II-model some turbulent wedges could be seen on the infrared images of the boundary layer flow (Fig. 3), which origins have to be detected by measuring the distribution of the suction velocity.

The Hot Wire Measurements

To measure suction velocities in the absence of a free stream, hot wire anemometry was used in the constant-temperature-mode. The probe is surrounded by a cylinder with a diameter of 7 mm and a height of 8 mm (Fig. 4). The cylinder is used to prevent unwanted disturbances caused by parallel components of the velocity from influencing wallnormal velocity measurements. As suggested by Tewari and Jaluria [7] the probe is calibrated in a DISA calibration-wind-tunnel in the same position as it is placed over the surface of the actual chamber. The calibration speed is varied from 0 to 0.5 m/s. The calibration curve (Fig. 5) is in accordance to Hoffmeister [8] and Bruun [9].

The probe was mounted in a computer aided traverse with a resolution of 0.01 mm. The traverse was levelled parallel to the surface and to the actual chamber (Fig. 6). The radius of operation is about 3000 mm. The probe was moved automatically in steps of 10 mm while the surrounding cylinder was kept in contact with the perforated titanium surface. A tent was set up around the wing model to minimize the influence of air circulation in the surrounding room. The room was air-conditioned, but the ventilation was reduced to minimum.

The average suction velocity was set to about 0.2 m/s with the help of a mass flow meter in the suction tube located outside the model. The inner chamber pressure was measured and recorded.

Results

The measured local suction velocities were mapped in a surface diagram (Fig. 7 and 8). It is possible to reproduce specific integral velocity distributions from the data as they were used in the S1 wind tunnel experiment. However, discontinuities along the flow direction may be stronger than in the shown presentation with a constant mean suction velocity for all compartments.

It is obvious that there are variations of the suction velocity along most of the compartments. It is also remarkable for the downstream chambers that there are gradients from the root of the wing, where the suction tubes are mounted inside the model. The length of the chambers leads to additional pressure losses. One can see in the leading edge area that there are locally high suction velocities at a few positions where the suction supplies are placed. There are 2 or 3 supplies for each chamber for the upstream compartments.

The static pressure taps attract attention at half of the panel length by reduced suction velocity. The installation of the taps has blocked the perforation. Some areas are remarkable for up to ±70 % diverging flow rate from average. Some of these areas are visible on the lower side of the wing. Less number of discontinuities can be seen on the upper side. The evaluation of the data in comparison to the infrared images is still in progress, but obviously areas of high suction velocities on the lower half of the wing model are related to a higher number of turbulent wedges on the infrared images, although the suction rates are in the usual range of laminar flow experiments. Otherwise the root region of the wing was favoured for particle impact. Detailed analysis does not correlate origins of turbulent wedges to specific local excess or discontinuities of the suction velocity except the positions at both spanwise ends of the perforation and the blocked areas at the pressure taps.

Conclusions

The distribution of the suction velocity in the absence of a free stream has been measured. The goal was to determine the origins of turbulent wedges in discontinuities of the suction distribution, what is still difficult. Additional investigations have to be carried out. Obviously there are regions of varying suction velocities in the order of 70% on average. A correlation between the distribution of the suction velocities and discontinuities and the positions of the turbulent wedges are not clearly visible. Some of the larger areas of different suction velocity hint at variations of the drilling process of the holes. For the moment, it has to be taken into account that there were additional influences, such as single roughness elements from impacts of dust, that initiated boundary layer disturbances during the test phase. Nevertheless, the results can be helpful to identify different influences on receptivity of hybrid laminar flow at high speed.

References

[1] Fasel, H.F.; Joslin, R.D.: *Numerical Simulation of Receptivity and Transition in a Boundary Layer on a Flat Plate with Suction Hole*, Final Report, NASA-CR-195789, 1994.

[2] Fasel, H.F.; Meitz, H.L.: *Navier-Stokes Simulations of the Effects of Suction Holes on a Flat Plate Boundary Layer*, Application of Direct and Large Eddy Simulation to transition and Turbulence, AGARD Conference Proceedings 551, 1994.

[3] Deyhle, H.: *Experimentelle Untersuchungen zu Stabilität und Transition in dreidimnesionalen Grenzschichten mit Absaugung*, DLR, IB 223-94 C 21, 1994.

[4] Reimann, T.: *Untersuchungen zu den Eigenschaften des Absaugesystems an einem Modell eines Hybrid-Laminarflügels*, Diplomarbeit, RWTH Aachen, 1992.

[5] Lay, S.; Fletcher, S.; Poll, D.I.A.: *Report of Progress of Task 3E ELFIN II, Period Aug. 93 - Aug. 94*, Department of Engineering, University of Manchester, Internal Report.

[6] MacManus, D.; Eaton, J.: *Micro-Scale Three-Dimensional Navier-Stokes Investigation Laminar Flow Control Suction Hole Configurations*, AIAA 96-0544, 1996.

[7] Tewari, S.S.; Jaluria, Y.: *Calibration of constant-temperature hot-wire anemometers for very low velocities in air*, Rev. Sci. Instrum. 61 (12), pp. 3834-3845, December 1990.

[8] Hoffmeister, M.: *Methoden der Thermoanemometrie, Teil 1 - 3*, Akademie der Wissenschaften der DDR, Institut der Mechnaik, 1989.

[9] Bruun, H.H.: *Hot-Wire Anemometry, Principles and Signal Analysis*, pp. 103-109, Oxford University Press, 1995.

Fig. 1 The ELFIN II wing model in the ONERA S1 wind tunnel

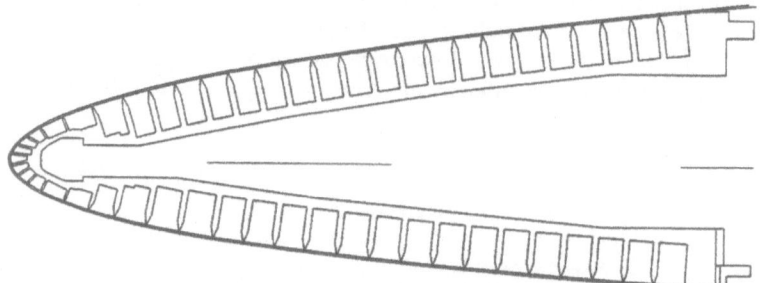

Fig. 2 Sectional plane of the ELFIN II suction system

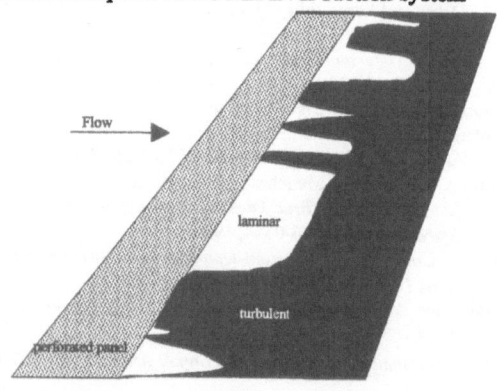

Fig. 3 Schematic view of an IR-image with transition region comparable to typical results from the ELFIN II wind tunnel test

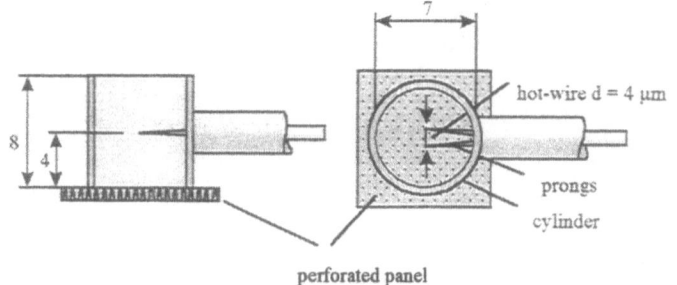

perforated panel

Fig. 4 Hot-wire probe for measuring of the suction velocity

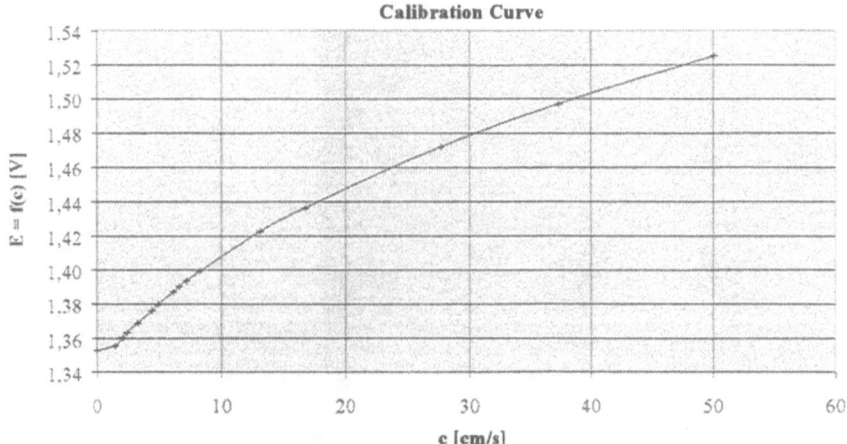

Fig. 5 Calibration at low velocity

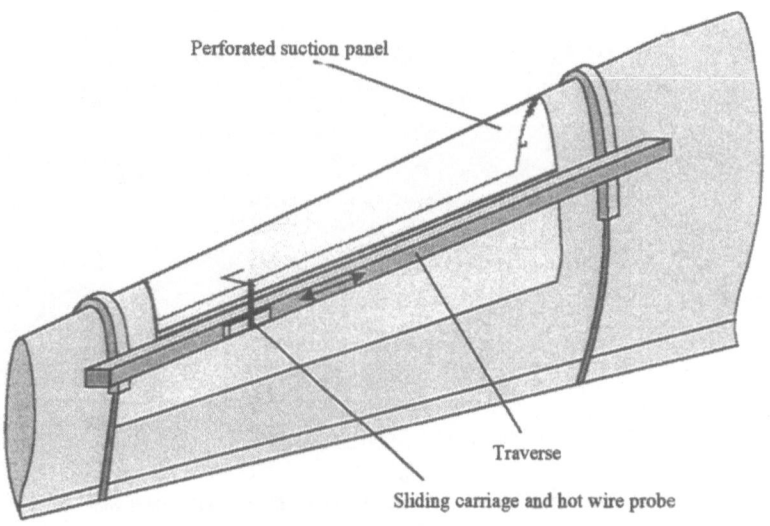

Fig. 6 ELFIN II-wing model with hot-wire mounted on a traverse

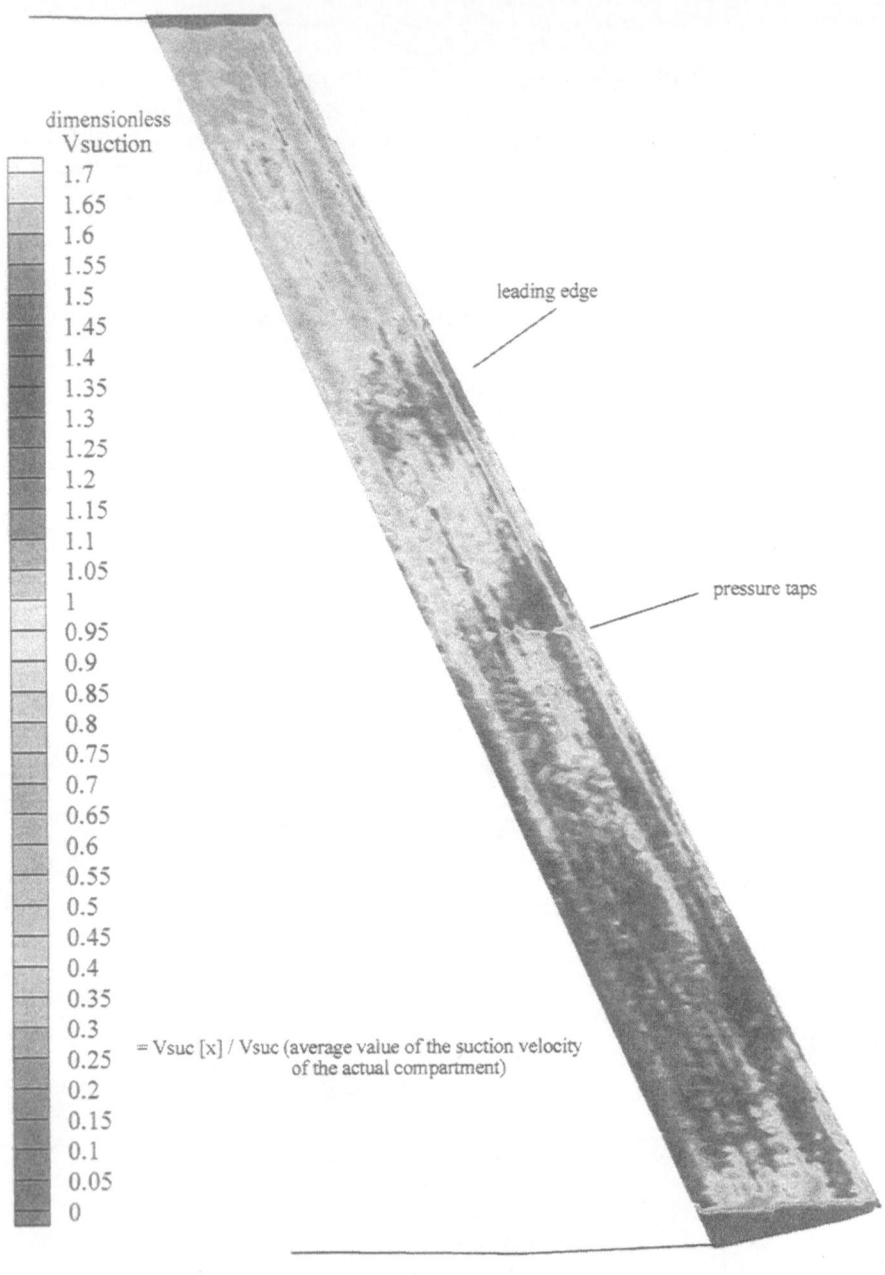

dimensionless
Vsuction

1.7
1.65
1.6
1.55
1.5
1.45
1.4
1.35
1.3
1.25
1.2
1.15
1.1
1.05
1
0.95
0.9
0.85
0.8
0.75
0.7
0.65
0.6
0.55
0.5
0.45
0.4
0.35
0.3
0.25
0.2
0.15
0.1
0.05
0

leading edge

pressure taps

= Vsuc [x] / Vsuc (average value of the suction velocity
of the actual compartment)

Fig. 7 Suction distribution on the upper side of the ELFIN II wing model

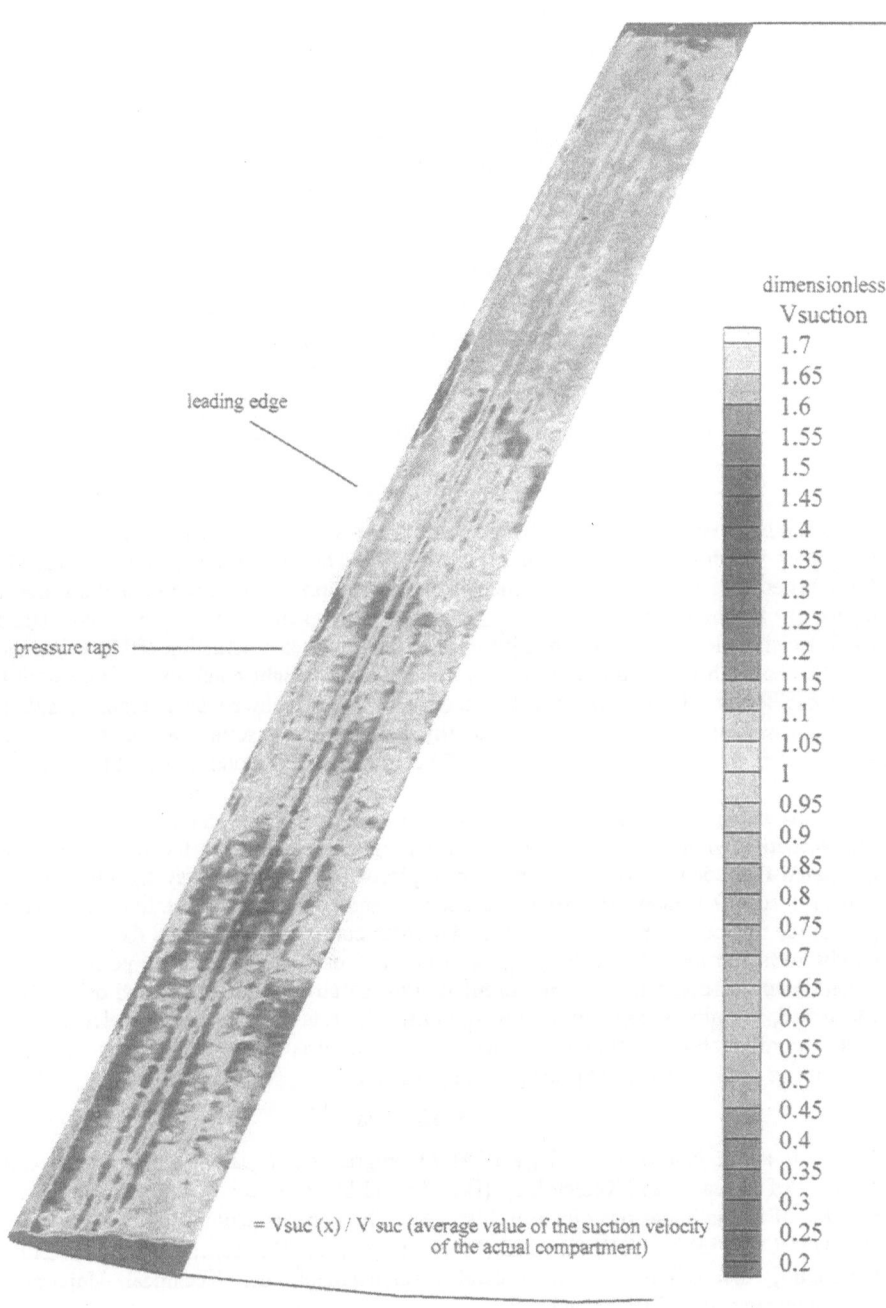

leading edge

pressure taps

dimensionless
Vsuction
1.7
1.65
1.6
1.55
1.5
1.45
1.4
1.35
1.3
1.25
1.2
1.15
1.1
1.05
1
0.95
0.9
0.85
0.8
0.75
0.7
0.65
0.6
0.55
0.5
0.45
0.4
0.35
0.3
0.25
0.2

= Vsuc (x) / V suc (average value of the suction velocity
of the actual compartment)

Fig. 8 Suction distribution on the lower side of the ELFIN II wing model

Active Rotor Control by Servo-Flap and Blade Root Control Recent Results from Flight Test and Wind Tunnel

D. Schimke
EUROCOPTER DEUTSCHLAND GmbH
Willy-Messerschmitt-Straße, 85521 Ottobrunn, Germany

Uwe T.P. Arnold
ZF Luftfahrttechnik GmbH
Kassel Flughafen, 34379 Calden, Germany

W. Geißler, R. Kube, W. R. Splettstösser
Deutsches Zentrum für Luft- und Raumfahrt e.V.
37073 Göttingen, Bunsenstraße 10, Germany
38108 Braunschweig, Lilienthalplatz 7, Germany

Summary

The inflight evaluation of an active rotor control system was performed on a BO105 helicopter, which was equipped with an individual blade root control system provided by ZF Luftfahrttechnik. A powerful experimental system comprises of a high resolution measuring equipment for the recording and a fast preprocessing capability. To capture rotor data these were decisive elements for the analysis of the blade vortex interaction (BVI) phenomenon. For the approach flight condition, the typical BVI noise flight condition, a sinusoidal blade pitch oscillation with two times rotor frequency (2/rev) showed encouraging results: The blade pressure measurements clearly identify BVI for the baseline case and non optimum 2/rev phases. For a phase between 60 and 90 degree this BVI condition could be completely suppressed and the noise on ground was noticeably reduced. In addition, the cockpit vibrations were reduced by 50 to 80% for a similar range of 2/rev phases.

The activities within the second part of the programme comprise the development of an integrated flap control unit on the basis of a piezoelectric actuation system, the integration into a full scale 2-D model of a rotor blade section and a wind tunnel test. In the two week test period, the flap control principle was tested in three configurations: In the fixed configuration steady measurements with steady flap deflection and oscillating flap were performed. In the second configuration a forced 1/rev oscillation of the airfoil with steady and oscillating flap was tested, in order to compare steady and unsteady tests with theoretical calculations. In a further configuration a torsional degree of freedom was used. During this test, general phenomena of the flap control principle were evaluated.

Introduction

The rotor active control technology (RACT) programme, equally shared by the German Ministry of Research and Technology (BMBF) and by resources of the partners, forms the baseline of the activities described in this paper. The partners within the RACT programme are EUROCOPTER DEUTSCHLAND (Project leader), the Deutsches Zentrum für Luft- und Raumfahrt, the Daimler Benz research establishment, the Technical University of Braunschweig and ZF Luftfahrttechnik.

The programme has two main objectives, which require quite different levels of technology for the test and evaluation:

- Inflight evaluation of individual blade control laws using an existing hydraulic blade root control system

- Wind tunnel demonstration of a servo flap control unit using a piezoelectric actuation system.

From both parts of the programme interesting results have been already gathered. Some of these recent results, which refer to interesting aerodynamic phenomena will be presented in this paper. Further literature about the RACT programme itself and results from both parts of the programme can be found in [1, 2, 3, 4, 5, 6, 7].

The Experimental Helicopter and Results from Flight Test

Individual Blade Root Control System (IBC)

The implemented IBC system on the BO105 S1 (Fig. 1) is an enhanced version of the ZFL-built hardware which was used during the 1990/91 flight test campaign [8]. This purely experimental system has a modular structure and can be divided into two parts, one in the rotating system and another one in the fixed frame. The rotating subsystem consists of:
- four servo-hydraulic actuators replacing the conventional pitch link rods
- separate controller electronics for each actuator
- electrical slipring
- hydraulic slipring and hub adapter.

The Test Equipment and Flight Test Procedure

The BVI noise emissions and the effects of IBC control input were evaluated by a large measurement installation on ground and on the BO105 test aircraft. The main components of the test equipment on board was:
- Comprehensive sensing equipment of the rotor blade. One rotor blade was equipped with 20 pressure transducers to detect impulsive pressure changes due to blade vortex interaction (Fig. 2). Five of these signals at the leading edge were processed and analysed in real time by BVI identification algorithms [5].
- Rotor measurement equipment in the rotating frame providing the high resolution and transmission rate needed for the sensors
- Fast data recording and preprocessing in the fixed system.
- Microphones were installed on the helicopter landing skids for a fast evaluation of noise reduction by IBC inputs.

On the ground, a microphone array installed along a line of 300m to both sides of the flight path showed the effects of IBC inputs on neighbourhood noise. The microphone array comprised of 11 ground microphones and 3 microphones installed on 1.2 m tripods at the certification positions -150 m, 0 m and 150 m. The noise data measured by these tripod microphones should contribute to an analysis in accordance to the ICAO noise certification rules. The increased density of the ground microphones on the advancing side should resolve the BVI noise emission in a more accurate way. For a precise approach flight condition a pulse light allocation system was installed on ground. The main part of the investigations was performed at an approach flight condition of 6° glide slope angle and 65 kts also in accordance with the ICAO noise certification rules.

Reduction of BVI Noise and Cabin Vibration

A specific annoying noise is radiated by a helicopter if the blade tip vortex collides with a following blade. The BVI noise is primarily radiated during landing approach, when the helicopter is descending into its own rotor wake. Looking at the measured blade pressure data gathered during the flight test, the most probable explanation of BVI noise. is the local change of the rotor downwash, which influences the miss-distance between vortex and blade. In wind tunnel experiments, the concept of higher harmonic control were proved to be very effective for BVI noise reduction [4, 9, 10]. Fig. 3 illustrates the noise measured below the rotor for a 2/rev phase sweep [4]. These wind tunnel results demonstrate, that a remarkable noise reduction could be achieved for a 2/rev phase between 60° and 90° and at 270°.

The decisive question for the inflight evaluation was therefore, whether these results can be confirmed in a real flight test condition. Although the comparison of flight and wind tunnel test has to take into account differences with regard to the test set up, possibly different trim condition and other differences [4], the flight test results in Fig. 4 confirm the wind tunnel test results qualitatively. Unfortunately, the flight test results for phase angles between 180° to 300° exhibit a high scatter and the second optimum cannot be confirmed with confidence. For the 60° phase angle, the highest number of test points was acquired and the scatter of flight test noise data is surprisingly low. Additionally, a decisive reduction of cabin vibration' was observed between 0° and 90°, whereas the vibrations strongly increase beyond 120°. Fig. 7 shows the intrusion index for the 2/rev phase sweep, which is derived from the weighted average value of the x-, y-, z- cabin vibration (4/rev).

The optimum phase angle range of 60° to 90° could also be confirmed by the skid microphone data. The measured noise reduction correlates very well with the results from the microphones on the ground (Fig. 6). The location of body microphone installation does not have a great influence on the measured noise reduction, although the microphone mounted on top of the skid seems to deliver the best reproduction of the ground noise characteristics.

An important additional sensor information, which was not available during the wind tunnel tests shown in Fig. 3 were the blade pressure sensors (Fig. 2). The leading edge blade pressure transducers installed on one rotor blade of the test helicopter allow for detailed analysis of the source for BVI noise reductions. In Fig. 5, the azimuthal distribution of the blade pressures measured at 3 % chord on three radius stations are presented. For IBC not activated, the impulsive pressure changes at all three radius stations clearly show the occurence of BVI. Since these pressure pulses are all located at the same azimuth range of 60° to 90°, the interacting vortex is positioned parallel to the rotor blade. This so-called 'parallel BVI' situation is supposed to be the one with the most annoying noise emission. With IBC engaged, the pressure level in the parallel BVI region is nearly completely suppressed and the impulsive pressure peaks are smoothed out resulting in the noise level reductions shown in the figures before. On the retreating side, the characteristic of the leading edge blade pressure distribution does not change significantly due to IBC input.

Wind Tunnel Tests with Full Scale Piezoelectric Flap Control Unit

Tests with Trailing Edge Flap in the Transonic Wind Tunnel Göttingen (TWG)

The wind tunnel tests were conducted in the transonic wind tunnel of DLR in Göttingen. The 2-dimensional wing model with a chord length of 0.3 m, a span of 1 m and an OA312 airfoil is attached at both ends to a cross spring and on each end a torsional exciter is used, providing a fully symmetrical set-up. Between the outer torsional exciters and the inner cross spring a piezoelectric balance is mounted in order to measure steady and unsteady forces. On each side the balance consists of four multi-component force transducers. The main feature of such wind tunnel balances is their high stiffness. Thus piezoelectric balances are particularly suitable for dynamic measurements [11, 12]. In addition the pressure distribution is measured by 50 pressure transducers for one cross section, especially for the analysis and the validation of the dynamic measurements. Four laser triangulators determine the pitching angle $\alpha(t)$ of the wing itself. This angle is different from the adjusted angle at the torsional exciters, if the

388

torsional spring is installed. Figure 8 shows the test set up for the different suspensions providing the configurations used during the evaluation tests.

Tests with Torsional Degree of Freedom (DOF)

For a better understanding of the aerodynamic conditions of a flap controlled rotor, a torsional DOF was integrated in the wind tunnel suspension using a torsional spring. The dynamic layout simulates the first torsional mode of the rotor blade at about 28Hz (4/rev). For the analysis of the total lift in relation to the dynamic and aerodynamic design parameters the distinction between direct lift due to the flap deflection and lift due to torsional DOF (Servo-effect) is important. The direct lift is dominating, if the total lift in Fig. 11-14 follows the time history of the flap motion. The servo effect lift is the change of lift due to the elastic twist of the blade. It is dominating, if the total lift follows the oscillation of the blade angle of attack. However, this simulation of the torsional DOF can only take into account the basic flap control principle of the torsional DOF of a rotor blade.

Fig. 13 summarises the results of Fig. 9 to 12 and shows, that for the quite simple test configuration a quite complex transfer characteristic is obtained at Ma=0.33. For a steady flap deflection and low flap control frequency, the direct lift dominates the total lift (Fig. 9). Some loss is induced by the low torsional stiffness of the suspension (Servo effect). This loss is increased with higher flap control frequency and consequently a frequency can be adjusted, where the two effects cancel each other. For the selected design parameters this frequency is slightly above 21 Hz. With a control frequency of 21 Hz (Fig. 10) this condition is nearly reached. Maximum transmissibility between flap deflection and lift is reached close to the resonance frequency due to maximum blade torsion. The time history of the lift curve shows clearly, that it follows the torsional oscillation (angle of attack). This indicates, that the servo effect dominates the lift in this condition (Fig. 11). Beyond the resonance frequency two effects define the lift due to flap deflection. First, the servo effect is strongly reduced with increasing frequency and second, the phase change of about 180° influences the sense of this effect (Fig. 12): Servo effect and direct lift act in the same sense. This phenomenon leads to the quite good transfer characteristic beyond the resonance frequency. However, for a controller design the changed phase of the servo effect has to be taken into account.

Dynamic Measurement (Forced Oscillations)

Fig 14 shows calculated (Time-accurate Navier-Stokes Code, [13]) and measured lift- drag- and pitching moment distributions versus azimuth (left) corresponding to the time dependent incidence and flap deflection of the model (right).

The 1/rev incidence variation is combined with a 4/rev flap deflection. The calculated beta-deflection is assumed to be sinusoidal. The experimental beta-curve shows some differences from the sinus curve. These differences can be detected in the moment distributions (c_M) as well, which show also a 4/rev time-dependency. The correspondance between calculation and experiment is very satisfactory. In addition the calculated hinge moment cr is included in Fig 14. The measured value is not included here, because this sensor signal includes also other effects, which are not included in this calculation (e.g. friction of hinge).

Some deviations can be found in the lift distributions (c_L): The cl curves show a 1/rev variation with a 4/rev modulation. The peak values show some differences between calculation and experiment, which must be attributed to unsteady wind tunnel wall interference effects (only steady wall corrections have been applied).

The drag distribution show an offset of the measured data. Different to measured c_L and c_M which are integrated from pressures (Kulites) the drag was measured by piezo-electric balance, which is quite precise for dynamic changes but not suited for static measurements.

Therefore the static offset was expected. However, the measured and calculated data show definitely the same time dependencies.

Conclusions

The flight test results showed the potential of the individual blade control strategy with regard to noise and vibration reduction. The analysis of the blade pressure sensors provided a more detailed understanding of the BVI phenomenon and possible control strategies in the future:

- The blade pressure vs rotor azimuth shows clearly the blade vortex interaction (BVI) for the baseline helicopter and non optimum 2/rev phases.
- For a phase of 60 and 90 degree this BVI condition could be completely suppressed and the noise on ground was noticeably reduced.
- In addition, the cockpit vibrations were reduced by 50 to 80% in the three axes for the minimum noise condition.

The wind tunnel evaluation has many aspects because of the different types of suspension systems tested and the first demonstration of a full scale piezoelectric flap control unit. The results encourage a continuance of the flap development towards whirl tower and flight test:

- The piezoelectric flap control unit was very reliable and achieved the expected deflections needed for the envisaged control strategies.
- The steady and unsteady measurements with fixed and oscillating flap give a good data base for the code validation and was a test for the actuator performance in steady and unsteady conditions up to dynamic stall and up to a maximum Mach number of 0.74.
- The tests with a torsional DOF demonstrated the transfer characteristic of the lift due to flap control inputs below, near and above the eigenfrequency of the torsional DOF, derived from a typical rotor blade design. The test showed some interesting effects due to the superimposed direct lift and servo effect.

Future Activities

The next step in the development of a flap control unit will be the testing on a centrifugal test bench. For these tests an optimisation process for the design has already been started. End of this year the complete unit will be tested at about 1000g. After these tests, the integration of the flap control unit in full scale rotor blades is planned together with the testing on the whirl tower. This milestone will be the last important step before testing the flap in flight. An important design philosophy of EUROCOPTER for the development and test of the flap control principle was the beginning of the whole development process in a full-scale realisation, because this has a decisive influence on important feasibility aspects. The evaluation of bench and wind tunnel tests presented in this paper are the first step in this process. The results encourage to continue the development.

References

1. Schimke, D.,Jänker, P., Blaas, A., Kube, R., Schewe, G., Keßler, Ch., "Individual Blade Root Control by Servo-Flap and Blade Root Control - A Collaborative Research and Development Programme", 23rd European Rotorcraft Forum, Dresden, Germany, September 1997.
2. Schimke, D., Arnold, U., Kube, R., „Individual Blade Root Control Demonstration - Evaluation of Recent Flight Tests", American Helicopter Society, 54th Annual Forum, Washington, D.C., May 1998.
3. Schimke, D., Jänker, P., Wendt, V., Junker, B., "Wind Tunnel Evaluation of a Full Scale Piezoelectric Flap Control Unit", 24th European Rotorcraft Forum, Marseille, France, September 1998.
4. Schöll, E., Gembler, W., Bebesel, M., Splettstößer, W., Kube, R., Pongratz, R., "Noise Reduction by Blade Root Actuation - Validation and Analysis of Flight and Wind Tunnel Tests", 24th European Rotorcraft Forum, Marseille, France, September 1998.
5. Honert, H., van der Wall, B., Fritsche, M., Niesl, G., "Realtime BVI Noise Identification from Blade Pressure Data", 24th European Rotorcraft Forum, Marseille, France, September 1998.

6. Splettstoesser, W. R., Schultz, K.-J, van der Wall, B., Buchholz, H., Gembler, W., Niesl, G., "The Effect of Individual Blade Pitch Control on BVI Noise - Comparison of Flight Test and Simulation Results", 24th European Rotorcraft Forum, Marseille, France, September 1998.

7. Morbitzer, D., Arnold, U.T.P., Müller, M., "Vibration and Noise Reduction Through Individual Blade Control - Experimental and Theoretical Results", 24th European Rotorcraft Forum, Marseille, France, September 1998.

8. Richter, P., Eisbrecher, H.-D., „Design and First Tests of Individual Blade Control Actuators", 16th European Rotorcraft Forum, Glasgow, September 1990.

9. W.R. Splettstößer, K.J. Schultz, R. Kube, T.F. Brooks, E.R. Booth jr., G. Niesl, and O. Streby, "BVI Impulsive Noise Reduction by Higher Harmonic Pitch Control: Results of a Scaled Model Rotor Experiment in the DNW", Seventeenth European Rotorcraft Forum, Berlin, Germany, 1991.

10. S.A. Jacklin, A. Blaas, D. Teves, R. Kube, "Reduction of Helicopter BVI Noise, Vibration, Power Consumption Through Individual Blade Control", 51st Annual Forum of the American Helicopter Society, Fort Worth, USA 1995.

11. Schewe, G., "Force Measurements in Aero-elasticity Using Piezoelectric Multicomponent Transducers", Int. Forum on Aeroelasticity and Structural Dynamics, Aachen, June 3-6, 1991, DGLR-Bericht 91-06 (1991), 91-071.

12. Schewe, G., "Beispiele für Kraftmessungen im Windkanal mit piezoelektrischen Mehrkomponenten-meßelementen", ZFW 14 (1990), pp. 32-37.

13. Geissler, W., Sobieczky, H., Volmers, H., „Numerical Study of the Unsteady Flow on Pitching Airfoil with Oscillating Flap", 24th European Rotorcraft Forum, September 1998, Marseille, France.

Figures

Fig. 1: IBC Demonstrator BO105 S1

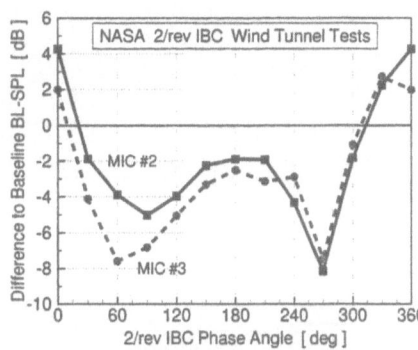

Fig. 3: BVI Noise Reduction vs Phase Angle
(1° Amplitude, Full Scale Test)

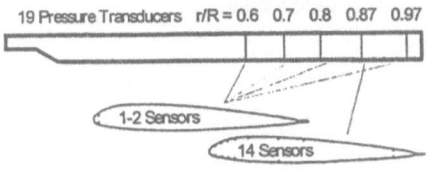

Fig. 2: Pressure Transducers Integrated in Rotor Blade

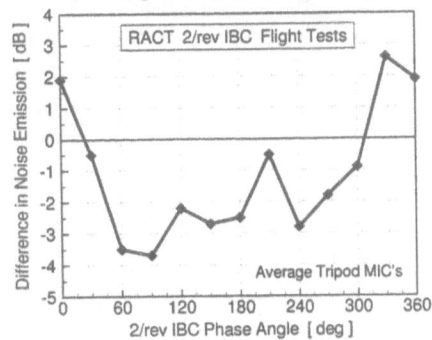

Fig. 4: BVI Noise Reduction vs Phase Angle
(1° Amplitude, Flight Test)

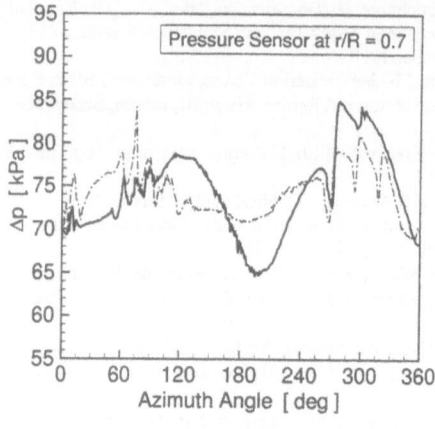

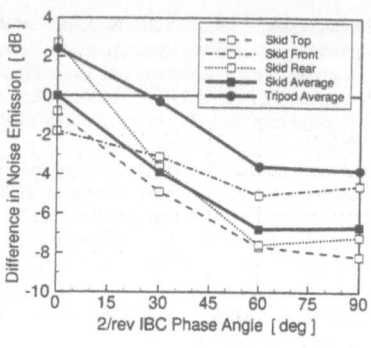

Fig. 6: Noise Reduction vs Phase Angle as Measured by the Landing Skid

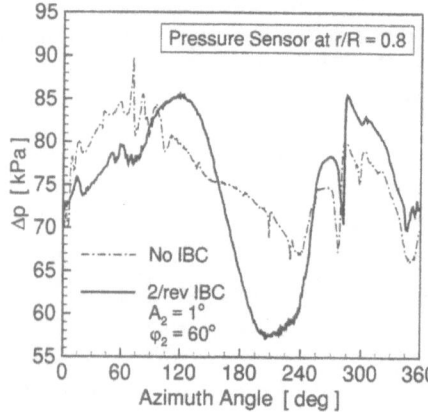

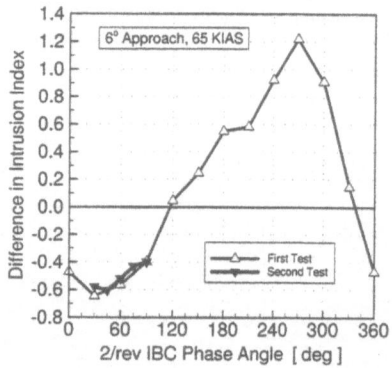

Fig. 7: Reduction of Intrusion Index (Vibration) by 2/rev during Approach Flight Condition

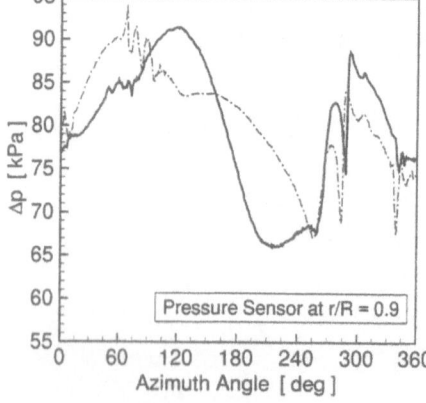

Fig. 5: Leading Edge Blade Pressure Distribution vs Azimuth Angle

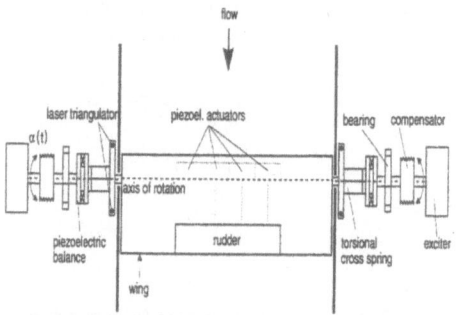

Fig. 8: Schematic sketch of the test set-up

392

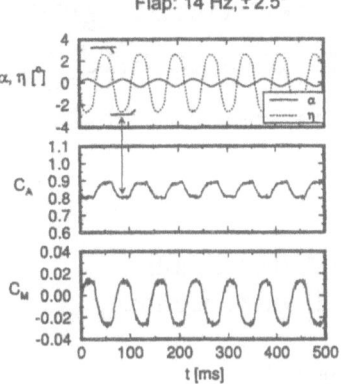

Fig.9: Direct lift effect dominates the total lift at low flap control frequency.

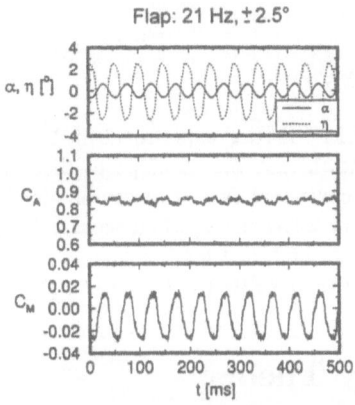

Fig.10: Cancelling of direct lift effect and servo effect at increased flap control frequency

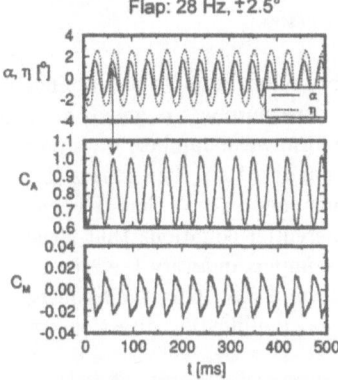

Fig. 11: Maximum effeciency at the resonance condition (servo effect)

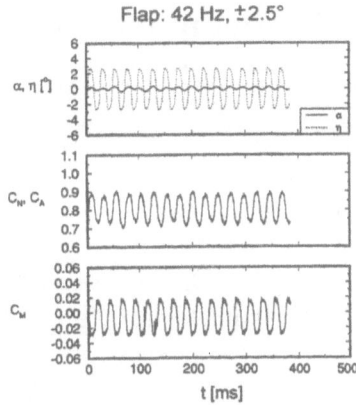

Fig.12: Transfer Characteristic beyond the resonance condition

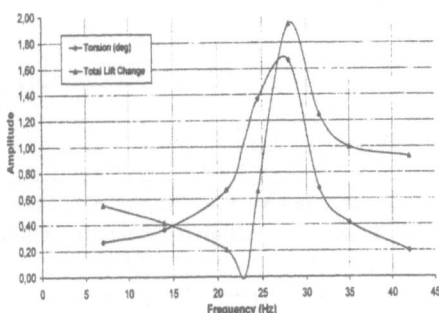

Fig. 13: Transfer Characteristic of Total Lift at different Flap Control Frequencies

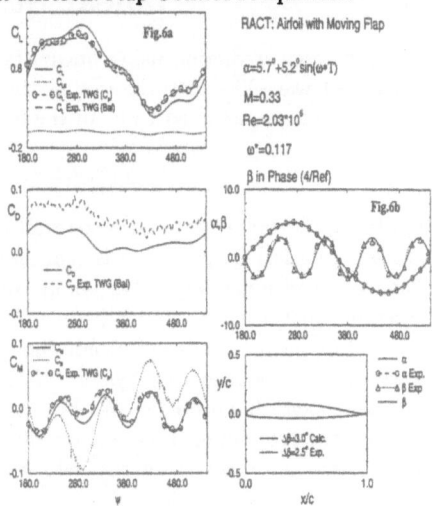

Fig. 14: Airfoil in Motion (1/rev, 7 Hz), Flap in Motion (4/rev, 28 Hz)

393

Comparison of N-Factor Strategies using Wind Tunnel Experiments and Flight Tests

G. Schrauf

Daimler-Benz Aerospace Airbus GmbH

Hünefeldstraße 1-5, D-28199 Bremen, Germany

W. Schröder

Deutsches Zentrum für Luft- und Raumfahrt

Lilienthalplatz 7, D-38108 Braunschweig, Germany

Summary

This paper reports on the evaluation of the ELFIN II HLF experiment in the ONERA S1MA wind tunnel with the e^N-method. It is shown that the envelope method can fail for swept wings with leading edge suction. More consistent N-factors are obtained if the amplification of Tollmien-Schlichting and cross-flow waves is described separately by two N-factors. The new results fit well to previous ones obtained for the same wind tunnel. They are, furthermore, compared to the results of the ELFIN I Fokker 100 flight tests.

Linear Stability Theory

Transition prediction using linear stability theory is based on computing the amplification of wave-like disturbances in the boundary layer. In local stability theory, the disturbances are assumed to be of the form

$$q'(x, y, z, t) = \widehat{q}(z)\, e^{i\,(\alpha x + \beta y - \omega t)}, \tag{1}$$

x, y being the tangential and z the wall-normal coordinate. For spatial amplification the (circumferential) frequency ω is real and the two wave numbers α and β are complex. Their real and imaginary parts can be combined to form wave number, wave propagation direction, spatial amplification rate, and amplification direction. The ansatz (1) contains five real quantities. The amplitude function depends only on z so that the conservation equations are reduced to a system of ordinary differential equations that constitutes a complex eigenvalue problem. We can compute one complex quantity (i.e. two real quantities) if the remaining three real quantities are known. Prescribing the frequency of the disturbance and using the group velocity direction as amplification direction, we need to choose one additional quantity to close the problem. Possible choices are the propagation direction, the wave length, or the spanwise wave number. Another possibility is to choose the mode with maximal amplification. Thus, there are four different approaches to calculate the local amplification rate of a disturbance at one station in the boundary layer, giving rise to four different N-factor integration strategies. They can be used to establish two different transition prediction methods: first, two methods based on two N-factors, one for cross-flow and a second one for Tollmien-Schlichting amplification, and, second, the envelope method.

The details of linear local stability theory and N-factor integration are described in [1, 5], and [6, 9], respectively.

Envelope Method and Suction

In contemporary HLF-design, suction is applied in the leading edge region to dampen the cross-flow vortices. By varying the suction rate, we influence cross-flow amplification. If transition occurs downstream of the suction panel and is primarily caused by Tollmien-Schlichting waves, it is only slightly influenced by a variation in cross-flow amplification. To demonstrate this, we present two cases from the ELFIN II S1 experiment, case 403CP5U ($M_\infty = 0.82$, $Re = 22 \times 10^6$, $\varphi_{LE} = 32^o$, transition at $X/C = 57\%$) and case 405CP5U with the same flow parameters, but with reduced suction. In spite of the smaller suction rate, transition is observed between 55% and 60% chord (i.e. nearly the same location as in the first case). As the envelope method searches for the wave with maximal local amplification at each station in the boundary layer cross-flow amplification rates are used for the N-factor integration near the leading edge of a swept wing and Tollmien-Schlichting rates further downstream. Thus, the envelope method adds the amplification of Tollmien-Schlichting waves to the previously encountered amplification of cross-flow waves. Because only cross-flow amplification is changed by a variation of the suction rates, we correlate different N-factors for the two cases using the envelope method (cf. Figure 1). For the first case we obtain a value of 11 and, for the second one, values between 14 and 14.5. This constitutes a difference of nearly 30%.

Transition Prediction using Two N-factors

To predict transition with two N-factors, we use the "prescribed frequency/prescribed propagation direction" integration strategy for Tollmien-Schlichting (TS) waves and the "prescribed frequency/prescribed wave length" integration strategy for cross-flow (CF) waves. It was shown in [7] that for Tollmien-Schlichting amplification on transonic wings with high aspect ratio it is sufficient to consider the 0°-direction even though this direction is generally not the direction of largest TS-amplification. This is due to the fact that the TS-amplification rates vary slowly with the propagation direction. Their rates have a relatively flat maximum, whereas the CF-rates form a sharp peak if plotted versus propagation direction.

In a low-turbulence environment, stationary cross-flow waves, or cross-flow vortices, dominate the transition process [3, 4]. Therefore, we consider only the $0Hz$ - modes for the evaluation of flight experiments. It is not known whether the restriction to stationary cross-flow modes is justified for the evaluation of wind-tunnel experiments, even though all design and evaluation work with the two N-factor transition prediction method that has been carried out to date is based on stationary cross-flow vortices. Therefore, we present in this paper only results for stationary cross-flow vortices.

The N_{TS}-factors obtained with incompressible and compressible stability theory for case 403CP5U are plotted in Figure 2. TS-amplification starts behind the suction region, ending at 20% chord. For this high Mach number case, compressible stability theory gives much smaller N_{TS}-factors than does incompressible theory. Stationary cross-flow vortices are less affected by compressibility as can be seen in Figure 3.

The N_β-factors obtained with the "prescribed frequency/prescribed spanwise wave number" integration strategy are shown in Figure 4. We see that the N_{CF}-factors and the N_β-factors reach nearly the same level at transition at 57%. This behavior is encountered for all forty-six cases from the ELFIN II experiment. The correlated N_{CF}-factors agree with the correlated N_β-factors if transition occurs far enough downstream of the leading edge, because, behind the leading edge region modes with prescribed wave length have nearly constant spanwise wave number. The correlated (N_{CF}/N_{TS})-factor pairs as well as the (N_β/N_{TS})-factor pairs, obtained with incompressible stability theory in both cases, are shown in Figure 5. As for the VFW614/ATTAS [6] and the Fokker 100 [9] flight tests, transition prediction using the N_{TS}-factor together with either the N_{CF} or the N_β-factor are equivalent.

The (N_{CF}/N_{TS})-factor pairs obtained with compressible and incompressible stability theory are shown in Figure 6. The correlated N-factor pairs obtained with compressible theory lie inside those obtained with incompressible theory and form a somewhat narrower band.

The pairs obtained with incompressible theory lie outside of an imaginary convex curve. They compare well with the earlier ELFIN I results [2]. The N-factor pairs from both HLF-experiments fit well together [8] even though the older ELFIN I HLF experiment used a different type of pressure distribution, namely one especially designed for N-factor correlation. This pressure distribution produces a linear increase of the N-factor envelopes with chordwise position and avoids interaction of Tollmien-Schlichting and cross-flow waves as much as possible. In addition to the differences in the pressure distributions, different boundary layer and stability codes were used for the stability analysis experiments. Because only an incompressible analysis was performed, no comparisons using compressible theory can be made.

Wind Tunnel to Free-flight Extrapolation

Comparisons of the wind tunnel results with the previously obtained results from the ELFIN I Fokker 100 flight tests are shown in Figures 7 and 8. It is worth noting that, with incompressible as well as compressible instability theory, the correlated wind tunnel N-factor pairs lie inside the band of N-factor pairs obtained from the flight experiment. It is generally believed that this behavior is caused by the larger disturbances in the oncoming free-stream encountered in the wind tunnel. These, in turn, generate instability waves with larger initial amplitudes. These waves have to be amplified less to reach an amplitude necessary to cause transition. Noise lowers the correlated N_{TS}-factors, and the vorticity generated by the wind tunnel grids reduces the correlated N_{CF}-factors. By comparing Figures 7 and 8 we notice that compressibility reduces the difference between the limiting N-factors much more for the N_{TS}-factors than for the N_{CF}-factors. Does this mean that a correlation using compressible theory is better than one using incompressible theory and that noise plays only a secondary role? Furthermore, we have to keep in mind that we are comparing an NLF flight test with an HLF wind tunnel experiment. In the HLF experiment, the suction panel acts like a rough surface causing larger initial amplitudes for the stationary cross-flow vortices. Therefore, we cannot decide whether the difference in the N_{CF}-factors is caused by the difference in the environment or by the larger surface roughness in the HLF experiment.

References

[1] Arnal, D.: *Boundary Layer Transition: Predictions based on Linear Theory.* In: Special Course on Progress in Transition Modelling, AGARD Report N0. 793, pp. 2-1 – 2-63, 1994.

[2] Bieler, H., Ponsin, J.: *LISW-COAST Stability Analysis of the ELFIN I S1 Test.* DA Technical Report EF 2021-94, July 1994.

[3] Bippes, H.: *Instability Features Appearing on Swept Wing Configurations.* Laminar-Turbulent Transition, IUTAM Symposium Toulouse/France, September 11-15, 1989. Springer-Verlag, Berlin/Heidelberg/New York, 1990.

[4] Deyhle, H., Bippes, H.: *Disturbance growth in an unstable three-dimensional boundary layer and its dependence on environmental conditions.* JFM, Vol. 316. pp. 73-113, 1996.

[5] Mack, L. M.: *Boundary-Layer Linear Stability Theory.* Special Course on Stability and Transition of Laminar Flow. AGARD Report N0. 709, pp. 3-1 - 3-81, 1984.

[6] Schrauf, G.: *Transition prediction using different linear stability analysis strategies.* AIAA-Paper 94-1848, 12th AIAA Applied Aerodynamics Conference, Colorado Springs, CO, USA, June 1994.

[7] Schrauf, G., Perraud, J., Vitiello, D., Lam, F., Stock, H. W., Abbas, A.: *Transition Prediction with Linear Stability Theory. Lessons Learned from the ELFIN F100 Flight Demonstrator.* 2nd European Forum on Laminar Flow Technology, 10.-12. 6. 1996, Bordeaux.

[8] Schrauf, G.: *Linear stability theory applied to wind tunnel and flight experiments.* Computational Fluid Dynamics '98 (ECCOMAS), John Wiley & Sons, Ltd., Vol 2. pp. 126-131. 1998.

[9] Schrauf, G., Perraud, J., Vitiello, D., Lam, F.: *Comparison of boundary layer transition predictions using flight test data.* AIAA Journal of Aircraft Vol. 53, No. 6. November-December 1998, pp. 891-897.

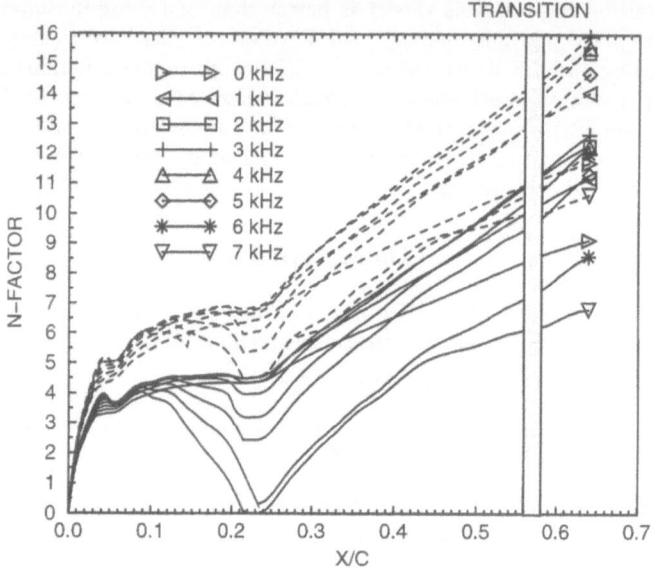

Figure 1: Envelope N-factor curves for two different suction rates

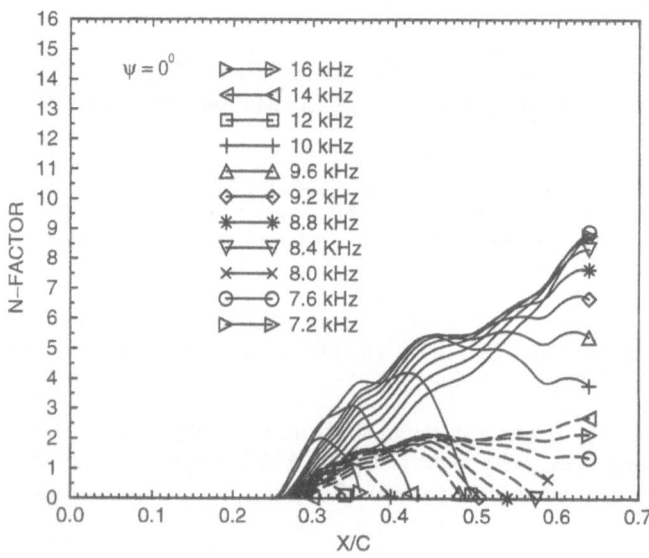

Figure 2: N_{TS}-factor curves obtained with incompressible (solid lines) and compressible (dashed lines) theory

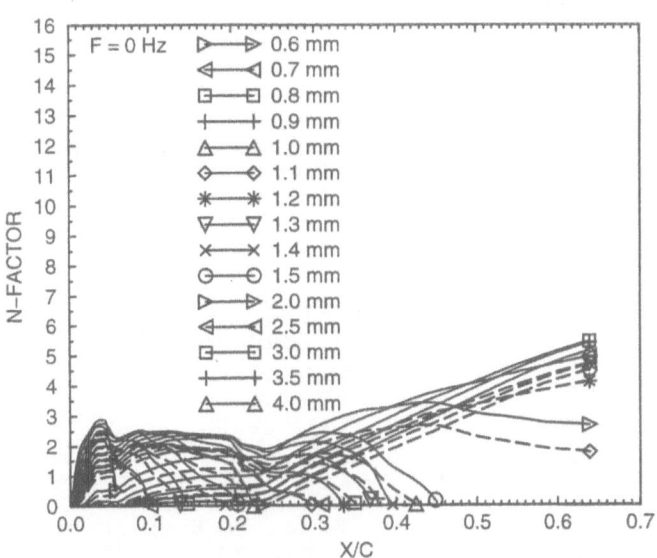

Figure 3: N_{CF}-factor curves obtained with incompressible (solid lines) and compressible (dashed lines) theory

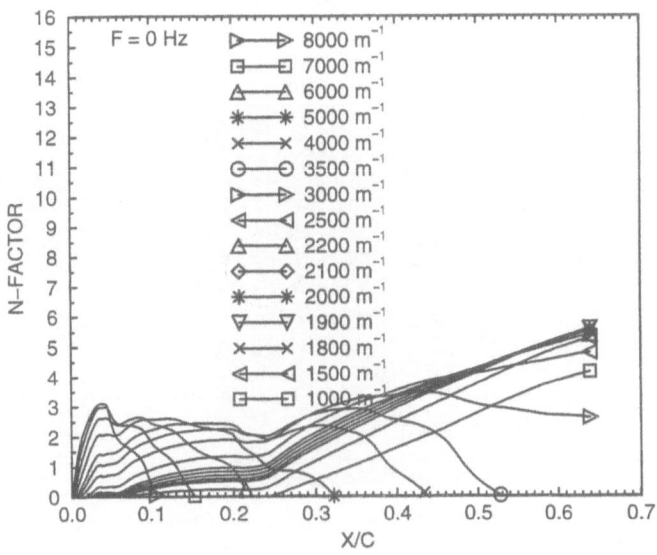

Figure 4: N_{β}-factor curves obtained with incompressible theory

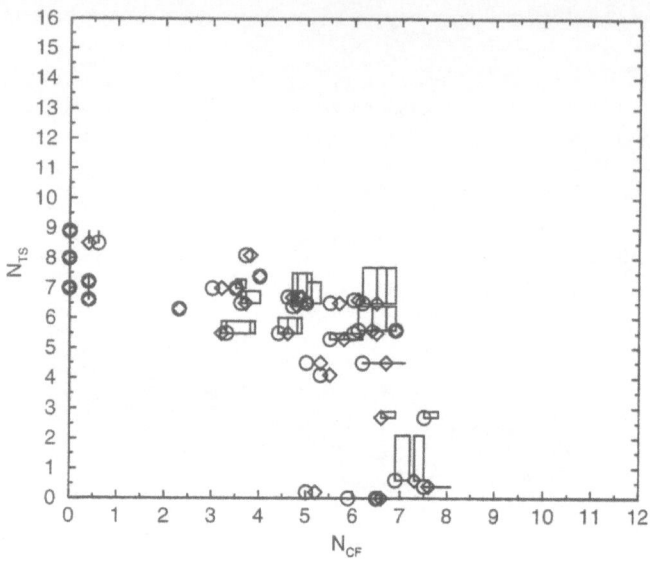

Figure 5: Correlated (N_{CF}, N_{TS})-factor pairs $\circ$ and (N_β, N_{TS})-factor pairs $\diamond$ obtained with incompressible theory

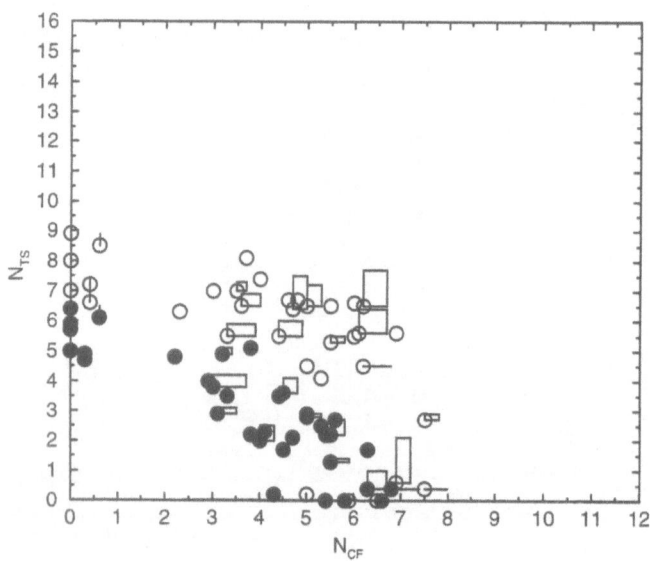

Figure 6: Correlated (N_{CF}, N_{TS})-factor pairs obtained with incompressible $\circ$ and compressible $\bullet$ theory

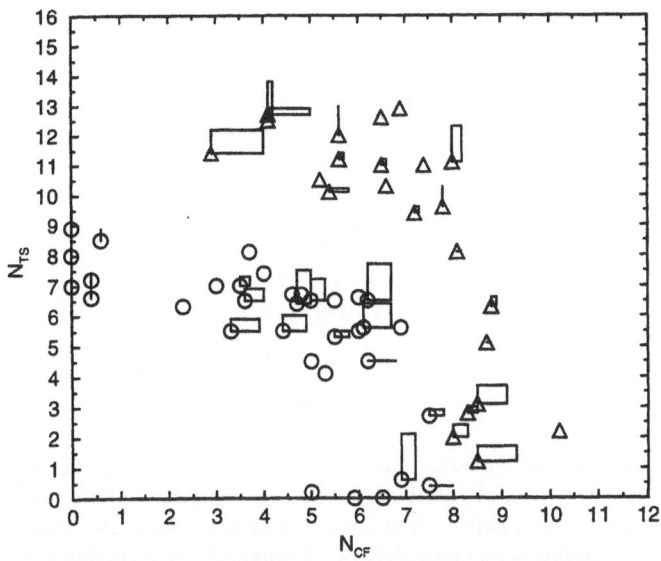

Figure 7: Correlated (N_{CF}, N_{TS})-factor pairs for the ELFIN II S1 experiment ○ and the ELFIN I Fokker 100 flight tests △ obtained with incompressible theory

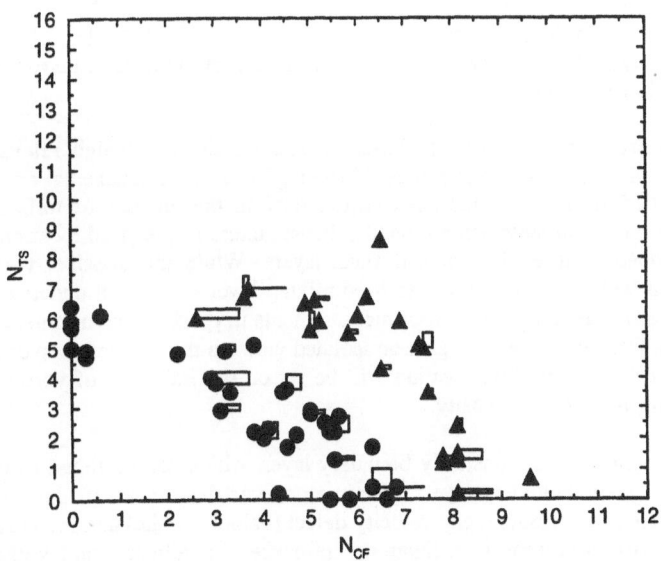

Figure 8: Correlated (N_{CF}, N_{TS})-factor pairs for the ELFIN II S1 experiment ● and the ELFIN I Fokker 100 flight test ▲ obtained with compressible theory

COMBINED TEST FACILITY FOR THREE-DIMENSIONAL MIXING LAYER AND UNSTEADY BOUNDARY LAYER RESEARCH

N. Schröder, W. Lou, J. Hourmouziadis

Jet Propulsion Laboratory
Aerospace Institute F1
Berlin University of Technology
Marchstr. 12, D-10587 Berlin, Germany

Summary

This paper describes the design concept of two low speed test facilities driven by a common set of two integrated blowers. The facilities can be used separately for the study of 3D-mixing layers and unsteady flow, particularly unsteady boundary layers. The overall flow quality obtained in the test sections was investigated. The preliminary measurements of 3D-mixing layers and unsteady boundary layers showed that the present design, which is on some points unusual, was successful. The combined test facility is an example of cost-saving and space-saving integration of two experimental tasks by using a common air supply.

Introduction

The design concept of the new low-speed test facilities at the Jet Propulsion Laboratory of the Berlin University of Technology realizes the diverging requirements of investigations on two fundamental flow phenomena:

One of them are three-dimensional mixing layers, which are of high relevance in many engineering applications. The plane, skewed mixing layer is a fundamental flow case, which can be identified as trailing edge shed vortex flow in the annulus of turbomachinery. In particular, the mixing process behind lobed exhaust mixers is enhanced by strong streamwise vorticity bounded in three-dimensional shear-layers. While the classical two-dimensional mixing layer is well investigated, the skewed mixing layer is not well understood in general and in detail. Since the interpretation of measurements in practical environment is difficult, a low-speed test facility was built to give an isolated view on the free shear layer development. The focus of experimental investigation will be set on the influence of vorticity on mixing processes at low turbulence intensity.

The other phenomena is the unsteady boundary layer, which can be found in blade rows of turbomachines. These boundary layers contain a periodic disturbance (wake) of the previous blade row. There is a periodic mean velocity defect (velocity wake) and increased turbulence (turbulence wake). In order to investigate the influence of a velocity wake without increased turbulence on the development of boundary layers, it is necessary to build a flow facility which can produce an oscillatory main flow. Such an unsteady boundary layer can also be found on helicopter blades. In order to represent the flow over a helicopter blade in a wind tunnel it is necessary to combine pitching oscillations of the airfoil with streamwise oscillations of the main flow, representing the relative motion of the rotor blade.

Design

The combined test facility is driven by two centrifugal blowers of the same size and maximum power consumption of 48.9 kW. For optimum use of available space and best accessibility, the blowers are mounted on top of one other. They are independently controlled by adjustable pre-swirl vanes as well as by electronic speed controls. The complete facility can be functionally divided in a Double Stream Wind Tunnel at both pressure sides of the blowers and an Unsteady Suction Tunnel at the suction side of the lower blower. To allow an independent use of the wind tunnel parts and to avoid undesirable interactions, a bypass system was installed at both sides of the blowers (see Fig. 1). The components of the complete facility were constructed in sections, which can be interchanged and removed so that various different tunnel combinations are possible.

Fig. 1 View on the test facility

Double Stream Wind Tunnel

The design of the Double Stream Wind Tunnel is that of a typical low-speed blower tunnel with two particular characteristics, the overall modularity and the two segment 3D/2D-nozzle. The wind tunnel build of wood consists of two separate legs divided by a splitter plate downstream of the two wide-angle diffusers. The wide-angle diffusers are attached to the bypass tubes via flexible sleeves. Following *Mehta* [1], the boundary layer control is provided by two coarse-mesh screens in the middle and at the exit of each diffuser. The settling chamber has a cross section area of twice 1 m x 2 m and incorporates a honeycomb straightener and two fine-mesh screens in 0.5 m distance. The nozzle is of special design and has an overall contraction ratio of 6.67. It consists of two structurally separate segments, a primary 3D-contraction (contraction ratio 4.0) and a secondary 2D-contraction (contraction ratio 1.67). This solution avoids having to build huge nozzles for every new experimental setup, since the primary nozzle ends in the vertical direction at straight wall with an angle of 30°, which allows a further flow acceleration with different secondary nozzles. At the exit of the nozzles, which are 1.0 m high and 0.3 m wide each, a maximum flow speed of about 40 m/s can be reached. To realize different skewing angles between the upper and the lower flow, a special deflection module was built (see Fig. 1). The deflection module is based on wooden frames with flexible metal sheets as side walls, while the deflection angle is given by wedge-shaped top- and bottom-plates. By interchanging the top- and bottom-plates the skewing angle can be varied in steps of 5° up to total 40°. The overall modular design of the test facility permits the replacement of individual parts for special experiments. This includes the removal of the splitter plate downstream of the honeycombs, transforming the facility into a large single flow facility. In combination with an adapted 2D-nozzle, a single wind tunnel leg can be used as a separate wind tunnel. A more detailed description can be found in [2].

Flow Quality

The static pressure distribution along the wind tunnel walls was measured to check the efficiency of the screens and the flow straightener and the quality of the internal flow. The streamwise pressure evolution of both wind tunnel legs was found to be practically identical and quite smooth. As expected, the first screens produce the highest pressure losses, because of the about 2.7 times higher flow velocity compared to the other screens. In Fig. 2 the pressure distribution of the nozzle section is plotted against the axial coordinate normalized by the length of the primary nozzle. Deviations from the predicted one-dimensional pressure distribution can be seen on both the horizontal and vertical contour. The strong over-accelerations indicate the wall curvature effects of the asymmetric nozzle. Without a relaxation section the region of static pressures below ambient ends directly at the exit of the 2D-nozzle. A resulting non-homogenous exit flow with a strong vertical gradient can be expected, as shown by the hot-wire measurement of Fig. 3a. With a short rectangular relaxation section the region of over-acceleration moves upstream and a nearly constant ambient pressure is reached downstream of x/L(3D-nozzle) = 1.6. The total equalization of pressure gradients leads to a homogenous flow, with non-uniformities of mean velocity below ± 1% (see Fig. 3b). The turbulence level of the mean flow is 0.4% ± 0.1%, with slightly lower levels at higher velocities.

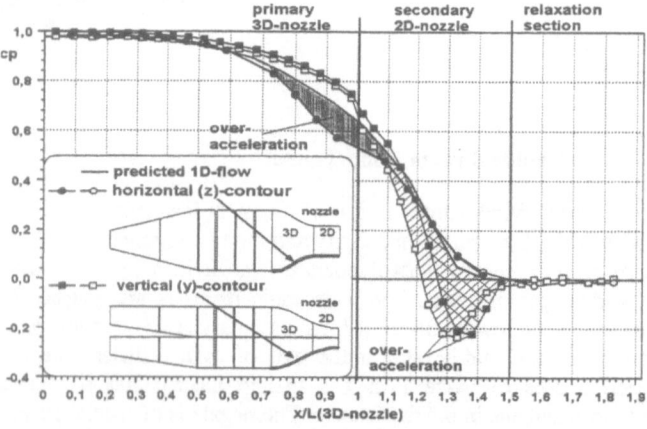

Fig. 2 Pressure distribution at the side wall and lower wall of the nozzle, mean flow
velocity u = 30 m/s, filled symbols without, empty symbols with relaxation section

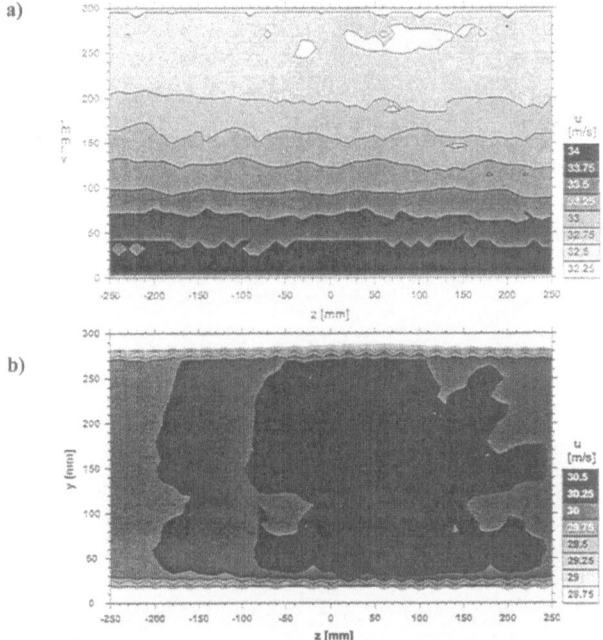

a)

b)

Fig.3 Mean velocity distribution in the middle section of the lower leg
    a) 2D-nozzle exit and b) relaxation section exit

Preliminary Results

The first X-wire measurements and smoke visualizations showed the capability of the deflection module to generate a skewed mixing layer with good flow quality in both streams. Fig. 4 shows an example of X-wire measurements at a velocity ratio of $C_1/C_2 = 2/1$ and the maximum skewing angle of 40°. The probe was traversed in a x,y-plane along the symmetry axis of the test facility. The boundary layer at the trailing edge of the splitter plate was found to be laminar, with a momentum thickness of the wake directly behind the splitter plate of 5.4 mm. In Fig. 4 the normalized flow components $(u-U_1)/\Delta U$, $(w-W_1)/\Delta W$ and the flow angle θ are plotted for different x-positions. It can be seen that the velocity defect in the u-component, resulting from the wake of the splitter plate, is not removed up to a development distance of $x \geq 400$ mm. This corresponds with measurements of the reference case of 0° skewing angle (2D-mixing layer), where a good agreement with theoretical prediction for the 2D-shear layers is met about 400 mm downstream . The predictions for self-similarity of a 2D-mixing layer in [4] give a development length of minimum 4.05 m, which can not be reached by this free stream test facility. The w-component has an overshoot in velocity decreasing with development length, which is shown more clearly in the flow angle plot. This indicates the presence of strong streamwise vorticity-components. All plots show the expected asymmetric spreading of the mixing layer to the low-speed side. A very homogenous velocity distribution at constant flow angle can be recognized outside of the mixing layer. In a short distance away from the mixing layer, the turbulence intensity showed no gradient in y-direction at a constant level of 0.32 %. The upstream influence of the streamwise vortex-patterns was found to be negligible.

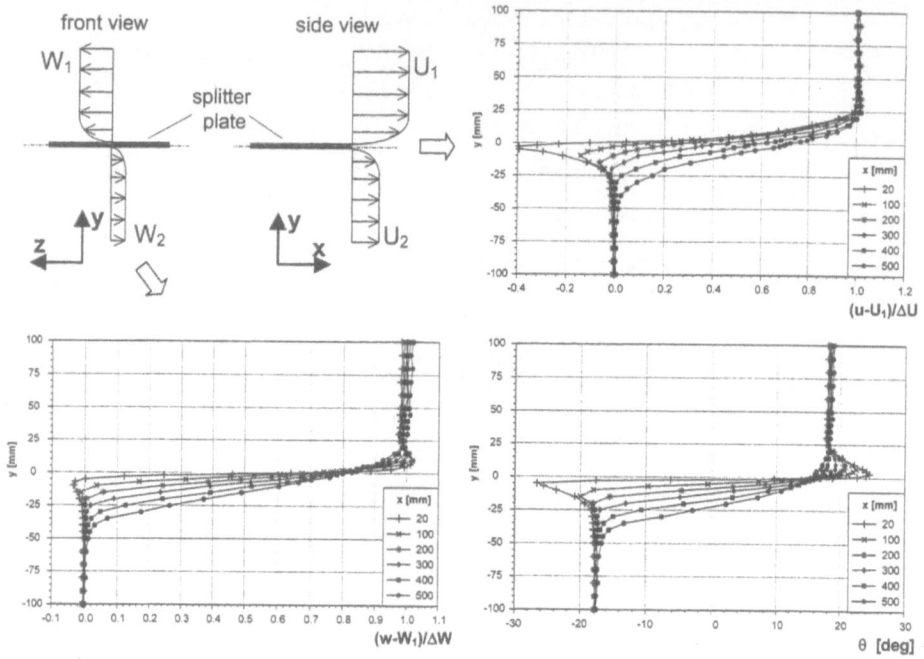

Fig. 4 Mean flow components and flow angle at a velocity ratio of $C_1/C_2 = 2/1$ and a "symmetrical" skewing angle of total 40°, C_1 = ca. 30 m/s

Unsteady Suction Tunnel

The unsteady low speed wind tunnel is an open circuit wind tunnel of suction type. It incorporates an inlet, a settling chamber fitted with a honeycomb straightener, a square 9:1 contraction and a square test section of 0.4 m width, 0.4 m height and 1.5 m length. The test section is followed by a rotating flap and a diffuser. The oscillating flow is generated by the rotating flap between the test section and the diffuser. The flap closes the passage twice every cycle of rotation. The rotating flap is made of sheet metal plate and can be varied in width so that in the limit case the duct is completely closed. It is driven at constant speed from 0 to 50 cycles per second, giving a frequency range of flow oscillation up to 100Hz. The generation of static pressure oscillations poses some problems for the combined test facilities, since it makes the use of screens impossible and requires a solid frame and a good basement. To protect the screens of the Double Stream Wind Tunnel from damage a sealed off bypass system is required.

A plate of 1 m length is located in the test section, which is shown in Fig. 5. Particular attention was paid to surface finish giving excellent flatness and smoothness. The leading edge was made of a NACA-profile form to avoid leading edge separation. A contoured wall opposite the test plate was designed to generate the required pressure distribution. The 1D potential flow analysis was used to design the contour to produce a pressure distribution with minimum surface friction. This method was recommended by *Stratford* [3] to optimize the turbulent boundary to reach a maximum diffusion. This controlled diffusion concept is widely used in turbomachinery blade design, especially in compressor blade design.

The two major aerodynamic parameters: overall Reynolds number and Strouhal number can be varied from 10^5 to 10^6 and from 0 to 2.5, respectively, which are the relevant values for turbomachinery.

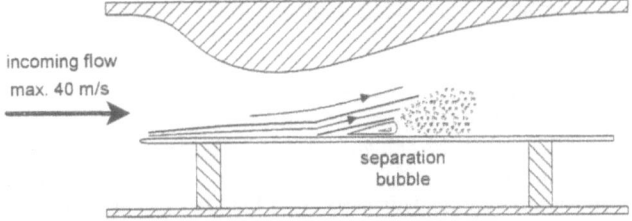

Fig. 5 Test section

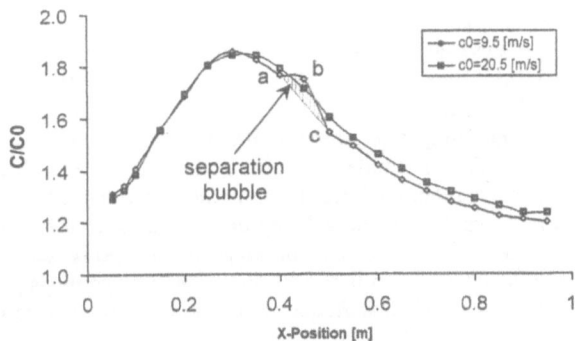

Fig. 6 Velocity distribution over the flat plate

Flow Quality
Under steady flow conditions the static pressure distribution along the walls of the wind tunnel and the flow field at the inlet of the test section were measured for the global assessment of the flow quality. The static pressure distribution along the wall of the wind tunnel was satisfactorily smooth. The maximum pressure loss can be found in the diffuser. The flow velocity measurements were made using a TSI 1212-60 hot film probe. The deviation of the mean flow velocity across the test section was below 0.75%. The turbulence level of the mean flow across the tunnel is about 0.5%.

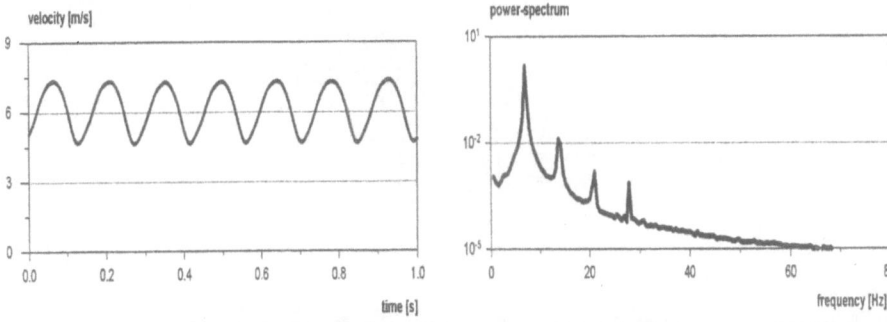

Fig. 7 Main flow velocity oscillations

In unsteady operation the wind tunnel was run for a range of different flow speeds, flap sizes and flap frequencies for the setting of different Reynolds number and Strouhal number combinations. Fig. 7 shows an example of the velocity signal of the mean flow and its power-spectrum. It can be seen that the flow contains a fundamental frequency (equal to twice the flap frequency) and a number of harmonics, which decrease in amplitude rapidly.

Preliminary Results

After the installation of the test plate and the contoured wall, the static pressure distributions on the plate were measured. The normalized velocities over the plate derived from the static pressure measurements for two typical cases are shown in Fig. 6, where c_0 is the incoming flow velocity. It can be seen that in the case of $c_0 = 9.5$ m/s (low Reynolds number) a separation bubble (zone a-b-c in Fig. 6) occurs. This separation bubble can not be found at $c_0 = 20.5$ m/s (high Reynolds number).

The characteristic of periodic-unsteady transition was determined by hot-wire measurements. Fig. 8 shows the effects of Strouhal number on the transition of boundary layers without pressure gradient. The main flow and boundary layer velocity signals measured at the same x-position with two hot-wires are plotted against time. The measurements were made with the same rotating flap, which means the time-averaged flow velocity and the flow velocity amplitude are constant. For a low Strouhal number of 0.2, the upper diagram of Fig. 8 shows that the turbulent zone begins immediately after the maximum velocity and ends before reaching the next velocity rise zone. The intermittency factor is about 0.5. The effects of a more than three times higher Strouhal number 0.65 are shown in the lower diagram of Fig. 8. The turbulent zone stretches from the middle of the velocity decrease zone to the middle of the velocity rise zone, while the intermittency factor is the same. It can also be seen in this figure that the switching between the laminar and turbulent part in both diagrams is periodic and its fundamental frequency is equal to the first frequency of the main flow. This results agree well with the measurements made by *Miller and Fejer* [5].

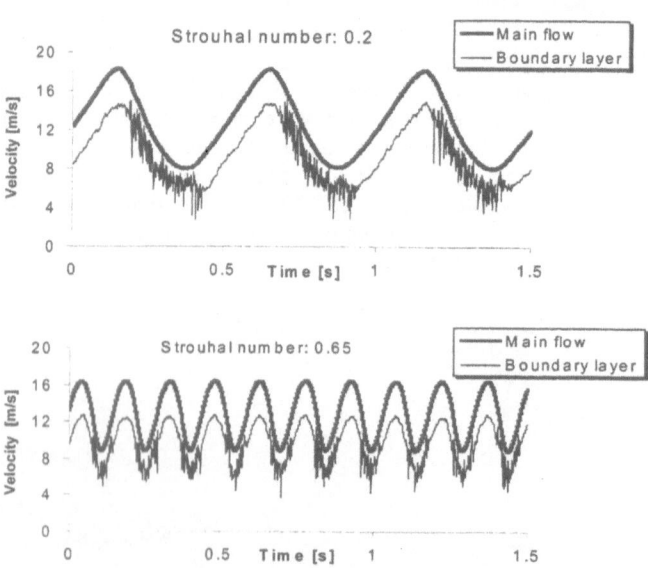

Fig. 8 Strouhal-number effect on unsteady transition

Concluding Remarks

A combined low speed test facility for the investigation of skewed mixing layers and unsteady boundary layers with pressure gradients was built at the Jet Propulsion Laboratory of the Berlin University of Technology. A special arrangement of two independently controlled radial blowers drives both wind tunnel parts of the test facility: the double stream blower tunnel and the unsteady suction tunnel. A bypass system allows the independent operation of the two wind tunnel parts. The design of the wind tunnel legs resulted in an overall classic open-wind tunnel design, with some special features like the extensive modularity of the wind tunnel segments. The flow quality of both parts of the facility is characterized by a non-uniformity within ±1% and a turbulence level below 0.5 %.

The Double Stream Wind Tunnel has a specially designed nozzle with an attached deflection module to generate a plane, skewed mixing layer. For better adaptability and easier handling, the nozzle consists of two segments: a universal primary 3D-nozzle and an adaptable secondary 2D-nozzle. The deflection of the two streams is provided by an adjustable deflection module attached on the 2-D-nozzle. Preliminary free stream X-wire measurements show a very uniform velocity distribution at a constant flow angle normal to the mixing layer. This mixing layer facility is expected to provide new insights into the structure and mixing properties of skewed mixing layers.

The oscillating flow of the Unsteady Suction Tunnel is generated by a downstream rotating flap driven by an electronically controlled motor. Flaps of different width can be used and the rotational speed can be varied to set the amplitude and frequency of the oscillating flow. The Reynolds number and Strouhal number of turbomachinery boundary layers can be simulated by this arrangement. Using a specially designed contoured wall opposite the test plate at low velocities, a separation bubble, which occurs in turbomachines at low Reynolds number, can be generated on the flat plate. This phenomenon has not been investigated in unsteady flow. Based on first results, this powerful test facility offers possibilities to investigate unsteady boundary layers with pressure gradients, unsteady separation bubbles as well as unsteady film cooling.

References

[1]   R. D. Mehta
      „The aerodynamic design of blower tunnels with wide-angle diffusers",
      Pro. Aerospace Sci., Vol. 18, pp. 59 -120, 1977.

[2]   N. Schröder, J. Hourmouziadis
      „Design and Calibration of a Double Stream Wind Tunnel", CEAS European Forum „Wind Tunnels and Wind Tunnel Test Techniques", Cambridge, P49.1-P49.8, 1997.

[3]   B. S. Stratford
      "The Prediction of Separation of the Turbulent Boundary Layer", J. Fluid Mech. Vol. 5,
      pp. 1 -16, 1959.

[4]   B. Dziomba, H. E. Fiedler
      „Effect of intial conditions on two-dimensional free shear layers", J. Fluid Mech, vol. 152,
      pp. 419-442, 1985.

[5]   J. A. Miller, A. A. Fejer
      "Transition phenomena in oscillation boundary flows", J. Fluid Mech. Vol. 18, part 3,
      pp. 438-449, 1964 .

Data preparation for the stability analysis of the ELFIN II HLF wing measurements in S1Ma wind tunnel

W. Schröder

Deutsches Zentrum für Luft- und Raumfahrt (DLR) e. V.
Institut für Entwurfsaerodynamik
Lilienthalplatz 7, D-38108 Braunschweig, Germany

G. Schrauf

Daimler-Benz Aerospace Airbus
EFV
D-28183 Bremen, Germany

Summary

For the stability analysis of the ELFIN II HLF half wing measurements which were conducted in the S1Ma wind tunnel the input data for 55 selected test runs have been prepared.

The following steps have been treated:
- Selection of the test cases
- Smoothing and interpolation of measured wing coordinates and pressures in the nose section
- Determination of transition locations (span- and chordwise) from IR images
- Interpolation of contour pressures
- Determination of the stagnation point conditions (location, pressure and effective sweep)
- Determination of the suction velocity distribution by the measurement of the suction skin perforation characteristics and the interpolation of the pressure difference between wing surface and suction chambers of the experiments.

With these prepared geometry, wall pressures and suction velocities the boundary layer parameters were calculated as a common input for the individual stability analysis of the different European partners within the HYLDA program (task 3.2). First results are presented in [1].

Introduction

One of the objectives of the European laminar flow investigations is to build up design criteria for laminar flow wings. Herein an important objective are criteria for the laminar/turbulent transition. In case of NLF (natural laminar flow) a wide experience base already exists, both for wind tunnel as well as for flight tests (for example F100 flight tests in the ELFIN program). The further studies are concerned with widening the knowledge of HLF (hybrid laminar flow) wing design.

Thus in the framework of the European ELFIN II program tests with an HLF half wing model have been done in the S1Ma wind tunnel of ONERA in Nov. 96. The tests were carried out with the model at fixed sweep angles of 32 and 36° and incidences between ± 4°, at a Mach number range from 0.3 to 0.84 and at Reynolds numbers between 9 and 19 millions. The measurement of all wind tunnel and model data was with ONERA as the wind tunnel operator [2]. The final data were collected, prepared and distributed by DASA on a CD-ROM to the partners for further assessment [3]. The infrared images of both inner and outer wing surfaces have been taken by DLR and ONERA and are available on CD as well. The stability analysis of the laminar boundary layer for selected tests are subject of task 3.2 of the subsequent HYLDA program. The data preparation for the boundary layer calculations are partly shared between the HYLDA partners. The boundary layer calculations as final input for the stability analysis have been carried out at DLR by the method of [4].

Data preparation

Selection of the test cases

Based on the measured wing surface pressure distribution and on the IR images the partner have defined suitable candidate test cases [5].

Based on these recommendation 45 HLF and 10 NLF cases have finally been selected for the stability analysis [6]. The selection was mainly directed by the quality of the IR images. The selection covers the whole range of model and tunnel parameter i.e. leading edge sweep, Mach number, incidence and suction velocity.

Preparation of the wing geometry and pressure close to the nose

Fig. 1 shows a sketch of the wing. Surface pressure measurements were done at 25, 50 and 75% of half span. The actual wing geometry including the pressure tap locations had been measured at DASA. Due to dense and spanwise displaced locations of pressure taps in the vicinity of the wing leading edge, the transfer of these locations onto the three ideal sections of constant span resulted in wavy positions of the taps (fig. 2) and the related pressure values. To smooth the data the partner have agreed on:

- reducing the coordinate data points to one point resulting from the pressure distribution.
- coordinates which deviated remarkably from the expected elliptical nose contour were omitted, for example the data points 1 on both wing surfaces as well as 3-5 on the upper side.

Transition locations

The transition locations have been determined from the IR images [5]. The images show turbulent wedges at the sections of the pressure tap rows. This can be seen in fig. 3 for the outer section DV3 on the upper wing surface.

All of the six sections CP1 – CP6 on the HLF part of the wing, see fig. 1, have been identified as candidates for the HLF stability analysis. In addition section CP7 on the NLF reference panel just outside the suction panel has been selected for the NLF stability calculations. Thus the chordwise transition locations of one wind tunnel run are defined for a particular spanwise section on a particular wing surface.

Interpolation of the contour pressures

Taking into account the results of Navier Stokes calculations based on the measured wing geometry, including the deformations due to aerodynamic loads, the following spanwise interpolation procedures have been agreed between the HYLDA partners:

- linear interpolation between pressure sections DV1 and DV2
- constant pressure between sections DV2 and DV3 i.e. cp of section DV2
- pressure in section CP7 being identical to that of section DV3.

Determination of stagnation line conditions and effective sweep

For the conical wing boundary layer calculation precise information on the stagnation line and its pressure distribution is needed. Fig. 4 shows the measured contour pressure in the vicinity of the nose being densely interpolated in the nose area. The maximum of a parabolic curve fit yields the pressure and the location of the stagnation line. The effective sweep angle is derived from the velocity along the stagnation line. It amounts to about 1-1.5° larger than the geometric leading edge sweep depending on Mach number [7].

Suction skin loss characteristics

The loss characteristics of the suction panel has been measured in the six sections CP1 to CP6 on both wing surfaces at ambient conditions [8]. Fig. 5 shows layout details of the suction nose. In each of the six spanwise sections the skin losses have been calibrated in the centre of each of the 51 suction chambers. Fig. 6 shows the principles of the measurement system. The laminar flow meter probe with a suction diameter of 8 mm is shown in more details in fig. 7.

The complete equipment is mobile, fast, independent and energy efficient in comparison to previous ones which needed the complete wing suction system for just local loss data. At the mid position of each suction chamber the flow is pressed through the perforated skin for preselected pressure differences. The suction chamber side is the ambient pressure reference.

Fig. 8 shows the calibration curves for all 51 chambers of section CP4. The bandwith reflects directly the inhomogenity of the laser drilled suction surface. The density scaling of the loss curves is illustrated in fig. 9. The curve fit through the 5 measured data points is of 3rd order. The linear term represents the laminar skin friction in the suction holes in analogy to the laminar pipe flow behaviour. Thus this term is independent on density and Re number respectively. The higher order terms represent the stagnation pressure dependent losses which needs a linear correction of density. The figure 9 demonstrates the scaling for a suction pressure level of 0.5 bar. For a constant pressure jump across the skin, 9% higher suction velocities are obtained.

Calculation of suction velocities

At each suction chamber center the surface as well as the chamber pressures are needed to obtain the suction velocities from the calibration charts. A span- and chordwise interpolation of the suction chamber pressures is needed. A pneumatic layout of the suction compartments is sketched in fig. 10. The spanwise locations of the pressure sections differ between the nose compartments 5 to 13 and the rear compartments 1 to 4 and 14 to 18. A similar behaviour can be found with the suction pipe connections. The rear compartments are sucked at the inner root, a monotonic spanwise increasing chamber pressure is expected. The suction chambers at the nose have 2 to 3 spanwise suction ports with different spanwise locations. A prediction of the spanwise pressures is difficult particularly because of local blockage of the surface pressure tubings inside these narrow suction chambers at the nose.

Fig. 11 gives an example of the internal pressure distribution inside the nose and rear compartments. The figure covers the mayor three data correction methods applied. In case a pressure tap has already been found blocked before or during tests (tap 404) its pressure has been set equal to that of a neighbour tap (tap 304). In such cases the wrong pressure has been overwritten by the linearly interpolated pressures of the two neighbour taps (304 and 504).

The chamber 14 showed for all wind tunnel runs large scatter of the central taps 314 and 414 which did not follow the tendency of the outer taps 114 and 514 at all and can not be explained. In this study these 2 pressures were corrected by a linear interpolation between the outer taps.

The correction applied to the pressures of the rear compartments seemed to be more reliable because a monotonic spanwise increasing pressure is expected. In case of compartment 2 tap reading 4015 is taken as reference to correct the readings of the taps 202 and 502 to get in line with the gradient of the adjacent compartments 1 and 3.

Based on the interpolated and corrected surface and chamber pressures, the suction velocities are calculated via the suction loss characteristics of each chamber of a particular spanwise section. An example from [9] is shown in fig. 12.

A typical suction velocity distribution is shown in fig. 13 for the outer section CP5. As comparison the mean suction velocities, based on overall flow meter readings of [3] are included. These mean reference suction data do not reflect the spanwise varying chamber pressures as well as the local suction skin characteristics.

Boundary Layer Calculation

Based on the prepared wing geometries, pressures and suction velocities the boundary layer profiles and parameter have been calculated with the method [4] for the flow conditions of each of the 55 selected cases. They were distributed to the HYLDA partners being engaged in the stability analysis.

References

[1] Schrauf, G.; Schröder, W.: Comparison of N-factor strategies using wind tunnel experiments and flight tests, Paper T12, 11. DGLR-Fach-Symp. "Strömungen mit Ablösung", TU Berlin, Nov. 10-12, 98.

[2] Garcon, F.: ELFIN II laminar wing tests in S1MA wind tunnel (Nov. 1996), ELFIN II TR 208, June 1997.

[3] Kotschote, J.: Compilation of test data [1], ELFIN II CD-ROM, May 1997.

[4] Horton, H.P.; Stock, H.W.: Computation of compressible, laminar boundary layers on swept, tapered wings, J. Aircraft, 32, 1402-1405, 1995.

[5] Schröder, W.: Determination of the laminar/turbulent transition stations of the ELFIN II HLF half wing S1MA wind tunnel experiment, HYLDA TR 5, March 1997.

[6] Schröder, W.: Definition of the first fifteen ELFIN II wing test cases for the boundary layer stability calculations, HYLDA TR 17, February 1998.

[7] Schrauf, G.: Preparation of the boundary layer input for the analysis of the ELFIN II HLF experiment for all 47 cases chosen for evaluation with linear stability theory, HYLDA TR 7, Oct. 1998.

[8] Müller, R.; Trenz, Th.: Measurement of the suction surface characteristics of the ELFIN II half wing, HYLDA TR 22, in preparation.

[9] Schröder, W.: Determination of the suction velocities for the boundary layer stability analysis of the first fifteen ELFIN II wing test cases, HYLDA TR 23, Febr. 1998.

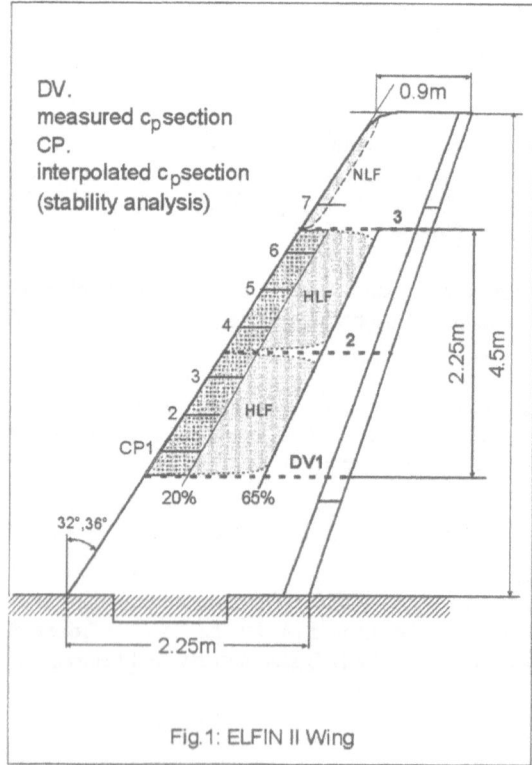

Fig.1: ELFIN II Wing

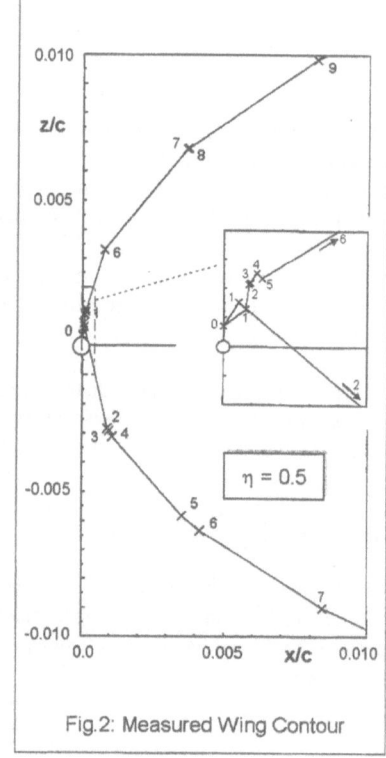

Fig.2: Measured Wing Contour

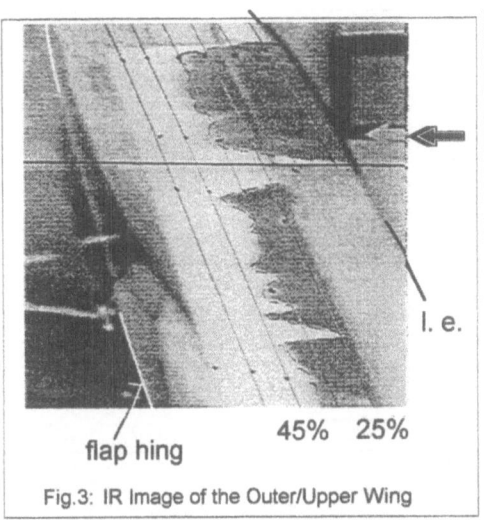

Fig.3: IR Image of the Outer/Upper Wing

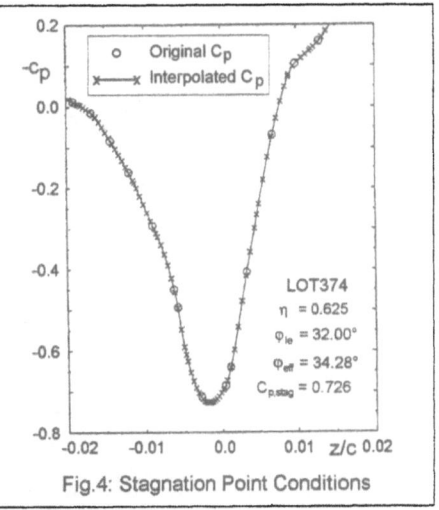

Fig.4: Stagnation Point Conditions

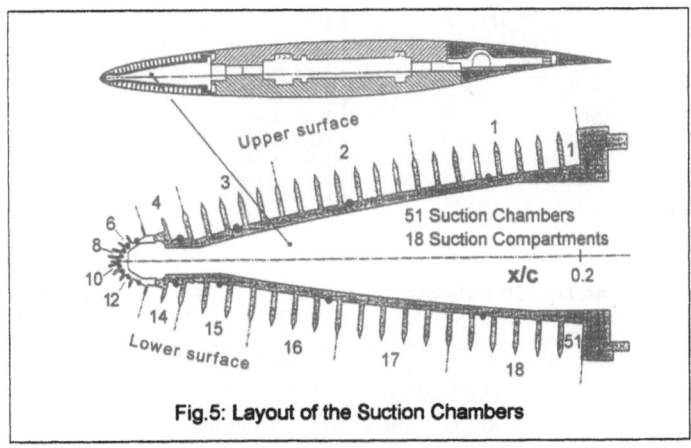

Fig.5: Layout of the Suction Chambers

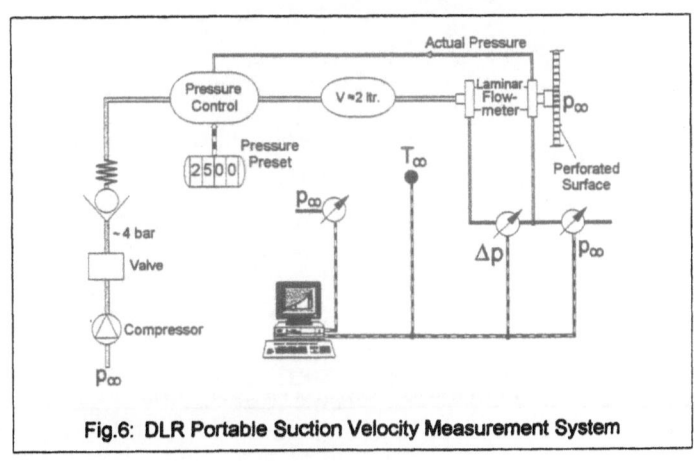

Fig.6: DLR Portable Suction Velocity Measurement System

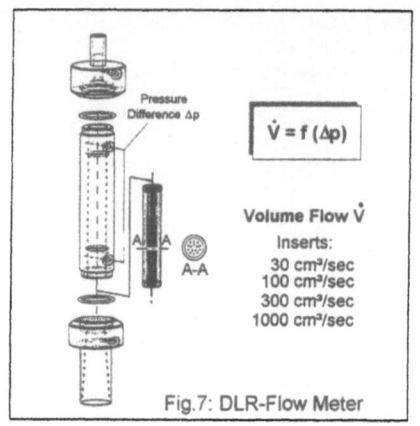

$$\dot{V} = f\,(\Delta p)$$

Volume Flow $\dot{V}$

Inserts:
30 cm³/sec
100 cm³/sec
300 cm³/sec
1000 cm³/sec

Fig.7: DLR-Flow Meter

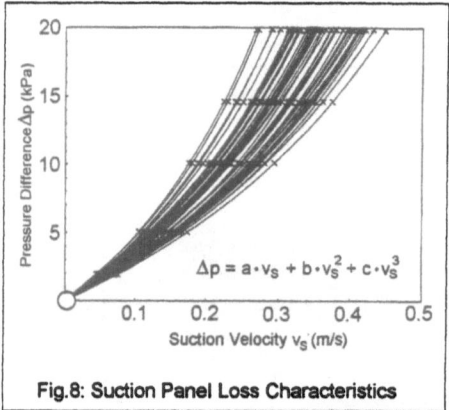

$$\Delta p = a \cdot v_S + b \cdot v_S^2 + c \cdot v_S^3$$

Fig.8: Suction Panel Loss Characteristics

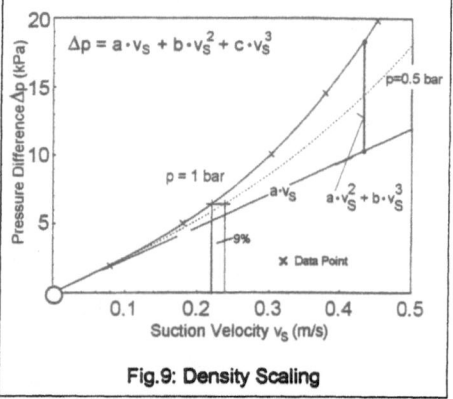

$$\Delta p = a \cdot v_S + b \cdot v_S^2 + c \cdot v_S^3$$

Fig.9: Density Scaling

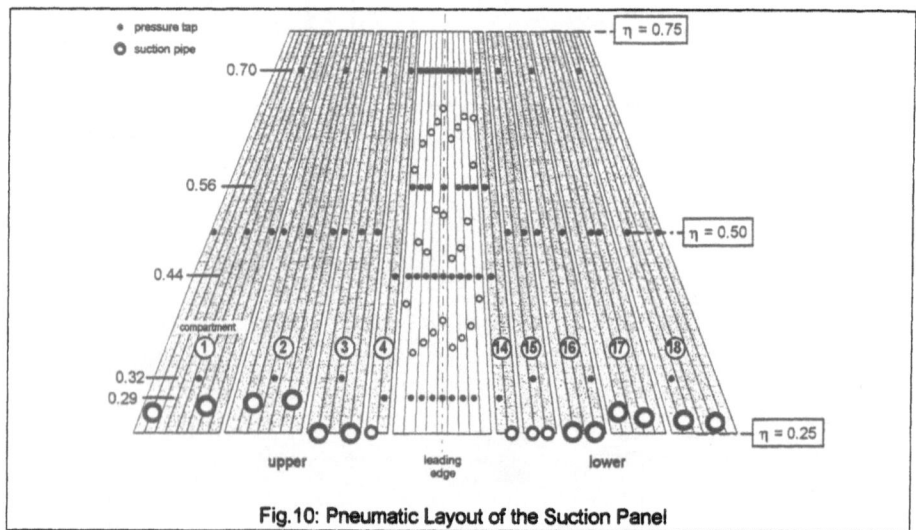

Fig.10: Pneumatic Layout of the Suction Panel

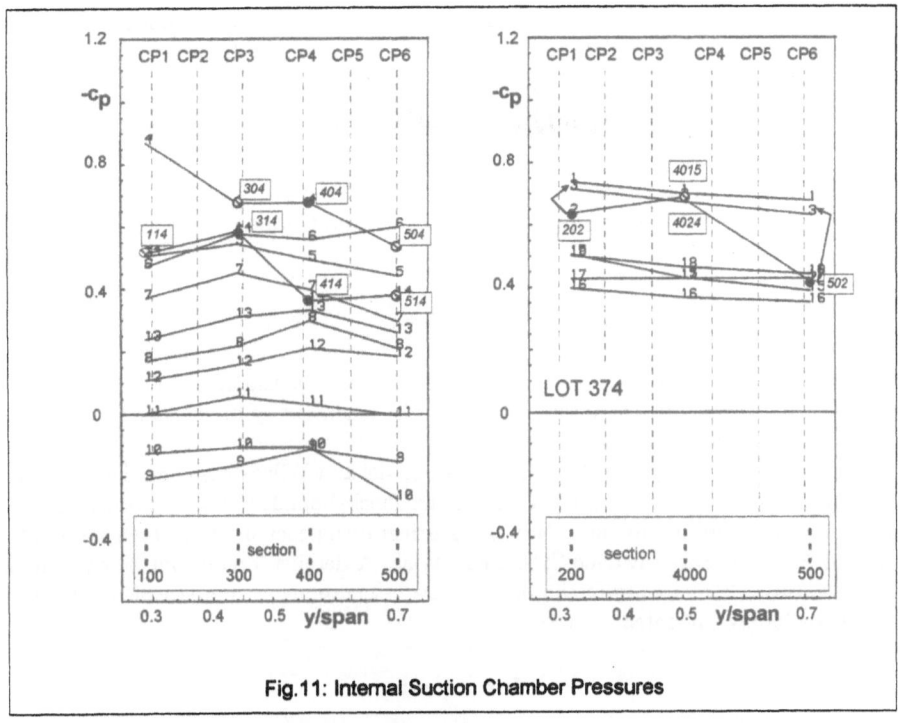

Fig.11: Internal Suction Chamber Pressures

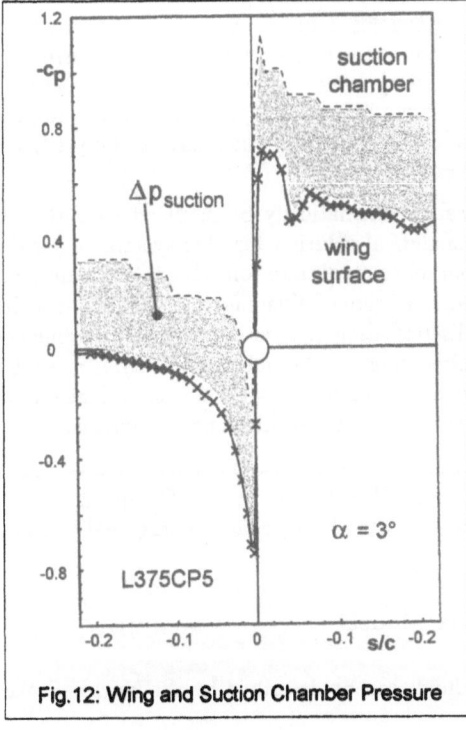

Fig.12: Wing and Suction Chamber Pressure

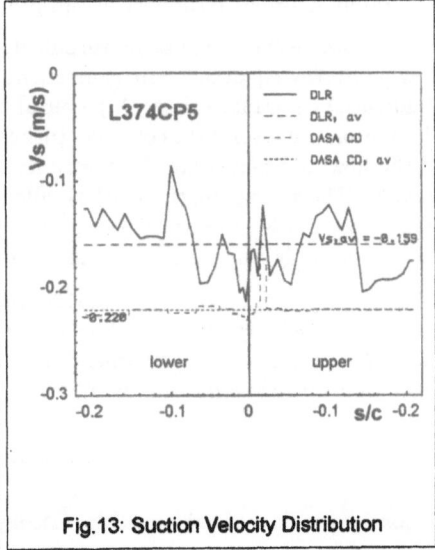

Fig.13: Suction Velocity Distribution

Numerical Investigation of Boundary Layer Separation Caused by Turbulent Shock/Boundary Layer Interaction

A. Schubert, T. Coratekin, J. Ballmann

Lehr- und Forschungsgebiet für Mechanik der RWTH Aachen,
Templergraben 64, 52062 Aachen, Germany

Summary

This work focuses on the numerical simulation of two–dimensional supersonic turbulent flows close to solid walls. The interaction of shocks generated by discontinuous changes of the contour slope with turbulent boundary layers is of particular interest. Two different turbulent models, namely the Baldwin–Lomax model and Wilcox's k–ω model, are tested regarding their prediction quality in simulating shock/boundary layer interactions with separation. They belong to the group of statistical turbulence models which are used to close the Favre–averaged Navier–Stokes equations. A detailed description of each model as well as two different compression corner flow calculations in comparison with experimental results are presented.

Introduction

High speed flows characterized by very high Reynolds numbers are dominated by turbulence. Due to the intensive exchange of momentum and energy caused by turbulent fluctuations within the fluid, turbulent boundary layers separate later than laminar. Shock/boundary layer interactions can cause boundary layer separation and are responsible for high mechanical and thermal loads on the structure, which have to be predicted accurately in order to support design processes.

Plane flows, either laminar or turbulent, can mathematically be modeled with the 2D Navier–Stokes equations. In general, no analytical solution for this system of partial differential equations exists. A powerful tool for approximate solutions is the numerical simulation. To resolve the complete turbulent spectrum of different time and length scales of the flow, one can either use a very fine discretization or apply appropriate turbulence models. The first approach, which is called direct numerical simulation (DNS), is presently limited to detailed investigations for low Reynolds number flows. Thus, turbulence modeling is the only way for calculating technically relevant, turbulent flow problems.

This investigation concentrates on statistical turbulence models, in particular on two eddy viscosity models: the Baldwin–Lomax model and Wilcox's k–ω model. Computations for two supersonic compression corner flows with boundary layer separation are analyzed and compared with experimental data.

Turbulence Models

A continuum flow of compressible, viscous fluids can be described by the Navier–Stokes equations. The conservation equations of mass, momentum and total energy plus an additional caloric and thermal equation of state and appropriate material laws form a closed set of governing equations for laminar and turbulent flows.

One possibility for the numerical treatment of turbulent flows is statistical turbulence modeling. The key is to write all flow quantities as the sum of a time averaged mean value $\overline{f}$ and a fluctuation about the mean value f'. For compressible flows, a combination of Reynolds–averaging eqn. (1) and Favre–averaging eqn. (2) is used [8].

Reynolds–averaging:
$$
\begin{cases}
f(\mathbf{x}, t) = \overline{f(\mathbf{x})} + f'(\mathbf{x}, t) \quad \text{with} \\[2mm]
\overline{f(\mathbf{x}, t)} = \frac{1}{\delta t} \int\limits_{t}^{t+\delta t} f(\mathbf{x}, t) dt \ , \quad \overline{f'(\mathbf{x}, t)} = 0 \ .
\end{cases}
\tag{1}
$$

Favre–averaging:
$$
\begin{cases}
f(\mathbf{x}, t) = \tilde{f}(\mathbf{x}, t) + f''(\mathbf{x}, t) \quad \text{with} \\[2mm]
\tilde{f}(\mathbf{x}, t) = \dfrac{\overline{\rho(\mathbf{x}, t) f(\mathbf{x}, t)}}{\overline{\rho(\mathbf{x}, t)}} \ , \quad \overline{\rho(\mathbf{x}, t) f''(\mathbf{x}, t)} = 0 \ .
\end{cases}
\tag{2}
$$

The statistical turbulence modeling for compressible flows yields additional terms in the so called Favre–averaged Navier-Stokes equations eqn. (3)–(5), which are framed with boxes [3],[6],[8]:

$$
\frac{\partial \overline{\rho}}{\partial t} + \frac{\partial (\overline{\rho} \tilde{v}_\gamma)}{\partial x_\gamma} = 0 \ ,
\tag{3}
$$

$$
\frac{\partial (\overline{\rho} \tilde{v}_j)}{\partial t} + \frac{\partial (\overline{\rho} \tilde{v}_j \tilde{v}_\gamma)}{\partial x_\gamma} = -\frac{\partial \overline{p}}{\partial x_j} + \frac{\partial}{\partial x_\gamma} \left(\overline{\tau_{j\gamma}} - \boxed{\overline{\rho v_j'' v_\gamma''}} \right) \ ,
\tag{4}
$$

$$
\frac{\partial (\overline{\rho} \tilde{e})}{\partial t} + \frac{\partial}{\partial x_\gamma} \left(\tilde{v}_\gamma (\overline{\rho} \tilde{e} + \overline{p}) \right) = \frac{\partial}{\partial x_\gamma} \left(k \frac{\partial \overline{T}}{\partial x_\gamma} - \boxed{c_p \overline{\rho v_\gamma'' T''}} - \boxed{\tfrac{1}{2} \overline{\rho v_\lambda'' v_\lambda'' v_\gamma''}} \right) +
$$
$$
\frac{\partial}{\partial x_\gamma} \left(\overline{v_\lambda \tau_{\lambda\gamma}} - \boxed{\tilde{v}_\lambda \overline{\rho v_\lambda'' v_\gamma''}} \right) \ .
\tag{5}
$$

These new terms now have to be adequately modeled in order to close the set of equations. Besides the turbulent heat flux $c_p \overline{\rho v_i'' T''}$ and the triple correlation $\tfrac{1}{2} \overline{\rho v_\lambda'' v_\lambda'' v_i''}$, the main difficulty is the calculation of the Reynolds stress tensor $(\tau_{ij})_{turb} = -\overline{\rho v_i'' v_j''}$. Two different approaches are proposed in this investigation.

The approximation of Boussinesq is widely accepted and forms the basis for all eddy viscosity models. The Reynolds stress tensor is described according to the ansatz of Stokes for the molecular stress tensor:

$$
-\overline{\rho v_j'' v_k''} = \mu_T \left(\frac{\partial \tilde{v}_j}{\partial x_k} + \frac{\partial \tilde{v}_k}{\partial x_j} \right) - \delta_{jk} \frac{2}{3} \mu_T \frac{\partial}{\partial x_\gamma} (\tilde{v}_\gamma) - \delta_{jk} \frac{2}{3} \overline{\rho} k \ .
\tag{6}
$$

Two new variables are introduced: the turbulent viscosity μ_T and the turbulent kinetic energy $k=\frac{1}{2}\widetilde{v''_\lambda v''_\lambda}$. Both can be obtained with an eddy viscosity model, e.g. an algebraic turbulence model or a two-equation turbulence model. The present research uses a Baldwin–Lomax model [1] and the k–ω model proposed by Wilcox [8] and Haidinger [3]. A major deficiency of eddy viscosity models is that they account only for isotropic turbulence.

The components of the Reynolds stress tensor can be determined in a more sophisticated way with the Reynolds stress transport equations (RSE). In this approach each component of the Reynolds stress tensor has its own transport equation. As the turbulent stress tensor is symmetric, three (2D) or six (3D) additional differential equations have to be solved. The use of an algebraic Reynolds stress model (ARSM) based on the RSE offers a reduction in computation time. It calculates the Reynolds stresses with algebraic expressions derived from the RSE but still requires a two–equation turbulence model. According to the authors' experiences it has not been possible to achieve satisfying results of the presented compression corner flows with an ARSM [4],[6].

The Baldwin–Lomax Model

Algebraic turbulence models neglect the turbulent pressure term $-\delta_{jk}\frac{2}{3}\bar{\rho}k$ in eqn. (6). Hence, the only variable left to be determined is the turbulent viscosity μ_T. The Baldwin–Lomax model is a two–layer model, in which the turbulent viscosity is computed separately in each of them [1]:

$$\mu_T = \begin{cases} \mu_{T_{inner}} & \text{for} \quad y \leq y_{crossover} \\ \mu_{T_{outer}} & \text{for} \quad y > y_{crossover} \end{cases} ,$$

where $y_{crossover}$ is the smallest value of y for which $\mu_{T_{inner}} = \mu_{T_{outer}}$ holds. The expressions for μ_T include known values of the mean flow variables and several closure coefficients. For wall bounded flows with y_s being the normal distance of the wall, it gives:

$$\mu_{T_{inner}} = \bar{\rho}l^2|\omega| \tag{7}$$

with

$$|\omega| = \left|\frac{\partial u}{\partial y} - \frac{\partial v}{\partial x}\right|, \; l = ky_s[1 - \exp(-y^+/A^+)], \; y^+ = y_s\frac{\sqrt{\rho_w \tau_w}}{\mu_w}, \; \tau_w = \mu_w\left|\frac{\partial \bar{v}_1}{\partial x_2} + \frac{\partial \bar{v}_2}{\partial x_1}\right|_w$$

and

$$\mu_{T_{outer}} = K\,C_{cp}\,\rho\,F_{wake}\,F_{kleb}(y) \tag{8}$$

with

$$F_{wake} = \min\left(y_{max}F_{max}, \frac{C_{wk}y_{max}u^2_{diff}}{F_{max}}\right), \quad F_{kleb}(y) = \left(1 + 5.5\left(\frac{C_{kleb}\,y}{y_{max}}\right)^6\right)^{-1}.$$

F_{max} is the maximum value of $F(y) = y|\omega|[1 - \exp(-y^+/A^+)]$ at the position y_{max}. u_{diff} describes the difference between the maximum and minimum velocity within the boundary layer: $u_{diff} = (\sqrt{\mathbf{vv}})_{max} - (\sqrt{\mathbf{vv}})_{min}$.

The remaining constants have the following values:

$$
\begin{array}{llll}
A^+ & = & 26, & k & = & 0.4, \\
C_{cp} & = & 1.6, & K & = & 0.0168, \\
C_{wk} & = & 0.25, & C_{kleb} & = & 0.3 \; .
\end{array}
$$

The turbulent heat flux is calculated in a similar way as the molecular heat flux according to Fourier:

$$
c_p \overline{\rho v_i'' T''} = -k_T \frac{\partial \tilde{T}}{\partial x_i} \; . \tag{9}
$$

The newly introduced turbulent heat flux coefficient k_T is defined using a constant turbulent Prandtl number

$$
k_T = \frac{c_p \mu_T}{Pr_T} \quad \text{with} \quad Pr_T = 0.9 \; . \tag{10}
$$

Finally, the triple correlation $\frac{1}{2}\overline{\rho v_\lambda'' v_\lambda'' v_i''}$ in the energy equation is simply neglected. This model is easy to implement, robust and computational costs are moderate. It yields good estimates for the mean flow field but is, like any statistical turbulence model, not capable to predict turbulent structures.

The k–ω Model

All two–equation turbulence models introduce two additional transport equations for two turbulent variables. In this case these are k and ω. The specific turbulent dissipation rate ω is needed to calculate $\mu_t = \alpha^* \bar{\rho} k/\omega$. According to [3] and [8], the transport equations have the following form using the Einstein summation convention:

$$
\frac{\partial(\bar{\rho}k)}{\partial t} + \frac{\partial}{\partial x_\gamma}(\bar{\rho}k\tilde{v}_\gamma) = H_k + \frac{\partial}{\partial x_\gamma}\left[\left(\bar{\mu} + \frac{\mu_t}{Pr_k}\right)\frac{\partial k}{\partial x_\gamma}\right] , \tag{11}
$$

$$
\frac{\partial(\bar{\rho}\omega)}{\partial t} + \frac{\partial}{\partial x_\gamma}(\bar{\rho}\omega\tilde{v}_\gamma) = H_\omega + \frac{\partial}{\partial x_\gamma}\left[\left(\bar{\mu} + \frac{\mu_t}{Pr_\omega}\right)\frac{\partial \omega}{\partial x_\gamma}\right] \tag{12}
$$

with

$$
H_k = (\tau_{\lambda\gamma})_{turb}\frac{\partial \tilde{v}_\lambda}{\partial x_\gamma} - \left(\beta^* - \alpha_2 M_T^2\right)\bar{\rho}k\omega,
$$

$$
H_\omega = \hat{\alpha}(\tau_{\lambda\gamma})_{turb}\frac{\partial \tilde{v}_\lambda}{\partial x_\gamma}\frac{\omega}{k} - \left(\beta + \alpha_2 M_T^2\right)\bar{\rho}\omega^2 - \underbrace{(C_{\epsilon2} - 1)\beta^*\sqrt{\frac{1}{2}\left(\frac{\partial \tilde{v}_\lambda}{\partial x_\gamma} - \frac{\partial \tilde{v}_\gamma}{\partial x_\lambda}\right)^2}\rho\omega}_{\text{vorticity term } vc}
$$

and $Pr_k = Pr_\omega = 2$, $M_T = \sqrt{2k}/c$ with $c = $ speed of sound.

For free stream turbulence the following high–Reynolds number values are used:

$$
\hat{\alpha} = (C_{\epsilon1} - 1), \quad \beta^* = C_\mu = 0.09, \quad \beta = (C_{\epsilon2} - 1)\beta^*,
$$

$$
C_{\epsilon1} = 1.56, \quad C_{\epsilon2} = 1.83, \quad \alpha^* = 1, \quad \alpha_2 = 0.2 \; .
$$

These coefficients are corrected for low–Reynolds number flows with $Re_T = \bar{\rho}k/\bar{\mu}\omega$ as follows:

$$
\begin{aligned}
\hat{\alpha} &= \frac{(C_{\epsilon1} - 1)}{\alpha^*}\frac{0.1 + Re_T/2.7}{1 + Re_T/2.7} \quad \text{without } VC, \\
&= \frac{36(C_{\epsilon1} - 1)}{25}\frac{1}{\alpha^*}\frac{0.1 + Re_T/2.7}{1 + Re_T/2.7} \quad \text{with } VC,
\end{aligned}
$$

$$\beta^* = C_\mu \frac{5/18 + (Re_T/8)^4}{1 + (Re_T/8)^4} \quad \text{and} \quad \alpha^* = \frac{0.15 + Re_T}{6 + Re_T}.$$

Wall bounded flows also need a special treatment of ω. Thus, an asymptotic solution ω_a is implemented for ω in near wall regions ($y^+ < 2.5$) [3],[8], where the closest cell to the wall should be at $y^+ < 1$ and the value of ω at the wall ($y_s=0$) is extrapolated.

$$\omega_a = \frac{6\mu_w}{(C_{\epsilon 2} - 1)C_\mu \rho_w y_s^2}.$$

The turbulent heat flux and the triple correlation are treated in the same way as in the Baldwin–Lomax model.

Solving eqn. (11) and eqn. (12) requires more CPU time than using the Baldwin–Lomax model, but the accuracy provided by additional physics strongly encourages this approach.

Numerical Method

The numerical algorithm used to calculate approximate solutions of the Favre–averaged Navier–Stokes equations and the transport equations for k and ω utilizes a Finite–Volume–Method to discretize all spatial derivatives appearing in the balance equations. Upwind discretization is applied to the hyperbolic part of the equations, while centered differences are used for the diffusive part. The upwind discretization is handled with a second-order-in-space Godunov–Type method. Time integration is performed by means of an implicit scheme. The resulting system of nonlinear equations is linearized with Newton iterations. The obtained systems of linear equations are solved with a relaxation scheme based on conjugate gradient methods [2]. The Navier–Stokes equations and the k–ω equations are solved in a loosely coupled scheme with the same numerical algorithm. This means that between each Newton step of the mean flow computation the k–ω equations are solved with constant mean flow variables. Similarly, the equations of conservation for the mean flow field are integrated in time using frozen values of k and ω from the previous iteration step. This method is applicable if one is interested in the steady state solution.

Results

The computations of two different supersonic flows passing a compression corner with a deflection angle of 24° serve as test cases. This kind of shock/boundary layer interaction can be observed for example at the deflected flap of a hypersonic airplane. In both computations the solid wall boundary was supposed to be adiabatic and air was modeled as a gas consisting of 23% O_2 and 77% N_2 with the following free stream conditions:

	$V_\infty \, [m/s]$	$p_\infty \, [Pa]$	$\rho_\infty \, [kg/m^3]$	$T_\infty \, [K]$	$Re_{x\infty} \, [1/m]$	$M_\infty \, [-]$
Exp. a)	579	23940	0.8068	103	$6.7 \cdot 10^6$	2.84
Exp. b)	2377	8666	0.11797	255	$1.70 \cdot 10^7$	7.41

The first experiment was performed by Settles [5] with a fully turbulent incoming boundary layer. The second example, which was investigated at the shock tunnel of the RWTH Aachen [7], is characterized by a much higher Mach number. Here, the boundary layer was also supposed to be turbulent. The numerical simulation, however, suggested the existence of laminar flow and motivated further experiments which confirmed that the flow in fact was laminar.

Both flow situations show similar physical behavior. The pressure rise following the deflection causes a flow separation close to the kink which can clearly be identified in Fig. 1 and 2. As the size of the separation bubble is much smaller in the second case, it can hardly be seen in Fig. 3. One can observe in both cases that a shock starts upstream of the corner, close to the separation point. The shock angle decreases with increasing free stream Mach number. On the deflected part of the ramp, the flow later reattaches. Qualitatively, the calculated contour plots show good agreement with the experimental results (Fig. 1, 2 and 3). For the wall pressure distribution p_w/p_∞ (Fig. 4) as well as for the first part of the skin friction coefficient c_f (Fig. 5) we found close agreement with the measurements, but for both quantities, the k-ω results are closer to the experimental data than the Baldwin–Lomax calculation.

In the second test case, the flow, which was assumed to be turbulent, was first calculated with the k-ω model. As the values of the kinetic turbulent energy were far too low for the existence of turbulence, a second calculation was performed with a laminar program version. The turbulent and the laminar results agree reasonably well with the experimental data as can be seen in Fig. 6. Due to this result the incoming boundary layer was investigated again in further experiments and was finally characterized as laminar.

Conclusion

A study was performed to investigate the prediction quality of the Baldwin–Lomax and the k-ω turbulence model in order to calculate separated supersonic boundary layers. Besides the description of the eddy viscosity models, the results of two 2D compression corner flows, each with a deflection angle of 24°, were presented. The first case was a turbulent shock/boundary interaction and both turbulence models were used to calculate the flow. The predicted flowfields showed all the expected features of the physical problem. Quantitative comparison of the wall pressure and the skin friction coefficient with experimental data showed only minor discrepancies. The best agreement was obtained with the k-ω model, but a surprisingly good result was also achieved with the Baldwin–Lomax model. The second flow case was a shock/boundary interaction. The numerical results could clarify the laminar character of the flow in a previously unclear experimental situation.

Acknowledgments

The numerical results were calculated on computers of the supercomputing centers of the RWTH Aachen and the Forschungszentrum Jülich. We also appreciate the fruitful cooperation with the RWTH Aachen Shock Wave Laboratory and the financial support of the Deutsche Forschungsgemeinschaft under Grant No. GRK 5/2.

References

[1] Baldwin B.S., Lomax H. (1978): *Thin–Layer Approximation and Algebraic Model for Separated Turbulent Flows*. AIAA Paper 78-257, Huntsville, Al, USA.

[2] Grotowsky I.M.G. (1996): *A numerical algorithm for calculating flows in hypersonic inlets.* ZFW **20**, pp. 95-104.

[3] Haidinger F.A. (1992): *Numerische Untersuchung turbulenter Stoß/Grenzschicht-Wechselwirkungen.* Dissertation TU Munich, Germany.

[4] Rizzetta D.P. (1998): *Evaluation of Explicit Algebraic Reynolds-Stress Models for Separated Supersonic Flows.* AIAA-Journal, **36**, Nr. 1, pp. 24-30.

[5] Settles G.S., Fitzpatrick T.J., Bogdonoff S.M. (1979): *Detailed Study of Attached and Separated Compression Corner Flowfields in High Reynolds Number Supersonic Flow.* AIAA Journal, Vol. 17, pp. 579-585.

[6] Schubert A. (1998): *Grundlagen für die numerische Simulation von laminaren und turbulenten Hyperschallströmungen.* Dissertation RWTH Aachen, Germany.

[7] Schulte-Rödding J.-H., Olivier H. (1998): *Experimental Investigations on Hypersonic Inlet Flows.* AIAA-Paper 98-1528.

[8] Wilcox D.C. (1994): *Turbulence Modeling for CFD.* DCW Industries, Inc., La Cañada, Ca, USA.

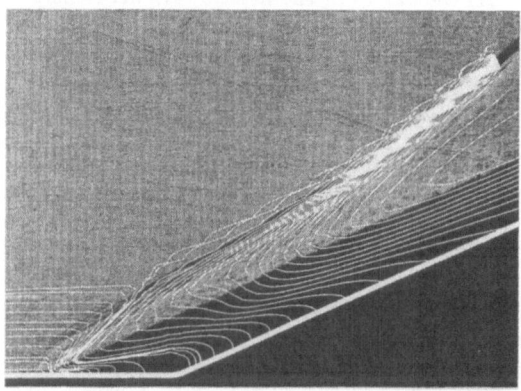

Figure 1. Exp. a) Shadowgraph with density contour plot (k-ω) [6]; flow from left to right.

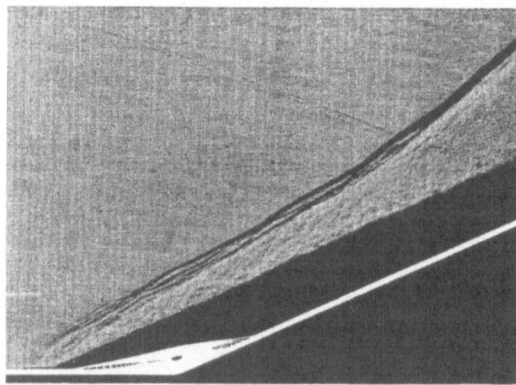

Figure 2. Exp. a) Shadowgraph with streamtraces inside the separation bubble (k-ω) [6].

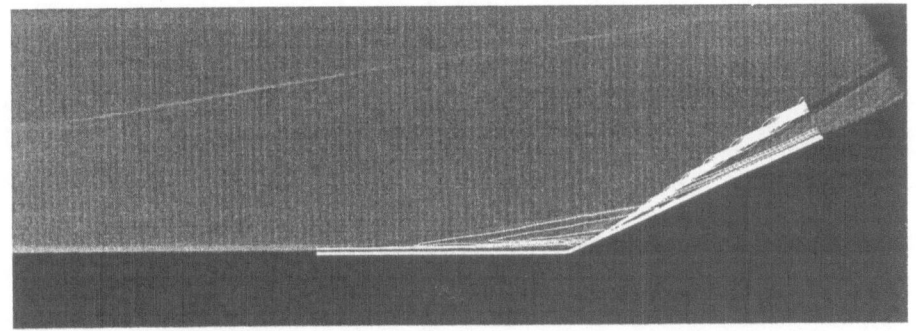

Figure 3. Exp. b) Color Schlieren photography with pressure contour plot (laminar) [6]; flow from left to right.

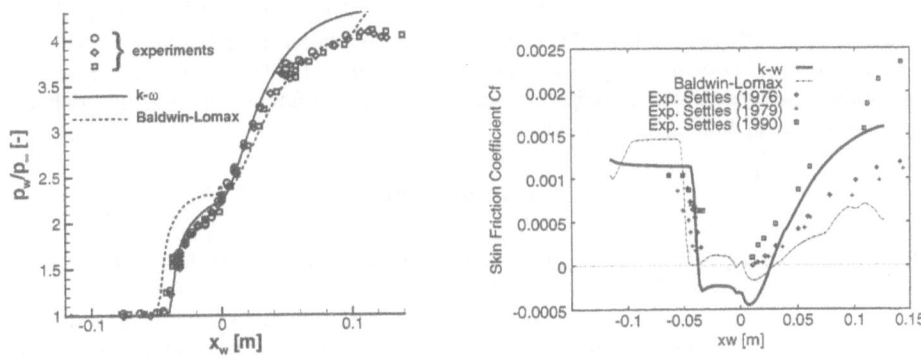

Figure 4. Exp. a) Wall pressure distributions [6] compared with experimental results [5].

Figure 5. Exp. a) Friction coefficients [6] compared with experimental results [5].

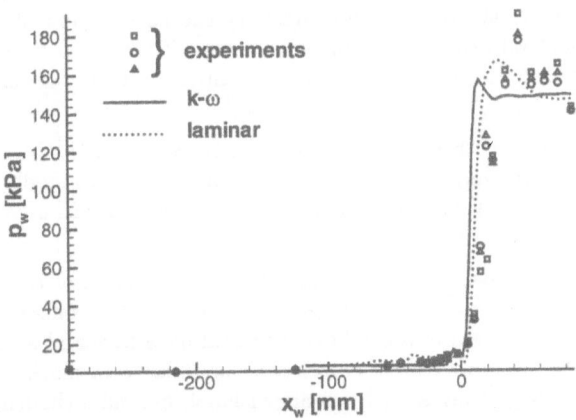

Figure 6. Exp. b) Wall pressure distributions [6] compared with experimental results [7].

425

On the Validation of the DLR-TAU Code

D. SCHWAMBORN, T. GERHOLD, V. HANNEMANN

Deutsches Zentrum für Luft- und Raumfahrt e.V.
Institut für Strömungsmechanik
Bunsenstraße 10, D-37073 Göttingen, Germany
Fax: +49 551 709 2416
email: Dieter.Schwamborn@dlr.de

Summary

This paper deals with the validation of turbulence models in the frame work of the DLR-TAU code, a Navier-Stokes solver for hybrid, unstructured grids. Examples discussed here include the transonic RAE2822 airfoil (cases 9 and 10)and the ONERA Bump C. The turbulence models used in the TAU code are the one-equation model of Spalart-Allmaras and a variety of two-equation models of kω-type. These models are evaluated and the results are compared with experiments and earlier results obtained by a structured code with algebraic or half-equation models.

Introduction

In recent years CFD has become increasingly important in the development of new or advanced aircraft, since it allows in principle for an improved cost effectiveness of the design cycles and it provides information details of which are otherwise difficult to obtain, e.g. in a wind tunnel. Although the last decade has seen a tremendous progress in numerical algorithms, mesh generation, flow solvers and last but not least the computer power, the accuracy of numerical solutions is still very much dependent on the grid spacing and the physical modelling of transition and turbulence. Thus the validation of codes including the assessment of these models is an important pre-requisite for the successful, i.e. accurate simulation of complex flow fields, as it is needed in the aircraft design and development. Due to the increase in complexity the airframe industry has become more interested in one and two-equation turbulence models which allow for a more local modelling of the turbulence terms which is especially important for the use of unstructured grids.

During the development of DLR's adaptive Navier-Stokes solver TAU [5] for unstructured hybrid grids this led to the implementation of the Spalart-Allmaras model in its variant due to Edwards [4] and several two-equation models, like the kω model of Wilcox [12] and it's modification by Menter [10].

This paper deals with the basic steps necessary for the validation of the latter models and is based on experience gained in European projects [7],[8] , where the test cases were chosen such that there exists a good experimental data base and a broad range of other numerical results. Here we restrict ourselves to two flows with shock boundary-layer interaction which, although they are all two-dimensional, are quite challenging. All the necessary information on these cases can be found in [7]. For further interesting test cases and a concise description of numerous turbulence models the reader is also refered to [8].

Numerical Methods and Turbulence Models employed

The DLR TAU code is described in [5],[6]. This code is a 3D adaptive Navier-Stokes solver for hybrid unstructured meshes based on a finite volume method operating on a dual grid. The code is of cell-vertex type, i.e. of cell-centered type with respect to the dual grid cells and employs various central and upwind schemes for the convective fluxes and an agglomeration multigrid approach for convergence acceleration. The adaptation feature of the code was, however, not used to obtain most of the results presented here, since the results for different models should be compared for the same mesh, if possible. The structured grid method which is used in comparison for the calculations with the algebraic or half-equation models employs a very similar but cell centered approach [9].

The turbulence models used in our calculations are not described here in detail, instead the reader is referred to the original papers or to [8], which gives a very concise, yet relatively complete overview over many turbulence models from algebraic over one- and two-equation to Reynolds-stress models. Here we only briefly discuss the modifications, if any, that we have made to the models or their constants.

The Baldwin Lomax model (BL) is used with two modifications: To arrive at a more stable convergence the original $C_{wk} = 0.25$ is replaced by $C_{wk} = 1.$, resulting in that mainly the formulation $f_{wake} = y_{max} f_{max}$ is used in the outer viscosity formula. This leads, however, sometimes to a larger discrepancy between experimental and numerical shock location than the original value. Additionally, $C_{Kleb} = 0.52$ is used instead of 0.3, taking the findings of [9] into account . The Granville modification to the BL model (GR) results mainly in a pressure gradient dependent evaluation of C_{CP} and C_{Kleb} which are constants in the BL model (see [7],[8]).

The Johnson-King model (JK) is a half-equation model which allows for some history effects in the boundary layer based on the solution of a differential equation for the maximum shear stress. Here we use a time-dependent approach to the equations of the original JK model and employ the BL model for the outer eddy viscosity formulation instead of the Cebeci-Smith model in the original approach. The Spalart-Allmaras model (SA) is used in the modifation of [4], while the kω models without (kω) and with SST correction (kω+SST) are implemented as suggested in [12] and [10].

RAE2822 Airfoil

The RAE2822 airfoil test cases 9 and 10 [11] are certainly among the most often calculated transonic flows with a shock strength that leads the turbulent boundary layer almost to separation or to incipient separation, respectively. Especially, case 10 is known to be difficult to compute with respect to the correct shock location and thus is a challenge for turbulence model validation. The calculations discussed in the following utilised a structured grid of 257 by 65 nodes or a hybrid grid of 13937 points based on the same structured grid near the airfoil and triangles further away, i.e. in this case outside of the boundary layer. A smallest grid size normal to the wall of $y^+ = O(1)$, was ensured in all grids. Although an usually advisable mesh convergence study is not discussed here, we looked into the influence of adaptation in the case of the hybrid grid. For the influence of mesh size on the results, see [7] ,[13] . Moreover it should be mentioned that we used the camber-corrected version of the airfoil shape and the correspondingly corrected flow parameters [5]. This, unfortunately leads generally to a somewhat too high pressure near the suction peak of all calculations of case 9 and 10 (compare also [7]).

Figures 1 and 2 show the pressure and skin-friction distribution, respectively, for case 9 indicating that the BL model as used here is not able to predict a correct shock location, while the GR and the SA models perform much better as does the JK model, not shown here. It should be mentioned, however, that the BL model will do a better job, when the numerically less stable original version with $c_{wk} = .25$ is employed. Both two-equation models predict a shock position which is farther down stream of the experiment than SA or GR, with the $k\omega$+SST coming closer to the shock, as one might expect from the SST correction. With respect to the transiton to turbulence it is noticed in both figure 2 and 4, that the immediate switching on of the algebraic models from one position to the other leads to a peak in the skin friction, which is only a little smaller in the case of the two-equation models, while the SA model shows a more reasonable smooth transition, although there is no experimental evidence whether this is physical. With respect to the skin friction downstream of the shock it is not clear from the few experimental data whether the one- or the two-equation model is more correct. The latter, however, will become a little clearer when we come to the discussion of the velocity profiles.

For the more challenging case 10, depicted in figures 3 and 4 the scatter between the models becomes larger and the BL model is far off from the experimental shock location. The GR model predicts the shock position somewhat better and even better than the two-equation models. The latter is an effect also noticed for many other two-equation models in [7], although it is not clear why the $k\omega$+SST does not produce better results as often mentioned in the literature. Only the SA comes closer to the correct shock location, but at the same time exhibits a somewhat lower pressure plateau.

To check the influence of the mesh to a certain extent the solution obtained for the $k\omega$ model was three times adapted by TAU's adaptation tool . The resulting grid was then additionally adapted normal to the wall in the structured boundary layer part. All this had almost no influence on the pressure. The changes in skin friction provided in figure 8 indicate that the refinement influences mainly the areas of steep gradients, while the adaptation in the separation resulted finally in some local unsteadyness in that area.

Since the DLR-TAU code allows the use of a variety of discretisations of the convective fluxes and even different flux formulations for different equations and /or on different multi-grid levels, we also looked into the influence of the flux formulation used for the turbulence equation(s). In figures 5 and 6 the influence of a central or a Roe-upwind flux formulation is presented. Additionally the influence of using only a first order formulation in the Roe-case for the turbulence equation only is shown. Here it can be seen that the changes are minor and neglectable compared to the use of different turbulence models. The influence of the discretisation for the two-equation models is similar, with the largest influence seen in the case of $k\omega$+SST in the area of separation, figure 7.

The influence of the adaptation is finally shown in figure 8. There is not much changed in the pressure distribution except a somewhat steeper shock also recognized in the skin friction, where additionally the transition and the separation are influenced. Already the threefold grid refinement brings out most changes, while the additional adaptation of the structured sublayer resolves the separation somewhat more, leading to a local unsteadyness there at te same time.

The velocity distributions at different locations normal to the upper airfoil surface (fig. 9) provide us with some more information on the model performance. Since all models used here produce a shock location which is at least a little bit too far downstream, none of the models matches the profile closest to the experimental shock position (a), while all models perform very good upstream of that position (not shown). Farther downstream (b) the agreement between the experiment and the SA, JK and GR models becomes good again except close to the wall where SA and JK predict already a little separation, i.e. GR is best, but as we have seen with the wrong shock positon. It should, however, be mentioned that there is also some possible error in the experimental data, because skin friction could not be measured at positions a), b) and c) of figure 9. The two-equation models and the BL model are far away from the experiment which reflects the too far downstream shock position. At the next position (c) the discrepancies between the models become smaller again, but unfortunately none is close to the experiment, where the tendency to separation is even more pronounced without taking place. The same discrepancy was also found in the EUROVAL project [7]. At the final station (d) the agreement between simulation and experiment is better again, with the SA result being farthest away from the experiment indicating that the SA model has some difficulty to recover from the separated flow situation.

ONERA Bump C

Although being an internal flow (fig. 10), the Onera bump C flow case [3] offers a very good opportunity to validate models also for external flow. Unlike in the airfoil cases it is possible here by a careful adjustment of the downstream pressure to obtain the same shock position (in the inviscid part of the flow) in all numerical calculations as found in the experiment, without influencing the flow far upstream of the shock. Thus all models start from the same accelerated boundary layer which eases the comparison quite a bit. The mesh for the calculations presented here is the structured mesh from [7] with a dimension of 193 by 97.

The strong shock boundary-layer interaction of this flow leads to separation as can be seen from the λ-shock in the Mach isolines in figure 10. In figures 11 and 12 the isentropic Mach number along the upper wall and lower wall is presented. The first discrepancy attracting the eye is certainly that in the level of the isentropic Mach number far downstream of the shock. This, however, can not be attributed to the flow simulation, but to some three-dimensional effects in the experiment farther downstream: there exists a pressure gradient across the chanel far downstream from the bump. Therefore the exit pressure in the calculations is set such that the same shock position is obtained. Taking this necessary correction into account, it seems that the one and two-equation models reflect the experiment well with some slight advantage for SA. The BL and to a lesser extent the GR model are unable to simulate the λ-shock along the lower wall, while the JK tries to produce a λ-shock also at the upper wall.

To gain more insight we briefly discuss the velocity profiles at some location just before and after the end of the bump (L=1.0). The good results of the one and two-equation models are found again in figure 13a) (somewhat downstream of the separation point) and figure 13b) (about the middle of the separation region). The JK model is not to far away from the experiment and the discrepancy becomes larger for GR and especially BL. In figure 13c) (near the end of the separation), however, we find what we have already seen in the RAE2822 results: the SA model predicts still backflow and thus a much longer separation region. The other models except BL recover just in the right moment. Thus from the velocity profiles alone one would be tempted to judge the JK better than the SA, but one should not forget the tendency of the JK to predict a too strong shock boundary-layer interaction (λ-shock) which it not found in the experiment.

Conclusion

As we have seen in this part of the validation process, the BL model is not really worthwhile to be considered for computations in cases of non-equilibrium turbulence, like strong shock boundary-layer interaction, although it can perform well in near equilibrium cases of turbulence. The Granville modification is usually considerably better, easy to implement and computationally cheap. Unfortunately, this model is restricted to mainly two-dimensional flow. Comparing the performance of the algebraic models with that of the half-equation model one might be tempted to prefer the latter in strong interaction cases. Its complexity, however, is considerablly higher, and with respect to complex geometries it is as restricted as an algebraic model. This is especially true for unstructured grids as they are used in the TAU code.

The one and two-equation models are the better choice here anyway, because they can be used in complex flow situations and are not much more complex or costly then the half-equation model. Nevertheless also these models have their deficiencies as we have seen, either w.r.t. recovery of the flow after separation or w.r.t. shock location. This is, of course, not an issue of the code the models are used in, since other methods as used e.g. in [7] showed the same behaviour. One question, however, seems to be open: whether the implementation of the SST correction in TAU is correct, because its influence has been reported to be much more rewarding w.r.t. to shock boundary-layer interaction than is observed here.

References

[1] Abid, R., Vatsa, V.N., Johnson, D.A., Wedan, B.W., 'Prediction of Separated Transonic Wing Flows with a Non-equilibrium Algebraic model', AIAA paper 89-0558, 1989.

[2] Cook, P.H., McDonald, M.A., Firmin, M. C. P., 'Airfoil RAE2822 - Pressure distributions, boundary layer and wake measurements', AGARD-AR-138, 1997.

[3] Delery, J., 'Investigation of strong shock turbulent boundary layer interaction in 2D flows with emphasis on turbulence phenomena', AIAA paper 81-1245, 1981.

[4] Edwards, J.R., Chandra, S., 'Comparison of Eddy Viscosity-Transport Turbulence Models for Three-Dimensional, Shock-Separated Flowfields', AIAA J., Vol.34, pp. 756-763, 1996.

[5] Gerhold, T., Galle, M., Friedrich, O., Evans, J., 'Calculation of Complex Three-Dimensional Configutrations Employing the DLR-TAU code', AIAA paper 97-0167, 1997.

[6] Galle, M., Gerhold, T., Evans, J., 'Technical Documentation of the DLR-TAU code', DLR IB 223-97 A-43, 1997.

[7] Haase, W., Brandsma, F., Elsholz, E., Leschziner, M. A., Schwamborn, D., 'EUROVAL - An European Initiative on Validation of CFD Codes', Notes on Num. Fluid Mech., Vol. 42, 1993.

[8] Haase, W., Chaput, E., Elsholz, E., Leschziner, M. A., Müller, U. R., 'Validation of CFD Codes and Assessment of Turbulence Models', Notes on Num. Fluid Mech., Vol. 58, 1997.

[9] Kloppmann, Ch., Schwamborn, D., Singh, J.P., 'Multigrid Solution of the 2-D Navier-Stokes Equations for Transonic Internal and External Flows', in Contributions to Multigrid, CWI Tracts 103, CWI, Netherlands, 1994.

[10] Menter, F.R., Rumsey, C.L., 'Assessment of Two-Equation Turbulence Models for Transonic Flows'. AIAA paper 94-2343, 1994.

[11] Stock, H.W., Haase, W., 'The Determination of Turbulent Length Scales in Algebraic Turbulence Models for the Navier-Stokes Method', AIAA J., Vol.27, No.1, pp.5-14, 1987.

[12] Wilcox, D.C., 'Reassessment of the Scale-Determining Equation for Advanced Turbulence Models', AIAA J. Vol. 26,pp. 1299-1310, Nov. 1988.

[13] Williams, B.R., 1994, Computations of 2D Navier-Stokes equations. GARTEUR/TP-067.

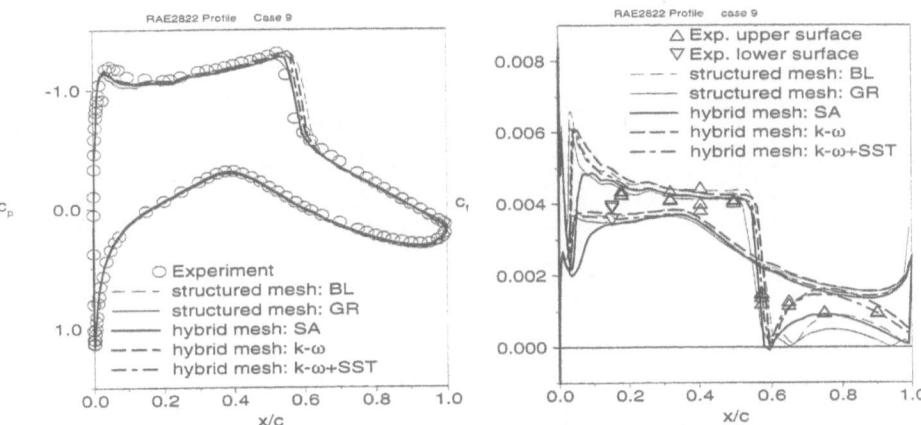

Figure 1: C_p-distribution for case 9.   Figure 2: C_f-distribution for case 9.

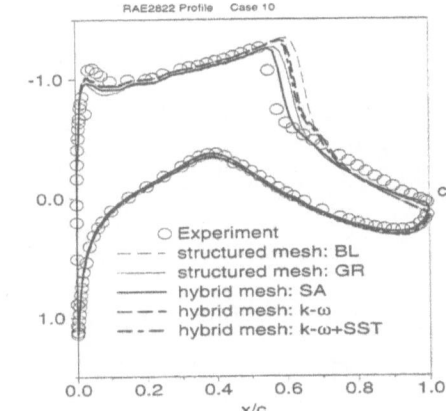
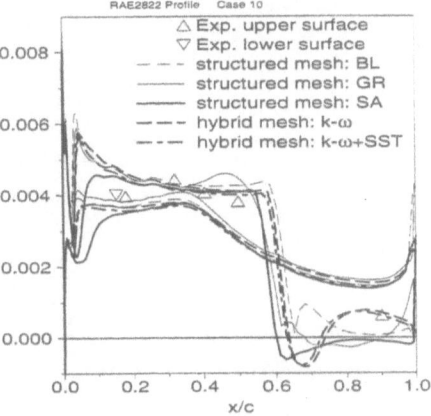

Figure 3: C_p-distribution for case 10.   Figure 4: C_f-distribution for case 10.

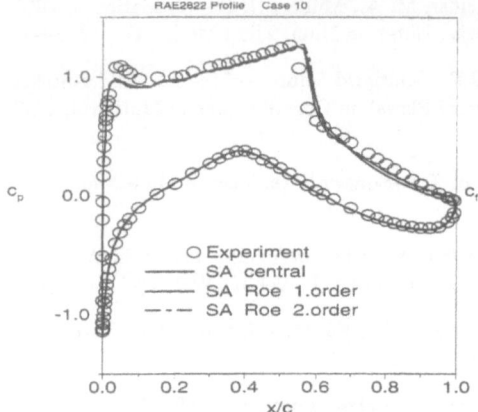

Figure 5: Influence of discretisation on C_p.

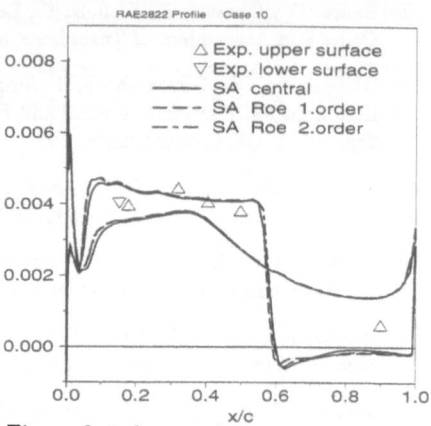

Figure 6: Influence of discretisation on C_f.

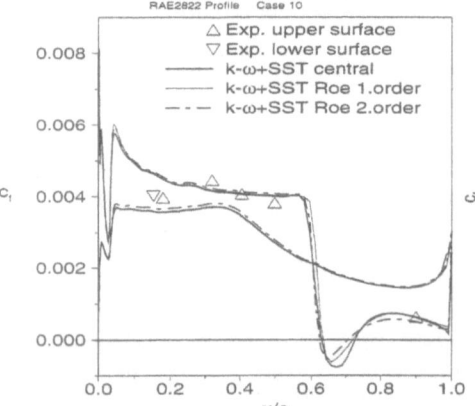

Figure 7: Influence of discretisation on C_f.

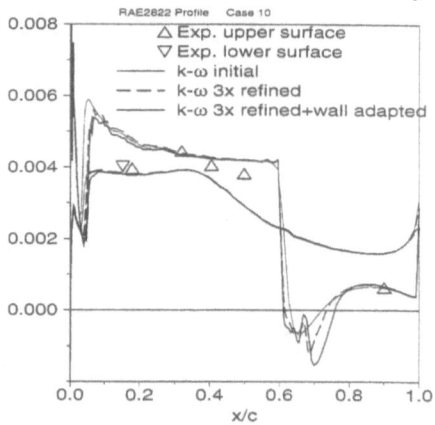

Figure 8: Influence of adaptation on C_f (wall adapted solution unsteady in separation).

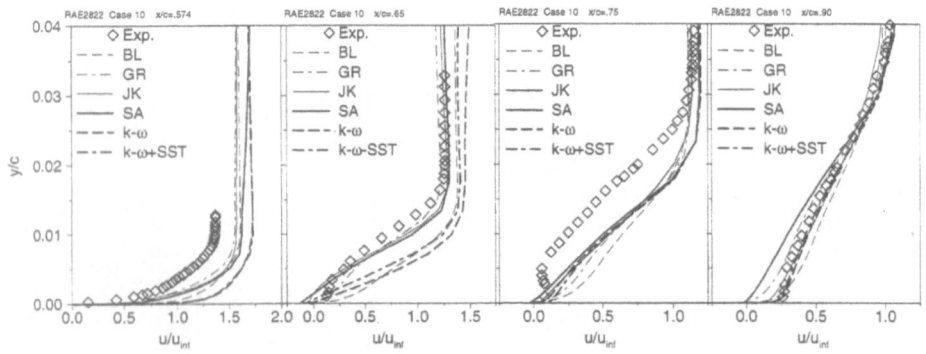

Figure 9: Velocity profiles from Rae2882 case 10

a) x/L = 0.574   b) x/L = 0.65   c) x/L = 0.75   d) x/L = 0.90

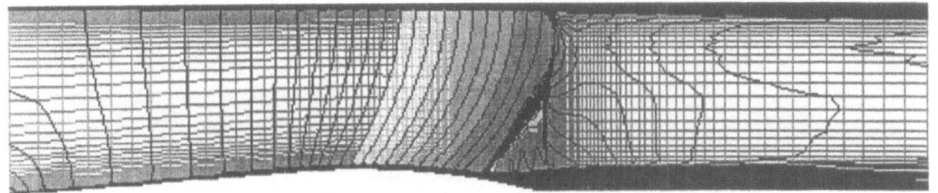

Figure 10: Bump C: Isolines of the Mach number for the bump C flow (computed with the TAU code and the SA model).

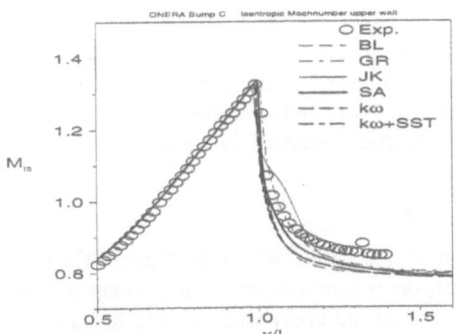

Figure 11: Bump C: Isentropic Mach number along upper wall.

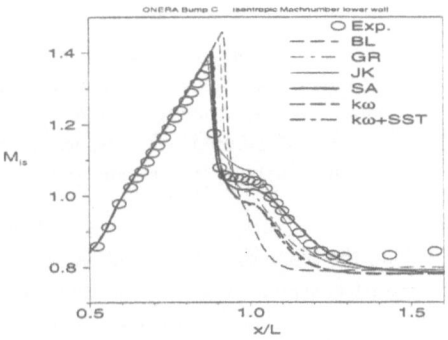

Figure 12: Bump C: Isentropic Mach number along lower wall.

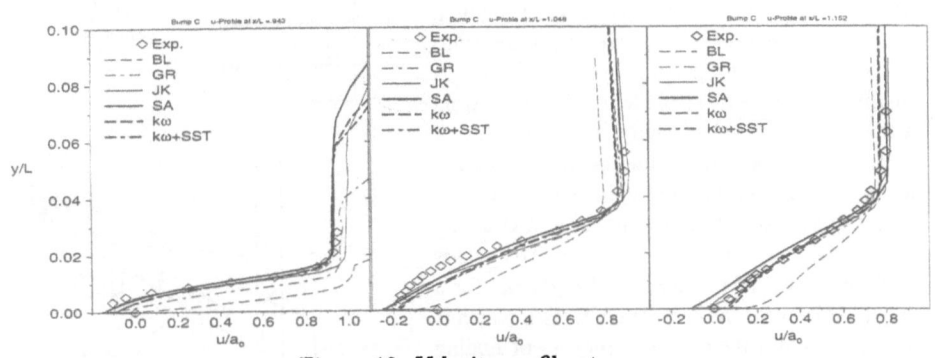

Figure 13: Velocity profile at

a) $x/L = 0.940$, b) $x/L = 1.048$, c) $x/L = 1.152$.

Effects of a Contour Bump on the Unsteady Transonic Aerodynamic Forces of the AMP Wing

W. Send, R. Voß

Deutsches Zentrum für Luft- und Raumfahrt (DLR) e.V.
Institut für Aeroelastik
Bunsenstraße 10, D-37073 Göttingen, Germany

Summary

The effects of a contour bump on steady and unsteady aerodynamics are investigated for both a 2D profile section and a 3D wing, e.g. the AMP model. The supercritical airfoil NLR7301 has been chosen as the 2D case, for which the pressure distributions in steady and unsteady transonic flow demonstrate differences between cases with and without bump. For the AMP wing, a 3D bump is shaped in such manner that it traces the shock area of the elastically deformed wing. Steady and unsteady pressure coefficients are shown for both configurations. The unsteady transonic aerodynamic forces of the AMP wing are subject to flutter calculations. A first comparison of the flutter boundaries with and without bump indicates that the presence of a bump, with its desired effect of drag reduction in transonic flow, also reduces flutter stability for the same physical reasons.

Introduction

Within the scope of the concept „Adaptiver Flügel" (Adaptive Wing) by Daimler-Chrysler Aerospace, Daimler-Chrysler Forschung and DLR, the research on the *contour bump* pursues the aim of reducing drag by adapting a small region of the upper wing contour to transonic flow conditions. Research also comprises studies on side effects like altered aeroelastic stability as a consequence of adaptive measures. To study the effects on 3D surfaces, the AMP wing (Aeroelastic Model Program) has been chosen as an appropriate model (Fig. 1) for modern transport aircraft wings. The 3D bump traces the shock area of the elastically deformed wing. The suitable bump shape is derived from 2D numerical and experimental investigations on drag reduction. In the presented results, the deflections are still taken from experimental data. A simplified finite element model of the AMP wing has been constructed to study the techniques of coupling transonic flow around aircraft wings and their structural response on the basis of a certified finite element tool. First results from a coupling computation in steady flow are shown. The goal is the development of the *numerical flight test* for a wing (and later for the complete aircraft), which is intended to cover the complete process of finding the steady deformation according to the respective flight level and the unsteady loads from gusts or forced excitations.

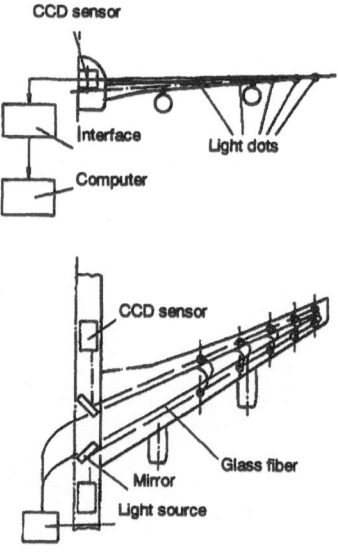

Fig. 1 The AMP windtunnel model

Results from the 2D Airfoil NLR7301

Within the project, the NLR7301 airfoil was chosen because it is a standard AGARD airfoil. Unsteady aerodynamic investigations have been carried out for this airfoil both in windtunnel tests and by CFD. The CFD computations are carried out using the DLR steady/unsteady Full Potential code with boundary layer coupling VST2D/VUN2D, which had previously been validated for the NLR7301 airfoil [1]. Currently, detailed windtunnel tests for this airfoil with contour bump are under work in the DLR transonic windtunnel TWG in Göttingen. Results of flow calculations for the two configurations with and without bump are shown in Fig. 4. The flow conditions for steady flow are described in the following.

Design condition: For the Reynolds number $Re = 2.6 \ 10^6$ in the windtunnel test, the flow parameters $Ma = 0.725$ and angle of incidence $\alpha = 0.6°$ yield a nearly shock-free pressure distribution with $c_L = 0.514$ and $c_D = 0.0105$.

Off-design condition: A small change in the flow parameters to $Ma = 0.730$ and $\alpha = 0.7°$ yields an increase in lift $c_L = 0.5515$ together with the development of a pronounced shock wave and a corresponding increase of drag by 10% to $c_D = 0.0116$.

Bump condition: The contour bump has been designed for the off-design flow parameters. The bump does not change the lift, but locally alters the pressure distribution. The shock wave is changed to a smooth recompression and, correspondingly, the drag is decreased by 5%.

The shape of the bump is sketched in Fig. 4. Its height h is added to the clean airfoil contour on the upper surface and is analytically defined:

$$h(x) = h_0 (2 f(x)^3 - 3 f(x)^2 + 1) \quad with \quad f(x) = \begin{cases} (x_0 - x)/(x_0 - x_1) & for \quad x_1 < x < x_0 \\ (x - x_0)/(x_2 - x_0) & for \quad x_0 < x < x_2 . \end{cases} \quad (1)$$

The lengths are scaled with the airfoil chord, and the leading and trailing edges of the airfoil are at $x = 0$ and 1, respectively. The bump parameters are:

$h_0 = 0.0015$	Maximum height
$x_0 = 0.56$	Position of maximum height
$L = 0.20$	Length
$x_b = 0.70$	Relative backward position of max. height
$x_1 = x_0 - x_b L$	Upstream edge
$x_2 = x_1 + L$	Downstream edge

Unsteady pressure distributions induced by pitching oscillations about an axis at 40% chord are computed with the unsteady CFD code VUN2D. The results in Fig. 6 show that the unsteady flow behavior is significantly changed by the bump, although the pitching amplitude of 0.5° is relatively small. While the shock wave of the clean airfoil oscillates harmonically about its mean position with increasing and decreasing shock strength, the shock motion is more complicated in presence of the bump. A well-pronounced shock wave appears during longest part of the oscillatory motion, and even double shocks develop and vanish. The smooth recompression of the steady flow appears for only a very short time. In Figures 4 and 6, the bump height is scaled by a greater factor. A CFD grid with 261 points in circumferential direction, 41 points normal to the surface, and with 14 cells within the bump length was used for the computations.

Simplified Aeroelastic Model of the AMP Wing

A simplified finite element model has been constructed to study the techniques for coupling transonic flow around transport aircraft wings and their structural response. The design of the

FE model (Fig. 2) refers to the windtunnel model in a series of experiments named AMP, which were carried out between 1987-1991. The results were presented by *Deutsche Airbus* [2], among others.

Based on a total of 1.500 nodes and 2.200 elements, the static analysis - without the flow field computation - and the evaluation of modal characteristics require a moderate computing time of approximately two minutes on a modern workstation (SGI Indigo[2] ; R10000 processor).

On the surface of the wing, two grids used for different purposes are active: The aerodynamic grid is the computational basis for the flow calculation, whereas the structural grid describes the nodes of the FE model. The position of the deflected surface for recalculating the deformed aerodynamic CFD grid is reconstructed from the respective structural net by a two-dimensional B-spline interpolation. This method provides a very accurate reconstruction. Figure 3 shows a typical result from the iteration process [3]. Aerodynamic lift is compared to the total reaction force of the whole structure, which

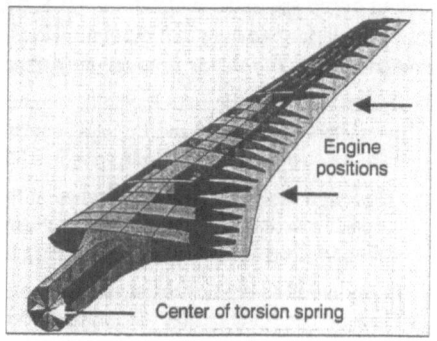

Fig. 2 Interior of the simplified FE model for the AMP wing suspended from a torsion spring.

acts along the center line of the torsion spring. To achieve a steady convergence, the calculated deflection is increased stepwise to its full value within the first ten steps. A detailed

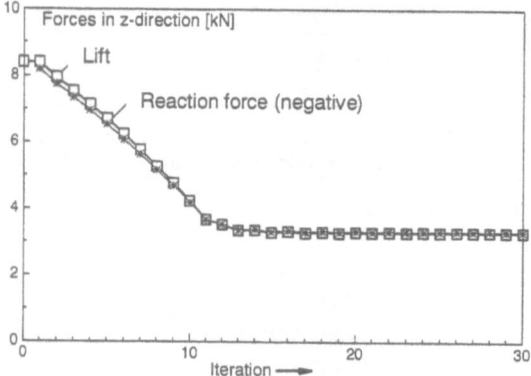

Fig. 3 Convergence of a typical static loading (AMP FE model)
Ma=0.82, Re=3.57 10[6] p_0=0.9 bar, c_L=0.265.

description of the model and its elastomechanical properties is given in [3]. An overview of the static behavior is shown in Fig. 11, in which the $c_L(\alpha)$ function of the simplified model and of the windtunnel experiments are compared with each other. The aerodynamics are computed with VST3D [4].

As is well known, the frequencies of the first bending mode and of the first torsion mode of a typical swept wing are far apart from each other by a ratio of about one to seven or even higher. In modern transport aircraft, flutter occurs as a complex phenomenon of coupled fuselage and engine nacelle modes. To achieve flutter with a simple half model and for a reasonable Mach number of about Ma = 0.8, the actuator for the AMP windtunnel model is equipped with a built-in torsion spring (Fig. 2). The spring introduces an adjustable torsional rigid-body mode which couples with the bending mode at the desired operating conditions. Time-dependent coupling is controlled by integrating the simplified dynamic equation

$$M\ddot{q} + Kq = -f_c \left[Q_S(\dot{q},q)\dot{q} + Q_R(\dot{q},q)q \right] \qquad (2)$$

with **M** being the mass matrix and **K** the stiffness matrix. The aerodynamic forces are given as matrices $\mathbf{Q_R}$ and $\mathbf{Q_S}$ proportional to displacements **q** and velocities $\dot{\mathbf{q}}$. All nonlinearities of the flow field are assumed to be included in these matrices. Test computations in subsonic flow show that amplitude ratios and phase shifts are well preserved in comparison with the corresponding flutter computation in the frequency domain. With $\mathbf{q} = [\, h(t)/c_r,\ \alpha(t)\,]$ (bending and torsion modes) and $\mathbf{Q_X}$ (the aerodynamic coefficients) being dimensionless quantities, the factor f_c is given as

$$f_c = \frac{1}{2}\rho_\infty u_0^2\, S/c_r \quad [\text{N/m}]. \tag{3}$$

The reference area of the AMP wing is given as $S = 0.289$ m². The root chord length reflects the 1:25 scale of an Airbus A340 wing: $c_r = 0.4221$ m. Flutter results from the unsteady computations are given below.

Unsteady Investigations for the AMP Wing

The AMP wing and the computation of its static aeroelastic deformation has been described above. A 3D contour bump is designed stripwise as a sequence of 2D bumps. The shape is constant along the wing span, but position x_0 is adapted to the shock position, and length and height are scaled with local chord. Corresponding with the varying shock strength, the bump height is varied between 0.15% and 0.10%, and the bump vanishes for the wing tip region; see Fig. 5. One of the investigated flow conditions of the AMP windtunnel model was chosen to investigate the effect of a bump. It is significant only if strong shocks appear, and thus data $Ma = 0.820$, $c_L = 0.40$ ($\alpha = 1.55°$), $Re = 3.57\ 10^6$ was chosen as the reference test case. Figure 7 shows a comparison of the steady pressure distribution obtained by CFD with the test results. For these parameters the flow is transonic and the shocks are quite strong. Three-dimensional CFD computations were carried out with the DLR steady/unsteady 3D Full Potential code with boundary layer VST3D/VUN3D [4]. A grid with 139*35*31 (circumferential, normal to the surface, spanwise direction) points was used. This is coarser than that of the 2D case, but there are still 7 grid cells along the length of the bump. The pressure distribution obtained for the wing with 3D bump is shown in Fig. 8. The shocks are less strong but the total drag is reduced by only 1 %.

Unsteady flow simulations were carried out for the AMP wing. The wing oscillated harmonically in two degrees of freedom around the steady flow conditions of Fig. 8, both with bump and without bump. The investigated oscillation modes were rigid heaving and rigid pitching around an axis located at a position of midchord wing root and swept back by 30 degrees. The first harmonic component of unsteady pressure distribution is plotted as the real and imaginary part. The pressure coefficients are divided by the amplitude, which is one degree for pitching (Fig. 9) and 2 % root chord for heaving (Fig. 10). The results are shown at seven wing span positions. The unsteady pressure

Tab. 1	Aerodnamic coefficients applied in Eq. (2)	
Without bump, alf0 = 0.5 deg		
0.5371◇	1.2573<	CQ11
5.37115◇	−0.92818<	CQ12
−0.0024◇	−0.0781<	CQ21
−0.3966◇	−0.1149<	CQ22
Without bump, alf0 = 1.0 deg		
0.5371◇	1.25725<	CQ11
5.090 ◇	−0.9957<	CQ12
−0.0024◇	−0.0781<	CQ21
−0.2796◇	−0.0833<	CQ22
With bump, alf0 = 0.5 deg		
0.5260 ◇	1.3149 <	CQ11
5.11756◇	−0.63028<	CQ12
+0.00635◇	−0.0842 <	CQ21
−0.29227◇	−0.13548<	CQ22
With bump, alf0 = 1.0 deg		
0.5260 ◇	1.3149 <	CQ11
5.5627◇	−0.9446<	CQ12
+0.00635◇	−0.0842 <	CQ21
−0.3960◇	−0.1620<	CQ22

The coefficients of the original computation for the heaving motion are divided by 2, which better reflects the bending mode of the AMP wing.

values differ significantly between the cases with and without bump, but only in the shock region. They are identical far upstream and downstream of the shock and near the wing tip. All data is given in Tab. 1.

Influence on Aeroelastic Stability

The applied *quick-look* method concerning the influence of the bump on aeroelastic stability provides a rough estimate of the effects. This first look is far from being a comprehensive survey on the basis of a sufficiently large number of aerodynamic calculations, which still needs to be established. The aerodynamic forces taken from time integration are decomposed into their Fourier components. The first harmonics are thought to be the determining parts for predicting aeroelastic stability. Their values are put into Eq. (2), in which the structural data is taken from the FE model. In the first step, a time integration of this equation with the IMSL routine DIVPAG [5] for various forces f_c shows the flutter point for the configuration *without bump*. In the next step, the aerodynamic coefficients are replaced by the ones from the configuration *with bump*. If f_c has to be diminished to reach a stable situation, the clear consequence is that the bump reduces the aeroelastic stability.

Calculations have been made for $\rho_\infty = 1.0733\,\text{kg/m}^3$, $c_S = 344.89\,\text{ms}^{-1}$, $Ma = 0.82$, yielding $f_c = 29387$ N/m. The preceding value f_c may also be obtained from $p_\infty \cdot 0.5\gamma\,Ma^2\,S\,/\,c_r$ with $p_\infty = 91197$ N/m² ($\gamma = 1.4$). The reduced frequency $\omega^* = \omega\,c_r\,/u_0$, for which the unsteady airloads have been computed, is $\omega^* = 0.5$ in all cases. This slightly higher value compared with the resulting value of 0.28 from the preceding data affects the level of f_c, for which flutter occurs. However, the relations among the results do not change. Finally, f_c has to be multiplied with S^*/S, where $S^* = 0.243$ m² is the effective wing area.

The flutter analysis shows the following overall behavior, indicated by the value of f_c, for which the wing reaches the stability limit:

Configuration	Pitch amplitude 0.5°	Pitch amplitude 1.0°
Without bump	$f_c = 21500 \pm 500$ N/m	$f_c = 25500 \pm 500$ N/m
With bump	$f_c = 16600 \pm 500$ N/m	$f_c = 18500 \pm 500$ N/m

Obviously, a larger pitch amplitude results in nonlinear stabilizing effects. The values for f_c are shown in Fig. 12, recalculated for the respective windtunnel pressure. In Fig. 13, the time functions $h(t)$ and $\alpha(t)$ are plotted for the stable wing with bump (pitch amplitude 1°). In Fig 14, the time functions for the wing without bump are drawn for the same value of $f_c = 18500$, showing the large gain in stability for the given initial deflections.

Conclusions and Outlook

Further computations are needed to establish a reliable assessment of the influence of a bump on aeroelastic stability. The results presented in this paper indicate a destabilizing effect. However, to which extent the reduction of flutter speed will occur is still uncertain for two reasons: The rigid modes assumed in the present investigation exaggerate the influence, and the bump area will probably be smaller than is assumed here for the AMP wing. A third uncertainty deals with the mode shapes in effect for flutter. The combination of the simple torsion and bending motions is not a very realistic assumption for a large transport aircraft where modes with wing/fuselage interaction are relevant for flutter prediction.

Currently, the approach is extended to include the three-dimensional elastic modes from the FE model. The final goal is the coupling of fluid and structure for the 3D case in the time domain using the CFD codes interacting with the deflected FE model.

References

[1] Müller,U.R., Henke, H., Le Balleur, J.C.: *Computation of Transonic Steady and Unsteady Flow About the NLR 7301 Airfoil* in : Notes on Numerical Fluid Mechanics Vol. 58, Vieweg Verlag (1997).

[2] Zimmermann, H., Vogel, S., Henke, H., Schulze, B.: *Computation of Flutter Boundaries in the Time and Frequency Domain*, AGARD Structures and Materials Panel, Specialists' Meeting on Transonic Unsteady Aerodynamics and Aeroelasticity, San Diego, CA, October 9-11, 1991.

[3] Send, W., *Kopplung von Fluid und Struktur bei Tragflächen* (Coupling of Fluid and Structure for Wings), Paper DGLR JT98-214, Deutscher Luft- und Raumfahrtkongreß 1998, Oct. 5-8, Bremen, Germany.

[4] Lu, Z., Voß, R.: *FP + VII Code Improvements and Computation of Unsteady Transonic Flows*. DLR IB232-96J03 (1996).

[5] Internat. Math. and Stat. Library, Version 2.0, 1991, distributed by Visual Numerics, Inc..

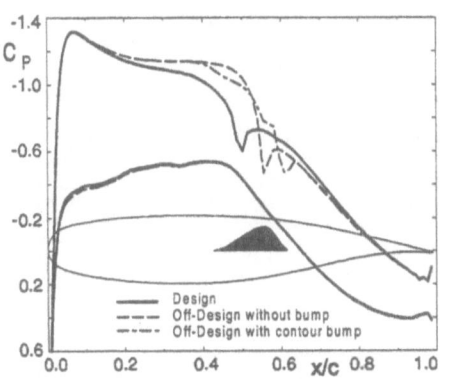

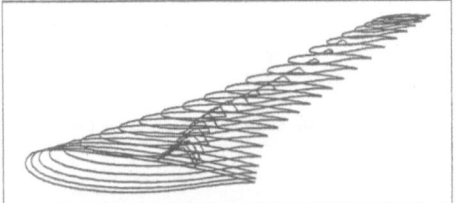

Fig. 5 Geometry of contour bump for AMP wing; Bump height is exaggerated (above)

Fig. 4 NLR 7301 – Contour bump geometry and steady pressure distribution for various conditions

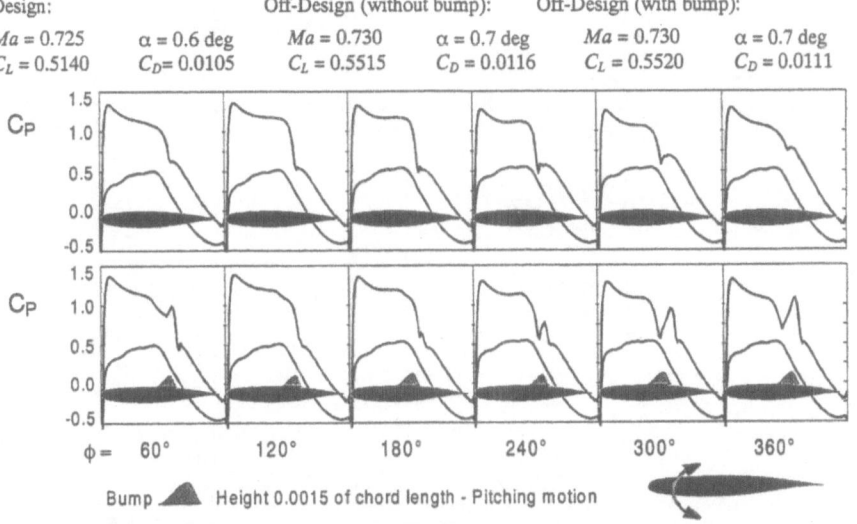

Design:

$Ma = 0.725$ $\alpha = 0.6$ deg
$C_L = 0.5140$ $C_D = 0.0105$

Off-Design (without bump):

$Ma = 0.730$ $\alpha = 0.7$ deg
$C_L = 0.5515$ $C_D = 0.0116$

Off-Design (with bump):

$Ma = 0.730$ $\alpha = 0.7$ deg
$C_L = 0.5520$ $C_D = 0.0111$

Bump ◢ Height 0.0015 of chord length - Pitching motion

Fig. 6 NLR7301 - Unsteady pressure distributions, $Ma = 0.730$, $\alpha_s = 0.7°$, $\alpha_0 = 0.5°$, $\omega^* = 0.5$

439

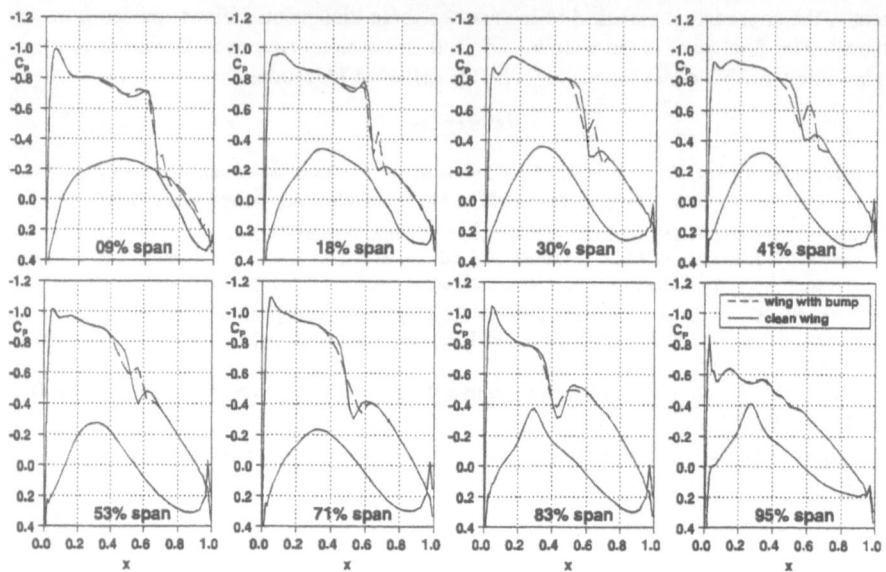

Fig. 8 AMP wing deformed, steady pressure distribution; Ma = 0.82, α_S = 1.55°, Re = 3.57 10^6 .

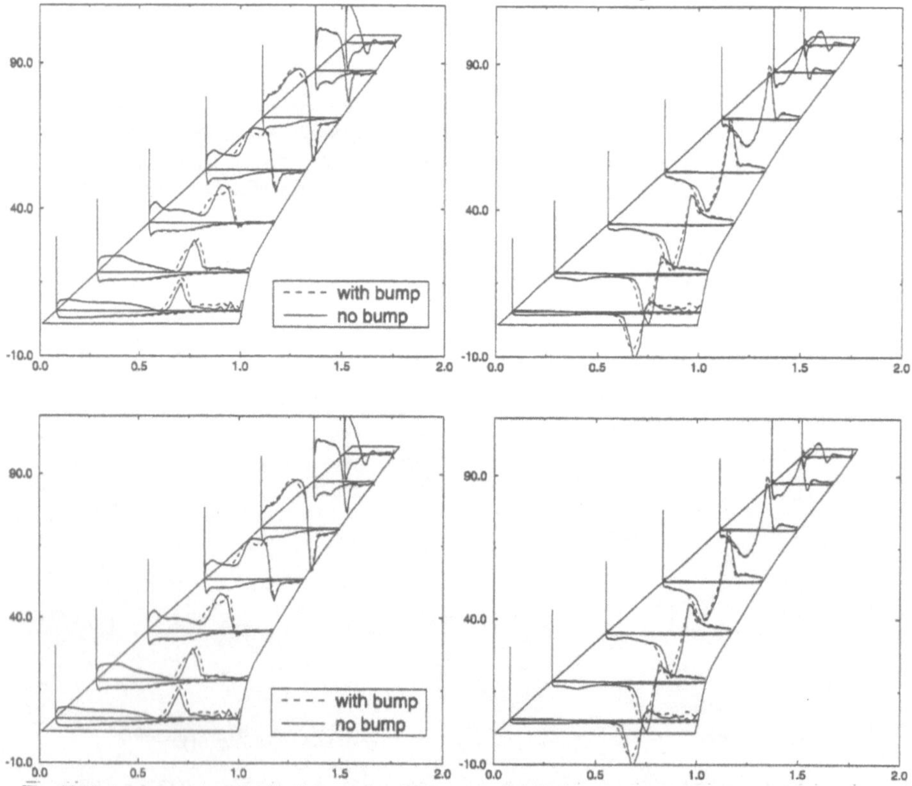

Fig. 10 Unsteady pressure distributions, real and imag. part (left/right), amplitude 0.02 root chord, heaving.

440

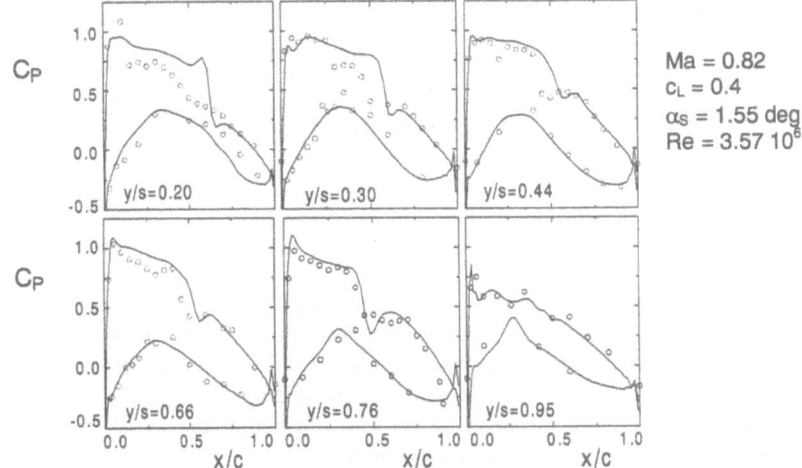

Fig. 7 Steady pressure distribution for elastic deformation in comparison to windtunnel experiment

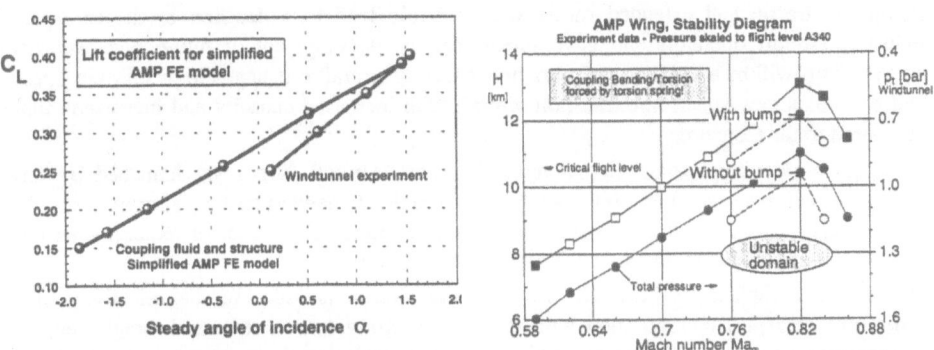

Fig. 11 Steady lift coefficient for simplified FE model

Fig. 12 Effect of the bump on aeroelastic stability

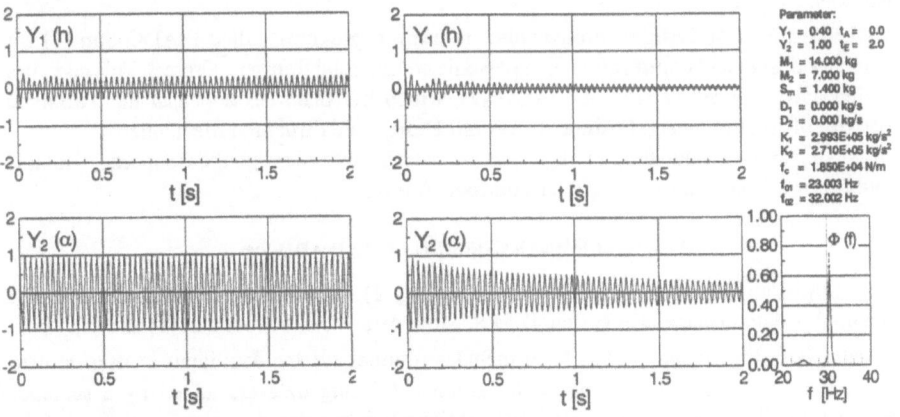

Fig. 13 Stable time functions for wing with bump **Fig. 14** Damped motion for wing without bump, same structure

Validation of the Preliminary Aircraft Design and Optimization Program for Supersonic Commercial Transport Aircraft PrADO-Sup

Rainer Seubert

DLR, Institute of Design Aerodynamics

Lilienthalplatz 7, D-38108 Braunschweig, Germany

SUMMARY

This paper presents the validation of the multi-disciplinary design code PrADO-Sup for supersonic commercial transport aircraft. For this purpose the only flying SCT, the Aerospatiale/BAe Concorde was redesigned. A typical SCT mission was flown and the results for the redesigned Concorde are compared to the real aircraft. Some further improvements to the aerodynamic model of PrADO-Sup are discussed and the resulting more accurate supersonic drag polars for the Concorde are shown.

INTRODUCTION

Rising air traffic and extended routes will probably lead to a decision of the aerospace industry, aircraft operators and governments for the development of future high capacity aircraft. This will be a conventional ultra high capacity aircraft cruising at high subsonic speed and perhaps a new supersonic transport aircraft with increased capacity and increased range compared to the Concorde.

Although a new Supersonic Commercial Transport Aircraft is not yet launched by any aerospace industry several feasibility studies confirm the demand of such an aircraft.

A new supersonic aircraft has to fulfill current and future environmental requirements and must be economically competitive against subsonic aircraft.

Due to the complexity of a totally revised new aircraft and due to the sensitivity of a supersonic transport configuration with respect to aerodynamics, structural weight, engine performance and environmental constraints a multidisciplinary design and optimization tool has to be used to get the most economic result fitting within all the requirements and constraints.

The DLR-Institute of Design Aerodynamics is using a program, called PrADO-Sup [1] for preliminary design and optimization of supersonic commercial transport aircraft. This code is a modified and extended version of PrADO [2], which has been developed at the Technical University of Braunschweig, Institute of Aircraft Design and Structural Mechanics.

In a first step and for validation, PrADO-Sup has been used to redesign the Concorde, which is the only flying Supersonic Commercial Transport Aircraft.

THE PRELIMINARY DESIGN CODE PrADO-Sup

PrADO-Sup is a modular code. It's structure (see Fig. 1) shows 4 levels which communicate by a complex data management system D.M.S. More details can be found in [1] and [2].

1. Level - Data Input/Output: This level includes routines for the data input of user defined baseline data, mission requirements or constraints. Missing data are added by a statistics subroutine. Finally all data of the converged configuration are stored.

2. Level - Optimization Loop: A given objective function is minimized varying user defined parameters and checking the constraints.

3. Level - Multidisciplinary Design Process: In this level all the modules for the interdisciplinary design process are run, the aircraft and the powerplant system are sized for the specified mission.

4. Level - Design Program Libraries: The program libraries include all subroutines necessary for the different disciplines called by Level 3.

REDESIGN OF THE CONCORDE AIRCRAFT

For the redesign of the Concorde Aircraft, PrADO-Sup was run with only a few input parameters: single floor fuselage, 144 passengers at nearest pitch, ogive wing, wing loading 500 kg/m^2, no horizontal tail, turbojet engines with afterburner, cruise at 16 km with a cruise mach number of 2.0, range 6000 km.

A circular fuselage with cylindrical mid section was selected providing the required planform area for the integration of the whole equipment, e.g. floors, seats, cockpit. For the wing an ogive planform with trailing edge flaps, a 3% symmetrical airfoil, no twist, no dihedral was selected. Only two nacelles were allowed and installed at the wing lower side, picking up all the required engines. One vertical fin was defined to ensure directional stability.

Within the next step the aerodynamics was calculated. PrADO-Sup used the panel code HISSS [3] to provide all the required data for sub- and supersonic onset flow conditions. The flow in the transonic regime was interpolated. For the drag breakdown of the configuration the viscous drag for all main components was added to the surface pressure drag of the wing-fuselage combination from HISSS. For more details see [1].

With the aerodynamic data achieved, the engines were scaled according to the FAR25 requirements, such for take off, single engine climb, cruise and go around. It was also checked, that the plane can pass the sound barrier with that engines. The engine performance, e.g. thrust, fuel consumption, NO$_x$ emission was calculated within a thermodynamic cycle for adaptable engines [4].

With that initial virtual aircraft a 6000 km flight was simulated including take-off, climb, cruise at constant lift coefficient, decent, approach and landing. The equations of motion were solved for the aircraft at each point of the flight envelope with the mass of the aircraft initially estimated by statistics and continuously updated within the iteration process. The aircraft was trimmed iteratively by setting the flaps until the resulting pitch moment was zero. For the first calculations the static longitudinal stability was set constant. But trim drag due to the flap settings was not yet added to the total drag.

A more detailed mass calculation for the different aircraft components followed. For the calculation of the fuselage mass a semi-empirical method developed at the University of Braunschweig [2] has been applied. For the wing, engines, nacelles, inlets, nozzles also an empirical method [5] was used. The masses of empennage, landing gear and operational equipment were calculated using the WAATS method [6] and added to the component weights to get the total empty weight. For the calculation of the take-off weight the mass of passengers, payload and fuel was added according to [2].

The final result within each iteration was used to rerun the code iteratively until the take-off weight and the required thrust converged.

RESULTS

Fig. 2 shows the geometry of the redesigned Concorde compared to the real aircraft. The main dimensions, such as wing span, wing area, fuselage length, diameter, fin and nacelles are correctly predicted by PrADO-Sup. The predicted weight data is the following:

	PrADO-Sup	Literature [8],[9]
Max. Taxi Weight	183.0 t	178.3 t
Basic Operating Weight	85.2 t	78.2 t
Max. Landing Weight	112.1 t	111.3 t
Max. Fuel Capacity	85.6 t	93.4 t
Max. TO-Thrust (AB on)	178.5 kN	175.3 kN
Number of Engines	4	4
Services:	4.2%	4.5%
Furnishings:	6.3%	6.8%
Systems:	8.0%	13.6%
Landing Gear:	7.9%	9.1%
Fuselage:	24.0%	14.8%
Tail:	2.2%	4.1%
Wing:	17.8%	17.3%
Inlets:	6.4%	6.4%
Nozzles:	6.4%	6.4%
Engines:	16.8%	16.8%

It can be seen that there are still some discrepancies in the weight prediction, especially in the calculation of the systems and fuselage weight. Further investigations have to improve the weight prediction. For that purpose it is planned that a finite element code for structure analysis of wings and fuselages will be integrated into PrADO-Sup. This code is currently developed at the Technical University of Braunschweig, Institute of Aircraft Design and Structural Mechanics. The pressure distribution from HISSS is used to define the aerodynamic loading, the weight prediction of PrADO-Sup for the engines, nacelles, intakes, nozzles and the landing gear is used to simulate external stores. Finally the aircraft will be considered as totally flexible and the interaction of aerodynamics and structure will be investigated.

IMPROVEMENTS TO THE AERODYNAMIC MODEL

During the interpretation of the PrADO-Sup results for the redesigned concorde a lack in the aerodynamic prediction model was found. In supersonic flow the zero lift drag calculated by PrADO-Sup was less than the published one [10]. On the other hand, the lift depending drag (wave drag plus induced drag), derived from the surface pressure integration with HISSS (coarse grid), was overestimated. Fortunately, at the design lift coefficient the predicted drag corresponded well with the measured drag so that the Concorde aircraft could be redesigned!

To achieve a higher accuracy of the aerodynamic prediction especially for the supersonic cruise conditon the following modifications are used: the influence of the vertical tail and the nacelles on the zero wave drag are added running the Harris-Code (Mach-Cuts, slender body theory [12]). A new HISSS-Version from DASA-M calculates the induced drag in the Trefftz-plane, the lift depending wave drag is calculated by far field integration (momentum theory [11]).

Following the slender body theory as described by Harris the wave drag can be calculated by:

$$\left(\frac{D}{q}\right)_{Pressure} = -\frac{1}{4\pi^2} \int_0^{2\pi} d\Theta \int_{-\infty}^{\infty} \int_{-\infty}^{\infty} A''(x;\Theta)A_1''(x_1;\Theta)\ln|x-x_1|\,dxdx_1$$

with: $A''(x;\Theta) = S''(x;\Theta) - \frac{\beta}{2}L'(x;\Theta)$

444

where S is the cross section of the Mach cuts and L the corresponding lift. For the zero lift wave drag calculation only the distribution of the cross sections is used (see Fig. 3). The geometry is successively cut by Mach planes and for each roll angle Θ the equivalent cross section distribution is calculated. The wave drag is integrated using a Fourier series representation and finally averaged [12].

Wave drag can also be calculated by far field integration using momentum theory (see Fig. 4):

$$\frac{D}{q} = -2 \int\int_{S_2} \phi_x \phi_r dS_2 + \int\int_{S_3} \left(\phi_y^2 + \phi_z^2\right) dS_3 \quad \text{(Wave Drag + Induced Drag)}$$

where ϕ is the pertubation velocity potential and $\phi_{x,y,z,r}$ the components parallel to the coordinate axes and in radial direction respectively. S describes the surface of the control cylinder and q is the dynamic pressure. As the integration of the induced drag is already added to the latest HISSS version, only the wave drag is calculated using the momentum theory.

Fig. 5 shows the result for the zero lift wave drag of the redesigned Concorde aircraft by the Harris code. The result is splitted into increments for the fuselage, wing, fin and nacelles. The result compares well to Van der Velden [13].

Fig 6. shows the network definition for the panel code HISSS to perform the data on the cylinder for the field momentum integration. A fine grid was only used between the intersection of the front, rear Mach cone and the control cylinder. According to the theory this is the only region of importance. Results of the farfield calculation are depicted in Fig.7. It can be clearly seen that there is only a change in pressure between the front and the rear Mach cone.

Finally the drag of the Concorde aircraft is summed up using the pressure drag (far field integration) and the induced drag (Trefftz) of the wing-fuselage from HISSS, the drag increments for the vertical tail and nacelle from the Harris code and the friction drag for the main components from the flat plate analogy. The results are multiplied by a factor of 1.1 for interference, excrescent and trim drag according to the following formula:

$$C_D = [C_{D,P} \text{ (Wing + Fuselage)} + \Delta C_{D,P} \text{ (Vert. Tail + Nacelle)} + C_{D,F} \text{ (Main Components)}] * 1.1 .$$

Fig. 8 shows the revised drag polars of the concorde aircraft for Mach Numbers of 1.1, 1.4, 1.6 and the design Mach number of 2.0. The results compare well to publications of the flying aircraft [12] while the surface pressure integration overestimates the lift depending drag.

CONCLUSIONS

The Concorde aircraft was redesigned running the Preliminary Design Code for Supersonic Commercial Transport Aircraft PrADO-Sup developed at the Institute of Design Aerodynamics at DLR Braunschweig. It was shown that the code predicts the aircraft reasonable but some further investigations concerning the weight prediction have to be performed. Furthermore the aircraft has to be trimmed correctly by flap setting and fuel transfer. Fin and nacelles must be added to the surface panel grid to achieve consistent wave drag results only from the panel code HISSS. PrADO-Sup will then be used for the design of future SCT-Configurations and for sensitivity investigations.

REFERENCES

[1]   Seubert, R.: The Preliminary Aircraft Design and Optimization Program for Supersonic Commercial Transport Aircraft PrADO-Sup, in: H. Körner, R. Hilbig: New Results in Numerical and Experimental Fluid Mechanics, Contributions to the 10th AG STAB/DGLR Symposium Braunschweig, Germany 1996, NNFM Vol. 60, Vieweg 1997, pp. 311-318.

[2]   Heinze, W.: Ein Beitrag zur quantitativen Analyse der technischen und wirtschaftlichen Auslegungsgrenzen verschiedener Flugzeugkonzepte für den Transport großer Nutzlasten, Dissertation, ZLR-Forschungsbericht 94-01, 1994.

[3]   Fornasier, L.: HISSS - A Higher - Order Subsonic/Supersonic Singularity Method for Calculating Linearized Potential Flow, AIAA-Paper 84-1646, 1984.

[4]   Deidewig, F.: Ermittlung der Schadstoffemissionen im Unter- und Überschallflug, DLR-FB 98-10, DLR Köln 1998.

[5]   Nicolai, L. M.: Fundamentals of Aircraft design, University of Dayton, Ohio, 1975.

[6]   Glatt, C. R.: WAATS - A Computer Program for Weight Analysis of Advanced Transportation Systems, NASA CR-1420, 1974.

[7]   Rech, J. and Leyman, C. S.: A Case Study by Aerospatiale and British Aerospace on the Concorde, AIAA Professional Study Series, 1980.

[8]   Barfield, N.: Aerospatiale/BAe Concorde, Aircraft Profiles, Profile Publications, Coburg House, 1973.

[9]   Mizuno, H; Hagiwara, S.; Hanai, T.; Takami, H.: Feasibility Study on the Second Generation SST, AIAA Paper 91-3104, 1991.

[10]   Döpelheuer, A.: Abschätzung des Brennstoffverbrauchs und der NO_x-Emission von Überschallverkehrsflugzeugen, DLR IB-325-13-94, DLR Köln, 1994.

[11]   Leyman, C. S. and Markham, T.: Prediction of Supersonic Aircraft Aerodynamic Characteristics, AGARD LS-67, 1974.

[12]   Harris, R. V., Jr.: An Analysis and Correlation of Aircraft Wave Drag, NASA TM X 947, 1964.

[13]   Van der Velden, A. J. M.: Aerodynamic Design and Synthesis of the Oblique Flying Wing Supersonic Transport, Dissertation, Stanford University, 1992.

FIGURES

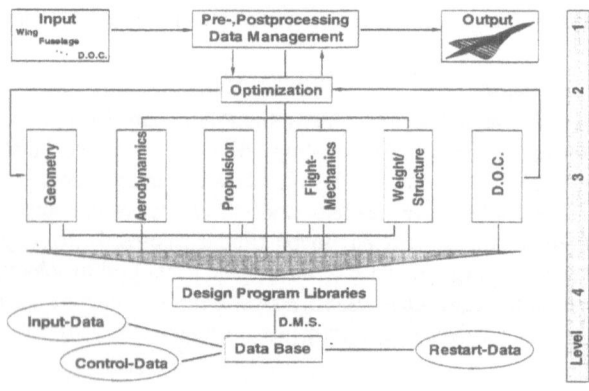

Fig. 1:   The Structure of Prado-Sup

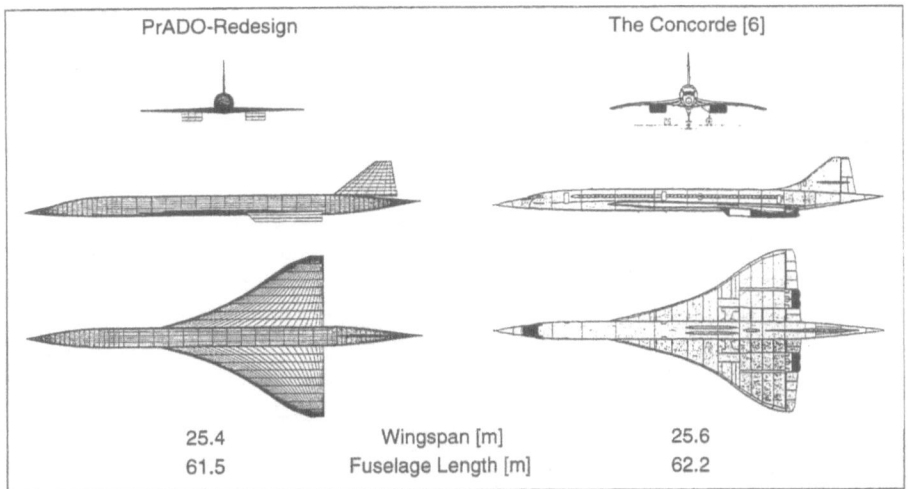

Fig. 2: The Redesigned Concorde

PrADO-Redesign		The Concorde [6]
25.4	Wingspan [m]	25.6
61.5	Fuselage Length [m]	62.2

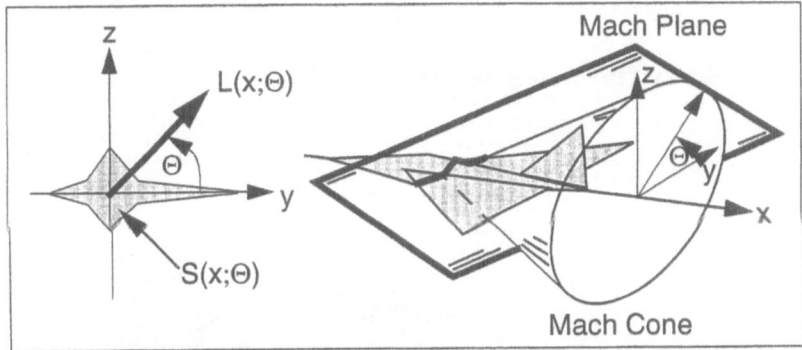

Fig. 3: The Mach Cut Analogy

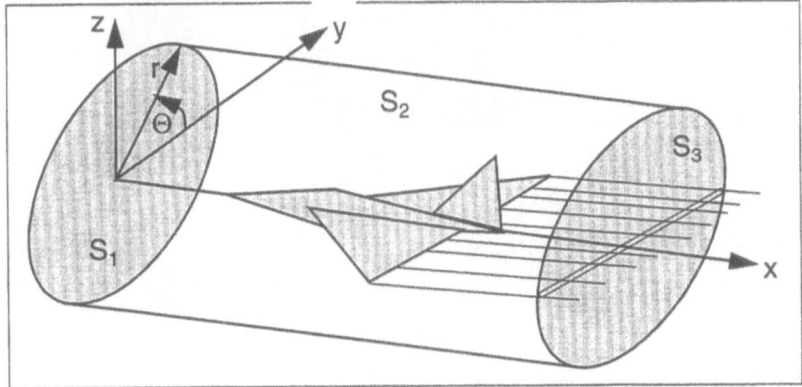

Fig. 4: Farfield Drag Integration by Momentum Theory

447

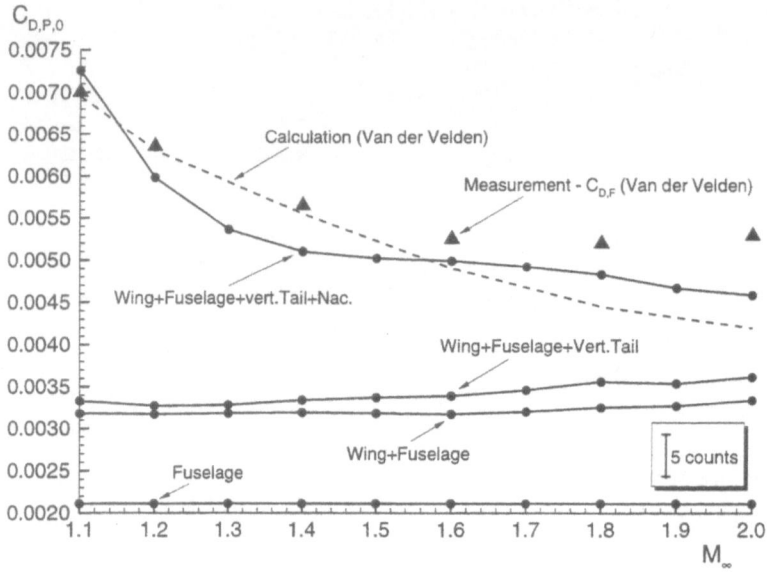

Fig. 5: Concorde Zero Lift Wave Drag from HARRIS Correlation

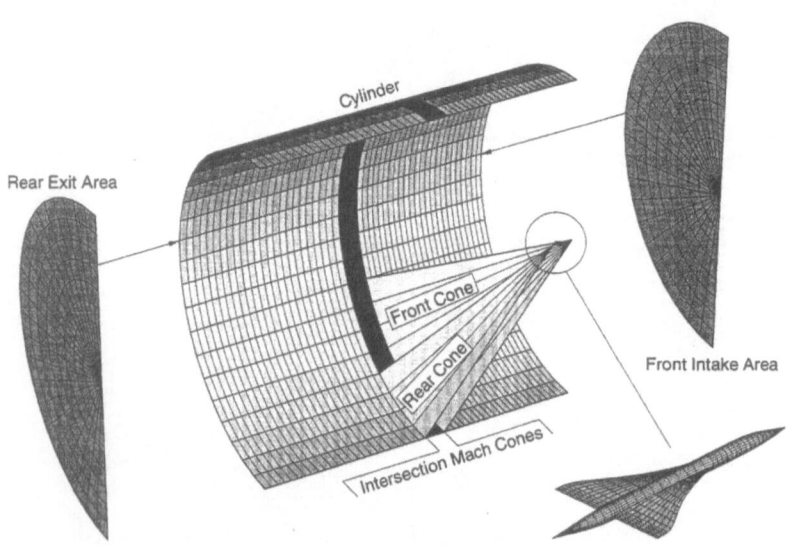

Fig. 6: Control Surfaces for HISSS Cylinder Integration

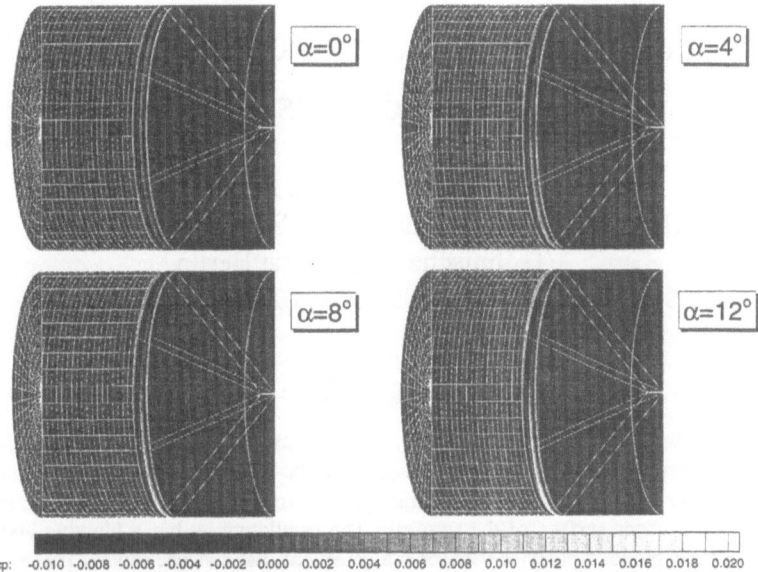

Fig. 7: Pressure Signature from HISSS for the Concorde Wing-Body at M_∞=1.3

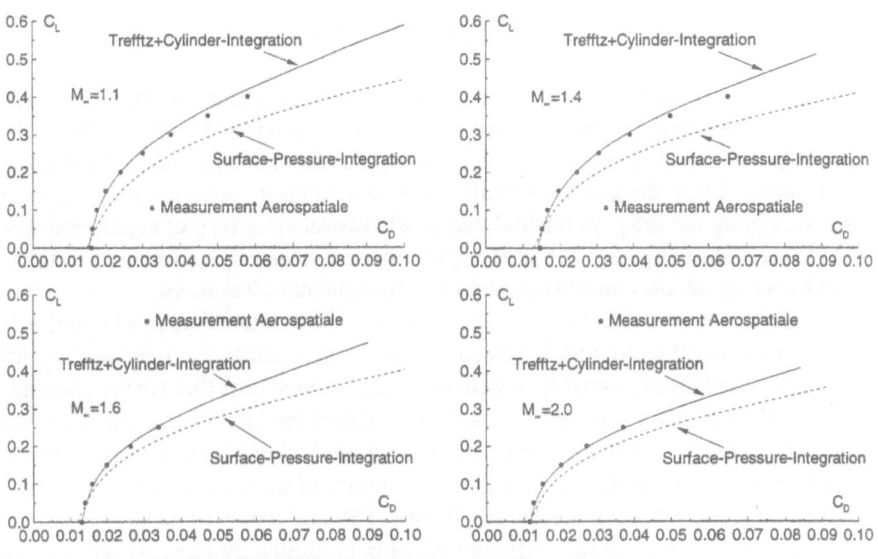

Fig. 8: Revised Drag Polars of Redesigned Concorde

Active Separation Control on an Aerofoil by a periodic cross-flow

H.A. Siller and H.H. Fernholz
Hermann-Föttinger-Institut für Strömungsmechanik
Technische Universität Berlin,
Straße des 17. Juni 135, D–10623 Berlin, Germany

Summary

Wind-tunnel experiments were performed using an HQ-17 laminar aerofoil with active control of the flow separation in the recompression region. A loud-speaker system generates an oscillating, two-dimensional jet in a spanwise slot on the upper surface of the aerofoil. The oscillating jet has a blowing and a suction phase. Spanwise vortices are generated that enhance mixing processes and increase the lift coefficient c_l. The main parameters affecting the lift gain are the position of the separation line relative to the slot, the frequency, and the amplitude of the oscillating jet.

1. Introduction

Boundary layer separation at high angles of attack limits the maximum lift of aerofoils. The flow begins to separate at moderate angles of attack at the trailing edge. With increasing angles of attack, the separation region moves forward towards the leading edge of the aerofoil, creating an area of constant pressure, reducing the lift and increasing the drag. A method that would prevent this type of separation would enable aerofoils to operate at higher angles of attack. Aircraft could land and take off with lower speed, they would consume less fuel and emit less noise.

A review on the subject of separation control [1] states that "the most popular flow separation control technique has been to add momentum to the near-wall region". This can be achieved passively or actively: Passive methods, like vortex generators, stabilise the boundary layer very effectively, but they tend to increase the drag forces under off-design conditions. An active control method can be turned on when it is needed and would ideally not alter the performance of the aerofoil otherwise. Active methods include vibrating mechanical devices [2], and steady or oscillatory blowing from slots [3]. A variant of oscillatory blowing is oscillatory blowing with zero net mass-flux, i.e. periodically alternating blowing and suction. Investigations of separation flow by periodic suction and blowing are reported in [4] and [5]. The aim of the present study is to investigate the potential of this method of separation control on the trailing edge separation of an aerofoil.

2. Experimental setup

A section of a laminar glider profile *HQ-17* is mounted in the test-section of the wind-tunnel on top of a six-component force balance. (see figure 1). The chord length of the aerofoil is $c = 0.65$ m, the span 1.55 m. Its support is isolated from the flow by double walls and the angle of attack can be adjusted by a motor drive. The wind tunnel can run with a maximum velocity of $U_\infty \approx 40$ m/s. The maximum Reynolds number based on the aerofoil chord is $Re_c = 1.7 \times 10^6$.

A slot in the spanwise direction with a length of 1.05 m is cut into the upper surface of the aerofoil at a chordwise position of $x/c = 0.66$. Underneath the slot, there is a pressure chamber with an array of loudspeakers. The sound-pressure level can be monitored with a microphone. The loudspeakers pump a periodically alternating jet and sink flow at the slot exit (see figure 2). The loudspeaker input signal is a sine wave, the frequency range is $10 \leq f \leq 300$ Hz.

The surface of the aerofoil has 95 pressure taps. They are connected to two Scanivalves (TM) inside the aerofoil. Two differential pressure transducers measure the pressure difference relative to the reference static pressure. Data acquisition is controlled by a personal computer. No wind-tunnel corrections are applied to the data, because the main concern are the relative changes imposed by the acoustic excitation.

3. Separation control

The periodic flow at the slot exit generates spanwise vortices that are convected downstream by the free-stream velocity. They mix the flow in the near-wall region with the potential flow very efficiently. Depending on the state of the flow, it is possible to either suppress boundary layer separation or induce early reattachment of an already separated flow.

One has to keep in mind that two systems interact in this process: i) the source of the excitation (the system consisting of the loudspeakers, the pressure chamber and the slot), and ii) the boundary layer or the separation region on the aerofoil that interacts with the spanwise vortices. The greatest effects can be achieved, if an inherent instability of the separation zone can be exploited [4].

The efficiency of the source depends on its aero-acoustic properties. The interaction of the spanwise vortices with the boundary layer (or the separation bubble) depends on the flow speed, the angle of attack, the relative distance of the separation line to the slot where the vortices are formed, the frequency, and the amplitude of the excitation.

In order to optimise the method, first the frequencies and amplitudes have to be determined which most effectively delay or suppress boundary layer separation in a specified situation. Then, the system consisting of the loudspeakers, the pressure chamber and the slot has to be designed in such a way that it generates the required frequency and amplitude.

3.1 The source of the excitation

Figure 2 shows phase averaged velocity measurements at a position located 0.3 mm above the slot. The data were measured using a laser-Doppler-anemometer at a wind-tunnel velocity of zero. The measuring volume had a length of 1.4 mm and was aligned perpendicular to the slot, thus averaging the velocity over the width of the slot. Figure 2 shows the vertical velocity component as a function of the phase angle. The loudspeakers act as pumps: during the upward stroke, a jet is produced in the slot, and during the downward stroke, the slot acts as a sink, drawing in fluid from all sides. The velocities measured above the slot are therefore much lower during the suction phase than during the ejection phase of the cycle.

3.2 The effect of excitation on the lift forces

The lift of the aerofoil depends on the flow speed U_∞, ρ and ν, the angle of attack α, the position of the slot x_s, and the frequency f and the amplitude of the excitation. In non-dimensional terms this relationship can be expressed as

$$c_l = f(Re_c, \alpha, x_s/c, St, L_p) .$$

A variation of the Reynolds number has little effect on c_l once a certain threshold Reynolds-number is exceeded. The angle of attack α influences the position of the separation line and the lift and drag forces on the aerofoil.

Figure 3 shows the variation of the position of separation on the aerofoil versus the angle of attack. The flow separates at the position of the slot at an angle of $\alpha = 13°$. If separation occurs downstream of the slot, the vortices generated at the slot are weakened by dissipation before they reach the separation region. If the slot is already inside the separation region, the disturbance cannot mix fluid from above sufficiently.

Figure 4 shows polars of the lift and drag coefficients of the aerofoil at a Reynolds number of 6.0×10^5. c_l rises linearly with α until the onset of separation from the trailing edge at $\alpha \approx 7°$.

When the flow is forced at a frequency of $f = 50.1$ Hz with a sound-pressure level of $L_{p1} = 129$ dB, c_l is increased for $\alpha > 7°$. The maximum lift gain of 12% is achieved at $\alpha = 12°$, when the unforced flow separates just a small distance downstream of the slot. The forcing loses its effectiveness when the slot is situated inside the separation bubble: figure 4 shows a rapid decline of c_l for angles above $\alpha = 13°$ in the excited case.

A small hysteresis effect occurs: For increasing α, the excitation delays separation up to $\alpha = 13°$. When the flow is already separated and the angle of attack is decreased, the excitation cannot induce reattachment until $\alpha \approx 12°$.

The Strouhal number influences the efficiency of the excitation, since the wavelength of the vortices generated has to match with the length scales of the separation region. Figure 5 shows the maximum of the lift coefficient measured at different forcing frequencies scaled with the corresponding value without forcing. The frequency is scaled with the distance from the slot to the trailing edge L and the free stream velocity U_∞, yielding the Strouhal number $St = (fL)/U_\infty$. The velocity measured at the slot exit had a constant value of $v_{rms} = 9.0$ m/s for all frequencies. The maximum lift increase is achieved at a Strouhal number $St = 0.6$, a second relative minimum occurs at the subharmonic $St = 0.3$.

The sound pressure level L_p inside the pressure chamber can be used as a measure for the amplitude of the excitation. Measurements of c_l and c_D at a fixed $\alpha = 10°$ and varying sound-pressure level L_p are shown in figure 6. The lift and drag forces are nonlinear functions of the amplitude and they are not coupled: For sound-pressure levels below 116 dB, both lift and drag increase, between 116 and 126 dB, c_l increases while c_D is reduced. For even higher sound-pressure levels, there is a steeper lift increase, coupled with an increase of the drag. Due to the limited power of the loudspeakers, a saturation point could not be reached.

To illustrate the effect of the forcing on the flow around the aerofoil, figure 7 shows pressure distributions measured around the aerofoil. Measurements were performed for the unforced and the forced case at an angle of attack of $\alpha = 10°$ and a Reynolds number $Re_c = 6.0 \times 10^5$. At the pressure tap situated immediately downstream of the slot (at $x_s/c = 0.66$), the disturbance has a strong local effect. While the flow separates at $x/c \approx 0.76$ in the reference case, it remains fully attached with the excitation turned on.

4 Conclusions

The lift on an aerofoil at high angles of attack can be increased by acoustic excitation. Boundary layer separation is delayed or a separated boundary layer can be induced to reattach. The lift increase depends on the Strouhal number, the amplitude and the position of the slot. The main problem is to match the properties of the source of the excitation with the frequencies and amplitudes required to control the separation most efficiently.

Further experiments are planned to study the influence of the different parameters. Especially questions regarding the optimisation of the source of the excitation, the system consisting of the slot, the loudspeakers and the pressure-chamber, demand further study.

Acknowledgements

The project was financed by the German ministry for education, science and technology. The authors would like to thank the Italian loudspeaker manufacturers AU-DYON for their cooperation.

References

[1] M. Gad-el Hak and D. M. Bushnell. Separation control: Review. *Journal of Fluids Engineering*, 113:5–30, 1991.

[2] A. Seifert, D. Eliahu, and D. Greenblatt. Use of piezoelectric actuators for airfoil separation control. *AIAA Journal*, 36(8):1535–1537, 1998.

[3] B. Nishri and I. Wygnanski. Effects of periodic excitation on turbulent flow separation from a flap. *AIAA Journal*, 36(4):547–556, 1998.

[4] L. W. Sigurdson. The structure and control of a turbulent reattaching flow. *Journal of Fluid Mechanics*, 298:139–165, 1995.

[5] P.P. Erk. *Separation control on a post-stall airfoil using acoustically generated pertubations*. Number 328 in Fortschrittberichte, Reihe 7: Strömungstechnik. VDI-Verlag, 1997.

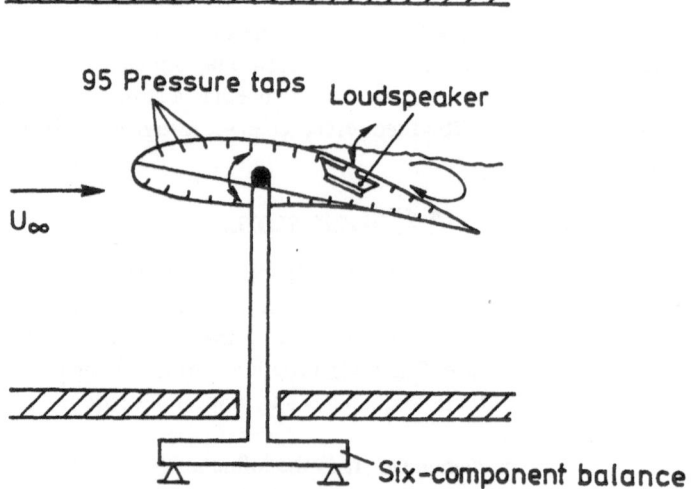

Figure 1: Experimental setup

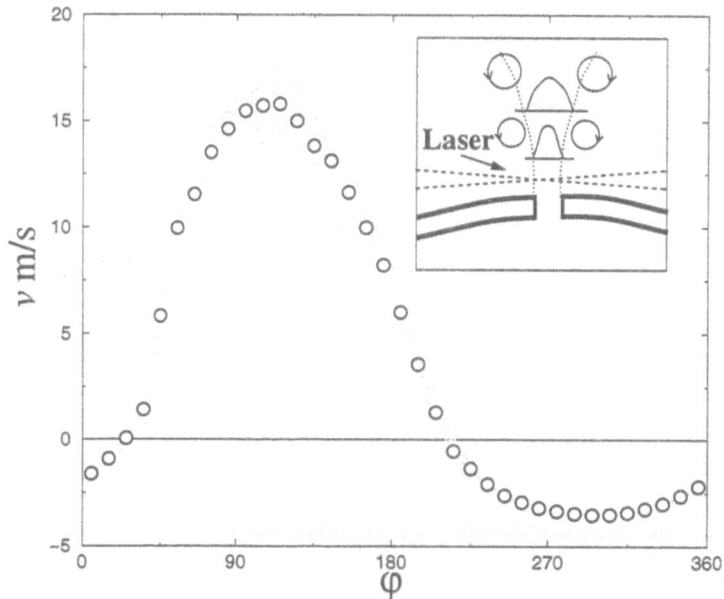

Figure 2: Velocity measurements above the slot exit

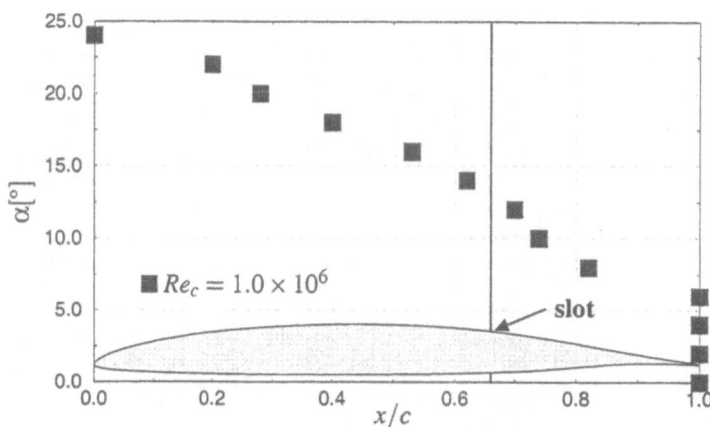

Figure 3: Position of the separation line as a function of the angle of attack

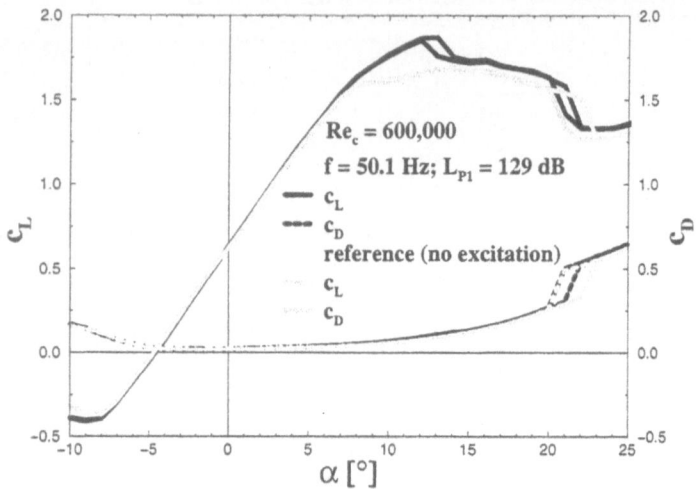

Figure 4: Lift and drag coefficients for a Reynolds number of 6.0×10^5

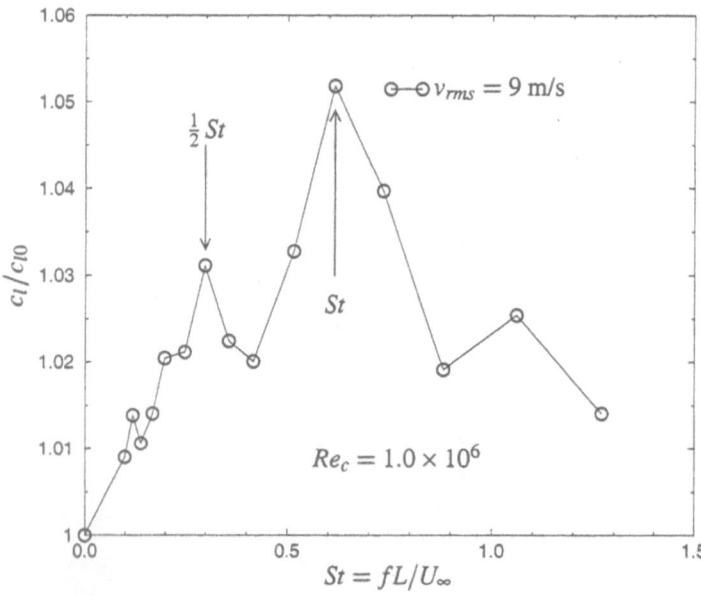

Figure 5: Maximum lift increase as a function of the forcing Strouhal number

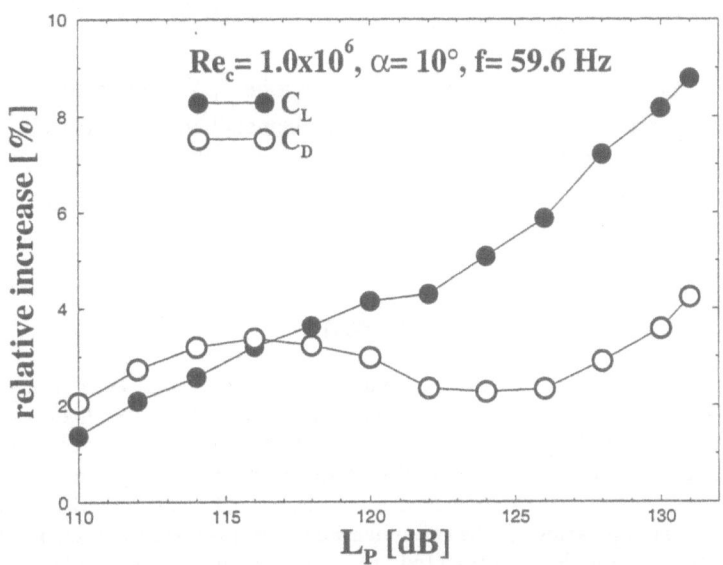

Figure 6: Lift and drag increase with rising sound-pressure level

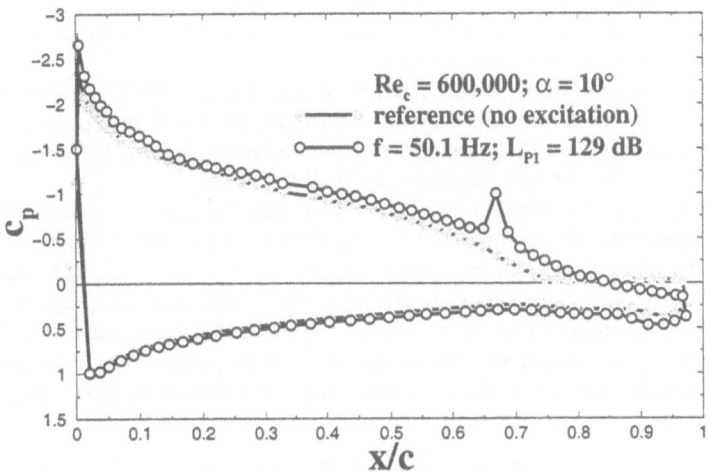

Figure 7: Pressure distribution around the aerofoil

Point-source induced transition in free flight

C. Stemmer[1], J. Suttan[2], M. Kloker[1], W. Nitsche[2]

[1]Universität Stuttgart, Institut für Aerodynamik und Gasdynamik
Pfaffenwaldring 21, 70550 Stuttgart, Germany

[2]Technische Universität Berlin, Institut für Luft- und Raumfahrt
Marchstr. 14, 10587 Berlin, Germany

Summary

This paper reports on experimental and numerical results, which were obtained during the course of work of the joint university research group "Natürliche Transition". Main target of the investigations was the downstream development of an artificially induced point source disturbance in a decelerated transitional boundary layer in free flight. Experimental results obtained through in-flight measurements and numerical results from Direct Numerical Simulations of the twodimensional adverse pressure gradient flow over a wing glove on a motorglider are presented and compared qualitatively and quantitatively.

Introduction

Laminar turbulent transition experiments in wind tunnels have been undertaken extensively in the last decades. From the many insights gained through valuable measurements one question remains: "Do the results represent the conditions in free flight ?" A joint research effort has been undertaken to measure transition on the wing of a motor glider in free flight. These extensive experimental efforts are accompanied by a numerical investigation of the flow using Direct Numerical Simulation (DNS). The research effort encompasses various measuring techniques, which are described in [1]. In this paper, a closer comparison between the piezofoil-array measurements by the TU Berlin group and the DNS by the Universität Stuttgart group shall be made.

As a mutual test case, the twodimensional boundary layer flow on a wing glove mounted on the right wing of the motor glider was chosen. All experiments were conducted with controlled disturbances in the area of adverse pressure gradient. The disturbances are introduced by a harmonic point source at a chord-wise position of $x/c = 0.27$. The frequency of the harmonic disturbance source ($f = 900Hz$) is chosen such that the excited disturbance waves experience maximum amplification as predicted by a Linear Stability Theory analysis in the presumed area of transition (downstream of $x/c = 0.40$).

Numerical Method

The numerical model is based on the complete Navier-Stokes equations for incompressible, 3-D unsteady flow in the vorticity-velocity formulation (for a comprehensive description of the numerical method see [2, 3]) The equations are solved numerically in disturbance formulation with finite differences of fourth-order accuracy in space and time. The span-wise direction is discretized with Fourier modes. The resolution of the calculation which was used to obtain the results presented in this paper is $1410\times145\times93$ points in the streamwise, wall-normal and spanwise direction, respectively, corresponding to a domain of $0.3972 \leq x/c \leq 0.51; 0 \leq y \leq 4.5mm; -124mm \leq z \leq 124mm$. The inflow boundary for this high-resolution simulation has been shifted downstream compared to a first run. Unsteady boundary conditions have been used in the second run at the inflow bound-ary extracted from the first computation within the domain. The unsteady pressure is calculated from the complete velocity field with the modified Poisson's equation:

$$\tilde{\Delta} p = \frac{2}{Re} \left\{ \frac{\partial u}{\partial x} \left\{ \frac{\partial v}{\partial y} + \frac{\partial w}{\partial z} \right\} + \frac{\partial v}{\partial y} \frac{\partial w}{\partial z} - \frac{\partial u}{\partial y} \frac{\partial v}{\partial x} - \frac{\partial u}{\partial z} \frac{\partial w}{\partial x} - \frac{\partial v}{\partial z} \frac{\partial w}{\partial y} \right\}. \qquad (1)$$

The calculation of the eigenvalue for the λ_2-criterion follows the procedure described in [4].

Measurement System

For the in-flight measurements, a laminar wing glove equiped with a multi-sensor PVDF (Polyvinylidenfluorid)-array consisting of 192 discrete sensors (Figure 1) is mounted on the wing of a motorglider. The wing glove contains a digital multichannel measurement sys-tem, which is able to record the signals of all 192 sensors simultaneously with a sampling frequency of 20 kHz per channel. A PVDF-sensor measures dynamic pressure, wall shear stress and temperature fluctuations in the near-wall flowfield. During the recent experi-ments, a small temperature gradient ($\Delta T \approx 1K$) between surface and flow was used to amplify the pyroelectric capabilities of the PVDF-foil. The pyroelectric sensitivity is some orders of magnitude higher than the piezoelectric sensitivity, so the temperature gradient leads to higher signal amplitudes and thus, higher signal-to-noise ratios [5, 6].

Since all of the above mentioned fluctuations are correlated with the near-wall stream-wise velocity fluctuations u' in a boundary layer, a PVDF-sensorarray is a good means to gather qualitative measurement data with a high spatial and temporal resolution. As a disadvantage, a calibration of the signal amplitudes in terms of pressure or wall shear stress is difficult for this kind of sensors.

Results

The character of the flow on an airfoil is very sensitive to the local pressure gradient. Slight changes in the pressure gradient can result in large changes of the stability properties of the flow. Therefore, it is essential to maintain a reproducible pressure gradient distribution over the entire chord length. Pressure sensors have been distributed over the entire wing glove to deliver the pressure distribution on the surface of the glove in flight [7].

The agreement of the numerical and the experimental conditions has been verified through hot-wire measurements of the root-mean-square (RMS) value of the streamwise velocity u'_{rms} at an arbitrary x/c-location of 0.375 at the center position $z = 0$ (see Figure 2). The calculated u'_{rms} profile over the wall-normal direction y agrees very well with the three measurements obtained in different flights. The downstream position of the comparison is still in the area where the linear stability theory can be applied ($u' < 1\%$).

The fact that the PVDF-sensors not only pick up the pressure fluctuations but also wall shear stresses, rises the question of the appropriate comparison with computed data. Figures 3 and 4 show that a qualitative agreement between the structures that can be identified through the wall shear stress ($\sim \omega_z$, Figure 3) and the pressure fluctuation at the wall (p', Figure 4) is given. The comparison of the measured data with DNS results will be made by using the pressure fluctuation at the wall.

At two time instances one half fundamental time period apart (Figure 4) the wall pressure fluctuation is shown in a streamwise-spanwise plane (x-z plane). One half of the spanwise extension of the integration domain is shown. The data is symmetric with respect to $z = 0$. The wave train originating from the point source can be identified between $z = 0$ and $\sim 20mm$. Shown is the pressure fluctuation, e.g. the steady part of the calculated pressure is removed in this representation. This is necessary since the PVDF-array is sensitive to fluctuations only, not regarding the pressure gradient of the mean flow. Due to the strong dynamics of the pressure fluctuation, the development downstream of $x/c = 0.43$ exceeds the chosen scale. The pressure is normalized with $p = \tilde{p}/\tilde{\rho}\tilde{U}_\infty^2$, where the tilde denotes dimensional values. Due to nonlinear interaction of the waves building the wave train, the center part of the wave train flattens after exhibiting the expected half moon shape in the linear region. This straightening can be seen at $x/c = 0.405$ in Figure 4a). One half fundamental time period later (Figure 4b), the area of pressure below the mean intensifies at $x/c = 0.412$ and a distinctive dent at $z = 8mm$ starts to develop. This dent becomes very pronounced at $x/c = 0.42$ and $z = 8mm$ (Figure 4a). The formed localized bow shape represents the "*footprint*" of a Λ-vortex which is shown in Figure 12 and described later.

The amplification of the downstream traveling disturbance wave pattern is also clearly visible in the snapshots of signal amplitudes from the PVDF-sensors for similar time levels (Figure 5). The grey scale in this case was chosen as an expression of the relation between the actual and the maximum signal amplitude E/E_{max}. The overall pattern matches the results from the DNS very well, though smaller structures are not as clearly visible due to the grid resolution of 6 x 6 mm.

Figure 6 compares the measurements in the transitional region with the DNS results. Figure 6a) shows the p'-RMS values at the wall. The presented plot resolution for the DNS data reflects the spacing of the sensors on the glove. Three distinct features stand out that can be identified through the measurements. The maximum RMS value in the measurements can be found at location **A'** (Figure 6b). The first sensor row off centerline ($\Delta z = 6mm$) picks up the maximum at $x/c = 0.448$, in comparison with the DNS that shows a maximum at $x/c = 0.452$. The spanwise deviation of this maximum suggests that the airplane flew with a slight yaw angle ($\approx 1.5°$). The streamwise deviation can result from a slightly different basic pressure gradient as was taken for the computation. Other features of the p'-rms can be found in comparable distribution when the spanwise deviation is taken into account. The line **M-M'** should represent the centerline in the case of a small yaw angle present on the glove. One might argue that the flow becomes three-dimensional in this case, but it will be shown that the relevant results compare astonishingly well. The maximum **A** corresponds to the position of the Λ-vortex to be described later. At location **B**, a local maximum appears well before the global maximum **A** shows up. This is clearly off center and is an expression of the fact, that the maximum of the spanwise distribution of the pressure fluctuations is not on centerline as might be expected. This local maximum **B** appears where the outward "end" of the wave train passes by. This end wraps up and remains visible much longer than the entire wave train, which is disintegrating rapidly as breakdown occurs (compare [8]). Another local maximum at location **C** is identified which lies at the border of the measurement region. This is not identified yet to correspond to a certain structure or property of the flow. The measured RMS-value decreases as the flow becomes more and more turbulent ($x/c > 0.46$). This might have several reasons. The most likely one is the limitation of the sampling mechanism for the measurement equipment towards high frequencies that attain considerable amplitudes in the breakdown and turbulent region.

The p'-rms distribution using the full DNS data resolution in the described $x - z$ region is presented in Figure 7. The features described above still hold although they are almost over-detailed and pronounced.

The spatial N-factor distribution of the p'-rms of the wall pressure fluctuation is shown in Figure 8. The reference for the calculation of the N-factor A_0 is taken at $x/c = 0.3975$ and $z = 0$. A comparison with the corresponding measurement data (in this case $N_{rel.} = \ln(A/A_{min})$, A_{min} at $x/c = 0.30$) also shows a satisfying agreement (Figure 9).

The centerline N-factor curves (figure 10) show a similar increase of the N-factor for $0.35 \leq x/c \leq 0.45$. Upstream of $x/c = 0.35$, the resolution of the piezofoil measurement is reached whereas the DNS can resolve much smaller amplitudes (negative N-factors here). The maximum N-factor reached for a reference value $A_0(x = 0.35)$ is about 4 for the DNS and the measurement.

The measured time traces (solid lines) for three consecutive fundamental time periods at six different downstream positions are compared to the corresponding DNS results (dashed lines) in Figure 11. For the first measurement position $x/c = 0.40$, the DNS data show the dominance of the fundamental frequency through the sinusoidal shape of the time traces. Each pair of curves ($\Delta x/c = 0.005$) is shifted up by $p'/p_{ref} = 0.01$ for clarification. For $x/c = 0.41$, the sinusoidal shape changes, and higher (harmonic) frequencies attain reasonable amplitudes. For $x/c = 0.425$, the shape of the DNS time traces are highly nonlinearly deformed, whereas the measured time traces show only the fundamental frequency. The phase-velocity of the disturbances are captured through the down shift of the wave-maxima with increasing x/c. In this respect, the measurement and the DNS match very well.

The λ_2-criterion ($\lambda_2 < 0$) corresponds to the detection of local inflection points in the pressure distribution. The velocity-gradient-tensor is related to the pressure tensor through the Navier-Stokes equation and the eigenvalues of the latter are determined. Negative eigenvalues are an indication for the presence of a vortex. The iso-surface level of $\lambda_2 = -500$ is shown in Figure 12. At the location **A** of the p'-rms in Figure 6, a Λ-vortex is clearly visible. It is interesting to note that the development of the vortex as well as very high rms-values for the wall pressure appear off centerline and not on the centerline as might have been expected. The flow is far from being "two-dimensional". The flattening of the wave crest (compare Figure 4) in the middle part of the wave train is the result of nonlinear wave interaction of various three-dimensional waves. The development of the wave train is dominated by the three-dimensional waves (ref. to [8]).

Conclusions

This work shows that a comparison of DNS data with data obtained through piezofoil measurements for free flight experiments can deliver valuable insight into the investigated adverse pressure gradient flow. The instantaneous results compare qualitatively well as well as the time averaged and RMS results. The wall signals from the measurements can be interpreted in detail by means of the high resolution DNS results. A maximum at the wall of the piezofoil signals can be directly correlated with the Λ-Vortex found in the DNS.

Acknowledgments

This research has been financially supported by the German Research Council (DFG) under contract number Ni 282/7-4 and Be 1192/5-4.

References

[1] Suttan, J., Baumann, M., Fühling, S., Erb, P., Becker, S. and Stemmer, C., In-Flight Research on Boundary Layer Transition – Works of the DFG-University Research Group. In: H. Körner and R. Hilbig (eds.), *Notes on Numerical Fluid Mechanics*, vol. 60, pp. 343–350. Vieweg Verlag (1997), 10^{th}Stab Symposium 96, Braunschweig.

[2] Rist, U. and Fasel, H., Direct numerical simulation of controlled transition in a flat-plate boundary layer. *J. Fluid Mech.*, 298, 211–248 (1995).

[3] Kloker, M., *Direkte Numerische Simulation des laminar-turbulenten Strömungsumschlages in einer stark verzögerten Grenzschicht.* Dissertation, Universität Stuttgart (1993).

[4] Jeong, J. and Hussain, F., On the identification of a Vortex. *J. Fluid Mech.*, 285, 69–94 (1995).

[5] Sturzebecher, D. and Nitsche, W., Visualization of the Spatial-Temporal Instability Wave Development in a Laminar Bouindary Layer by Means of a Heated PVDF Sensor Array. In: H. Körner and R. Hilbig (eds.), *Notes on Numerical Fluid Mechanics*, vol. 60, pp. 335–342. Vieweg Verlag (1997), 10^{th}Stab Symposium 96, Braunschweig.

[6] Nitsche, W., Bose, S., Haselbach, F. and Suttan, J., Capabilities of Surface Measurement Techniques and their Impact on Modern Wing-Design and Assessment. ICAS-98-3.3.4 (1998), 21st Congress of the International Council of the Aeronautical Sciences, Melbourne, Australia.

[7] Erb, P., Ewald, B. and Roth, M., Flight experiment guidance technique for research on transition with Grob G109 aircraft of the Technische Hochschule Darmstadt. In: H. Körner and R. Hilbig (eds.), *Notes on Numerical Fluid Mechanics*, vol. 60, pp. 143–150. Vieweg Verlag (1997), 10^{th}Stab Symposium 96, Braunschweig.

[8] Stemmer, C., Kloker, M. and Wagner, S., DNS of Harmonic Point Source Disturbances in an Airfoil Boundary Layer. AIAA Paper 98-2436 (1998), 29th AIAA Fluid Dynamics Conference, Albuquerque, NM.

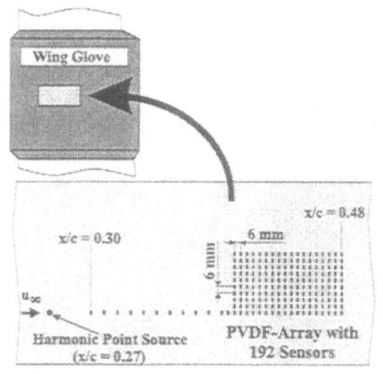

Fig. 1: Wing glove with PVDF-array setup

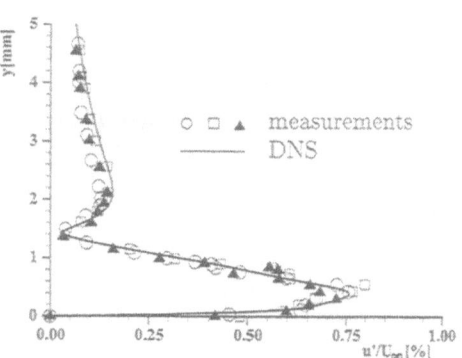

Fig. 2: u'-rms amplitudes at $x/c = 0.375$ on centerline

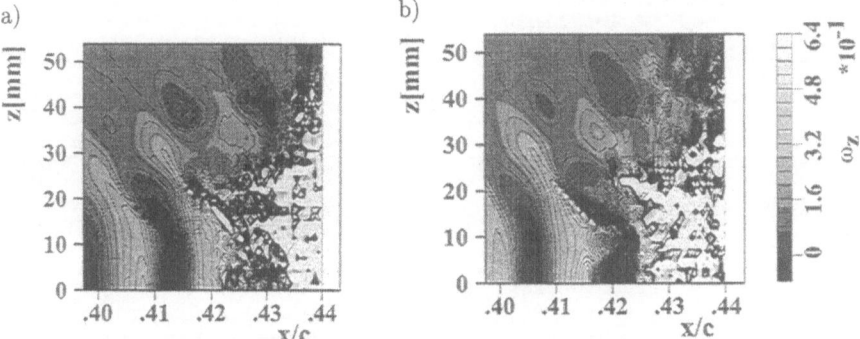

Fig. 3: Instantaneous total spanwise vorticity ω_z at the wall from DNS a) at $t=t_0$ and b) at $t=t_0+T/2$

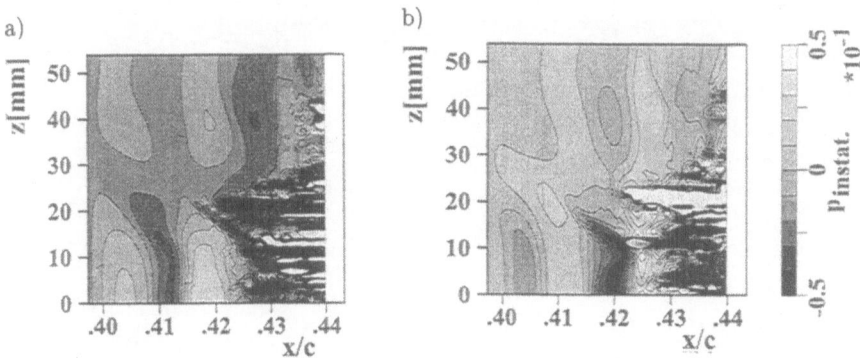

Fig. 4: Instantaneous wall pressure fluctuations p' from DNS a) at $t=t_0$ and b) at $t=t_0+T/2$

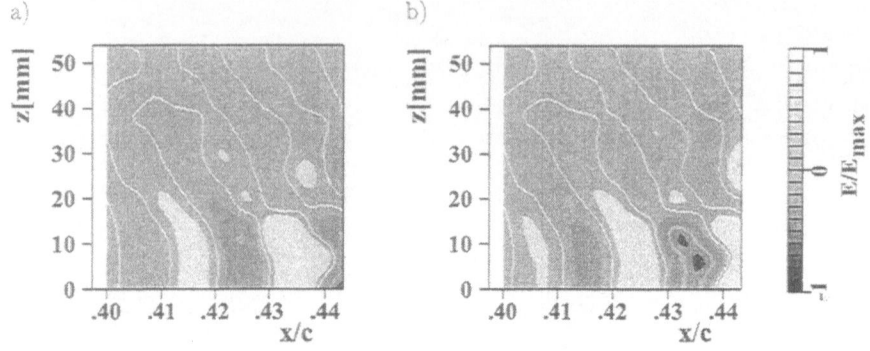

Fig. 5: Instantaneous piezofoil-array signals from experiment a) at $t=t_0$ and b) at $t=t_0+T/2$

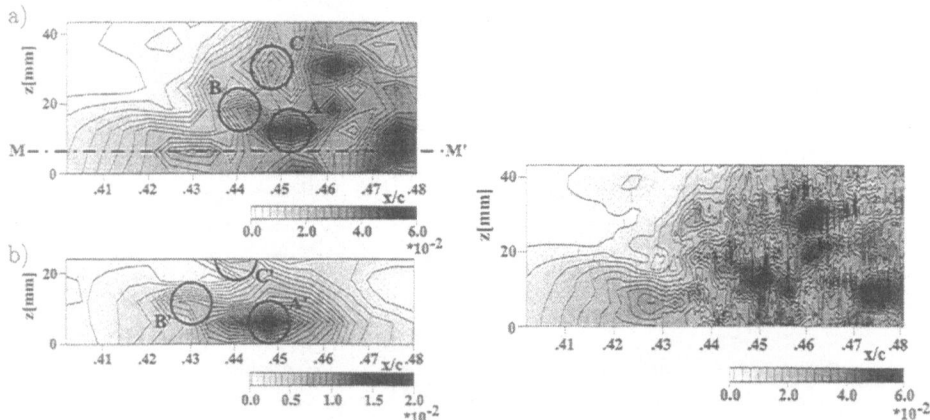

Fig. 6: p'-rms values on the glove surface; measurement-resolution (18×5 points in x-z; to scale) a) DNS b) measurement

Fig. 7: p'-rms values on the glove surface; DNS-resolution (241×14 points in x-z; to scale)

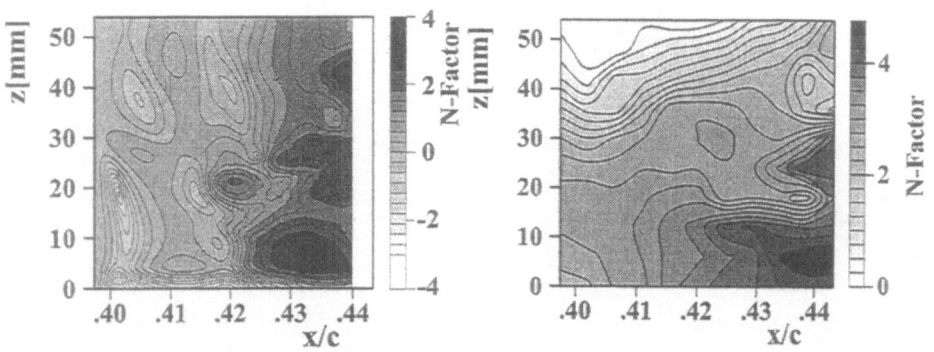

Fig. 8: Spatial N-Factor distribution of the wall pressure fluctuation from DNS; $N = \ln(A/A_0(x = 0.3975, z = 0))$; not to scale

Fig. 9: Spatial N-Factor distribution of the piezofoil-array signals from experiment; $N = \ln(A/A_0(x = 0.30, z = 0))$; not to scale

464

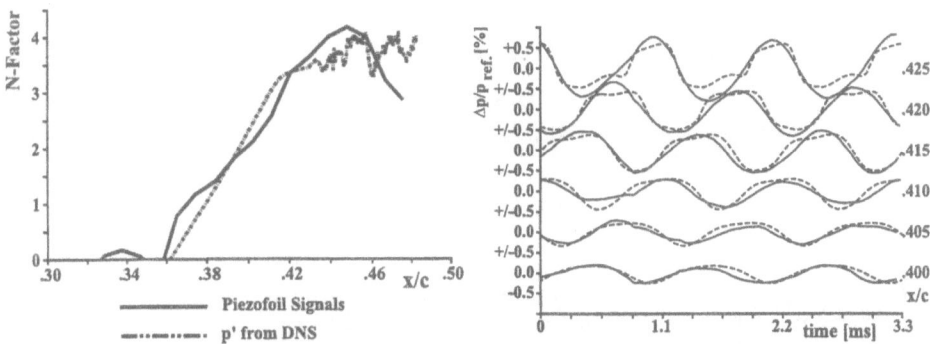

Fig. 10: N-Factor of the wall pressure on centerline

Fig. 11: unsteady p' for six different downstream positions on centerline; $0.40 < x/c < 0.425, \Delta x/c = 0.005$; solid: measurements, dashed: DNS.

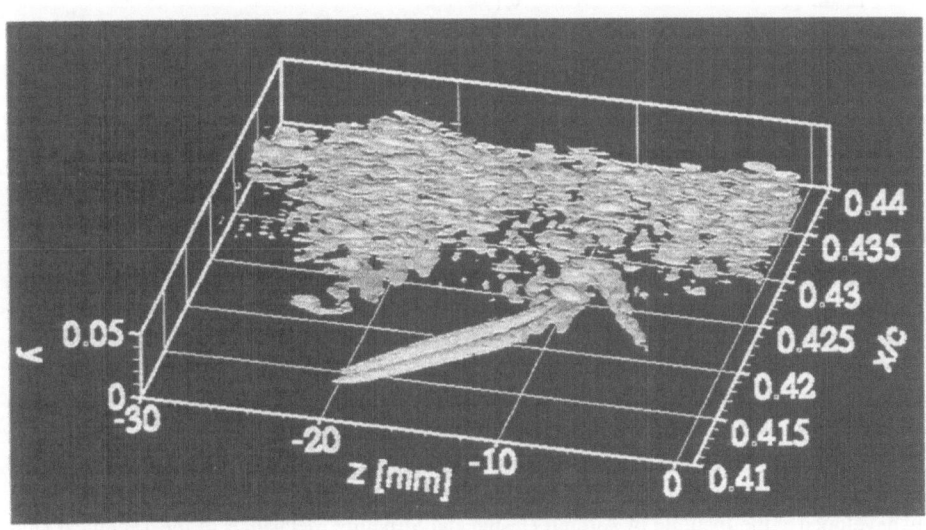

Fig. 12: Flow visualization using the λ_2-criterion for the DNS data; $t=t_0$ (iso-surface of λ_2=-500)

465

Extension of the Predesign Code PrADO for the Evaluation of a Three Surface Aircraft

D. Strohmeyer

Deutsches Zentrum für Luft- und Raumfahrt (DLR) e.V.
Institut für Entwurfsaerodynamik
Lilienthalplatz 7, D-38108 Braunschweig, Germany

Summary

For the analysis of a three surface Megaliner configuration a preliminary aircraft design code was extended by a higher-order panel method, a transonic data base and a trim routine. Preliminary results of the design studies show that the integration of the canard as lifting control device promises a significant reduction of the drag (≈ 3 %) and mission fuel (≈ 2.3 %).

Nomenclature

				Indices	
A	aspect ratio	M_∞	freestream Mach number		
B	span	R	range	A.C.	aerodynamic center
C_L	lift coefficient	S	reference area	C	canard
C_M	pitching-moment coefficient	x,y,z	cartesian coordinates	C.G.	center of gravity
C_p	pressure coefficient	α	angle of attack	cr	cruise
c	chord length	ϵ	trim angle	loc	local
H	flight altitude	φ_{25}	quarter-chord sweep angle	req	required
HTP	horizontal tailplane	γ	circulation	tr	trimmed
h	height	λ	bypass ratio		
L	length	η	span coordinate, 2y/B		
L/D	lift-over-drag ratio	τ	taper ratio		

Introduction

Analyses of the International Air Traffic Association (IATA) predict an annual increase of the worldwide air traffic between 5 % and 7 % for the next decades (Fig. 1). Based on these predictions the large aircraft companies and national aeronautical research institutes think about alternative concepts such as the Megaliner (A3XX), which go beyond the conventional stretching of the existing wide bodies (e.g. B747, A330/A340).

At the Institute of Design Aerodynamics, DLR Braunschweig, special attention is drawn to the evaluation of advanced technologies and unconventional concepts for the improvement of the aerodynamic efficiency of a Megaliner configuration. Within the DLR-project 'Dreiflächen-Flugzeug - 3FF' (Three Surface Aircraft, TSA) a second lifting control device in the nose region, a so-called canard, is designed and its effects with respect to aerodynamics, flight mechanics and control as well as aeroelastics are respectively will be studied.

The final evaluation of this concept is based on integrated predesign studies with an improved version of the **Preliminary Aircraft Design and Optimization Program** PrADO [1]-[3], originally developed at the Institute of Aircraft Design and Structure Mechanics of the Technical University Braunschweig. For this purpose, the aerodynamics module of PrADO was improved by the implementation of the **Higher-Order Subsonic / Supersonic Singularity Method** HISSS [4], provided by Daimler-Benz Aerospace (DASA), Munich. In addition, a transonic data base was included for wave drag calculation and the flight mechanics module of PrADO was extended to allow the simulation of trimmed missions for TSA configurations.

It is the aim of this contribution to show the present state of development, validation and calibration of PrADO, drawing special attention to the aerodynamic part of the design process, and discuss first preliminary results obtained for the Megaliner canard design.

PrADO - Structure and Analysis Models

The structure of the design code PrADO [1]-[3], which has been used by university and industry for the assessment of various flight vehicle and engine related technologies and concepts, is shown in Fig. 2. PrADO is subdivided into four levels, which comprise: a) data input/output routines on the first level, building up a data base to start the iterative design process and to save the result, b) an optimization loop on the second level for design optimization, including gradient and evolution methods, c) a sequential multidisciplinary design process on the third level, covering all disciplines relevant for the integrated aircraft predesign and d) program libraries on the fourth level, which include the physical models and form the kernel of the program. The modular set-up coupled with a clever data management system (DMS) guarantees a high flexibility and easy extension of the program system.

In the canard studies, the basic geometry (wing planform, span, body length, horizontal and vertical tail volume etc.) of the current A3XX configuration with dimensions according to [5] was frozen in the first step. Instead of the proposed elliptical fuselage cross section a circular cross section was assumed, leading to a 9-abreast upper-deck and a 10-abreast lower-deck layout. Hence, the canard was integrated into a geometrically fixed configuration.

The aerodynamic analysis of the original PrADO version is based on lifting-surface theory, using engineering methods [6] in the transonic flow regime. This certainly leads to good results for conventional aircraft configurations, Fig. 3, but it is not suitable for the simulation of the aerodynamic interactions between canard, wing and tailplane, for instance. For this reason the higher-order panel code HISSS was implemented and the aircraft aerodynamics simulated for three Mach numbers: $M_\infty = 0.3$, 0.6 and 0.85. Between the three Mach numbers, the aerodynamic characteristics were interpolated and kept constant below $M_\infty = 0.3$. Since both control devices, horizontal tailplane and canard, are used for trimming, nine different combinations of canard and tailplane trim angle were simulated for each Mach number to determine the interaction effects on the aerodynamic behavior. In transonic flow, wing wave drag was added, taking into account the distribution of the local lift coefficient in spanwise direction. It is based on two-dimensional wind-tunnel measurements of a typical supercritical airfoil [7]. Friction drag was added according to the friction on a flat plate in turbulent flow.

In the following propulsion module thrust and fuel consumption characteristics of existing (e.g. CFM56-5) or projected engines (e.g. CRISP propfan) are available. Alternatively, a scalable turboprop-, turbojet- or turbofan 'rubber engine' can be chosen, modeling the thermodynamic cycle for the calculation of thrust and fuel consumption. For the investigations a high bypass turbofan rubber engine with a bypass ratio of $\lambda = 6$ was chosen and validated using thrust and fuel consumption data of modern high thrust / high bypass ratio engines, provided by BMW Rolls-Royce AeroEngines. The necessary maximum take-off thrust for the engine layout is determined according to FAR 25 by the highest of the following demands: Meeting the maximum take-off distance, minimum climb potential at cruise begin, in the second segment and during approach with one engine inoperative as well as the minimum landing climb potential to ensure safe wave-off after a balked landing.

After the engine layout the mission is simulated in the flight performance module. A trimmed mission of 14,200 km (7650 nm) with 656 passengers and no additional payload is prescribed assuming a cruise flight with $M_{\infty,\,cr} = 0.85$ = const. and $C_{L,\,cr} = 0.5$ = const. To determine the trim conditions an iterative algorithm was integrated, analyzing the relevant combinations of α, ε_{HTP} and ε_C which guarantee $C_L = C_{L,\,req}$ and $C_{M,\,C.G.} = 0$ at the lowest drag. Static stability $\partial C_{M,\,C.G.} / \partial C_L < 0$ is not required, but the neutral point position along the mission is provided in the output data to determine whether the configuration is stable or not.

467

The design of the aircraft structure is based on a finite element method for wing and tailplane. These components are modeled with 2-knot beam elements having a multi-cell cross section. According to JAR / FAR regulations three critical load cases are taken into account for the dimensioning of the wing: rolling on the runway at take-off, flight with maximum manoeuver load and maximum gust load. Additional load cases are analyzed for the layout of tailplane and canard. The canard is treated in the same way as the horizontal tailplane. For the calculation of the fuselage structure a semi-empirical method [8] is used, which takes into account structure loads coming from wing, tailplane and canard as well as the fuselage pressure difference under cruise conditions. The mass of the propulsion unit is determined according to [9], the landing gear mass as well as the masses of all systems, equipment and operational items are based on the WAATS program [10].

In the final step of the iterative design process the tailplane is sized to guarantee sufficient control along the flight envelope from take-off to landing for the entire range of the center of gravity position.

The convergence of the iterative aircraft design process is checked at the end of the stability / control module. A design is considered converged, if the variation of prescribed relevant design parameters (e.g. maximum take-off thrust and take-off weight, mission fuel) falls below a prescribed absolute and relative limit (e.g. maximum take-off weight variation below 0.5% and 50 kg). Larger variations of the chosen parameters lead to a return to the beginning of the design procedure. Finally, the converged design may be assessed based on a defined object function, which is especially relevant for the control of the optimization algorithm.

HISSS - Code, Grid Generation and Validation

The panel code HISSS [4] is a higher-order singularity method for the solution of linear potential flow around arbitrary three-dimensional configurations at subsonic and supersonic speeds. Composite source / doublet panels are used on the surface of the configuration, the wake panels have a doublet distribution to carry downstream the vorticity generated over the surface of the configuration. The Kutta condition at trailing edges is fulfilled without additional Kutta panels.

The application of HISSS requires a panel grid consisting of the surface grid and wake networks. For this purpose the grid generation routine PrADO-Grid was developed. It provides suitable panel grids for conventional and canard configurations using a C-H topology to guarantee regular grids in the intersection between fuselage and wing (tailplane, canard), as shown in Fig. 4. Since HISSS also runs in the optimization mode of PrADO, where non-viable designs may be proposed by the optimization algorithm, Fig. 4, special importance was attached to the robustness of the grid generation method. Up to now, pylon, nacelle and engine are not modeled for the panel calculations. Furthermore, only one half of the configuration is calculated for reason of symmetry; sideslip conditions are not taken into account in order to reduce the numerical effort.

Detailed analyses of the influence of the grid fineness on the aerodynamic coefficients showed a strong dependency of the induced drag on the grid fineness, if it is calculated via surface integration of the pressure coefficient, as illustrated in Fig. 5 for the elliptical wing with NACA-0012 profile. Even fine grids do not provide a proper result. However, the influence of the grid fineness on lift and pitching moment coefficient is neglectable. To guarantee proper drag calculations even for coarse grids, which have to be applied in the predesign studies to reduce iteration time, the drag coefficient is calculated based on the circulation distribution in spanwise direction and the downwash in the Trefftz plane. This method provides accurate drag coefficients almost independent of the fineness of the surface grid.

The validation of the panel code is illustrated exemplarily in Fig. 6, showing the wing surface pressure distribution of the ALVAST configuration for a typical subsonic testcase. The differences between the result of HISSS and the Euler code are small. However, viscous effects, such as the reduction of the rear loading due to the boundary layer influence on the effective camber of the profile cannot be predicted, neither from the panel method nor from the Euler code.

Results

Before the design of the Megaliner configuration with and without canard, PrADO was calibrated on the basis of a redesign of the A340-300 as an existing wide body. The configuration, mission data and results of the calibration are given in Fig. 7. The differences concerning operational empty weight, fuel weight and maximum take-off weight are marginal.

In the second step, a Megaliner configuration corresponding to the geometry, performance and mission data of the A3XX-200 was designed. It is called in the following the baseline configuration. For the cruise flight at $M_{\infty, cr} = 0.85$ a lift coefficient of $C_{L, cr} = 0.5$ was chosen. The required mission data and dimensions as well as the configuration and weight breakdown of the design are given in Fig. 8 and are compared with published data [5]. With respect to the relevant weights the differences between PrADO design and predicted weights in literature are below 6 %. This is comparable to the order of accuracy which was obtained with the original version of PrADO for conventional transport aircraft, Fig. 3.

In the final step, three canard configurations with different canard span, $B_C = 20$ m, 25 m and 30 m were designed, based on the A3XX-200 layout. The geometric characteristics of the canards as well as the planform shape of the configurations are given in Fig. 9a. To illustrate the fineness of the grids used in the panel calculations within the design process, the surface grid of the configuration with $B_C = 25$ m is given in Fig. 9b. The result of the integrated predesign is summarized in Tab. 1 and compared with the A3XX-200 design without canard.

For all four configurations the same cruise lift coefficient was chosen, therefore a direct comparison of drag and aerodynamic efficiency L/D is possible. With respect to the induced drag, the introduction of a canard provides a significant reduction with a maximum of 15 counts at $B_C = 25$ m. It is primarily based on the trimming with the lifting canard instead of the horizontal tailplane download as indicated in Fig. 10a (note that the lift of each component was calculated via C_p integration, they do not yield 100 % due to the missing contribution of the body). Compared with the baseline configuration, the canard provides between 13.9 % and 20.5 % of the total lift, reducing the wing contribution down to 64.9 % at $B_C = 30$ m. In addition, also the horizontal tailplane provides lift for all canard configurations to fulfill the trim condition $C_{M, C.G.} = 0$. The second effect of the canard is an improvement of the lift distribution in spanwise direction, shown in Fig. 10b. The lift decrease of the baseline design in the inner part, based on the download of the horizontal tailplane, is compensated by the canard lift, which leads to a more elliptical distribution of the circulation in spanwise direction. On the other hand, these positive effects of the canard are partially compensated by an additional friction drag which increases from 105 counts of the baseline design by up to 11 counts for $B_C = 30$ m. Since wave drag and interference drag are almost the same for all configurations and also one order of magnitude lower, the combination of induced and friction drag determines the resulting aerodynamic efficiency of the configuration. The results show a maximum of L/D = 21.2 for $B_C = 25$ m, which is almost 3% higher compared with the baseline configuration.

A comparison of the weight analysis shows the expected increase of the operational empty weight with increasing canard span, primarily due to the introduction of the canard. However, in the case of the B_C = 20 m and 25 m the additional canard weight is overcompensated by the reduced mission fuel, leading to a lower maximum take-off weight compared with the baseline configuration. With respect to the static stability, both, center of gravity and neutral point shift to the nose with increasing canard span. Due to the stronger gradient of the neutral point shift, the static margin under cruise conditions (based on the reference chord length of 11.7 m) is reduced from 43.8 % of the baseline design over 30.6 % (B_C = 20 m) and 20.9 % (B_C = 25 m) down to 11.0 % of the canard configuration with B_C = 30 m.

Conlusions

In the present paper the improvement of the integrated aircraft predesign code PrADO and its application to the design of a three surface Megaliner configuration is reported. For the simulation of aerodynamic interference effects between wing, tailplane, canard and body the higher-order panel method HISSS was integrated and extended by a transonic data base for wave drag calculation. Furthermore, a trim routine for three surface aircraft was added.

The application of the calibrated predesign code to a conventional Megaliner configuration leads to differences between published and predicted design weights below 6 %. Preliminary results of the canard studies show that the integration of a canard may provide an increase of the aerodynamic efficiency L/D of almost 3 %. This drag reduction is based on the trimming with the canard as lifting control device and leads to a mission fuel redcution of up to 2.3 %.

Further sensitivity studies including a new integral finite element model for wing, body, canard and tailplane as well as the consideration of center of gravity control via tailplane trim tanks are in progress to corroborate the results.

Acknowledgements

The author thanks L. Fornasier, Daimler-Benz Aerospace (DASA), Munich, for providing the panel code HISSS and his support. W. Heinze of the Institute of Aircraft Design and Structure Mechanics, TU Braunschweig, is gratefully acknowledged for the helpful discussions with respect to the application of PrADO. Further acknowledgement is given to J. Stilla, BMW Rolls-Royce AeroEngines, Berlin, for providing engine data for validation purpose.

References

1. Kossira, H.; Heinze, W.; "Untersuchung über die Auslegungsgrenzen zukünftiger Transportflugzeuge", DGLR-Jahrbuch I (1990), pp. 35-49.

2. Hertel, J.; Albers, M.; "The Impact of Engine and Aircraft Design Interrelations on DOC and its Application to Engine Design Optimization", AIAA Paper 93-3930, 1993.

3. Heinze, W.; Bardenhagen, A.; "Waverider Aerodynamics and Preliminary Design for Two-Stage-to-Orbit Missions, Part 2", Journal of Spacecraft and Rockets, Vol. 35, No. 4, 1998, pp. 450-458.

4. Fornasier, L.; "HISSS - A Higher-Order Subsonic/Supersonic Singularity Method for Calculating Linearized Potential Flow", AIAA Paper 84-1646, 1984.

5. Jane's All The World's Aircraft, 1997-98.

6. Hoerner, S.F.; "Fluid-Dynamic Drag", Hoerner Fluid Dynamics, New York, 1965.

7. Knauer, A.; "Die Leistungsverbesserung transsonischer Profile durch Konturmodifikation im Stoßbereich", Dissertation Universität Hannover, 1998.

8. Simpson, D.M.; "Fuselage Structure Weight Prediction", SAWE Paper 981, 1973.

9. Waters, M.H.; Schairer, E.T.; "Analysis of Turbofan Propulsion System Weight and Dimensions", NASA TM-73195, 1977.

10. Glatt, C.R.; "WAATS - A Computer Program for Weight Analysis of Advanced Transportation Systems", NASA CR-2420, 1974.

Figures

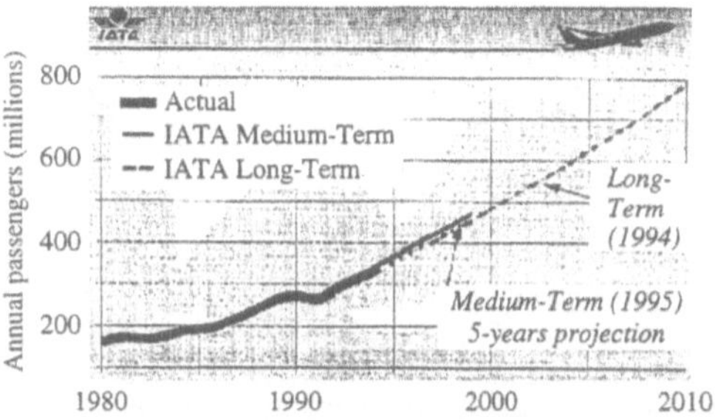

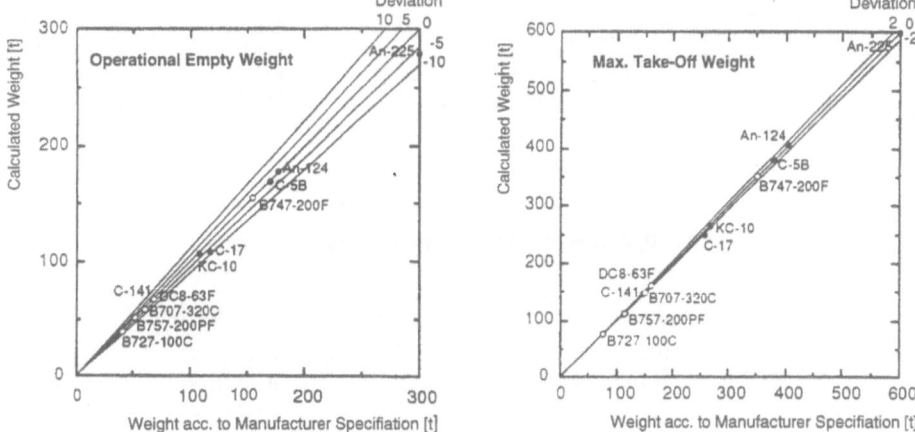

Fig. 1 IATA projection of world international scheduled passenger traffic.

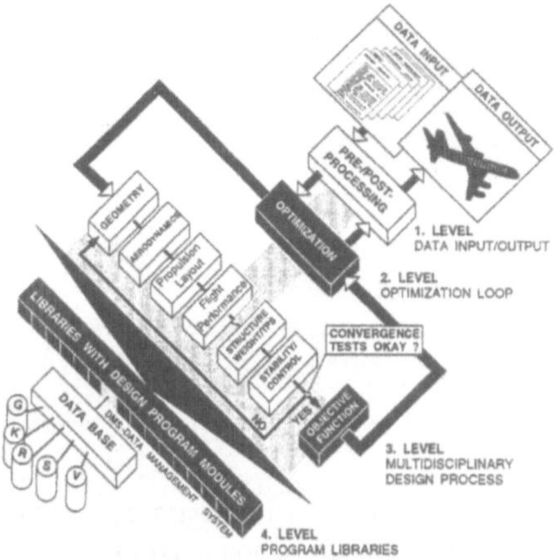

Fig. 2 Structure of the integrated aircraft design program PrADO [1].

Fig. 3 Results obtained with PrADO (original version) for conventional cargo aircraft [1].

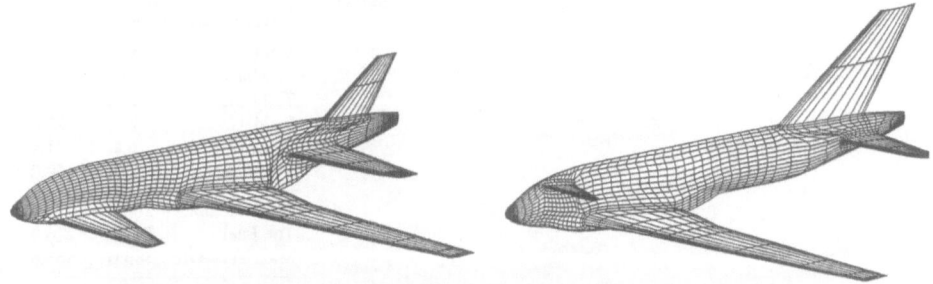

Fig. 4 PrADO-Grid results for arbitrary three surface configurations.

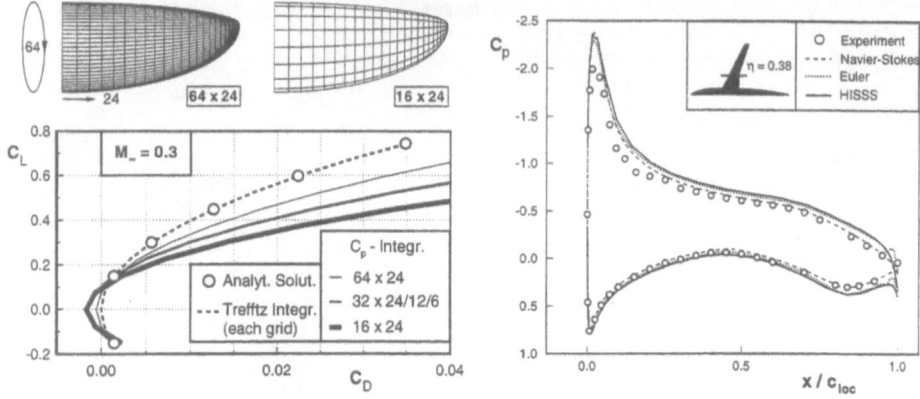

Fig. 5 Influence of grid fineness and method on drag calculation in HISSS.

Fig. 6 HISSS validation using the ALVAST configuration ($M_\infty = 0.27$, $\alpha = 4°$).

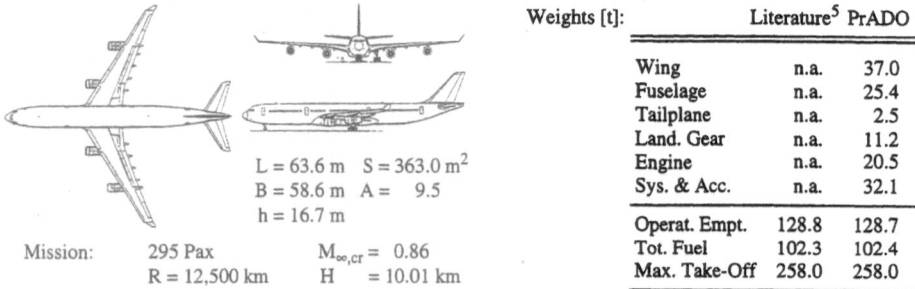

Mission: 295 Pax $M_{\infty,cr} = 0.86$
          R = 12,500 km   H = 10.01 km

L = 63.6 m   S = 363.0 m^2
B = 58.6 m   A = 9.5
h = 16.7 m

Weights [t]:	Literature[5]	PrADO
Wing	n.a.	37.0
Fuselage	n.a.	25.4
Tailplane	n.a.	2.5
Land. Gear	n.a.	11.2
Engine	n.a.	20.5
Sys. & Acc.	n.a.	32.1
Operat. Empt.	128.8	128.7
Tot. Fuel	102.3	102.4
Max. Take-Off	258.0	258.0

Fig. 7 Redesign of A340-300 for calibration of PrADO.

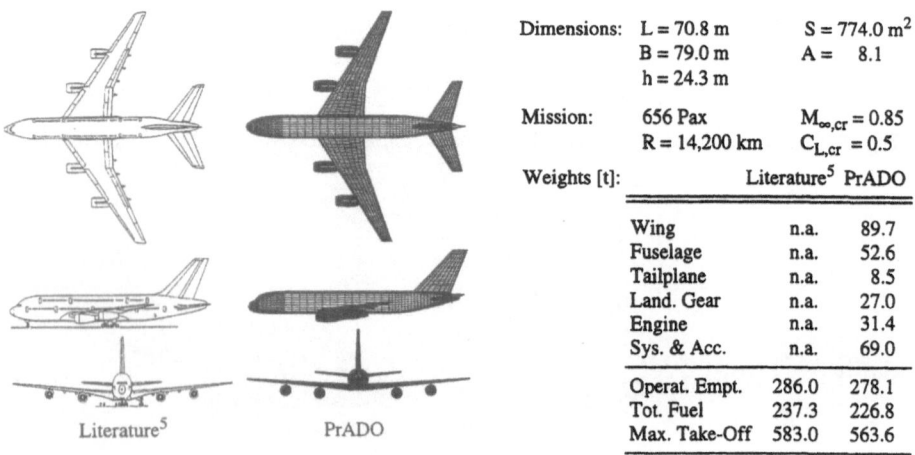

Literature[5]   PrADO

Dimensions: L = 70.8 m   S = 774.0 m^2
             B = 79.0 m   A = 8.1
             h = 24.3 m

Mission: 656 Pax $M_{\infty,cr} = 0.85$
             R = 14,200 km $C_{L,cr} = 0.5$

Weights [t]:	Literature[5]	PrADO
Wing	n.a.	89.7
Fuselage	n.a.	52.6
Tailplane	n.a.	8.5
Land. Gear	n.a.	27.0
Engine	n.a.	31.4
Sys. & Acc.	n.a.	69.0
Operat. Empt.	286.0	278.1
Tot. Fuel	237.3	226.8
Max. Take-Off	583.0	563.6

Fig. 8 Design of A3XX-200 with PrADO.

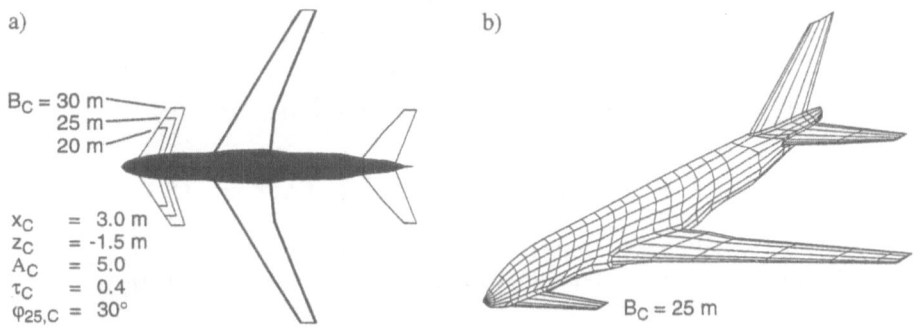

Fig. 9　Analyzed canard configurations (a) and panel grid for the HISSS calculations (b).

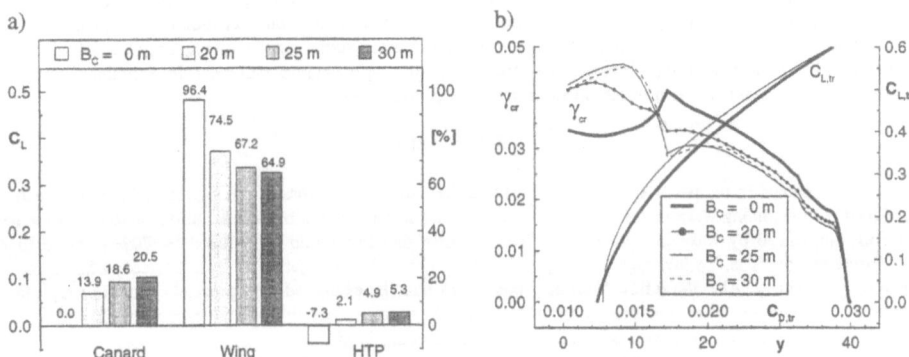

Fig. 10　Contribution of canard, wing and horizontal tailplane to the cruise lift coefficient (a), spanwise circulation distribution and trimmed drag polar for cruise conditions (b).

Tab. 1　Preliminary results of the canard analyses.

Configuration:	canard span [m]	0.	20.	25.	30.
Aerodynamics:	lift	0.5	0.5	0.5	0.5
(cruise)	drag: induced	0.0125	0.0115	0.0110	0.0111
	friction	0.0105	0.0110	0.0113	0.0116
	wave	0.0002	0.0002	0.0002	0.0002
	interference	0.0011	0.0011	0.0011	0.0011
	total	0.0243	0.0238	0.0236	0.0240
	L / D	20.6	21.1	21.2	20.8
Weights:	canard [t]	0.0	2.1	3.5	5.3
	operat. empty [t]	278.1	280.8	282.4	285.9
	tot. fuel [t]	226.8	221.6	221.6	228.5
	max. take-off [t]	563.6	561.7	563.2	573.5
Stability:	x-pos. center of gravity [m]	34.83	34.53	34.30	34.06
(cruise)	x-pos. neutral point [m]	39.96	38.11	36.75	35.55

Experimental Investigation of Boundary Layer Transition in Compressor Cascades at Unsteady Periodic Flow Conditions

M. Swoboda[*], R. Teusch[**], V. Gümmer[*], L. Fottner[**], U. Wenger[*]

[*] BMW Rolls Royce GmbH, Eschenweg 11, 15827 Dahlewitz, Germany
[**] ISA, Univ. BW Munich, Werner-Heisenberg-Weg 39, 85577 Neubiberg, Germany

Abstract

The boundary layer transition process on compressor blades in a multi-row environment is mainly influenced by incoming periodic wakes generated by the upstream blade row. To simulate such conditions experimentally, a moving bar wake generator device was used which is arranged in front of a compressor cascade in a high speed wind tunnel. A Hot-Film-Array measuring technique was used to investigate the transition process on compressor blades with unsteady inlet flow conditions. This paper discusses some aspects of unsteady transition on the suction side of the compressor blades. The structure of the wake and in particular its turbulence intensity distribution is shown to cause different kinds of multi-mode transition.

Introduction

It has been shown in the past (Adamczyk, 1996) that the total pressure loss of a compressor blade row can be reduced if an incoming wake mixes out after rather then before a following blade row. In that case the blade row is fully influenced by wake passing effects. It was shown that loss could be reduced by 70% relative to an undisturbed flow case. To investigate this global effect a detailed investigation of local unsteady effects on the blades is necessary. Also the inflow boundary conditions must be examined very precisely because they play a dominant role in the transition behaviour under periodic unsteady conditions and especially the turbulence quantities are needed for unsteady code validation, Delimar et al., 1998.

The process of boundary layer transition in compressor blade rows depends on one side upon the adverse pressure gradient on the blades, Reynolds number or Turbulence intensity of the inlet flow but on the other side will be especially influenced by unsteady effects of incoming wakes of previous blade rows. Under this periodic disturbance the boundary layer development is mainly influenced by wake-passing effects resulting in a multi-mode transition process. The boundary layer behaviour under these conditions was investigated in the past mainly in low speed wind tunnel flows using rotating rods, Schröder, 1984, or in low speed rig compressors, Halstead et. al, 1995. These investigations show that the incoming wake induces a turbulent spot on the blade suction side followed by a calmed region similar to laminar flow but more stable and thus which is able to survive up to the trailing edge of a blade. The existence of a calmed region behind a turbulent spot was already reported by Schubauer and Klebanoff in 1955. Significance gains in understanding the processes leading to the development of this region and the role of the calmed region in the loss generation process have been made in recent years.

Experimental Apparatus and Cascade Instrumentation

The experimental investigations were carried out at the High Speed Cascade Wind Tunnel of the University of Federal Armed Forces in Munich/Germany. The wind tunnel is built in a large pressurized tank and thus offers the possibility of varying the Mach and Reynolds Number in the test section independently and in a wide range which is relevant to simulate turbomachinary flows, Sturm et al., 1985. Recently, this facility was equipped with a wake generator to simulate unsteady, periodic wakes and thus effects of rotor-stator interaction, Acton, 1998. This moving bar device consists of cylindrical rods which are moved in front of the cascade at velocities up to 40m/s. In Figure 1 the moving bar device is shown which is mounted on the wind tunnel nozzle in front of the compressor cascade. The cylindrical bars are mounted in rubber belts (minimal pitch = 10mm) which are driven by an electric motor. More details of this device can be found in Acton, 1998.

The inlet Mach number is measured by means of a pitot probe (60mm in front perpendicular to the cascade) and an average of three static wall pressures measured in the cascade inlet plane in the middle of a blade passage. In addition to the measurement of these steady quantities, a single hot-wire probe was used to measure the velocity and turbulence profiles of the moving bar wakes. These measurements were performed without the cascade but are representative to the periodic and unsteady inlet conditions to the cascade. The hot-wire was calibrated prior to every measurement to account for different tank pressures at varying Reynolds number. Because only a single wire was used the values measured are mean velocity and isotropic turbulence intensity. No statement can be made in that case about the fluctuating velocity components or length scales of turbulence.

The CDA (Controlled Diffusion Airfoil) compressor cascade V110 was used in this investigation. This airfoil was conventionally designed via camberline and thickness distribution by means of a 2D blade-to-blade code to give a prescribed velocity distribution for inlet conditions of Table 1. For the application in the Cascade Wind Tunnel, the airfoil was scaled up to give a chord length of 100mm. The aerodynamic details of the cascade are summarised in Table 1 for the design point.

Table 1: Design data for the V110 cascade

BLADE NUMBER	7
Inlet Mach Number	0.77
Reynolds Number	584000
Stagger Angle β_s	122.55°
Flow Inlet Angle β_1	136.43°
Flow Exit Angle β_2	114.53°
Relative Pitch t/l	0.611
Relative Thickness d/l	0.049
Chord Length l	100mm

The instrumentation of the cascade consists of the measurement of static pressure distribution on two blades on suction and pressure side respectively. However, pressure measurement results are not presented in this paper because only the influence of wakes on the suction side boundary layer behaviour was of interest here. Another airfoil was equipped with a Hot-Film-Array on the suction side to analyse the unsteady effects caused by the incoming wakes. The glue-on Hot-Film-Array consists of 32 separate sensors arranged in the mid span of the airfoil from 9 to 84% of axial chord. The thickness of the individual sensors is 0.3μm and the distance between the sensors is 2.5mm. The sensors are operated simultaneously in blocks of 12 sensors in CTA (Constant Temperature Anemometry) mode. In this mode the steady signal output is proportional to the local wall shear stress according to Reynolds analogy. However, a calibration of the signal was not carried out because the raw signals allow an assessment of the state of the boundary layer flow, i.e. whether it is laminar, transitional, turbulent or separation and to what extent.

Cascade Flow in Presence of Wakes Generated by Cylindrical Rods

The simulation of periodic unsteady conditions representative for turbomachinery flow can be carried out by means of a moving bar device assuming that the wake of the cylinder in its far field is similar to a wake of an upstream blade row having similar loss, Pfeil et al., 1975. The arrangement of the moving bar device and the V110-cascade is shown schematically in Figure 2, as well as the position of the measuring devices. In the case of a compressor the movement direction of the cylinders is from upper left to lower right (in turbine case the direction is opposite). It can be seen that the bars are moved with a component in flow direction. Assuming that there will be a small amount of work done on the flow by viscous forces the incidence angle to the blades will also increase by a small amount. Another aspect which should be mentioned here is the direction of the wakes relative to the blades. When the bar speed is approximately the same as the flow speed the wakes will enter the blade passage in a direction almost rectangular to the blades which is representative to the flow in a real machine. In compressible flow at high Mach numbers of approx. M=0.7 the maximum bar speed of 40m/s is to low to simulate that wake direction. In this case the wake is almost parallel to the blades and enters the passage while covering almost the whole blade. This is shown schematically by velocity triangles for the two cases and the accompanying wake direction. However, to overcome this problem also measurements were performed at very low Mach number of M=0.15 to show the incompressible transition effects similar to previous publications, Cumpsty et al., 1995, Halstead et al., 1995.

Analysis Methods

In the case of the moving bar measurement the flow signal contains as well periodic unsteady shares as turbulent fluctuations. To separate these shares the entire unsteady measured data was processed by a so called Phase Locked Ensemble Average Technique proposed by Lakshminarayana et al., 1974. The periodic share will be extracted from the raw data by means of averaging over a large variety of measured signals. In the case of the presented measurements the signals were recorded 300 times over 5 bar passages triggered to the same 5 bars. As well the hot-wire as the hot-film-array measurements were treated in this way.

The ensemble averaged hot-film signals can be analysed in different ways. The quasi wall shear stress $(E-E_0)/E_0$ can be derived directly from the DC signal of the Wheatstone anemometer bridge. The Root Mean Square (RMS) values are derived from the AC signal of distinct hot-films and can be used to analyse the state of the boundary layer. Low RMS values characterise generally laminar flow regions while a maximum of the RMS distribution is typical for the transition point. Local maximum stream up of the absolute RMS_{MAX} is typical for separation bubbles. Another statistical analysis method which is used here is the skewness function which represents the deviation of the signal from the normal GAUSS distribution. The skewness is zero in laminar and fully turbulent flow whereas it is positive in front of the transition point showing turbulent spots and negative behind the transition point. The sign change gives then the transition point or 50% intermittence point.

Results

Wakes of the Moving Bars

For analysis of the flow on the blade suction side exact knowledge of the inflow conditions is very important. Figure 3 shows the ensemble averaged wake profiles versus bar passing period for different Mach numbers (compressible and incompressible conditions) and for different bar pitch (40 and 80mm) measured by the single hot wire. It can be clearly seen on the velocity profiles that there is almost no existing core flow in the case of the small bar pitch for both Mach numbers. In that case the cascade flow is fully wake influenced. The respective distributions of the measured turbulence intensity show a very interesting behaviour. Although both measurements are performed at the same blade Reynolds number and thus also the same bar Re number the turbulence intensity shows different distributions. In the case of the low Mach number (M=0.15) the turbulence intensity increases from 2% free stream level up to approx. 5.5% in the wake. In the case of the high Mach number (M=0.7) the free stream level is lower (~1%) and in contrary to the M=0.15 case the turbulence distribution shows distinctive double peaks indicating vortex shedding behind the bars. The intensity is smaller on the forerunning side of the bars because of the non uniform inflow conditions to the bars, i.e. a superposition of flow around the bars with flow caused by bar movement and thus mutual influence of the potential flow fields. To explain the occurrence of the double peak Figure 4 shows the measured mean turbulence level in the test section which decreases for given Reynolds Number with increasing Mach number. This can also be seen on the core flow turbulence level in Figure 4 (Tu=0.5-1% for M=0.7 and Tu=2% for M=0.15). The higher turbulence intensity of approx. 2% in the case of the lower Mach number causes a dumping of the vortex shedding behind the bars and thus only one discrete maximum is visible.

Time Space Diagrams

Figure 5 shows the simultaneously recorded raw hot-film traces (AC voltage signal) for sensors in the rear part of the airfoil's suction side (from 43.3 to 83.4% axial chord length) for five wake passing periods measured at the low Mach number (M=0.15). The effects of wake induced transition can clearly be seen in the first period of the raw traces. A turbulent spot caused by the bar wake is visible on the high voltage peak which travels with typical leading and trailing edge speeds of $0.9v_-$ and $0.5v_-$ respectively. The region within the turbulent spot develops through a transition to fully turbulent flow. In the region behind the turbulent spot or between the wakes the transition is suppressed and a calmed region develops a long way down the blade chord. Not every wake causes the described effect. In the second and third wake passing period the turbulent spot is not visible and thus no calmed region can delay the transition process. In this case the boundary layer transition takes place at about 56% axial chord in a bypass manner. Figure 5 clearly shows the necessity of the ensemble averaging technique to analyse the periodic unsteady effects. In the raw signals the periodic and the turbulent effects can not be separated. For the case of the low Mach number Figure 6 shows the so called Time-Space diagrams of quasi wall shear stress (same axial chord range as in Figure 5), RMS and skewness distribution in form of contour plots of the ensemble averaged quantities. The periodic effects in all presented passages are very pronounced and the stochastic turbulent effects are suppressed. Easily to identify are the calmed regions indicated by the low voltage values in Figure 6a. The effects of wake induced boundary layer transition can be explained with the help of the RMS distribution, Figure 6b. The transition develops along two paths: the wake induced path and the path between the wakes. The wake induced path develops along laminar region A, transitional region B with highest RMS values, increasing voltage and a distinct sign change in the skewness distribution. The wake induced turbulent region can be identified as region C. The path between the wakes begins again with a laminar region followed by the calmed region with very low pseudo wall shear stress (voltage) and ends up with the transitional region E which is extended almost to 84% of the airfoil. Although the quasi wall shear stress in the calmed region is very low, it is quite higher then in a fully laminar region and thus the calmed region is able to suppress flow separation and delaying transition.

The Time-Space diagrams for the high Mach number case are shown in Figure 7 in form of RMS contour plots for different Reynolds numbers. For comparison reasons the RMS and voltage distribution for the undisturbed case (bar speed=0) is also shown in the lower part of the Figure. As can be clearly seen the wakes are much more disturbed in that cases. There are in both cases two maximums per wake passage, a smaller one on the forerunning side of the bars. This corresponds to the vortex shedding effects which result in the double peak in the turbulence intensity as described earlier. Thus it can be stated that the turbulence intensity in the wake is, to a great extend, responsible for the different transition behaviour on the suction side of compressor blades.

Comparing both distributions the effects of different Reynolds number are visible. In the case of the higher Re number no calmed region can develop because of very high turbulence intensity in the described double peaks. The highest RMS value is therefore located further upstream in comparison to the undisturbed flow. In the case of the low Re number a calmed region develops again between the wakes. In this case the maximum of RMS is located nearly at the same streamwise position as in the undisturbed flow case. The transition process is extended to almost 70% of axial chord length whereas in the case of the high Re number this process ends at about 55% axial chord length because no calmed region is present.

Finally, Figure 8 shows exemplary the total pressure losses versus bar speed in the compressible case (M=0.7) for the lower Re number and for different bar pitches. The losses are calculated from total pressure traverses and are related to the undisturbed flow case (bar speed=0). As can be seen in the case of the small bar pitch (40mm) the loss decreases to approx. 82% related to the undisturbed case (100%) and increases then again. It can be stated that the effect of calming the suction side flow by suppressing the development of separation bubble has a optimum for a given speed of the wakes. At higher speeds the losses are increasing again because there is probably less time between the wakes to suppress the bubble and thus delay the transition process.

Conclusions

The following conclusions about unsteady periodic boundary layer development and transition process can be drawn from the presented measurements:

1. The transition process on compressor blades follows two paths, a wake induced path and a path between wakes which follows the wake induced path and contains the calmed region.

2. The calmed region is able to suppress flow separation and delay the transition onset due to the presence of higher wall shear stress compared with a fully laminar region.

3. The level of turbulence intensity in the wake is very important for the development of different structures in the boundary layer on the suction side of compressor blades. This level can trigger the transition process, especially in the case when vortex shedding on the trailing edge of previous blade row is generated.

Acknowledgements

This reported work was performed within a research project TurboTech II that is part of the national research corporation „AG TURBO". The project has been supported by the German Ministry of Education, Science, Research and Technology (BMBF) and the BMW Rolls Royce GmbH, Dahlewitz. The permission for publication is gratefully acknowledged.

References

- Adamczyk, J. J., Wake Mixing in Axial Compressors, ASME 96-GT-29, 1996.
- Acton, P.: Untersuchung des Grenzschichtumschlages an einem hochbelasteten Turbinengitter unter inhomogenen und instationären Zuströmbedingungen, PhD Thesis, Univ. BW Munich, 1998.
- Cumpsty, N. A., Dong, Y., Li, Y. S.: Compressor Blade Boundary Layers in the Presence of Wakes, ASME Paper 95-GT-443, 1995.
- Delimar, D., Swoboba, M., Lötzerich, M.: Numerical Analysis of an Unsteady Flow Field in a Compressor Cascade Under Periodically Changing Inflow Conditions, to be presented on 11. DGLR Symposium Strömungen mit Ablösung (AG STAB), TU-Berlin, 10.-12. Nov. 1998.
- Halstead, D. E., Wisler, D. C., Okiishi, T. H., Walker, G. J., Hodson, H. P., Shin, H. W., Boundary Layer Developmentin in Axial Compressors and Turbines: Part 2 of 4: Compressors, ASME Paper 95-GT-462, 1995.
- Lakshminarayana, B., Poncet, A.: A Method of Measuring Three-Dimensional Rotating Wakes Behind Turbomachinery Rotors, J. of Fluid Engineering, 1974.

- Schröder, Th.: Entwicklung des instationären Nachlaufs hinter quer zur Strömungsrichtung bewegten Zylindern und dessen Einfluß auf das Umschlagverhalten von ebenen Grenzschichten stromabwärts angeordneter Versuchskörper, PhD Thesis, TH Darmstadt, 1985.
- Schubauer, G. B., Klebanoff, P. S.: Contributions on the Mechanics of Boundary Layer Transition, NACA TN 3489, 1955.
- Sturm, W.,Fottner, L.: The High-Speed Cascade Wind Tunnel of the German Armed Forces University Munich, Proc. 8th Symposium on Measuring Techniques for Transonic and Supersonic Flows in Cascades and Turbomachines, Genoa, 1985.

Figures

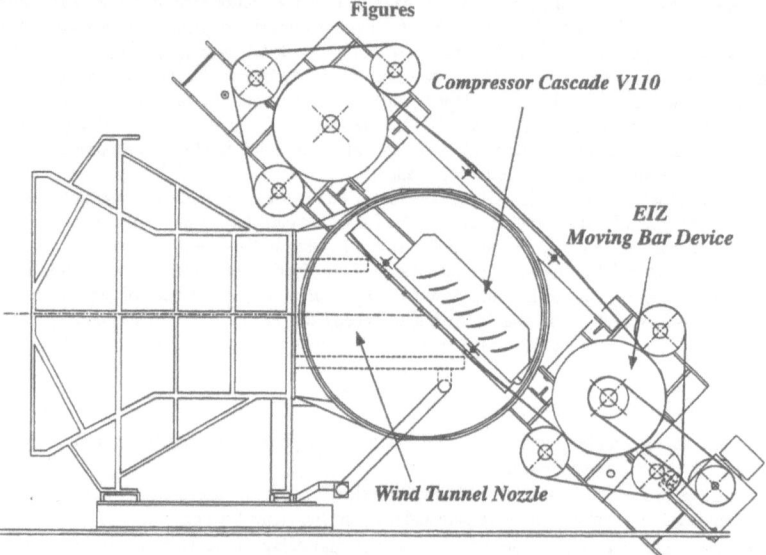

Figure 1: Arrangement of the Moving Bar Device Within the Wind Tunnel

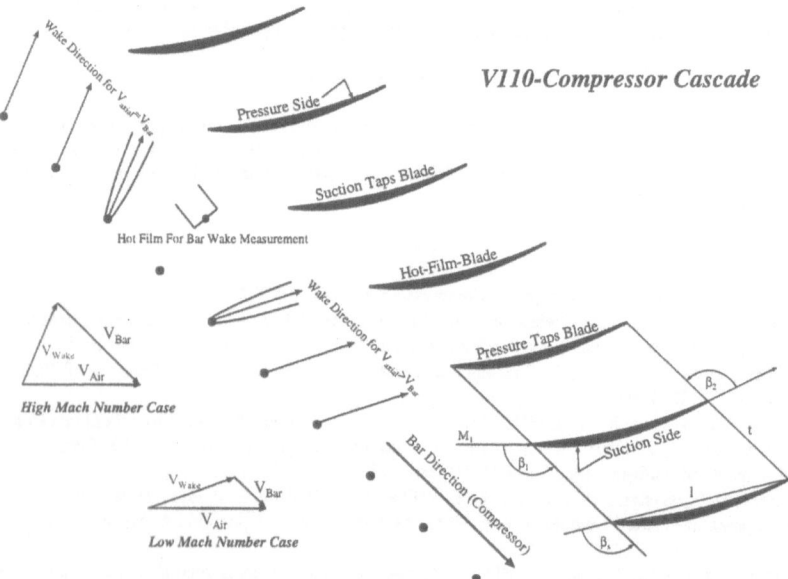

Figure 2: Principal Sketch of the V110-Cascade with Moving Bar Arrangement, Wake Direction at Different Speeds

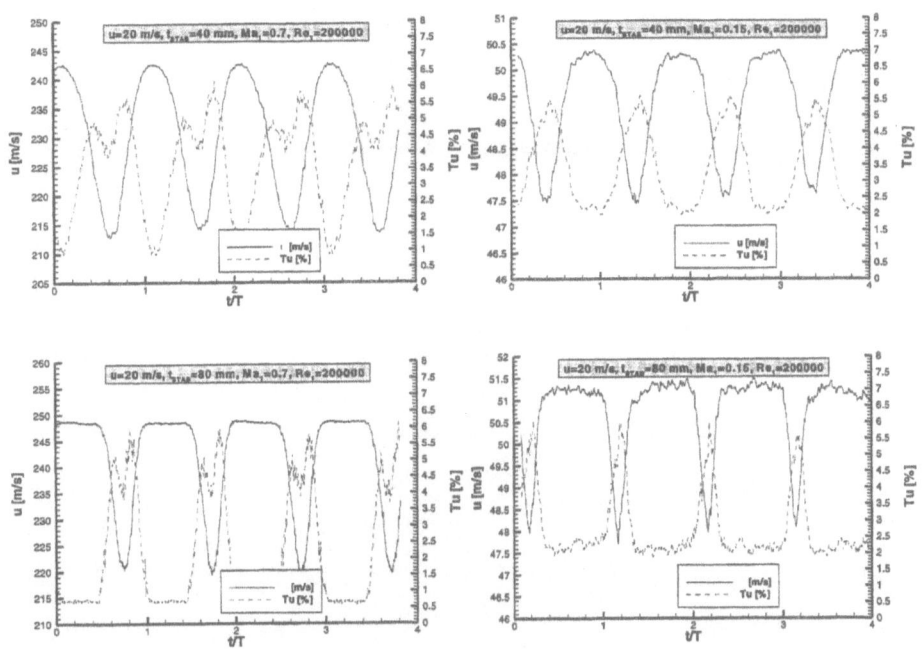

Figure 3: Velocity And Turbulence Intensity Behind Moving Bars for Different Mach Numbers and Bar Pitch

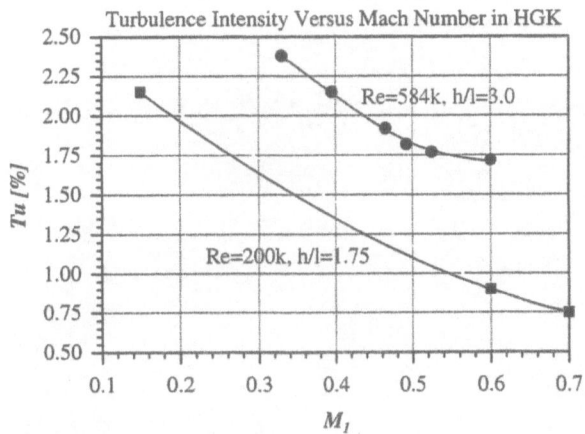

Figure 4: Turbulence Intensity versus Mach Number in the HGK for different Reynolds Numbers and for different test Sections

479

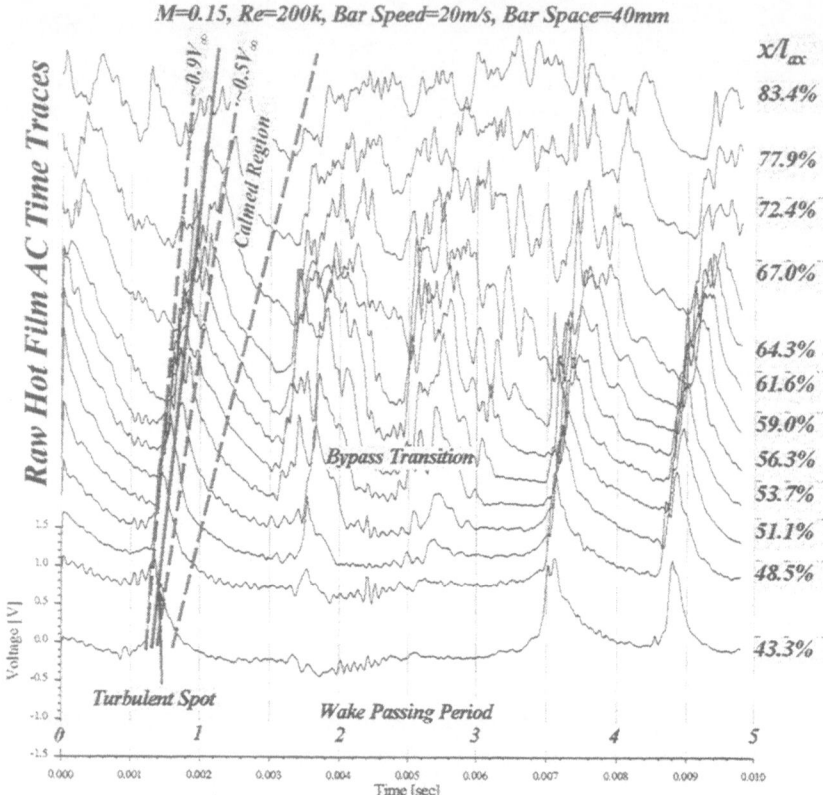

Figure 5: Raw Hot-Film_Array time traces for 12 simultaneously recorded sensors between 43.3% and 83.4% axial chord length

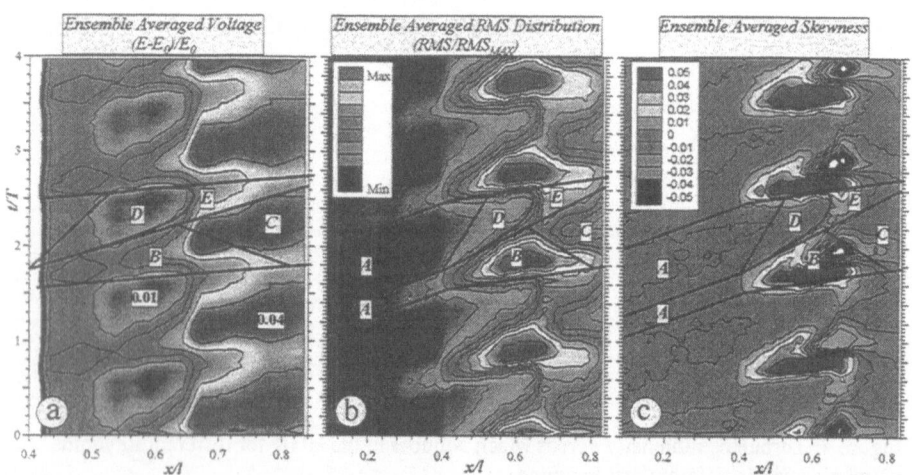

Figure 6: Time-Space diagrams of ensemble averaged Hot-Film Signals for M=0.15, Re=200000, Bar Speed=20m/s, Bar Space=40mm; Voltage (Quasi Wall Shear Stress), RMS and Skewness distribution (Note: same axial chord length in voltage distr. as in Figure 6)

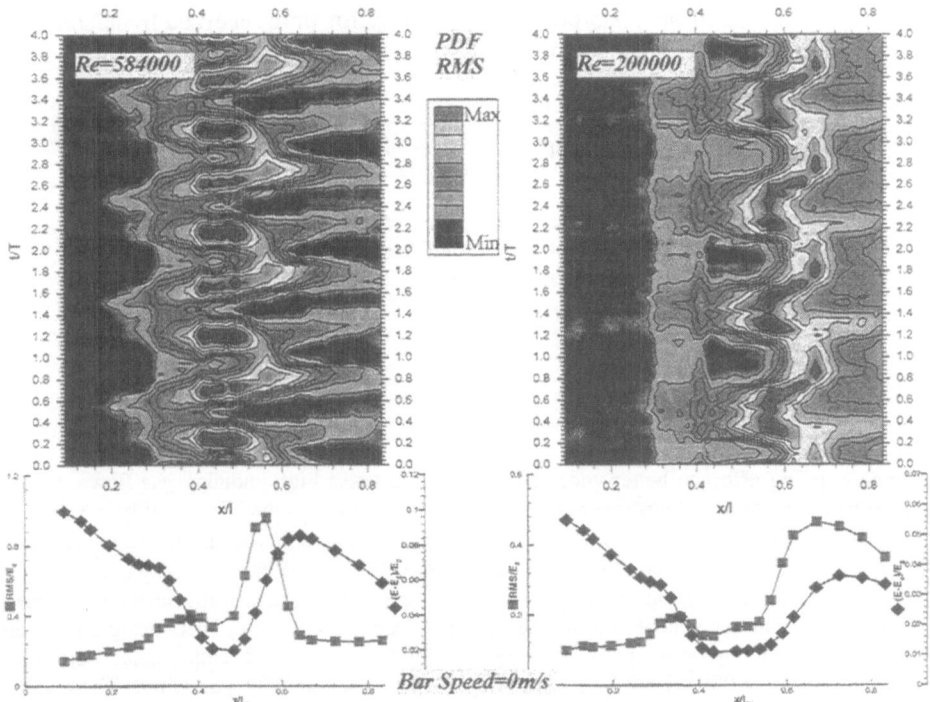

Figure 7: Time-Space Diagrams of RMS Distribution for Different Re Numbers in Compressible Flow (M=0.7), Bar Speed=20m/s, Bar Space=40mm, Comparison With Undisturbed Flow: Bar Speed=0

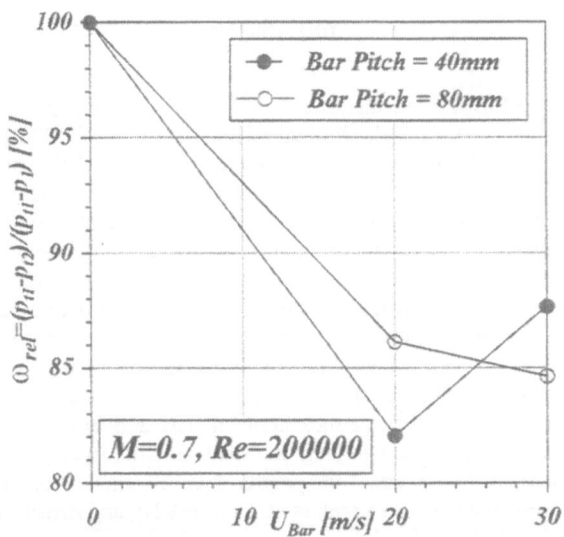

Figure 8: Total Pressure Losses Related to the Undisturbed Flow Case Versus Bar Speed for M=0.7, Re=200000; Variation of Bar Pitch

Application of the absorption spectroscopy for time resolved hypersonic flowfield measurements using the D$_2$-transition of seeded atomic rubidium

O. Trinks, W. H. Beck

German Aerospace Center (DLR)
Institute of Fluid Mechanics
Bunsenstr. 10, D-37073 Göttingen

Summary

In an ongoing effort to better understand and characterise high enthalpy gas flows, a diode laser based absorption lineshape diagnostic has been applied to the HEG facility in which the atomic rubidium $5S_{1/2} \rightarrow 5P_{3/2}$ (780.2 nm) resonance transition was used. Atomic rubidium (Rb) was still present in the flow after seeding $RbNO_3$ (2-3 µg) into the test gas at the shock tube end. The absorption technique has been used to determine the change in kinetic temperature and gas velocity during the first millisecond test time in HEG using a lineshape analysis. Depending on the particular flow condition, the measured temperatures were found to be in the range 900 – 1400 K and the velocities in the range 3900 – 6200 m/s. These values supply important information for a better understanding of the free stream and model gas flows produced in HEG. It has been shown that this technique is a reliable instrument for performing optical measurements in the free stream flow of high enthalpy facilities.

Introduction

The characterization of hypersonic airflows produced in high enthalpy shock tunnels such as HEG (High Enthalpy Shock Tunnel, Göttingen [Beck et al. (Sep. 1992)]) is necessary for a better understanding of the free stream and model gas flows. The extreme environments produced in the HEG facility require special considerations for diagnostic techniques to enable measurements of critical parameters such as temperature, velocity and species concentration. Presently available measurements of the test flow include heat transfer and bulk pressure measurements made by several probes placed in the free stream [Krek et al. (1994)]. While these standard measurement techniques are suitable for measurement of surface properties, they do not provide any details of the thermodynamic state and chemical composition of the gas, which are needed to validate the codes used in CFD calculations. Modern laser diagnostic techniques, however, provide non-intrusive, often species-specific and thermodynamically state-sensitive measurement data and have a sufficiently fast time response which is necessary for studying transient flowfields generated in HEG. Some preliminary estimates of the rotational temperatures are available from LIF (Laser Induced Fluorescence) measurements on NO [Beck et al. (Jan. 1992)], and density field measurements are available from HI (Holographic Interometry) [Kastell (1997)], but the kinetic temperature and flow velocity are among the parameters still to be measured with laser diagnostic techniques.

A diode laser-based absorption lineshape diagnostic has been applied to the HEG facility in which the atomic rubidium $5S_{1/2} \rightarrow 5P_{3/2}$ (780.2 nm) resonance transition is used. Since no rubidium is naturally present in the flow this species has to be seeded into the test gas. This diagnostic technique provides line-of-sight measurements in the flow and can be used to measure time resolved flow velocities in addition to kinetic temperatures. The technique has been used already to examine NO absorption in a high enthalpy wind tunnel [Rosier et al. (1993)], and to measure O atom absorption in the heated gas behind the reflected shock in a small shock tube [Chang et al. (1992)].

Experimental Method

The inherent tunability of diode lasers make them well-suited for laser absorption measurements in that the lasing frequency of a diode laser can be tuned by either temperature or injection current variations. The possibility, then, of setting the wavelength of the laser to an absorption line by controlling its temperature and rapidly scanning across the absorption profile by ramping the injection current presents an elegant opportunity to collect transient absorption lineshape data. Furthermore, since the shape of an absorption line is dependent on the absorbing specie's environment, the lineshape can, in principle, be used to deduce information about the environment in cases where the line broadening mechanisms are sufficiently understood. In particular, the measurement of Doppler broadened lineshapes with rapidly tuned diode lasers has been used successfully in determining gaseous kinetic temperatures in environments where Doppler broadening dominates [Chang et al. (1995)].

The experimental setup for diode laser absorption measurements at HEG is shown in Figure 1. A single mode diode laser beam (GaAlAs-laser) is tuned with a 15 kHz current ramp across the atomic rubidium $5S_{1/2} \rightarrow 5P_{3/2}$ (780.2 nm) resonance transition[1] and passed through the test section where it is detected by a photodiode and recorded on a transient recorder giving a time-dependent absorption signal which contains highly resolved spectral lineshape information. Since the shape of a spectral line is in part dependent on the kinetic temperature of the emitting or absorbing species, the shape of spectral lines can be used to deduce the local temperature in cases where the line broadening due to the temperature effects (Doppler-broadening) dominates. This line broadening, $\Delta \bar{v}_D$, is given by

$$\Delta \bar{v}_D (cm^{-1}) = \frac{2\sqrt{2R \cdot \ln 2}}{c} \cdot \bar{v}_0 \cdot \sqrt{\frac{T}{M}} \qquad (1)$$

where R is the gas constant equal to $8.314 \cdot 10^7$ *erg deg* $^{-1}$ *mol* $^{-1}$, $\bar{v}_0$ is the wave number at the center of the absorption line, T is the gas temperature in K and M is the atomic weight.

The free stream test flow of HEG is of sufficiently high temperature and low density that Doppler broadening dominates, making the freestream flow a good candidate for this type of measurement. The Doppler-linewidth for the rubidium transition is of the order of 0.05 *cm* $^{-1}$ [Trinks (1997)]. By comparison, the low pressures present in the test chamber (0.5 – 1.3 kPa) would lead to a collisional broadening component closer to 0.001 *cm* $^{-1}$. Since the laser linewidth itself is only $5 \cdot 10^{-4}$ *cm* $^{-1}$, the Doppler-lineshape and the shifting of the absorption frequency can be fully resolved [Trinks (1997)]. The use of high repetition rate (15 kHz) linear scans of the wavelength allows an application of the technique to high speed flows

---

[1] D_2-line of Rb.

generated in HEG. The flow velocity has also been determined in the freestream by observing the Doppler-shift of an absorption feature as measured at an angle of 53° to the flow direction. The magnitude of the shift was determined accurately by comparison with absorption from a second beam orthogonal to the flow direction.

A portion of the beam is sent through a Rb-reference cell where atomic rubidium is produced for a reference signal. An interferometer is also used to monitor the relative frequency of the laser output during each scan. The observed absorption signal was interpreted by fitting a model of the line broadening and shifting mechanisms, leading to a kinetic temperature from the Doppler-broadened lineshape and gas velocity from the Doppler-shift of the absorption frequency.

In addition to the absorption measurements extinction data has been collected from HEG tests in parallel to the system preparations to evaluate a problem with scattering of the beam by particles in the HEG flow. The high temperatures reached behind the reflected shock in the shock tube cause some ablation of the shock tube walls and nozzle throat. The ablated material is carried along with the test gases into the the test chamber, where it scatters light, causing extinction of light transmitted through the test section. A technique of overlapping two laser beams, one scanned across the Rb-absorption feature, the other away from any absorption (3mW HeNe-laser at 633 nm), has been used successfully to subtract out the scattering component of the signal in the free stream flow.

To ensure that the line-of-sight diagnostic technique provides only absorption features from the HEG core flow, the laser beam itself should pass through only the free stream core flow. For that reason sealed light-guiding pipes (see Figure 1; lower schematic) were mounted at the windows for the 53° laser beam and extended along the laser path into the flow through the boundary layer. Due to the construction of the HEG test section it was not possible to use light guiding pipes for the orthogonal beam which traverses the boundary layer and core flow providing mixed temperature values.

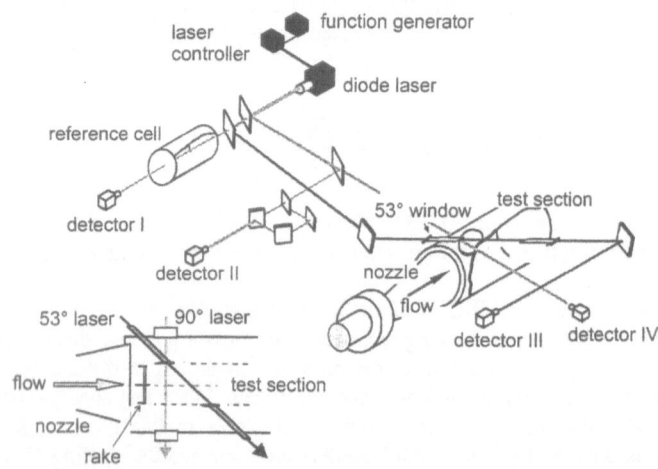

Figure 1: Experimental setup for diode laser absorption measurement at HEG using Rb. The lower schematic shows a detail picture of the test chamber with both laser beam directions, the light-guiding pipes and the rake position.

Results and Discussion

For evaluation of the measurement data, first, spectral lineshapes were obtained from the raw data set by dividing the transmitted intensity by the incident laser intensity. To convert the data from the time domain to the wavenumber domain the interferometer signal monitored simultaneously by a photodetector is used. By counting the fringes between the points in the interferometer signal which correspond to the appropriate points on the line profile in the absorption signal, a means for converting the data is provided. Finally Gaussian lineshape profiles were fitted to the absorption features for which Doppler-broadening dominated. Temperatures were then deduced from the fitted absorption linewidths. The gas velocity was determined from the Doppler shift of absorption observed by the 53° laser beam compared with the non-shifted signal from the orthogonal beam. The data reduction procedure is described in greater detail in [Trinks (1997)] and [Trinks (1998)]. An exact evaluation of the errors in the kinetic temperature and gas velocity obtained via the described data reduction method is quite difficult. Based on the goodness of fit, a (conservative) error estimate for velocity of ±5% was obtained, while a similar analysis led to ±20% for kinetic temperature.

The time development of temperature and flow velocity in the free stream, as determined from diode laser absorption measurements with Rb, is shown in Figure 2, middle and lower trace respectively; the typical error bars are also indicated. The measurements were performed for two experimental arrangements – with and without the presence of a model (viz. the measurement rake for calibration) in the test section, at a HEG test-condition which provides a high value for the flow enthalpy of about h_0 = 21.2 MJ/kg. The upper trace presents the pitot-pressure in the test chamber. The usual HEG test flow window is at the time 5.8 – 6.2 ms after shock initiation; additional theoretical values – calculated by a 1D Euler code (including chemistry) and a computer code using the standard shock relations – are plotted at this test window. Compared with the values observed without the rake in the free stream flow the measurements with the rake provided higher temperatures and lower velocities as a result of disturbances in the flow produced by the rake.

A summary of kinetic temperatures and flow velocities as determined from diode laser absorption measurements using Rb for different high enthalpy flows produced at HEG is given in Table 1. A comparison with calculated values leads to the conclusion that the measured velocity correlates somewhat better with numerical values whereas the measured temperature is in some cases much higher than its theoretical counterpart. But for a more quantitative statement more measurements without a model have to be performed [Trinks (1997)].

Table 1: Summary of measured kinetic temperatures ($T_{measured}$) and flow velocity ($u_{measured}$) for different flow conditions; each HEG condition is characterized by its test gas enthalpy h_0. The measured results are compared with the numerical values $T_{numerical}$ and $u_{numerical}$.

HEG-condition	h_o [MJ/kg]	$T_{measured}$ [K]	$T_{numerical}$ [K]	$u_{measured}$ [m/s]	$u_{numerical}$ [m/s]
I	21.2	1000 ± 200	910	5600 ± 280	5939
II	22.3	1410 ± 282	1036	5500 ± 275	6157
III	12.9	—	612	—	4757
IV	14.8	1100 ± 220	722	3900 ± 195	5148
V	10.8	1010 ± 202	503	4100 ± 205	4382
VI	11.6	910 ± 182	536	3850 ± 193	4582

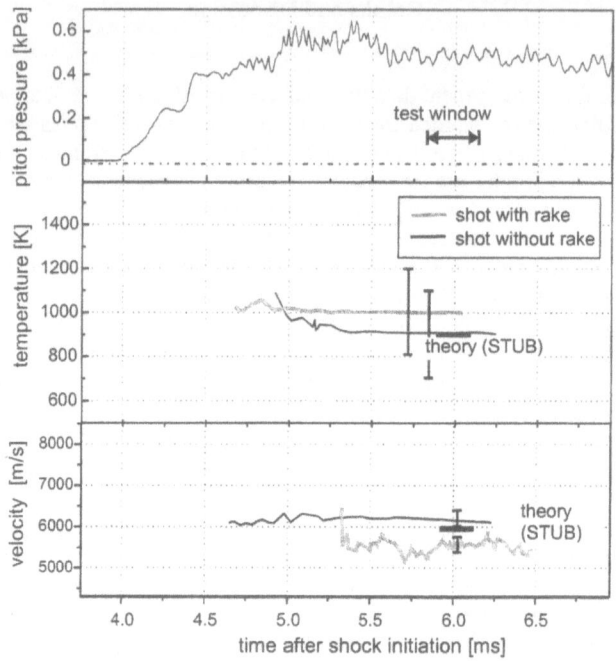

Figure 2: Time development of pitot-pressure, temperature and flow velocity, as determined from diode laser absorption measurements with Rb at HEG-condition I; =0 corresponds to the moment of shock initiation, each of the two data-traces belongs to measurements performed with and without the presence of a measurement rake in the test section.

Summary and Conclusions

A diode laser has been used to scan across the D_2-line of rubidium ($5S_{1/2} \rightarrow 5P_{3/2}$) to measure absorption lineshapes. The kinetic temperatures have been determined from lineshape analyses of the measured broadened spectral features; velocity is derived from the Doppler shift of the absorption profiles. The use of high repetition rate (15 kHz) linear scans of the wavelength region studied allows application of the technique to hypersonic flows generated in HEG to help in characterising the HEG free stream and to examine the gas flow in the HEG test section. For measurements carried out with the calibration-rake in position, the measured values of velocity are somewhat smaller than the theoretical values, whereas the measured temperature is much higher than determined by CFD. Depending from each HEG test condition the measured temperature and velocity ranged from 910 – 1410 K and 3900 – 6200 m/s, respectively. It has been shown that lineshape measurements using Rb absorption lines are a reliable instrument to perform optical diagnostics in free stream flows of high enthalpy facilities.

Acknowledgements

Significant contributions to the success of this effort is attributed to Dr. W. Gillespie[2]. The authors would also like to thank J. U. Frenzel and M. Schnieder (all from HEG) for their support.

References

[Beck et al. (1992)] W. H. Beck, C. Dankert, G. Eitelberg, G. Gundlach: *Preliminary laser-induced-fluorescence-measurements in several facilities in preparation for application to studies in the High Enthalpie Shock Tunnel Göttingen (HEG)*, AIAA paper 92-0143, 30th Aerospace Science Meeting & Exhibit, Jan. 6-9, 1992 / Reno, NV (1992).

[Beck et al. (Sep. 1992)] W. H. Beck, G. Eitelberg, D. Vennemann: *HEG – A new shock tunnel for high enthalpies*, Paper presented at European Forum on Wind Tunnels, Royal Astronautical Society of Southampton, Sep. 1992.

[Chang et al. (1992)] H. A. Chang, D. S. Baer, R. K. Hanson: *Semiconductor laser diagnostics of atomic oxygen for hypersonic flowfield measurements*, AIAA Paper 92-0628, (1992).

[Chang et al. (1995)] H. A. Chang, D. S. Baer, R. K. Hanson: *Semiconductor Laser Diagnostics for Simultaneous Determination of Kinetic and Population Temperatures in High-Enthalpy Flows*, Paper presented at 19th International Symposium on Shock Waves, July 6-9, Marseille / France, (1995).

[Kastell (1997)] D. Kastell, M. Carl, G. Eitelberg: *Phase step holographic interferometry applied to hypervelocity, non-equilibrium Cylinder Flow*, Experiments in Fluids, 22:57-66, (1997).

[Krek et al (1994)] R. M. Krek, G. Eitelberg: *Classical characterisation of HEG*, DLR-IB internal report 223-94 A 50, Göttingen (1994).

[Rosier et al. (1993)] B. Rosier, A. K. Mohamed, D. Henry, S. Juville: *Mesures par spectroscopie d'absorption diode laser a la soufflerie F4*, ONERA Final report 8/4383 PY, (1993).

[Trinks (1997)] O. Trinks: *Diodenlaser-Absorptionsspektroskopie am atomaren Rubidium im Hochenthalpiekanal HEG zur Bestimmung der Strömungsgeschwindigkeit und Gastemperatur*, Diplom thesis; DLR-IB 223-97 A31, (1997).

[Trinks (1998)] O. Trinks, W. H. Beck: *Application of a diode-laser absorption technique with the D_2 transition of atomic Rb for hypersonic flow-field measurements*, APPLIED OPTICS / Vol. 37, No. 30 , pg. 7070 – 7075 / 20 October 1998.

---

[2] Dynamics Technology, Inc.; Torrance.

Aerodynamic Excitation of Transonic Turbine Cascade; Description of the Experimental Method

Brigitte Urban, Heinz Stetter

Institute of Thermal Turbomachinery and Machinery Laboratory,
University of Stuttgart
Pfaffenwaldring 6, D-70550 Stuttgart
Germany

Nicolas Vortmeyer

Test Engineering WM TV
Siemens AG, KWU Group,
Wiesenstr. 35, D-45473 Muelheim
Germany

Summary

This paper presents the design of a cascade and the experimental method to investigate shock inducted flutter. To examine the interaction between a shock wave and blade oscillation at transonic flow, a cascade with 7 blades, each of them separately elastically mounted, has been developed so that only an aerodynamic coupling can occur in the system. The experimental investigations have been performed at a two-dimensional test rig with superheated steam as working fluid.

In order to study the interaction between the blade and shock wave both the blade vibration and the shock wave oscillation have to be measured simultaneously. Whereas well known measurement technique like strain gauges and Schlieren optic enables the observation of blade vibration, an optoelectronic measurement technique was developed for determining the shock wave oscillation frequency. This method is based on the different refraction of the laser light caused by different density fields.

Using an experimental test profile which is representative for the tip section of modern steam turbine last stages a shock induced flutter could be achieved in the range of $1{,}30 < Ma_{2is} < 1{,}35$. In this region the interaction between the blades and the fluid can be shown.

Nomenclature

List of symbols

c	chord length	
f	frequency	
l	length	
Ma	Mach number	
m	mass flow	
p	pressure	
Re	Reynolds number	
t	pitch	
T	temperature	
Tu	turbulence level	
β	flow angle	

subscripts

bi	bitangent
is	isentropic
t	total
0, 1	inlet
2	outlet

Introduction

High performance of advanced turbomachinery requires an increase of the flow velocities which was realised by the use of long and slender turbine blades with transonic profiles. As a result shock waves occur in a part of the stages. An interaction between blade and shock wave can develop into resonance, which can under unfavourable conditions be able to damage the structure. This interaction is an aeroelasticity problem, a shock flutter problem.

Self excited shock wave oscillation was already investigated at a fixed biconvex circular-arc airfoil by Yamamoto et al. [1].

Henne [2] examined a cascade with one forced excited blade. He systematically lists the results of transient flow measurements at different Mach- and Reynolds numbers. He found out that the oscillation and development of the shock wave depends on these two parameters (and in that way depends on the characteristics of the boundary layer).

Furthermore Araki et al. [3] showed at a similarly designed LP last stage tip section cascade that the shock wave and the blade oscillation frequency at the transient flow deviate.

Bölcs [4] designed an annular test facility to investigate blade flutter at steam and gas turbine blades.

Extensive investigation of steam and gas turbine blades of middle and tip sections are summarised in a report of Schläfi [5]. He excited the blades at an annular cascade [4] into controlled oscillations. The results of his investigations showed a blade excitation caused by shock wave oscillation, where coupled forces only occur at adjacent blades. In opposite to this the blades are aerodynamically stable at sub- and supersonic flow. Furthermore it is emphasised that at a overlapped is not influenced by the angle of incidence whereas at a weakly overlapped cascade (tip section) the angle of incidence has an important influence on the transient flow condition.

Moreover Nowinsky et al. [6] made investigations at the same annular cascade. They found out that the oscillation mode with reference to the inter blade phase angle has a significant influence.

Examination at an annular compressor blade cascade was done by Kobayashi et al. [7]. The blades were excited with variable frequencies and constant interblade phase angle. The result showed that the aerodynamic damping depends strongly on the blade's frequency.

This paper presents investigations of a two dimensional fully elastically attached cascade at transonic flow conditions. An elastic suspension system has been developed to get only aerodynamic coupling in the system. The investigations were done in a steam test rig with superheated steam. The cascade consists of seven test profiles which are representative for tip section of modern low pressure steam turbine.

Well known measurement techniques used up to now like high speed Schlieren technique and strain gauges enable an observation of the blade oscillation. In order to study the interaction between the blade and shock wave oscillation it additionally is necessary to know the shock wave oscillation simultaneously to the blade oscillation. For that reason an optoelectronic measurement technique was developed for determining the shock wave oscillation frequency. This method is based on the different refraction of the laser light caused by different density fields.

Based on the experimental results similarity scaling laws can be derived to design other turbine blades.

Test Facility

Test Rig

The experimental investigations were carried out at the steam test rig (STR) at the flow laboratory of Siemens' KWU Group Mülheim (Fig. 1). The test rig is suitable for long-term tests and is supplied with superheated steam or wet steam from a separate boiler. Table 1 shows the operating parameters of the test rig. The steam condition is superheated for this pressure and temperature range.

Because of the closed loop system of the steam supply the test rig enables an independent variation of Mach and Reynolds number.

Model Blade and Support

In order to be able to perform the experiments it was necessary to develop an elastic support system to attach the blades.

The airfoils used in this study are representative for the tip section of large last stage free standing low pressure steam turbine blades. The test profile is backward curved.

The important boundary conditions for the design of the support were:

1. Geometry of the Steam Test Rig of the Siemens' Flow Laboratory / Germany
2. The oscillation mode and resonance frequency of a typical industrial low pressure last stage blade had to be simulated
3. The support had to allow easily a variation of the fundamental natural frequency
4. The support had to permit a limited variation of the stagger angle
5. Between the blades only an aerodynamic coupling was allowed

Fig. 2 shows the model blade and support. The height of the model blade was scaled up to 148 mm to be adapted to the width of the test chamber. The beams were electronic-beam-welded with the model blade. Furthermore the beams were adapted to the blades contour to permit the use of optical measurement techniques at the blades surface.

To simulate the first mode, a bending mode, and the resonance frequency of the original blade the beams were designed as bending springs. An FEM-prediction and pre-tests were used to design the geometry of the beams. The results from the FEM prediction and holographic measurement are shown in Fig. 3 and Fig. 4. The natural frequency of the model was determined by a geometrical scaling factor from the original blade. To obtain a natural frequency of 180 Hz of the whole model system the height and length of the bending springs were chosen with a differential equation of mass affected springs by Dunkerley.

For the tuning and mistuning of the system the torsion springs were used. Furthermore the torsion springs were divided into a cylindrical and a cone part. This geometry enables a variation of the angle of stagger. The cone is designed as self jamming and is used to attach the blade-beam system at the test rig.

Cascade

The investigated cascade consists of seven prismatic blades with a chord length of 89 mm, a pitch-to-chord length ratio of $t/c=0,8988$ and a stagger angle of $\beta_{Bi}=14,1°$.

As mentioned above each blade is suspended by an elastic spring system which allows the respective blade to vibrate similar to the real blade's first bending mode.

Fig. 5 shows the cascade. Windows at the test rig's wall were used to observe the shock wave motion with optical measurement techniques like laser and Schlieren optic.

In order to ensure that only an aerodynamic coupling can occur one blade was excited and the response from the other blades with a strain gauge installation was checked. None of the blades showed a response when a adjacent blade was excited. So it could be proven that there is no mechanical coupling. The blade oscillations during the tests were caused only by the shock wave oscillation.

Measurement Technique

Measurement Equipment

At Fig. 6 the measurement equipment is shown. The first part is the measurement equipment of the test rig like pressure transducer and thermocouples at the inlet and outlet of the cascade to control the flow conditions like Mach- and Reynolds number.

To observe the blade oscillation strain gauges were used. At each blade one strain gauge was installed.

The third part of the measurement equipment is optically. It is necessary to observe the motion of the shock wave range at the blade's surface and to inquire the frequency of the shock wave oscillation. For getting the range of the shock wave motion a high speed schlieren optic was installed. The camera is a 16 mm film camera which is able to take up to then thousand photographs per second. For the presented investigations a picture frequency of 4500 pictures per second was used.

To investigate the frequency range of the shock wave a laser measurement technique was developed.

All data except the Schlieren pictures were recorded parallel at a digital audio tape.

Strain Gauges

To examine the blade oscillation under various isentropic outlet Mach numbers strain gauges were used. For the selection of the optimal place various pre-tests with a mechanical excited model blade in it's difference natural frequencies were done. At Fig. 7 the model-blade-beam-system is presented. At each beam there are applicated 5 strain gauges. The Diagram 1 shows the qualitative trend of the strain gauges elongation at the blades first bending mode. The maximum is detected where the blade is attached to the torsion springs. Furthermore the elongation decreases with the beam length. Another maximum is at the blade's nose. At this point the welding joint ended. A destructive test showed that the model system broke at strain gauge positions 2 and 4. Because of this results one strain gauge was applicated at position 2 at each blade system (Fig. 7). The indicated data from the strain gauges at the torsion springs during these test are neglectably in comparison to the beam strain gauges.

Laser Measurement Technique

The structure of this measurement technique is shown at Fig. 8. The measurement equipment consists of a laser, a special optic, an aperture and a photodetector.

The measurement principle is as following. If there is no shock wave in the subsonic range at the blade channel the laser beam is focused to the photodetector. During the test a shock wave motion at the blade channel deflects the laser beam because of the pressure gradient of the shock wave. So the photodetector gets a negative signal. All data were analysed with an FFT-analyser and the result is the frequency of the shock wave oscillation.

To investigate the complete shock wave range at the blades' suction side the laser and detector were traversed simultaneously over a range of 50 mm.

An aperture in front of the photodetector is able to confine the sensitive field of the detector.

Results

Strain Gauge Data

The investigations were done at different isentropic outlet Mach numbers at the transonic flow range. The blade oscillation began at a certain outlet Mach number. With further increase of the Mach number the amplitudes increase to a maximum. Fig. 9 contains a plot of the FFT signals during the fully developed oscillation. All blades of the mistuned cascade oscillated at this Mach number with the same frequency of 183 Hz, which is near to the natural frequency of the blades. A further frequency with a smaller amplitude could be detected at 309 Hz. A third frequency with a very small amplitude at 366 Hz could also be detected. Both frequencies are not natural frequencies of the model blade system.

Schlieren Optics

In Fig. 10 and Fig. 11 the results of the high speed schlieren optic measurement are presented. An important result from the schlieren pictures and the strain gauge data is that the blades oscillated counterphase with the main frequency of 183 Hz.

Furthermore the high speed film showed that the shock waves move nearly over 33% of the blade's suction side length.

It could be seen that the shock waves move upstream if the blades build a minimum blade channel width and downstream if the blades build a maximum blade channel width. While the blade oscillation is a continuously one the slow motion of the film showed that the shock wave motion is not continuously. On the basis of this result it was assumed that there is more than one typical shock wave frequency. This result was confirmed by the results of the laser measurements.

Laser Measurements

The laser detector method is an optical method and measured only the shock wave frequency. Fig.12 shows the results of the laser measurement at $Ma_{2is}=1,3$. The data were recorded at the same time as the strain gauges data at Fig. 9.

The FFT of the detector data shows the same frequencies of 183 Hz and 309 Hz as at the strain gauge data. This study clearly indicates, that there must be an interaction between the shock wave and blade oscillation. The nature of this interaction will be the focus at further studies.

Conclusions

This paper presents a fully elastically attached cascade and a measurement technique to investigate the aeroelasticity problem of shock flutter. The two dimensional fully elastically attached cascade was developed in a way that only an aerodynamic coupling occurs between the blades.

The blade oscillation was determined by strain gauges and observed by a high speed schlieren optic. At the same way the shock wave oscillation was examined. Additionally the frequency of the shock wave was investigated with a laser detector measurement technique, which was developed because of this problem.

The results showed that there is an interaction between the blade and shock wave oscillation.

To get an answer about the excitation mechanism further investigations on the blade oscillation due to the shock wave boundary layer interaction will be performed

Acknowledgements

This research was supported by BMBF AG TURBO TURBOTECH II, 1.221.

References

[1]   Yamamoto, K., Tandida, Y., 1990, „Self-Excited Oscillation of Transonic Flow Around an Airfoil in Two-Dimensional Channels",
      Journal of Turbomachinery, October 1990, Vol. 112.

[2]   Henne, J.M., 1989, „Instationäre Stoß- und Grenzschichtphänomene an Einzelprofilen und in einem ebenen Gitter bei transsonischer Strömung",
      Dissertation RWTH Aachen 1989.

[3]   Araki, T., Okamoto, Y., Ohtomo, F., 1981, „Self-Excited flow oscillation in the low pressure steam Turbine Cascade",
      Proceeding of the 2nd International Symposium on „Aeroelasticity in Turbomachines", H.P. Suter, Juris Verlag, Zürich 1981.

[4]   Bölcs, A., 1983, „A Test Facility for the Investigation of steady and Unsteady Transonic Flows in Annular Cascade", ASME-Paper 83-GT-34.

[5]   Schläfli, D., 1989, „Experimentelle Untersuchungen der instationären Strömung in oszillierenden Ringgittern",
      Communication du Laboratoire de Thermique Appliquée et de Turbomachines, EPFL, Nr. 17, Lausanne 1989.

[6]   Nowinski, M., Panovsky, J., 1998, „Flutter Mechanisms in Low Pressure Turbine Blades"
      ASEM Paper 98-GT-573, June 1998, Stockholm / Sweden.

[7]   Kobayashi H., Oinuma, H., Araki, T., 1995, „Shock Wave Behaviour of Annular Blade Row Oscillating in Torsional Mode with Interblade Phase Angle"
      Unsteady Aerodynamics and Aeroelasticity of Turbomachines, 7th International Symposium, 1995.

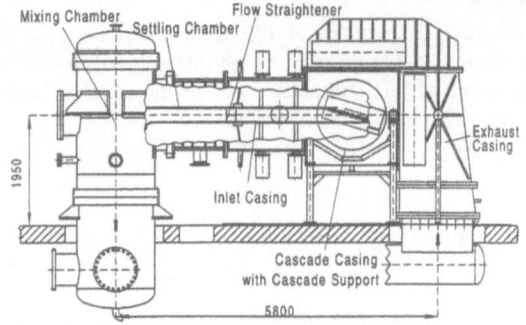

Fig. 1: Steam Test Rig (STR)

p_{t0}:	$= 0.35 \cdot 10^5 \text{N/m}^2$
T_{t0}:	$\approx 420 \text{ K}$
Tu_0:	$\leq 1\%$
Ma_2:	$\leq 1,5$
Re_2:	$\leq 1 \cdot 10^6$
m:	$\leq 2.7 \text{ kg/s}$

Table 1: Operating parameters

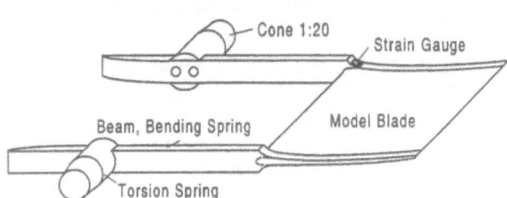

Fig. 2: Model Blade and Support

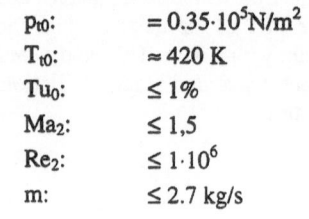

Fig. 3: FEM-Prediction

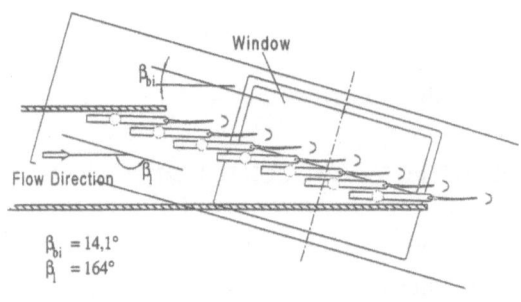

$\beta_{bi} = 14,1°$
$\beta_1 = 164°$

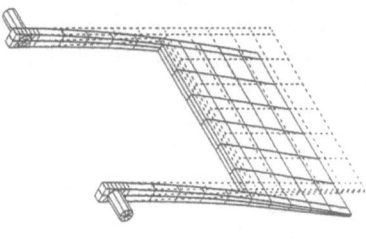

Fig. 4: Holographic Measurement

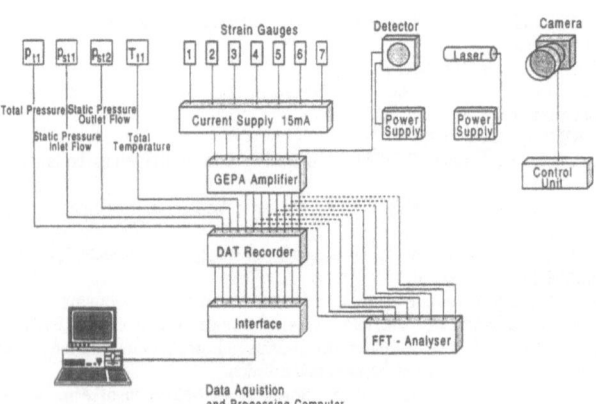

Fig. 6: Measurement Equipment

494

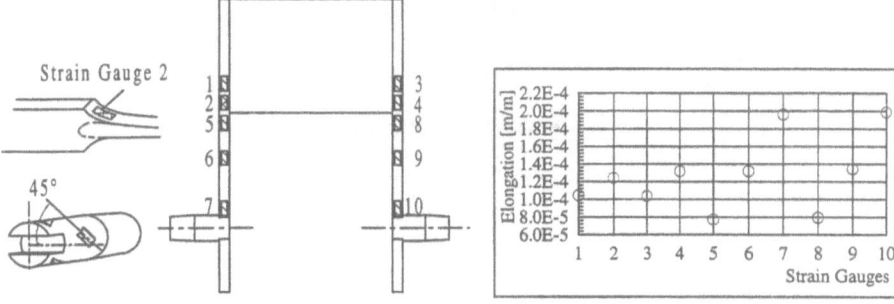

Fig.7: Pri-Test Spring System with Strain Gauges   Diagramm 1

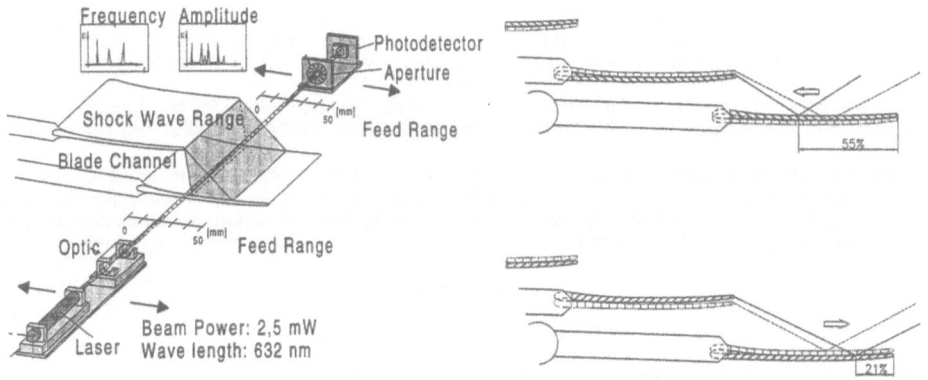

Fig. 8: Laser Measurement Technique

Figure 11: Maximum blade channel width

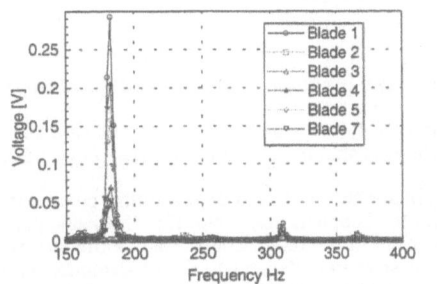

Ma$_{2is}$= 1,30; f=183; 309 and 366 Hz

Figure 9: Blade frequencies at Ma$_{2is}$=1,30

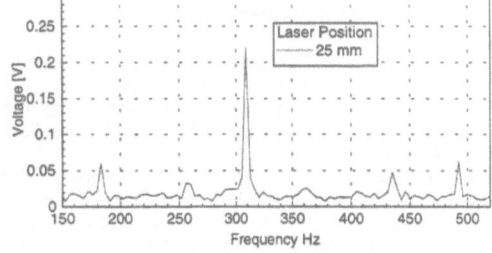

Ma$_{2is}$= 1,30; f=183; 309; 435 and 492 Hz

Figure 12: Shock wave frequencies at Ma$_{2is}$=1,30

Experimental Analysis of Unsteady Mixing in an RQL Combustor Segment aimed for Validation of PDF Models in the TRUST Code

Peter Voigt, Peter Theisen

Institute of Propulsion Technology
German Aerospace Center
Linder Hoehe
D-51170 Cologne, Germany.

Summary

Each newly developed flow simulation code requires the experimental validation of underlying models. One fundamental flow parameter used for comparison is the mixing of mass flows. At DLR Cologne an efficient measurement technique has been established, producing a large number of data sets for comparison with numerical data.

In this paper some aspects of the QLS measurement technique are presented. QLS has been developed as a planar laser optical method for the analysis of mass concentration in mixing flows. Based on experimental results of mixing in an RQL combustor a numerical model for the prediction of chemistry turbulence interaction has been successfully validated. The test rig has been supplied by MTU Munich in the frame of a co-operation project.

QLS has been applied successfully in a number of mixing configurations for the analysis of time averaged mixing under isothermal conditions (see *Hassa et al. 1998*) one typical test arrangement is shown is Fig. 1.

Particular aspects of proficient light sheet generation have been presented in *Voigt and Schodl 1997, non-linear* aspects have been treated in *Voigt 1998*. In this paper the measurement signal is examined and the extension of QLS aimed for time resolved data extraction as well as comparisons with numerical results are presented.

Introduction

Up to now QLS has been utilized exclusively for the time averaged analysis of mixing fields. However, it is possible to 'freeze' the physical process of mixing by shortening the camera integration time, yielding a deeper comprehension of mixing that can be integrated in numerical models. In the first part of this paper the QLS signal is explored. After that the experimental setup and the data processing are introduced. Then, the experimental results are analyzed and on the basis of this analysis statements on improved numerical models are shown.

These improved models will be integrated in coming versions of the TRUST code (see *Schütz et al. 1997*).

1. The QLS Measurement Signal

1.1 Scattered light signal

In Fig.2 a typical flow arrangement for QLS measurements is shown. Two separated flow components are combined in the mixing zone.

Most of the examined flow arrangements were of the type "jet in crossflow", where one flow component is injected perpendicularly into a crossflow through sets of injection holes (see *Perry 1993* for details).

One of the flow constituents is seeded with small particles that are used as scattering centers for the laser light sheet, used for visualization. The separated supply is inevitable. If both mixing partners are seeded, only an unstructured fog will originate from the test section. Most of the incident laser light is transmitted by the particles. Only a part is scattered, depending on the complex index of refraction of the particle, the laser line wavelength and the particle diameter. A part of the scattered light is recorded in the far field by a CCD camera. The scattered light is the raw signal of the QLS concentration measurement. By appropriate experimental setups and digital image processing the scattering amplitude can be transformed into concentration values. Now it will be shown, which physical property is determined by QLS and under which conditions the measurement signal matches the mass concentration.

Basically, the power of the scattered light, P_{sca}, is proportional to the number of affected scattering particles in the probe volume. The signal amplitude depends linearly on the power of the incident laser light P_L. The light sheet height $h(x)$ and thickness $\Delta z(x)$ spread a cross section $A_L(x)$. The energy flux per cross section is defined as incident intensity I_L. In Fig. 3 the light sheet height is constant. This is a special arrangement. In general a variable cross section $A_L(x)$ must be taken into account.

The scattered light is given by the product of I_L and the effective projection cross section of the particle. Unlike in the case of geometrical optics, the geometrical cross section cannot be considered. The particle diameter in QLS experiments is in the same order of magnitude as the wavelength of the incident laser light. Consequently the **Mie-Theory** must be used for the description of the scattering process. In the range of Mie-Scattering for each particle the effective scattering cross section c_{sca} must be computed.

In the Mie scattering range the scattered light signal increases with the fourth power of the particle diameter as can be deduced from Fig. 4. The scattered light from a single particle is then:

$$P_{sca} = c_{sca} \cdot I_L .\tag{1}$$

Since Mie scattering is dipole scattering, the signal amplitude depends largely on the scattering angle. The so called Mie-Function $i(\vartheta,\varphi)$ considers the directional dependence. The practically transmitted scattering flux depends also on the receiving optics' aperture angle Ω, given by the camera lens diameter d_{OL} and the distance r_S between camera and radiation source. It corresponds to the aspect ratio of the lens area related to a sphere shell with radius r_S. Since polarization and scattering angle differ on distinct optical paths between radiation source and detector, the scattered power must be integrated over differential angle elements $d\Omega$. The integration over all angle elements results in the integral energy flux P_{sca} as follows:

$$P_{sca}(x,y,d_P,\Omega) = \int_{\Omega} i(\vartheta(x,y),\varphi(x,y),d_P) \cdot d\Omega \cdot I_L .\tag{2}$$

In order to receive sufficient signal amplitude it will be necessary in QLS experiments to use particle ensembles. As long as effects as obscuration and energy exchange between adjacent particles can be neglected, the power of the ensemble will be the sum over all single scattering events. In the case of monodisperse diameter distribution[1], the integral scattering power is given by:

$$P_{sca,\Sigma} = N_P \cdot P_{sca} .\tag{3}$$

The sigma index in Equation (3) indicates that the sum over a particle ensemble is meant. The local particle density $c_N(x,y,z,d_P)$ can be seen as a probability density function over the three space coordinates as well as the distribution over particle diameters d_P and is defined as:

$$c_N = \frac{N_P}{\rho \cdot dxdydz} .\tag{4}$$

The total number of scattering particles in a mass element is then:

$$N_P = \int_{d_P} \left[\int_z^{z+dz} \int_y^{y+dy} \int_x^{x+dx} c_N(x,y,z,d_P) \cdot \rho(x,y,z) \cdot dxdydz \right] dd_P .\tag{5}$$

In the case of a constant volume element with a known particle ensemble the integral scattered power is then:

$$P_{sca,\Sigma}(x,y,d_P,\Omega) = \int_{d_P,min}^{d_P,max} \left[dN_{P,V}(d_P) \cdot \int_{\Omega} i(\varphi(x,y),\vartheta(x,y)) \cdot d\Omega \right] dd_P \cdot I_L .\tag{6}$$

The scattered power $P_{sca,\Sigma}$ is integrated during the integration time τ_B and yields- generated by the inner photoelectric effect- the scattered light energy.

1.2 The mass concentration c_m

The mass concentration is defined as the mass ratio of one mixing partner related to the integral mass m:

$$c_{m_i} = \frac{m_i}{m} .\tag{7}$$

If the combination is setup of two flow constituents, for component one the following equation holds:

---

[1] monodisperse particle distribution indicates that all particles have the same diameter

$$c_{m_1} = \frac{m_1}{m} = \left(\frac{p_1 \cdot V_1}{R_1 \cdot T_1}\right) \cdot \left(\frac{R_M \cdot T_M}{p_0 \cdot V_0}\right). \tag{8}$$

In Equation (8) the perfect gas law is assumed as well as the fact that the partners mix in the same volume V. In equilibrium a mixture temperature T_m and an average specific gas constant R_M will arise. The pressure is given by the sum of all partial pressures (Dalton's law):

$$p_0 = \frac{R_M}{R_1} \cdot p_1 + \frac{R_M}{R_2} \cdot p_2. \tag{9}$$

The mass concentration can then be expressed as:

$$c_{m_1} = \frac{p_1}{p_0} \cdot \frac{R_M}{R_1} \cdot \frac{T_M}{T_1}. \tag{10}$$

For the case of identical gases in both flows ($R_M = R_1 = R_2$) follows:

$$c_{m_1}\big|_{R=const} = \frac{p_1}{p_0} \cdot \frac{T_M}{T_1}. \tag{11}$$

If both flows are supplied isothermally ($T_M = T_1 = T_2$), Equation (11) can be further simplified to:

$$c_{m_1}\Big|_{R=const}^{T=const} = \frac{p_1}{p_0}. \tag{12}$$

The scattered power is proportional to the total number of particles N_P in the measurement volume $P_{sca,\Sigma} \propto c_N$. Therefore the following expression holds:

$$I \propto N_P = c_{N_1} \cdot \frac{p_1 V}{R_1 T_M}. \tag{13}$$

Since not all transmission coefficients are known in the experiment and since some of the post processing steps are of iterative nature, the QLS technique has been setup as a relative method, meaning that all mixing states in the measurement volume are related to the location of the injection jets, where the seeded particle flow penetrates the cross flow. This reference point is used as 100 per cent jet flow concentration. The intensity I_0 can be extracted at any one of the injection orifices. At the jet core location the total pressure is equal to the partial pressure inside the jet, thus the pressure p_1 can be replaced by the pressure p_0:

$$I_0 = c_{N_1} \cdot \frac{p_0 \cdot V}{R_1 T_1}. \tag{14}$$

The intensity ratio I versus I_0 is then:

$$\frac{I(x,y)}{I_0} = \frac{T_1(x,y)}{T_M} \cdot \frac{p_1}{p_0}. \tag{15}$$

In the isothermal case $T_1 = T_2 = T_M$ the intensity ratio is equal to the ratio of partial pressures:

$$\frac{I(x,y)}{I_0}\bigg|_{T=const} = \frac{p_1}{p_0}. \tag{16}$$

The comparison with Equation (12) yields in this case:

$$\frac{I}{I_0}\bigg|_{c_{N_2}=0}^{R=const,T=const} = c_{m_1}. \tag{17}$$

This shows that **exclusively in the case of isothermal mixing of identical gas components QLS can measure the mass concentration c_m.**

2 Experimental Realization

2.1 Test rig setup

QLS measurements have been performed in an RQL combustor segment that has been designed specifically for isothermal mixing investigations. Nevertheless the mixer geometry is in accordance with a geometry that is under investigation simultaneously in experiments with reaction. The light sheet is generated from downstream of the mixing duct as can be seen in Fig. 5. The part of the image that has been used for comparison between experiment and numerical prediction, is marked with a rectangle.

2.2 Light Sheet Generation and Detection Optics

The light sheet was generated with a polygon scanner with 20 facets. It was thus possible to scan the full combustor height within a time interval of 70 µs. The detection optics consisted of an intensified, Peltier-cooled CCD camera, type La Vision Flame Star, with a resolution of 384*286 pixels and a 12 bit dynamic range. Background radiation could be largely suppressed by the application of a laser line filtering technique which is explained in *Voigt and Schodl 1997*.

2.3 Image Acquisition and Post Processing

For the statistical data extraction a large number of time resolved images had to be recorded. Three campaigns with 10.000 images each have been executed and analyzed. The different campaigns only differed in the size of the imaged light sheet section. In the part that has been taken as a reference for comparison with numerical data, the jet flow downstream of the center swirl nozzle at the axial position of the first row of injection orifices, was imaged to the CCD chip, yielding an image with lengths of 12*8 mm (see Fig. 6). Considering the pixel resolution this results in a spatial resolution of 31 µm. The section was limited in vertical direction by the proximity of the outer burner liner, in axial direction the jet core was imaged to the sensor center.

After that, the images were corrected for the influences of Mie scattering like remaining background radiation, scattering direction, polarization, light sheet intensity distribution, etc. Details of the correction chain can be found in *Voigt 1999*.

2.4 Statistical Data Extraction in Form of PDF's

A novelty in the QLS data extraction chain is the generation of probability density functions (PDF) for each pixel location. The PDF consists of 64 linearly scaled bins of the mass concentration. For each data point in each image it is checked, which concentration interval is appropriate. Since the dynamic range of c_m is limited by zero and one, the division into 64 data arrays produces an interval width of 1.56 per cent. 384*286 data points result in 109 824 local PDF's. For the purpose of data reduction 4*4 data points have been merged, still yielding a data set of 32 MB for each PDF that was used for comparison.

2.5 Experimental Results

The first result is an image of the averaged concentration distribution, obtained by numerical averaging over 10 000 images. Images of this kind represent typically the result of time averaged QLS measurements. Figs. 7 and 8 display an overview over the total mixing field, but only the marked area has been considered for comparison.

In Fig. 8 the field of concentration fluctuations is shown (RMS distribution). The large area of high fluctuation level related to the jet shear layer region strikes especially.

The last presented resulting image shows the image section used for comparison. It represents one single time resolved image.

3 Utilization of the Experimental Results in Numerical Simulation

To calculate the chemical sources for simulation of technical combustion the mixture fraction is one of the most important parameters. If the diffusion coefficients of all species are equal, the mixture fraction in a system of two input streams yields the atomic concentration in the control volume. The mixture fraction is frequently defined by the mass fraction of the input stream containing the fuel versus the total mass flow. Consequently, it is equivalent to the physical property that is measured by QLS experiments (see 1.2).

The next topic is the aspect of modeling the *Probability Density Function (PDF)* of mixture fractions. Within the computation process the PDF has to be integrated more than 10^8 times, an extremely time consuming part of the chain. Therefore, it is essential to check, what kind of PDF is the most appropriate providing the best results. Since this averaging procedure signifies a dominant part of the total computation time, a large potential for acceleration and code improvement is assumed.

The chemical reaction rate is a nonlinear function of the atomic concentration in the control volume. All possible values of the mixture fraction have to be used in order to calculate the chemical sources φ. One solution is to solve a transport equation for every point of the PDF of mixture fractions. In practical combustion cases, this operation would largely exceed acceptable computation time.

To reduce the number of necessary transport equations, the PDF is approximated by an assumed two parameter PDF. In the computation only the equation for mean values and the variances have to be solved, in order to know the form of the PDF. The chemical sources are calculated by integrating the product of the PDF value and the source term, as shown in Equation 18. These averages are determined as follows:

$$\overline{\phi} = \int_0^1 \phi(f) \cdot P(f) df .$$ (18)

There are several possibilities to replace P(f) by an assumed PDF. Using the results of time resolved QLS we are able to decide, which of these functions yields the best results for certain applications and we are able to estimate the error caused by a two parameter model.

The Gaussian function is defined as:

$$P(f) = \frac{1}{f' \sqrt{2\pi}} e^{\frac{(f-\overline{f})}{2 \cdot f'^2}} .$$ (19)

The double delta function is described as:

$$P(f) = A \cdot \delta\left[f - \left(\overline{f} - \overline{f'}\right)\right] + B \cdot \delta\left[f - \left(\overline{f} - \overline{f'}\right)\right].$$ (20)

One of the most approved PDF's is the Gaussian function. The mixture fraction is only defined within the interval [0,1]. The values of the Gaussian function outside this interval are integrated and handled as Dirac peaks at 0 and 1. This type of Gaussian function is known as *clipped Gaussian function*.

Another possible form of two parameter PDF's is the double delta function, whose greatest convenience is the simple numeric treatment in integrating this function. To integrate the Gaussian function one needs at least ten table lookups, whereas the double delta approach only requires two of them. Fig.9 shows three PDF's, one double delta PDF, one Gaussian PDF and the measured PDF.

The Gaussian function is in good agreement with the measured PDF. The double delta function usually yields good results as well. In the following, we check, based on comparison with experimental data, in which cases one can use the double delta function in order to decrease the computation time and in which cases one has to use the Gaussian function to decrease the numeric error.

The comparison between measured PDF and Gaussian PDF is performed by using the correlation coefficient, which is defined as:

$$r_{xy} = \sqrt{\frac{S_{xy}}{S_x \cdot S_y}} .$$ (21)

s_{xy} is the covariance of x with y, s_x is the variance of x and s_y is the variance of y. If the correlation coefficient is near unity, a linear dependence between the two functions can be stated. Because of the normalizing procedure of the two PDF's a correlation coefficient of one means that the two PDF's, are equal. However, if the correlation coefficient tends against zero, no correlation between the data pairs can be assumed.
Fig. 10 shows the distribution of the correlation coefficient between measured PDF and the Gaussian PDF. The location is the region of the first jet, marked in Figs. 7 and 8.

In regions where the PDF is relevant for the computation of chemical source terms, the correlation coefficient is greater than 98 per cent. In regions where the correlation coefficient ist lower than 90 per cent the chemical source terms are linear with the mixture fraction and the integration would yield the same result as taking only the value at mean mixture fraction.

To understand the reason for the low coefficient in the critical area of the middle of the jet the typical measured PDF is show in Fig. 11.

Here we can see that the measured PDF is bimodal. Measurements of acoustic pressure of the combustion chamber have confirmed two peaks in frequency produced by the swirl nozzle and amplified by the mixing jets. In the jet injection area the same frequency peaks appeared. This implies a break off of the jets, triggered by the swirl nozzle.

This result is very important for the interpretation of the mean values in the inflow region of the jets, because this location is used to normalize the intensities of the QLS values. One can see that the value one is never reached as mean value, so that one has to scale with the maximum values and not with the maximum of mean values. For the calculation of the mean chemical source terms the numerical error produced during integration must be known.

The minimum number of data points, necessary for an accurate integration of the product of the Gaussian function with the chemical sources, is ten. During the computation process this product will be integrated more than 10^8 times. For higher level chemical models one has two integrate the Gaussian function in mixture fraction direction and in the direction of several progress variables as well. This is not realistic in practical setups. So we have to confirm, in which cases we can replace the Gaussian function by the double delta function aiming to decrease the computation time.

The result of the averaging process does not only depend on the PDF shape. The function of the chemical source terms is important, too. In the following we use a hypothetical function for temperature and NO production rate, which is a result of a realistic RQL experiment, intending to compare the measured PDF with the double delta PDF. The boundary conditions are shown in the table.

Mixture fraction=0	Mixture fraction=1
Equilibrium composition of kerosene at $\lambda=0.61$	Air
T = 2050 K	T = 700 K

The source terms correspond to the equilibrium values. The measured PDF yields a mean value and a variance. These two values are taken to construct a double delta function. In Fig. 12 the difference of temperature between an average of equilibrium values with the QLS–PDF and an averaging with the double-delta values is shown.

The difference is small in stoechiometric regions, i.e. one can get the double delta function to compute the mean heat release. In the region of the jet inflow the error is slightly larger. The reason for this is the bimodal distribution, explained above. However, this region is not crucial for the calculation and yet poses no problems.

The production of thermal NO is a nonlinear function of temperature and a linear function of the oxygen radical. Fig. 13 shows the same analysis, this time performed with temperature for thermal NO as described by Zeldovic et al 1994. One can see a larger difference in stoechiometric regions. The reason is the large gradient of o-radicals near stoechiometry. The overlapping of three nonlinear functions cannot be predicted with only two peaks.

The result shows that it is possible to average the chemical source terms with the double delta function in order to calculate the average heat release. Because of the low coupling of other chemical parameters to the flow field and the large coupling with temperature one can use it to reach the stationary point in flow field computation. **This operation decreases the computation time from 10 to 1000 times.**

For the calculation of NO the double delta function is not appropriate. Because of the low coupling of NO with the flow field it can be calculated within the post processing. Therefore, the Gaussian function is very effective.

Higher level chemistry models use higher dimension PDF's. The nonlinearly of the other input values as CO_2 or H_2O is smaller so that one can use the double delta function to compute the heat release also. **In this case the computation time can be decreased by a factor 1000.**

4 Conclusion and Outlook

It could be demonstrated that the QLS measurement technique is an appropriate means for the time resolved analysis of mixing fields. The gathered measurement data sets could be used for a better understanding of the mixing process in problems of type jet in crossflow. The experimental PDF data could be compared to numerically assumed PDF's. On the basis of these comparisons the numerical model for the prediction of chemistry turbulence interaction can be essentially accelerated.

References

A.E. Perry, R.M. Kelso and T.T. Linn
'Topological structure of a jet in a cross flow', AGARD Conference 534, *Computational and Experimental Assessment of Jets in Cross Flow*, Winchester, UK, 1993.

H. Schütz, H. Eickhoff, P. Theisen, C. Hassa, J. Koopman
'Analysis of the Mixing Zone of an Air Staged Combustor', ISABE-97 7225, Chattanooga, Tennessee, 1997.

P. Voigt, R. Schodl
'Using the Laser Light Sheet Technique in Combustion Research', 90th AGARD PEP-Symposium on Non-Intrusive Measurement Techniques for Propulsion Engines, Brussels, Belgium, 1997.
P. Voigt
'2D-Streulichtverfahren zur Konzentrationsbestimmung in Mischungsströmungen', 8. Workshop der Arbeitsgemeinschaft *Strömungen mit Ablösung (STAB)*, Göttingen, 1997.

C. Hassa, C.E.S.S. Migueis, P. Voigt
'Design Principles for the Quench Zone of Rich-Quench-Lean Combustors', AGARD PEP-Symposium on *Design Principles and Methods for Aircraft Gas Turbine Engines,* Toulouse, 11.-15.05.1998.

P. Voigt
'Non-linear effects in planar scattering techniques: proof of existence, simulations and numerical corrections of extinction and multiple scattering', 9th International Symposium on Applications of Laser Techniques to Fluid Mechanics, Lisbon, Portugal, 1998.

P. Voigt
'Entwicklung und Einsatz eines Laserlichtschnittverfahrens zur quantitativen Konzentrationsmessung bei Mischungsprozessen', Dissertation Ruhr-Universität Bochum, 1999.

Ya. B. Zeldovich, P. Ya. Sadovnikov and Frank- D.A. Kamenetskii
'Oxidation of nitrogen in combustion' (Translation by M. Shelef). Academy of Sciences of USSR, (Institute of Chemical Physics), Moscow, 1974.

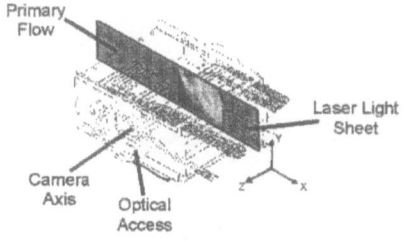

Fig 1 Typical QLS setup for an RQL combustor test rig

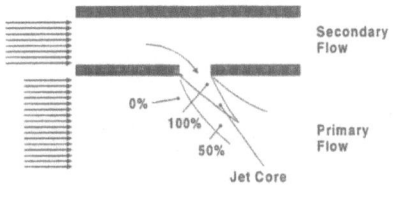

Fig.2 Principle of *Jet in Crossflow*

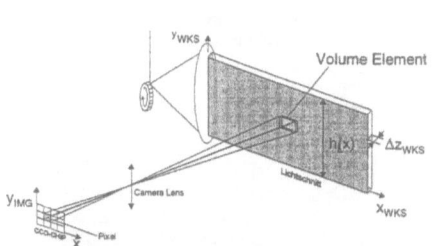

Fig. 3 Imaging the light sheet onto the CCD camera chip

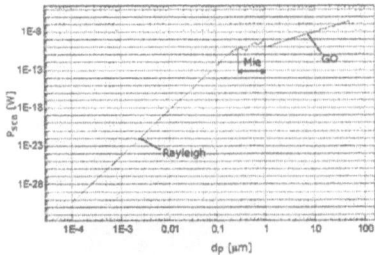

Fig. 4 Scattering ranges - power of scattered light of a single particle, l=514.5 nm, n=1.45+0*i

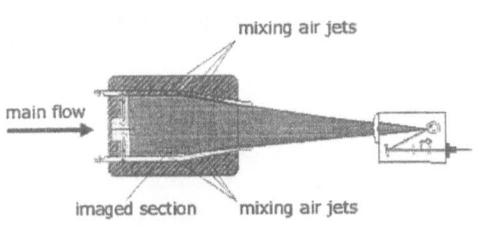

Fig. 5 RQL combustor geometry and light sheet setup

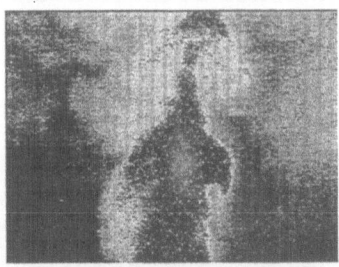

Fig. 6 Pseudo colored image of time resolved concentration distribution

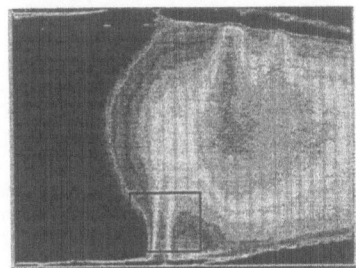

Fig. 7 Concentration distribution, averaged numerically over 10 000 images

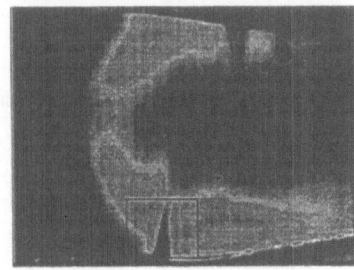

Fig. 8 RMS field of the mass concentration, computed over 10 000 images

503

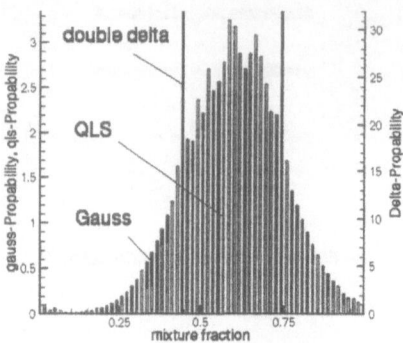

Fig. 9 Comparison of PDF shapes

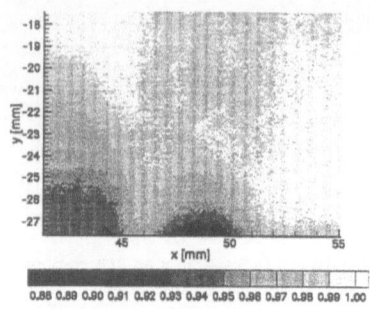

Fig. 10 Correlation coefficient between Gaussian and measured PDF

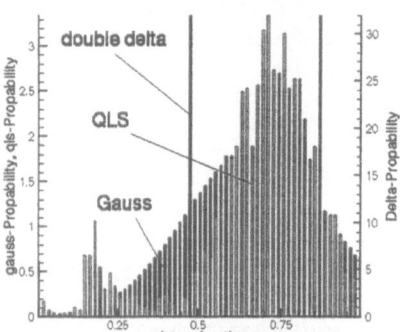

Fig. 11 Comparison of PDF shapes

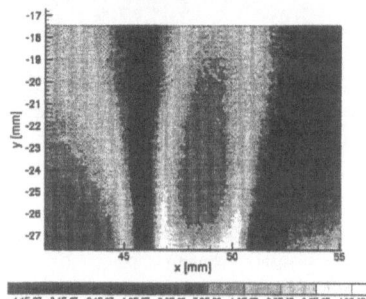

Fig. 12 Absolute temperature differences

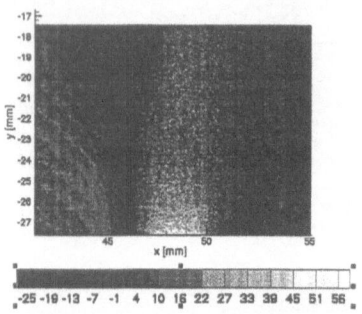

Fig. 13 Absolute differences of NO production

A Direct Navier-Stokes solver for turbulent flows over round steps

C. Wagner and U.Ch. Dallmann
DLR, Institute of Fluid Mechanics,
Bunsenstr. 10, 37073 Göttingen, Germany

Summary

Direct Numerical Simulations (DNS) provide an increasing data source to improve our understanding of turbulent flows. Such DNS are especially required (together with generic experiments) to support flow physical modeling of separated, two- and three-dimensional vortex flows. In perspective to improve the prediction of such flows (either by new turbulence models or Large Eddy Simulations (LES)) data fields are required which resolve turbulence and unsteady separated flow features. This is still a challenging task.

However, a fully conservative, second order accurate, boundary correction method for Cartesian grids has been developed and implemented in an existing direct Navier-Stokes (NS) solver. For the case of a round backward facing channel step, laminar flow results generated with the proposed method agree well with those calculated on curvilinear, colocated grids. A Direct Numerical Simulation (DNS) of the turbulent flow in a minimal channel domain is performed for a Reynolds number based on friction velocity and channel height of $Re_\tau = 265$ for an efficiency analysis. Additionally the turbulent flows in a channel and a round backward facing channel step are calculated by means of DNS. The former simulation provides inflow conditions for the latter. The upstream Reynolds is $Re_\tau = 360$. Good agreement is obtained comparing statistically averaged variables of the turbulent channel flow with those of Kim et al. [3]. A mean recirculation zone, with mean separation/reattachment points being located 2.0/7.5 step heights downstream the step entrance, is computed in the DNS through the round backward facing channel step. Instantaneous and statistical flow variables are presented. They give an impression of the complex flow dynamics in the free shear layer, the recirculation and reattachment regions.

Introduction

Direct numerical simulations have proven to be a valuable tool not only for fundamental researchers but also for turbulence modelers who need DNS results for calibriation and validation purposes. Despite of the tremendeous efforts paid to numerous international turbulence model validation projects the frustration increased due to permanent failures in predicting separated turbulent flows or three-dimensional turbulent vortex flows. Hence, there is an increasing necessity and a fundamental interest in a better physical understanding of the multi-scale spatio-temporal structures of turbulence; either to develop new statistical turbulence models or to support subgrid scale modeling required for LES. Due to the substantial resources which are required for DNS in terms of the large number of gridpoints and time steps most of the work in this area has been reported on turbulent flows in or past simple geometries. So far only spectral discretization methods as used by Kim et al. [3] to simulate the turbulent channel flow and finite volume methods with second order central differencing on staggered grids (see Eggels et al. [5]) have proven to be accurate enough to simulate turbulent flows realistically. The latter method was used to study turbulent flow separation in Le and Moin [4] for the case of a backward facing step domain and in Wagner et al. [7] to simulate the turbulent flow through a sudden pipe expansion.

For domains with curved boundaries Navier-Stokes (NS) solvers of different algorithmic nature have been developed in the past. Second order finite differences on curvilinear, colocated grids are used by Peric et al. [12] and Deng et al. [10] to discretize the NS equations in complex domains. The disadvantages are the not fully conservative formulation and high computational costs to either store or calculate the geometrical quantities. An application of this approach is the massive parallel computation of the turbulent flow around a sphere by Seidl et al. [9]. A fully conservative NS-solver based on staggered boundary fitted non-orthogonal grids was developed by Wesseling et al. [11]. Due to the immense computational costs this method was not yet applied to turbulent flows in complex geometries.

One alternative is the use of a boundary correction method in a Cartesian grid, as investigated by Gullbrand et al. [13]. They apply higher order Lagrange interpolations on a Cartesian grid to accurately realize curved boundaries. The resulting system of algebraic equations are solved using a multigrid Gauss-Seidel method.

It is the aim of this work to develop a direct, second order accurate, conservative, boundary correction method and implement this method in an existing DNS code (see Wagner [6]) to simulate the turbulent flow through a round backward facing channel step.

Numerical approach

Incompressible, isothermal turbulent flows can be completely decribed by solutions of the Navier-Stokes equations

$$\frac{\partial \vec{u}}{\partial t} + \nabla \cdot (\vec{u}\vec{u}) = -\frac{1}{\rho}\nabla p + \nu\nabla^2\vec{u} \tag{1}$$

$$\nabla \cdot \vec{u} = 0 \quad , \tag{2}$$

where $\vec{u} = (u, v, w)$ respresents the velocity vector, p the pressure, ρ the density and ν the dynamic viscosity, and suitable boundary conditions. The presented NS-solver FLOWSI integrates equations 1 and 2 in Cartesian (x, y, z)-coordinates using staggered grids and second order central differences in space. An explicit, essentially second order accurate, Euler-Leapfrog time-integration scheme serves to advance the solution in time. The maximum permissible time-step is obtained from a linear stability analysis.

The coupling between the pressure and velocity fields is provided by a fractional step approach, which leads to a 3D Poisson equation for the pressure. Considering the flow in a straight channel with two periodic directions this is done by means of FFT's in x- and y-directions and a standard tridiagonal matrix algorithm for the resulting set of 1D Helmholtz problems. For the channel step, depicted in Fig. 1, FFT's are used in the periodic y-direction. The remaining set of 2D Helmholtz problems in irregular domains are treated with a cyclic reduction algorithm together with the influence matrix technique (see Schumann [1]). More details on the numerical method are given in Wagner [6].

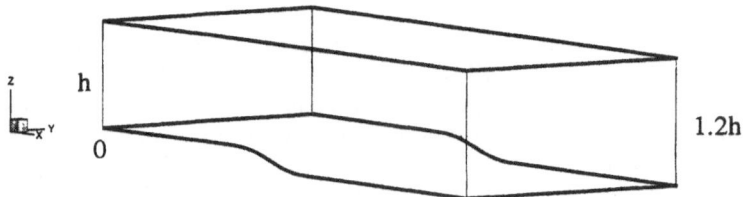

Figure 1: Computational domain of the round backward facing channel step

Laminar flow results

A computational domain of 6 channel heights h length, πh width and a lower wall following

$$z_0(x) = 20h_s(x - x_s)^7 - 70h_s(x - x_s)^6 + 84h_s(x - x_s)^5 - 35h_s(x - x_s)^4 + h_s, \quad (3)$$

where $h_s = 0.2h$ defines the step height and $x_s = 1.975$ the streamwise location of step entrance, was chosen to simulate the laminar flow for a Reynolds number based on the bulk velocity u_b and channel height h upstream of $Re_b = u_b h / \nu = 1000$. Simulations were performed on three different grids with (60×30), (80×45) and (120×60) gridpoints in (x, z)-directions and 8 gridpoints in spanwise direction. Additionally this flow was computed on a grid with 120×60 curvilinear colocated cells using the stationary, 2D multigrid solver by Peric et al. [12].

Boundary correction method

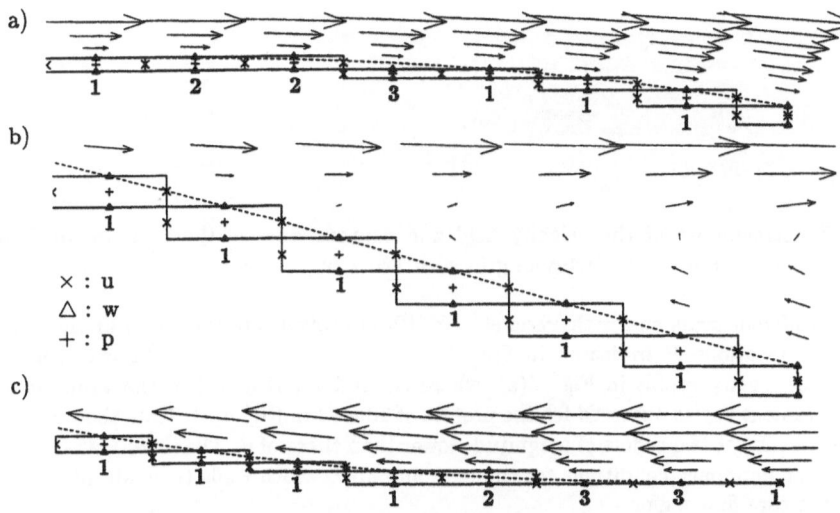

Figure 2: Curved boundary oriented control volumes and neighbouring velocity vectors of the laminar flow calculation on the fine grid for a Reynolds number of $Re_b = 1000$.

In order to correctly model the curved boundary, grid refinement in lateral direction is applied such that the boundary values of the lateral velocity component w are defined

507

on the curved wall. For uniform gridspacing in x-direction the staggered streamwise velocity component u is then, within second order accuracy, defined on the wall as well. The majority of these boundary cells (indicated by 1 in Fig. 2(a)-(c)) reflect reasonable stretching factors of less than 15 per cent. In regions with high boundary curvature, as shown in Fig. 2a and c, this approach would lead to excessive grid stretching. To overcome this two or more of these fine cells are substituted by a coarser cell, leading to boundary values of w which formally have to be defined either inside or outside flow domain as indicated by 2 or 3 respectively. In order obtain a conservative formulation for cells of type 2 and 3, auxiliary control volumes are assumed such that the finite difference expressions realize the no-slip and impermeability conditions of the curved wall. The boundary neighbouring velocity vectors, which have been calculated on the fine grid, are included in Fig. 2. The orientation of the velocity vectors follows the boundary contour. No influence of the underlying 'rough' Cartesian control volumes is observed.

Validation

Streamlines of the velocity field and pressure contour lines are presented in Fig. 3. A pressure minimum in the vincinity of the step entrance reflects the acceleration and deceleration of the flow leading to separation shortly downstream. A recirculation zone, extending 7.5 step heights in streamwise direction, is obtained. Even streamline next to the curved wall are not influenced by the Cartesian grid. The inflow field, which is the analytical profil a fully developed laminar channel flow, is shown to be undisturbed. This also holds for the outflow field, where a convective boundary condition as discussed in Wagner [6] is applied. In Fig. 4(a) and (b) velocity profiles, calculated with FLOWSI

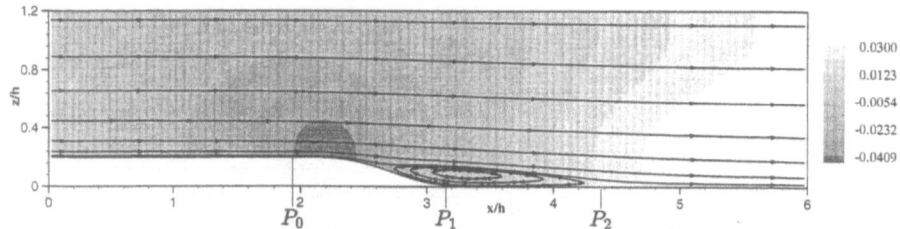

Figure 3: Streamlines of the velocity field and contour lines of the pressure field for $Re_b = 1000$, calculated on a Cartesian grid with 120×60 cells in (x, z)-directions.

on three different grids and with Peric et al.'s [12] NS-solver, are compared at the three streamwise locations P_l indicated in Fig. 3. Excellent agreement is obtained for the streamwise velocity profils in Fig. 4(a), where the index l is added to the value of u . Good agreement is also observed for the profiles of $10w + l$ in Fig. 4(b). Note that the values of w are almost two orders of magnitude lower than those of u. The investigated codes use different boundary conditions for the outgoing flow, which leads to small differences of w in the core flow region.

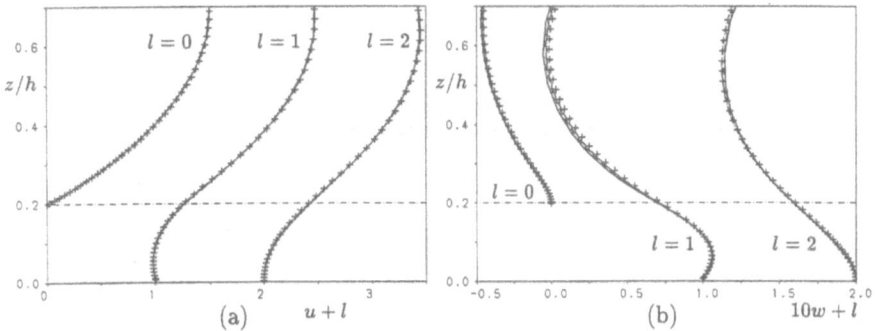

Figure 4: Laminar flow profiles of velocity components u and w in (a) and (b) at the streamwise locations P_l. P_0 : $x = 1.98$, P_1 : $x = 3.18$ and P_2 : $x = 4.38$. ——— :(60 × 30), – – – – :(80 × 45) and –·–·–· :(120 × 60) gridpoints. + : Peric et al.

Turbulent channel flow

Efficiency Analysis

Two codes of different algorithmic nature have been selected for the efficiency analysis. The first one is a Cartesian and therefore fully conservative DNS version of the 'colocated curvilinear' code named HORUS by Deng et al. [10]. The second, fully implicit, code DeFT is discretized on staggered, non-orthogonal grids. It was developed by Wesseling et al. [11] to solve the Navier-Stokes equations in geometrically complicated flow domains using boundary fitted coordinates. In HORUS the CPU-time consuming solution of the Poisson equation is obtained by an algebraic multi-grid solver, while DeFT uses a GMRES method with one multigrid V-cycle as a preconditioner.

The performance of the above described NS-solvers was investigated by Manhart et al. [8] simulating the turbulent flow in a minimal channel unit on a Fujitsu VPP700 for a Reynolds number based on friction velocity u_τ and channel height h of $Re_\tau = 265$ as defined by Jimenez and Moin [2].

The superior performance of FLOWSI is demonstrated in Table 1 where the CPU-time and storage requirements the three different codes are summarized. Using less than 1/10 of Random Access Memory, FLOWSI is 3 times faster than HORUS and 358 times faster than DeFT. It must be noted that the curvilinear version of HORUS either needs approximatly twice the memory or additional CPU-time in order to store or calculate geometrical coefficients.

Table 1: CPU-time and memory requirements to simulate the turbulent minimal channel flow on a Fujitsu VPP700 at $\approx 400 MFlops$

Code	FLOWSI	HORUS (Cartesian)	DeFT
CPU-sec per Δt and gridpoint	$1.48 \cdot 10^{-6}$	$4.43 \cdot 10^{-6}$	$530.1 \cdot 10^{-6}$
Byte per gridpoint	60	720	1950

Generation of inflow conditions

The computational domain, a straight channel of length $5.12h$ extending $2.4h$ in spanwise direction, is resolved by 256×128 equidistant grid points in (x, y)-directions and 85 points in lateral direction which are clustered close to the wall. In terms of wall units, ν/u_τ, the grid spacing is:

$$\Delta x^+ = 7.2, \quad \Delta y^+ = 6.75, \quad \Delta z^+ = 0.47 \div 7.85 \tag{4}$$

Non-dimensionalizing equations 1 and 2 with the friction velocity u_τ and channel height h a Reynolds number $Re_\tau = u_\tau h/\nu = 360$ and mean pressure gradient $\partial \langle p \rangle / \partial x = 2$ was prescribed for this simulation. In order to demonstrate the accuracy of FLOWSI, profiles

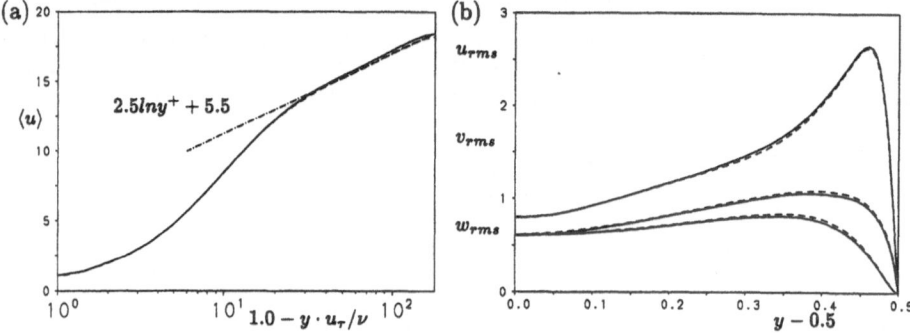

Figure 5: Mean streamwise velocity component in (a) and profiles of rms-velocity fluctuations (b). ———— : FLOWSI, – – – – : Kim et al.

of the statistically averaged streamwise velocity $\langle u \rangle$ and the rms fluctuating velocity vector $\vec{u}_{rms} = \sqrt{\langle \vec{u}^2 \rangle - \langle \vec{u} \rangle^2}$ are compared in Fig. 5 to those of Kim et al. [3]. A overall good agreement is obtained. Deviations of the mean velocity profile in the logarithmic region are due to the rather coarse resolution of the core flow in lateral direction.

At each time step of this simulation a 3D time-dependent velocity vector field of a certain cross section is transfered to the inflow plane of the round backward facing channel step.

Turbulent flow in a round backward facing channel step

In order to avoid disturbances in the vincinity of the inflow plane the Reynolds number $Re_\tau = 360$, the grid resolution and the time step of the turbulent channel flow DNS are adopted. For a domain of length $4.32h$ and width $2.4h$ this leads to 216×128 equidistantly spaced grid points in (x, y)-directions. In addition to the 85 lateral distributed gridpoints of the channel domain 44 gridpoints are used to discretize the expanded part of the round backward facing channel step. Applying the contour oriented grid generation on the round step following eq. 3 with $x_s = 1.19$, the grid spacing (in dimensionless wall units) varies between $\Delta z^+ = 0.94$ and $\Delta z^+ = 2.13$.

Instantaneous flow fields

Streamlines of the instantaneous velocity field projected into a (x, z)-plane and contour lines of the fluctuating pressure field $p' = p - \langle p \rangle$ are shown in Fig. 6. Recirculating

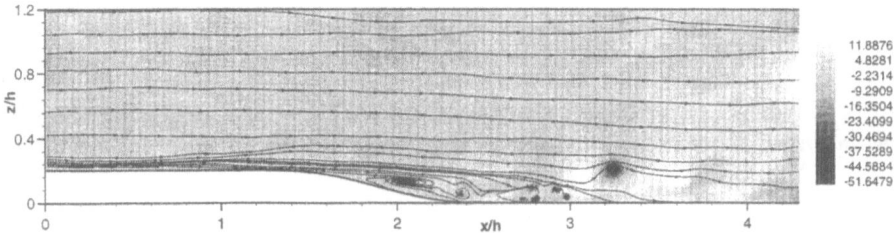

Figure 6: Streamlines of the instantaneous velocity field and contours of the fluctuating pressure field.

streamlines downstream the convex part of the step reflect the generation of a stationary vortex. Complex streamline patterns and intense minima in the fluctuating pressure field are observed within the mixing layer.

Statistical averaged flow fields

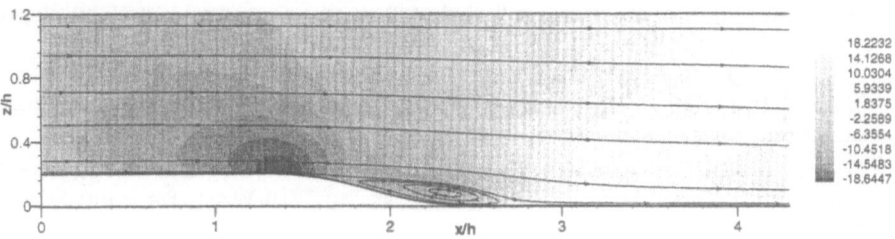

Figure 7: Streamlines of the mean velocity field superposed on contour lines of the mean pressure field.

In order to obtain statistically averaged flow fields more than 800 realizations were averaged in spanwise direction and in time. In Fig. 7 streamlines of the mean velocity field are superimposed on contour lines of the mean pressure field. Starting at $x/h = 1.49$ the mean recirculation zone extends 5.5 step heights downstream. Again the streamlines are not affected by the Cartesian grid. Similar to the laminar flow field, a pressure minimum in the vincinity of the step entrance reflects flow acceleration/deceleration induced by high boundary curvature.

Conclusions

Motivated by the immense computational costs for DNS of turbulent flows in complex geometries using boundary fitted grids a boundary correction method was developed and implemented in a Cartesian DNS code. The accuracy of this method was underlined comparing laminar flow results with those computed on curvilinear colocated cells. An efficiency analysis performed for the turbulent flow in a minimal channel unit revealed the superior behaviour of the proposed method compared to Navier-Stokes solvers which are either based on collocated, curvilinear grids or non-orthoganal staggered grids. The reliability of the finite voloume method for DNS of turbulent flows was demonstrated comparing turbulent channel flow results to Kim et al.'s [3] data. Using the flow field of the latter DNS to prescribe instantaneous inflow conditions a DNS of the turbulent flow in a round backward facing channel step was performed for a Reynolds number of $Re_\tau = 360$.

References

[1] Schumann, U. - Fast elliptic solvers and their application in fluid dynamics, In: Computational Fluid Dynamics, W.Kollmann (Ed.), Washington: Hemisphere, pp.402-430, (1980).

[2] Jimenez, J. and Moin, P. - The minimal flow unit in near wall turbulence, J. Fluid Mech., vol. 255, pp. 213-240, (1991).

[3] Kim, J., Moin, P., Moser, R., - Turbulence statistics in fully developed channel flow at low Reynolds number, J. Fluid Mech., vol. 177, pp. 133-166, (1987).

[4] Le, H., Moin, P., Moser, R., - Direct numerical simulation of turbulent flow over a backward-facing step, In; Proc. 9th Symp. on Turbulent Shear Flows, August 16-18, Kyoto, Japan, P13/2/1-P13/2/6, (1993).

[5] Eggels, J.G.M., Unger, F., Weiss, M.H., Westerweel, J., Adrian, R.J., Friedrich, R. and Nieuwstadt, F.T.M. - Fully developed turbulent pipe flow: a comparison between direct numerical simulation and experiment. J. Fluid Mech. 268, 175-209 (1994).

[6] Wagner, C. - Direkte numerische Simulation turbulenter Strömungen in einer Rohrerweiterung, Ph.D. Thesis, Munich University of Technology, VDI-Verlag, Fortschrittsberichte Reihe 7, Nr. 283, (1996).

[7] Wagner, C., Friedrich, F. - Direct numerical simulation of turbulent flow in a sudden pipe expansion, -In: AGARD Conference Proceedings of the 74th Fluid Dynamics Symposium on Application of direct and large eddy simulation to transition and turbulence, April 1994, Chania, Crete, Greece, pp. 6.1-6.11 (1994).

[8] Manhart, M., Deng, G. B., Hüttl, T. J., Tremblay, F., Segal, A., Friedrich, R., Piquet, J., Wesseling, P. - The minimal turbulent flow unit as a test case for three different computer codes. Notes on Numerical Fluid Mechanics, Vol. 66, Vieweg Verlag, Braunschweig (1998).

[9] Seidl, V., Muzaferija, S. and Peric, M. - Parallel Computation of Unsteady Separated Flows Using Unstructured, Locally Refined Grids. Notes on Numerical Fluid Mechanics, Vol. 64, Vieweg Verlag, Braunschweig (1998).

[10] Deng, G.B., Piquet, J., Queutey, P. and Visonneau, M. - Three Dimensional Full Navier-Stokes Solvers for Incompressible Flows Past Arbitary Geometries. Int. J. for Num. Method in Eng., vol. 31, pp. 1427-1451, (1991).

[11] Wesseling, P., Segal, A., van Kan, J.J.I.M., Oosterlee, C.W. and Kassels, C.G.M. - Finite volume discretization of the incompressible Navier-Stokes equations in general coordinates on staggered grids. Comp. Fluid Dynamics Journal, vol. 1, pp. 27-33, (1992)

[12] Peric, M. Kessler, R. and Scheurer, G. - Comparison of finite-volume numerical methods with staggered and colocated grids. Comp. and Fluids, vol. 16, pp. 389-403, (1988).

[13] Gullbrand, J., Bai, X.S. and Fuchs, L. - High Order Boundary Corrections for Computation of Turbulent Flows. -In: Numerical methods in laminar and turbulent flow, C. Taylor, J. Cross (eds.), Pineridge Press, Swansea UK, Vol. 10, pp. 191-202 (1997).

Direct Numerical Simulation of the Development and Control of Boundary-Layer Crossflow Vortices

P. Wassermann, M. Kloker

Institut für Aerodynamik und Gasdynamik
Universität Stuttgart
Pfaffenwaldring 21, D-70550 Stuttgart, Germany

Summary

The development and control of stationary crossflow-vortex packets in a 3-D flat-plate boundary-layer flow is investigated by means of spatial direct numerical simulations based on the complete incompressible 3-D Navier-Stokes equations and a combined 6th-order compact Finite-Difference / spanwise Fourier-spectral scheme. The considered spanwise-invariant laminar base flow is a model of the boundary-layer flow in the front region of a swept wing, with favorable and ensuing adverse chordwise pressure gradient (range of approximate local Hartree parameter gained from potential flow: $\beta_H = 0.99 \sim -0.42$). The crossflow-vortex packets are excited by imposing different steady wall-normal velocity distributions within a spanwise disturbance strip at the wall to represent point-like disturbances. Their downstream growth and nonlinear interaction as well as their control by a downstream control strip and different control strategies is investigated in detail.

Introduction

On a swept-back airfoil the chordwise acceleration of the potential flow induces an inboard-oriented crossflow component inside the boundary layer perpendicular to the meanflow direction. The crossflow velocity profile $W_s(y)$, y being the wall-normal coordinate, is inflectional and causes a strong primary instability of the flow with respect to so-called crossflow (CF) modes. The wave number vectors of both instationary and stationary CF modes - the latter referred to as CF vortices (CFVs) - are oriented approximately parallel to the crossflow direction. At low freestream-turbulence conditions, CFVs excited by surface nonuniformity are found to be dominant (see e.g. [1] and [6] for experimental and [3] for numerical evidence). Upon downstream vortex saturation, local inflectional streamwise-velocity profiles $U_s(y)$ are formed and strong subsequent (secondary) instability mechanisms that are not yet fully understood typically lead to transition. Under uncontrolled experimental conditions, a spanwise CFV wavelength is observed that is related to the most unstable mode as predicted by (primary) Linear Stability Theory; the role of other primarily unstable stationary modes in the nonlinear process is unclear.

The region of favorable pressure gradient on the airfoil is followed by increasing pressure, giving rise to a directional change of the crossflow due to the sign change of the external streamline curvature. Downstream of the pressure minimum, W_s may change sign within the base boundary layer, and, in case of still transitional flow, this influence on the grown CFVs has not yet been under investigation. They may decay or strongly influence the ongoing transition that now seems to be dominated by strong Tollmien-Schlichting (TS)-type instability (with the respective wave number vectors approximately aligned with the streamlines) caused by the adverse pressure gradient (APG).

Our investigations focus on these open issues, in this paper by considering exclusively stationary CF modes to elucidate specific basic mechanisms. CFV packets consisting of four modes are chosen to enable investigations of a realistic development and control of superimposed CFV disturbances at still feasible identification of mode interactions.

Governing Equations and Numerical Method

The numerical model is based on the complete 3-D Navier-Stokes equations for incompressible unsteady flows in a vorticity-velocity formulation. All variables are nondimensionalized by the reference length $\bar{L} = 0.05m$, the freestream velocity $\bar{U}_\infty = 30m/s$ and the Reynolds number $Re = \bar{U}_\infty \bar{L}/\bar{\nu} = 100000$, where ¯ denotes dimensional variables and $\bar{\nu}$ is the kinematic viscosity (see also [4]):

$$x = \bar{x}/\bar{L}, \quad y = \sqrt{Re} \cdot \bar{y}/\bar{L}, \quad z = \bar{z}/\bar{L}, \quad t = \bar{t} \cdot \bar{U}_\infty/\bar{L}, \tag{1}$$

$$u = \bar{u}/\bar{U}_\infty, \quad v = \sqrt{Re} \cdot \bar{v}/\bar{U}_\infty, \quad w = \bar{w}/\bar{U}_\infty \quad .$$

With the vectors of vorticity $\vec{\omega} = \{\omega_x, \omega_y, \omega_z\}^T$ and velocity $\vec{u} = \{u, v, w\}^T$, where u denotes the velocity in chordwise (x-), v in wall-normal (y-) and w in spanwise (z-) direction, the equations are:

$$\frac{\partial \omega_x}{\partial t} + \frac{\partial}{\partial y}(v\omega_x - u\omega_y) - \frac{\partial}{\partial z}(u\omega_z - w\omega_x) = \tilde{\Delta}\omega_x, \tag{2}$$

$$\frac{\partial \omega_y}{\partial t} - \frac{\partial}{\partial x}(v\omega_x - u\omega_y) + \frac{\partial}{\partial z}(w\omega_y - v\omega_z) = \tilde{\Delta}\omega_y, \tag{3}$$

$$\frac{\partial \omega_z}{\partial t} + \frac{\partial}{\partial x}(u\omega_z - w\omega_x) - \frac{\partial}{\partial y}(w\omega_y - v\omega_z) = \tilde{\Delta}\omega_z, \tag{4}$$

$$\frac{\partial^2 u}{\partial x^2} + \frac{\partial^2 u}{\partial z^2} = -\frac{\partial \omega_y}{\partial z} - \frac{\partial^2 v}{\partial x \partial y}, \tag{5}$$

$$\tilde{\Delta}v = \frac{\partial \omega_x}{\partial z} - \frac{\partial \omega_z}{\partial x}, \tag{6}$$

$$\frac{\partial^2 w}{\partial x^2} + \frac{\partial^2 w}{\partial z^2} = \frac{\partial \omega_y}{\partial x} - \frac{\partial^2 v}{\partial y \partial z}, \tag{7}$$

wherein $\tilde{\Delta}$ denotes the modified Laplace operator

$$\tilde{\Delta} = \frac{1}{Re}\frac{\partial^2}{\partial x^2} + \frac{\partial^2}{\partial y^2} + \frac{1}{Re}\frac{\partial^2}{\partial z^2} \quad . \tag{8}$$

The three vorticity components are defined as

$$\omega_x = \frac{1}{Re}\frac{\partial v}{\partial z} - \frac{\partial w}{\partial y}, \quad \omega_y = \frac{\partial w}{\partial x} - \frac{\partial u}{\partial z}, \quad \omega_z = \frac{\partial u}{\partial y} - \frac{1}{Re}\frac{\partial v}{\partial x}, \tag{9}$$

and at the wall the following equations are used:

$$\frac{\partial^2 \omega_x}{\partial x^2} + \frac{\partial^2 \omega_x}{\partial z^2} = -\frac{\partial^2 \omega_y}{\partial x \partial y} + \frac{\partial}{\partial z}\tilde{\Delta}v, \tag{10}$$

$$\frac{\partial \omega_z}{\partial x} = \frac{\partial \omega_x}{\partial z} - \tilde{\Delta}v. \tag{11}$$

We use a disturbance formulation, i.e. each flow variable is split into the steady 3-D base flow part (index B) and the unsteady 3-D disturbance flow part (denoted by a prime).

$$f = f_B + f' \quad \text{with} \quad f \in \{u, v, w, \omega_x, \omega_y, \omega_z\}. \tag{12}$$

The simulation is carried out in a rectangular integration domain (Fig. 1). First, the steady base flow is calculated; next, defined disturbances are introduced in a disturbance strip at the wall and the unsteady disturbance flow is calculated.

Calculation of the base flow

The calculation of the 3-D base flow relies strongly on the assumption of infinite span, i.e. all quantities are independent of the spanwise coordinate z, but there is a velocity component w_B in spanwise direction. The equations are obtained from (2)-(7) by neglecting all z-derivatives and are treated by high-order finite differencing. The vorticity equations are solved by a semi-implicit pseudo-temporal technique, and the Poisson equations by a vectorizable stripe-pattern LSOR-technique (iteratively in x-direction).

The boundary conditions are no-slip condition at the wall and vanishing vorticity and a prescribed chordwise velocity distribution $u_e(x)$ at the upper boundary. At the outflow boundary the equations are solved by neglecting the second x-derivative terms following usual boundary-layer-theory assumptions, and u_B and w_B are calculated from

$$\frac{\partial^2 u_B}{\partial y^2} = \frac{\partial \omega_{z,B}}{\partial y} \quad , \quad \frac{\partial^2 w_B}{\partial y^2} = \frac{1}{Re} \frac{\partial \omega_{y,B}}{\partial x} - \frac{\partial \omega_{x,B}}{\partial y}. \tag{13}$$

In this work, $u_e(x)$ at the upper boundary is gained from the potential velocity distribution $u_{p0}(x)$, valid at $y = 0$ in inviscid flow, by a complex flow function for potential flow. This function is obtained by integrating the analytical distribution $u_{p0}(x)$ (see below). Thus, an effect of the integration-domain height on the actual $u_\delta(x)$ ($\approx u_{p0}(x)$) and integral boundary-layer parameters is excluded. At the inflow boundary a local Falkner-Skan-Cooke solution appropriate to the local $u_e(x)$ is used.

Calculation of the disturbance flow

The equations for the disturbance quantities are derived from the equations (2)-(7) with the decomposition (12) leaving out the zero sum of all pure base-flow terms.

In spanwise direction the numerical method uses a complex Fourier spectral representation (the assumption of infinite span yields periodic boundary conditions) to calculate the nonsymmetric 3-D flow , i.e all variables are decomposed as

$$f'(x, y, z, t) = \sum_{k=-K}^{K} F_k(x, y, t) \cdot e^{ik\gamma z}, \tag{14}$$

where γ is the basic spanwise wavenumber, related to the width λ_z of the integration domain by $\lambda_z = 2\pi/\gamma$. The F_{-k} are the complex conjugates of the F_k and do not have to be computed. In x- and y-direction, a Finite-Difference (FD) discretisation is used, based on a blockwise equidistant rectangular grid with a special wall zone, where the stepsize Δy is halved. Principally, sixth-order compact FDs are used. The nonlinear terms in the vorticity transport equations are computed pseudospectrally and their x-derivatives are differenced with a special split-type method with inherent damping (see

[2]); the time integration is done by a 4-step fourth-order Runge-Kutta (RK) method. Other RK schemes modified specifically to faster achieve the steady state in theory (e.g. by maximizing the amplitude damping or the stability limit) were not successful with respect to reducing computation time.

The boundary conditions at the upper boundary are vanishing vorticity and exponential decay of the wallnormal velocity v'. At the inflow boundary all disturbance quantities are set to zero and at the wall the no-slip condition is satisfied for u' and w'. Also the velocity v' is zero at the wall, except in the disturbance strip, where the disturbances are forced, with momentum input but no net mass flow:

$$v'(x, y = 0, z) = \sum_{k=1}^{K} A_k \cdot f_v(x) \cdot \cos(k\gamma z - \Theta_k). \tag{15}$$

As for the outflow boundary, a well tested buffer-domain technique is applied, wherein the vorticity disturbance vector is slowly forced to zero upstream the outflow boundary. A more detailed description of the numerical method is given in [3] and for the main FD stencils see [2].

The grid used for the simulations contains 2626 gridpoints in x-, 257 points in y- (including 33 points in the wall zone) and 16 gridpoints in z-direction, corresponding to $K = 8$ Fourier spectral modes in (14). The problems were solved on a NEC SX4/32 vector supercomputer. The code has been parallelized in spanwise direction with a speedup of about 8.25 with the use of 9 processors. The calculation of a single time step took about 3.6 μs per gridpoint on a single processor corresponding to roughly 1 GFLOPS.

Numerical Results

Base flow

The base flow is designed to resemble the flow in the front region of a swept wing. The streamwise edge velocity is defined analytically (see [5]) by

$$u_{p0}(x) = \frac{3}{2\pi} \left(\arctan \left(\frac{x - a}{b} \right) + \arctan \left(\frac{x + a}{b} \right) \right) - cx \tag{16}$$

with $a = 0.2611, \quad b = 0.41015, \quad c = 0.056$.

The integration domain starts at $x_0 = 0.25$ close to the leading edge (local Hartree $\beta_H(x_0) = 0.99$) and extends over a large distance downstream (outflow boundary at $x_E = 4.96$). The sweep angle is $\varphi_\infty = 45°$ and the local angle of the external streamline varies from $\varphi_e(x_0) = 68.1°$ to the minimum value of $\varphi_e = 39.7°$ at $x = 2.65$ and $\varphi_e(x_E) = 41.1°$. The Reynolds number based on the chordwise displacement thickness δ_1 rises from $Re_{\delta 1}(x_0) = 67$ to $Re_{\delta 1}(x_E) = 1630$ and the displacement thickness increases by a factor of five within the domain. The maximum crossflow amplitude is 13.3%, and for $x \geq 3.5$ the crossflow profiles $w_{B,s}(y)$ are S-shaped (see Figs. 2, 3).

Linear-stability-theory analysis

To start with, the base flow has been analysed by means of spatial linear stability theory (LST) to get an overview of relevant instability modes. The stability diagram for stationary modes is shown in Figure 4 ($\alpha_i < 0$: amplification). As a result, $\gamma_1 = 45 \ (= \bar{\gamma}_1 \cdot \bar{L})$ was chosen for the subsequent DNS, so that the CF-mode with the largest local amplification rate ($\gamma = 135 = 3 \cdot \gamma_1$) as well as the mode with nearly the largest integral amplification ($\gamma = 90 = 2 \cdot \gamma_1$) can be investigated introducing the packet of the stationary modes $(0, k), k = 1 - 4$ in the disturbance simulation (the discrete waves are presented in the frequency-spanwise wavenumber spectrum (h,k), so the mode (0,k) denotes a stationary crossflow mode with the spanwise wavenumber $k \cdot \gamma_1$).

Quasilinear development

In case 1 the CFV packet (as mentioned above) is forced with small amplitude ($A_k = 1.0 \cdot 10^{-7}$ for $k = 1 - 4$) and its quasilinear downstream development has been calculated to study the disturbance growth and the propagating direction especially in the APG region. It can be observed (Fig. 5), that all CFV-modes are strongly amplified first and that the modes (0,1) and (0,2) are still growing in the APG region, where the mode (0,2) attains the largest amplitude. The mode (0,3) behaves nearly neutral in this region, while only the mode (0,4) is damped. The amplification rates are larger than predicted by LST throughout the whole domain. The vortex-axis directions are similar for all modes and are roughly aligned with the direction of the inviscid streamline (Fig. 11).

Nonlinear development

For the investigation of non-linear phenomena, the CFV packet is introduced with larger amplitudes ($A_k = 5.0 \cdot 10^{-3}$ for $k = 1-4$) and the z-phase shift Θ_k (eqn. 15) of each mode is chosen in order to model point-like perturbations, with momentum but no net-mass input. Two different cases have been considered, case 2 with local "blowing" ($\Theta_{1,3} = \pi, \Theta_{2,4} = 0$) and case 3 with local "suction" ($\Theta_{1,3} = 0, \Theta_{2,4} = \pi$) (see Figs. 6,7, respectively).
The downstream spectral amplitude development (Figs. 8, 9) reveals differences between the two cases, beginning with slightly different receptivity at the disturbance strip. The disturbances are strongly amplified first and saturate at $x \approx 2.8$ with different amplitudes. In both cases the mode (0,2) has the highest amplitude, as expected from the quasilinear development (case 1). This mode attains a maximum amplitude of 17.8% in case 3 and 20.3% in case 2, where it clearly dominates the scenario by suppressing the development of the modes (0,1) and (0,3). In both cases the mode (0,4) is nonlinearly generated by and coupled to (0,2). Despite the distinctly different spectral amplitudes, the total rms-disturbance amplitudes (not shown here) of case 2 and 3 approximately coincide. The significant influence of the z-phase relation between the spectral modes can also be seen in the nonlinear disturbance profiles $u'(y)$ (Fig. 10), especially of the dominated modes.

Control of crossflow vortices

Considering the dominant role of stationary crossflow vortices for crossflow-induced transition, one idea to delay transition is to reduce the amplitude of the vortices in early stages by the introduction of an antiphase disturbance. For this purpose the CFV packet of case 2 (local blowing) has been excited with small amplitude and for its control by a second disturbance strip at the wall (see Fig. 1, $x_1 = 0.668, x_2 = 0.693, x_1' = 1.207, x_2' = 1.232$) several strategies have been considered.
First, localized disturbances with the same z-phase correlation between the spectral modes as in case 2, 3 have been used, since in the early stages all vortices have nearly the same direction downstream (see Fig. 11, control of a "hole-induced" disturbance by another

"hole-induced" disturbance). In this case the amplitude and z-phase shift of the control forcing is oriented at the amplitude and phase development of the dominating mode (0,2). Two different simulations have been performed, one with local "suction" (case A) and one with local "blowing" in the second strip (case B). Second, all modes have been individually controlled (case C), i.e. for each mode a different amplitude and z-phase shift has been calculated considering the downstream development of each mode separately (modal control). In Figure 12 the resulting v'-velocity distribution in spanwise direction in the control strip and in Figure 13 the u'_{rms}-amplitudes of the simulations with the different control strategies are shown. A striking reduction of the amplitudes is only achieved by modal control, although the control forcing of case B and case C are very similar (recall that the control amplitude and z-phase shift of (0,2) is identical for all cases). It turns out that the exact determination of the z-phase shift of all spectral control disturbances is very important to substantially reduce the total disturbance energy of the CFV packet. Nevertheless, the growth rates after control in case A, B are diminished since they are given by the less unstable remaining modes. For control in a fully nonlinear case difficulties arise by altered receptivity and nonlinear correlation between the downstream sensor point and the control strip, as for amplitudes and phases.

Conclusions

The nonlinear downstream development of a crossflow-vortex (CFV) packet depends on the initial spanwise (z-)phase relations of the spectral modes it is composed of. In case of localized positive momentum input (blowing-like disturbance) the development is clearly dominated by the primarily most unstable component, that attains first amplitudes higher than approximately 10% and suppresses the growth of the other modes. In the suction-like case, however, the dominance is only weak. In the region of adverse pressure gradient no significant principal decay of the CFV modes is observed. Thus, any laminar breakdown downstream will be strongly influenced by the CFV modes.
Due to the sensitivity of the controlled linear CFV packet with respect to the correct z-phase shift of each control-disturbance mode, an effective total amplitude reduction can only be achieved by modal control. The effectiveness of modal control in the nonlinear case still has to be investigated.

Acknowledgements

The financial support by the German Research Council (DFG) under contract number Kl 890/2-2 is gratefully acknowledged.

References

[1] Bippes, H.: Environmental conditions and transition prediction in 3-D BL. AIAA-97-1906, 1997.

[2] Kloker, M.: A robust high-resolution split-type compact FD scheme for spatial direct numerical simulation of boundary-layer transition. *Applied Scientific Research*, 59 (4), pp. 353-377. Kluwer Acad. Publishers, NL, 1998.

[3] Bonfigli, G.; Kloker, M.: Spatial Navier-Stokes simulation of crossflow-induced transition in a 3-D boundary layer. In Nitsche, W.G.; Hilbig, R., (eds.): *New results in numerical and experimental fluid dynamics*. 11. STAB/DGLR Symposium, November 1998. NNFM, Vieweg Verlag, Berlin.

[4] Müller, W.: *Numerische Untersuchung räumlicher Umschlagvorgänge in dreidimensionalen Grenz-schichtströmungen.* Diss. , Institut für Aerodynamik und Gasdynamik, Universität Stuttgart, 1995.

[5] Spalart, P.R.; Crouch, J.D.; Ng, L.L.: Numerical study of realistic perturbations in 3-D boundary layers. In *Proc. AGARD Conf.: Application of Direct and Large Eddy Simulation to Transition and Turbulence*, AGARD-CP-551, pp. 30.1 – 30.10., Chania, Crete, Greece, 1994.

[6] Reibert, M.S.; Saric, W.S.: Review of Swept-Wing Transition. AIAA-97-1816, 1997.

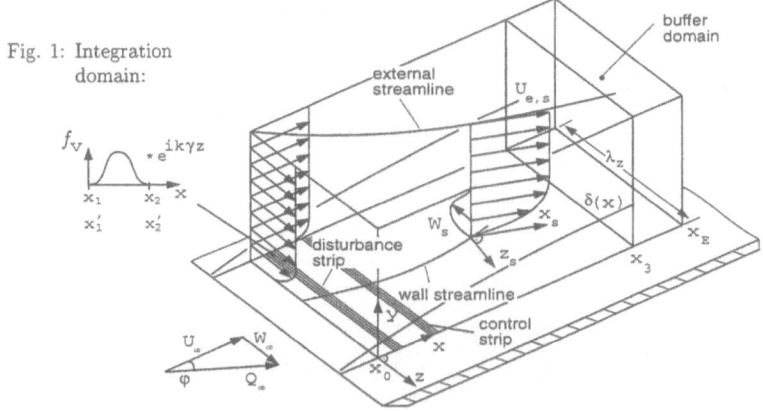

Fig. 1: Integration domain:

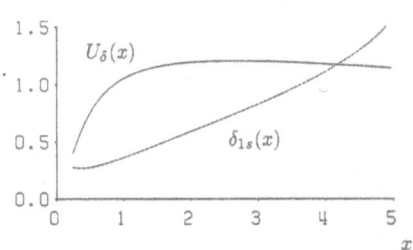

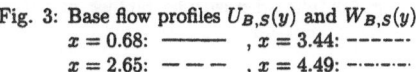

Fig. 2: Chordwise edge velocity $U_\delta(x)$, streamwise displacement thickness stretched by 0.35

Fig. 3: Base flow profiles $U_{B,S}(y)$ and $W_{B,S}(y)$
$x = 0.68$: ———— , $x = 3.44$: ------
$x = 2.65$: — — — , $x = 4.49$: —·—·—·

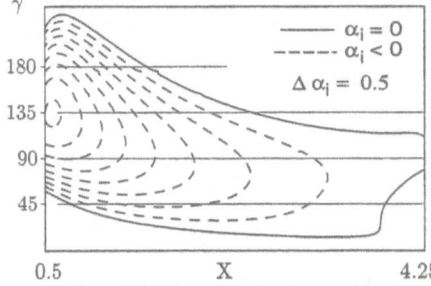

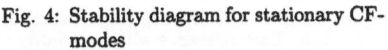

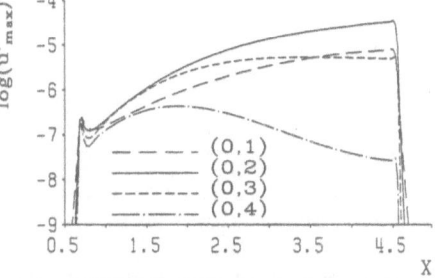

Fig. 4: Stability diagram for stationary CF-modes

Fig. 5: Quasilinear development of stationary CF-modes (u'_{max}: y-maximum)

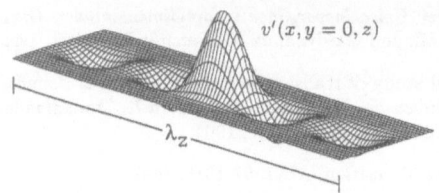

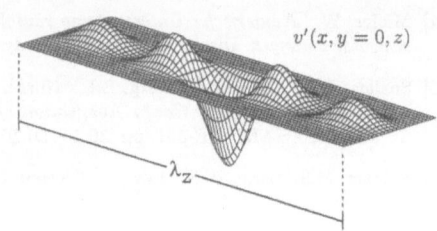

Fig. 6: v'-velocity distribution in the disturbance strip, case 2, "blowing"

Fig. 7: v'-velocity distribution in the disturbance strip, case 3, "suction"

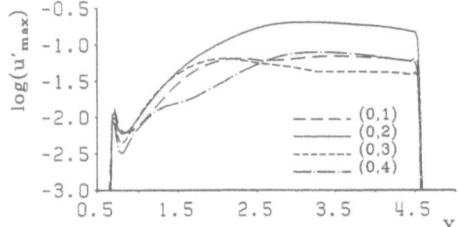

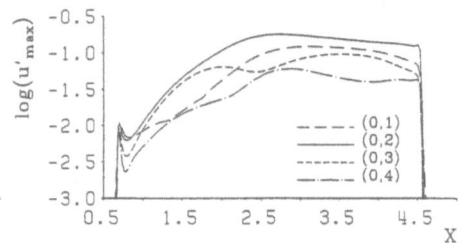

Fig. 8: Amplitude development, case 2

Fig. 9: Amplitude development, case 3

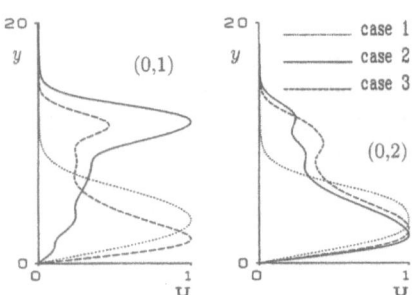

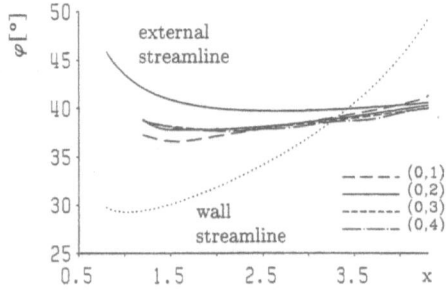

Fig. 10: $u'(y)$-disturbance profiles at
$x = 4.0$

Fig. 11: Angle of the vortex axes versus x-axis,
case 1

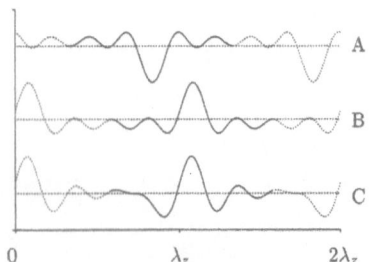

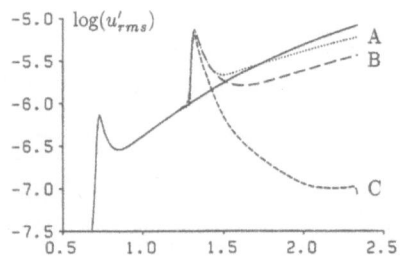

Fig. 12: Normalized spanwise v'-velocity distribution in the x-center of the control
strip; cases A, B and C

Fig. 13: u'_{rms}-amplitude development for
various controlled cases (A, B, C);
solid line: reference without control

Euler calculations for the flow
around a delta wing during flap oscillations

C. Weishäupl, E. Kreiselmaier, G. Heller, B. Laschka

Lehrstuhl für Fluidmechanik, Technische Universität München
Boltzmannstraße 15, 85748 Garching, Germany

Summary

For maneuverability of fighter aircraft flap efficiency over the whole flight envelope is necessary. In this context unsteady Euler calculations are performed for a delta wing with 53° leading–edge sweep in the transonic and supersonic region for different angles of attack. The wing is fitted with two trailing–edge flaps, an inboard and an outboard flap. The main features of the Euler code and grid generator used are described. For interpretation a Fourier analysis of the time–dependent results is performed. The influence of angle of attack and incoming Mach number on the flap efficiency is discussed. In the supersonic region especially the flow physics is regarded.

Nomenclature

c_p	pressure coefficient	s	wing half span
$c_{A\eta}, c_{M\eta},$		α	angle of attack
$c_{A\dot\eta}, c_{M\dot\eta}$	flap derivatives	η_0	mean flap angle
f	frequency [Hz]	η_1	amplitude of the flap angle
M	Mach number		

Introduction

Flap efficiency in the whole flight envelope is necessary to guarantee the maneuverability of fighter aircraft. Experiments for delta wings show, that the flap efficiency is influenced significantly by the angle of attack. Up to a certain angle of attack flap efficiency is nearly constant. A further increase of the angle of attack leads to a dramatic loss in efficiency. The aim of the performed numerical simulations is to show, whether this effect can be reproduced with an Euler code. The second aspect is the flow topology and the flap efficiency in the supersonic region. Therefore simulations for a delta wing with oscillating flaps are performed in the transonic and supersonic region for different angles of attack, documented in detail in Refs. 1–3.

In order to decide whether an Euler code is suitable for the simulations, the flow parameters are compared with the Miller–Wood–diagram [4], Fig. 1. There, eight different topologies for a delta wing are classified, depending on the angle of attack α, the leading–edge sweep Φ_V and the incoming Mach number M_∞ with $\tan\alpha_N = \tan\alpha/\cos\Phi_V$ and $M_N = M_\infty\sqrt{1 - \cos^2\alpha\sin^2\Phi_V}$. The regions 5–8 are dominated by viscous flow phenomena, for which an Euler code is not suitable. In region 1 a leading–edge separation can be modelled due to numerical viscosity. In the regions 2–4 viscous effects play only a minor role and the Euler equations can be used. Therefore an Euler code is appropriate for the simulated cases, marked in the diagram as dots and squares.

Numerical method

Inviscid compressible flow is assumed. In this case the governing equations are the Euler equations. The used Euler code is a finite–volume code with TVD–feature. Second order accuracy in space and time is guaranteed. The time integration can be performed explicitly or implicitly with the LU–SSOR (Lower-Upper-Symmetric-Successive-Overrelaxation) scheme [5,6]. With shock capturing an adequate shock resolution can be obtained. For interpretation a Fourier analysis of the time–dependent results is performed. In this way the important stability derivatives $c_{A\eta}$, $c_{M\eta}$, $c_{A\dot\eta}$ and $c_{M\dot\eta}$ can be determined directly from the real and imaginary part of the complex formulation for the aerodynamic forces [3]. $c_{A\dot\eta}$ and $c_{m\dot\eta}$, corresponding to the imaginary part, reflect the unsteadiness of the flow. For discretization structured multiblock grids are used, which allow flexible grid generation for complex geometries. Based on a suitable start grid elliptic smoothing of the grid inside the blocks and over block boundaries is performed. For unsteady simulations, here the flap oscillations, the grid has to be adapted to the actual wing geometry during the motion. The deformation of the grid should be kept small, realized again with elliptic smoothing. Details concerning the Euler code and grid generator are given in [3,7,8,9].

Geometry and flow parameters

The simulations are performed for a delta wing with 53° leading–edge sweep and −3° trailing–edge sweep, Fig. 2. The wing is fitted with two trailing–edge flaps, an inboard and an outboard flap. Furtheron the wing has a round nose and a wash–out at the wing tips. For grid generation the wing is subdivided into 6 segments, whereby 2 segments define the region of the flaps, Fig. 2a. The grid is smoothed inside the segments and over the boundaries of the segments, too, if the segment boundary does not define the boundary to a movable segment, Fig. 2b. The wing is discretized as a continuous surface, i. e. gaps at the flap leading– and side–edges are not modelled. For the simulations in the supersonic region a grid with CH–topology, in the transonic cases a 5–block–topology is used. The parameters of the performed simulations are given in Tab. 1.

Tab.1 Parameters of the performed simulations.

	M	α	f[Hz]	η_0	η_1
transonic region	0.8	$0 - 25°$	6	0°	0.4°
supersonic region	1.6	$0 - 10°$	6	0°	1.0°

Flap efficiency in the transonic regime

The flap efficiency in the transonic regime is given in Fig. 3 with $c_{A\eta}$, $c_{A\dot\eta}$, $c_{M\eta}$ and $c_{M\dot\eta}$ as function of angle of attack. The inboard flap shows a higher efficiency than the outboard flap. This effect is caused by the following mechanism: with increasing angle of attack leading–edge vortices develop at the wing, which induce crossflow velocities in the flap region, Fig. 3. This velocity component reduces the flap efficiency. Because of higher velocities in the region of the outboard flap, its efficiency is more reduced than that of the inboard flap. Furtheron the flap efficiency is reduced with increasing angle of attack, as expected from experimental results. The axis of the leading–edge vortices is shifted upwards and inboard with increasing angle of attack. At the same time the vortex strength is increased and leads to higher crossflow velocities, reducing flap efficiency with increasing angle of attack.

Flow topology in the supersonic regime

Case I ($M = 1.6$, $\alpha = 0°$): Fig. 4a shows the distribution of the pressure coefficient in 5 sections normal to the incoming flow on the lower wing side. At the leading–edge a separation bubble exists leading to a distinct suction peak. Extension and intensity of the region with negative pressure grow downstream, corresponding to the wash–out at the wing tips. The isolines show only a small compression behind the suction peak, i. e. a shockfree recompression occurs. In the plane $x/s = 1.6$ a weak tip vortex can be identified from the crossflow velocities, Fig. 4b. Its orientation corresponds to a negative lift. The local Mach number is supersonic over the whole wing.

Case II ($M = 1.6$, $\alpha = 10°$): The distribution of the pressure coefficient on the upper wing side, Fig. 5a, is characterized by a significant pressure minimum at $60 - 70\%$ of the local half span. Behind this region a recompression area occurs corresponding to a shock depicted by the compression of the isolines. In the plane $x/s = 1.6$, Fig. 5b, a distinct vortex comes up. The local Mach number, Fig. 6, is again supersonic over the whole wing. On the lower side, Fig. 6a, only small changes occur, the flow is accelerated downstream. On the upper side, Fig. 6b, the Mach–contour–plot shows the region of negative pressure with increasing extension downstream. First, the streamlines are nearly parallel to the leading–edge, then the following acceleration leads to a deflection of the streamlines. The kink in the streamlines corresponds to the oblique shock.

For further analysis the steady pressure distributions in 5 sections along the span with 2 sections in the region of the inboard and 2 in the region of the outboard flap are presented in Fig. 7. The kink near the trailing–edge results from the discretization in this region: the wing has a finite thickness at the trailing–edge, which is reduced to zero in the last cell. This kink in the contour leads to a kink in the pressure distribution. As mentioned before, for $\alpha = 0°$ a suction peak on the lower side results. For $\alpha = 10°$ the suction peak occurs on the upper side and is increased in comparison to $\alpha = 0°$. Due to the supersonic Mach number there is no flow around the trailing–edge, the pressure coefficient is nearly constant in this region. The case $\alpha = 0°$ is characterized by a negative lift coefficient and a positive pitch moment coefficient and the case $\alpha = 10°$ by a positive lift coefficient and a negative pitch moment coefficient.

Flap efficiency in the supersonic regime

To determine the flap efficiency the first harmonic of the pressure distribution in the relevant sections is given in Fig. 8 for the inboard flap and in Fig. 9 for the outboard flap. The sections at 35% and 50% of the span are in the region of the inboard flap, the sections at 75% and 90% in the region of the outboard flap. For both flaps the real and imaginary part show values not equal zero only at the moving flap because of the supersonic local Mach number over the whole wing. Against that in a subsonic case the moving flap causes a singularity at the flap axis and influences the whole wing. The imaginary part is much lower than the real part, so that only small unsteady effects exist. Changing the angle of attack from 0° to 10°, the pressure distribution on the lower and upper side is shifted in the same direction and therefore flap efficiency depends hardly from the angle of attack. The real parts for the inboard and the outboard flap differ only slightly. The imaginary part for the outboard flap is smaller than for the inboard flap. The pressure distributions result in the flap derivatives $c_{A\eta}$, $c_{M\eta}$, $c_{A\dot\eta}$ and $c_{M\dot\eta}$ as function of the angle of attack, Fig. 10. As in the transonic region the inboard flap shows a higher efficiency than the outboard flap. Now this is caused by the bigger surface, not by the crossflow velocities arising from the leading–edge vortices. The angle of attack has only small influence on flap efficiency. Corresponding to the low values of the imaginary part, only small unsteady effects occur. The values of the derivatives are considerably reduced against the transonic ones, Fig. 3, due to the local limited effect of the flaps. Compared to the transonic case an efficiency of only 25% is obtained.

Conclusions

For a delta wing with 53° leading–edge sweep the flap efficiency of an inboard and an outboard flap was investigated numerically with an Euler code. Transonic and supersonic conditions were regarded for different angles of attack. In the subsonic region the inboard flap shows a higher efficiency than the outboard flap and the efficiency is reduced with increasing angle of attack. In the supersonic region again the inboard flap is more efficient than the outboard flap, but the difference is smaller than in the transonic case. Furtheron the supersonic efficiency is considerably reduced against the transonic case and only small unsteady effects occur.

Acknowledgements

This work was supported by the Daimler Chrysler Aerospace AG (DASA). The authors thank J. Becker for his kind assistance in the research project.

References

[1] Heller, G., Kreiselmaier, E., " Eulerluftkräfte für Klappenschwingungen." Technical report, Technische Universität München, TUM–FLM–96/05, 1996.

[2] Kreiselmaier, E., " Eulerluftkräfte für Klappenschwingungen – Phase II." Technical report, Technische Universität München, TUM–FLM–97/01, 1997.

[3] Kreiselmaier, E., " Eulerluftkräfte für Klappenschwingungen – Phase III." Technical report, Technische Universität München, TUM–FLM–97/20, 1997.

[4] Rom, J., " High Angle of Attack Aerodynamics," Springer–Verlag, New York, 1992.

[5] Jameson, A., Turkel, E., " Implicit Schemes and LU–Decompositions," Mathematics of Computation, 37:385–397, 1981.

[6] Blazek, J., " Investigations of the Implicit LU–SSOR Scheme," DLR–FB 93-51, 1993.

[7] Heller, G., " Aerodynamik von Deltaflügelkonfigurationen bei Schieben und Gieren," Dissertation, Lehrstuhl für Fluidmechanik, Technische Universität München, 1997.

[8] Kreiselmaier, E., " Berechnung instationärer Tragflügelumströmungen auf der Basis der zeitlinearisierten Eulergleichungen," Dissertation, Lehrstuhl für Fluidmechanik, Technische Universität München, 1998.

[9] Weishäupl, C., " CFD Methods Applied to Non–standard Unsteady Problems in Aircraft–aerodynamics." in Proceedings of the 3rd International Symposium on Advanced and Aerospace Science & Technology in Indonesia, ISASTI 98, Jakarta, Aug. 31–Sept. 3, 1998.

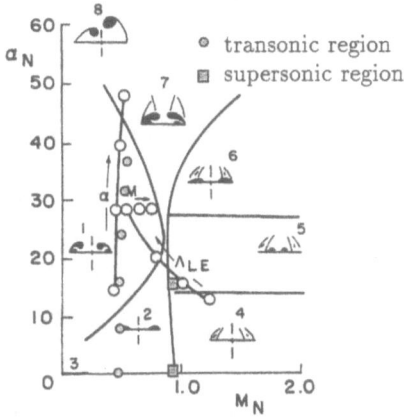

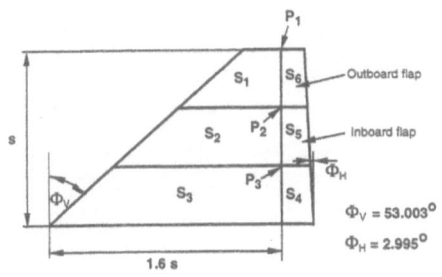

○ transonic region
◻ supersonic region

1 leading–edge vortex

2 separation bubble with no shock

3 no shock / no separation

4 shock with no separation

5 shock–induced separation

6 separation bubble with shock

7 leading–edge vortex with shock

8 asymmetric vortex separation

Fig. 1 Miller–Wood–diagram.

a) segments of the surface

$\Phi_V = 53.003^{\circ}$
$\Phi_H = 2.995^{\circ}$

b) surface grid

Fig. 2 Surface grid of the delta wing.

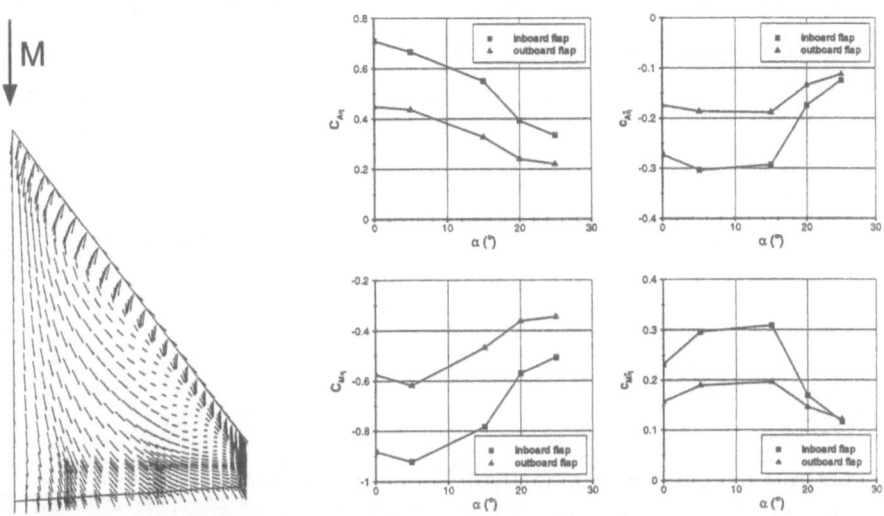

Fig. 3 Flap efficiency in the transonic region for $M = 0.8$ over the angle of attack.

525

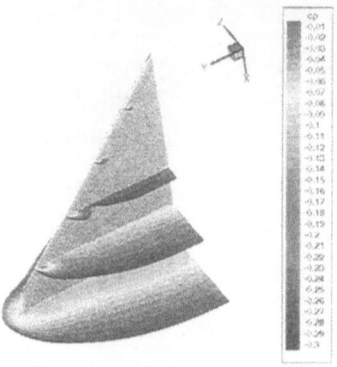

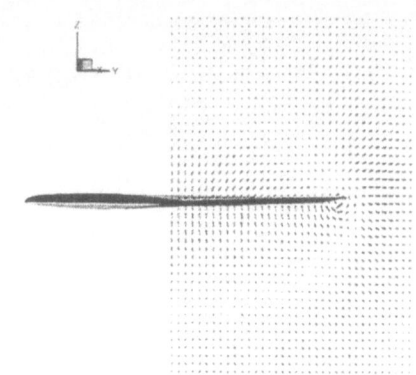

a) c_p–contour–plot of the lower wing side

b) crossflow–velocity in the section $x/s = 1.6$

Fig. 4 Flow topology for $M = 1.6$, $\alpha = 0°$.

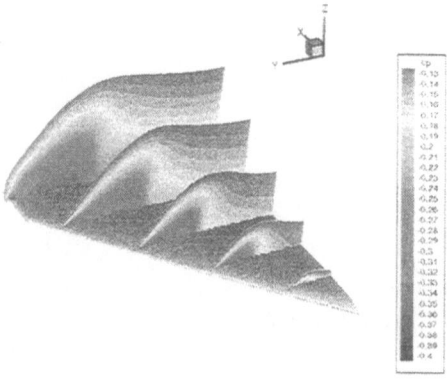

a) c_p–contour–plot of the upper wing side

b) crossflow–velocity in the section $x/s = 1.6$

Fig. 5 Flow topology for $M = 1.6$, $\alpha = 10°$.

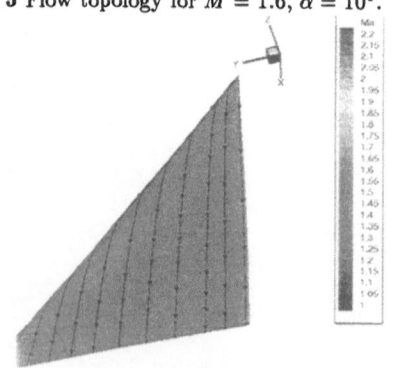

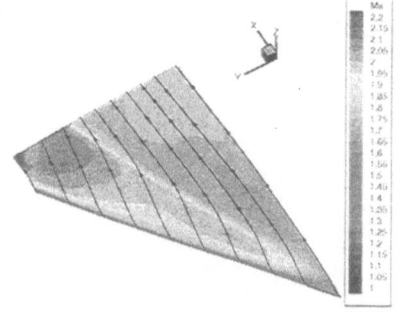

a) M–contour–plot of the lower wing side

b) M–contour–plot of the upper wing side

Fig. 6 Distribution of Mach number on the wing surface for $M = 1.6$, $\alpha = 10°$.

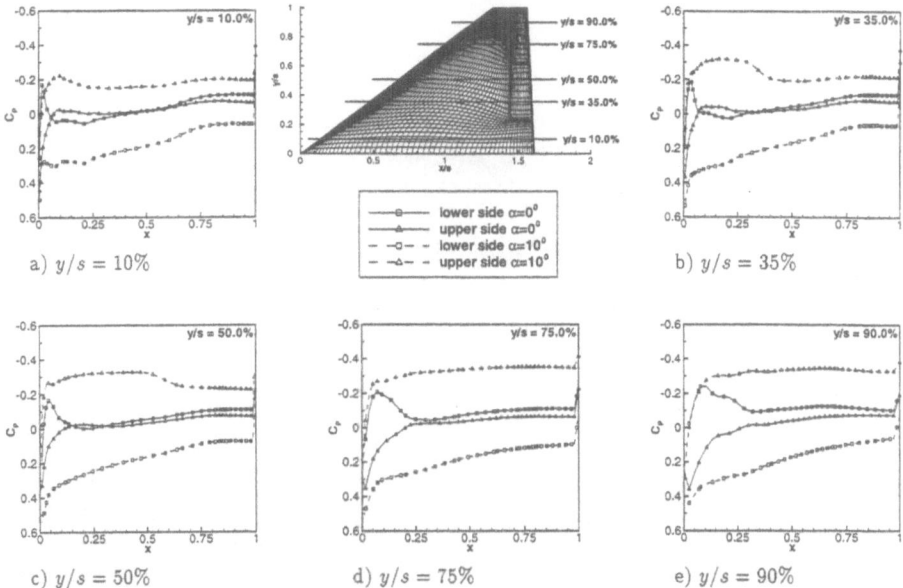

Fig. 7 Steady pressure distributions for $M = 1.6$, $\alpha = 0°, 10°$.

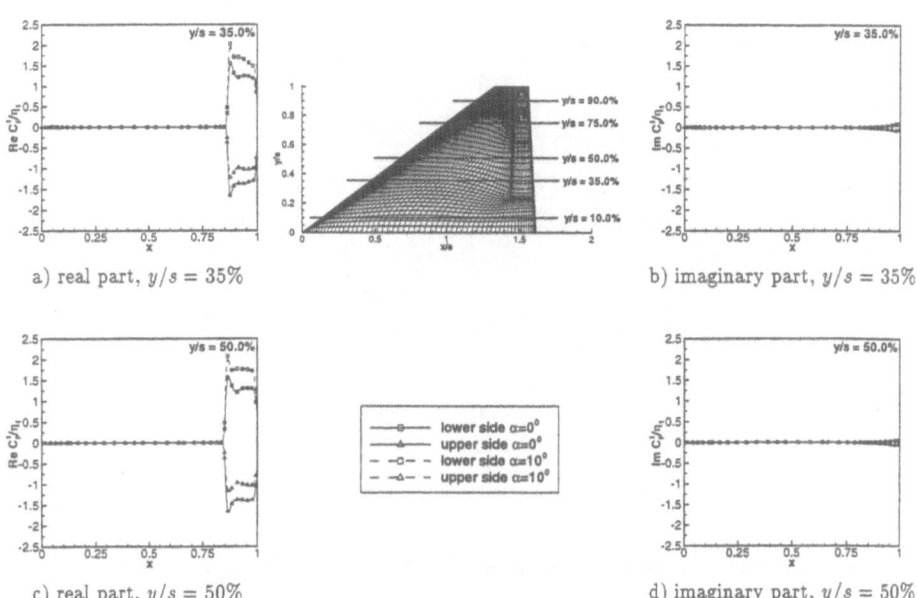

Fig. 8 Flap efficiency of the inboard flap for $M = 1.6$, $\alpha = 0°, 10°$.

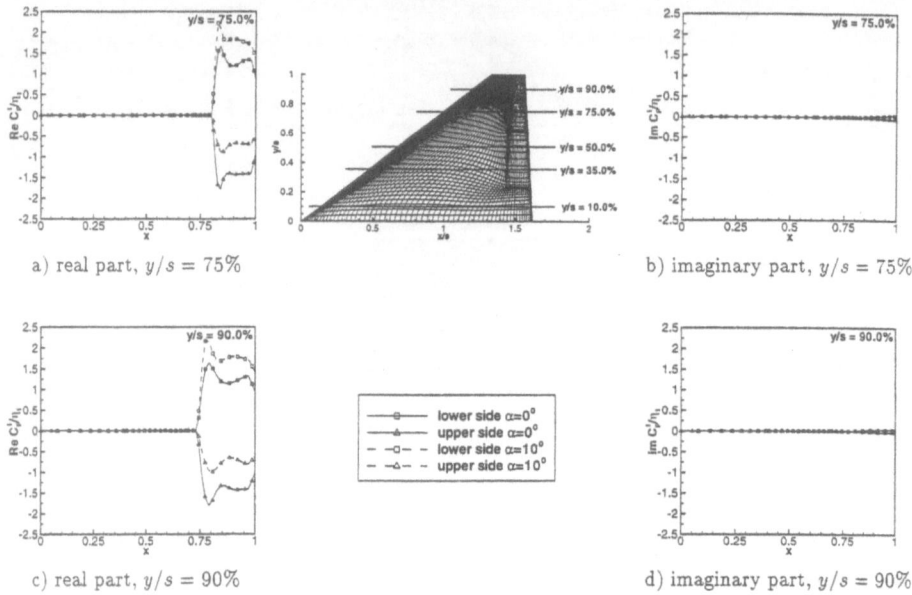

a) real part, $y/s = 75\%$

b) imaginary part, $y/s = 75\%$

c) real part, $y/s = 90\%$

d) imaginary part, $y/s = 90\%$

Fig. 9 Flap efficiency of the outboard flap for $M = 1.6$, $\alpha = 0°, 10°$.

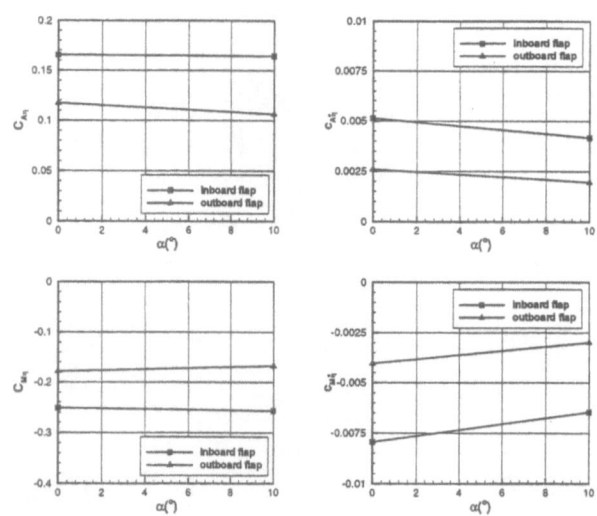

Fig. 10 Flap efficiency in the supersonic region for $M = 1.6$ over the angle of attack.

528

Experimental Investigation on 3D Acoustic Receptivity of a Laminar Boundary Layer in the Presence of Surface Non-Uniformities

W. Würz, S. Herr, S. Wagner

Institut für Aerodynamik und Gasdynamik
Pfaffenwaldring 21, D-70550 Stuttgart, Germany

Y. S. Kachanov

Institute for Applied and Theoretical Mechanics
Russian Academy of Science, 630090 Novosibirsk, Russia

Summary

Hot-wire measurements were performed under controlled conditions in the Laminar Wind Tunnel on an airfoil with pressure gradient. A quantitatively known acoustic field was established in the test section. The receptivity of the 2-dimensional laminar boundary layer in the presence of a localized 3D surface roughness was studied. The development of the generated Tollmien-Schlichting-waves was measured at different streamwise stations as cuts in spanwise direction. The quasi steady 3D roughness was modeled by a vibrating source driven at a frequency two orders below the acoustic one. The amplitude of the TS-waves was measured at combination frequencies. Fourier decomposition in time and space leads to wave number spectra which can be compared to linear stability theory. The receptivity function was evaluated by upstream extrapolation to the initial distributions at the source position.

Introduction

The problem of the boundary layer transition from laminar to turbulent flow still attracts much attention because of the fundamental importance to the study of fluid motion. It has three main parts. First, there is the laminar flow receptivity to external perturbations. The second part is the region where the boundary layer instabilities grow according to linear stability theory. And finally there is the nonlinear flow breakdown to turbulence.

The receptivity is defined by the ratio of the generated amplitude in the boundary layer to the amplitude of the external perturbation. In case of receptivity due to surface distortions this definition can be extended by normalization with the height of the involved roughness element [3], [7]. When linear receptivity is studied, the initial TS-amplitudes are too low to be measured directly. Therefore we used the boundary layer as a selective amplifier. We measured TS-amplitude distributions downstream of the roughness element and in the linear stage of the transition development. Linear stability theory was used to extrapolate the data to the position of the receptivity element.

This paper is devoted to 3D acoustic receptivity at a localized roughness element which is poorly studied yet [7], [11]. A single acoustic frequency (f_{ac}) similar to the most unstable TS-frequency was chosen as the external perturbation. The roughness was simulated by a quasi steady vibrating source driven at frequency $f_v = \frac{1}{64} \cdot f_{ac}$. This leads to a modulation in amplitude of the generated TS-wave, resulting in frequency space in distinct combination frequencies ($f_{CF,1/2} = f_{ac} \pm f_v$) which can be measured instead of the center frequency (f_{ac}). This allows to separate the acoustic amplitude from the amplitude of the TS-wave.

529

The TS-waves develop downstream as a wave train consisting of waves with different propagation angles and phase speeds.

Therefore, the measured spanwise distributions of the hot-wire signal are Fourier transformed to decomposite them into oblique modes [5], [6], [9]. The complex wave number spectra are extrapolated to the position of the source. Together with the double Fourier transform of the shape of the source and the dispersion characteristics of the boundary layer the complex receptivity function can be evaluated.

Investigation Technique

The experiments were carried out in the Laminar Wind Tunnel of the Institute of Aerodynamics and Gasdynamics [12]. The Laminar Wind Tunnel is an open return tunnel of the Eiffel design. The rectangular test section measures $0.73 \times 2.73 m^2$ and is $3.15m$ in length. The 2D airfoil models span the short distance of the test section. The high contraction ratio of 100:1 as well as five screens and filters result in a very low turbulence level of less than $2 \cdot 10^{-4}$.

The boundary layer measurements were performed on a symmetrical airfoil section with 15% thickness (XIS40MOD). It was specially designed to have a long instability ramp at zero angle of attack and to meet all requirements for boundary layer measurements. The Reynoldsnumber was chosen to $1.2 \cdot 10^6$ based on the arclength $s_{max} = 0.615m$ measured from leading edge. The velocity distribution (fig.5) was measured on a similar model, which is equipped with pressure orifices, and is in good agreement with a distribution calculated with XFOIL [4]. Based on the latter distribution the boundary layer profiles were calculated by using a finite difference scheme [2]. The stability diagram

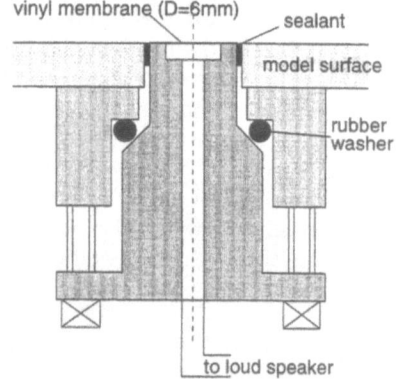

Fig. 1: Vibrating source

(fig.5) was finally evaluated according to linear stability theory. The position of the vibrating source $(s/s_{max} = 0.2)$ was selected to the instability point for the acoustic frequency $(f_{ac} = 1088Hz)$ at branch I of the neutral curve.

The acoustic wave was generated by a 1020 McCauley loud speaker. It was mounted inside a low drag housing at the centerline and about $4.4m$ downstream of the test section. The introduced sound wave had a sound pressure level of approximately $100dB$, that is $16dB$ higher than the natural sound pressure level for the frequency range $10Hz - 5000Hz$ of the wind tunnel itself.

Vibrating Source

The 3D roughness element was modeled by the inflection of the membrane of a vibrating source [8]. The active diameter ($6mm$) was chosen to approximately one half of the TS-wavelength (at f_{ac}) to allow a significant amount of amplitude at higher spanwise wave numbers. The body of the source (fig.1) was manufactured as a lathe work. A vinyl membrane was pasted with rapid glue to the smaller side of the cone. The other side was connected to a loud speaker. The source body was mounted inside the airfoil model. With three screws it was possible to adjust the body to set up the membrane flush-mounted with the model surface. A sealant was applied to the gap between the body and the model surface to smooth it completely.

To avoid a deflection of the membrane due to static pressure differences, the loud speaker was mounted in a box which was connected to a static pressure orifice at the same stream-wise station.

The hump of the membrane produced by the pressure oscillations of the loud speaker was measured with a laser triangulation device. A maximum height of $28\mu m$ RMS was used, which is about 8.5 % of the displacement thickness $\delta_{1,ref}$ at the source position.

The shape of the hump was double Fourier transformed in streamwise and in spanwise direction in order to find the complex wave number spectrum of the surface vibration at the vibrational frequency:

$$\tilde{A}_v(\alpha_r, \beta) = \frac{1}{2\pi^2} \int_{-\infty}^{\infty} \int_{-\infty}^{\infty} \tilde{A}_m(s, z) \cdot e^{-i(\alpha_r s + \beta z)} ds dz .$$

Measurement Procedure

For the hot-wire measurements a modified DISA55P15 boundary layer probe with $1mm$ wire in length was used together with a DISA55M10 bridge. A small static probe with $1mm$ in diameter yields as a velocity reference at the boundary layer edge.

The probes are mounted on a traversing mechanism which allows computer controlled scans in normal to wall and in spanwise direction. A resolution of $5\mu m$ in wall distance was achieved by a high precision rack-and pinion drive together with an optical encoder. The distance in spanwise direction was measured inductively with a resolution of $0.1mm$.

The DC-output of the hot-wire anemometer was integrated with a $1Hz$ low-pass filter. The AC-output was high-pass filtered with a first order low-noise filter with $100Hz$ cut-off frequency. A programmable amplifier was used to fit the signal always optimal to the input range of the 12 bit AD-converter of the control PC. Prior to sampling the signal was low-pass filtered at $4400Hz$ to omit aliasing problems connec-

Fig. 2: Experimental setup

ted with the sampling frequency of $8704Hz$. The acoustic frequency as well as the vibrational frequency and the sampling trigger were generated by the DA-converter of an additional PC. The ratios between the frequencies were chosen as integer powers of two $(f_{source} : f_{ac} : f_{samplerate} = 1 : 64 : 8)$ and the frequencies were strictly phase locked because they were subdivided from a single quartz based clock. Therefore it was possible to use a Fast Fourier Transform for the hot-wire signal where the acoustic frequency, the source frequency and the combination frequencies were represented exactly by a Fourier coefficient.

The data acquisition was started with fixed phase relation to the signal for the vibrating source. Five sets of 4096 points were collected and averaged in time domain. The FFT analysis was performed and the Fourier coefficients were corrected in amplitude for the influence of the filters. All results were monitored online. The complete set up is sketched in fig.6.

The direction of the acoustic field was checked by a single slanted hot-wire in the vicinity of the vibrating source. The angle between the direction of the acoustic field and the mean flow direction is below 5 degree. Amplitude and phase of the acoustic wave were measured before and after every spanwise scan by positioning the hot-wire at the s-position of the vibrating source but with $20mm$ offset in spanwise direction and outside the Stokes layer. To control the vibrational amplitude and phase, the hot-wire was positioned at a repeatable reference point over the source, which gave a fixed relation to the surface displacement measurements at the start and the end of the main set measurements.

Results of Measurements

Mean Flow Characteristics

Mean flow velocity profiles (fig.7) were taken at three downstream positions: The source position ($s = 122mm$), about three TS-wavelength downstream ($s = 162mm$) and about six TS-wavelength downstream ($s = 200mm$). The symbols mark the results from different measurements. The solid lines represent the calculated profiles based on the velocity distribution described before. The velocity values have been normalized by the velocity at the boundary layer edge U_δ and the y coordinate (normal to wall) has been normalized by the local displacement thickness δ_1. The calculated and the measured profiles agree well. The displacement thickness $\delta_{1,ref}$ at the source position was then used to normalize the length scales in the remainder of this paper.

For the spanwise traverses we had to decide for a constant wall distance which allows to measure the TS-amplitude close to their maximum of the eigenfunction. Linear stability calculations for different wave angles show that a good compromise is given for $y/\delta_2 = \text{const.} = 2.2$ (fig.8). This position was adjusted during the experiment by keeping U/U_δ at the necessary corresponding value.

Previous Measurements

In this paper we will focus on linear receptivity. A criterion for the linearity of the present problem is the independence of the spatial distributions of phases and normalized amplitudes from absolute values. In fig.9 the normal to wall profiles of streamwise velocity fluctuations for three different levels of excitation are compared with each other. The first was taken for an acoustic amplitude of $u'_{ac,RMS} = 0.00567\frac{m}{s}$ ($\simeq 101dB$) and a vibrational amplitude of $A_{v,RMS} = 33\mu m$. A spanwise distribution at $s = 200mm$ and a normal to wall profile at $s = 200mm$ $z = 0mm$ were measured. Then the hump height was kept constant and the acoustic amplitude was cut to $u'_{ac,RMS} = 0.00249\frac{m}{s}$ ($\simeq 93dB$). Finally the hump height was reduced to $A_{v,RMS} = 16\mu m$ and the acoustic amplitude was increased again to $u'_{ac,RMS} = 0.00535\frac{m}{s}$. The three measurements agree well. Additionally the hot-wire signal was checked carefully for the appearance of additional combination frequencies. The maximum TS-amplitude at the end of the measurement section was $\frac{u'_{RMS}}{U_\delta} \leq 0.02\%$ for the combination frequencies and $\frac{u'_{RMS}}{U_\delta} \leq 0.15\%$ for the center frequency.

During preliminary measurements it turned out that the acoustic wave excites the membrane of the source at the acoustic frequency and also at the combination frequencies ($f_{CF,1/2} = f_{ac} \pm f_v$). This produces TS-waves at combination frequencies resulting from receptivity due to surface vibrations. To estimate the magnitude of these TS-waves, we measured the vibrational amplitude at combination frequencies. Then the source was driven with a similar amplitude at acoustic frequency. Without sound field we measured the generated TS-amplitudes at a reference station downstream. The shape of the y- and z-profiles was comparable to those generated by acoustic receptivity but we found an amplitude 18.5 times lower. This additional amplitude is neglected in the post processing of the data.

We performed some scans in streamwise direction to look for the extension of the near field of the source. We found no significant near field influence at the combination frequencies (fig.10).

The quality of the installation of the source in respect to the model surface was estimated by measurements of the natural transition position. The influence was small even if we only turned on both the vibrating source or the acoustic.

Main set of data

A main set of data consists of 8 to 9 spanwise scans covering the whole width of the wave train generated downstream of the vibrating source (fig.5). The time signals at every measurement point were analyzed with a FFT and the complex amplitude

$$\tilde{B}_{CF,1/2,raw}(s_i, z) = B_{CF,1/2,raw}(s_i, z) \cdot e^{i\phi_{CF,1/2}(s_i,z)}$$

formed by the amplitude part ($B_{CF,1/2,raw}$) and the phase part ($\phi_{CF,1/2}$) for the combination frequencies was used for further processing. The amplitudes were normalized by the local free stream velocity U_δ. The phases were corrected in streamwise and spanwise direction by adding times of 360°.

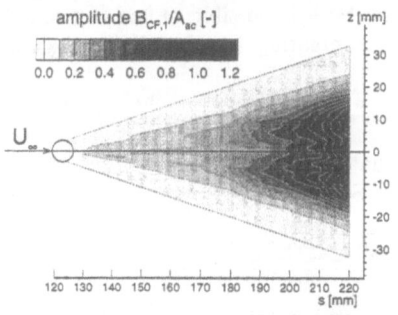

Fig. 3: Amplitude part of the wave train   Fig. 4: Phase part of the wave train

In order to correct changes in the acoustic field during the measurements the 'raw' complex values were also normalized by the complex amplitude $\tilde{A}_{ac} = A_{ac} \cdot e^{i\phi_{ac}}$ of the acoustic field:

$$\tilde{B}_{CF,1/2,ac} = \frac{B_{CF,1/2,raw}}{A_{ac}} \cdot e^{i(\phi_{CF,1/2}-\phi_{ac})} .$$

The resulting amplitude and phase of the wave train are shown in fig.3 and fig.4.

Wave characteristics

The complex values of the wave train in physical space were mapped for each spanwise cut to wave number spectra by the complex Fourier transform

$$\tilde{B}_{TS}(s_i, \beta) = \frac{1}{2\pi} \int_{-\infty}^{\infty} \tilde{B}_{CF}(s_i, z) \cdot e^{-i\beta z} dz .$$

After this decomposition the downstream development of waves with different spanwise wave numbers can be followed separately. The development of the amplitudes for the left combination frequency for waves with three different wave numbers is shown in fig.11. The wave numbers were chosen such that they correspond to propagation angles of 0°, 25° and 60°. The solid lines in fig.11 denote the amplification as it is calculated by linear stability theory. It can be seen that the measurements for zero propagation angle coincide very well with the calculated curves. When the propagation angle increases, the agreement gets poorer, probably as a result of non parallel effects [1] which are not included in linear theory.

The downstream development of the associated phases is shown in fig.12. The points can easily be fit by a straight line. The gradient of this line can be interpreted as the real part α_r of the streamwise wave number [10].

By this procedure we found the dispersion characteristic $\alpha_r = \alpha_r(\beta)$ that gives a fixed relation between the streamwise and spanwise wave number (fig.13). Also the propagation angle $\theta = \theta(\beta) = \arctan \frac{\beta}{\alpha_r}$ was found as a function of the spanwise wave number β. The wavelength in downstream direction is defined as $\lambda_{CF} = \frac{2\pi}{\alpha_r}$. Together with the connected frequency the 'virtual' propagation speed in streamwise direction $C_{sCF} = \lambda_{CF} \cdot f_{CF}$ can be calculated. This completes the necessary knowledge about the stability characteristics in our experiment.

Receptivity Function

The downstream behavior of the amplitudes of the wave train is consistent with the linear stability theory for Tollmien-Schlichting waves. We have also seen that the development of the phases can be represented by a straight line. So it is possible to find the complex initial amplitudes of the generated TS-waves by extrapolating, based on linear theory, to the position of the vibrating source.

From the double Fourier decomposition of the shape of the membrane the complex wave number spectrum of the surface vibration in spanwise and streamwise direction is known. From this spectrum the vibrational amplitudes, that generate physically reasonable TS-waves, can be found along the dispersion function. So the complex receptivity function

$$\tilde{G}_{av}(\alpha_r, \beta) = G_{av}(\alpha_r, \beta) \cdot e^{i\phi_{av}(\alpha_r,\beta)} = \frac{\tilde{B}_{inTS}(\alpha_r, \beta)}{\tilde{A}_{ac} \cdot \tilde{A}_v(\alpha_r, \beta)}$$

can be evaluated by calculating its amplitude part and its phase part separately. The values of the amplitude and phase part of the complex receptivity function in dependence on the propagation angle of the generated TS-wave are shown in fig.14. We must denote here, that the presented values for the receptivity function are preliminary results because the post processing of the data is still in progress.

Conclusion

The results of an experiment on 3D acoustic receptivity in a 2D laminar boundary layer in the presence of a localized quasi steady surface non uniformity were presented. Hot-wire measurements were performed as spanwise scans of the wave train downstream from the vibrating source. The results of these spanwise scans were Fourier transformed in order to decomposite the wave train into oblique modes. The downstream development of these modes is in good agreement with the linear stability theory. So it was possible to extrapolate the values to the initial ones at the vibrating source. Together with the knowledge about the acoustic field and the wave number spectrum of the surface vibration in spanwise and streamwise direction the complex receptivity function could be evaluated.

Acknowledgment

This work was performed under grant of the German research council (DFG) and is a contribution to the research program 'Transition'.

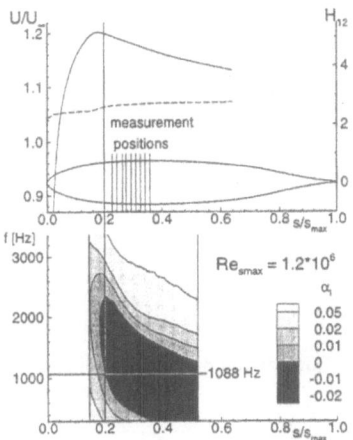

Fig. 5: Pressure distribution and position of the source in stability diagram

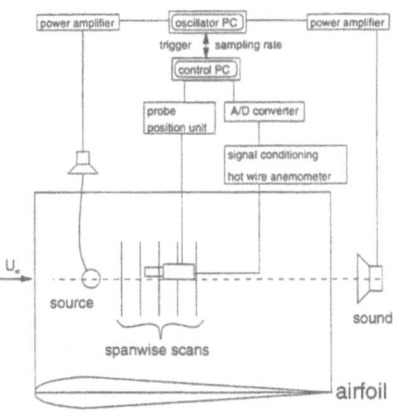

Fig. 6: Sketch of the experimental setup and data acquisition

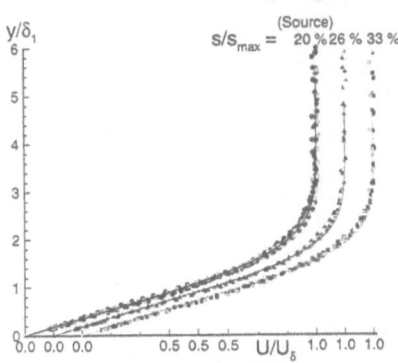

Fig. 7: Mean velocity profiles

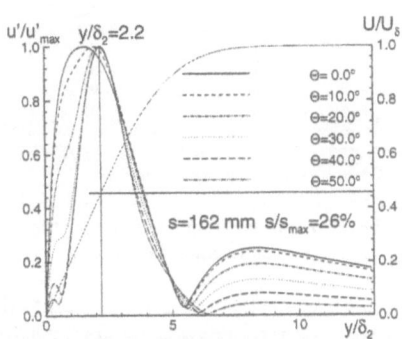

Fig. 8: Eigenfunction of TS-waves with different propagation angles at $s = 162mm$

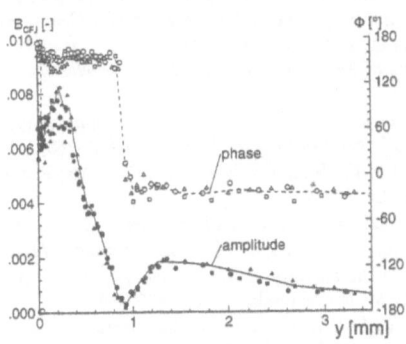

Fig. 9: Normalized y-profiles of amplitude and phase

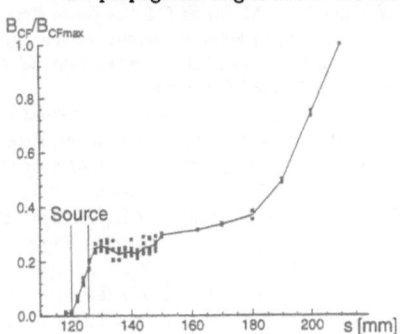

Fig. 10: Nearfield

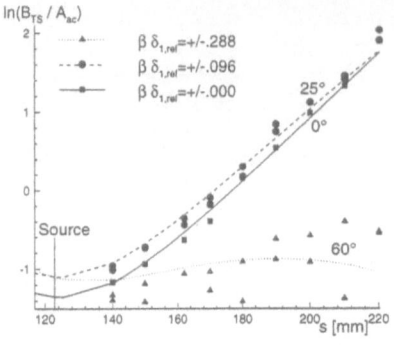

Fig. 11: Downstream development of amplitudes

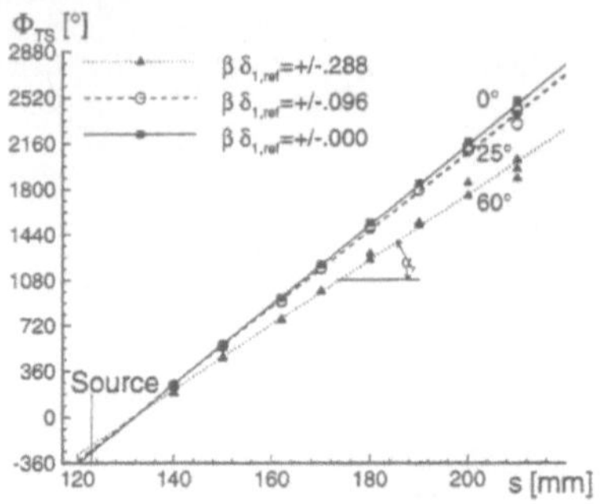

Fig. 12: Downstream development of phases

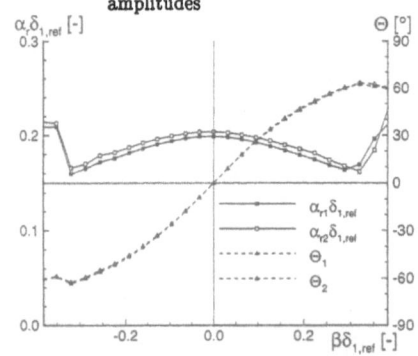

Fig. 13: Dispersion characteristics for the two combination frequencies

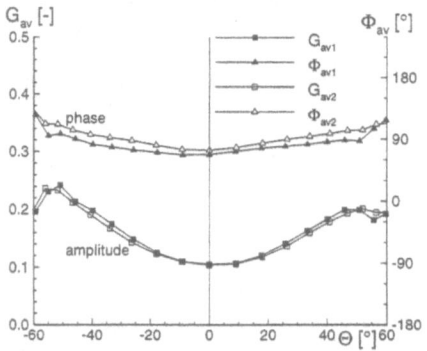

Fig. 14: Amplitude and phase part of the complex receptivity function

References

[1] Bertolotti F.P.: *Linear and nonlinear stability of boundary layers with streamwise varying properties*; PhD thesis, The Ohio State University USA, 1991.

[2] Cebeci T., Smith A.M.O.: *Analysis of Turbulent Boundary Layers*; Academic Press, NY, 1974.

[3] Choudhari, M. Street C.L.: *A finite Reynolds-number approach for the prediction of boundary layer receptivity in localized regions*; Physics in Fluids 11, 1992, pp. 2495-2514.

[4] Drela M., Giles M.B.: *Viscous-Inviscid Analysis of Transonic and Low Reynolds Number Airfoils*; AIAA-86-1786-CP, 1986.

[5] Gaster M., Grant I.: *An experimental investigation of the formation and development of a wave packet in a laminar boundary layer*; Proc. Royal Society of London A; 347, pp. 253-269, 1975.

[6] Gaster M.: *A theoretical model of a wave packet in the boundary layer on a flat plate*; Proceedings of the Royal Society of London A; 347, pp. 271-289, 1975.

[7] Ivanov A.V., Kachanov Y.S., Koptsev D.: *An exp. investigation of inst. wave excitation in 3D boundary layer at acoustic scattering on a vibrator*; Thermophysics & Aeromechanics 4/4, 1997.

[8] Ivanov A.V., Kachanov Y.S.: *A method of study the stability of 3D boundary layers using a new disturbance generator*; ICMAR Proc. Part 1, ITAM, Novosibirsk, pp. 125-130, 1994.

[9] Kachanov Y.S., Obolentseva T.G.: *Development of 3D disturbances in the blasius boundary layer I. Wave-trains*; Thermophysics and Aeromechanics, Vol.3, No.3, 1996.

[10] Kachanov Y.S., Michalke A.: *Three-dimensional instability of flat-plate boundary layers: Theory and experiment*; European Journal of Mechanics B/Fluids, 13, No.4, pp. 401-422, 1994.

[11] Saric W.S., Radetzky J.A., Hoos R.H. jr: *Boundary layer receptivity of sound with roughness*; Boundary Layer Stability and Transition, FED-Vol. 114, Eds. Reda, Reed, Kobayashi, ASME, 1991.

[12] Wortmann F.X., Althaus D.: *Der Laminarwindkanal des Instituts für Aerodynamik und Gasdynamik der Technischen Hochschule Stuttgart*; Zeitschrift für Flugwissenschaften Nr.12 Heft 4.

Brief Instruction for Authors

Manuscripts should have well over 100 pages. As they will be reproduced photomechanically they should be produced with utmost care according to the guidelines, which will be supplied on request. Figures and diagrams should be lettered accordingly so as to produce letters not smaller than 2 mm in print. The same is valid for handwritten formulae. Manuscripts (in English) or proposals should be sent to the general editor, Prof. Dr. E. H. Hirschel, Herzog-Heinrich-Weg 6, D-85604 Zorneding.